Abiotic Stresses

Plant Resistance Through Breeding and Molecular Approaches

Crop Science
Amarjit S. Basra, PhD
Senior Editor

Handbook of Formulas and Software for Plant Geneticists and Breeders edited by Manjit S. Kang

Postharvest Oxidative Stress in Horticultural Crops edited by D. M. Hodges

Encyclopedic Dictionary of Plant Breeding and Related Subjects by Rolf H. G. Schlegel

Handbook of Processes and Modeling in the Soil-Plant System edited by D. K. Benbi and R. Nieder

The Lowland Maya Area: Three Millennia at the Human-Wildland Interface edited by A. Gómez-Pompa, M. F. Allen, S. Fedick, and J. J. Jiménez-Osornio

Biodiversity and Pest Management in Agroecosystems, Second Edition by Miguel A. Altieri and Clara I. Nicholls

Plant-Derived Antimycotics: Current Trends and Future Prospects edited by Mahendra Rai and Donatella Mares

Concise Encyclopedia of Temperate Tree Fruit edited by Tara Auxt Baugher and Suman Singha

Landscape Agroecology by Paul A. Wojkowski

Concise Encyclopedia of Plant Pathology by P. Vidhyasekaran

Molecular Genetics and Breeding of Forest Trees edited by Sandeep Kumar and Matthias Fladung

Testing of Genetically Modified Organisms in Foods edited by Farid E. Ahmed

Agrometeorology: Principles and Applications of Climate Studies in Agriculture by Harpal S. Mavi and Graeme J. Tupper

Concise Encyclopedia of Bioresource Technology edited by Ashok Pandey

Genetically Modified Crops: Their Development, Uses, and Risks edited by G. H. Liang and D. Z. Skinner

Plant Functional Genomics edited by Dario Leister

Immunology in Plant Health and Its Impact on Food Safety by P. Narayanasamy

Abiotic Stresses: Plant Resistance Through Breeding and Molecular Approaches edited by Muhammad Ashraf and Philip John Charles Harris

Multinational Agribusinesses edited by Ruth Rama

Crops and Environmental Change: An Introduction to Effects of Global Warming, Increasing Atmospheric CO_2 and O_3 Concentrations, and Soil Salinization on Crop Physiology and Yield by Seth G. Pritchard and Jeffrey S. Amthor

Teaching in the Sciences: Learner-Centered Approaches edited by Catherine McLoughlin and Acram Taji

Durum Wheat Breeding: Current Approaches and Future Strategies edited by Conxita Royo, M. M. Nachit, N. Di Fonzo, J. L. Araus, W. H. Pfeiffer, and G. A. Slafer

Abiotic Stresses
Plant Resistance Through Breeding and Molecular Approaches

M. Ashraf, PhD
P. J. C. Harris, PhD
Editors

CRC Press
Taylor & Francis Group
Boca Raton London New York

CRC Press is an imprint of the
Taylor & Francis Group, an informa business

Reprinted 2010 by CRC Press

CRC Press
6000 Broken Sound Parkway, NW
Suite 300, Boca Raton, FL 33487
270 Madison Avenue
New York, NY 10016
2 Park Square, Milton Park
Abingdon, Oxon OX14 4RN, UK

For more information on this book or to order, visit
http://www.haworthpress.com/store/product.asp?sku=5288

or call 1-800-HAWORTH (800-429-6784) in the United States
and Canada or + (607) 722-5857 outside the United States and Canada

or contact orders@HaworthPress.com

Published by

Food Products Press®, an imprint of The Haworth Press, Inc., 10 Alice Street, Binghamton, NY 13904-1580.

Cover design by Kerry E. Mack.

Library of Congress Cataloging-in-Publication Data

Abiotic stresses : plant resistance through breeding and molecular approaches / M. Ashraf, P.J.C. Harris, editors.
p. cm.
Includes bibliographical references and index.
ISBN 1-56022-964-0 (hard : alk. paper) — ISBN 1-56022-965-9 (soft : alk. paper)
1. Crops—Effect of stress on. 2. Plants—Disease and pest resistance—Genetic aspects. 3. Plant breeding. 4. Plant molecular genetics. I. Ashraf, M. (Muhammad), 1953- II. Harris, P. J. C. (Phil J. C.)
SB112.5.A255 2004
632'.9—dc22

2004015042

CONTENTS

ABOUT THE EDITORS

Muhammad Ashraf, PhD, is Professor and Head of Botany at the University of Agriculture in Faisalabad, Pakistan. Dr. Ashraf received his PhD in botany from the University of Liverpool, UK, and undertook postdoctoral work as a Fulbright Scholar at the University of Arizona. His research is focused on the improvement of stress tolerance in plants using breeding and physiological approaches. He has published over 200 scientific papers and reviews. Dr. Ashraf has earned several prestigious awards and honors for his outstanding contributions in the fields of agriculture and biology, including two Gold Medals from the Pakistan Academy of Sciences, the Salam Prize, the National Book Foundation of Pakistan Awards, and the Presidential Awards Izaz-e-Fazeelat and Pride of Performance. He was inducted as a Fellow of the Pakistan Academy of Sciences in 2000, and was elected a Fellow of the Third World Academy of Sciences (TWAS), Italy, in 2003.

Phil Harris, PhD, is Professor of Plant Science in the School of Science and the Environment at Coventry University in the United Kingdom. He received his PhD in botany from the University of Glasgow. For the past 28 years, his research interests have included plant physiology and propagation related to environmental stress as well as research applied to tropical crop development, organic and sustainable agriculture, forestry, and agroforestry. He has published over 140 scientific papers and books. Professor Harris has extensive agricultural consultancy experience in Africa, Asia, and Latin America. He has acted as an advisor to the UK Department for International Development and several other international agencies including the International Foundation for Science in Stockholm. He is editor of the international journal *Biological Agriculture & Horticulture*. He was elected a Fellow of the Institute of Biology in 1995.

CONTRIBUTORS

Thomas Blake, PhD, is director of the Germplasm Program, International Center for Agricultural Research in the Dry Areas (ICARDA), Aleppo, Syria.

Daryl T. Bowman, is cotton breeder, Crop Science Department, North Carolina State University, Raleigh, North Carolina.

Ana Butrón, is maize breeder, Misión Biológica de Galacia, Spanish Council for Scientific Research, Pontevedra, Spain.

Gloria S. Cabuslay, PhD, is postdoctoral fellow (plant physiologist), Plant Breeding, Genetics, and Biochemistry Division, International Rice Research Institute, Los Banos, Philippines.

M. Elena Cartea, is *Brassica* crops breeder, Misión Biológica de Galacia, Spanish Council for Scientific Research, Pontevedra, Spain.

Viswanathan Chinnusamy, PhD, is scientist (plant physiology), Stress Physiology Laboratory, Water Technology Centre, Indian Agricultural Research Institute, New Delhi, India.

David A. Dierig, PhD, is research plant geneticist, USDA, ARS, U.S. Water Conservation Laboratory, Phoenix, Arizona.

Wilfried H. O. Ernst, PhD, is Emeritus Professor of plant science, Department of Ecology and Physiology of Plants, Faculty of Earth and Life Science, Vrije Universiteit, Amsterdam, The Netherlands.

Majid R. Foolad, PhD, is Professor of plant genetics, Department of Horticulture and Intercollege Graduate Degree Program in Genetics, Pennsylvania State University, Pennsylvania.

Glen B. Gregorio, PhD, is international research fellow (plant breeder), Plant Breeding, Genetics, and Biochemistry Division, International Rice Research Institute, Los Banos, Philippines.

Catherine J. Howarth, PhD, is research scientist, Plant Genetics and Breeding Department, Institute of Grassland and Environmental Research, Aberystwyth, Wales, United Kingdom.

Bernhard Huchzermeyer, PhD, Botany Institute, Hanover University, Hanover, Germany.

Mervyn O. Humphreys, PhD, is head of plant genetics and breeding, Institute of Grassland and Environmental Research, Aberystwyth, Wales, United Kingdom.

Michael W. Humphreys, PhD, is project leader and European project coordinator, Plant Genetics and Breeding Department, Institute of Grassland and Environmental Research, Aberystwyth, Wales, United Kingdom.

Sakti Jana, PhD, is population geneticist and Professor of plant sciences, University of Saskatchewan, Saskatoon, Saskatchewan, Canada.

Hans-Werner Koyro, PhD, is Assistant Professor of botany and plant ecology, Institute for Plant Ecology, Justus-Liebig-University, Giessen, Germany.

D. Kumar, PhD, is principal scientist and project coordinator (AICRP on arid legumes), Central Arid Zone Research Institute, Jodhpur, India.

Rajinder S. Malhotra, PhD, is principal chickpea breeder, International Center for Agricultural Research in the Dry Areas (ICARDA), Aleppo, Syria.

Rosa Ana Malvar, is maize breeder, Misión Biológica de Galacia, Spanish Council for Scientific Research, Pontevedra, Spain.

Amando Ordás, is maize and *Brassica* crops breeder, Misión Biológica de Galacia, Spanish Council for Scientific Research, Pontevedra, Spain.

Ivan Ortiz-Monasterio, PhD, is senior scientist, agronomist, and project coordinator, Wheat Program, International Maize and Wheat Improvement Center (CIMMYT), Ciudad Obregon, Sonora, Mexico.

Andrew H. Paterson, PhD, is director, Plant Genome Mapping Laboratory, University of Georgia, Georgia.

Wolfgang H. Pfeiffer, PhD, is principal scientist, durum wheat breeder, and global project coordinator, Wheat Program, International Maize and Wheat Improvement Center (CIMMYT), El Batán, Mexico.

Sanjaya Rajaram, PhD, is distinguished scientist and director of the Wheat Program, International Maize and Wheat Improvement Center (CIMMYT), El Batán, Mexico.

Zed Rengel, PhD, is professorial fellow, soil science and plant nutrition, University of Western Australia, Crawley, Australia.

Pedro Revilla, is maize breeder, Misión Biológica de Galacia, Spanish Council for Scientific Research, Pontevedra, Spain.

Theerayuth Toojinda, PhD, is rice molecular breeder, Rice Gene Discovery, National Center for Genetic Engineering and Biotechnology (BIOTEC), Kasetsart University, Thailand.

Somvong Tragoonrung, PhD, is molecular geneticist, DNA Technology Laboratory, National Science and Technology Development Agency (NSTDA), National Center for Genetic Engineering and Biotechnology (BIOTEC), Kasetsart University, Thailand.

Richard M. Trethowan, PhD, is senior scientist and spring bread wheat breeder (marginal environments), Wheat Program, International Maize and Wheat Improvement Center (CIMMYT), El Batán, Mexico.

Shafiq-ur-Rehman, PhD, is postdoctoral fellow, Faculty of Bioresources Science, College of Agriculture, Chonbuk National University, Chonju, South Korea.

Apichart Vanavichit, PhD, is Associate Professor in rice genomics, Department of Agronomy, Kasetsart University and Rice Gene Discovery, Thailand.

Maarten van Ginkel, PhD, is principal scientist, spring bread wheat breeder (optimal environments) and global project coordinator, Wheat Program, International Maize and Wheat Improvement Center (CIMMYT), El Batán, Mexico.

Nguyen Thi Vinh, PhD, is researcher, Department of Genetics and Hybrid Rice Technology, Institute of Agricultural Genetics, Ministry of Agriculture and Rural Development, Hanoi, Vietnam.

Randy Wells, is cotton physiologist, Crop Science Department, North Carolina State University, Raleigh, North Carolina.

Ron W. Wilen, PhD, is oil and cereal biotechnologist, Department of Sciences and College of Biotechnology, University of Saskatchewan, Saskatoon, Saskatchewan, Canada.

Liming Xiong, PhD, is Professor, Donald Danforth Plant Science Center, St. Louis, Missouri.

Jian-Kang Zhu, PhD, is Professor, Department of Plant Sciences, University of Arizona, Tucson, Arizona.

Preface

The human population is expected to increase to over 8 billion by the year 2020, generating considerable concern that current food inequalities will worsen over the next 20 years. It is predicted that in developing countries, two-thirds of the required increase in food output will come from rising crop yields; the rest will be achieved through expansion of the arable area into marginal and degraded lands and from increased cropping intensity.

Some abiotic stresses, such as drought, flooding, salinity, mineral deficiency, and adverse pH, are soil-based and amenable to physical improvement by irrigation, drainage, and chemical amelioration. However, physical approaches can be expensive and often do not provide permanent solutions. Other stresses, principally high and low temperature, are more difficult to counter by physical means and demand a biological approach. Plant breeding is an attractive approach to the problem of abiotic stress because it is seen as an efficient and economic means to increase crop yield. It also holds the promise of increasing overall productivity by extending potential growing seasons, allowing marginal land to be brought into cultivation, and allowing more productive crops to be chosen for difficult environments.

The model upon which such an approach has been based derives from the extensive experimental evidence of the natural evolution of populations of many plant species that are highly tolerant of environmental stresses. Over the centuries, and especially in the past fifty years, plant breeding has proved highly effective in selecting for desirable agronomic traits, and for pest and disease resistance in crop species, through the exploitation of natural variability. Using conventional techniques, plant breeders also have been successful in developing cultivars with at least some tolerance to a number of abiotic stresses. Since the 1990s, molecular sciences have been used to investigate the fundamental structure and actions of plant genomes and to begin to assemble an understanding of the genetic and biochemical bases of abiotic stress tolerance. These recent studies form an excellent basis for modern breeding approaches and advancement in producing stress-tolerant crop varieties may now accelerate more rapidly than previously thought possible.

Information in this book is presented in two main sections. Part I presents a general overview of the subject and the main techniques available to the plant breeder. There are general chapters on the principles of breeding for stress resistance and on molecular approaches to breeding. Detailed coverage is provided of current and potential progress in breeding for resistance to the main abiotic stresses of salinity, drought, flooding, soil metal content, low soil nutrient status, and high and low temperatures. Part II contains chapters by acknowledged experts covering work done in relation to breeding for different stresses in many of the most important world food crops.

It is no exaggeration to say that we are at the threshold of major new developments in plant breeding for abiotic stress tolerance. This book sets out to present and interpret the knowledge that has brought us to this point. In celebrating the achievements to date and considering the future prospects of conventional and molecular approaches to breeding for stress tolerance, this book is also intended to inspire the next generation of plant breeders.

PART I: GENERAL OVERVIEW

Chapter 1

Stress Environments and Their Impact on Crop Production

Shafiq-ur-Rehman
P. J. C. Harris
M. Ashraf

INTRODUCTION

In common with all living organisms, plants acquire the materials they need for energy and growth from their surroundings. In the case of plants, this acquisition usually has to be made against a concentration gradient. For land plants, tissue water must be retained against gradients of atmospheric and soil water potential, which can vary from small in humid tropical forests to very large in hot, dry deserts. The concentrations of many essential ions in plants are often a thousand-fold greater than those in the soil solution from which their roots extract them. Since plants invest considerable amounts of metabolic energy to acquire and retain external resources in the face of usually limited supply, it could be said that stress is a normal condition of plants even in relatively benign environments.

Stress has a strictly defined physical science definition describing the force per unit area acting upon a material, inducing strain and leading to dimensional change. More generally, it is used to describe the impact of adverse forces, and this is how it is usually applied to biological systems. In the widest biological sense, stress can be any factor that may produce an adverse effect in individual organisms, populations, or communities. Biologically, stress also has been defined as the overpowering pressure that affects the normal functions of individual life or the conditions in which plants are prevented from fully expressing their genetic potential for growth, development, and reproduction (Levitt, 1980; Ernst, 1993). In the agricultural context, stress has been defined as a phenomenon that limits crop productivity or destroys biomass (Grime, 1979).

Abiotic stress management is one of the most important challenges facing agriculture. Clear evidence of increasing stress problems in many parts of the world exists, with the worst-affected areas often in the poorest countries. Abiotic stress can persistently limit choice of crops and agricultural production over large areas, and extreme events can lead to total crop failures. Abiotic stresses adversely affect the livelihoods of individual farmers and their families as well as national economies and food security.

BIOTIC AND ABIOTIC STRESSES

It has become traditional for ecologists, physiologists, and agronomists to divide stresses experienced by plants into two major categories: biotic and abiotic. Biotic stresses originate through interactions between organisms, while abiotic stresses are those that depend on the interaction between organisms and the physical environment. Strictly speaking, biotic stresses result from competition between organisms for resources, from predation and parasitism, and from the actions of allelopathic chemicals released by one organism and affecting another (Fitter and Hay, 1987). Abiotic stresses include potentially adverse effects of salinity, drought, flooding, metal toxicity, nutrient deficiency, high temperature, and low temperature, all elements covered in detail in this book. In addition, abiotic stresses can include shade, UV exposure, photoinhibition, air pollution, wind, hail, and gaseous deficiency, factors that are not covered in detail in this book and which are often sporadic and highly localized in occurrence.

Clearly, from these definitions, plants can experience abiotic stress resulting from the shortage of an essential resource or from the presence of high concentrations of a toxic or antagonistic substance. In some cases, such as the supply of water, too little (drought) or too much (flooding) can both impose stress on plants. In reality, abiotic and biotic stresses are often inextricably linked. Thus, shade is an abiotic stress, but is most important when imposed by the competing canopy of other plants. The competition between plants can also lead to stresses through depletion of mineral or water resources. Conversely, the presence of abiotic stress can predispose plants to biotic stress, making them more susceptible, though in some cases less susceptible, to competition, predation, and parasitism. However, this book is primarily concerned with abiotic stresses that arise from the inherent physical nature of the environment in which plants, and crops in particular, are grown, rather than from abiotic stresses generated by biotic processes.

MULTIPLE AND VARIABLE STRESSES AND TOLERANCE

Stressful environments are often characterized by the simultaneous or sequential occurrence of more than one stress. For example, salinity is often associated with drought or with waterlogging, and drought is often associated with high temperature. Perennial crops in higher latitudes may experience summer droughts and winter cold. The occurrence, nature, and intensity of stress can vary with time, whether from hour to hour, day to day, or season to season. The tolerance of plants to stress has been widely shown to vary with physiological growth stage, developmental phase, and size of plants. There is also growing evidence of multiple tolerances to stress in plants, with plants showing tolerance to more than one stress. All of these factors complicate the management of abiotic stress in an agricultural situation and make the jobs of physiologists, molecular biologists, and plant breeders more difficult as they attempt to understand the genetic and physiological bases of stress response and stress tolerance as part of crop improvement programs.

ABIOTIC STRESS SYMPTOMS

With a wide range of abiotic stresses able to adversely affect plants through a multitude of mechanisms, there are no universal symptoms of stress and it is sometimes, as with certain mineral deficiencies, possible to use the visible symptoms of stress to discern the cause. The symptoms of stress, of course, vary with its severity, from the subtle to the catastrophic. Taking a wide ecological perspective, the symptoms of stress have previously been grouped into the following categories:

1. invisible but measurable injuries, e.g., change in biomembranes or enzyme activities;
2. visible injury of plant parts, e.g., necrosis, chlorosis, and discoloration;
3. retarded growth, development, and reproduction;
4. changes in the interference caused by heterotrophs;
5. changes in the genotypes of which populations are composed; and
6. change in species composition of an ecosystem (Ernst, 1993; Lerner, 1999).

From the point of view of crop productivity in managed agricultural ecosystems it is the occurrence of categories 1 and 2, with their resulting impact on category 3, that are the symptoms of immediate economic importance.

MAJOR ABIOTIC STRESSES LIMITING CROP YIELD

Salinity

"All plants can be killed by salt if the concentration is high enough. The ability of plants to survive salinity stress is important for natural distribution of plant species and to agriculture" (Flowers and Yeo, 1989). Generally, salinity is defined as the presence of excessive amounts of soluble salts that hinder or affect the normal functions of plant growth. It is measured in terms of electrical conductivity (ECe), with the exchangeable sodium percentage (ESP) or sodium adsorption ratio (SAR) and pH of a saturated soil paste extract. Therefore, saline soils are those that have saturated soil paste extracts with an ECe of more than 4 $d{\cdot}Sm^{-1}$, ESP less than 15 percent, and pH below 8.5 (Waisel, 1972; Abrol, 1986; Szabolcs, 1994). Saline soils have a mixture of salts of chloride, sulfate, sodium, magnesium, and calcium ions, with sodium chloride often dominant.

There are two main sources of salinity: primary or natural sources resulting from weathering of minerals and the soils developed/derived from saline parent rocks (Sposito, 1989; Ashraf, 1994), and secondary salinization caused by human factors such as irrigation, deforestation, overgrazing, or intensive cropping (Sposito, 1989; Ashraf, 1994).

Salinization affects the chemical properties of soil by changing the cation exchange capacity (CEC) and altering the physical properties. Soil structure is damaged by defloculation of clay particles and hydraulic conductivity is decreased, resulting in a slow movement of irrigation water. Soil salinity also affects the soil microflora that play important roles in the improvement of soil structure, the decomposition of organic matter, and the nitrogen and sulfur cycles (Waisel, 1972; Lal and Khanna, 1994).

One of the most common forms of land degradation results from soil salinization. Almost all of the continents have saline soils (Pessarakli, 1999). However, salinity is predominantly a problem of arid and semiarid regions of the world, where the potential for evapotranspiration exceeds rainfall and there is insufficient rain to leach away soluble salts from the root zone (Abrol, 1986; Miller and Donahue, 1990). The impact of salinity on the economic exploitation of land for agriculture and forestry is very severe (Singh and Singh, 1995).

According to an early Food and Agriculture Organization (FAO) study (Massoud, 1974), salt-affected soils occupy 7 percent of the world land area, and salinity is also a problem that is increasing rapidly (Pessarakli, 1999), having more than doubled in the past two decades. It is estimated that the world is losing at least 10 ha of arable land every minute and 3 ha of this is lost by salinization. In this way, 1.6 Mha of arable land is lost every year (Ghassemi et al., 1995). Secondary salinization is widely considered a key process leading to degradation and desertification of the world's dry lands (Thomas and Middleton, 1993). Ironically, much of the loss of land due to salinization is in fact caused by agriculture itself through the adverse effects of irrigation, this occurring in up to 20 percent of irrigated land (Flowers and Yeo, 1995).

Drought

Water is essential for plant growth and makes up 75 to 95 percent of plant tissue. A vast amount of water moves throughout the plant daily. Plants use water and carbon dioxide to form sugars and more complex carbohydrates, and water also acts as a carrier of nutrients and a cooling agent, and provides an element of support through turgor and as an intercellular reaction medium.

Many attempts to define drought objectively have been made, with the result that different definitions exist depending on the perspectives of the user. Most generally, drought can be defined as an extended period of deficient rainfall relative to the statistical mean for a region. Drought can also be defined according to meteorological, hydrological, or agricultural criteria. Meteorological drought is qualified by any significant deficit of precipitation, hydrological drought is manifest in noticeably reduced river and stream flow and critically low groundwater tables, and agricultural drought indicates an extended dry period that results in crop stress and harvest reduction. The impact of drought on agriculture is due to a deficit of moisture in the soil, when the moisture in the soil is no longer sufficient to meet the needs of growing crops. This results from a lack of input of moisture from rainfall or irrigation for an extended period.

It is impossible to specify a period of time without rain as an agricultural drought, as the soil moisture deficit depends on rate of loss as well as rate of input. Furthermore, the severity of stress imposed on crops also depends on the susceptibility of different crops during different stages of their development. When soil moisture is lacking, crop establishment may be reduced, growth limited, normal development patterns disrupted, and, eventually, final yields lowered.

Among the environmental stress factors, one of the most widely limiting for crop production on a global basis is water. According to one estimate, around 28 percent of the world's land is too dry to support vegetation (Kramer and Boyer, 1995). On a global basis, water is a paramount factor in determining the distribution of species, and the responses and adaptation of species to water stress are critical for their success in any environmental niche and for their use and productivity in agricultural ecosystems. Severe droughts occur periodically in several major food-producing countries, having far-reaching impacts on global food production and supply. The global production of grain has, in some years, been reduced by 5 percent or more as a result of severe droughts in key countries (Wisner and Chase, 1984). It has been estimated that drought causes an average annual yield loss of 17 percent in the tropics (Edmeades et al., 1992), but losses can be much more severe and total crop failures are not unknown.

Flooding

At its simplest, flooding may be defined as any situation of excess water. Flooding can be catastrophic, with flash floods causing major soil erosion and direct physical destruction of crops. Sudden inundation following high rainfall events also poses a severe physiological stress on crops. The gradual inundation of crop lands that occurs in a more regular cycle of seasonal changes in river levels and associated gradual flooding of crop lands poses a different, but equally challenging, flooding environment to which plants must adapt. As a result, some plants, such as rice, evolve a semiaquatic habit.

Flooding stress in terrestrial species is referred to as waterlogging and the damage symptoms caused are primarily due to the prolonged exposure of the plants to hypoxia. The effect of waterlogging of roots and lower stems are apparent as a range of symptoms on the shoots, including rapid wilting and severe physiological disruption.

Vast areas of rainfed crops, particularly in South and Southeast Asia, are annually affected by flooding. In Southeast Asia alone, about 15 percent of the total maize growing area is affected by flooding (Chapter 7, this volume).

Metal Toxicity

The adverse effects of metals on plants is closely linked to soil acidity and the effect that the latter parameter has on metal form and availability. For example, aluminum is highly soluble at pH values <5 and this can cause severe aluminum toxicity in acid soils (Kochian, 1995). Soil acidity in some areas is actually increasing as a result of industrial release of sulfur dioxide, ammonium, and other soil-acidifying compounds. This has exacerbated the problems of metal toxicity in many areas, with effects spreading widely from the point sources of pollution.

Some soils naturally contain high levels of heavy metals (Ernst, 1974) to the point that this may be the dominant feature determining the natural flora of an area. In addition, a number of human activities have resulted in increasing levels of metal contamination of land, including prime agricultural land. Such activities include the use of metal-based pesticides, lead-containing fuels, and metal-contaminated organic and inorganic fertilizers.

Metal toxicity in soils is increasingly becoming a major problem in some areas and is often associated with high soil acidity. This is most serious in the tropics, where 1.6 billion ha of land are acidic, including more than 50 percent of soils in tropical America and nearly 40 percent of land in tropical Africa (Sanchez and Salinas, 1981).

Mineral Nutrient Deficiency

Stress resulting from mineral nutrient deficiency occurs when the availability of soil nutrients and/or the ability of the crop to acquire nutrients is below that required for the optimum growth and activity of the plant. As with many other stresses, the requirement for mineral nutrients varies with species and growth stage.

Mineral nutrient deficiency may result from a low nutrient status of the soil, but also from low mobility or solubility of nutrients within the soil, often under the control of some other chemical or physical property of soil such as water content, adsorption capacity, or pH. There are also very close links between the supply of essential mineral elements and the activity of soil microorganisms, either directly through symbiotic associations or indirectly through the mineralizing role of free living microorganisms.

With 14 mineral elements essential for all plants, any one of these can theoretically limit plant growth and hence impose stress. The form of a particular element may also affect its availability and value to a plant. In the case of some mineral nutrients, such as P and K, deficiency is more often due to low availability rather than to low total levels in the soil, whereas total N is often limiting. The essential mineral elements have a multitude of specific roles in plant structure and function and mineral stress often manifests itself with symptoms characteristic of particular deficient elements.

Many soils are inherently low in one or more essential plant nutrients to the extent that crop growth is severely limited. Dudal (1976) estimated that about 20 percent of arable land is under some kind of mineral stress. The objective of crop cultivation is to maximize yield and the availability of mineral nutrients rarely if ever match the levels required for optimum plant production. In the sense that stress may be defined as the general under-performance of a crop, deficiency in one or more mineral nutrients is likely to be the norm rather than the exception in unamended soils.

High Temperature

Levitt (1980) classified plants into psychrophiles, mesophiles, and thermophiles according to whether or not they tolerate low, medium, or high temperatures. Psychrophiles are those plants whose high temperature threshold is 15 to 20°C, mesophiles are those plants whose high temperature threshold is 35 to 45°C, and thermophiles are those plants whose high temperature threshold ranges from 45 to 100°C. Levitt (1980) proposed that the high temperature injury process progresses from a direct reversible strain, i.e., excess respiration over photosynthesis due to elevated temperatures, to an indirect strain, i.e., loss of reserves, or to a direct or indirect injury, i.e., starvation injury.

High temperatures may be experienced by plants on a daily or seasonal basis. There is also growing evidence of long-term climatic changes leading to both higher average temperatures, widening the geographic range where high temperatures become routinely limiting to crop production, and increasing the frequency and severity of extreme temperature events. As with most other stresses, the growth stage of the crop is a major factor determining the extent of damage caused by exposure, with early and late stages of the crop growth generally particularly sensitive to temperature extremes. Furthermore, plants may be as affected by exposure to prolonged periods of moderately high temperature as to short periods of extreme temperature, though the mechanisms for coping with these stresses may differ. Heat stress affects grain quality and yield in 40 percent of the irrigated wheat-growing areas of the world (Fischer and Byerlee, 1991).

Low Temperature

As an ecological factor, low temperature exerts an extremely important influence on the global distribution of organisms and is no less important in determining the suitability and performance of agricultural crops. Decreasing temperature is also a feature of height above sea level. Therefore, temperature acts as a major factor influencing the distribution of natural vegetation and the performance of agricultural crops by both altitude and latitude. The lowland tropics are the only regions where low temperatures rarely influence the choice, growth, and yield of crops. Humidity, air movement, evaporation, and intensity of the ultraviolet and visible radiation of the sun can all affect cold resistance in plants. Low-temperature exposure can be both a daily and seasonal factor to which plants must adapt, including, in some regions, prolonged periods of low temperature lasting many months. Although temperature extremes are common and unpredictable, the diurnal and seasonal pattern of temperature change is sufficiently settled for plant adaptation to have occurred. Selection and management of crops within the annual seasonal temperature cycle has been refined over a long time, as settled agriculture has developed.

Greaves (1996, p. 308) defines suboptimal temperature stress as

> any reduction in growth or induced metabolic, cellular or tissue injury that results in limitations to the genetically determined yield potential, caused as a direct result of exposure to temperatures below the thermal thresholds for optimal biochemical and physiological activity or morphological development.

Low temperatures can damage plants both by a chilling effect, leading to physiological and developmental abnormalities, and by freezing, causing cellular damage directly or via cellular dehydration. Lyons (1973) described many symptoms of low-temperature injury. Some physiological processes such as flowering in rice are extremely sensitive to low temperatures and damage may occur at temperatures as high as 20°C. Commonly visible symptoms of low-temperature injury to the leaves include wilting, bleaching due to photooxidation of pigments, waterlogging of the intercellular spaces, browning, and eventually leaf necrosis and plant death (Levitt, 1980; Witt and Barfield, 1982).

Dudal (1976) estimated that 15 percent of arable land is affected by freezing stress. Low temperatures can reduce crop yields in several ways. Chilling and freezing injury can directly affect crop growth by causing physical damage or by interfering with normal biochemical and physiological functions, thus reducing yield. More subtly, low temperatures reduce potential agricultural productivity by limiting the crops or varieties that can be grown in a particular area, with cold-tolerant species/cultivars often not those with the highest potential yield.

CROP PRODUCTION IN STRESSFUL CONDITIONS

Salinity

Amelioration and utilization of salt-affected soils have been the focus of research for over half a century. Problems of salinity may be addressed by using technical (water and soil management) and/or biological approaches.

Reclamation of saline areas by physical or chemical means is not only expensive but may also itself raise environmental concerns. Current practices used for reclamation of saline soils often ignore the natural environment and may even result in further ecological deterioration. Therefore, traditional engineering approaches to salinity problems may no longer be adequate or sustainable (Epstein et al., 1980).

There is a need for a biological approach, and the first step in this direction is to screen and develop different plant species for their salt tolerance and ability to grow in saline areas. Biological approaches to coping with saline soils include both the use of plants to ameliorate the effects of salt and the selection and breeding of crops to tolerate prevailing conditions. Biological amelioration of soils represents an economic alternative to other costly and extensive land-reclamation methods, especially in third-world countries. Farrington and Salama (1996) examined tree and shrub plantations on saline soils and suggested them to be one of the most favored and best long-term options for controlling salinity. Chhabra (1996) summarized the beneficial effects of vegetation on saline soils, including the effect of trees and grasses in lowering the salt content of soil by absorbing and thus biologically harvesting the salts, preventing erosion, and conserving rainwater by increasing the infiltration rate. Physiochemical and biological properties of the soil are improved by the increase in organic matter content of the soil from leaf litter; salt accumulation can be shifted from the soil surface to lower layers by the removal of water from lower layers, thus minimizing salt injury to shallow-rooted crops.

There is also no doubt that breeding and selection of salt-tolerant species, explored in detail in this book, will have immense significance in bringing saline areas under cultivation/utilization. Thus, there is a need for physiologists and biochemists to pinpoint the critical mechanisms and responses of plants to saline conditions.

Drought and Flooding

The rate of growth of plants is generally in direct proportion to the available water, although too much water may also be harmful. Therefore, under most plant growing conditions, an increase in the amount of available soil water is desirable. This can be achieved by increasing total water input to the system by irrigation, by increasing the effective use of the water that is available as precipitation or through irrigation, and by minimizing the losses of water from the system. From a technical point of view, contour plowing, contour planting, strip cropping, crop rotation, terracing, conservation tillage, and cover crops can all reduce runoff of water and increase infiltration. Even the act of plowing will break the crust sealing the surface of soil and improve water entry and water storage of soil. Water retention can be improved by incorporating crop residues or green or animal manures into the soil, or by mulching the soil with crop residues, manure, hay, straw, or artificial plastic mulches.

In a sense, the problems of flooding and consequent waterlogging are the converse of drought. There is considerable scope in some terrestrial situations for reducing the incidence of flooding though engineering works to maintain water flow in rivers, direct flow away from cropland, and to provide flood barriers and defenses. Similarly, drainage schemes at both landscape and field scales can reduce the localized flooding impact of excess rainfall. There are, however, limits to the extent to which fluctuations in water levels in rivers and flood plains can be managed. The huge seasonal variation in water availability in some areas is too large to be controlled, affects land that is too valuable to be ignored, and can even add nutrients to seasonally flooded areas. The alternative to technical approaches and, in many cases, an addition to these approaches is the selection of appropriate drought or flooding-tolerant crop species and the development through selection and breeding of drought- or flooding-tolerant varieties.

Metal Toxicity

Heavy metals are stable elements and relatively immobile in soil, so crop production problems caused by the presence of excess metal in soils are difficult to correct directly. They may, however, be ameliorated by modification of the soil pH. Even this can be difficult and expensive as a recurrent cost. There is thus considerable interest in biological approaches to solving these problems, both by the use of bioaccumulating plants to remove excess metals and in the breeding of metal-tolerant crop varieties, in many cases by exploiting the natural tolerance that is present in wild relatives of existing crops.

Mineral Nutrient Deficiency

Various strategies can be adopted to overcome mineral nutrient deficiencies in agricultural soils and traditionally have been of two main types: increasing the total amount and improving availability of nutrients. Increases in the amount of essential mineral nutrients are achieved by the addition of organic and inorganic fertilizers and by agronomic practices that minimize loss by leaching or volatilization. Improved availability of mineral nutrients is achieved by amelioration of soil properties that limit supply, such as pH or soil microbial biomass activity. Despite the fact that these two approaches, often employed together, are routine and almost universal in modern agriculture, they place a high demand on resources, are expensive, and can lead to environmentally damaging side effects of pollution. Increasing the level of inputs to supply more nutrients or improve the physical or biological properties of soils is also currently beyond the financial capability of many farmers in poorer countries. The alternative approach is to cultivate crops and select and breed cultivars that are more efficient at obtaining adequate supplies of nutrients from soils with relatively low "available" reserves.

High Temperature

The opportunities for significantly altering the ambient temperatures of field-crop-growing areas are limited. The selection of crops adapted to high temperatures and the breeding of tolerant crops is thus the best option for dealing with the direct stress effects of high temperature. However, high temperature and water stress are often inextricably linked with water loss. The latter is both a potential means of cooling plants exposed to high temperatures and the cause of secondary stress when it increases excessively with increasing temperature or humidity gradients between the leaf and the atmosphere. Thus, one means of reducing high-temperature stress is to increase the availability of soil moisture or reduce the evaporative demand on the plants. Increasing the availability of moisture is the process most commonly used. Irrigation, timing of planting, and the use of mulches are all techniques that accomplish this. Evaporative demand can be reduced by cooling the air, using sprinkler wetting of crop surfaces, providing crop covers, and using wind breaks.

Low Temperature

A number of crop management and environmental modification practices are available to eliminate or alleviate low-temperature stress. Much effort by researchers and growers has gone into developing methods of protecting valuable crops. Climatic modifications on a micro scale include heating, overhead irrigation, and wind machines to modify the climate by additional energy input, making effective use of heat in the system. Unfortunately, many of these approaches are by definition energy demanding and thus expensive and are often inappropriate for all but high-value protected cropping. Low-temperature tolerance has, therefore, long been a target for crop selection and breeding.

FUTURE PROSPECTS

Over the centuries, and especially in the past fifty years, plant breeding has proved highly effective in selecting for desirable agronomic traits and for pest and disease resistance in crop species through the exploitation of natural variability. Using conventional techniques, plant breeders have also been able to successfully develop cultivars with at least some tolerance of a number of abiotic stresses. The source of tolerance used in conventional breeding has been the variation that exists among present varieties and land races. This has often resulted from long-term farmer selection or from wild relatives of crop plants that have evolved abiotic stress tolerance as a means to allow colonization of marginal and extreme habitats. There have, however, been limits to the extent and rate of progress of conventional breeding for abiotic stress tolerance. Many desirable traits are controlled by several genes, the techniques employed are time consuming and consequently expensive, and conventional breeding possibilities are limited by reproductive barriers between more distantly related plant groups.

Over the past decade or so, molecular sciences have been used to investigate the fundamental structure and actions of plant genomes and to begin to assemble an understanding of the genetic and biochemical basis of abiotic stress tolerance. Probably the most conclusive, if not the most welcome or unexpected, message coming out of recent work is that the genetics and molecular biology of stress tolerance are complex. Multiple genes may control a particular response, one gene may be involved in more than one response, and there is a complex interaction between genes, all adding to the complexity. Nevertheless, recent studies, many reported in this book, form an excellent basis for modern breeding approaches and advancement in producing stress-tolerant crop varieties and may now accelerate more rapidly than previously thought possible. Not least has been the development of new approaches to plant breeding using marker technologies, which are described in detail in several chapters of this book and which offer the prospect of revolutionizing the often laborious process of selection.

The advent of plant biotechnology has also provided a range of techniques to increase the genetic variation, including for stress tolerance, available to plant breeders. Somaclonal variation generated by regeneration from cell and tissue cultures and the generation of somatic hybrids by protoplast fusion can provide material for breeding programs, and techniques of haploid production and selection can provide shortcuts to the exploitation of existing variation. The laboratory production of novel genetic combinations using plant transformation techniques is now routine for most important crops and offers a future route to the production of stress-tolerant crop cultivars. Among the early targets for the genetic engineering approach to abiotic stress tolerance are a number of fundamental cellular properties and processes that give tolerance of a variety of stresses. Thus, engineering plants for accumulation of compatible solutes, induction of stress proteins, and improvements in membrane stability show some promise as generic approaches to improving stress tolerance.

The future production of stress-tolerant varieties by genetic modification will no doubt benefit from the detailed understanding of the molecular and genetic bases of plant stress responses, and plant transformation, in turn, is proving a powerful tool in explaining gene function. However, genetic modification for stress tolerance will remain hampered by the obvious complexity of the systems.

In the long term, as techniques are developed and refined, it is very likely that breeding strategies will focus more specifically on crops adapted to specific climates. As discussed previously, most of the abiotic stresses discussed in this book are not single, uniform stresses but vary in cause, severity, and duration. Although we may often take a rather simplistic view of stress tolerance in crops, it is clear that the characteristics, and hence varieties, required for crops resistant, for example, to prolonged periods of drought may be significantly different from those required for crops exposed to daily, growth-limiting water deficits.

REFERENCES

Abrol, I.P. (1986). Salt-affected soils: An overview. In *Approaches for Incorporating Drought and Salinity Resistance in Crop Plants,* V.L. Chopra and S.L. Paroda (Eds.). New Delhi, India: Oxford and IBH Publishing Company, pp. 1-23.

Ashraf, M. (1994). Breeding for salinity tolerance in plants. *Critical Reviews in Plant Sciences* 13:17-42.

Chhabra, R. (1996). *Soil Salinity and Water Quality*. Rotterdam, The Netherlands: Swets and Zeitlinger Publishers.

Dudal, R. (1976). Inventory of the major soils of the world with special reference to mineral stress hazards. In *Plant Adaptation to Mineral Stress in Problem Soils,* M.J. Wright (Ed.). Ithaca, NY: Cornell University Agricultural Experiment Station, pp. 3-14.

Edmeades, G.O., J. Bolanos, and H.R. Lafitte (1992). Progress in breeding for drought tolerance in maize. In *Proceedings of 47th Annual Corn and Sorghum Industry Research Conference.* Washington, DC: American Seed Trade Association, pp. 93-111.

Epstein, S., J.D Norlyn, D.W. Rush, R.W. Kingsbury, D.B. Kelley, G.A. Cunningham, and A.F. Wrona (1980). Saline culture of crops: A genetic approach. *Science* 210:399-404.

Ernst, W.H.O. (1974). *Heavy Metal Vegetation of the World* (in German). Stuttgart, Germany: G. Fischer.

Ernst, W.H.O. (1993). Population dynamics, evolution and environment: Adaptation to environmental stress. In *Plant Adaptation to Environmental Stress,* L. Fowden, T. Mansfield, and J. Stoddart (Eds.). London, UK: Chapman and Hall Publishing, pp. 19-44.

Farrington, P. and R.B. Salama (1996). Controlling dryland salinity by planting trees in the best hydrological setting. *Land Degradation and Development* 7:183-204.

Fischer, R.A. and D.R. Byerlee (1991). Trends of wheat production in the warmer areas: major issues and economic considerations. In *Wheat for Nontraditional, Warm Areas,* D.A. Saunders (Ed.). Mexico, DF: CIMMYT, pp. 3-27.

Fitter, A.H. and R.K.M. Hay (1987). *Environmental Physiology of Plants,* Second Edition. New York: Academic Press.

Flowers, T.J. and A.R. Yeo (1989). Effect of salinity on plant growth and crop yields. In *Environmental Stress in Plants: Biochemical and Physiological Mechanisms,* J.H. Cherry (Ed.). Berlin, Germany: Springer-Verlag, pp. 101-119.

Flowers, T.J. and A.R. Yeo (1995). Breeding for salinity resistance in crop plants. Where next? *Australian Journal of Plant Physiology* 22:875-884.

Ghassemi, F., A.J. Jakeman, and H.A. Nix (1995). *Salinization of Lands and Water Resources: Human Causes, Extent, Management and Case Studies.* Sydney, Australia: UNSW Press and Wallingford, UK: CAB International.

Greaves, J.A. (1996). Improving suboptimal temperature tolerance in maize: The search for variation. *Journal of Experimental Botany* 47:307-323.

Grime, J.P. (1979). *Plant Strategies and Vegetation Processes.* New York: John Wiley.

Kochian, L.V. (1995). Cellular mechanisms of aluminum toxicity and resistance in plants. *Annual Review of Plant Physiology and Plant Molecular Biology* 46:237-260.

Kramer, P.J. and J.S. Boyer (1995). *Water Relations of Plants and Soils.* San Diego, CA: Academic Press.

Lal, B. and S. Khanna (1994). Selection of salt-tolerant rhizobium isolates of *Acacia nilotica. World Journal of Microbiology and Biotechnology* 10:637-639.

Lerner, H.R. (1999). Introduction to the response of plants to environmental stress. In *Plant Responses to Environmental Stress,* H.R. Lerner (Ed.). New York: Marcel Dekker, pp. 1-27.

Levitt, J. (1980). *Responses of Plants to Environmental Stress.* New York: Academic Press.

Lyons, J.M. (1973). Chilling injury in plants. *Annual Review of Plant Physiology* 24:445-566.

Massoud, F.I. (1974). *Salinity and Alkalinity As Soil Degradation Hazards.* Rome, Italy: FAO-UNESCO Publications.

Miller, R.W. and R.L. Donahue (1990). *Soils: An Introduction to Soils and Plant Growth,* Sixth Edition. Englewood Cliffs, NJ: Prentice-Hall.

Pessarakli, M. (1999). *Handbook of Plant and Crop Stress.* New York: Marcel Dekker.

Sanchez, P.A. and J.G. Salinas (1981). Low-input technology for managing oxisol and ultisols in tropical America. *Advances in Agronomy* 34:279-406.

Singh, J. and J.P. Singh (1995). Land degradation and economic sustainability. *Ecological Economics* 15:77-86.

Sposito, G. (1989). *The Chemistry of Soils*. New York: Oxford University Press.

Szabolcs, I. (1994). Prospects of soil salinity for the 21st century. Paper presented at 15th World Congress of Soil Science, Acapulco, Mexico, July 10-16 (Volume I, pp. 123-141).

Thomas, D.S.G. and N.J. Middleton (1993). Salinization: New perspectives on a major desertification issue. *Journal of Arid Environments* 24:95-105.

Waisel, Y. (1972). *Biology of Halophytes*. New York: Academic Press.

Wisner, R.N. and C.A. Chase (1984). *World Food Trade and US Agriculture 1960-1983*. Ames, IA: World Food Institute, Iowa State University.

Witt, M. and B.J. Barfield (1982). Environmental stress and plant productivity. In *Handbook of Agricultural Productivity*, M. Rechcigl Jr. (Ed.). Boca Raton, FL: CRC Press, pp. 347-374.

Chapter 2

Breeding for Stress Resistance: General Principles

Mervyn O. Humphreys
Michael W. Humphreys

INTRODUCTION

Plant productivity is fundamentally affected by abiotic stress. Climate and soil determine many adaptations in plants and the ecogeographical distribution of species and ecotypes demonstrate differences in physiology and developmental patterns that provide good evidence of adaptive mechanisms. Plants respond to environmental change as individuals through phenotypic plasticity and in populations through selection and associated evolutionary processes. It is not always easy to determine the genetics underlying adaptive processes because environmental factors may be complex or not clearly defined. However, extreme environmental pressures such as heavy metal contamination of soils (Humphreys and Bradshaw, 1977) or severe winter conditions (Helgadóttir et al., 2001) may produce detectable genetic shifts. Several genes may be responsible for a response to a given factor or the same gene(s) may be involved in different adaptive responses. Specific gene interactions may be in a state of flux or become "fixed," restricting opportunities for further evolution. Phenotypic plasticity serves as a buffer to avoid excessive genetic flux in response to short-term changes. Traditional methods of evaluating phenotypic plasticity in terms of G × E (genotype × environment) interactions (e.g., Finlay and Wilkinson, 1963) equate "environment" to the mean performance of all genotypes in response to a particular site, management, or year. Yield of individual varieties is plotted against the measure of environment and yield potential may be defined as the highest yield in the best environment. Yield stability of a particular genotype should be measured across environments and may be equated to stress resistance (Thomas, 1997). Traits desirable in temperate European grasslands for resistance to periods of summer drought and cool winters are outlined in Table 2.1 and can be seen to differ among areas where grassland

TABLE 2.1. Evaluation of traits relevant to stress resistance of perennial forage grasses in Europe

		Importance in Different Climatic Zones of Europe[a,b]		
Trait	**Strategy[c]**	**Marit**	**Cont**	**Med**
Ontogenetic and Developmental				
Rapid establishment	DR	+	+	+
Early flowering	DA		+	+
	WH		–	
Concentrated flowering period	DA	+	+	+
Summer quiescence	DA	– –	–	++
Survival of axillary buds	DR WH	+	+	++
Ability to regrow rapidly when stress relieved	DR	±	+	++
	WH		–	+
Winter dormancy	DR	–	+	– –
	WH	–	–	– –
Winter growth	DR	+	–	++
	WH	±	–	++
Roots				
Dense surface roots	DD	+	+	+
At least some deep roots	DA DD	+	+	++
Ability to extract nutrients from soil	DR WH	+	+	+
Leaves				
Maintenance of high RWC	DD	+	+	+?
Leaf rolling at high RWC	DA	–	–	+
Leaf rolling at low RWC	DD	+	+	+
Continued leaf growth in drought	DR	+	?	–
Maintenance of green leaf area	DR	+	?	?
	WH	+	+	+
Leaf shedding or rapid senescence	DA DD	–	–	+
High abaxial stomatal density (allied with leaf rolling)	DD	+	+	+
Stomata close at high RWC	DD	–	–	–
	DA			–
Stomata close at low RWC	DD	+	+	+
	DT		–[d]	–[d]
Metabolism/Anatomy				
C3 metabolism	DR	+	±	±
	WH	+	+	+

C4 metabolism	DR WH	–	±	±
Osmotic adjustment, hydration, solute accumulation	DD DT WH	+?	+	+
Osmoprotectants, antioxidants	DT WH	+	+	+
Membrane acclimation and stability	DT WH	+	+	+
Survival of green leaf function: robustness of assimilation	DA DT	+	+	+[e]
	WH	+	+	+
Cell size, cell walls, rheology	DT WH	+	+	+
Storage of WSC and N reserves in sheaths	DA DT WH	+	+	++
Translocation of reserves to meristems	DA DT WH	+	+	++

[a]The maritime climate (Marit) is characterized by moderate temperatures and drought, and risk of dehardening in mild spells. The continental climate (Cont) has more extreme summer and winter temperatures and greater risk of drought. The Mediterranean climate (Med) has generally mild winters, although frost can damage young growth in winter, and severe but not necessarily fatal summer drought.

[b]Symbols: + advantageous, – deleterious, ± dependent on circumstances.

[c]Drought resistance (DR) traits may be grouped into strategies: DA drought avoidance, DD desiccation delay (or avoidance), DT desiccation tolerance. Winter hardiness traits are indicated by WH.

[d]If closure causes overheating.

[e]More important during early rather than later stages of drought.

agriculture is practiced in Europe. Blum (1988) reviewed general aspects of crop improvement for stress environments. This chapter aims to build on this and give an overview of the principles underlying breeding for abiotic stress in an era of rapidly advancing genetic information and technical development.

BREEDING FOR IMPROVED STRESS RESISTANCE

Although technologies such as tissue cultures to create somatic hybrids or gene transformation systems may become effective procedures for improving plant adaptation to abiotic stresses, conventional breeding through sexual hybridization is still the principal route for the development of stress-resistant varieties. Conventional plant breeding has had considerable success in improving the world's major crops (Humphreys, 1999). It depends on phenotypic selection assembling optimum combinations of large numbers of genes and seldom targets individual genes. However, modern conventional plant breeding is undergoing revolutionary changes that embrace new marker technologies and a more profound understanding of the mechanisms that constitute complex traits. Terms such as "designer breeding" and "breeder's toolkits" have been used to describe new genotypes that contain genes of known function whose presence confers adaptations to specific stress conditions. Genetic markers are assigned to the genes of interest to enable their selection in subsequent generations.

A need to broaden the gene pools of our major crop species and to use adaptations that have evolved in their wild relatives has led to major efforts in intergeneric and interspecific plant breeding. Domesticated crops are often derived from a very narrow genetic base and modern breeding techniques can further restrict the available variation. Related wild species may extend the range of variation that can be used in crop improvement programs, but this depends on their accessibility to the breeder. In the case of some crop plants such as maize, identification of a progenitor is still not clear. One method of combining desirable traits from related species is amphiploidy, in which complete parental genomes are combined and homologous chromosome pairing and disomic inheritance is encouraged by chromosome doubling. The disadvantages of this procedure are the frequent instability of such hybrids (Thomas and Humphreys, 1991) and the presence of deleterious gene combinations that may reduce crop yields and quality and negate benefits accrued from improved stress resistance. An alternative procedure is introgression, where a limited number of genes are transferred from a wild relative to an agricultural crop species. While the amphiploid approach requires homologous chromosome pairing, introgression requires widespread homoeologous chromosome pairing and recombination for gene transfer. Addition lines of wheat were used to introgress genes for salt-tolerance from the grass *Thinopyrum bessarabicum* (William and Mujeeb-Kazi, 1995). The alien chromosome and the subsequent location of introgressed genes from *Thinopyrum* were detected using genomic in situ hybridization (GISH) (King et al., 1993). In other cases, the detection of alien introgressions in wheat has proven difficult, since those species that recombine frequently with wheat retain too much structural homology with wheat for their DNA to be distinguished using GISH. For example, it is possible to introduce genes from *T. urartu* (donor of the A genome) or *Aegilops squarrosa* (donor of the D genome) (Gale and Miller, 1987) into wheat, but these genes cannot be physically mapped using GISH. More distantly related species do not recombine with wheat because of the presence of the *Ph1* pairing control gene on chromosome 5B. In the absence of the 5B chromosome, recombination can occur albeit at fairly low frequencies; for example, in wheat/rye at 9.5 percent (Fernández-Calvín et al., 1995).

Some major advances in breeding for stress resistance have been achieved in the forage grasses within the *Lolium/Festuca* complex of the Poaceae (Humphreys and Thomas, 1993). *Lolium* species provide high-quality nutritious forage but lack the stress resistance normally found in closely related ecotypes of *Festuca* species. The hexaploid species *F. arundinacea,* like wheat, has disomic inheritance (Lewis et al., 1980) and probably derives from three ancestral species, although only two thus far have been identified: *F. pratensis* (2x) and *F. glaucescens* (4x) (Humphreys et al., 1995). Hybrids between *Lolium* species and *F. arundinacea* and its progenitors differ from wheat and all other known crops in that intergeneric chromosome pairing and recombination is very frequent, but chromosomes of *Lolium* and *Festuca* can be distinguished by GISH. This has allowed a novel approach to breeding for stress resistance, termed introgression mapping, whereby a range of different *Festuca* genes has been introduced into *Lolium,* thereby "dissecting" complex QTL (quantitative trait loci). GISH can determine the location of alien *Festuca* genes and genetic markers ascribed to the targeted sequence (Humphreys et al., 1998). *Lolium* lines carrying different *Festuca* genes that convey a range of adaptations to abiotic stresses are currently being developed alongside breeder's toolkits to aid their commercial exploitation. The most advanced introgression line is a *L. multiflorum* genotype with genes on chromosome 2 for drought resistance derived from *F. arundinacea* (Humphreys and Pasakinskiene, 1996).

THE GENETICS OF ABIOTIC STRESS TOLERANCE

Successful breeding depends on a broad understanding of the genetic architecture of relevant traits. Genes with major effects and genes contributing to the expression of quantitative traits both have a role in controlling abiotic stress tolerance. Traditionally, major gene and polygenic variation has been analyzed in different ways, but the use of new DNA markers and techniques of QTL analysis now allow a more integrated approach in dissecting complex traits and assessing gene effects. Genetic fingerprinting of *Hordeum spontaneum* has revealed genetic marker associations with ecogeographic factors and experimentally imposed stresses (Forster et al., 2000). Examples of QTL associated with valuable traits are increasing in a wide range of crops including rice, wheat, maize, sorghum, barley, forage and turf grasses, brassica, soybean, pea, alfalfa, beans, tomato, beet, and potato (McCouch and Doerge, 1995; Mohan et al., 1997; Duncan and Carrow, 1999).

Useful information on the genetic basis of abiotic stress tolerance has also been obtained by moving genes between plants of the same or closely related species. Gene introgression achieved by conventional cross-pollination means that the range of genetic variation available to understand and manipulate genetic adaptation to environmental change is greatly enhanced. A concern is that much of the genetic variation for improving abiotic stress tolerance has been lost during domestication and selection in modern breeding, leaving only pleiotropic effects of a relatively small sample of genes for adaptation. In barley, greater variation to abiotic stresses exists in primitive landraces and gene pools in related wild species (Forster et al., 2000), which are easily hybridized with cultivated barley to provide extended sources of variation. Drought and cold tolerance has been improved by hybridization within the *Lolium/Festuca* species complex (Humphreys et al., 1998). White clover plants have been produced with varying resource allocation in stolons and rhizomes, which also affects tolerance to stresses such as cold and drought (Marshall et al., 2001).

Compared with many diploid crops, the buffering-effect of gene duplications in the genomes of polyploids such as wheat offers some advantages for mapping traits. Chromosome addition and substitution lines have been created in wheat that have enabled traits to be assigned to chromosome arms. For example, it has been demonstrated that most chromosome pairs are implicated in freezing tolerance (Thomashow, 1990), especially QTL associated with chromosomes 5A, 7A, and 4D (Sutka and Snape, 1989). Wheat chromosome substitution lines indicate that the accumulation of antifreeze proteins, freezing tolerance, and the dehydrin-like protein WSC120 are all regulated by chromosome 5 (Griffith et al., 1997). Five vernalization loci were identified on chromosomes 5A, 5B, 5D (Xin et al., 1988), 2B, and 7B (Law, 1996), and the *Vrn1* gene is a major determinant of growth habit (Brule-Babel and Fowler, 1988). Winter wheat has recessive *vrn1* alleles and requires vernalization to promote floral development while spring wheat carries dominant *Vrn1* alleles, allowing floral development without vernalization. Winter wheat is more freezing tolerant than spring wheat, and there is evidence to show close linkage of the *Vrn1* gene to a gene *Fr1,* which has a major effect on freezing tolerance (Galiba et al., 1995).

Mutants are also a promising way to detect genes involved in response to environmental stress. The model plant species *Arabidopsis* is particularly amenable for dissecting the genetics and molecular mechanisms underlying physiological responses and a wide variety of mutants are available. For example, mutant plants with impaired hormone biosynthesis (deficient mutants) or signal transduction pathways (responsive mutants) provide valuable tools for understanding drought tolerance (Vartanian, 1996). In particular, abscisic acid (ABA) mutants affect developmental processes during the desiccation phase of seed maturation and/or in response to drought in vegetative tissues.

Advances in understanding the effectiveness of stress responses are also being made using transgenic plant analyses (Hasegawa et al., 2000). A complex set of stress-responsive transcription factors is emerging. The imminent availability of genomic DNA sequences and global and cell-specific transcript expression data, combined with determinant identification based on gain- and loss-of-function molecular genetics provides the means of dissecting adaptive traits. Genetic activation and suppression screens will lead to an understanding of the interrelationships of the multiple signaling systems that control stress-adaptive responses in plants. For example, osmotic stress plays a role in drought, salt, and cold stress, and is often associated with the synthesis of novel polypeptides and/or osmolytes. A number of stress-related genes have been characterized (Bartels and Nelson, 1994). After transforming plants with these genes, the effect of the encoded proteins on altered stress behavior can be examined. Using the approaches outlined previously, the genetic control of tolerance to a range of abiotic stresses is becoming clearer.

Drought

Various morphological, physiological, and biochemical characters confer drought resistance and exhibit different patterns of inheritance (monogenic and polygenic) and gene action (additive and nonadditive) (Mitra, 2001). Genetic improvement of adaptation to drought is often based on selecting for yield and its stability over locations and years. Dhanda and Sethi (1998) found that additive gene action played a major role in determining the inheritance of excised-leaf water loss and relative water content in bread wheat. Because of low heritabilities in field trials, selection programs are expensive and slow in attaining progress. Molecular markers linked to root traits and osmotic adjustment are being identified in rice, which should speed up selection response (Nguyen et al., 1997). In sorghum, molecular markers have been used to identify regions of the genome specifically associated with preflowering drought tolerance (Tuinstra et al., 1996). Several loci were associated with the expression of drought tolerance under both mild and severe drought-stress conditions.

Dormancy is a drought avoidance mechanism and is generally governed by many genes, which in a few cases have been mapped to specific chromosome regions. Genes that are differentially expressed between dormant and nondormant seeds have been isolated, although their identity and function remain obscure (Li and Foley, 1997). Other genes and QTL linked to drought-tolerance traits are listed in Duncan and Carrow (1999).

Flooding

The adaptation of deepwater rice to flooding is attributed to two mechanisms, submergence tolerance and plant elongation. Sripongpangkul and colleagues (2000) identified some key genes/QTL for these traits. The most important gene was *QIne1,* which mapped near sd-1 on chromosome 1. The second locus was a major gene, *subI(t),* mapped to chromosome 9, which contributed to submergence tolerance only and *QIne4,* which mapped to chromosome 4, which had a large effect on internal elongation. An additional locus that interacted strongly with both *QIne1* and *QIne4* appeared near RG403 on chromosome 5, suggesting a complex epistatic relationship among the three loci. Several QTL with relatively small effects on plant elongation and submergence tolerance were also identified. Setter and colleagues (1997) suggested that there is one dominant gene for submergence tolerance that is present in three out of four of the world's most tolerant rice cultivars. This may be associated with the high levels of carbohydrate required to maintain energy supply partly through fast rates of alcoholic fermentation. Mishra and colleagues (1997) also suggested that a single dominant gene governed submergence tolerance.

Salt

Salinity in soil affects about 7 percent of the land's surface and about 5 percent of cultivated land. About 20 percent of irrigated land has suffered from secondary salinization and 50 percent of irrigation schemes are also affected. Thus, the development of salt-resistant crops is of high priority, but progress is hampered by the complex effects that salt has on plants (Flowers et al., 1997). Salt tolerance depends on a range of physiological adaptations, including ion compartmentation and the production of compatible solutes. However, new genetic and biochemical approaches to studying plant responses to abiotic stresses have recently began to bear fruit (Hasegawa et al., 2000). A total of 38 traits related to different tissue or whole-plant Na^+ and/or Cl^- accumulation was analyzed in salinized and nonsalinized BC1 progeny clones from a *Poncirus* × *Citrus* cross (Tozlu et al., 1999). QTL were mapped for traits associated with salinity and it was possible to identify 17 regions of the citrus genome of interest, of which eight contain QTL of large effect. Analysis of salt overly sensitive (sos)

Arabidopsis mutants has revealed a novel calcium-regulated protein kinase pathway for response to the ionic aspect of salt stress (Zhu, 2001). Other specific genes controlling osmoprotectants, such as glycine betaine, dehydrins, and K-Na selectivity are listed in Duncan and Carrow (1999).

Aluminium, Boron, and Heavy Metals

Soil mineral stresses are increasingly becoming growth-limiting factors for crop plants in many parts of the world. These problems are often not easily correctable by conventional agricultural practices but, fortunately, some crop plants and their wild relatives possess considerable genotypic variation in tolerance. Plants have both constitutive (present in most phenotypes) and adaptive (present only in tolerant phenotypes) mechanisms for coping with elevated metal concentrations. Where studies on the mechanisms of tolerance fall down is in their failure to integrate tolerance mechanisms within cell or whole-plant function by not relating adaptive mechanisms to constitutive mechanisms. Consideration of the response to metals at a cellular rather than a biochemical level will lead to a greater understanding of mechanisms to withstand elevated levels of metals in both contaminated and uncontaminated environments. Genetics also has a role in unraveling tolerance mechanisms (Meharg, 1994). Examples of major genes for tolerance to aluminium, arsenic, boron, cadmium, copper, and manganese have been reported, although polygenic inheritance is often apparent, including various modifiers (Smith and Macnair, 1998). Manyowa and Miller (1991) identified the chromosomal locations of factors conferring tolerance to aluminium, excess manganese, boron, and copper in the Triticeae. Duncan and Carrow (1999) list genes relevant to the genetic engineering of acid soil/aluminium tolerance and alkaline soil-based iron deficiency. They also list genes associated with the efficient use of nitrogen, phosphate, and potassium, but soil nutrient deficiencies are not considered further in this chapter.

Acid soil and associated aluminium toxicity are considered the major abiotic factors limiting crop production. Over 2 billion hectares of acid soils exist worldwide, both in tropical and moderate climatic zones. Diallel analysis of aluminium tolerance in rice (Wu et al., 1997) indicated greater additive compared to nonadditive genetic control. A molecular linkage map was used to identify QTL for relative root length (RRL) in a rice population derived from a cross between an Al-sensitive lowland indica variety and an Al-tolerant upland japonica variety (Wu et al., 2000). Two QTL for RRL were detected on chromosomes 1 and 12. Additive gene effects predominantly controlled tolerance in younger seedlings, while an epistatic effect was also important for tolerance in older seedlings. The QTL may be multiple allelic loci for Al tolerance and phosphorus-uptake efficiency, or for Al and Fe^{2+} tolerance. Combining ability analysis of genetic differences in the aluminium tolerance of soybeans also indicated predominantly additive gene effects (Spehar and Galwey, 1996). Inheritance of tolerance to high concentrations of soil boron in pea (*Pisum sativum* L.) was explained in terms of two major gene loci interacting in an additive manner with incomplete dominance at each locus (Bagheri et al., 1996).

The degree to which tolerance to one metal may often confer a degree of tolerance to some other metals is variable (Tilstone et al., 1997; Schat and Vooijs, 1997). Very little is known about the molecular genetics of tolerance and the mechanisms of tolerance to most metals. Common major genes for tolerance among different geographically isolated populations appears to be the result of independent parallel evolution in local nontolerant ancestral populations (Schat et al., 1996). The possible role of phytochelatins and metallothionein-like proteins in metal tolerance have been discussed (Macnair, 1993).

Cold

Freezing tolerance is a multigenic trait; freezing injury in most plants and tissues results largely from the severe cellular dehydration that occurs upon ice formation, and the cellular membrane systems are a primary site of freeze-induced injury. Cold acclimation is a critical part of freezing tolerance, and response may differ between autumn and spring. The ability to resist cold damage in different seasons is influenced by genotype and by various environmental cues. In Douglas fir, spring cold hardiness is under strong genetic control and is responsive to temperature, whereas autumn cold hardiness is less heritable and is more strongly influenced by photoperiod, available moisture, and other environmental factors (Jermstad et al., 2001). Eleven QTL controlling autumn cold hardiness were detected on four linkage groups, and 15 QTL controlling spring cold hardiness were detected on four linkage groups in Douglas fir. *Arabidopsis* is used as a model to study freezing tolerance, and genes associated with cold acclimation have been identified and their modes of action determined (Thomashow, 2001). Low-temperature signaling and regulatory pathways involved in activating the cold acclimation response are being investigated and novel approaches have been suggested to improve the environmental stress tolerance of plants. A comprehensive list of genes linked to low-temperature stress tolerance is presented in Duncan and Carrow (1999). Genes may encode responses to low-temperature acclimation, cold acclimation, or freezing tolerance. They include antifreeze protein genes, COR (cold-regulated) protein genes, and cDNA clones associated with response to low temperature, drought, and ABA.

Heat

High-temperature stress protection may involve heat stability of photosynthetic systems and other enzymes, proteins, and membranes. Heat absorption may be reduced by changes in leaf morphology, and appropriate carbohydrate resource allocation may also be critical, such as between roots and shoots and between cell walls (structural) and cell contents (nonstructural).

During high temperatures, plants synthesize an abundant amount of low-molecular-weight heat shock proteins, which have been correlated with acquired thermal tolerance. A genetic analysis of heat shock protein (HSP) synthesis (Jorgensen and Nguyen, 1995) revealed both qualitative and quantitative polymorphisms and inheritance patterns corresponding to dominant gene expression. There was evidence for at least four unlinked gene loci, and a low-molecular-weight Class I heat shock protein cDNA has been isolated from maize (Jorgensen and Nguyen, 1994). A comprehensive list of genes associated with high-temperature stress tolerance is also presented in Duncan and Carrow (1999).

Photoinhibition and UV Stress

Abiotic stress often increases production of active oxygen species (AOS) such as superoxide, hydrogen peroxide, and the hydroxide radical that can inhibit photosynthesis and cause loss of integrity of cell membranes (Foyer et al., 1998). Increasing the activity of antioxidant enzymes such as superoxidase dismutase (SOD), catalase, ascorbate peroxidase (APX), and glutathione reductase, can increase protection against stresses. Chilling-induced photoinhibition of maize may be reduced by the overexpression of antioxidant enzymes such as SOD (Foyer et al., 1998).

In plants, UV-B (ultraviolet-B radiation) can potentially interfere with growth, development, photosynthesis, flowering, pollination, and transpiration (Rozema et al., 1997; Jansen et al., 1998). Jansen and colleagues (2001) studied the mechanism of UV protection in duckweed by exploiting the UV sensitivity of photosystem II as an in situ sensor for radiation stress. A UV-tolerant ecotype had significantly higher indole-3-acetic acid (IAA) levels and increased activity of a specific peroxidase isozyme compared to a UV-sensitive ecotype. It appears that peroxidase genes contribute to UV protection as well as controlling leaf and plant architecture. Smith and colleagues (2000) found that biomass accumulation was correlated with UV-B sensitivity when a variety of vegetable crop plants were screened. Plants that accumulated more biomass over a two-week period were more likely to be UV-B sensitive, which suggests that a rapid growth rate is disadvantageous. An *Arabidopsis* mutant that has increased sensitivity to UV-B was produced by chemical mutagenesis (Lois and Buchanan, 1994). The UV-sensitive mutation proved to be recessive and to segregate as a single Mendelian locus. This single gene defect was shown to lead to a block in the synthesis of a group of flavonoids, which normally accumulate in *Arabidopsis,* particularly when exposed to UV radiation. One of these compounds was identified as a rhamnosylated derivative of the flavonol kaempferol and provides evidence that flavonoids play an important role in the protection of plants from the damaging effects of UV-B.

MECHANISMS OF STRESS RESISTANCE

Although it is often convenient to consider components of abiotic stress tolerance in isolation, it must be remembered that resistance to abiotic stresses usually requires complex gene interactions. Many stress-tolerance traits may overlap (see Figure 2.1) and their separation as components of

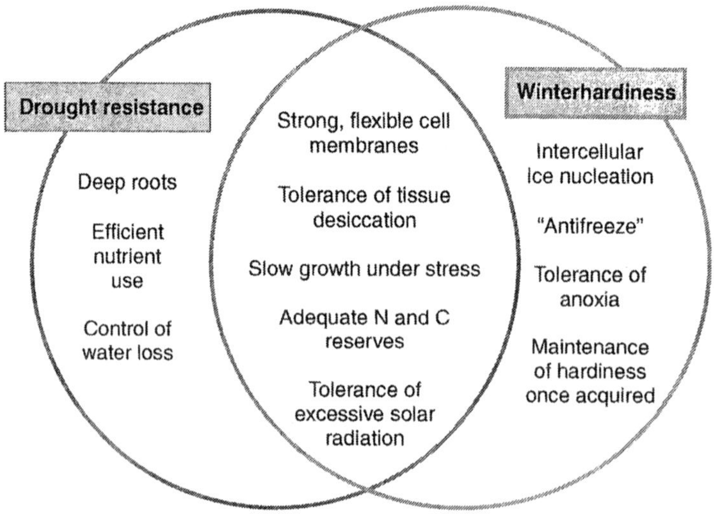

FIGURE 2.1. Coincidence of traits that contribute to drought resistance and winterhardiness.

drought resistance, winter hardiness, and so on may well confuse overall understanding of mechanisms. For example, signaling drought is a complex phenomenon and involves not only a shortage of water but also high transpiration rates, supraoptimal temperatures, photooxidation, mineral deficiency, and soil impenetrability. Winterhardiness is similarly complex and will be considered in more detail to illustrate the interactions involved.

Winterhardiness comprises tolerance of subzero temperatures, ice encasement, anoxia, low light inputs, especially near the arctic circle, and disease attack. Winterhardiness may be lost more rapidly than it is acquired, and persistence over winter may be worse in relatively mild climates, which can induce dehardening, leading to more frost damage, than in colder climates in which hardiness is maintained. Winter dormancy can help prevent dehardening. To survive the winter, a plant must evolve mechanisms whereby freezing-sensitive tissues can avoid freezing or undergo freezing tolerance compatible with the normal variations of the local climate, coordinate the induction of the tolerance at the appropriate time, maintain adequate tolerance during times of risk, and properly time the loss of tolerance and resumption of growth when the risk of freezing has passed (Guy, 1990).

Low-Temperature Stress

There are two types of injuries a plant can sustain through exposure to low temperatures (Stushnoff et al., 1984). The first is chilling injury that occurs from approximately 20 to 10°C. The resultant injuries may include a variety of physiological disruptions in germination, flower and fruit development, yield, and storage life. Minor chilling stress at nonlethal temperatures is normally reversible. Exposure to gradually decreasing temperatures above the critical range also can result in hardening of plants, which may reduce or eliminate injury during subsequent exposure to low temperatures. The second type of injury is called freezing injury. This type of injury occurs when the external temperature drops below the freezing point of water. Some varieties of plants that are susceptible to chilling injury can be killed by the first frost. At the other extreme, many plants that are native to cold climates can survive extremely low temperatures without injury (Levitt, 1980). Winter-hardy species require a period of cold acclimation (CA) at low, nonfreezing temperatures in order to survive subsequent freezing and other winter stresses. Development of winterhardiness requires exposure of plants to low, nonfreezing temperatures typically 0 to 10°C and shortened photoperiod (Levitt, 1980). Many physiological and biochemical changes occur during CA, including slowed or arrested growth; reduced tissue water content; altered cell pH, protoplasm viscosity, and photosynthetic pigments; reduced adenosine triphosphate (ATP) levels (Levitt, 1980); transient increases in ABA (Chen et al., 1983); changes in membrane lipids (Uemura and Steponkus, 1994); accumulation of compatible solutes including proline, betaine, polyols, and soluble sugars; and accumulation of antioxidants (Tao et al., 1998).

Some tree species show extreme freezing tolerance. Nonacclimated (NA) birch and dogwood are injured at −10°C but after CA can survive −40 to −50°C under natural conditions. NA wheat and rye are killed at −5 to −10°C, but after CA wheat can survive to −15°C and rye to −30°C (Thomashow, 1990). The plasma membrane is the primary site of freezing injury and must respond to freeze-thaw–induced changes in osmotic gradient, allowing the passage of water, retaining solutes, and preventing ice-nucleation of the cell from the extracellular space (Steponkus, 1990). Freeze-induced cell dehydration is such that >80 percent of the osmotically active water is removed at −5°C and >90 percent at −10°C.

The rate and extent of dehardening is a critical factor in winter survival. Overwintering plants are particularly susceptible to freezing damage in the spring if the deacclimation process occurs prematurely or too rapidly, or if unpredictable temperature fluctuations occur (Levitt, 1980; Gay and Eagles, 1991). Eagles (1989) suggested that the nature of an adaptive CA process would vary with the stability and predictability of winter conditions in a particular environment. In stable and predictably cold continental climates, where the onset of freezing temperatures is rapid, a photoperiod-triggered and rapid acclimation process is desirable, while in the more variable and less severe conditions of a maritime climate, a temperature-dependent response might enable plants to exploit a mild autumn or spring by continuing to grow. However, in cultivars adapted to maritime climates, deacclimation may occur in response to fluctuations in winter and spring temperature with a risk of damage by subsequent frosts (Eagles, 1994).

Anoxia, Flooding, and Ice Encasement

Possibly the most important component of winterhardiness in Northern Europe is the ability to withstand periods of ice-encasement. Ice-encasement tolerance expressed as LD_{50} (lethal dose for 50 percent of plants) varied from 50 days at −1°C in Berings hairgrass *(Deschampsia berengenis)* to two days in a cultivar of orchardgrass *(Dactylis glomerata)*. Ice encasement, waterlogging, and soil compaction all include a component of hypoxic or anoxic stress. In waterlogged soils, respiration by roots and soil microflora depletes dissolved oxygen, and toxic by-products of anaerobic respiration accumulate (Pulli, 1989). In colder climates, soil and root respiration is slowed, but freezing and ice encasement may completely seal soil and plant surfaces against penetration by oxygen. Injury to cells may result from ethanol self-poisoning, cytoplasmic acidosis, insufficient ATP generation, and metabolic lesions caused by reentry of oxygen (Vartapetian and Jackson, 1997). Metabolic adaptations are important in short-term survival of anaerobic stress (Vartapetian and Jackson, 1997). Increased activity in both the glycolytic and fermentation pathways is observed, along with increased expression of enzymes in this pathway. Avoidance of cytoplasmic acidosis by early switching from production of acidic lactate to neutral ethanol may underlie root tolerance of reduced oxygen (Davies, 1980). Anatomical adaptations in roots may promote survival by allowing avoidance of anoxia (Jackson, 1990), and includes root aerenchyma, which transports oxygen to, and remove volatiles from, the roots. Dormancy is exhibited by shoot tissue of many species exposed to prolonged anaerobic conditions, as in the rhizomes of wetland species (Brandle, 1991). Stomatal closure, epinastic leaf curvature, slowed leaf expansion, and enhanced leaf senescence are also triggered by soil waterlogging, as a means of reducing transpiration (Else et al., 1995). Hypoxic acclimation enhances fermentation rate and promotes subsequent survival of anoxia in maize (Saglio et al., 1988) and wheat (Waters et al., 1991), and a flooding pretreatment enhanced fermentation rate and survival of anoxic ice encasement in winter wheat and rye (Andrews and Pomeroy, 1983).

Antifreeze Proteins

Antifreeze proteins (AFPs) were first isolated from Antarctic fish and were shown to depress the freezing point of blood serum to –2°C, one degree below that at which the serum melts (Duman and Olsen, 1993). This noncolligative effect, termed thermal hysteresis, is evident in other aqueous solutions and appears to result from the adsorption of the proteins to the surface of ice crystals and inhibition of further crystal growth. In general, plant AFPs exhibit low levels of thermal hysteresis and are present at much lower concentrations than fish AFPs (Duman and Olsen, 1993). In overwintering plants, ice does form in intercellular spaces and plant AFPs are thought to modify rather than prevent ice formation (Raymond and DeVries, 1977). During CA, proteins accumulate in the apoplast of winter rye, and extracts from the apoplast modify ice (Griffith et al., 1992). The accumulation of rye AFPs during CA correlates with the development of freezing tolerance (Marentes et al., 1993). Very recently, a heat-stable antifreeze protein has been isolated from *Lolium perenne* (Sidebottom et al., 2000). The *Lolium* AFP causes only slight thermal hysteresis (0.1°C) but is very efficient at inhibiting ice recrystallization.

Sugars

Carbohydrates also accumulate at low temperature, and species and varieties that grow more slowly during CA accumulate more soluble carbohydrate and develop more freezing tolerance (Eagles, 1967). However, carbohydrate accumulation may not have a specific role in CA but rather results from the conjunction of slowed growth and continuing photosynthesis at low temperature. Temperate grasses store fructans, a soluble polymer capable of rapid polymerization and depolymerization (Olien and Lester, 1985; Pontis and Del Campillo, 1985). The partitioning of solutes may be important as survival of the plant depends on survival of apices and meristems rather than mature leaf tissue (Eagles et al., 1993).

The Potential Role of Photoreceptors in Plant Responses to Cold

Modulations in light intensity, wavelength, or daylength affect a large number of developmental and physiological processes in plants, including germination, greening, phototropism, floral initiation, and development of circadian rhythms (Ahmad, 1999). The photosystem II (PSII) reaction center is the key site for regulation of light energy and also the main site of photoinhibitory damage. PSII comprises the oxygen-evolving complex (OEC or water-splitting system), a light-harvesting chlorophyll protein complex (LHCII), and a reaction center. The D1/D2 protein dimer at the core of PSII appears to be crucial in maintaining the integrity of the complex (Mattoo et al., 1989). Photoinhibition is related to the redox state of PSII, expressed as excitation pressure. The redox state of PSII reflects fluctuations in the photosynthetic energy balance and so acts as a sensor of any environmental stresses that disturb that balance. Change to the redox state of PSII, triggered by a low-temperature shift, has been proposed to be one of several potential temperature-sensing mechanisms involved in cold acclimation (Huner et al., 1996). Conditions of high PSII excitation pressure occur at high light or when Calvin cycle activity has been restricted by low temperature or other stress factors. Plants grown under elevated PSII excitation pressure have been shown to have a more compact morphology with elongation growth inhibited. Changes occur to the photosynthetic apparatus during cold acclimation (Gray et al., 1997). During CA there is a general increase in the activities of Calvin cycle enzymes, including RuBisCO and a selective increase in the activity of enzymes for sucrose synthesis. Such changes are necessary to avoid low-temperature photoinhibition and to provide energy necessary for cold acclimation (Hurry et al., 1995).

Cessation of elongation growth during cold acclimation (Levitt, 1980) and also compact plant morphology (Andersson and Olsson, 1961) have long been considered one of the main requirements for proper hardening and overwintering. Elongation growth is considered to be competition for cold acclimation because it causes an increase in water content and water potential in cells, facilitating ice formation and cell death during freezing events. Changes related to growth under elevated PSII excitation pressure are detectable in herbaceous plants in early autumn, and this phase of plant acclimation was named prehardening (Rapacz, 1998). Since changes in the redox state of PSII seem to control changes in plant morphology and elongation growth rate during prehardening and cold acclimation, PSII may be a temperature sensor involved during cold deacclimation.

Gene Expression During Cold Acclimation

The first step at the onset of a stress factor is the perception by a plant of that stress and the relay of relevant information via various signal transduction pathways that lead eventually to the triggering of various physiological defense mechanisms. It is now clear that there is considerable interaction of these pathways with, for instance, many genes inducible by one particular stimulus. An understanding of stress resistance requires as its first challenge, the identification and a measure of the efficacy of the sensing system. This might be located at the point of the initial stress perception. Knight and Knight (2001, and references therein) cite an example involving a putative osmosensor *AtHK1,* a transmembrane histidine kinase considered to relay changes in osmotic potential outside the cell to elicit drought-inducible gene expression within the cell. If signaling pathways stimulated by more than one stress converge, this has been termed "cross-talk" (Knight and Knight, 2001). These can be instances where different pathways achieve the same end or where different pathways interact to achieve an outcome. Cross talk includes instances where a signal may be initiated in one plant cell or organ but achieve a response in another region of the plant. Other signal transduction pathways are clearer to understand and may be considered specific to one stress stimulus. Most abiotic stresses have been shown to stimulate increased levels of free calcium ions in the cytosol and to involve kinases and protein phosphatases. Elevation of calcium levels is considered to be the principal stimulus-sensing event for the onset of several abiotic stresses. The stimulus may be able to regulate response appropriate to the degree of stress. For example, the timing and closure of stomatal guard cells may correspond directly with oscillations in calcium ion levels. While elevation of calcium ions is considered a response common to the onset of various stresses, its siting may differ with the stress. Calcium levels are increased throughout the plant at the onset of cold stress but only in the root at the onset of drought. Tissue-specific responses to elevated calcium may differ with the nature of the stress, for example, drought, cold, or salt, and may in each case induce production of different stress proteins relevant to the specific stress. Elevated calcium levels achieve control of various processes via calcium sensor proteins. These include calmodulin, calcium-dependent protein kinases, and calcium-regulated phosphatases. There are many different types of calcium-sensor proteins, and their production may be specific to the stress. *Arabidopsis* has around 40 isoforms of calcium-dependent protein kinase. For example, *OsCDPK7* is specific to the induction of responses to salt and drought, but not to cold stress. Calcium sensors may also act as nodes at which cross talk can occur. Calcium ion elevations derived from one stress stimulus may regulate phosphatases and kinases involved in the transduction of another stimulus. The AB11 and AB12 proteins have possible roles in the early stages of the phytohormone ABA signal transduction. These proteins are potential nodes for cross talk between different signaling pathways involving ABA, for example, cold and drought.

Cold Response Genes in **Arabidopsis**

Many genes induced during CA in *Arabidopsis* are homologues of late embryogenesis-abundant (LEA) genes (Thomashow, 1998). These proteins are extremely hydrophilic, tend to have a simple amino acid structure using few amino acids, repeat motifs, and are boiling soluble. Although the precise function of both novel and LEA-like genes in *Arabidopsis* remains unknown, many are induced in seedlings by water deficit and ABA (Thomashow, 1994). LEA proteins are synthesized late in embryogenesis just before seed desiccation (Ingram and Bartels, 1996), suggesting that the cold-induced genes in *Arabidopsis* may reduce the effects of freeze-induced dehydration (Thomashow, 1998). The most highly expressed cold-responsive genes of *Arabidopsis* are the four *COR* gene families, each gene family comprising two genes linked in tandem array (Thomashow, 1994). *COR78, COR15,* and *COR6.6 (kin1)* encode novel polypeptides, and *COR47* encodes an LEA group II (dehydrin) homologue. Although cognates of some genes have been isolated from different species, very few homologues of the cold-responsive genes isolated from *Arabidopsis* have been isolated in monocots (Hughes and Dunn, 1996). This may be because the CA response is very diverse or may reflect differences in starting material. Most of the barley genes were isolated from shoot meristem tissue, while *Arabidopsis* genes have been isolated from whole-plant tissue (Hughes and Dunn, 1996). Cold-regulated expression of the *Arabidopsis RAB18* and *LT165* genes is severely impaired by either *aba1* or *abi1* mutations, and the promoters of these genes have ABA-responsive elements. Both are only weakly induced by cold, but are strongly expressed in response to drought and ABA consistent with the much higher levels of ABA generated in drought-stressed plants than in cold-treated plants (Thomashow, 1999, and references therein). Thomashow (1999) concluded that the direct role of ABA in induction of cold-responsive genes is probably a limited one.

THE ROLE OF GENETIC ENGINEERING IN BREEDING: IMPLICATIONS AND LIMITATIONS

Genetic engineering provides opportunities to create new genetic variation when natural allelic variation may be limited or inappropriate. A disadvantage is that, unlike conventional plant breeding, it does not deal with large numbers of genes simultaneously and multigene interactions are often ignored. Environmental stresses can promote a cascade of events, and it may be difficult to identify the most appropriate target for genetic engineering. Synergistic genetic relationships in multiple stresses are well-documented (Duncan and Carrow, 1999) and make it difficult to identify the most effective genes to target. Further information is required concerning gene dosage effects, gene silencing and transmission through ovules and pollen, and the potential value of targeting genes to specific tissues. Specific management practices may be required for the successful cultivation of altered phenotypes, and evaluation systems must be sufficiently flexible to recognize the value of novel material. The products of new technology must not endanger the environment and should have public acceptance. Further research is required to identify ways of controlling gene flow and understanding other factors that govern the spread of new genes. Risk assessments and long-term monitoring must be properly carried out for specific GMO (genetically modified organisms) releases and any associated management practices. Public awareness of the benefits of new technology should be increased while at the same time recognizing and dealing with public concerns. Examples where genetic engineering are or may be applied to improving abiotic stress tolerance in crops are emerging.

Promising strategies for improving drought resistance lie in molecular technologies that allow genes to be tagged, isolated, and expressed in transgenic plants (Zhang et al., 1999). Drought tolerance could be improved by targeting component mechanisms associated with cellular protection, for example, membrane stabilization, antioxidants, osmotic adjustment, cell-wall elasticity and lipid modifications, and/or cellular repair/recovery, including dehydration repair, membrane reassembly, lipid and protein synthesis of dehydrins and rehydrins, and good RNA stability (Duncan and Carrow, 2001). In model studies, a gene *(codA)* for choline oxidase, which catalyzes the production of betaine from choline, was cloned from a soil bacterium and introduced into *Arabidopsis*. Transformed plants performed better under salt, cold, heat, and drought stress (Hayashi et al., 1997; Alia et al., 1998). Transgenic plants having tolerance to water deficit and osmotic stresses, through genetic engineering of osmoprotectants, such as proline and glycine betaine, have been reported in rice (Nguyen et al., 1997). The barley *hval* gene responsible for LEA proteins has been transferred to rice to produce drought-resistant transgenics (Mitra, 2001).

Arabidopsis mutants defective in elements of stress-signal transduction pathways provide the infrastructure for functional physiological dissection of salt tolerance (Hasegawa et al., 2000). Cell membranes are sensitive to cold damage but can be protected by a process of acclimation, which appears to be controlled by many genes. Constitutive expression of one such gene, *COR15a*, enhanced freezing tolerance in transgenic *Arabidopsis* (Thomashow et al., 2001). Other possible targets for genetic engineering of cold tolerance include expression of fatty acid desaturase genes (Kodama et al., 1994).

Transgenic crop plants that secrete citric acid from their roots are better adapted to growth in highly acidic tropical soils (Barinaga, 1997). Transgenic tobacco and papaya expressing a citrate synthase gene from *Pseudomonas* demonstrated improved aluminium tolerance (De la Fuente et al., 1997). Enhanced citrate efflux also enabled the tobacco plants to access normally unavailable phosphate from calcium phosphate (Lopez-Bucio et al., 2000). Other ways of increasing phosphate availability through organic acid exudation include overexpression of phosphoenolpyruvate carboxylase (PEPC), which catalyzes conversion of phosphoenolpyruvate and CO_2 to oxalocetic acid (OAA), and malic dehydrogenase (MDH), which catalyzes conversion of OAA to malic acid (Raghothama, 1999). Targets for enhancing heavy-metal tolerance by transgenesis include genes involved in the production of metallothioneins and phytochelatins. For example, expression of the α-domain of metallothionione IA in transgenic tobacco plants has been shown to improve cadmium tolerance (Pan et al., 1994).

Increased production of potentially damaging AOS such as superoxide, hydrogen peroxide, and the hydroxide radical is an intrinsic feature of stress metabolism (Foyer et al., 1998). Reducing AOS concentrations by increasing antioxidant enzyme activity has the potential to increase abiotic stress protection. Target enzymes for transformation include superoxide dismutase

(SOD), catalase, ascorbate peroxidase (APX), monodehydroascorbate reductase (MDHAR), dehydroascorbate reductase (DHAR), and glutathione reductase (GR), together with low-molecular-weight antioxidants such as α-tocopherol, ascorbate, and glutathione. Overexpression of GR in poplar produced increases in the total glutathione pool and the redox state of the pool and enhanced stress tolerance (Foyer et al., 1998). However, it seems unlikely that increase in the activity of single enzymes will consistently lead to improved stress tolerance. Elevated SOD activity in a range of transgenic plants produced very variable responses to a range of stress conditions such as exposure to paraquat herbicide, ozone, drought, and salt. Jansen and colleagues (2001) studied the role of peroxidases in UV protection using genetically modified tobacco. It was found that increased activity of the anionic peroxidase correlated with increased tolerance to UV radiation as well as decreased levels of free auxin.

CONCLUSION

Breeding for stress resistance is a very cost-effective way of adapting crops to abiotic stresses. It can provide long-term solutions to problems that otherwise require short-term and often expensive solutions through temporary environmental alleviation. Increasing genetic knowledge enables breeding targets to be more clearly defined. However, emerging information continues to demonstrate the complexities inherent in mechanisms involved in stress resistance and the interactions that occur in plant-growth response to different stresses. Although genetic engineering may have a role in specifically creating novel variation, it is likely that this can be used only as part of a multivariate approach based on conventional breeding that incorporates new marker-assisted selection strategies. A changing climate, both socioeconomic and environmental, requires from the plant breeder more than increased crop yields. It requires responses that can achieve more sustainable and environment-friendly crop production. It is fortunate that the technologies cited herein, that currently are becoming available to the plant breeder, will assist in rising to these new challenges.

REFERENCES

Ahmad, M. (1999). Seeing the world in red and blue: Insight into plant vision and photoreceptors. *Current Opinions in Plant Biology* 2:230-235.

Alia, H., H. Hayashi, A. Sakamoto, and N. Murata (1998). Enhancement of the tolerance of *Arabidopsis* to high temperature by genetic engineering of the synthesis of glycinebetaine. *Plant Journal* 16:155-161.

Andersson, G. and G. Olsson (1961). Cruciferous oilseeds. In *Manual of Plant Breeding,* Volume 5, T. Römer and W. Rudorf (Eds.). Berlin: Paul Barey, pp. 1-66.

Andrews, C.J. and M.K. Pomeroy (1983). The influence of flooding pretreatment on metabolic changes in winter cereal seedlings during ice-encasement. *Canadian Journal of Botany* 61:142-147.

Bagheri, A., J.G. Paull, and A.J. Rathjen (1996). Genetics of tolerance to high concentrations of soil boron in peas (*Pisum sativum* L.). *Euphytica* 87:69-75.

Barinaga, M. (1997). Making plants aluminium tolerant. *Science* 276:1497.

Bartels, D. and D. Nelson (1994). Approaches to improve stress tolerance using molecular-genetics. *Plant Cell and Environment* 17:659-667.

Blum, A. (1988). *Plant Breeding for Stress Environments.* Boca Raton, FL: CRC Press.

Brandle, R.A. (1991). Flooding resistance of rhizomatous amphibious plants. In *Plant Life Under Oxygen Deprivation: Ecology, Physiology and Biochemistry,* M.B. Jackson, D.D. Davies, and H. Lambers (Eds.). The Hague, The Netherlands: SPB Academic, pp. 35-46.

Brule-Babel, A.L. and D.B. Fowler (1988). Genetic control of cold hardiness and vernalization requirement in winter wheat. *Crop Science* 28:879-884.

Chen, H.-H., P.H. Li, and M.L. Brenner (1983). Involvement of abscisic acid in potato cold acclimation. *Plant Physiology* 71:362-365.

Davies, D.D. (1980). Anaerobic metabolism and the production of organic acids. In *The Biochemistry of Plants,* Volume 2, D.D. Davies (Ed.). New York: Academic Press, pp. 581-611.

De la Fuente, J.M., V. Ramirez-Rodriguez, J.L. Cabrera-Ponce, and L. Herrera-Estrella (1997). Aluminium tolerance in transgenic plants by alteration of citrate synthesis. *Science* 276:1566-1568.

Dhanda, S.S. and G.S. Sethi (1998). Inheritance of excised-leaf water loss and relative water content in bread wheat (*Triticum aestivum*). *Euphytica* 104:39-47.

Duman, J.G. and M.T. Olsen (1993). Thermal hysteresis protein activity in bacteria, fungi and phylogenetically diverse plants. *Cryobiology* 30:322-328.

Duncan, R.R. and R.N. Carrow (1999). Turfgrass molecular genetic improvement for abiotic/edaphic stress resistance. *Advances in Agronomy* 67:233-305.

Duncan, R.R. and R.N. Carrow (2001). Molecular breeding for tolerance to abiotic/edaphic stresses in forage and turfgrass. In *Molecular Breeding of Forage Crops* (Developments in Plant Breeding, Volume 10), G. Spangenberg (Ed.). Dordrecht, The Netherlands: Kluwer Academic, pp. 251-260.

Eagles, C.F. (1967). Variation in the soluble carbohydrate content of climatic races of *Dactylis glomerata* (cocksfoot) at different temperatures. *Annals of Botany* 31:645-651.

Eagles, C.F. (1989). Temperature-induced changes in cold tolerance of *Lolium perenne. Journal of Agricultural Science* 113:339-347.

Eagles, C.F. (1994). Temperature, photoperiod and dehardening of forage grasses and legumes. In *Crop Adaptation to Cool Climates,* Report of Cost 814 EC Workshop, K. Dorffling, B. Brettschneider, H. Tantau, and K. Pithan (Eds.). Hamburg, Germany: Tyskland, pp. 75-82.

Eagles, C.F., J. Williams, and D.V. Louis (1993). Recovery after freezing in *Avena sativa* L., *Lolium perenne* L. and *L. multiflorum* Lam. *New Phytologist* 123:477-483.

Else, M.A., W.J. Davies, M. Malone, and M.B. Jackson (1995). A negative hydraulic message from oxygen-deficient roots of tomato plants: Influence of soil flooding on leaf water potential, leaf expansion and synchrony between stomatal conductance and root hydraulic conductivity. *Plant Physiology* 109:1017-1024.

Fernández-Calvín, B., E. Benavente, and J. Orellana (1995). Meiotic pairing in wheat-rye derivatives detected by genomic in situ hybridisation and C-banding: A comparative analysis. *Chromosoma* 103:554-558.

Finlay, K.W. and G.N. Wilkinson (1963). The analysis of adaptation in a plant breeding programme. *Australian Journal of Agricultural Research* 14:742-754.

Flowers, T.J., A. Garcia, M. Koyama, and A.R. Yeo (1997). Breeding for salt tolerance in crop plants: The role of molecular biology. *Acta Physiologia Plantarum* 19:427-443.

Forster, B.P., R.P. Ellis, W.T.B. Thomas, A.C. Newton, R. Tuberosa, D. This, R.A. El-Enein, M.H. Bahri, and M. Ben Salem (2000). The development and application of molecular markers for abiotic stress tolerance in barley. *Journal of Experimental Botany* 51:19-27.

Foyer, C.H., A.H. Kingston-Smith, J. Harbinson, A.-C.M. Arisi, L. Jouanin, and G. Noctor (1998). The use of transformed plants in the assessment of physiological stress responses. In *Responses of Plant Metabolism to Air Pollution and Global Change,* L.J. De Kok and I. Stulen (Eds.). Leiden, The Netherlands: Backhuys Publishers, pp. 251-261.

Gale, M.D. and T.E. Miller (1987). The introduction of genetic variation into wheat. In *Wheat Breeding: Its Scientific Basis,* F.G.H. Lupton (Ed.). London: Chapman and Hall, pp. 173-210.

Galiba, G., S.A. Quarrie, J. Sutka, A. Morgounov, and J.W. Snape (1995). RFLP mapping of the vernalization (Vrn1) and frost resistance (Fr1) genes on chromosome 5A of wheat. *Theoretical and Applied Genetics* 90:1174-1179.

Gay, A.P. and C.F. Eagles (1991). Quantitative analysis of cold hardening and dehardening in *Lolium. Annals of Botany* 67:339-345.

Gray, G.R., L.-P. Chauvin, F. Sarhan, and N.P.A. Huner (1997). Cold acclimation and freezing tolerance: A complex interaction of light and temperature. *Plant Physiology* 114:467-474.

Griffith, M., P. Ala, D.S.C. Yang, W.-C. Hon, and B.A. Moffatt (1992). Antifreeze protein produced endogenously in winter rye leaves. *Plant Physiology* 100:593-596.

Griffith, M., M. Antikainen, W.-C. Hon, K. Pihakaski-Maunsbach, X.-M. Yu, J.U. Chun, and D.S.C. Yang (1997). Antifreeze proteins in winter rye. *Physiologia Plantarum* 100:327-332.

Guy, C.L. (1990). Cold acclimation and freezing stress tolerance: Role of protein metabolism. *Annual Review of Plant Physiology and Plant Molecular Biology* 41:187-223.

Hasegawa, P.M., R.A. Bressan, J.K. Zhu, and H.J. Bohnert (2000). Plant cellular and molecular responses to high salinity. *Annual Review of Plant Physiology and Plant Molecular Biology* 51:463-499.

Hayashi, H., H. Alia, L. Mustardy, P. Deshnium, M. Ida, and N. Murata (1997). Transformation of *Arabidopsis thaliana* with the *codA* gene for choline oxidase: Accumulation of glycinebetaine and enhanced tolerance to salt and cold stress. *The Plant Journal* 12:133-142.

Helgadóttir, Á., S. Dalmannsdóttir, and R.P. Collins (2001). Adaptational changes in populations of contrasting white clover cultivars selected under Icelandic conditions. *Annals of Botany* 88:771-780.

Hughes, M.A. and M.A. Dunn (1996). The molecular biology of plant acclimation to low temperature. *Journal of Experimental Botany* 47(296):291-305.

Humphreys, M.O. (1999). The contribution of conventional plant breeding to forage crop improvement. In *Proceedings of the XVIII International Grassland Congress,* Volume 3, J.G. Buchanan-Smith, L.D. Bailey, and P. McCaughey (Eds.). Winnipeg and Saskatoon, Canada: Canadian Forage Council, Canadian Society of Agronomy, Canadian Society of Animal Science, pp.71-77.

Humphreys, M.O. and A.D. Bradshaw (1977). Genetic potentials of solving problems of soil mineral stress: Heavy metal toxicity. In *Proceedings of Workshop on the Adaptation of Plants to Mineral Stress in Problem Soils*, M.J. Wright (Ed.). Beltsville, MD: USDA Plant Stress Laboratory, pp. 95-105.

Humphreys, M.W. and I. Pasakinskiene (1996). Chromosome painting to locate genes for drought resistance transferred from *Festuca arundinacea* into *Lolium multiflorum. Heredity* 77:530-534.

Humphreys, M.W., I. Pasakinskiene, A.R. James, and H. Thomas (1998). Physically mapping quantitative traits for stress-resistance in the forage grasses. *Journal of Experimental Botany* 49:1611-1618.

Humphreys, M.W. and H. Thomas (1993). Improved drought resistance in introgression lines derived from *Lolium multiflorum* × *Festuca arundinacea* hybrids. *Plant Breeding* 111:166-161.

Humphreys M.W., H.M. Thomas, W.G. Morgan, M.R. Meredith, J.A. Harper, H. Thomas, Z. Zwierzykowski, and M. Ghesquiére (1995). Discriminating the ancestral progenitors of hexaploid *F. arundinacea* using genomic in situ hybridisation. *Heredity* 75:171-174.

Huner, N.P.A., D.P. Maxwell, G.R. Gray, L.V. Savitch, M. Krol, A.G. Ivanov, and S. Falk (1996). Sensing environmental temperature change through imbalances between energy supply and energy consumption: Redox state of photosystem II. *Physiologia Plantarum* 98:358-364.

Hurry, V.M., O. Keerberg, T. Parnik, P. Gardestrom, and G. Oquist (1995). Cold-hardening results in increased activity of enzymes involved in carbon metabolism in leaves of winter rye (*Secale cereale* L.). *Planta* 195:554-562.

Ingram, J. and D. Bartels (1996). The molecular basis of dehydration tolerance in plants. *Annual Review of Plant Physiology and Plant Molecular Biology* 47:377-403.

Jackson, M.B. (1990). Hormones and developmental change in plants subjected to submergence or soil waterlogging. *Aquatic Botany* 38:49-72.

Jansen, M.A.K., V. Gaba, and B.M. Greenberg (1998). Higher plants and UV-B radiation: Balancing damage, repair and acclimation. *Trends in Plant Science* 3:131-135.

Jansen, M.A.K., R.E. van den Noort, M.Y.A. Tan, E. Prinsen, L.M. Lagrimini, and R.N.F. Thorneley (2001). Phenol-oxidizing peroxidases contribute to the protection of plants from ultraviolet radiation stress. *Plant Physiology* 126:1012-1023.

Jermstad, K.D., D.L. Bassoni, N.C. Wheeler, T.S. Anekonda, S.N. Aitken, W.T. Adams, and D.B. Neale (2001). Mapping of quantitative trait loci controlling adaptive traits in coastal Douglas-fir. II. Spring and fall cold-hardiness. *Theoretical and Applied Genetics* 102:1152-1158.

Jorgensen, J.A. and H.T. Nguyen (1994). Isolation, sequence and expression of a cDNA-encoding a class-I heat-shock protein (Hsp17.2) in maize. *Plant Science* 97:169-175.

Jorgensen, J.A. and H.T. Nguyen (1995). Genetic-analysis of heat-shock proteins in maize. *Theoretical and Applied Genetics* 91:38-46.

King, I.P., K.A. Purdie, S.E. Orford, S.M. Reader, and T.E. Miller (1993). Detection of homoeologous chiasma formation in *Triticum durum* × *Thinipyrum besssarabicum* hybrids using genomic in situ hybridisation. *Heredity* 71:369-372.

Knight, H. and M.R. Knight (2001). Abiotic stress signaling pathways: Specificity and cross-talk. *Trends in Plant Science* 6:262-267.

Kodama, H., T. Hamada, G. Horiguchi, M. Nishimura, and K. Iba (1994). Genetic enhancement of cold tolerance by expression of the gene for chloroplast w-3 fatty acid desaturase in transgenic tobacco. *Plant Physiology* 105:601-605.

Law, C.N. (1996). The location of genetic factors affecting a quantitative character in wheat. *Genetics* 53:487-498.

Levitt, J. (1980). *Responses of Plants to Environmental Stress,* Second Edition, Volume I, *Chilling, Freezing and High Temperature Stresses.* New York: Academic Press.

Lewis, E.J., M.W. Humphreys, and M.P. Caton (1980). Disomic inheritance in *Festuca arundinacea* Schreb. *Zeitschrift für Pflanzenzüchtung* 84:335-341.

Li, B.L. and M.E. Foley (1997). Genetic and molecular control of seed dormancy. *Trends in Plant Science* 2:384-389.

Lois, R. and B.B. Buchanan (1994). Severe sensitivity to ultraviolet-radiation in an *Arabidopsis* mutant deficient in flavonoid accumulation. 2. Mechanisms of UV-resistance in *Arabidopsis. Planta* 194:504-509.

Lopez-Bucio, J., O.M. de la Vega, A. Guevara-Garcia, and L. Herrera-Estrella (2000). Enhanced phosphorus uptake in transgenic plants that overproduce citrate. *Nature Biotechnology* 18:450-453.

Macnair, M.R. (1993). The genetics of metal tolerance in vascular plants. *New Phytologist* 124:541-559.

Manyowa, N.M. and T.E. Miller (1991). The genetics of tolerance to high mineral concentrations in the tribe Triticeae: A review and update. *Euphytica* 57:175-185.

Marentes, E., M. Griffith, A. Mlynarz, and R.A. Brush (1993). Proteins accumulate in the apoplast of winter rye leaves during cold acclimation. *Physiologia Plantarum* 87:499-507.

Marshall, A.H., C. Rascle, M.T. Abberton, T.P.T. Michaelson-Yeates, and I. Rhodes (2001). Introgression as a route to improved drought tolerance in white clover (*Trifolium repens* L.) *Journal of Agronomy and Crop Science* 187:11-18.

Mattoo, A.K., J.B. Marder, and M. Edelman (1989). Dynamics of the photosystem II reaction centre. *Cell* 56:241-246.

McCouch, S.R. and R.W. Doerge (1995). QTL mapping in rice. *Trends in Genetics* 11:482-487.

Meharg, A.A. (1994). Integrated tolerance mechanisms: Constitutive and adaptive plant-responses to elevated metal concentrations in the environment. *Plant Cell and Environment* 17:989-993.

Mishra, S.B., D. Senadhira, and N.L. Manigbas (1997). Genetics of submergence tolerance in rice (*Oryza sativa* L). *Field Crops Research* 46:77-181.

Mitra, J. (2001). Genetics and genetic improvement of drought resistance in crop plants. *Current Science* 80:758-763.

Mohan, M., S. Nair, A. Bhagwat, T.G. Krishna, M. Yano, C.R. Bhatia, and T. Sasaki (1997). Genome mapping, molecular markers and marker-assisted selection in crop plants. *Molecular Breeding* 3:87-103.

Nguyen, H.T., R.C. Babu, and A. Blum (1997). Breeding for drought resistance in rice: Physiology and molecular genetics considerations. *Crop Science* 37:1426-1434.

Olien, C.R. and G.E. Lester (1985). Freeze-induced changes in soluble carbohydrates of rye. *Crop Science* 25:288-290.

Pan, A., F. Tie, Z. Duau, M. Yang, Z. Wang, L. Li, Z. Chen, and B. Ru (1994). Alpha domain of human metallothionein IA can bind to metals in transgenic tobacco plants. *Molecular and General Genetics* 242:666-674.

Pontis, H.G. and E. Del Campillo (1985). Fructans. In *Biochemistry of Storage Carbohydrates in Green Plants,* P.M. Dey and R.A. Dixon (Eds.). New York: Academic Press, pp. 205-227.

Pulli, S.K. (1989). Metabolic effects of flooding in red clover and brome grass during growth and hardening. *Icelandic Agricultural Sciences* 2:75-85.

Raghothama, K.G. (1999). Phosphate acquisition. *Annual Review of Plant Physiology and Plant Molecular Biology* 50:665-693.

Rapacz, M. (1998). Physiological effects of winter rape (*Brassica napus* var. oleifera) prehardening to frost. II. Growth, energy partitioning and water status during cold acclimation. *Journal of Agronomy and Crop Science* 181:81-87.

Raymond, J.A. and A.L. DeVries (1977). Adsorption inhibition as a mechanism of freezing resistance in polar fishes. *Proceedings of the National Academy of Sciences, USA* 74:2589-2593.

Rozema, J., J. van de Staaij, L.O. Björn, and M. Caldwell (1997). UV-B as an environmental factor in plant life: Stress and regulation. *Trends in Ecology and Evolution* 12:22-28.

Saglio, P.H., M.C. Drew, and A. Pradet (1988). Metabolic acclimation to anoxia induced by low (2-4kPa) partial pressure oxygen pretreatment (hypoxia) in root tips of *Zea mays. Plant Physiology* 86:61-66.

Schat, H. and R. Vooijs (1997). Multiple tolerance and co-tolerance to heavy metals in *Silene vulgaris:* A co-segregation analysis. *New Phytologist* 136:489-496.

Schat, H., R. Vooijs, and E. Kuiper (1996). Identical major gene loci for heavy metal tolerances that have independently evolved in different local populations and subspecies of *Silene vulgaris*. *Evolution* 50:1888-1895.

Setter, T.L., M. Ellis, E.V. Laureles, E.S. Ella, D. Senadhira, S.B. Mishra, S. Sarkarung, and S. Datta (1997). Physiology and genetics of submergence tolerance in rice. *Annals of Botany* 79:67-77.

Sidebottom, C., S. Buckley, P. Pudney, S. Twigg, C. Jarman, C. Holt, J. Telford, A. McArthur, D. Worrall, R. Hubbard, and P. Lillford (2000). Heat-stable antifreeze protein from grass. *Nature* 406:256-257.

Smith, J.L., D.J. Burritt, and P. Bannister (2000). Shoot dry weight, chlorophyll and UV-B-absorbing compounds as indicators of a plant's sensitivity to UV-B radiation. *Annals of Botany* 86:1057-1063.

Smith, S.E. and M.R. Macnair (1998). Hypostatic modifiers cause variation in degree of copper tolerance in *Mimulus guttatus*. *Heredity* 80:760-768.

Spehar, C.R. and N.W. Galwey (1996). Diallel analysis for aluminium tolerance in tropical soybeans (*Glycine max* [L] Merrill). *Theoretical and Applied Genetics* 92:267-272.

Sripongpangkul, K., G.B.T. Posa, D.W. Senadhira, D. Brar, N. Huang, G.S. Khush, and Z.K. Li (2000). Genes/QTLs affecting flood tolerance in rice. *Theoretical and Applied Genetics* 101:1074-1081.

Steponkus, P.L. (1990). Cold acclimation and freezing injury from a perspective of the plasma membrane. In *Environmental Injury to Plants*, F. Katterman (Ed.). San Diego, CA: Academic Press, pp.1-16.

Stushnoff, C., D.B. Fowler, and A. Brule-Babel (1984). Breeding and selection for resistance to low temperature. In *Plant Breeding: A Contemporary Basis*, P.B. Vose (Ed.). Oxford, UK: Pergamon Press, pp. 115-136.

Sutka, J. and J.W. Snape (1989). Location of a gene for frost resistance on chromosome 5A of wheat. *Euphytica* 42:41-44.

Tao, D.L., G. Oquist, and G. Wingsle (1998). Active oxygen scavengers during cold acclimation of Scots pine seedlings in relation to freezing tolerance. *Cryobiology* 37:38-45.

Thomas, H. (1997). Drought resistance in plants. In *Mechanisms of Environmental Stress Resistance in Plants*, S. Basra and R.K. Basra (Eds.). Amsterdam, The Netherlands: Harwood Academic Publishers, pp. 1-42.

Thomas, H. and M.O. Humphreys (1991). Progress and potential of interspecific hybrids of *Lolium* and *Festuca*. *Journal of Agricultural Sciences* 117:1-8.

Thomashow, M.F. (1990). Molecular genetics of cold acclimation in higher plants. *Advances in Molecular Genetics* 28:99-131.

Thomashow, M.F. (1994). *Arabidopsis thaliana* as a model for studying mechanisms of plant cold tolerance. In *Arabidopsis*, E. Meyerowitz and C. Somerville (Eds.). Cold Spring Harbor, NY: Cold Spring Harbor Laboratory Press, pp. 807-834.

Thomashow, M.F. (1998). Role of cold-responsive genes in plant freezing tolerance. *Plant Physiology* 118:1-7.

Thomashow, M.F. (1999). Plant cold acclimation: Freezing tolerance genes and regulatory mechanisms. *Annual Review of Plant Physiology and Plant Molecular Biology* 50:571-599.

Thomashow, M.F. (2001). So what's new in the field of plant cold acclimation? Lots! *Plant Physiology* 125:89-93.

Thomashow, M.F., S.J. Gilmour, E.J. Stockinger, K.R. Jaglo-Ottosen, and D.G. Zarka (2001). Role of the *Arabidopsis* CBF transcriptional activators in cold acclimation. *Physiologia Plantarum* 112:171-175.

Tilstone, G.H., M.R. Macnair, and S.E. Smith (1997). Does copper tolerance give cadmium tolerance in *Mimulus guttatus? Heredity* 79:445-452.

Tozlu, I., C.L. Guy, and G.A. Moore (1999). QTL analysis of Na^+ and Cl^- accumulation-related traits in an intergeneric BC1 progeny of Citrus and Poncirus under saline and nonsaline environments. *Genome* 42:692-705.

Tuinstra, M.R., E.M. Grote, P.B. Goldsbrough, and G. Ejeta (1996). Identification of quantitative trait loci associated with pre-flowering drought tolerance in sorghum. *Crop Science* 36:1337-1344.

Uemura, M. and P.L. Steponkus (1994). A contrast of the plasma membrane lipid composition of oat and rye leaves in relation to freezing tolerance. *Plant Physiology* 104:479-496.

Vartanian, N. (1996). Mutants as tools to understand cellular and molecular drought tolerance mechanisms. *Plant Growth Regulation* 20:125-134.

Vartapetian, B.B. and M.B. Jackson (1997). Plant adaptations to anaerobic stress. *Annals of Botany* 79(Supplement A):3-20.

Waters, I., S. Morrell, H. Greenway, and T.D. Colmer (1991). Effects of anoxia on wheat seedlings. 2. Influence of O_2 supply prior to anoxia on tolerance to anoxia, alcoholic fermentation and sugar levels. *Journal of Experimental Botany* 42:1437-1447.

William, M.D.H.M. and A. Mujeeb-Kazi (1995). Biochemical and molecular diagnostics of *Thinopyrum bessarabicum* chromosomes in *Triticum aestivum* germplasm. *Theoretical and Applied Genetics* 90:952-956.

Wu, P., C.Y. Liao, B. Hu, K.K. Yi, W.Z. Jin, J.J. Ni, and C. He (2000). QTLs and epistasis for aluminum tolerance in rice (*Oryza sativa* L.) at different seedling stages. *Theoretical and Applied Genetics* 100:1295-1303.

Wu, P., B. Zhao, J. Yan, A. Luo, and D. Senadihra (1997). Genetic control of seedling tolerance to aluminum toxicity in rice. *Euphytica* 97:289-293.

Xin, Z.Y., C.N. Law, and A.J. Worland (1988). Studies on the effects of the vernalization-responsive genes on the chromosomes of homeologous group 5 of wheat. In *Proceedings of the Seventh International Wheat Genetics Symposium*, Volume 1, T.E. Miller and R.M.D. Koebner (Eds.). Cambridge, UK: Institute of Plant Science Research, pp. 675-680.

Zhang, J.X., H.T. Nguyen, and A. Blum (1999). Genetic analysis of osmotic adjustment in crop plants. *Journal of Experimental Botany* 50:291-302.

Zhu, J.K. (2001). Cell signalling under salt, water and cold stresses. *Current Opinion in Plant Biology* 4:401-406.

Chapter 3

Use of Genetic Engineering and Molecular Biology Approaches for Crop Improvement for Stress Environments

Viswanathan Chinnusamy
Liming Xiong
Jian-Kang Zhu

INTRODUCTION

Drought, salinity, and extreme temperatures are major environmental stresses that limit crop productivity and quality. Drought is a problem in 45 percent of the world's geographical area, where 38 percent of the world's human population lives (Bot et al., 2000). In rainfed agricultural ecosystems, crops experience either continuous drought stress or intermittent cycles of water availability and stress. Fresh water is increasingly becoming a scarce resource, and plants account for about 65 percent of global fresh water use (Postel et al., 1996). Soil salinity limits crop production in about 20 percent of irrigated land (Flowers and Yeo, 1995). The problem of salinity is aggravated in dry land as salts accumulate in the root zone due to high evaporative demand and insufficient leaching of ions due to inadequate rainfall. Temperature extremes (temperatures lower or higher than the physiological optimum of a crop plant) exacerbate drought and salinity stresses and are also a limiting factor of production even in irrigated environments. Temperate crops need freezing tolerance for their survival and production. Tropical and subtropical crops such as rice, maize, soybean, cotton, and tomato are susceptible to chilling injury when ambient temperatures fall below a nonfreezing critical threshold (0 to 15°C). High-temperature stress is one of the major limiting factors of crop production in the subtropic and tropics. Heat stress affects grain quality and yield in 40 percent of the irrigated wheat-growing areas of the world (Fischer and Byerlee, 1991).

Among these abiotic stresses, drought is highly unpredictable in terms of occurrence, severity, duration, and its timing in relation to crop phenology.

Resistance to abiotic stresses is very complex at whole-plant and crop levels because of the interaction of stress with various levels of molecular, biochemical, and physiological processes at different stages of plant development, such as leaf area development, floral initiation, and fertilization. Development of crop plants tolerant to various abiotic stresses is essential to meet the growing food demand and sustainable agriculture. The objective of plant breeding for stress environments is to accumulate favorable alleles that contribute to stress tolerance in a plant genome. The sources of genes that confer stress resistance are generally the wild relatives of crop plants or the organisms that have evolved and currently live in habitats of drought, extreme temperature, and salinity. Conventional breeding has not been effective in breeding for abiotic stress resistance because (1) it is time and labor consuming, (2) undesirable genes are transferred along with desirable traits, and (3) reproductive barriers limit transfer of favorable alleles from interspecific and intergeneric sources. Genetic engineering offers a way to achieve gene transfer in an accurate way across reproductive barriers. One approach to engineering stress-resistant crop plants would be to transfer the genes from those organisms that are adapted to live in stressful environments to crop plants. In the past two decades, forward and reverse genetic approaches to studying abiotic stress responses have provided considerable insight into abiotic stress tolerance at the molecular level.

Although stress-tolerance mechanisms vary from species to species, basic cellular responses to abiotic stresses appear to be conserved among all plants (reviewed by Zhu, 2001a,b, 2002). Diverse environmental stresses induce osmotic stress, oxidative stress, and protein denaturation and thus often elicit similar cellular adaptive responses such as accumulation of compatible solutes, induction of stress proteins, and up-regulation of reactive-oxygen species' scavenging systems. Cellular membrane stability is affected by abiotic stresses either directly or through production of reactive-oxygen species. In this chapter, genetic engineering approaches for osmolytes/osmoprotectants, antioxidant systems, cellular membrane stability, and stress proteins are discussed in general for drought, salt, and temperature stresses, while engineering for water-use efficiency and ion homeostasis are discussed as strategies for drought and salinity tolerance, respectively. In addition to the genetic engineering approaches, quantitative trait loci-based marker-assisted selection breeding is also briefly discussed.

OSMOLYTES/OSMOPROTECTANTS

Positive turgor is essential for expansion growth of cells and stomatal opening in plants. Decrease in soil water availability under drought or decrease in water potential of soil solution under salinity cause osmotic stress that leads to decreased water uptake and loss of turgor. Low-temperature stress is often accompanied by osmotic stress, as low temperature may limit water uptake by roots, while freezing-induced ice formation in the apoplast leads to movement of water from the symplast into the apoplast, resulting in severe cellular dehydration. Osmotic adjustment is one of the vital cellular responses to water deficit created by drought, salinity, and low temperatures, conserved in both halophytic and glycophytic plants. Osmotic adjustment refers to the lowering of osmotic potential due to net accumulation of solutes in response to water deficit and thus helps to maintain turgor at low water potentials. Osmotic stress may induce ion uptake (K^+), compartmentalization (Na^+ into vacuole), or synthesis of compatible solutes such as proline, betaine, polyols, and soluble sugars. Compatible solutes are low-molecular-weight, highly soluble compounds that are nontoxic at high cellular concentrations. These organic-compatible solutes protect plants from stress by (1) osmotic adjustment which helps in turgor maintenance, (2) detoxification of radical oxygen species, and (3) stabilization of the quaternary structure of proteins (Yancey et al., 1982; Bohnert and Jensen, 1996). Among these solutes, polyols and proline also possess antioxidant ability (Smirnoff and Cumbes, 1989). The compatible solutes are synthesized only when water deficit develops gradually, not as a shock. Osmotic adjustment has been associated with increase in crop yield under drought in pea (*Pisum sativum* L.), chickpea (*Cicer arietinum* L.), *Brassica*, wheat, and sorghum (Zhang et al., 1999). These studies provide impetus to engineer higher plants that overproduce compatible solutes either constitutively or in response to stress. Several transgenic plants engineered to overproduce osmolytes/osmoprotectants showed enhanced abiotic stress tolerance (Tables 3.1 to 3.5).

Glycine Betaine

Biosynthesis of glycine betaine, a quaternary ammonium compound, is induced under environmental stresses and its concentration is correlated with the level of stress tolerance (Rhodes and Hanson, 1993). In higher plants, glycine betaine is synthesized from choline in chloroplasts. Choline is converted to betaine aldehyde by choline monooxygenase (CMO) and betaine aldehyde is then converted to glycine betaine by betaine aldehyde

TABLE 3.1. Glycine betaine overproducing transgenics.

Gene	Plant	Stress tolerance	Remarks	Reference
CodA (choline oxidase) from *Arthrobacter globiformis*	*Arabidopsis*	Tolerance to germination at 0°C and seedling growth at 10 to 15°C		Alia, Hayashi, et al., 1998
CodA (choline oxidase) from *Arthrobacter globiformis*	*Arabidopsis*	Constitutive freezing tolerance in photosystem II activity (–5°C)	Constitutive freezing tolerance was not due to the induction of *COR* genes.	Sakamoto et al., 2000
CodA (choline oxidase) from *Arthrobacter globiformis*	*Arabidopsis*	Better germination at 300 mM NaCl, growth at 200 mM NaCl, Low-temperature tolerance	Retention of 50 percent of photosystem II activity at 400 mM NaCl	Hayashi et al., 1997
CodA (choline oxidase) from *Arthrobacter globiformis*, targeted to the chloroplasts or to the cytosol	Rice (*Oryza sativa* L.)	150 mM NaCl inhibited the growth but recovery was faster	ChlCOD plants were more tolerant than CytCOD plants to photo inhibition under salt stress and low-temperature stresses	Sakamoto et al., 1998
CodA (choline oxidase) from *Arthrobacter globiformis*	*Brassica juncea*	Better germination in 100-150 mM NaCl and seedling growth in 200 mM NaCl		Prasad et al., 2000

COX (choline oxidase) gene *Arthrobacter pascens*	*Arabidopsis, Brassica napus,* tobacco	Drought tolerance increased only in *B. napus.* Modest improvement of freezing tolerance only in *Arabidopsis.*	One of the tobacco transgenic lines showed better tolerance in shoot weight, but stomatal closure was highly sensitive to salinity.	Huang et al., 2000
E. coli choline dehydrogenase *(betA)* and betaine aldehyde dehydrogenase *(betB)* genes	Tobacco	Increased tolerance to salt stress as measured by biomass production of greenhouse-grown, intact plants	Recovery from photo inhibition under high light, salt stress, and cold stresses	Holmström et al., 2000
Betaine aldehyde dehydrogenase *(BADH)* gene of *Atriplex hortensis* driven by maize ubiquitin promoter	Wheat (*Triticum aestivum* L.)	Seedling growth in 0.7 percent NaCl		Guo et al., 2000
peroxisomal betaine aldehyde dehydrogenase of barley	Rice	Tolerance to salt stress (chlorophyll fluorescence; 100 mM), Seedling growth at low (15°C) or high (37/32°C) temperature	Transgenic rice plants accumulated fewer Na+ and Cl– ions and more K+ ions	Kishitani et al., 2000
codA (choline oxidase) from *Arthrobacter globiformis.*	*Arabidopsis*	Better germination after treatment at 40-55°C for 60 min during imbibition. Better seedling growth at 35°C	HSP70 expression was less in transgenic under heat stress, which indicates alleviation of heat stress by betaine	Alia, Sakamoto, et al., 1998

TABLE 3.2. Trehalose overproducing transgenic plants.

Gene	Plant	Stress tolerance	Remarks	Reference
Yeast *TPS1* gene driven by CaMV 35S promoter	Tobacco	Improved drought tolerance	Stunted growth, lancet-shaped leaves, reduced sucrose content	Romero et al., 1997
otsA Trehalose-6-phosphatesynthase and *otsB* Trehalose-6-phosphate synthase from *E. coli*	Tobacco	Improved drought tolerance	Increase in leaf size and altered stem morphology	Pilon-Smits et al., 1998
Yeast trehalose-6-phosphate synthase *(TPS1)* gene driven by the promoter of *rbcS* gene	Tobacco	Improved drought tolerance	30 to 50 percent reduction in growth rate compared to control plants	Holmström et al., 1996
Yeast *TPS1* gene driven by promoter of RD29A	Tobacco	Improved drought tolerance	Dwarf and fine shoot, lancet-shaped leaves, and vigorous auxiliary buds	Zhao et al., 2000

TABLE 3.3. Fructan overproducing transgenic plants.

Gene	Plant	Stress phenotype	Remarks	Reference
SacB gene (levansucrase) from *Bacillus subtilis subtilis* fused to the vacuole-sorting signal of carboxypeptidase Y from yeast driven by CaMV 35S promoter	Tobacco	Transgenics produced 59 percent greater dry weight than wild-type plants under drought.	Polyethylene glycol-mediated drought stress. Root growth was enhanced in transgenic.	Pilon-Smits et al., 1995
SacB gene (levansucrase) from *Bacillus subtilis subtilis* fused to the vacuole-sorting signal of carboxypeptidase Y from yeast driven by CaMV 35S promoter	Sugar beet (*Beta vulgaris* L.)	The beet storage roots were 15 to 31 percent heavier in transgenic plants than in their wild-type under drought.	Higher biomass due to increased production of leaves and fibrous roots	Pilon-Smits et al., 1999

TABLE 3.4. Engineering for polyol overproduction.

Gene	Plant	Stress tolerance	Remarks	Reference
E. coli mt1D mannitol-1-phosphate dehydrogenase driven by CaMV 35S promoter	Tobacco	Increased fresh weight, plant height, and flowering under salinity stress	250 mM NaCl for 30 days; new root and leaf formation in transgenic	Tarczynski et al., 1993
E. coli mt1D mannitol-1-phosphate dehydrogenase with amino-terminal transit peptide targeted to chloroplast	Tobacco	Increased resistance to methyl viologen (MV)-induced oxidative stress	Accumulated 100 mM mannitol in chloroplasts; helped in radical-scavenging	Shen et al., 1997
E. coli mt1D mannitol-1-phosphate dehydrogenase	*Arabidopsis*	Increased germination under salinity stress	Mannitol-expressing seeds were able to germinate in up to 400 mM NaCl, whereas control seeds ceased to germinate at 100 mM NaCl.	Thomas et al., 1995

Mannitol-1-phosphate dehydrogenase gene from *E. coli*	Tobacco	Salt stress tolerance	Nonstressed transgenic plants were 20 to 25 percent smaller. Mannitol contributed only 30 to 40 percent of the osmotic adjustment	Karakas et al., 1997
IMT1 myo-inositol O-methyl transferase from common ice plant	Tobacco	Drought and salinity stress	Photosynthetic CO_2 fixation was slightly better	Sheveleva et al., 1997
Stpd1 (sorbitol-6-phosphate dehydrogenase) from apple, driven by CaMV 35S promoter	Japanese persimmon	Tolerance in Fv/Fm ratio under NaCl stress		Gao et al., 2001

TABLE 3.5. Engineering for oxidative stress tolerance.

Gene	Plant	Stress tolerance	Remarks	Reference
Mn-SOD targeted chloroplast or mitochondria	Tobacco	Reduced cellular damage under oxidative stress	Leaf discs used in assay	Bowler et al., 1991
Pea *Cu/Zn SOD* targeted to chloroplast	Tobacco	Retained 90 percent photosynthesis under chilling and high light stress	Short duration (4 h) chilling stress	Gupta et al., 1993
Tobacco *Mn-SOD* driven by CaMV 35S promoter	Alfalfa	Increased tolerance to freezing stress, tolerance to water deficit under field conditions	Three-year field-testing	McKersie et al., 1993, 1996
Mn-SOD targeted to the mitochondria or to the chloroplast	Alfalfa	Field trial—survival and yield; twice the herbage yield of the control plants	No apparent difference in primary freezing injury	McKersie et al., 1999
Arabidopsis Fe-SOD, driven by CaMV 35S promoter, targeted to chloroplast	Alfalfa	Field trial—winter survival and yield	No difference between control and transgenic plants when stressed with methyl viologen	McKersie et al., 2000
Arabidopsis FeSOD under the control of CaMV 35S promoter targeted to chloroplast with Rubisco small subunit transit peptide	Maize	Enhanced tolerance to methyl viologen stress and chilling temperature		Van Breusegem, Slooten, Stassart, Moens, et al., 1999
Tobacco mitochondrial *MnSOD* targeted to chloroplast	Maize	Maintenance of stomatal conductance under methyl viologen stress; better growth under chilling stress	Mn-SOD in transgenic plant is mainly expressed in bundle sheath cells	Van Breusegem, Slooten, Stassart, Botterman, et al., 1999

Arabidopsis Apx3 gene encoding a peroxisomal ascorbate peroxidase	Tobacco	Protection against oxidative stress		Wang et al., 1999
Barley *HvAPX1* encoding a peroxisomal ascorbate peroxidase overexpression	*Arabidopsis*	Heat tolerance	Thermotolerance is not positively correlated to the *HvAPX1* transcript level in the transgenic lines	Shi et al., 2001
Tobacco *Nt107* gene (encoding enzyme with both *GST/GPX* activity) overexpression	Tobacco	Higher seedling growth rate under chilling or salt stress		Roxas et al., 1997
Tobacco *Nt107* gene encoding enzyme with both *GST/GPX* targeted to chloroplast with Rubisco small subunit transit peptide overexpression	Tobacco	Higher seedling growth under cold and salinity stress	Also enhanced the activity of MDHAR; prevention of lipid peroxidation	Roxas et al., 2000
Alfalfa ferritin	Tobacco	Photosynthetic tolerance to iron excess or paraquat		Deak et al., 1999

dehydrogenase (BADH). In *E. coli,* choline dehydrogenase (CDH) produces betaine aldehyde from choline and then betaine aldehyde is converted to glycine betaine by BADH. In *Arthrobacter,* glycine betaine is synthesized from choline by a single enzyme, choline oxidase. In halophilic microorganisms such as *Actinopolyspora halophilia* and *Ectothiorhodospira halochloris,* glycine is methylated to produce glycine betaine by glycine sarcosine methyltransferase (GSMT) and sarcosine dimethylglycine methyltransferase (SDMT) (Nyyssola et al., 2000). Crop plants such as barley, sugar beet, maize, and sorghum, and halophytic plants such as *Atriplex* naturally accumulate glycine betaine under stress (Rhodes and Hanson, 1993). Plants such as rice, *Brassica juncea, Arabidopsis,* and tobacco do not produce glycine betaine naturally, and hence they have been engineered to accumulate glycine betaine, using genes from either microbes or plants. These transgenic plants showed enhanced accumulation of glycine betaine and tolerance to salt, cold, heat, and drought stresses when compared with the control plants (Table 3.1). However, glycine betaine accumulation in transgenic plants was very low compared with that of naturally occurring glycine betaine–accumulating plants under stresses (Rhodes and Hanson, 1993). Since in the transgenic plants only the glycine betaine-synthesizing genes are overexpressed, the limitation for overproduction of glycine betaine appears to be the availability of substrate, i.e., choline or transport of choline, into the chloroplast, which is the cellular site of glycine betaine biosynthesis. Choline is synthesized by the cytosolic enzyme phosphoethanolamine *N*-methyltransferase (PEAMT) that successively methylates phosphoethanolamine to produce phosphocholine. PEAMT is induced by salt stress. Since PEAMT exerts major control over choline synthesis, this enzyme was targeted for overexpression to enhance choline synthesis and thus glycine betaine synthesis in plants. The *PEAMT* gene from spinach was overexpressed in tobacco plants, which were already expressing transgenes encoding chloroplast enzymes for glycine betaine synthesis, CMO, and BADH. The transgenic plants accumulated fiftyfold higher free choline and accumulated thirtyfold more glycine betaine than the control plant (McNeil et al., 2001). However, this concentration of glycine betaine is still very low. Serine is decarboxylated to ethanolamine, which is a precursor for choline biosynthesis in plants. It is not known whether the availability of ethanlamine limits choline production. Hence, availability of substrates and metabolic flux must be considered while engineering for overproduction of osmolytes such as glycine betaine.

Glycine betaine–overproducing transgenic plants showed enhanced tolerance to drought, chilling, freezing, heat, and salt stresses in different plants. However, the mechanism of protection is still not clear. The possibility of glycine betaine as a compatible solute in osmotic adjustment is ruled out as the amount of glycine betaine accumulation in transgenic plants was too low (<5 $\mu mol \cdot g^{-1}$ fresh weight) for osmotic adjustment. At high salt concentrations or at extreme temperatures, in vitro, glycine betaine could stabilize the quaternary structures of proteins and highly ordered state of membranes (Papageorgiou and Murata, 1995). Glycine betaine lowered the phase-transition temperature of an artificial membrane system (Rudolph and Goins, 1991). Hence it is proposed that glycine betaine might stabilize the quaternary structures of proteins and maintain the highly ordered state of membranes under abiotic stresses (Chen and Murata, 2002). Also, glycine betaine appears to improve stress tolerance by protection of photosystem II under stresses (Hayashi et al., 1997; Sakamoto et al., 1998, 2000). Transgenic rice plants overexpressing peroxisomal betaine aldehyde dehydrogenase showed increased ion selectivity under salt stress, i.e., they accumulated more K^+ and less Na^+ and Cl^-. It is not clear how glycine betaine overproduction increases K^+/Na^+ ratio, which is regulated by ion transporters/channels. The effect of glycine betaine accumulation on stress tolerance also varies from species to species. For example, transgenic *Arabidopsis* overexpressing the *CodA* gene could germinate even at 300 mM NaCl stress (Hayashi et al., 1997), but transgenic *Brassica* could germinate only up to 150 mM NaCl stress (Prasad et al., 2000). When transgenic *Arabidopsis, Brassica napus,* and tobacco were compared for stress tolerance, drought tolerance was enhanced only in *Brassica,* while freezing tolerance was enhanced only in *Arabidopsis* (Huang et al., 2000). Hence, the species-specific effect of glycine betaine accumulation on abiotic stress tolerance needs to be understood. In glycine betaine–overproducing transgenics, stress tolerance was assessed at the seedling stage under short durations of stress. *Arabidopsis* transgenics were assessed for heat tolerance at germination by imposing a very short duration (60 min) heat stress (Alia, Sakamoto, and Murata, 1998). However, under field conditions, heat stress persists for longer durations; for example, soil temperature during sowing and seedling establishment of wheat and maize may be up to 45 and 55°C, respectively (Wardlaw and Wrigley, 1994). Hence, testing for short duration tolerances in germination may not be sufficient to allow extrapolation to the long-duration stress tolerance required for crop plants in the field. Crop plants such as rice, wheat, *Brassica juncea,* and *Brassica napus* were engineered to overproduce glycine betaine, but their stress tolerances in yield need to be assessed under field conditions.

Trehalose

Trehalose is a nonreducing disaccharide composed of two alpha-glycosidically linked glucose units. Trehalose is synthesized from UDP glucose and glucose-6-phosphate in two reactions by trehalose-6-phosphate synthase and trehalose-6-phosphate phosphatase. In *E. coli,* trehalose-6-phosphatesynthase (otsA) catalyzes the synthesis of trehalose-6-phosphate from UDP glucose and glucose-6-phosphate, and trehalose-6-phosphate synthase (otsB) catalyzes trehalose synthesis from trehalose-6-phosphate, while in yeast trehalose-6-phosphate synthase (TPS1) catalyzes both the steps of trehalose synthesis. In yeast, trehalose is a regulator of the carbohydrate allocation to glycolytic flux and stress survival strategies. Trehalose levels correlate with cellular stress resistance to heat and desiccation in yeast and desert resurrection plants such as *Selaginella lepidophylla* and *Myrothamnus flabellifolius* (Hottiger et al., 1987; Müller et al., 1995). Transgenic tobacco plants engineered to overexpress yeast trehalose-6-phosphate synthase *(TPS1)* gene under the control of a CaMV 35S or rbcS or RD29A promoter accumulated high concentrations of trehalose up to 0.17 $mg \cdot g^{-1}$ fresh weight compared with 0.02 $mg \cdot g^{-1}$ fresh weight in control plants. These transgenics exhibited multiple phenotypic alterations, including stunted growth, although they were drought tolerant (Table 3.2). Transgenic tobacco plants expressing *E. coli otsA* and *otsB* genes did not exhibit growth defects. Under drought stress, these transgenic tobacco plants produced 28 to 39 percent higher dry weight, mainly due to increased leaf production, and more efficient photosynthesis than control plants. Dehydration tolerance of detached leaves and more negative leaf osmotic potentials of transgenics indicated better osmotic adjustment in transgenic plants than in the control plants. Transgenic plants showed higher levels of glucose, fructose, and sucrose, which might have contributed to osmotic adjustment (Pilon-Smits et al., 1998). Transgenic potato plants overexpressing the yeast *TPS1* gene also showed enhanced drought tolerance (Yeo et al., 2000). All these trehalose-overproducing transgenics exhibited substantial changes in morphology and accumulated higher levels of nonstructural carbohydrates.

Fructans

Fructans are polyfructose molecules used as storage carbohydrates by many plants. Levansucrase synthesizes fructan from fructose. Tobacco plants engineered for fructan overproduction grows normally under nonstress conditions and showed enhanced stress tolerance (Table 3.3). Although the levansucrase gene was expressed under the control of the CaMV 35S promoter, fructan production in transgenic plants was enhanced sevenfold under drought compared with that of normal growth conditions. The fructan-producing plants accumulated 59 percent higher dry weight than the wild type under drought (Pilon-Smits et al., 1995). Transgenic sugar beet (*Beta vulgaris* L.) accumulated fructan to about 0.5 percent of their dry weight and performed better than wild-type plants under drought, as the beet storage roots yield was 15 to 31 percent higher in transgenics than in the wild-type plants (Pilon-Smits et al., 1999). Fructan overproduction also imparts freezing-stress tolerance (Konstantinova et al., 2002). Fructans enhanced the root branching and growth, which are important traits for increased water uptake under drought stress and also protected phosphatidylcholine vesicles/liposomes during freeze-drying (Hincha et al., 2000). Hence, it appears that fructans confer stress tolerance by protecting the integrity of biomembranes and enhancing root growth under stress.

Polyols

Polyols are sugar alcohols, such as glycerol, mannitol, sorbitol, and D-ononitol, that act as osmoprotectants in halophytic plants (Yancey et al., 1982). Mannitol is synthesized from fructose-6-P. Fructose-6-P is converted to mannitol-1-P by mannose-6-P isomerase. Mannitol-1-P is then reduced to mannitol by mannitol-1-phosphate dehydrogenase. Tarczynski and colleagues (1993) first demonstrated that genetic engineering for mannitol overproduction could improve stress tolerance. Polyol-overproducing transgenics showed enhanced tolerance to drought and salinity stresses (Table 3.4). Karakas and colleagues (1997) analyzed the contribution of mannitol to osmotic adjustment in tobacco plants transformed with a mannitol-1-phosphate dehydrogenase gene. Under drought stress, mannitol accumulation was enhanced but did not confer drought tolerance. These transgenic plants were 20 to 25 percent smaller than the wild type. Salt stress reduced dry weight by 44 percent in control plants, which did not show osmotic adjustment, while no significant weight reduction was observed in transgenic plants, which adjusted osmotically by 0.57 MPa. Since mannitol accounted for only 30 to 40 percent of the osmotic adjustment, mannitol might have indirectly enhanced osmotic adjustment and salt tolerance (Karakas et al., 1997).

In halophytes such as *Mesembryanthemum crystallinum*, myo-inositol serves as substrate for synthesis of osmoprotectants such as D-ononitol and D-pinitol under salt stress. D-ononitol is synthesized from myo-inositol by myo-inositol *O*-methyl transferase. Tobacco transgenics overexpressing myo-inositol *O*-methyl transferase accumulated a higher level of D-ononitol and showed improved photosynthesis under drought and salt stress, and also recovered faster than wild type (Table 3.4; Sheveleva et al., 1997).

Sorbitol is a primary photosynthetic product of some important fruit and nut crops such as *Malus, Prunus, Pyrus, Eriobotrya,* and *Rubus.* Glucose-6-P is converted into sorbitol-6-P by sorbitol-6-P dehydrogenase. Sorbitol is synthesized from sorbitol-6-P by sorbitol-6-P phosphatase. Japanese persimmon (*Diospyros kaki* Thunb.), a major tree-fruit crop in Japan, was engineered to overproduce sorbitol and showed higher photosynthetic tolerance as measured by leaf Fv/Fm ratio under NaCl stress (Table 3.4; Gao et al., 2001).

Polyol-overproducing transgenic plants showed growth reduction (Karakas et al., 1997). Transgenic tobacco plants expressing *Stpd1* (sorbitol-6-phosphate dehydrogenase) from apple under the control of a CaMV 35S promoter showed higher sorbitol accumulation (up to 130 $\mu mol \cdot g^{-1}$ fresh weight), causing necrotic lesions of leaves and infertility (Sheveleva et al., 1998). Transgenic Japanese persimmon plants accumulated 14.5 to 61.5 μmol sorbitol$\cdot g^{-1}$ fresh wt in leaves, did not show abnormal growth, but showed an Fv/Fm value less than that of wild type even under normal growth conditions (Gao et al., 2001). These growth abnormalities might be due to the alteration of flux toward myo-inositol biosynthesis, which in turn is a substrate for biosynthesis of inositol phosphates. Thus, change in the inositol pool might affect many signaling pathways directly, and it might indirectly affect the availability of free auxin as inositol is conjugated to auxins to regulate auxin action (Cohen and Bandurski, 1982). Under salt and drought stresses, transgenic tobacco overexpressing the myo-inositol *O*-methyl transferase gene accumulated D-ononitol in the cytosol up to 600 mM, but D-ononitol did not enter the vacuole (Sheveleva et al., 1997). Hence, it is essential to understand the metabolic flux and compartmentation of polyols among cell organelles in order to utilize engineered polyol overproduction for stress tolerance.

Proline

Plants accumulate free proline in response to abiotic stresses such as drought, salinity, and low temperature. In addition to the role of proline as an osmolyte for osmotic adjustment, it also stabilizes subcellular structures (membranes and proteins), scavenges free radicals, and buffers cellular redox potential under stress. Proline accumulation under stress has been correlated with stress tolerance of plants. Transgenic tobacco, *Arabidopsis,* and rice plants that overproduce proline confirmed the role of proline in stress tolerance of higher plants. Tobacco plants were engineered to overproduce proline by overexpression of moth bean (*Vigna aconitifolia* L.) [Δ^1]-pyrroline-5-carboxylate synthetase, a bifunctional enzyme, with activities of γ-glutamyl kinase and glutamyl-5-semialdehyde dehydrogenase, able to catalyze the conversion of glutamate to [Δ]-pyrroline-5-carboxylate, which is then reduced to proline. The transgenic plants produced ten- to eighteenfold more proline than the control plants and showed enhanced root biomass and flower development under drought stress. However, overproduction of proline did not contribute to osmotic adjustment, as the osmotic potentials of leaf sap from transgenic plants were less decreased under water stress than were those of control plants (Kishor et al., 1995). Rice plants were engineered to overaccumulate proline under stress by expressing moth bean [Δ^1]-pyrroline-5-carboxylate synthetase cDNA under the control of ABA-responsive element from the barley *Hva22* gene. Transgenic plants (eight-week-old) showed stress-inducible proline accumulation under water stress. The shoot biomass of transgenic plants was not significantly higher than that of wild-type plants, but transgenics produced significantly higher root biomass under water stress. Transgenic plants also showed faster recovery after a short period of salt stress (Zhu et al., 1998).

Proline biosynthesis is regulated by the end-product inhibition of γ-glutamyl kinase activity, which might limit the proline accumulation under stress. Feedback inhibition of proline on g-glutamyl kinase activity of moth bean [Δ^1]-pyrroline-5-carboxylate synthetase was eliminated through site-directed mutagenesis of [Δ^1]-pyrroline-5-carboxylate synthetase gene (Zhang et al., 1995). Tobacco transgenics overexpressing this mutated moth bean [Δ^1]-pyrroline-5-carboxylate synthetase gene accumulated two-fold more proline than the transgenics overexpressing the normal [Δ^1]-pyrroline-5-carboxylate synthetase gene. This higher accumulation of proline significantly reduced free radical levels and improved tolerance of the transgenic seedlings to 200 mM NaCl (Hong et al., 2000). Proline is metabolized to glutamine via pyrroline-5-carboxylate and glutamyl-5-semialdehyde, catalyzed by proline dehydrogenase and pyrroline-5-carboxylate dehydrogenase, respectively. *Arabidopsis* proline dehydrogenase gene is repressed by dehydration, but induced by both the application of exogenous proline and rehydration, indicating that the proline level is also regulated by degradation. Nanjo and colleagues (1999) engineered *Arabidopsis* plants with an antisense proline dehydrogenase cDNA that resulted in enhanced accumulation of proline. These antisense transgenics showed constitutive freezing tolerance (–7°C) and were more tolerant to high salinity (600 mM NaCl). The stress tolerance of these transgenic plants was also compared with *rd29A::DREB1A* transgenic plants. The *RD29A* promoter-driven *DREB1A* transgenic plants showed tolerance to drought, high salinity, and freezing (Kasuga et al., 1999). The freezing tolerance of the antiproline dehydrogenase transgenic plants was higher than that of the control plant but lower than that of the *rd29A::DREB1A* transgenic plants, while short duration (up to 22 min) NaCl stress tolerance was higher in antiproline dehydrogenase transgenic plants than in *rd29A::DREB1A* transgenic plants (Nanjo et al., 1999). Soybean plants expressing antisense [Δ^1]-pyrroline-5-carboxylate reductase gene under the control of inducible heat-shock promoter (IHSP) showed that induction of IHSP (37°C) results in decreased proline levels. The antisense plants failed to survive a six-day drought stress at 37°C while the wild-type plants survived. This study further supports the essential role of proline in osmotic stress tolerance (De Ronde et al., 2000). This evidence suggests that proline accumulation increases stress tolerance mainly by protection against reactive oxygen species rather than by increasing osmotic adjustment.

OXIDATIVE-STRESS TOLERANCE

Reactive oxygen species (ROS), namely superoxide radicals ($O_2^{\cdot-}$), hydrogen peroxide (H_2O_2), and hydroxyl radicals ($OH^{\cdot}$), are produced in aerobic cellular processes such as mitochondrial and chloroplast electron transport, and oxidation of glycolate (photorespiration), xanthine, and glucose. The ROS cause oxidative damage to membrane lipids, proteins, and nucleic acids. Hence, organisms have evolved various antioxidants and detoxifying enzymes to efficiently scavenge ROS. The antioxidants employed by plants are ascorbate, glutathione, α-tocopherol, and carotenoids, while detoxifying enzymes include superoxide dismutase (SOD), catalase, peroxidase, and enzymes of the ascarbate-glutothione cycle. Plant strategies for oxidative stress tolerance can be grouped into (1) avoidance of ROS production, (2) detoxification of ROS, and (3) repair of ROS-mediated damage. Under stress conditions, photosynthetic CO_2 reduction decreases, which results in accumulation of NADPH. This leads to decreased availability of nicotinamide adenine dinocleotide phosphate (NADP) to accept electrons from photosystem I. Under these conditions, reduced ferridoxin donates an electron to oxygen to produce O_{2-}, which is further converted into H_2O_2 by thylakoid/stromal SOD. The components of mitochondrial electron transport such as respiratory complexes I (NADH dehydrogenase) and III (ubiquinol-cytochrome bc1 reductase) can directly donate electrons to oxygen to produce ROS. The H_2O_2 produced from the photosynthetic/respiratory electron transport is scavenged by antioxidants such as ascorbate and glutathione primarily via the ascorbate/glutathione pathway. α-Tocopherol directly scavenges singlet oxygen and peroxides. SOD detoxifies superoxide to H_2O_2. H_2O_2 is detoxified to water and oxygen by catalase and/or by ascorbate-glutathione cycle. Based on the prosthetic group requirement, SOD is classified into Fe-SOD (localized in chloroplast), Mn-SOD (localized in mitochondria and peroxisomes), and Cu-Zn SOD (localized in chloroplast, cytosol, and apoplastic space). Biotic and abiotic stresses upregulate genes that encode detoxifying enzymes. Comparison between stress-tolerant and susceptible cultivars revealed a positive correlation between the capacity to scavenge ROS and stress tolerance. Alleviation of oxidative damage by scavenging ROS seems to be an important strategy of plants to tolerate stresses (Noctor and Foyer, 1998; Hasegawa et al., 2000; Møller, 2001; Zhu, 2002). Hence, several attempts have been made to engineer stress-tolerant plants by enhancing ROS scavenging capacity (Table 3.5).

Transgenic tobacco overexpressing chloroplast targeted to pea Cu-Zn SOD showed resistance to environmental stress in terms of photosynthesis. Transgenic plants were subjected to chilling stress at moderate (500 μmol of quanta·m^{-2}·s^{-1}) or high (1500 μmol of quanta·m^{-2}·s^{-1}) light stress. Transgenic SOD plants maintained nearly 90 percent of their photosynthetic capacity following exposure to chilling at high light intensity for 4 h (Gupta et al., 1993). Transgenic maize overexpressing Mn-SOD or Fe-SOD targeted to chloroplasts showed tolerance to oxidative stress and chilling stress (Van Breusegem, Slooten, Stassart, Botterman, et al., 1999; Van Breusegem, Slooten, Stassart, Moens, et al., 1999). Oxidative stress tolerance is an essential component of winter survival of alfalfa. McKersie and colleagues (1993, 1996, 1999, 2000) engineered alfalfa plants that overexpress different forms of SODs targeted to the mitochondria or chloroplasts. Transgenic alfalfa showed enhanced tolerance to oxidative stress generated by diphenyl ether herbicide (acifluorfen), increased regrowth after freezing stress, and drought tolerance as determined by chlorophyll fluorescence, electrolyte leakage, and regrowth from crowns after relieving the drought stress. Field trials also confirmed the tolerance of transgenic plants to oxidative stress in field drought environments (McKersie et al., 1996). Comparison of 33 primary transgenics and control plants for winter survival and yield showed that transgenic plants had higher survival rates and foliage yield (McKersie et al., 1999). Alfalfa plants engineered to overexpress *Arabidopsis* Fe-SOD in chloroplasts showed better winter survival. However, no association was found between Fe-SOD activity under stress and total shoot dry matter production in these transgenics (McKersie et al., 2000).

Hydrogen peroxide is detoxified by catalase or by the ascorbate-glutathione cycle. The activity of the key enzyme of the ascorbate-glutathione cycle, ascorbate peroxidase (APX), is increased under drought (Smirnoff and Colombe, 1988; Mittler and Zilinskas, 1994), salt (Lopez et al., 1996), heat stress, and ABA (Sato et al., 2001; Shi et al., 2001).

Arabidopsis transgenic plants overexpressing a peroxisomal type *APX* from barley *(HvAPX1)*, showed heat tolerance (35°C with humidity of 70 percent in growth chamber for five days), but the expression level of the *APX* gene did not correlate with heat tolerance (Shi et al., 2001). Transgenic tobacco overexpressing *Arabidopsis APX3* gene showed oxidative stress tolerance when the oxidative stress was created by aminotriazole (inhibitor of glyoxysomal and peroxisomal catalase activity) while it did not show oxidative stress tolerance when the oxidative stress was created by treatment with PS II inhibitor, methyl viologen (Wang et al., 1999). Hence, it is important to target enzymes at the site where the stress-induced ROS production is essential for detoxification and stress tolerance. One of the key enzymes of ascorbate-glutathione cycle, glutathione S-transferase/glutathione peroxidase (GST/GPX) was overexpressed in tobacco. These transgenic plants were tolerant to chilling and salt stresses (Roxas et al., 1997, 2000). Iron reacts with oxygen to produce very reactive hydroxyl radicals through a Fenton reaction. Transgenic tobacco plants expressing alfalfa ferritin in chloroplasts or in the cytosol-retained photosynthesis under oxidative stress generated by iron excess or paraquat treatment (Deak et al., 1999).

McKersie and colleagues (1999, 2000) found no correlation between SOD activities of individual transgenic plants in the greenhouse and improved field performance. Also, when the freezing tolerance of field-acclimated plants were examined in leaves or roots by tetrazolium staining, which measures cell viability, no difference was observed between transgenic and control plants. Hence, the better performance of transgenic alfalfa was not due to a difference in primary freezing injury. Some investigations showed that engineered overexpression of SOD did not contribute to stress tolerance. Transgenic tobacco overexpressing chloroplastic Cu-Zn SOD (Tepperman and Dunsmuir, 1990; Pitcher et al., 1991) or Fe-SOD with chloroplast targeting sequence (Van Camp et al., 1996) and transgenic cotton overexpressing Mn-SOD (Payton et al., 1997) did not show significant improvement in oxidative stress tolerance. However, many other transgenic plants engineered for enhanced ROS scavenging capacity showed abiotic stress tolerance. This variation in the results from transgenic plants that express SOD might be due to the fate of H_2O_2 produced by the SOD. Since $O_2^{\cdot -}$ scavenging is dependent on catalazes and ascorbate glutathione cycle, engineered alteration in SOD expression and activity might lead to accumulation of H_2O_2, which can affect plant growth by inducing oxidative damage or stress signaling. H_2O_2 acts as a signaling molecule in biotic and abiotic stress signaling. In plant-pathogen interactions, H_2O_2 signaling induces an orchestrated sequence of reactions leading to cell death or plant immunity (Alvarez et al., 1998). H_2O_2 is also an intermediate in ABA signaling in guard cells (Pei et al., 2000). In field environments, crop plants often experience more than one biotic and abiotic stress. Hence, careful examination of the effect of engineered alterations in the antioxidant system on crop productivity in normal environments as well as under multiple stress environments under field conditions is needed.

STRESS PROTEINS

Abiotic stresses induce and up-regulate many proteins, and the expression levels of some of these proteins are positively correlated with the abiotic stress tolerance of plants. Efforts have been made to understand the function of these stress proteins using transgenic plants.

Dehydration-Stress-Induced Proteins

In higher plants, dehydration stress induces several proteins in vegetative tissues, which are related to late-embryogenesis-abundant (LEA) proteins. The correlation between LEA protein accumulation in vegetative tissues and stress tolerance in various plant species indicated its protective role under dehydration stress (reviewed by Ingram and Bartels, 1996). For example, in salt-tolerant indica rice varieties, LEA proteins are induced to higher levels by salt or ABA, while salt-sensitive varieties do not show high accumulation (Moons et al., 1995). LEA proteins are rich in hydrophilic amino acids and hence might help in dehydration tolerance by binding water so as to maintain the minimum cellular water requirement, and by binding to other proteins to prevent denaturation and to protect cytosolic structures (reviewed by Ingram and Bartels, 1996). To investigate the function of LEA proteins in dehydration tolerance, Xu and colleagues (1996) engineered rice plants with a barley *LEA* gene, *HVA1*, fused to rice actin 1 promoter. T2 generation of transgenic rice plants accumulated a high level of HVA1 protein (0.3 to 2.5 percent of total soluble protein) constitutively in both leaves and roots of transgenic rice plants. Water stress was imposed on three-week-old rice seedlings grown in a tray. Transgenic plants showed better stress tolerance and faster recovery once the stress was relieved. Similarly, under 200 mM NaCl stress, transgenic plants also performed better in terms of seedling growth. Stress symptoms such as wilting, dying of old leaves, and necrosis of young leaves were delayed in transgenic rice under both salt and water stress (Xu et al., 1996). Spring wheat (*Triticum aestivum* L.) engineered with the barley *HVA1* gene under the control of maize ubiquitin *(ubi1)* promoter, showed high levels of expression of the *HVA1* gene. T3 progenies of selected transgenic wheat were grown in pots under greenhouse conditions and tested under moderate water deficit. Two transgenic lines that showed high HVA1 expression had significantly higher water-use efficiency (WUE, 0.66 to 0.68 $g \cdot kg^{-1}$), as compared to that of non-expressing transgenic (0.57 $g \cdot kg^{-1}$) and nontransgenic controls (0.53 $g \cdot kg^{-1}$). Transgenic plants, which showed higher WUE, also produced significantly higher total biomass when compared to the control plants under soil-water deficit conditions (Sivamani et al., 2000). Molecular analysis of protein turnover, activity of some essential proteins, and water relations under dehydration stress in these transgenics may shed some light on the precise mode of action of HVA1 in the protection against dehydration stress.

Heat-Shock Proteins

Being poikilothermic, the temperature of land plants changes with the ambient temperature, although a few plants can control their temperatures by several degrees above (through alternate oxidase respiration) or below (through transpirational cooling) the ambient temperature. Hence, survival and productivity of plants depends upon their environmental temperatures. Synthesis of heat-shock proteins (HSPs) in response to heat shock is a highly conserved cellular response among all organisms. The induction temperature of HSPs is dependent on the temperature at which each species ordinarily grows. Induction of HSPs is associated with acquired thermotolerance in almost all organisms examined. Some HSP homologs are synthesized even during normal growth conditions and serve as molecular chaperones. Under heat-stress conditions HSPs prevent protein denaturation, prevent the aggregation of denatured proteins, and repair the denatured proteins. HSPs are classified into five classes based on their molecular weight (HSP100, HSP90, HSP70, HSP60, and low-molecular weight HSPs). Heat-shock activated transcription factors (HSFs) form homo-trimers upon exposure to heat stress and then bind to the heat-shock elements of HSP genes to induce HSP expression. Transgenic *Arabidopsis* expressing HSFs or HSF-glucuronidase translational fusion proteins showed derepression of HSPs under normal growth temperatures. The level of constitutive thermotolerance was enhanced in the transgenic plants (Lee et al., 1995; Prändl et al., 1998). Transgenic carrot cell lines expressing the *Hsp17.7* gene driven by the *CaMV 35S* promoter and an antisense of the *Hsp17.7* gene driven by *HSP17.7* promoter were produced to examine the role of small-molecular-weight HSPs in plants. Transgenic plants over-expressing *HSP17.7* showed enhanced thermotolerance at 50°C for 3 h as determined by electrolyte leakage, while the antisense plants showed reduced thermotolerance (Malik et al., 1999). Transgenic *Arabidopsis* expressing antisense *HSP70* driven by the *HSP70* promoter (Lee and Schoffl, 1996) or the *CaMV 35S* promoter (Queitsch et al., 2000) showed decrease in thermotolerance. Antisense suppression of the *HSP101* gene also resulted in loss of basal thermotolerance during seed germination. Transgenic *Arabidopsis* overexpressing *HSP101* driven by the *CaMV 35S* promoter showed enhanced thermotolerance (recovery after 45°C heat shock for 60 min) without compromising growth at normal temperatures (Queitsch et al., 2000). Thus, engineering for enhanced HSP production may be a promising approach to improve the thermotolerance of plants.

Salt-Stress-Induced Proteins

The *Alfin1* gene of *Medicago sativa* encodes a member of the zinc-finger family transcription factors, and its expression is correlated with NaCl tolerance. In vitro, *Alfin1* binds to the promoter of *MsPRP2,* which encodes a root-specific cell-wall protein induced by NaCl. The *Alfin1* gene appears to be conserved in alfalfa, rice, and *Arabidopsis* (Winicov and Bastola, 1997). The role of Alfin1 in salt-stress tolerance was examined in transgenic alfalfa expressing *Alfin1* driven by the *CaMV 35S* promoter in the sense and antisense orientations. Although overexpression lines did not show any growth defect, the antisense transgenic plants grew poorly in soil in a normal environment, demonstrating that *Alfin1* expression is essential for normal plant development. *Alfin1* overexpression enhanced the root growth significantly under both normal and saline conditions, while the antisense plants showed poor root growth (Winicov and Bastola, 1999; Winicov, 2000). The tobacco-stress-induced-gene 1 *(Tsi1)* encodes a DNA-binding protein with EREBP/AP2 DNA-binding motif, which is involved in defense- and drought-responsive gene expression. *Tsi1* gene expression was rapidly induced by salt stress, but not by drought or ABA. Overexpression of *Tsi1* in tobacco enhanced retention of chlorophyll content when leaves were floated in 400 mM NaCl solution for 48 or 72 h (Park et al., 2001). Further detailed studies are needed to assess the role of Alfin1 and TSI1 in salt-stress tolerance as it is not clear at present whether these proteins and their targets are involved in ion/osmotic homeostasis or detoxification.

Cold-Stress Proteins

Tropical and subtropical plants are incapable of cold acclimation. Most temperate plants can acquire tolerance to freezing temperatures by the prior exposure to low nonfreezing temperatures, a process called cold acclimation. This is achieved by altering the expression of many genes, changes in membrane lipid composition, accumulation of compatible osmolytes (proline, betaine, polyols, and soluble sugars), a transient rise in ABA, and reduction or cessation of plant growth (Levitt, 1980). Cold stress induces genes that encode fatty acid desaturases, chaperones, lipid transfer proteins, proteins involved in osmoprotectant biosynthesis, antifreeze proteins, and signal transduction components such as transcription factors, kinases, and phosphatases (reviewed by Viswanathan and Zhu, 2002; Xiong et al., 2002). A change in membrane fluidity is one of the immediate effects of cold stress and hence membranes may act as the primary sensor of low temperature (Levitt, 1980). Cold acclimation is accomplished by the production of stress proteins similar to LEA proteins. In *Arabidopsis,* the genes that encode cold-stress proteins are called *rd* (responsive to dehydration), *erd* (early responsive to dehydration), *lti* (low-temperature induced), *kin* (cold-induced), and *cor* (cold-regulated). These genes can be collectively called cold-responsive *(COR)* genes, although many of them are also induced by dehydration (due to water deficit or high salt) and ABA. Analysis of the promoter regions of *COR* genes showed that many of them contain dehydration-responsive elements (DRE) or C-repeat (CRT), as well as ABA-responsive elements or ABREs. Genetic evidences from cold-signaling mutants showed that cold-responsive genes may be regulated through both ABA-dependent and independent pathways (Ishitani et al., 1997). Transcription factors that regulate cold-responsive genes are called CBFs (C-repeat binding proteins) or dehydration-responsive element-binding proteins (DREBs). These DREBs are classified into two classes, DREB1 (DREB1A=CBF3, DREB1B=CBF1, and DREB1C=CBF2) and DREB2 (DREB2A and DREB2B). Some of the components of the cold-stress signaling pathway, which leads to the induction of *COR* genes, have been identified and engineered enhancement in their expression in transgenic-plants-conferred stress tolerance (Table 3.6).

In higher plants, cold-induced rigidification at microdomains in the plasma membrane may lead to cytoskeletal rearrangement, induction of stretch-sensitive Ca^{2+} channels, and increases in cytosolic Ca^{2+} that can trigger cold-induced gene expression and cold acclimation (Örvar et al., 2000; Sangwan et al., 2001). Increases in cytosolic Ca^{2+} during cold stress may also be brought out by activation of Ca^{2+} channels by inositol (1,4,5)-triphosphate (IP_3). Genetic evidences from *fiery1 (fry1)* mutant studies showed that IP_3 metabolism plays a crucial role in cold, ABA, salt, and dehydration signaling (Xiong et al., 2001). The *fry1* mutant showed enhanced induction of cold-regulated genes (*RD29A, KIN1, COR15A, COR47A,* and *ADH*) in response to cold, ABA, salt, or dehydration stress. FRY1 encodes a bifunctional enzyme with 3'(2'), 5'-bisphosphate nucleotidase and inositol polyphosphate 1-phosphatase activities. The FRY1 protein was found to be mainly acting as an inositol polyphosphate 1-phosphatase, which functions in the catabolism of IP_3. Thus, the *fry1* study provides the first genetic evidence showing the involvement of IP_3 in ABA and abiotic signaling in plants (Xiong et al., 2001).

Cold-stress–induced intracellular Ca^{2+} signatures are probably sensed by various Ca^{2+} binding proteins, including calcium-dependent protein kinases (CDPKs) (Monroy et al., 1993; Tähtiharju et al., 1997; Martín and Busconi, 2001) or protein phosphatases. Arabidopsis transgenic plants expressing a protein phosphatase 2C (*AtPP2CA)* in antisense showed superinduction of cold-responsive genes *(RAB18, RCI2A/LTI6, RD29A/LTI78)* during cold stress and conferred better freezing tolerance. Also, cold-responsive gene expression and cold acclimation were accelerated in the *AtPP2CA* antisense plants, i.e., less time of cold stress was required when compared to that of wild-type (WT). Since the expression pattern of *CBF1, CBF2, CBF3,* and *DREB2* was unaltered in *AtPP2CA* antisense plants, the enhanced expression of cold-responsive genes is not mediated through the CRT/DRE element but through ABRE elements (Tähtiharju and Palva, 2001). Hence, AtPP2CA is a negative regulator of cold-stress signaling that acts through ABA-dependent pathways (Tähtiharju and Palva, 2001).

In *Arabidopsis, COR15a* encodes a novel 15 kDa polypeptide that is targeted to the stromal compartment of the chloroplasts. Transgenic *Arabidopsis* overexpressing *COR15a* showed reduced freezing damage to photosystem II and enhanced freezing tolerance of both chloroplasts and protoplasts in vitro but showed no discernible effect on the freezing tolerance of the whole plant (Steponkus et al., 1998). Constitutive overexpression of *CBF*s strongly activated the expression of many *COR* genes and imparted constitutive freezing tolerance to transgenic *Arabidopsis* plants (Jaglo-Ottosen et al., 1998; Liu et al., 1998; Kasuga et al., 1999). Stress-induced expression of *CBF3* driven by the *RD29A* promoter resulted in a constitutive low-level expression of cold-regulated genes and enhanced expression under cold, dehydration, and salt stresses in transgenic *Arabidopsis* (Kasuga et al., 1999). *CBF3* overexpression in *Arabidopsis* resulted in elevated accumulation of proline and total soluble sugars, including sucrose, raffinose, glucose, and fructose. The increase in the proline levels is due to increased expression of the key proline biosynthetic enzyme Δ^1-pyrroline-5-carboxylate synthase (Gilmour et al., 2000). Constitutive overexpression of *Arabidopsis CBF1, CBF2,* and *CBF3* in transgenic *B. napus* also enhanced the freezing tolerance of acclimated and nonacclimated transgenic plants as measured by freezing-induced electrolyte leakage (Jaglo et al., 2001). Transgenic tomato overexpressing *CBF1* showed enhanced chilling tolerance but not freezing tolerance. The enhanced chilling tolerance was at least in part due to induction of catalase,

TABLE 3.6. Engineering cold-stress signaling components and their target genes.

Gene	Transgenic plant	Stress tolerance	Remarks	Reference
CBF1 (=DREB1B) of *Arabidopsis* under the control of CaMV 35S promoter	*Arabidopsis*	Constitutive freezing tolerance of seedlings; freezing-induced electrolyte leakage is less	No obvious effects on plant growth and development. Freezing tolerance is due to constitutive expression of *COR* genes.	Jaglo-Ottosen et al., 1998
CBF1 from *Arabidopsis* under the control of CaMV 35S promoter	Tomato	Chilling tolerance in survival rate, chlorophyll fluorescence value, and radical elongation	Expression of a catalase gene was induced in transgenic plant. The level of H_2O_2 was lower and tolerance to methyl viologen was higher in transgenic tomato; however, transgenic plants were not freezing tolerant.	Hsieh et al., 2002
CBF3 (=*DREB1A*) C-Repeat Binding Factor 3 under the control of CaMV 35S promoter or RD29A promoter	*Arabidopsis*	Constitutive freezing tolerance and tolerance to salinity and drought stress	Enhanced expression of *COR* genes and Δ^1-pyrroline-5-carboxylate synthase gene	Kasuga et al., 1999

CBF1, *CBF2*, and *CBF3* under the control of CaMV 35S promoter	*B. napus*	Increase in constitutive and acquired freezing tolerance as measured by electrolyte leakage	Higher expression of *Bn115* and *Bn28* (*COR* genes of *B. napus*) in transgenic both under normal and cold stress	Jaglo et al., 2001
SCOF1 (Soybean Cold-inducible Factor 1) driven by CaMV 35S	*Arabidopsis*	Constitutive freezing tolerance	Regulates ABA-mediated signaling	Kim, et al., 2001
ABI3, transcriptional activator driven by CaMV 35S promoter	*Arabidopsis*	Faster cold acclimation and significantly higher freezing tolerance	Enhanced expression of ABA/cold/drought-responsive genes such as *RAB18* and *LTI78*.	Tamminen et al., 2001
Antisense *AtPP2CA*, protein phosphatase 2C driven by CaMV 35S promoter	*Arabidopsis*	Freezing tolerance	Enhanced expression *COR* genes under stress	Tähtiharju and Palva, 2001
COR15a overexpression	*Arabidopsis*	Freezing tolerance of chloroplast (as measured by chlorophyll fluorescence) and protoplast	Prevents freeze-induced lamellar-to-hexagonal II phase transitions in cellular membranes	Steponkus et al., 1998

which prevents oxidative damage. It is interesting to note that the *CBF1* constitutive overexpression also resulted in a dwarf phenotype, which can be overcome by GA_3 (gibberellic acid) treatment (Hsieh et al., 2002). These studies provided functional evidence for the involvement of *CBF*s in cold signal transduction across species; they act as nodes of cross talk between chilling, freezing, dehydration, and salt-stress signaling pathways and also offer promising targets for engineering multistress-tolerant transgenic plants of agronomic value.

Kim and colleagues (2001) have cloned a novel cold-inducible zinc finger protein from soybean, SCOF1 (soybean cold-inducible factor 1), using the mRNA differential display technique. SCOF1 contains two C_2H_2-type zinc fingers. SCOF1 could also be induced by exogenous application of ABA. Overexpression of SCOF1 under the control of the *CaMV 35S* promoter in *Arabidopsis* resulted in constitutive expression of cold-responsive genes (*COR15a, COR47,* and *RD29B*) and constitutive freezing tolerance. However, SCOF1 does not bind directly to ABRE or DRE/CRT *cis*-elements. Transactivation experiments in *Arabidopsis* protoplast revealed that SCOF1 enhanced the DNA binding activity of SGBF1 (soybean G-box binding factor 1, a bZIP transcription factor). SGBF1 is cold and ABA inducible. Thus, SCOF1 interacts with SGBF1 to regulate the cold-responsive gene expression through activation of ABRE in the ABA-dependent pathway of cold-stress signal transduction (Kim et al., 2001). Further evidence for the role of ABA in cold acclimation came from transgenic *Arabidopsis* ectopically expressing the *ABI3* gene, which encodes a seed-specific transcriptional activator. Ectopic expression of *ABI3* increased ABA-induced accumulation of transcripts of ABA/cold/drought-responsive genes such as *RAB18* and *LTI78.* The *ABI3* transgenic plants acclimated faster than the wild-type plants and showed higher freezing tolerance than the wild-type plants (Tamminen et al., 2001).

A classical genetic study on freezing tolerance led to the identification of *sfr* (sensitive to freezing) mutants in *Arabidopsis* (Warren et al., 1996). In the *sfr6* mutant, the cold-induced expression of *KIN1, COR15a,* and *RD29A* was abolished and also osmotic stress and ABA-induced expression of *KIN1* was inhibited. However, the expression of *CBF1, CBF2, CBF3,* and *ATP5CS1* was not affected by the *sfr6* mutation. *Sfr6* appears to specifically affect the transactivation of *DRE/CRT* genes by *CBF*s (Knight et al., 1999). Identification of the *esk1* (*eskimo1,* constitutively freezing tolerant) mutant of *Arabidopsis* revealed proline accumulation as a possible mechanism of acquired freezing tolerance (Xin and Browse, 1998). The *esk1* mutant did not differ in its cold-responsive gene expression from WT, but maintained thirtyfold higher proline level due to higher expression of the *P5CS* gene (delta-pyrroline 5-carboxylate synthase) when compared with the WT plants in normal growing conditions. In *esk1,* the expression of *RD29A, COR47,* and *COR15a* were similar to that in WT, but *RAB18* expression was enhanced by three- or fourfold, and *COR6.6* expression was significantly reduced in the mutant. Further understanding of freezing tolerance in these mutants needs molecular cloning of *sfr6* and *esk1.*

Genetic analysis of mutants isolated from chemically mutagenized *Arabidopsis* expressing the *RD29A* promoter (contains both CRT/DRE and ABRE)-driven luciferase reporter gene provided evidence for the role of protein degradation in cold signaling. The *hos1* (high expression of osmotically responsive genes) mutation resulted in superinduction of *RD29A, COR47, COR15a, KIN1,* and their transacting factors (*CBF2* and *CBF3*) at 4°C (Ishitani et al., 1998). The expression of *CBF*s is transient in WT, while in the *hos1* mutant *CBF*s mRNA abundance was maintained at a much higher level, even up to 24 h during cold stress. Hence, HOS1 negatively regulates the cold-responsive genes by modulating the expression level of the CRT/DRE binding factors. Molecular cloning and characterization revealed that HOS1 encodes a ring finger protein, which has been implicated as an E3 ubiquitin-conjugating enzyme. HOS1 protein is present in the cytoplasm at normal growth temperatures and accumulates in the nucleus upon cold stress. The *hos1* mutation also affected the thermosensing mechanism, as evident from the fact that *RD29A* expression occurs at relatively warmer temperatures (Lee et al., 2001). The *los1-1* mutant showed enhanced expression of *CBF/DREB1*s but suppression of *COR* genes, which resulted in loss of freezing tolerance. The *LOS1* gene encodes a translation elongation factor 2-like protein (Guo et al., 2002).

Genetic analysis of *Arabidopsis* mutants, namely *srf6, hos1,* and *los1,* demonstrated that superinduction of either *CBF* or *COR* genes is not sufficient for enhanced freezing tolerance. However, overexpression of *CBF*s gave freezing-stress tolerance in *Arabidopsis* and *Brassica napus* (Liu et al., 1998; Kasuga et al., 1999; Gilmour et al., 2000; Jaglo et al., 2001). Tomato plants naturally lack the ability to acquire freezing tolerance. The *CBF1* overexpression in tomato also did not confer freezing tolerance, but did enhance chilling tolerance (Hsieh et al., 2002). Transgenic *Arabidopsis* plants overexpressing *CBF3 (DREB1a)* have a "dwarf" phenotype and produced significantly less biomass and seed yield (Liu et al., 1998; Kasuga et al., 1999; Gilmour et al., 2000; Jaglo et al., 2001). Although the *RD29A* promoter-driven *CBF* expression appears to be promising for engineering multiple stress-tolerant plants, the mechanism of freezing tolerance conferred by *CBF*s and stress tolerance in crop plants under field conditions needs further examination.

CELLULAR MEMBRANE STABILITY

Abiotic stresses induce deterioration of biological membranes, which results in ion leakage and deactivation of membrane proteins and thus loss of function of the cells. Membrane damage occurs mainly due to (1) stress-induced reactive oxygen species, which cause oxidation of lipids and proteins of the membrane, and (2) temperature/dehydration stress-induced change in phase transitions. Damages due to oxygen radicals could be alleviated by overexpression of antioxidant enzymes or some osmoprotectants. The temperature-dependent phase transition depends upon the membrane lipid composition and the level of saturation of membrane lipids. Hence, the stress tolerance of transgenic plants with engineered membrane lipid saturation levels was examined (Table 3.7). During cold acclimation, cold-tolerant plants increase their cellular membrane fluidity by increasing polyunsaturated fatty acids (Murata et al., 1982; Kodama et al., 1995; Nishida and Murata, 1996). Transgenic tobacco plants overexpressing a chloroplast [ω]-3 fatty acid desaturase gene from *Arabidopsis* showed increase in trienoic fatty acids and chilling tolerance (Kodama et al., 1994, 1995). Tobacco plants engineered to overexpress the *Anacystis nidulans* δ-9 desaturase gene had a highly reduced level of saturated fatty acid content in most membrane lipids and exhibited a significant increase in chilling resistance (Ishizaki-Nishizawa et al., 1996). In contrast to this evidence, transgenic tobacco plants overexpressing a microsomal ω-3 fatty acid desaturase gene did not show superior freezing or chilling tolerance when compared with that of control plants, although α-linolenic acid content was more in transgenic plants. The membrane fluidity of the transgenic and control plants was not different (Hamada et al., 1998). Phosphatidylglycerol of chloroplast membranes contains more saturated fatty acid in chilling-sensitive plants than in chilling-resistant plants. Also, phosphatidylglycerol from the chilling-sensitive plants showed a phase separation at approximately 30°C, whereas phosphatidylglycerol from the chilling-resistant plants went into the phase separation state at <15°C. The difference in the level of unsaturated/saturated fatty acids in phosphatidylglycerol depends upon substrate selectivity of glycerol-3-phosphate-acyltransferase, which initiates glycerolipid synthesis in plastids. To investigate whether glycerol-3-phosphate-acyltransferase can be engineered for stress tolerance, tobacco transgenic plants that overexpress the glycerol-3-phosphate-acyltransferase gene either from squash (chilling-sensitive plant) or *Arabidopsis* (chilling-resistant plant) were produced. Transgenic tobacco that overexpresses squash glycerol-3-phosphate-acyltransferase showed an increase in saturated phosphatidylglycerol and consequently increased chilling sensitivity (1°C for ten days). Conversely, transgenic tobacco that overexpresses *Arabidopsis*

TABLE 3.7. Genetic engineering for cellular membrane stability.

Gene	Plant	Stress phenotype	Remark	Reference
FAD7, chloroplast ω-3 fatty acid desaturase from *Arabidopsis*	Tobacco	Chilling stress at 1°C for 7 days, then recovery at 25°C	Transgenic plant showed less cold-induced leaf necrosis and better leaf growth	Kodama et al., 1994
Anacystis nidulans δ-9 desaturase gene, with the transit peptide of the pea RuBisCO small subunit, driven by CaMV 35S promoter	Tobacco	Seedlings grown at 1°C for 11 days, control plants showed leaf necrosis, while transgenic plant was normal.	When seeds germinated at 10°C and grown for 52 days at 10°C; also, transgenic plants showed better tolerance	Ishizaki-Nishizawa et al., 1996
Glycerol-3-phosphate-acyltransferase gene from squash or *Arabidopsis*	Tobacco	Transgenic plant expressing squash gene showed chilling sensitivity, while the transgenic plants expressing *Arabidopsis* gene were chilling tolerant		Murata et al., 1992
Glycerol-3-phosphate-acyltransferase gene from *Arabidopsis* driven by ubiquitin promoter	Rice	Chilling tolerance of photosynthesis	In leaf discs, O_2 evolution tolerance at 17°C was 1.36-fold higher in transgenic than in control plants; but at 11°C no significant difference was observed	Yokoi et al., 1998
Chloroplast ω-3 fatty acid desaturase gene (the *fad7* gene) from *Arabidopsis*	Tobacco gene silencing	High-temperature stress tolerance	Photosynthesis and seedling growth was higher in transgenic plants under heat stress.	Murakami et al., 2000

glycerol-3-phosphate-acyltransferase showed a significant decrease in saturated phosphatidylglycerol and consequently increased chilling tolerance (Murata et al., 1992). It is also interesting to note that transgenic *Arabidopsis* plants overexpressing the *E. coli* gene for glycerol-3-phosphate-acyltransferase (*plsB*, which shows substrate preference toward saturated acyl-ACP), produced more saturated phosphatidylglycerol in the chloroplast membranes and was more sensitive to chilling stress (Wolter et al., 1992).

In contrast to the increase in fatty acid desaturation under cold stress, in response to heat stress, plants increase fatty acid saturation to maintain the required membrane fluidity at higher temperatures. Mutants defective in fatty acid desaturase genes showed enhanced tolerance to high temperatures (Hugly et al., 1989; Kunst et al., 1989). Transgenic tobacco plants engineered to overexpress the chloroplast-localized ω-3 desaturase gene showed silencing of the chloroplast trienoic fatty acid synthetase gene. This resulted in reduction in trienoic fatty acid content in chloroplast membranes. In these transgenic plants, photosynthesis was measured over the range of temperatures between 25 and 55°C. Wild-type plants showed a significant decrease in photosynthesis at 40°C, while transgenic plants showed higher photosynthesis at 40°C than at the normal growing temperature of 25°C. Transgenic seedlings produced significantly higher biomass (ten times) than that of control plants when grown at 36°C for 45 days (Murakami et al., 2000). Although membrane lipid modifications in transgenic plants gave improved stress tolerance under short-duration stresses, it is essential to analyze the effect of membrane lipid modifications on plant metabolism and productivity during the entire life cycle of plants.

SALT-STRESS TOLERANCE: ION HOMEOSTASIS

Salinity causes both osmotic and ionic stresses. The cytotoxic ions in salinity stress are Na^+ and Cl^-. Hydration shells of proteins are essential to maintain their three-dimensional structures. Under salinity, ions such as Na^+ and Cl^- penetrate the hydration shells of proteins and interfere with the noncovalent interactions between amino acids of proteins. This leads to conformational changes and loss of function of proteins. Na^+ also competes with K^+, which is inimical to biochemical processes. Excess Na^+ competes with K^+ uptake, which results in a decrease in K^+/Na^+ ratios. Plant adaptive processes to salt tolerance can be grouped into (1) cellular homeostasis, which includes ion homeostasis and osmotic adjustment, (2) stress-damage control and repair, or detoxification, and (3) growth regulation (Zhu, 2001b). The contribution of osmolytes and antioxidant systems to salinity tolerance is discussed in the previous sections of this chapter. Cellular ion homeostasis under salinity is achieved by the following strategies: (1) exclusion Na^+ out of the cell by plasma membrane-bound Na^+/H^+ antiports or by limiting the Na^+ entry, (2) utilization of Na^+ for osmotic adjustment by compartmentation of Na^+ into the vacuole through tonoplast Na^+/H^+ antiporters, and (3) Na^+ secretion. Hence, regulation of ion transport systems is fundamental to plant salt tolerance.

Genetic analysis of salt overly sensitive *(sos)* mutants of *Arabidopsis* led to the identification of the SOS pathway that regulates cellular ion homeostasis and salt tolerance (Zhu, 2002). The SOS3, a myristoylated calcium binding-protein, senses salt stress-induced increases in cytosolic Ca^{2+} concentration (Liu and Zhu, 1998; Ishitani et al., 2000). SOS3 interacts with SOS2, a serine/threonine protein kinase, and activates SOS2 (Liu et al., 2000; Halfter et al., 2000; Quintero, et al., 2002). Activated SOS2 regulates the transport activity of SOS1, a plasma membrane Na^+-H^+ antiporter (Shi et al., 2000). Salt tolerance of yeast mutants lacking endogenous Na^+ transporters could be recovered by expression of SOS1. Coexpression of *SOS2* and *SOS3* significantly increased SOS1-dependent Na^+ tolerance of the yeast mutant (Quintero et al., 2002). Under salt stress (100 mM NaCl), Na^+ accumulation in shoots of *sos1* mutant plants was more than that of the wild type. Strong expression of SOS1 in cells bordering the xylem suggests that SOS1 is critical for controlling long-distance Na^+ transport from root to shoot (Shi et al., 2002). In addition to the positive control of exclusion of Na^+ from cytosol, the SOS pathway may also negatively regulate Na^+ influx systems. A suppressor genetic screen for the *sos3* mutation revealed that functional disruption of *AtHKT1*, a Na^+-K^+ cotransporter gene, could suppress the salt-sensitive phenotype of *sos3*. Expression of *AtHKT1* in *Xenopus laevis* oocytes and yeast showed that it could mediate Na^+ uptake. Also, *Athkt1* mutations alleviate the K^+-deficient phenotype of the *sos3* mutant (Rus, Yokoi, et al., 2001). Thus, the SOS pathway may coordinately regulate Na^+ efflux and influx to restore cellular ion homeostasis and salt tolerance of plants (Zhu, 2002). Transgenic tobacco expressing an activated form of yeast calcineurin (a Ca^{2+} and calmodulin-dependent protein phosphatase) showed enhanced salt-stress (200 mM NaCl) tolerance and recovery. Root growth was more tolerant than the shoot growth in transgenic plants (Pardo et al., 1998). Overexpression of the *SOS1* gene or an activated *SOS2* gene enhanced the salt-stress tolerance of *Arabidopsis* (unpublished data). Hence, genetic manipulation of the SOS pathway may help to engineer improved ion homeostasis under salt stress.

In addition to SOS3, salt-stress–induced calcium oscillations may also be perceived by Ca^{2+}-dependent protein kinases (CDPKs), which have both protein kinase and calmodulin-like domains in a single polypeptide. A rice

Ca^{2+}-dependent protein kinase, OsCDPK7, is a positive regulator involved in cold and salt/drought stress signaling. Rice transgenics overexpressing *OsCDPK7* under the control of the *CaMV 35S* promoter resulted in enhancement salt/drought-stress tolerance (Saijo et al., 2000). However, the targets of OsCDPK7 are not known.

The tonoplast Na^+/H^+ antiporters transport Na^+ into the vacuole by using electrochemical gradient of protons generated by the vacuolar H^+-adenosine triphosphatase (ATPase) and H^+-inorganic pyrophosphatase (PPase). Excess chloride ions are transported passively into the vacuole by anion channels. This vacuolar compartmentation prevents deleterious effects of Na^+ and Cl^- and also helps in osmotic adjustment. In *Arabidopsis*, the *AtNHX1* gene encodes a tonoplast Na^+/H^+ antiporter. Expression of *AtNHX1* in the yeast *nhx1* mutant suppressed some of the mutant phenotypes. *AtNHX1* is induced by salinity in *Arabidopsis* (Gaxiola et al., 1999), suggesting that NHX1 might play an important role in salt-stress tolerance in higher plants. Transgenic *Arabidopsis* plants that overexpress *AtNHX1* were able to grow in the presence of 200 mM NaCl, whereas the growth of wild-type plants was significantly reduced at 200 mM NaCl stress (Apse et al., 1999). Similarly, transgenic tomato plants overexpressing *AtNHX1* were also able to grow and produce fruit in the presence of 200 mM NaCl (Zhang and Blumwald, 2001). Both canola *(Brassica napus)* and tomato transgenics overexpressing *AtNHX1* accumulated high concentrations of sodium in leaves, but not in fruits/seeds. These transgenics were highly tolerant to salt stress, as the fruit/seed yield was not affected by salinity. Also, quality of the transgenic tomato fruits and canola oil was not affected by salt stress (Zhang and Blumwald, 2001; Zhang, Hodson, et al., 2001). Hence, it appears that significant improvement in salinity tolerance in crop plants may be achieved by engineering of a single gene. Vacuolar H^+-ATPase and H^+-PPiase energize the tonoplast Na^+/H^+ antiporters. To investigate the role of tonoplast H^+-PPase in salinity tolerance, *AVP1* gene (vacuolar H^+-pyrophosphatase) was overexpressed in *Arabidopsis*. The transgenics showed increased sequestration of Na^+ into the vacuole, maintained higher relative water content in leaves, and were more tolerant to salt and drought stress (Gaxiola et al., 2001).

In plants, Na^+ competes with K^+ for uptake under salinity. Hence, expression of cation transport systems that specifically transport K^+ into the cell might help in maintaining ionic balance. Overexpression of *AtHAL3a* (regulator of K^+ transport) in yeast and *Arabidopsis* conferred increased salt tolerance (Espinosa-Ruiz et al., 1999). Transgenic melon plants expressing the *HAL1* gene also showed salt tolerance in vitro (Bordás et al., 1997). To investigate the role of HAL1 in vivo, tomato plants were engineered to overexpress the yeast *HAL1* gene. Transgenics show better salt tolerance than the control plants (Gisbert et al., 2000). In another study, overexpression of *HAL1* in tomato increased the K^+ accumulation under NaCl stress. Tomato seedlings (30-day-old) were irrigated with 35mM NaCl till maturity. Salt stress decreased the fruit yield (fruit number and weight) by 57.5 percent in control plants while transgenic plants showed 24 to 42 percent decrease in fruit yield. However, under normal growing conditions the transgenic lines were less productive than the wild type (Rus, Estañ, et al., 2001). In rice, the contribution of carrier proteins to Na^+ uptake is much less significant when compared to the apoplastic pathway under salinity (Yadav et al., 1996). The deposition of silica in the endodermis and rhizodermis, and the polymerization of silicate via colloidal silica to silica gel or polysilicic acid throughout the root apoplast, appears to block Na^+ uptake through the apoplastic pathway in the roots of rice (Yeo et al., 1999). Quantitative trait loci (QTL) analysis in rice revealed that QTLs that control Na^+ uptake are different from those that control K^+ uptake. Hence, the uptake of Na^+ in rice under salinity is expected to be regulated by genes that affect root development and silicon uptake. However, in wheat, sodium/potassium selectivity by carrier proteins in the root appears to be a major determinant of salt tolerance (Gorham et al., 1997). Hence, the species-specific differences in salt tolerance mechanisms must be considered in order to engineer salt-tolerant crop plants.

WATER UPTAKE AND TRANSPIRATION

Drought-avoidance mechanisms help plants to maintain relatively high plant water potential (ψ) during drought by increasing the soil-water uptake and by decreasing transpirational water loss, while drought-tolerance mechanisms lead to maintenance of cell turgor and volume to help continue metabolic processes even under low plant-water potentials. Plants increase water uptake during drought by changing root characteristics such as length, depth, penetration ability, and xylem vessel diameter, and decrease water loss through stomatal closure, low stomatal density, increased cuticular resistance to water loss, leaf rolling, leaf area display, leaf area duration, decrease in leaf area, pubescence, and waxiness. At present, only a few of these traits are understood at the molecular level. Under drought stress, plants maintain water uptake by regulating their water potential through osmotic adjustment and maybe also by regulating the expression of water-channel proteins called aquaporins (AQPs). Plants prevent excessive transpirational water loss under drought stress primarily by stomatal regulation.

Aquaporins: Facilitators of Water Uptake

Water movement from the soil, through the plant and to the atmosphere, is described by Ohm's Law analog, flux = (driving force)/resistance. The driving force for water movement is the water potential gradient between soil to plant and plant to atmosphere. Hydraulic conductance and stomatal movement regulate the resistance to water flow through plant to atmosphere. Regulation of stomata in response to water deficit is well-understood at the molecular level (Schroeder et al., 2001). In contrast, water uptake by root at low water potential is poorly understood. Plants regulate their root growth and hydraulic conductance in order to maintain uptake of water and nutrients from soils with receding soil moisture during drought stress. Water movement through root cells to xylem or from xylem to other tissues occurs by the apoplastic pathway, symplastic pathway (cytoplasmic continuum of adjacent cells connected by plasmadesmata), and the transcellular (transport across the plasma membrane and tonoplast) pathway. The Casparian strip of the endodermal cells restricts entry of water from root cells into xylem and hence water transport to the xylem vessels is forced to transverse plasma membranes. Water diffusion is slow across lipid membranes. Moreover, under severe water deficit, root growth is slowed down and apoplastic barriers increase root hydraulic resistance. How does water transport occur quickly across biological membranes? How do plants regulate hydraulic conductance reversibly under stress? These questions were answered by the discovery of a class of water-channel proteins named aquaporins. During stress, aquaporins can reversibly increase the hydraulic conductivity and facilitate water uptake. Aquaporins belong to the family of major integral membrane proteins (MIB) with six transmembrane domains. *Arabidopsis* has about 30 distinct AQP genes. Hydraulic conductance through aquaporins can be modulated by their abundance (transcriptional or posttranscriptional regulation) or by their activity. The expression of aquaporin genes is regulated developmentally by hormones and by abiotic stresses such as drought and salinity (Schaffner, 1998; Johansson et al., 2000). The activity of some plant aquaporins is controlled by reversible phosphorylation. In spinach leaves, for instance, the phosphorylation level of a plasma-membrane aquaporin is dependent on the apoplastic water potential, thus establishing an intriguing link between driving force and conductance (Johansson et al., 1996).

Pharmacological studies showed that 20 to 80 percent of water transport might be under the control of aquaporins. Since the pharmacological agent used in these studies was a mercury derivative, which not only blocks aquaporins but also other cellular proteins, these results need to be taken with caution (Maurel and Chrispeels, 2001). In spinach leaf plasma membranes, aquaporins are activated by phosphorylation in response to changes in $Ca2^+$ and apoplastic water potential (Johansson et al., 1996). Aquaporins of pea (clone 7a, PIP1 subgroup), *Arabidopsis* (RD28, PIP2 subgroup), and tomato (TRAMP, a PIP1 homolog) are up-regulated by drought. In order to determine the role of aquaporins in plant-water transport, *Arabidopsis thaliana* was transformed with an antisense *PIP1b* gene (Kaldenhoff et al., 1998). Antisense lines showed reduced steady-state levels of PIP1b and the highly homologous PIP1a mRNA. The cell-membrane water permeability as measured by swelling of protoplasts indicated that the reduced expression of the specific aquaporins decreased the cellular membrane permeability to water. However, xylem pressure in antisense plants was similar to that of control plants. This is due to an increase of root mass of antisense plants, which compensated for the reduction in cellular water permeability. This study shows the importance of aquaporins for plant-water transport. Further studies on overexpression- or dehydration-responsive promoter-driven expression of aquaporins in different plants need to be examined to answer whether aquaporins play a significant role in water-stress tolerance.

Guard Cells: Major Regulators of Water-Use Efficiency

Water-use efficiency (the amount of organic matter per unit amount of water used by the plant) depends upon stomatal regulation. The rate of transpiration can be regulated by stomatal density and regulation of stomatal opening. Guard cells control the stomatal opening and closing and thus control the CO_2 intake and water loss from the plant. In response to drought stress, plants minimize water loss by ABA-mediated stomatal closure. ABA may be synthesized in guard cells or transported to guard cells from other leaf cells or root under stress. ABA-induced cytosolic Ca^{2+} elevations bring stomatal closure by activating anion efflux and K^+ efflux channels of the vacuole and plasma membrane. The signaling events that bring about stomatal closure have been investigated at the molecular level (Schroeder et al., 2001). The ABA-insensitive 1/2 (ABI1/2) protein serine/threonine phosphatase (PP2C) mutants are defective in drought-induced stomatal closure. The intragenic revertant of *abi1-1 (abi1-1R3)* displayed an enhanced drought tolerance compared with the wild-type (Gosti et al., 1999). Mutations in the farnesyltransferase gene *(ERA1)* cause hypersensitivity to ABA-mediated stomatal closure. Under drought stress *era1* mutants showed enhanced stomatal closure and thus reduced wilting (Pei et al., 1998). Transgenic analysis of guard-cell signaling components will shed light on how these genes can be utilized to improve drought tolerance of crop plants.

CELL-WALL ELASTICITY

Plants maintain their cell turgor under osmotic stress conditions by osmotic adjustment and/or by adjusting the elasticity of the cell wall. Although osmotic adjustment is well-understood in plants, only limited studies suggest that plant tissues can maintain turgor at reduced water volumes by physiological adjustment of their cell walls. In black spruce [*Picea mariana* (Mill) B.S.P.] and white spruce [*Picea glauca* (Moench) Voss.] increase in cell-wall elasticity significantly contributes to turgor maintenance under drought stress (Fan et al., 1994; Marshall and Dumbroff, 1999). Although the molecular basis of cell-wall adjustment during osmotic stresses is far from clear, genes that encode proteins of cell-wall metabolism are upregulated under osmotic stress conditions (Seki et al., 2002). A loss of function mutation in a gene encoding a cellulose synthase-like protein renders the mutant *Arabidopsis* plants more sensitive to osmotic stress (unpublished results). Turgor adjustment by modulation of cell-wall elasticity may involve less metabolic cost than would osmolyte synthesis for osmotic adjustment. In addition, osmotic adjustment can be useful only if extractable water is available in soil, but turgor maintenance is possible by elastic adjustment of cell wall, where soil moisture is not available for uptake by plants. Hence, further studies on turgor maintenance by cell-wall elasticity may be useful to further understand osmotic stress resistance.

MARKER-ASSISTED SELECTION

Since selection for yield under field-stress conditions across environments is time and labor consuming, crop breeders have identified several physiological traits that are linked to stress tolerance in yield. These traits include osmotic adjustment, leaf/xylem ABA content, leaf characters (waxiness, pubescence, and rolling), stomatal regulation, leaf expansion rate, leaf area duration (stay green in sorghum, maize, etc.), low stomatal density, canopy-temperature depression, cell-membrane stability, anthesis silking interval, root characters, tiller survival, seed size, early ground cover, spike photosynthesis, long coleoptile, and water-use efficiency, as determined by ^{13}C carbon discrimination or O18/O16 ratio. These physiological traits show continuous variation when they are examined in segregating populations. These types of traits are called continuous traits or quantitative traits, which are controlled by multiple genes. The regions of chromosomes or the loci controlling these traits are called quantitative trait loci (QTLs). In the QTL approach of plant breeding, parents displaying extreme phenotypes for a trait with potential value for breeding are crossed to produce progenies that segregate for the trait (e.g., F_2, backcross, recombinant inbred, or doubled haploid population). This population is phenotyped for that trait and screened for genetic polymorphism (RFLPs- or PCR-based markers). The construction of genetic map and identification of markers associated with the trait are achieved using computer programs. Identification of QTLs is the primary step for marker-assisted breeding and candidate gene cloning. Application of molecular markers to identify QTLs for physiological traits that confer stress tolerance has helped to dissect complex stress-tolerant traits into component traits (Table 3.8).

Cellular membrane stability (CMS) depends on the structure of phospholipid membranes, membrane proteins, tolerance to lipid peroxidation, activity of antioxidant enzymes, accumulation of osmolytes, and reconstitution of membranes after stress is relieved. Hence, CMS is a multigenic or quantitative trait. This is evident from genetic analysis of CMS under drought stress in rice (Tripathy et al., 2000). Therefore, manipulation of single genes in transgenic plants may have only small effects on CMS under stress. Breeding for heat-stress tolerance using CMS as a selection criterion, significant improvement in heat tolerance of wheat has been achieved (Saadalla et al., 1990; Shanahan et al., 1990). Martineau and colleagues (1979) found expected genetic gains for heat-tolerance selection by using CMS ranged from 16 to 18 percent in soybean. Heritability of CMS in maize was estimated as 73 percent (Ottaviano et al., 1991). Hence, better stress tolerance can be achieved by integrating the QTLs into desirable genotypes by marker-assisted selection breeding.

In crops such as sorghum, postflowering drought susceptibility is caused by premature leaf senescence. Premature leaf senescence results in reduced photoassimilate supply to the stalk and grain. Consequently, stalk carbohydrate reserves are remobilized to support grain development. This leads to stalk lodging. In addition to direct crop loss, lodging makes it difficult to employ mechanical harvesting. Resistance to premature leaf senescence is called "stay green," which confers drought tolerance in yield and lodging resistance. QTLs controlling the "stay green" trait have been identified in sorghum (Tuinstra et al., 1997; Crasta et al., 1999; Xu et al., 2000; Kebede et al., 2001) and maize (Beavis et al., 1994). Comparative analysis revealed that in maize and sorghum, orthologous QTL loci confer stay green and are also associated with preflowering drought tolerance, flowering time, and plant height (Kebede et al., 2001). Candidate gene cloning may help in identifying the genes involved in the stay-green trait, and can be useful in understanding the mechanism of drought-induced senescence. Integration of the stay-green trait into elite genotypes by marker-assisted selection breeding will help to improve postflowering drought tolerance.

TABLE 3.8. QTLs for physiological traits involved in abiotic stress tolerance.

Physiological trait	Crop	Stress	No. of QTLs	References
Cell-membrane stability	Rice	Drought (60 to 65 percent of leaf relative water content)	Nine QTLs (individual QTLs explained 13.4 percent to 42.1 percent of phenotypic variation)	Tripathy et al., 2000
Cell-membrane stability	Maize	High-temperature stress	Six QTLs accounting for 39 percent of phenotypic variation	Ottaviano et al., 1991
Root morphology	Rice	Hydroponically grown plants No stress	Root length controlled by four QTL (one QTL explained 30 percent of phenotypic variation). Adventitious root thickness controlled by 3 QTLs contributing 7 to 21 percent of phenotypic variation	Price and Tomos, 1997
Osmotic adjustment	Wheat	Drought	One major QTL	Morgan and Tan, 1996
Osmotic adjustment	Rice	Drought	One major QTL controlling 30 percent of phenotypic variation	Lilley et al., 1996
Osmotic adjustment	Rice	Drought	Five QTLs individually contributing 8 to 13 percent and together contributing 30 percent of phenotypic variation	Zhang, Zheng, et al., 2001

Osmotic adjustment	Barley	Drought (14 percent field capacity)	One QTL for osmotic adjustment contributing 17.7 percent of phenotypic variation	Teulat et al., 1998
Anthesis to silking interval	Maize	Drought in the field	Six QTLs together contribute to 47 percent of the phenotypic variance, of which four QTLs also contributed to grain yield	Ribaut et al., 1996, 1997
Stay green	Sorghum	Drought imposed just before anthesis in the field	Three QTLs each contributed up to 10 to 15 percent of phenotypic variation	Crasta et al., 1999. Kebede et al., 2001
Plant-water status	Barley	Drought (14 percent field capacity) in growth chambers	One QTL controlling 12 percent phenotypic variation for RWC	Teulat et al., 2001
Root traits: root penetration index, basal root thickness, penetrated root thickness, root pulling force, total root dry weight, penetrated root dry weight	Rice	Drought	36 QTLs identified on all 12 chromosomes; number for QTL for each trait varied from one to 11; contribution of individual QTL varied from 8 to 38 percent of phenotypic variance	Zhang, Zheng, et al., 2001

Stomatal characters	Rice	Excised leaf dehydration	Two QTL for stomatal resistance before leaf excision; one QTL for time for fastest stomatal closure after leaf excision	Price et al., 1997
Leaf rolling	Rice	Excised leaf dehydration	One QTL controlling 6.2 percent to the phenotypic variation	Price et al., 1997
ABA accumulation	Maize	Drought stress in the field	QTLs influencing L-ABA also harbored QTLs for stomatal conductance, a drought-sensitivity index, leaf temperature, leaf relative water content, anthesis-silking interval, and grain yield	Sanguineti et al., 1999
ABA accumulation	Maize	Drought stress in the field	Four QTLs accounting for 66 percent of the phenotypic variation and 76 percent of the genetic variation. These QTLs did not overlap the map position of mutants impaired in ABA biosynthesis	Tuberosa et al., 1998
ABA accumulation	Rice	Excised leaf dehydration	10 QTLs	Quarrie et al., 1997

In monoecious crops such as maize, abiotic stresses, such as drought and high temperatures, affect the timing of male and female flower production. In maize, male flowers are produced first, then female flowers are produced. Drought stress increases anthesis-silking interval (ASI), i.e., it delays silk emergence relative to the anthesis, which results in significant yield loss. Hence, in maize, stability of ASI is considered an important trait that contributes to yield stability under drought. Breeding for reduced ASI showed yield improvement under drought stress. Ribaut and colleagues (1996, 1997) have identified six QTLs for ASI under drought on chromosomes 1, 2, 5, 6, 8, and 10 in maize. QTLs for male flowering and/or female flowering under drought were detected on chromosomes 1, 2, 4, 5, 8, 9, and 10. Under drought, four QTLs were common for the expression of male flowering and female flowering, one for the expression of ASI and male flowering, and four for the expression of ASI and female flowering.

Comparison of QTLs for osmotic adjustment in cereals (rice, wheat, and barley) revealed that one QTL on chromosome 8 of rice (oa8.1) is homoeologous to that of wheat and barley. Another QTL for osmotic adjustment in rice (oa3.1) on chromosome 3 is homoeologous to a region in maize chromosome 1 that has the QTLs for xylem-leaf ABA concentration, stomatal conductance, and anthesis silking interval. Similarly, a QTL for osmotic adjustment in rice (oa9.1) on chromosome 9 and QTL for ABA content in wheat on chromosome 5A are homoeologous. Thus, it appears that some genomic regions for drought tolerance are conserved among cereals (Zhang, Zheng, et al., 2001), which can be used for marker-assisted selection breeding to improve stress tolerance and for candidate gene cloning in order to engineer stress tolerance. QTLs for physiological traits that confer stress tolerance have been identified in important food crops (Table 3.8). Each physiological trait involved in stress tolerance is controlled by multiple QTLs supporting that stress tolerance is a multigenic trait. Application of marker-assisted selection to QTLs of major effects should help in improving stress tolerance of crop plants significantly.

CONCLUSIONS AND FUTURE PERSPECTIVES

The transgenic approach demonstrated the possibilities of gene transfer across organisms and engineering abiotic stress tolerance by manipulation of a single gene or few genes. However, only a few from of a myriad of stress-tolerant traits found in nature have been unraveled today by application of molecular tools such as effective gene disruption methods and transgenic plants expressing the genes of interest in sense or antisense forms. Among these abiotic stresses, drought-stress signaling is still very poorly understood today. Genetic analysis of *Arabidopsis* mutants with the *RD29A* promoter::luciferase reporter transgene suggests complex signaling pathways in drought-, salt-, and cold-stress responses (Xiong et al., 2002; Zhu et al., 1998). A thorough understanding of stress signaling is essential for rational genetic engineering programs aimed at durable stress tolerance. The plant hormone ABA is a universal regulator of abiotic stress signaling and plant adaptation to abiotic stresses. However, ABA receptors have not yet been identified. Osmotic stress is imposed on cells by drought, salinity, and cold stresses. In bacteria, a two-component histidine kinase functions as osmosensor (Urao et al., 2000), but in higher plants osmosensors are yet to be identified. Hence, identification of ABA receptors and osmosensors in higher plants will give crucial knowledge for genetic engineering for abiotic stress tolerance. Transgenic assessment of genes controlling stress-specific signal components and transcription factors that control solute biosynthesis for osmotic adjustment and osmoprotectants will be of great use.

Abiotic stresses alter the duration of some developmental phases of crop plants. For example, drought and heat stresses shorten the spike development duration (during which the potential kernel number is determined) and grain development duration (during which the grain weight is determined) in wheat, while in rice drought delays anthesis. In addition to this, abiotic stresses affect the rate of biomass accumulation. Long duration of spike development and high spike weight at anthesis are positively correlated with final grain yield in wheat under drought and heat stresses (Warrington et al., 1977; Wardlaw et al., 1989; Bindraban et al., 1998). The molecular mechanisms involved in determination of the duration of phenological phases in normal and stress conditions need to be understood.

Analysis of metabolic flux is vital for designing efficient metabolic engineering. Polyol overaccumulating transgenic plants showed growth retardations, leaf necrosis, and infertility (Karakas et al., 1997; Sheveleva et al., 1998). These growth abnormalities might be due to the alteration of flux toward myo-inositol biosynthesis, which in turn is substrate for biosynthesis of inositol phosphates. Changes in the inositol pool might affect many signaling pathways directly, and it might indirectly affect the availability of free auxin as inositol is conjugated to auxins to regulate auxin action and hence results in growth abnormalities.

Engineered traits should be targeted to appropriate cellular locations. For example, rice transgenic plants that overexpress choline oxidase targeted to chloroplasts showed better tolerance to photoinhibition under salt and low-temperature stresses than did rice transgenic plants that overexpress choline oxidase targeted to cytosol (Sakamoto et al., 1998). Transgenic tobacco that overproduces proline (Kishor et al., 1995) or D-ononitol (Sheveleva et al.,

1997) did not show improved osmotic adjustment. It was found that D-ononitol did not enter the vacuole in transgenic tobacco that overproduced D-ononitol (Sheveleva et al., 1997). Compartmentation of osmolytes into the vacuoles is essential for osmotic adjustment. The transport of osmolytes into the vacuoles might need specific carrier proteins. Whenever a nonnative osmolyte is engineered into a plant, the problem of compartmentation may arise due to the lack of specific vacuolar transporters. When native osmolyte is engineered, then the limitation for the compartmentation may be expression of specific transporter genes, activity of the transporters, and the driving force required. Hence, in addition to overexpression of osmolytes/enzymes/stress proteins, their cellular targets should also be considered for better stress tolerance.

Today, the traits manipulated by genetic engineering to develop stress-tolerant transgenic plants are (1) signaling molecules, (2) transcription factors, and (3) tolerance effectors such as stress proteins and enzymes. Among these, the first two approaches appear to be more promising as they regulate a group of genes/regulons involved in stress response (for example, Jaglo-Ottosen et al., 1998; Liu et al., 1998; Kasuga et al., 1999; Tähtiharju and Palva, 2001). However, overexpression of metabolites, stress-responsive genes, or signaling components often results in reduction in plant size and other growth abnormalities even in normal growth conditions. Continuous diversion of energy and carbon sources toward the engineered pathway and the resulting alterations in the metabolism may be responsible for these growth defects. In addition, the engineered components may cross-talk with growth-regulation pathways (Zhu, 2002). Kasuga and colleagues (1999) have demonstrated that the use of a stress-responsive promoter can overcome this problem. Hence, selection of stress-responsive promoters for engineering stress-tolerance traits appears important.

Most of the studies discussed in this chapter are conducted using model plants such as *Arabidopsis* and tobacco; consequently, emphasis was not on long durations of stress, the effects of stresses in relation to plant ontogeny, or the effects of engineered traits on yield under nonstress and stress environments. In most of these studies, stress tolerance was assessed in controlled conditions for very short durations. Among all the transgenic plants listed in this chapter, the field performance was tested for only two transgenic plants, i.e., transgenic alfalfa overexpressing either Mn-SOD or Fe-SOD (McKersie et al., 1993, 1996, 1999). Hence, the real application of genetic engineering technology should be assessed by using crop plants. In nature, stresses develop gradually, occur either continuously or in cycles of stress and recovery, and more than one stress often occurs at the same time. Hence, it is imperative to keep in focus the environment for which the transgenic crop plants are developed.

If the manipulation of single genes can improve stress tolerance, then why is the response of crop plants to abiotic stresses always quantitative in a segregating population? QTL analysis reveals that multiple genes control each trait of stress tolerance. For example, the contribution of a single QTL for osmotic adjustment varies from 8 to 30 percent of the phenotypic variation (Table 3.8). Therefore, it is expected that only multiple gene transfers can achieve the engineering of osmotic adjustment. This is evident from the fact that plants engineered to overproduce osmolytes did not show significantly better osmotic adjustment. In only a few studies was the contribution of osmolytes to osmotic adjustment analyzed. Analysis of the contribution of mannitol to osmotic adjustment in tobacco plants that were engineered to overproduce mannitol showed that mannitol accounted for only 30 to 40 percent of the osmotic adjustment (Karakas et al., 1997). Kishor and colleagues (1995) analyzed osmotic adjustment in tobacco transgenics that overproduce proline and found that although the stress tolerance was significantly enhanced, the proline did not contribute to osmotic adjustment. Hence, the reason for stress tolerance should be better understood in order to extend these approaches to crop plants.

The use of tissue- and organ-specific promoters is probably essential for abiotic stress tolerance of crop plants. Unlike in the model plants, where survival is considered as stress tolerance, in crop plants survival and reproduction are equally important. The reproductive phase of crop plants is often more sensitive to abiotic stresses than the vegetative stage. Both low and high temperatures affect pollen germination and growth and thus the seed yield of crop plants. In maize, pollen germination and pollen growth are the most vulnerable processes for high-temperature stress. Temperatures >38°C decrease maize pollen germination and growth (Dupuis and Dumas, 1990). High vulnerability of pollen to heat stress may be due to the fact that pollen is unable to synthesize a full complement of HSPs (Mascarenhas and Crone, 1996). In the case of rice, low temperatures (below 15°C) induce spikelet sterility mainly due to the failure of pollens to germinate and grow on stigma (Satake and Koike, 1983). Overexpression of stress proteins such as HSPs, LEAs, and CBFs was shown to confer enhanced tolerance to vegetative tissues. It would be worthwhile to engineer stress proteins with pollen-specific promoters to evaluate their stress tolerance. Similarly, engineering for aquaporins, Na^+/H^+ antiporters to exclude sodium, and osmolyte accumulation may be more effective when their expression is driven by, for example, a root-specific promoter, because the root experiences the osmotic and ionic stresses first. Soil temperatures during wheat sowing may be >45°C, while during maize and sorghum sowing they may be >55°C. Antisense suppression of the *HSP101* gene in *Arabidopsis* resulted in loss of basal thermotolerance during seed germination (Queitsch et al., 2000), indicating the importance of this HSP in heat tolerance at germination. Overexpression of this HSP driven by a seed-specific promoter in developing seeds may help in the production of this protein in seeds and provide heat tolerance in environments where the temperature is very high during sowing. Thus, tissue/organ-specific promoter-driven transgenes may confer better stress tolerance and also save energy expenditure on the expression of the transgenes in tissues/organs where they are not needed.

In wheat, heat stress (>30°C) during grain development affects grain quality and grain weight. This is due to high sensitivity of starch biosynthesis, cross-linking of glutenin polymers and alterations in protein synthesis patterns, i.e., enhanced HSPs synthesis and gliadin synthesis; hence, the glutenin-to-gliadin ratio decreases. These changes in grains affect the bread-making quality and grain weight (Viswanathan and Khanna-Chopra, 2001, and references therein). In this case, genetic engineering should target appropriate physiological processes such as starch synthesis and protein synthesis, rather than traits such as antioxidants and HSPs. Thus, it is critical to consider the environment for which genetic engineering is aimed.

Overexpression of osmolytes and antioxidant systems appears to protect plants during stress by preventing oxidative damage. ROS production increases under stresses due to metabolic aberrations such as decrease in CO_2 fixation, and increase in photorespiration and mitochondrial defects. More effort is needed to identify the molecular basis of ROS production under stress so that those components can be engineered. This approach may be more beneficial, as it reduces the metabolic cost of scavenging ROS and repairing the damage caused by ROS. Engineering for antioxidant systems or polyol production may alter the pool size of H_2O_2 and inositol, respectively, which are involved in developmental and stress signaling.

Marker-assisted selection for QTLs of major effects on physiological traits that confer abiotic stress tolerance should be incorporated into breeding programs. QTL analysis has revealed that each component of stress tolerance such as CMS, osmotic adjustment, and so on, are multigenic. Clearly, successful genetic engineering for stress-tolerant crop plants needs engineering of many genes, so development of binary vectors or mini-chromosomes for multiple gene transfer (Goderis et al., 2002) and improvement in transformation-regeneration systems for crop plants are needed.

REFERENCES

Alia, H., T.H. Hayashi, H. Chen, and N. Murata (1998). Transformation with a gene for choline oxidase enhances the cold tolerance of *Arabidopsis* during germination and early growth. *Plant, Cell and Environment* 21:232-239.

Alia, H., A. Sakamoto, and N. Murata (1998). Enhancement of the tolerance of *Arabidopsis* to high temperatures by genetic engineering of the synthesis of glycinebetaine. *Plant Journal* 16:155-161.

Alvarez, M.E., R.I. Pennell, P.J. Meijer, A. Ishikawa, R.A. Dixon, and C. Lamb (1998). Reactive oxygen intermediates mediate a systemic signal network in the establishment of plant immunity. *Cell* 92:773-784.

Apse, M.P., G.S. Aharon, W.S. Snedden, and E. Blumwald (1999). Salt tolerance conferred by overexpression of a vacuolar Na^+/H^+ antiport in *Arabidopsis*. *Science* 285:1256-1258.

Beavis, W.D., O.S. Smith, D. Grant, and R. Fincher (1994). Identification of quantitative trait loci using a small sample of topcrossed and F4 progeny from maize. *Crop Science* 34:882-896.

Bindraban, P.S., K.D. Sayre, and E. Solis-Moya (1998). Identifying factors that determine grain number in wheat. *Field Crops Research* 58:223-234.

Bohnert, H.J. and R.G. Jensen (1996). Strategies for engineering water stress tolerance in plants. *Trends in Biotechnology* 14:89-97.

Bordás, M., C. Montesinos, M. Dabauza, A. Salvador, L.A. Roig, R. Serrano, and V. Moreno (1997). Transfer of the yeast salt tolerance gene *HAL1* to *Cucumis melo* L. cultivars and in vitro evaluation of salt tolerance. *Transgenic Research* 6: 41-50.

Bot, A.J., F.O. Nachtergaele, and A. Young (2000). *Land Resource Potential and Constraints at Regional and Country Levels*. World Soil Resources Report 90. Rome: FAO, Land and Water Development Division.

Bowler, C., L. Slooten, S. Vandenbranden, R.D. Rycke, J. Botterman, C. Sybesma, M.M. Van, and D. Inze (1991). Manganese superoxide dismutase can reduce cellular damage mediated by oxygen radicals in transgenic plants. *EMBO Journal* 10:1723-1732.

Chen, T.H.H. and N. Murata (2002). Enhancement of tolerance of abiotic stress by metabolic engineering of betaines and other compatible solutes. *Current Opinion in Plant Biology* 5:250-257.

Cohen, J.D. and R.S. Bandurski (1982). Chemistry and physiology of the bound auxins. *Annual Review of Plant Physiology* 33:403-430.

Crasta, O.R., W.W. Xu, D.T. Rosenow, J. Mullet, and H.T. Nguyen (1999). Mapping of post-flowering drought tolerance traits in grain sorghum: Association between QTLs influencing premature senescence and maturity. *Molecular and General Genetics* 262:579-588.

Deak, M., G.V. Horvath, S. Davletova, K. Torok, L. Sass, I. Vass, B. Barna, Z. Kiraly, and D. Dudits (1999). Plants ectopically expressing the iron-binding protein ferritin, are tolerant to oxidative damages and pathogens. *Nature Biotechnology* 17:192-196.

De Ronde, J.A., M.H. Spreeth, and W.A. Cress (2000). Effect of antisense [Δ^1]-pyrroline-5-carboxylate reductase transgenic soybean plants subjected to osmotic and drought stress. *Plant Growth Regulation* 32:13-26.

Dupuis, I. and C. Dumas (1990). Influence of temperature stress on in vitro fertilization and heat shock protein synthesis in maize (*Zea mays* L.) reproductive tissues. *Plant Physiology* 94:665-670.

Espinosa-Ruiz, A., J.M. Belles, R. Serrano, and F.A. Culianez-Macla (1999). *Arabidopsis thaliana AtHAL3:* A flavoprotein related to salt and osmotic tolerance and plant growth. *Plant Journal* 20:529-539.

Fan, S., T.J. Blake, and E. Blumwald (1994). The relative contribution of elastic and osmotic adjustments to turgor maintenance of woody species. *Physiologia Plantarum* 90:408-413.

Fischer, R.A. and D.R. Byerlee (1991). Trends of wheat production in the warmer areas: Major issues and economic considerations. In *Wheat for Nontraditional, Warm Areas,* D.A. Saunders (Ed.). Mexico, DF: CIMMYT, pp. 3-27.

Flowers, T.J. and A.R. Yeo (1995). Breeding for salinity resistance in crop plants. Where next? *Australian Journal of Plant Physiology* 22:875-884.

Gao, M., R. Tao, K. Miura, A.M. Dandekar, and A. Sugiura (2001). Transformation of Japanese persimmon (*Diospyros kaki* Thunb.) with apple cDNA encoding NADP-dependent sorbitol-6-phosphate dehydrogenase. *Plant Science* 160:837-845.

Gaxiola, R.A., J. Li, S. Undurraga, L.M. Dang, G.J. Allen, S.L. Alper, and G.R. Fink (2001). Drought- and salt-tolerant plants result from overexpression of the *AVP1* H^+-pump. *Proceedings of the National Academy of Sciences of the United States of America* 98:11444-11449.

Gaxiola, R.A., R. Rao, A. Sherman, P. Grisafi, S.L. Alper, and G.R. Fink (1999). The *Arabidopsis thaliana* proton transporters, AtNhx1 and Avp1, can function in cation detoxification in yeast. *Proceedings of the National Academy of Sciences of the United States of America* 96:1480-1485.

Gilmour, S.J., A.M. Sebolt, M.P. Salazar, J.D. Everard, and M.F. Thomashow (2000). Overexpression of the *Arabidopsis CBF3* transcriptional activator mimics multiple biochemical changes associated with cold acclimation. *Plant Physiology* 124:1854-1865.

Gisbert, C., A.M. Rus, M.C. Bolarín, J.M. López-Coronado, I. Arrillaga, C. Montesinos, M. Caro, R. Serrano, and V. Moreno (2000). The Yeast *HAL1* gene improves salt tolerance of transgenic tomato. *Plant Physiology* 123:393-402.

Goderis, I.J.W.M., M.F.C. De Bolle, I.E.J.A. François, P.F.J. Wouters, W.F. Broekaert, and B.P.A. Cammune (2002). A set of modular plant transformation vectors allowing flexible insertion of up to six expression units. *Plant Molecular Biology* 50:17-27.

Gorham, J., J. Bridges, J. Dubcovsky, J. Dvorak, P.A. Hollington, M.C. Luo, and J.A. Khan (1997). Genetic analysis and physiology of a trait for enhanced K^+/Na^+ discrimination in wheat. *New Phytologist* 137:109-116.

Gosti, F., N. Beaudoin, C. Serizet, A.R.R. Webb, N. Vartanian, and J. Giraudat (1999). ABI1 protein phosphatase 2C is a negative regulator of abscisic acid signaling. *Plant Cell* 11:1897-1910.

Guo, B.H., Y.M. Zhang, H.J. Li, L.Q. Du, Y.X. Li, J.S. Zhang, S.Y. Chen, and Z.Q. Zhu (2000). Transformation of wheat with a gene encoding for the betaine aldehyde dehydrogenase (BADH). *Acta Botanica Sinica* 42:279-283.

Guo, Y., L. Xiong, M. Ishitani, and J.K. Zhu (2002). An *Arabidopsis* mutation in translation elongation factor 2 causes superinduction of CBF/DREB1 transcription factor genes but blocks the induction of their downstream targets under low temperatures. *Proceedings of the National Academy of Sciences of the United States of America* 99:7786-7791.

Gupta, A.S., J.L. Heinen, A.S. Holaday, J.J. Burke, and R.D. Allen (1993). Increased resistance to oxidative stress in transgenic plants that overexpress chloroplastic Cu/Zn superoxide dismutase. *Proceedings of the National Academy of Sciences of the United States of America* 90:1629-1633.

Halfter, U., M. Ishitani, and J.K. Zhu (2000). The *Arabidopsis* SOS2 protein kinase physically interacts with and is activated by the calcium-binding protein SOS3. *Proceedings of the National Academy of Sciences of the United States of America* 97:3735-3740.

Hamada T., H. Kodama, K. Takeshita, H. Utsumi, and K. Iba (1998). Characterization of transgenic tobacco with an increased α-linolenic acid level. *Plant Physiology* 118:591-598.

Hasegawa, P.M., R.A. Bressan, J.K. Zhu, and H.J. Bohnert (2000). Plant cellular and molecular responses to high salinity. *Annual Review of Plant Physiology and Plant Molecular Biology* 51:463-499.

Hayashi H., H. Alia, L. Mustardy, P. Deshnium, M. Ida, and N. Murata (1997). Transformation of *Arabidopsis thaliana* with the *codA* gene for choline oxidase: Accumulation of glycinebetaine and enhanced tolerance to salt and cold stress. *Plant Journal* 12:133-142.

Hincha, D.K., E.M. Hellwege, A.G. Heyer, and J.H. Crowe (2000). Plant fructans stabilize phosphatidylcholine liposomes during freeze-drying. *European Journal of Biochemistry* 267:535-540.

Holmstrom K.O., S. Somersalo, A. Mandal, T.E. Palva, and B. Welin (2000). Improved tolerance to salinity and low temperature in transgenic tobacco producing glycine betaine. *Journal of Experimental Botany* 51:177-185.

Holmström, K.O., E. Mäntylä, B. Welin, A. Mandal, E.T. Palva, O.E. Tunnela, and J. Londesborough (1996). Drought tolerance in tobacco. *Nature* 379:683-684.

Hong, Z., K. Lakkineni, Z. Zhang, and D.P.S. Verma (2000). Removal of feedback inhibition of Δ^1-pyrroline-5-carboxylate synthetase results in increased proline accumulation and protection of plants from osmotic stress. *Plant Physiology* 122:1129-1136.

Hottiger T., T. Boller, and A. Wiemken (1987). Rapid changes of heat and desiccation tolerance correlated with changes of trehalose content in *Saccharomyces cerevisiae* cells subjected to temperature shifts. *FEBS Letters* 220:113-115.

Hsieh, T.H., L.T. Lee, P.T. Yang, L.H. Chiu, Y.Y. Charng, Y.C. Wang, and M.T. Chan (2002). Heterology expression of the *Arabidopsis C-Repeat/Dehydration Response Element Binding Factor 1* gene confers elevated tolerance to chilling and oxidative stresses in transgenic tomato. *Plant Physiology* 129:1086-1094.

Huang, J., R. Hirji, L. Adam, K.L. Rozwadowski, J.K. Hammerlindl, W.A. Keller, and G. Selvaraj (2000). Genetic engineering of glycinebetaine production toward enhancing stress tolerance in plants: Metabolic limitations. *Plant Physiology* 122:747-756.

Hugly, S., L. Kunst, J. Browse, and C. Somerville (1989). Enhanced thermal tolerance of photosynthesis and altered chloroplast ultrastructure in a mutant of *Arabidopsis* deficient in lipid desaturation. *Plant Physiology* 90:1134-1142.

Ingram, J. and D. Bartels (1996). The molecular basis of dehydration tolerance in plants. *Annual Review of Plant Physiology and Plant Molecular Biology* 47:377-403.

Ishitani, M., J. Liu, U. Halfter, C.S. Kim, W. Shi, and J.K. Zhu (2000). SOS3 function in plant salt tolerance requires N-myristoylation and calcium binding. *Plant Cell* 12:1667-1678.

Ishitani, M., L. Xiong, H. Lee, B. Stevenson, and J.K. Zhu (1998). *HOS1*, a genetic locus involved in cold-responsive gene expression in *Arabidopsis*. *Plant Cell* 10:1151-1161.

Ishitani, M., L. Xiong, B. Stevenson, and J.K. Zhu (1997). Genetic analysis of osmotic and cold stress signal transduction in *Arabidopsis:* Interactions and convergence of abscisic acid-dependent and abscisic acid-independent pathways. *Plant Cell* 9:1935-1949.

Ishizaki-Nishizawa, O., T. Fujii, M. Azuma, K. Sekiguchi, N. Murata, T. Ohtani, and T. Toguri (1996). Low-temperature resistance of higher plants is significantly enhanced by a nonspecific cyanobacterial desaturase. *Nature Biotechnology* 14:1003-1006.

Jaglo, K.R., S. Kleff, K.L. Amundsen, X. Zhang, V. Haake, J.Z. Zhang, T. Deits, and M.F. Thomashow (2001). Components of the *Arabidopsis* C-repeat/dehydration-responsive element binding factor cold-response pathway are conserved in *Brassica napus* and other plant species. *Plant Physiology* 127:910-917.

Jaglo-Ottosen, K.R., S.J. Gilmour, D.G. Zarka, O. Schabenberger, and M.F. Thomashow (1998). *Arabidopsis* CBF1 overexpression induces *cor* genes and enhances freezing tolerance. *Science* 280:104-106.

Johansson, I., M. Karlsson, U. Johanson, C. Larsson, and P. Kjellbom (2000). The role of aquaporins in cellular and whole plant water balance. *Biochimica et Biophysica Acta* 1465:324-342.

Johansson, I., C. Larsson, B. Ek, and P. Kjellbom (1996). The major integral proteins of spinach leaf plasma membranes are putative aquaporins and are phosphorylated in response to Ca^{2+} and apoplastic water potential. *Plant Cell* 8:1181-1191.

Kaldenhoff, R., K. Grote, J.J. Zhu, and U. Zimmermann (1998). Significance of plasmalemma aquaporins for watertransport in *Arabidopsis thaliana*. *Plant Journal* 14:121-128.

Karakas, B., P. Ozias-Akins, C. Stushnoff, M. Suefferheld, and M. Rieger (1997). Salinity and drought tolerance of mannitol-accumulating transgenic tobacco. *Plant, Cell and Environment* 20:609-616.

Kasuga, M., Q. Liu, S. Miura, K. Yamaguchi-Shinozaki, and K. Shinozaki (1999). Improving plant drought, salt, and freezing tolerance by gene transfer of a single stress-inducible transcription factor. *Nature Biotechnology* 17:287-291.

Kebede, H., P.K. Subudhi, D.T. Rosenow, and H.T. Nguyen (2001). Quantitative trait loci influencing drought tolerance in grain sorghum (*Sorghum bicolor* L. Moench). *Theoretical and Applied Genetics* 103:266-276.

Kim, J.C., S.H. Lee, Y.H. Cheng, C.M. Yoo, S.I. Lee, H.J. Chun, D.J. Yun, J.C. Hong, S.Y. Lee, C.O. Lim, and M.J. Cho (2001). A novel cold-inducible zinc finger protein from soybean, SCOF-1, enhances cold tolerance in transgenic plants. *Plant Journal* 25:247-259.

Kishitani, S., T. Takanami, M. Suzuki, M. Oikawa, S. Yokoi, M. Ishitani, A.M. Alvarez-Nakase, T. Takabe, and T. Takabe (2000). Compatibility of glycinebetaine in rice plants: Evaluation using transgenic rice plants with a gene for peroxisomal betaine aldehyde dehydrogenase from barley. *Plant, Cell and Environment* 23:107-114.

Kishor, P.B.K., Z. Hong, G.H. Miao, C.A.A. Hu, and D.P.S. Verma (1995). Overexpression of [delta]-pyrroline-5-carboxylate synthetase increases proline production and confers osmotolerance in transgenic plants. *Plant Physiology* 108:1387-1394.

Knight, H., E.L. Veale, G.J. Warren, and M.R. Knight (1999). The *sfr6* mutation in *Arabidopsis* suppresses low-temperature induction of gene dependent on the CRT/DRE sequence motif. *Plant Cell* 11:875-886.

Kodama, H., T. Hamada, G. Horiguchi, M. Nishimura, and K. Iba (1994). Genetic enhancement of cold tolerance by expression of a gene for chloroplast ω-3 fatty acid desaturase in transgenic tobacco. *Plant Physiology* 105:601-605.

Kodama, H., G. Horiguchi, T. Nishiuchi, M. Nishimura, and K. Iba (1995). Fatty acid desaturation during chilling acclimation is one of the factors involved in conferring low-temperature tolerance to young tobacco leaves. *Plant Physiology* 107:1177-1185.

Konstantinova, T., D. Parvanova, A. Atanassov, and D. Djilianov (2002). Freezing tolerant tobacco, transformed to accumulate osmoprotectants. *Plant Science* 163:157-164.

Kunst, L., J. Browse, and C. Somerville (1989). Enhanced thermal tolerance in a mutant of *Arabidopsis* deficient in palmitic acid unsaturation. *Plant Physiology* 91:401-408.

Lee, H., L. Xiong, Z. Gong, M. Ishitani, B. Stevenson, and J.K. Zhu (2001). The *Arabidopsis HOS1* gene negatively regulates cold signal transduction and encodes a RING finger protein that displays cold-regulated nucleo-cytoplasmic partitioning. *Genes and Development* 15:912-924.

Lee, J.H., A. Hubel, and F. Schoffl (1995). Derepression of the activity of genetically engineered heat shock factor causes constitutive synthesis of heat shock proteins and increased thermotolerance in transgenic *Arabidopsis. Plant Journal* 8:603-612.

Lee, J.H. and F. Schöffl (1996). An *Hsp70* antisense gene affects the expression of *HSP70/HSC70,* the regulation of *HSF,* and the acquisition of thermotolerance in transgenic *Arabidopsis thaliana. Molecular and General Genetics* 252:11-19.

Levitt, J. (1980). *Responses of Plants to Environmental Stress,* Volume I: *Chilling, Freezing, and High Temperature Stress.* New York: Academic Press.

Lilley, J.M., M.M. Ludlow, S.R. McCouch, and J.C. O'Toole (1996). Locating QTL for osmotic adjustment and dehydration tolerance in rice. *Journal of Experimental Botany* 47:1427-1436.

Liu, J., M. Ishitani, U. Halfter, C.S. Kim, and J.K. Zhu (2000). The *Arabidopsis thaliana SOS2* gene encodes a protein kinase that is required for salt tolerance. *Proceedings of the National Academy of Sciences of the United States of America* 97:3730-3734.

Liu, J. and J.K. Zhu (1998). A calcium sensor homolog required for plant salt tolerance. *Science* 280:1943-1945.

Liu, Q., Y. Sakuma, H. Abe, M. Kasuga, S. Miura, K. Yamaguchi-Shinozaki, and K. Shinozaki (1998). Two transcription factors, DREB1 and DREB2, with an EREBP/AP2 DNA binding domain, separate two cellular signal transduction pathways in drought- and low temperature-responsive gene expression, respectively, in *Arabidopsis. Plant Cell* 10:1391-1406.

Lopez, F., G. Vansuyt, F. Case-Delbart, and P. Fourcroy (1996). Ascorbate peroxidase activity, not the mRNA level, is enhanced in salt-stressed *Raphanus sativus* plants. *Physiologia Plantarum* 97:13-20.

Malik, M.K., J.P. Slovin, C.H. Hwang, and J.L. Zimmerman (1999). Modified expression of a carrot small heat shock protein gene, *Hsp17.7*, results in increased or decreased thermotolerance. *Plant Journal* 20:89-99.

Marshall, J.G. and E.B. Dumbroff (1999). Turgor regulation via cell wall adjustment in white spruce. *Plant Physiology* 119:313-319.

Martín, M.L. and L. Busconi (2001). A rice membrane-bound calcium-dependent protein kinase is activated in response to low temperature. *Plant Physiology* 125:1442-1449.

Martineau, J.R., J.H. Williams, and J.E. Specht (1979). Temperature tolerance in soybeans: II. Evaluation of segregating populations for membrane thermostability. *Crop Science* 19:79-81.

Mascarenhas, J.P. and D.E. Crone (1996). Pollen and the heat shock response. *Sexual Plant Reproduction* 9:370-374.

Maurel, C. and M.J. Chrispeels (2001). Aquaporins: A molecular entry into plant water relations. *Plant Physiology* 125:135-138.

McKersie, B.D., S.R. Bowley, and K.S. Jones (1999). Winter survival of transgenic alfalfa overexpressing superoxide dismutase. *Plant Physiology* 119:839-847.

McKersie, B.D., S.R. Bowley, E. Harjanto, and O. Leprince (1996). Water-deficit tolerance and field performance of transgenic alfalfa overexpressing superoxide dismutase. *Plant Physiology* 111:1177-1181.

McKersie, B.D., Y. Chen, M. Debens, S.R. Bowley, C. Bowler, D. Inze, K. D'Halluin, and J. Botterman (1993). Superoxide dismutase enhances tolerance of freezing stress in transgenic alfalfa (*Medicago sativa* L.). *Plant Physiology* 103:1155-1163.

McKersie, B.D., J. Murnaghan, K.S. Jones, and S.R. Bowley (2000). Iron-superoxide dismutase expression in transgenic alfalfa increases winter survival without a detectable increase in photosynthetic oxidative stress tolerance. *Plant Physiology* 122:1427-1438.

McNeil, S.D., M.L. Nuccio, M.J. Ziemak, and A.D. Hanson (2001). Enhanced synthesis of choline and glycine betaine in transgenic tobacco plants that overexpress phosphoethanolamine N-methyltransferase. *Proceedings of the National Academy of Sciences of the United States of America* 98:10001-10005.

Mittler, R. and B.A. Zilinskas (1994). Regulations of pea cytosolic ascorbate peroxidase and other antioxidant enzymes during the progression of drought stress and following recovery from drought. *Plant Journal* 5:397-405.

Møller, I.M. (2001). Plant mitochondria and oxidative stress: Electron transport, NADPH turnover, and metabolism of reactive oxygen species. *Annual Review of Plant Physiology and Plant Molecular Biology* 52:561-591.

Monroy, A.F., F. Sarhan, and R.S. Dhindsa (1993). Cold-induced changes in freezing tolerance, protein phosphorylation, and gene expression: Evidence for a role of calcium. *Plant Physiology* 102:1227-1235.

Moons, A., G. Bauw, E. Prinsen, M. Van Montagu, and D. Van Der Straeten (1995). Molecular and physiological responses to abscisic acid and salts in roots of salt-sensitive and salt-tolerant indica rice varieties. *Plant Physiology* 107:177-186.

Morgan, J.M. and M.K. Tan (1996). Chromosomal location of wheat osmoregulation gene using RFLP analysis. *Australian Journal of Plant Physiology* 23:803-806.

Müller, J., T. Boller, and A. Wiemken (1995). Trehalose and trehalase in plants: Recent developments. *Plant Sciences* 112:1-9.

Murakami, Y., M. Tsuyama, Y. Kobayashi, H. Kodama, and K. Iba (2000). Trienoic fatty acids and plant tolerance of high temperature. *Science* 287:476-479.

Murata, N., O. Ishizaki-Nishizawa, S. Higashi, H. Hayashi, Y. Tasaka, and I. Nishida (1992). Genetically engineered alteration in the chilling sensitivity of plants. *Nature* 356:710-713.

Murata, N., N. Sato, N. Takahashi, and Y. Hamazaki (1982). Compositions and positional distributions of fatty acids in phospholipids from leaves of chilling-sensitive and chilling-resistant plants. *Plant Cell Physiology* 23:1071-1079.

Nanjo, T., T.M. Kobayashi, Y. Yoshida, Y. Kakubari, K. Yamaguchi-Shinozaki, and K. Shinozaki (1999). Antisense suppression of proline degradation improves tolerance to freezing and salinity in *Arabidopsis thaliana*. *FEBS Letters* 461:205-210.

Nishida, I. and N. Murata (1996). Chilling sensitivity in plants and cyanobacteria: The crucial contribution of membrane lipids. *Annual Review of Plant Physiology and Plant Molecular Biology* 47:541-568.

Noctor, G. and C. H. Foyer (1998). Ascorbate and glutathione: Keeping active oxygen under control. *Annual Review of Plant Physiology and Plant Molecular Biology* 49:249-279.

Nyyssola, A., J. Kerovuo, P. Kaukinen, N. Von Weymarn, and T. Reinikainen (2000). Extreme halophiles synthesize betaine from glycine by methylation. *Journal of Biological Chemistry* 275:22196-22201.

Örvar, B.L., V. Sangwan, F. Omann, and R. Dhindsa (2000). Early steps in cold sensing by plant cells: The role of actin cytoskeleton and membrane fluidity. *Plant Journal* 23:785-794.

Ottaviano, E., M.S. Gorla, E. Pe, and C. Frova (1991). Molecular markers (RFLPs and HSPs) for the genetic dissection of thermotolerance in maize. *Theoretical and Applied Genetics* 81:713–719.

Papageorgiou, G.C. and N. Murata (1995). The unusually strong stabilizing effects of glycine betaine on the structure and function of the oxygen evolving photosystem II complex. *Photosynthesis Research* 44:243-252.

Pardo, J.M., M.P. Reddy, S. Yang, A. Maggio, G.H. Huh, T. Matsumoto, M.A. Coca, M. Paino-D'Urzo, H. Koiwa, D.J. Yun, et al. (1998). Stress signaling through Ca^{2+}/calmodulin-dependent protein phosphatase calcineurin mediates salt adaptation in plants. *Proceedings of the National Academy of Sciences of the United States of America* 95:9681-9686.

Park, J.M., C.J. Park, S.B. Lee, B.K. Ham, R. Shin, and K.H. Paek (2001). Overexpression of the tobacco *Tsi1* gene encoding an EREBP/AP2–type transcription factor enhances resistance against pathogen attack and osmotic stress in tobacco. *Plant Cell* 13:1035-1046.

Payton, P., R.D. Allen, N. Trolinder, and A.S. Holaday (1997). Over-expression of chloroplast-targeted Mn superoxide dismutase in cotton (*Gossypium hirsutum* L., cv. Coker 312) does not alter the reduction of photosynthesis after short exposures to low temperature and high light intensity. *Photosynthesis Research* 52:233-244.

Pei, Z.M., M. Ghassemian, C.M. Kwak, P. McCourt, and J.I. Schroeder (1998). Role of farnesyltransferase in ABA regulation of guard cell anion channels and plant water loss. *Science* 282:287-290.

Pei, Z.M., Y. Murata, G. Benning, S. Thomine, B. Klusener, G.J. Allen, E. Grill, and J.I. Schroeder (2000). Calcium channels activated by hydrogen peroxide mediate abscisic acid signaling in guard cells. *Nature* 406:731-734.

Pilon-Smits, E.A.H., M.J.M. Ebskamp, M.J. Paul, M.J.W. Jeuken, P.J. Weisbeek, and S.C.M. Smeekens (1995). Improved performance of transgenic fructan-accumulating tobacco under drought stress. *Plant Physiology* 107:125-130.

Pilon-Smits, E.A.H., N. Terry, T. Sears, H. Kim, A. Zayed, S. Hwang, K. van Dun, E. Voogd, T.C. Verwoerd, R.W.H.H. Krutwagen, and O.J.M. Gooddijn (1998). Trehalose-producing transgenic tobacco plants show improved growth performance under drought stress. *Journal of Plant Physiology* 152:525-532.

Pilon-Smits, E.A.H., N. Terry, T. Sears, and K. van Dun (1999). Enhanced drought resistance in fructan-producing sugar beet. *Plant Physiology and Biochemistry* 37:313-317.

Pitcher, L.H., E. Brennan, A. Hurley, P. Dunsmuir, J.M. Tepperman, and B.A. Zilinskas (1991). Overproduction of petunia chloroplastic copper/zinc superoxide dismutase does not confer ozone tolerance in transgenic tobacco. *Plant Physiology* 97:452-455.

Postel, S.L., G.C. Daily, and P.R. Ehrlich (1996). Human appropriation of renewable fresh water. *Science* 271:785-788.

Prändl, R., K. Hinderhofer, G. Eggers-Schumacher, and F. Schöffl (1998). HSF3, a new heat shock factor from *Arabidopsis thaliana,* derepresses the heat shock response and confers thermotolerance when overexpressed in transgenic plants. *Molecular and General Genetics* 258:269-278.

Prasad, K.V.S.K., P. Sharmila, P.A. Kumar, and P.P. Saradhi (2000). Transformation of *Brassica juncea* (L.) Czern with bacterial *codA* gene enhances its tolerance to salt stress. *Molecular Breeding* 6:489-499.

Price, A.H. and A.D. Tomos (1997). Genetic dissection of root growth in rice (*Oryza sativa* L.). II. Mapping quantitative trait loci using molecular markers. *Theoretical and Applied Genetics* 95:143-152.

Price, A.H., E.M. Young, and A.D. Tomos (1997). Quantitative trait loci associated with stomatal conductance, leaf rolling and heading date mapped in upland rice *(Oryza sativa). New Phytologist* 137:83-91.

Quarrie, S.A., D.A. Laurie, J. Zhu, C. Lebreton, A. Semikhodskii, A. Steed, H. Witsenboer, and C. Calestani (1997). QTL analysis to study the association between leaf size and abscisic acid accumulation in droughted rice leaves and comparisons across cereals. *Plant Molecular Biology* 35:155-165.

Queitsch, C., S.W. Hong, E. Vierling, and S. Lindquist (2000). Heat shock protein 101 plays a crucial role in thermotolerance in *Arabidopsis. Plant Cell* 12:479-492.

Quintero, F.J., M. Ohta, H. Shi, J.K. Zhu, and J.M. Pardo (2002). Reconstitution in yeast of the *Arabidopsis* SOS signaling pathway for Na^+ homeostasis. *Proceedings of the National Academy of Sciences of the United States of America* 99:9061-9066.

Rhodes, D. and A.D. Hanson (1993). Quaternary ammonium and tertiary sulfonium compounds in higher plants. *Annual Review of Plant Physiology and Plant Molecular Biology* 44:357-384.

Ribaut, J.M., D.A. Hoisington, J.A. Deutsch, C. Jiang, and D. Gonzalez-de-Leon (1996). Identification of quantitative trait loci under drought conditions in tropical maize. 1. Flowering parameters and the anthesis-silking interval. *Theoretical and Applied Genetics* 92:905-914.

Ribaut, J.M., C. Jiang, D. Gonzalez-de-Leon, G.O. Edmeades, and D.A. Hoisington (1997). Identification of quantitative trait loci under drought conditions in tropical maize. 2. Yield components and marker-assisted selection strategies. *Theoretical and Applied Genetics* 94:887-896.

Romero, C., J.M. Belles, J.L. Vaya, R. Serrano, and F.A. Culianez-Macia (1997). Expression of the yeast trehalose-6-phosphate synthase gene in transgenic tobacco plants: Pleiotropic phenotypes include drought tolerance. *Planta* 201:293-297.

Roxas Jr., V.P., R.K. Smith, E.R. Allen, and R.D. Allen (1997). Overexpression of glutathione S-transferase/glutathione peroxidase enhances the growth of transgenic tobacco seedlings during stress. *Nature Biotechnology* 15:988-991.

Roxas, V.P., S.A. Lodhi, D.K. Garrett, J.R. Mahan, and R.D. Allen (2000). Stress tolerance in transgenic tobacco seedlings that overexpress glutathione S-transferase/glutathione peroxidase. *Plant and Cell Physiology* 41:1229-1234.

Rudolph, A.S. and B. Goins (1991). The effect of hydration stress solutes on the phase-behavior of hydrated dipalmitoylphosphatidylcholine. *Biochimica et Biophysica Acta* 1066:90-94.

Rus, A., S. Yokoi, A. Sharkhuu, M. Reddy, B.-H. Lee, T.K. Matsumoto, H. Koiwa, J.K. Zhu, R.A. Bressan, and P.M. Hasegawa (2001). AtHKT1 is a salt tolerance determinant that controls Na^+ entry into plant roots. *Proceedings of the National Academy of Sciences of the United States of America* 98:14150-14155.

Rus, A.M., M.T. Estañ, C. Gisbert, B. Garcia-Sogo, R. Serrano, M. Caro, V. Moreno, and M.C. Bolarín (2001). Expressing the yeast *HAL1* gene in tomato increases fruit yield and enhances K^+ /Na^+ selectivity under salt stress. *Plant, Cell and Environment* 24:875-880.

Saadalla, M.M., J.S. Quick, and J.F. Shanahan (1990). Heat tolerance in winter wheat. II. Membrane thermostability and field performance. *Crop Science* 30:1248-1251.

Saijo, Y., S. Hata, J. Kyozuka, K. Shimamoto, and K. Izui (2000). Over-expression of a single Ca^{2+}-dependent protein kinase confers both cold and salt/drought tolerance on rice plants. *Plant Journal* 23:319-327.

Sakamoto, A., A. Murata, and N. Murata (1998). Metabolic engineering of rice leading to biosynthesis of glycinebetaine and tolerance to salt and cold. *Plant Molecular Biology* 38:1011-1019.

Sakamoto, A., R. Valverde, A. Murata, T.H.H. Chen, and N. Murata (2000). Transformation of *Arabidopsis* with the *codA* gene for choline oxidase enhances freezing tolerance of plants. *Plant Journal* 22:449-453.

Sanguineti, M.C., R. Tuberosa, P. Landi, S. Salvi, M. Maccaferri, E. Casarini, and S. Conti (1999). QTL analysis of drought related traits and grain yield in relation to genetic variation for leaf abscisic acid concentration in field-grown maize. *Journal of Experimental Botany* 50:1289-1297.

Sangwan, V., I. Foulds, J. Singh, and R.S. Dhindsa (2001). Cold-activation of *Brassica napus BN115* promoter is mediated by structural changes in membranes and cytoskeleton, and requires Ca^{2+} influx. *Plant Journal* 27:1-12.

Satake, T. and S. Koike (1983). Sterility caused by cooling treatment at the flowering stage in rice plants. *Japan Journal of Crop Science* 52:207-213.

Sato,Y., T. Murakami, H. Funatsuki, S. Matsuba, H. Saruyama, and M. Tanida (2001). Heat shock-mediated *APX* gene expression and protection against chilling injury in rice seedlings. *Journal of Experimental Botany* 52:145-151.

Schaffner, A.R. (1998). Aquaporin function, structure, and expression: Are there more surprises to surface in water relations? *Planta* 204:131-139.

Schroeder, J.I., G.J. Allen, V. Hugouvieux, J.M. Kwak, and D. Waner (2001). Guard cell signal transduction. *Annual Review of Plant Physiology and Plant Molecular Biology* 52:627-658.

Seki, M., M. Narusaka, J. Ishida, T. Nanjo, M. Fujita, Y. Oono, A. Kamiya, M. Nakajima, A. Enju, T. Sakurai, et al. (2002). Monitoring the expression profiles of 7000 *Arabidopsis* genes under drought, cold and high-salinity stresses using a full-length cDNA microarray. *Plant Journal* 31:279-292.

Shanahan, J.F., I.B. Edwards, J.S. Quick, and R.J. Fenwick (1990). Membrane thermostability and heat tolerance of spring wheat. *Crop Science* 30:247-251.

Shen, B., R.G. Jensen, and H.J. Bohnert (1997). Increased resistance to oxidative stress in transgenic plants by targeting mannitol biosynthesis to chloroplasts. *Plant Physiology* 113:1177-1183.

Sheveleva, E.V., W. Chmara, H.J. Bohnert, and R.G. Jensen (1997). Increased salt and drought tolerance by D-ononitol production in transgenic *Nicotiana tabacum* L. *Plant Physiology* 115:1211-1219.

Sheveleva, E.V., S. Marquez, W. Chmara, A. Zegeer, R.G. Jensen, and H.J. Bohnert (1998). Sorbitol-6-phosphate dehydrogenase expression in transgenic tobacco. High amounts of sorbitol lead to necrotic lesions. *Plant Physiology* 117:831-839.

Shi, H., M. Ishitani, C. Kim, and J.-K. Zhu (2000). The *Arabidopsis thaliana* salt tolerance gene *SOS1* encodes a putative Na^+/H^+ antiporter. *Proceedings of the National Academy of Sciences of the United States of America* 97:6896-6901.

Shi, H., F.J. Quintero, J.M. Pardo, and J.K. Zhu (2002). The putative plasma membrane Na^+/H^+ antiporter SOS1 controls long-distance Na^+ transport in plants. *Plant Cell* 14:465-477.

Shi, W.M., Y. Muramoto, A. Ueda, and T. Takabe (2001). Cloning of peroxisomal ascorbate peroxidase gene from barley and enhanced thermotolerance by overexpressing in *Arabidopsis thaliana. Gene* 273:23-27.

Sivamani, E., A. Bahieldin, J.M. Wraith, T. Al-Niemi, W.E. Dyer, T.H.D. Ho, and R.D. Qu (2000). Improved biomass productivity and water use efficiency under water deficit conditions in transgenic wheat constitutively expressing the barley *HVA1* gene. *Plant Science* 155:1-9.

Smirnoff, N. and S.V. Colombe (1988). Drought influences the activity of the enzymes of the chloroplast hydrogen peroxide scavenging system. *Journal of Experimental Botany* 39:1097-1108.

Smirnoff, N. and Q.J. Cumbes (1989). Hydroxyl radical scavenging activity of compatible solutes. *Phytochemistry* 28:1057-1060.

Steponkus P.L., M. Uemura, R.A. Joseph, S.J. Gilmour, and M.F. Thomashow (1998). Mode of action of the *COR15a* gene on the freezing tolerance of *Arabidopsis thaliana. Proceedings of the National Academy of Sciences of the United States of America* 95:14570-14575.

Tähtiharju, S. and T. Palva (2001). Antisense inhibition of protein phosphatase 2C accelerates cold acclimation in *Arabidopsis thaliana. Plant Journal* 26:461-470.

Tähtiharju, S., V. Sangwan, A.F. Monroy, R.S. Dhindsa, and M. Borg (1997). The induction of kin genes in cold-acclimating *Arabidopsis thaliana.* Evidence of a role for calcium. *Planta* 203: 442-447.

Tamminen, I., P. Mäkelä, P. Heino, and E.T. Palva (2001). Ectopic expression of *ABI3* gene enhances freezing tolerance in response to abscisic acid and low temperature in *Arabidopsis thaliana. Plant Journal* 25:1-8.

Tarczynski M.C., R.G. Jensen, and H.J. Bohnert (1993). Stress protection of transgenic tobacco by production of the osmolyte mannitol. *Science* 259:508-510.

Tepperman, J.M. and P. Dunsmuir (1990). Transformed plants with elevated levels of chloroplast SOD are not more resistant to superoxide toxicity. *Plant Molecular Biology* 14:501-511.

Teulat, B., C. Borries, and D. This (2001). New QTLs identified for plant water status, water-soluble carbohydrate and osmotic adjustment in a barley population grown in a growth-chamber under two water regimes. *Theoretical and Applied Genetics* 103:161-170.

Teulat, B., D. This, M. Khairallah, C. Borries, C. Ragot, P. Sourdille, P. Leroy, P. Monneveux, and A. Charrier (1998). Several QTLs involved in osmotic-adjustment trait variation in barley (*Hordeum vulgare* L.). *Theoretical and Applied Genetics* 96:688-698.

Thomas, J.C., M. Sepahi, B. Arendall, and H.J. Bohnert (1995). Enhancement of seed germination in high salinity by engineering mannitol expression in *Arabidopsis thaliana. Plant, Cell and Environment* 18:801-806.

Tripathy, J.N., J. Zhang, S. Robin, Th.T. Nguyen, and H.T. Nguyen (2000). QTLs for cell-membrane stability mapped in rice (*Oryza sativa* L.) under drought stress. *Theoretical and Applied Genetics* 100:1197-1202.

Tuberosa, R., M.C. Sanguineti, P. Landi, S. Salvi, E. Casarini, and S. Conti (1998). RFLP mapping of quantitative trait loci controlling abscisic acid concentration in leaves of drought-stressed maize (*Zea mays* L.). *Theoretical and Applied Genetics* 97:744-755.

Tuinstra, M.R., E.M. Grote, P.B. Goldsbrough, and G. Ejeta (1997). Genetic analysis of post-flowering drought tolerance and components of grain development in *Sorghum bicolor* (L.) Moench. *Molecular Breeding* 3:439-448.

Urao, T., K. Yamaguchi-Shinozaki, and K. Shinozaki (2000). Two-component systems in plant signal transduction. *Trends in Plant Science* 5:67-74.

Van Breusegem F., L. Slooten, J.M. Stassart, J. Botterman, T. Moens, M. Van Montagu, and D. Inze (1999). Effects of overproduction of tobacco MnSOD in maize chloroplasts on foliar tolerance to cold and oxidative stress. *Journal of Experimental Botany* 50:71-78.

Van Breusegem, F., L. Slooten, J.M. Stassart, T. Moens, J. Botterman, M. Van Montagu, and D. Inze (1999). Overproduction of *Arabidopsis thaliana* FeSOD confers oxidative stress tolerance to transgenic maize. *Plant Cell Physiology* 40:515-523.

Van Camp, W., K. Capiau, M. Van Montagu, D. Inze, and L. Slooten (1996). Enhancement of oxidative stress tolerance in transgenic tobacco plants overproducing Fe-superoxide dismutase in chloroplasts. *Plant Physiology* 112:1703-1714.

Viswanathan, C. and R. Khanna-Chopra (2001). Effect of heat stress on grain growth, starch synthesis and protein synthesis in grains of wheat (*Triticum aestivum* L.) varieties differing in grain weight stability. *Journal of Agronomy and Crop Science* 186:1-7.

Viswanathan, C. and J.K. Zhu (2002). Molecular genetic analysis of cold-regulated gene transcription. *Philosophical Transactions of Royal Society of London. B Biological Sciences* 357:877-886.

Wang, J., H. Zhang, and R.D. Allen (1999). Overexpression of an *Arabidopsis* peroxisomal ascorbate peroxidase gene in tobacco increases protection against oxidative stress. *Plant and Cell Physiology* 40:725-732.

Wardlaw, I.F., I.A. Dawson, and P. Munibi (1989). The tolerance of wheat to high temperature during reproductive growth. II. Grain development. *Australian Journal of Agricultural Research* 40:15-24.

Wardlaw, I.F. and C.W. Wrigley (1994). Heat tolerance in temperate cereals: An overview. *Australian Journal of Plant Physiology* 21:695-703.

Warren, G., R. McKown, A.L. Martin, and V. Teutonico (1996). Isolation of mutations affecting the development of freezing tolerance in *Arabidopsis thaliana* (L.) Heynh. *Plant Physiology* 111:1011-1019.

Warrington, I.J., R.L. Dunstone, and L.M. Green (1977). Temperature effects at three development stages on the yield of the wheat ear. *Australian Journal of Agricultural Research* 28:11-27.

Winicov, I. (2000). Alfin1 transcription factor overexpression enhances plant root growth under normal and saline conditions and improves salt tolerance in alfalfa. *Planta* 210:416-422.

Winicov, I. and D.R. Bastola (1997). Salt tolerance in crop plants: New approaches through tissue culture and gene regulation. *Acta Physiologiae Plantarum* 19: 435-449.

Winicov, I. and D.R. Bastola (1999). Transgenic overexpression of the transcription factor Alfin1 enhances expression of the endogenous, *MSPRP2* gene in alfalfa and improves salinity tolerance of the plant. *Plant Physiology* 120:473-480.

Wolter, F.P., R. Schmidt, and E. Heinz (1992). Chilling sensitivity of *Arabidopsis thaliana* with genetically engineered membrane lipids. *EMBO Journal* 11:4685-4692.

Xin, Z. and J. Browse (1998). *Eskimo1* mutants of Arabidopsis are constitutively freezing-tolerant. *Proceedings of the National Academy of Sciences of the United States of America* 95:7799-7804.

Xiong, L., B.-H. Lee, M. Ishitani, H. Lee, C. Zhang, and J.K. Zhu (2001). *FIERY1* encoding an inositol polyphosphate 1-phosphatase is a negative regulator of abscisic acid and stress signaling in *Arabidopsis*. *Genes and Development* 15:1971-1984.

Xiong, L., K.S. Schumaker, and J.-K. Zhu (2002). Cell signaling during cold, drought, and salt stress. *Plant Cell* 14:S165-183.

Xu, D., X. Duan, B. Wang, B. Hong, T.D. Ho, and R. Wu (1996). Expression of a late embryogenesis abundant protein gene, *HVA1*, from barley confers tolerance to water deficit and salt stress in transgenic rice. *Plant Physiology* 110:249-257.

Xu, W., P.K. Subudhi, O.R. Crasta, D.T. Rosenow, J.E. Mullet, and H.T. Nguyen (2000). Molecular mapping of QTLs conferring stay green in sorghum. *Genome* 43:461-469.

Yadav, R., T.J. Flowers, and A.R. Yeo (1996). The involvement of the transpirational bypass flow in sodium uptake by high- and low-sodium-transporting lines of rice developed through intravarietal selection. *Plant, Cell and Environment* 22:329-336.

Yancey, P.H., M.E. Clark, S.C. Hand, R.D. Bowles, and G.N. Somero (1982). Living with water stress: Evolution of osmolyte system. *Science* 217:1214-1222.

Yeo, A.R., S.A. Flowers, G. Rao, K. Welfare, N. Senanayake, and T.J. Flowers (1999). Silicon reduces sodium uptake in rice (*Oryza sativa* L.) in saline conditions and this is accounted for by a reduction in the transpirational bypass flow. *Plant, Cell and Environment* 22:559-565.

Yeo, E.T., H.B. Kwon, S.E. Han, J.T. Lee, J.C. Ryu, and M.O. Byu (2000). Genetic engineering of drought resistant potato plants by introduction of the trehalose-6-

phosphate synthase (TPS1) gene from Saccharomyces cerevisiae. *Molecules and Cells* 10:263-268.

Yokoi S., S. Higashi, S. Kishitani, N. Murata, and K. Toriyama (1998). Introduction of the cDNA for *Arabidopsis* glycerol-3-phosphate acyltransferase (GPAT) confers unsaturation of fatty acids and chilling tolerance of photosynthesis on rice. *Molecular Breeding* 4:269-275.

Zhang, C.S., Q. Lu, and D.P.S. Verma (1995). Removal of feedback inhibition of (delta1)-pyrroline-5-carboxylate synthetase, a bifunctional enzyme catalyzing the first two steps of proline biosynthesis in plants. *Journal of Biological Chemistry* 270:20491-20496.

Zhang, H.X. and E. Blumwald (2001). Transgenic salt-tolerant tomato plants accumulate salt in foliage but not in fruit. *Nature Biotechnology* 19:765-768.

Zhang, H.X., J.N. Hodson, J.P. Williams, and E. Blumwald (2001). Engineering salt-tolerant *Brassica* plants: Characterization of yield and seed oil quality in transgenic plants with increased vacuolar sodium accumulation. *Proceedings of the National Academy of Sciences of the United States of America* 98:12832-12836.

Zhang, J., H.T. Nguyen, and A. Blum (1999). Genetic analysis of osmotic adjustment in crop plants. *Journal of Experimental Botany* 50:291-302.

Zhang, J., H.G. Zheng, A. Aarti, G. Pantuwan, T.T. Nguyen, J.N. Tripathy, A.K. Sarial, S. Robin, R.C. Babu, B.D. Nguyen, et al. (2001). Locating genomic regions associated with components of drought resistance in rice: Comparative mapping within and across species. *Theoretical and Applied Genetics* 103:19-29.

Zhao, H.W., Y.J. Chen, Y.L. Hu, Y. Gao, and Z.P. Lin (2000). Construction of a trehalose-6-phosphate synthase gene driven by drought-responsive promoter and expression of drought-resistance in transgenic tobacco. *Acta Botanica Sinica* 42:616-619.

Zhu, B., J. Su, M.C. Chang, D.P.S. Verma, Y.L. Fan, and R. Wu (1998). Overexpression of a pyrroline-5-carboxylate synthetase gene and analysis of tolerance to water and salt stress in transgenic rice. *Plant Science* 139:41-48.

Zhu, J.K. (2001a). Cell signaling under salt, water and cold stresses. *Current Opinion in Plant Biology* 4:401-406.

Zhu, J.K. (2001b). Plant salt tolerance. *Trends in Plant Science* 6:66-71.

Zhu, J.K. (2002). Salt and drought stress signal transduction in plants. *Annual Review of Plant Physiology and Plant Molecular Biology* 53:247-273.

Chapter 4

Genome Mapping and Its Implications for Improving Stress Resistance in Plants

Nguyen Thi Vinh
Andrew H. Paterson

THE IMPACT OF DROUGHT, SALINITY, AND ACID SOIL CONDITIONS FOR WORLD AGRICULTURE

Abiotic stress is a highly probable occurrence for almost any plant growing under either natural or cultivated conditions. Some of the most important abiotic stresses that limit world crop production are drought, heat, salinity, freezing, and acidic soil (Blum, 1988). Mild stresses occur frequently, sometimes daily in the case of drought, and result in reduced growth, productivity, and quality of crops. Severe stresses can lead to catastrophic losses. Dudal (1976) estimated that only 10 percent of the world's arable land may be categorized as free of stress. About 20 percent of the land is under some kind of mineral stress, 26 percent is affected by drought stress, and 15 percent by freezing stress. Sanchez and Salinas (1981) estimated that approximately 55 percent of the soils in tropical America, 39 percent in tropical Africa, and 37 percent in tropical Asia are acidic, representing 1.6 billion hectares. A significant portion of the terrestrial environment is affected by high levels of salt. Including the major deserts, there are approximately 4 million km^2 of salt-affected land in the world (Flowers et al., 1977). Saline and sodic soils cover about 10 percent of the total arable land in over 100 countries of the world and, of the cultivated lands, about 23 percent are saline and 37 percent are sodic (Tanji, 1990). Loss of land to salt accumulation through irrigated agriculture has been estimated to be at least several hundred km^2 a year in India alone (Flowers et al., 1977). In California, 20 percent of irrigated agricultural land is affected to some extent by salt (Bernstein, 1975).

Plants respond to abiotic stresses at the molecular and cellular levels as well as at the physiological level. The physiology of drought resistance has been reviewed quite extensively (Fischer and Turner, 1978; Quisenberry,

1982; Castleberry, 1983; Dwyer and Stewart, 1986; Deng et al., 1990; Kuo et al., 1993; Volaire et al., 1998). Drought stress affects many physiological processes, including carbohydrate metabolism (Massacci et al., 1996; Wang et al., 1996; Escobar-Guti et al., 1998; Hashem et al., 1998), photosynthetic efficiency (Epron, 1997; Jagtap et al., 1998), and water relations (Perdomo et al., 1996; Paakkonen et al., 1998). Drought also causes oxidative stress (Zhang and Kirkham, 1994; Gogorcena et al., 1995; Olsson et al., 1996; Sairam et al., 1997; Kronfub et al., 1998) and protein alterations (Han and Kermode, 1996). Drought tolerance in crops usually appears to depend on one or more of the following components: (1) the ability of roots to exploit deep soil water to provide for evapotranspiration demand, (2) the capacity for osmotic adjustment (OA) that allows plants to retain turgor and protect meristems from extreme desiccation, and (3) control over nonstomatal water loss from leaves (Nguyen et al., 1997). Most mechanisms of tolerance fall into two categories in crop plants: (1) osmotic adjustment, a protective accumulation of solutes in the vacuoles of leaves; or (2) increased rooting capacity, allowing for more efficient moisture extraction from the soil. Both of the mechanisms can result in slower-than-normal wilting patterns under stress. Studies by Sloane and colleagues (1990) found that although soybean is not considered a species with great capacity for osmotic adjustment, there was some evidence for both of the mechanisms. As will be seen, in some respects the physiology of drought resistance is open to controversy (Blum, 1988). Mechanisms of plant tolerance to drought deserve further investigation, although numerous studies have been conducted in various species (Huang and Jiang, 2002). Identifying physiological traits involved in drought tolerance would expedite improvement of plant tolerance by conventional breeding or biotechnology approaches (Kang, 1998).

Widespread natural salinity has resulted in numerous species that show special adaptive mechanisms to growth in saline environments. Salinity is caused not only by NaCl but also by Na_2CO_3, $NaHCO_3$, and Na_2SO_4, and the relations of these salts to one another as well as to other nutrients such as K^+, Ca^{2+}, and Mg^{2+} are important and may differ greatly at different sites. The majority of plants, including almost all crop plants, are unable to tolerate permanently saline conditions in the soil (Kramer, 1984). In many environments, transient salinity stresses also occur when the concentration of salt in the soil solution fluctuates because of changes in water supply, drainage, and evaporation and transpiration.

Identification of plant genotypes capable of increased tolerance to salt and incorporation of these desirable traits into economically useful crop plants may reduce the effects of salinity on productivity (Epstein, 1980; Rains, 1981). Several mechanisms of salinity tolerance have been recognized. Roughly speaking, one can distinguish between salt excluders and salt includers (Collander, 1941). Salt excluders possess mechanisms that ensure that salt reaches the shoot only in very small amounts. This might be due to a very efficient selectivity toward K^+ during absorption, as in many grasses (Albert and Popp, 1977), or to reabsorption of Na^+ from the xylem sap in proximal parts of the root (Jacoby and Ratner, 1973) or shoot (Jacoby, 1965; Rains, 1969) followed by storage or retranslocation to the soil (Winter, 1982; Winter and Preston, 1982). In contrast, salt includers absorb salt and store it at high amounts in the stem and leaves. Salt-including species have special glands on the leaf surface that excrete salt at a high concentration (Thomson, 1975). Transport phenomena play a significant role in plant adaptation to salinity. Better knowledge of the structural features that are correlated with salt tolerance is essential for an understanding of the mechanisms of salt tolerance. It has not been possible to identify any one structure that fully explains the phenomenon of salt tolerance, instead requiring an integrative picture of different aspects of plant biology to understand how plants adapt to saline environments.

Crop productivity on acid soils is restricted by multiple abiotic stress factors. Since the forms of soil aluminum (Al) and their solubilities are high at a pH of 5 or less, aluminum toxicity becomes one of the major growth-limiting factors affecting plants on acid soil (Kochian, 1995). Symptoms of Al toxicity are not always easily identified; however, the initial and most dramatic symptom is inhibition of root elongation as a consequence of toxicity to the root apex (Wallace and Anderson 1984; Taylor, 1988, 1991; Ryan et al., 1992, 1993; Delhaize and Ryan, 1995; Kochian, 1995). Roots injured by high Al are usually stubby and thick, and become dark-colored, brittle, poorly branched, and suberized (Foy, 1983), with reduced root length and volume (Magnavaca and Bahia Filho, 1993).

The physiological and biochemical mechanisms of aluminum toxicity are a matter of controversy (Kochian, 1995). Because Al can interact with a number of extracellular and intracellular structures, many different mechanisms of Al toxicity have been hypothesized. Aluminum toxicity may inhibit shoot growth by limiting the supply of nutrients and water due to poorer subsoil penetration or lower root hydraulic conductivity. Shoot growth is also affected by Al toxicity, either due to reduced nutrient and water supply, or to limited supply of cytokinins from the roots (Pan et al., 1989). Massot and colleagues (1992) showed that scoring for Al tolerance using root elongation as a single criterion may avoid misclassification of genotypes that accumulate a large amount of Al in shoots.

Aluminum rhizotoxicity may be related to disruption of membrane function, probably due to changes in the structure and function of the root-cell plasmalemma (Zhao et al., 1987). Depending on the pH, aluminum can bind to membrane proteins and lipids (Campbell et al., 1994) and participate in the formation of cross-links between proteins and pectins within the cell wall, reducing membrane integrity (Foy, 1983). Disturbed mitotic processes may also contribute to abnormal root growth (Foy and Campbell, 1984; Morimura et al., 1978). Aluminum is particularly concentrated in the nucleus, and the cell cycle is inhibited, probably at the level of DNA replication (Foy et al., 1978; Foy, 1974). Although aluminum may bind to DNA (Matsumoto, 1991) of the root-cap cells in particular (Naidoo et al., 1978), inhibition of cell division is presumably an indirect effect. Rengel (1992) has hypothesized that Al^{3+} blockage of Ca^{2+} channels could prevent the formation of important cytoplasmic Ca^{2+} transients needed for cell division to occur. Al^{3+} is probably an effective cation channel blocker, since it blocks both Ca^{2+} and K^{+} channels in wheat root cell (Gassmann and Schroeder, 1994).

Many researchers have reported the identification of Al-tolerant genotypes in rice, wheat, maize, sorghum, and soybean (Armiger et al., 1968; Howeler and Cadavid, 1976; Fageria et al., 1988; Massot et al., 1992; Sivaguru et al., 1992; Sivaguru and Paliwal, 1993; Khatiwada et al., 1996; Urrea-Gomez et al., 1996; Sousa, 1998; Bushamuka and Zobel, 1998).

SELECTION OF STRESS-TOLERANT CROP CULTIVARS

Genetic variability within a species can be employed to understand the mechanisms of stress resistance, with the ultimate goal of producing stress-tolerant crop cultivars. Plants may respond to environmental stresses in a variety of ways. In the early stages of stress, plants may attempt to alleviate the damaging effect of stress by altering their own metabolism. Plants respond to and protect themselves from these stresses depending upon their genetic makeup (Nguyen et al., 1997).

The development of stress-tolerant crops has been hindered by lack of knowledge of more precise physiological parameters that reflect genetic potential for improved productivity under stress and by low heritability of indirect characters such as yield. Understanding the physiological, molecular, and biochemical effects of these stresses constitutes the first step toward the development of strategies for designing stress-resistant genotypes. The identification of DNA markers diagnostic of Al, drought, and salinity tolerance can accelerate the development of cultivars that can remain productive under stress and may be a starting point to identifying the specific genes responsible for differences in response of plant genotypes to toxic aluminum, drought, and saline levels.

APPLICATION OF DNA MARKERS TO IMPROVE STRESS TOLERANCE

Advances in molecular marker technology over the past decade have led to the development of detailed molecular linkage maps for many plant species. The important event is the discovery of DNA-based markers. DNA markers promise to accelerate progress in improvement of crops for stress tolerance in two specific ways. First, DNA markers provide an objective means to test the extent to which postulated physiological adaptations to stress account for genetic differences in productivity and quality under stress. Second, once the first goal is met, DNA markers provide a diagnostic tool to identify genotypes in large populations that combine superior genetic potential for productivity under stress with the attributes that are necessary to develop a commercially viable cultivar.

Because of the lack of more precise measures and consequent necessity to study indirect measures of stress tolerance, such as "yield," that integrate many aspects of growth and development throughout an organism's life cycle, the approach of QTL mapping has become central to the use of DNA markers in the improvement of stress tolerance. Quantitative trait locus, or QTL mapping, is a means to estimate the locations, numbers, magnitude of phenotypic effects, and modes of gene action of individual determinants that contribute to the inheritance of continuously variable traits. The fundamental task is to extricate genetic "signals" at an individual locus from the "noise" resulting from the collective effects of nongenetic factors, such as edaphic and microclimatic variations, as well as measurement error in assessment of continuous traits. Although QTL mapping is very helpful, a strong effort must be made to find additional methods to identify specific components that permit more direct assessment of stress tolerance.

Several different types of DNA markers and QTL mapping populations are in routine use. These will be briefly reviewed with regard to their relative strengths and weaknesses for study of the inheritance of abiotic stress tolerance.

Types of DNA Markers and Their Merits

Three general types of molecular tools figure prominently in the genetic toolbox for QTL mapping, each including a number of variations on a common theme.

Evolutionarily Conserved Loci Suitable for Establishing Orthology Across Taxa

The most widely used type of DNA marker in plant genomics has been the restriction fragment length polymorphism (RFLP) (Botstein et al., 1980). Detailed RFLP maps have taught us much about genome organization and were used in most of the early QTL mapping efforts in major crops. RFLPs remain especially important for learning about the comparative arrangement(s) of genes along the chromosomes of divergent plant taxa, as they are amenable to direct use of cDNAs or hypomethylated genomic DNA clones that tend to retain sufficient sequence similarity to hybridize to orthologous loci in different taxa. However, RFLPs are cumbersome, requiring large amounts of DNA, tedious blot hybridization, and autoradiographic methods, and detect only a small fraction of the DNA polymorphisms that occur between genotypes. A modification, the Cleaved Amplified Polymorphic Sequence (CAPS) method (Konieczny and Ausubel, 1993), uses locus-specific polymerase chain reaction (PCR) primers to amplify corresponding loci in different genotypes then cut the amplification product with a restriction enzyme, often with a 4-nt recognition sequence, retaining the locus-specificity of RFLP but reducing the amount of genomic DNA needed and eliminating autoradiography.

Rapidly Evolving Loci Suitable for Discerning Variation Among Closely-Related Genotypes

Simple-sequence repeat (SSR) markers have become popular in the past decade. SSRs are tandemly repeated arrays of simple sequence motifs, typically two to four nucleotides in length; for example . . . CACACA . . ., often comprising five to 50 tandem units of the motif. These elements rapidly evolve new "alleles" in the form of arrays with different numbers of units. By amplification of genomic DNA with PCR primers that immediately flank the arrays, alleles are visualized as size variants in acrylamide or agarose gels. By attaching fluorescent dyes to the primers, several loci can simultaneously be assayed on vertical acrylamide gels. Apparatuses for rapid size separation and laser detection of data are commercially available. Simple sequence repeats (SSRs) have been especially important in the detailed characterization of elite crop gene pools comprised of closely related individuals. However, SSRs are relatively costly to develop and have proven to detect only modest levels of DNA polymorphism in some recently formed polyploids such as groundnut (Hopkins et al., 1999).

Arbitrary or Semi-Arbitrary Sequence-Based PCR, Suitable for Rapid Assembly of Data with a Minimum of a priori Information

Large quantities of genomic segregation data can be accumulated with a minimum of a priori information, using any of several methods that rely upon PCR primers that are sufficiently short to occur by chance in a large number of locations in a higher-plant genome. While several such methods have been described, including AP-PCR (Welsh and McClelland, 1990) and RAPD (Williams et al., 1990), the most widely-used is presently the AFLP method (Vos et al., 1995). Some difficulties include: mapped loci tend to be unevenly distributed across the chromosomes, requiring more effort to obtain complete genome coverage; alleles tend to be population-specific, making comparative analysis difficult for different crosses and nearly impossible for different taxa; and the vast majority of alleles are dominant, failing to discern homozygotes from heterozygotes.

A recent development, transposon display, uses dispersed repetitive DNA to provide one primer and an arbitrary sequence as the second primer (Van den Broeck et al., 1998). A priori determination of the genomic distribution of individual transposable element (TE) families enables one to choose a sample that is likely to represent near-complete coverage of the genome, or possibly even to target specific chromosomes if warranted (Casa et al., 2000). This method has special appeal in that the priming sequence itself may be an important agent of genetic change, and in some cases is preferentially associated with genes (Zhang et al., 2000).

Comparative Mapping of Plant Genomes

Molecular markers have also provided a direct conduit for comparative genetic analysis of taxa that have been reproductively isolated for millions of years. Comparative mapping, the study of similarities and differences in gene order along the chromosomes of taxa that cannot be hybridized, has shown that surprisingly few major chromosomal rearrangements (inversions and/or translocations) often distinguish among many major crops and/or botanical models. Comparative maps permit information gathered during study of one taxon to be quickly applied to related taxa.

This capability has added new dimensions to comparative biology, with special significance for studying diverse crops. More than 200 different flowering plant taxa (angiosperms) sustain world agriculture, most tracing to a common ancestor thought to exist perhaps 150 to 250 million years ago. As a result of their shared heredity, gene arrangements and gene functions remain similar in the genomes of diverse plants (for example, see Paterson et al., 2000), and it is often possible to infer the locations and functions of genes in major crops based on seminal research in facile botanical models.

Different types of DNA markers are useful to different degrees for comparative mapping. Most of the comparative mapping studies published to date in plants have relied upon RFLPs, because this approach permits one to use genes (cDNAs) as probes. The relatively high degree of evolutionary conservation of gene sequences fosters their application to diverse taxa. Netic variations on RFLPs such as CAPS markers are similarly useful. At the opposite end of the spectrum, arbitrary-sequence markers such as RAPDs and AFLPs are of minimal utility for comparative mapping, as a result of their lack of specificity for single genetic loci. SSR markers are intermediate in their utility for comparative mapping, having the necessary single-locus specificity but often failing to hybridize to (amplify) genomic DNA from diverse taxa. Occasional SSRs that are found within the coding regions of genes are more successful.

Population Structure and Its Implications for QTL Mapping of Stress Tolerance

A wide range of population structures can be used for QTL mapping; the minimal requirement being the establishment of linkage disequilibrium between defined genotypes. Different population structures have different strengths and weaknesses with regard to the study of stress tolerance, in particular regarding the ability to conduct replicated studies that permit the study of response to different environments, such as stressed versus nonstressed, or simply different years. Plants that tolerate inbreeding enjoy especially great latitude for different population structures, including backcross F_2, recombinant inbred (RI) or single-seed descent (SSD), and double haploid (DH).

Backcross

Traditionally, genetic linkage mapping relied heavily upon intercrossing two homozygous genotypes to produce a heterozygous F_1 and crossing the F_1 back to one of its inbred parents to create 1:1 segregation for polymorphic alleles from the donor parent. This population structure remains useful for a number of purposes, especially introgression of exotic germplasm from a wild relative into a domesticated. The analysis of "advanced-backcross" or "backcross-self" populations in combination with molecular mapping has sometimes revealed cryptic alleles from wild relatives that exert unexpected desirable effects when uncoupled from background factors that suppress their statement (Paterson et al., 1988; Xiao et al., 1996; Tanksley et al., 1996). The backcross is not widely used in mapping of genes associated with stress tolerance because it is ephemeral and cannot easily be replicated. However, it is often used to transfer desirable genes to elite genotypes.

F_2

Selfing or intercrossing of heterozygous F_1s creates populations that segregate in the traditional 1:2:1 ratio, and enjoy the advantage of permitting the geneticist to see the consequences of all possible "dosages" of an allele, assuming bivalent pairing. This permits estimation of the mode of gene action: dominant, recessive, additive, or, most frequently, somewhere in between. Because of this merit and also the fact that it is quick to produce, the F_2 is often used in QTL mapping. A disadvantage in using it for study of stress tolerance is that, like the backcross, it is ephemeral and cannot easily be replicated. Some investigators have mitigated this problem by studying selfed progeny of F_2 individuals. Although there is uncertainty regarding the specific genotype(s) of individuals in such studies, the "breeding value" (average phenotype) of F_2-derived F_3 families has successfully been used to identify genes associated with plant response to water deficit (for example, see Saranga et al., 2001).

Recombinant Inbred (RI) or Single-Seed Descent (SSD)

These population structures are especially important for study of stress tolerance, in that they are comprised of homozygous genotypes that can be replicated in different environments or experimental treatments; for example, to quantify the "sensitivity" of individual QTLs to year-to-year fluctuations in climatic conditions, or to specific treatments such as an imposed stress. In plants, most RI or SSD populations are produced by selfing, and individuals are derived from two gametes that are identical by descent, so therefore offer only the information content of one gamete in, for example, an F_2 individual. However, this gamete has been through several, usually six to eight, generations of selfing, so there have been many additional opportunities for recombination in those regions of the genome that remained heterozygous. Since there is less heterozygosity with each passing generation, the result of this tradeoff is that RI or SSD individuals derived by selfing have similar "information content" to F_2 individuals, which comprise two different gametes that have been through only one meiotic cycle. A variation on these methods is the interjection of several generations of random intermating into the breeding scheme, followed by selfing, which permits one to obtain plant populations with higher information content, more like animal RI populations, but at a cost of time (Liu et al., 1996).

Doubled Haploid (DH)

This population structure combines the advantages of homozygosity with the speed at which an early-generation population can be made. A heterozygous F_1 is used to produce gametes that are artificially doubled in chromosome number, by means such as colchicine, and cultured to yield plants. DH populations exist for barley, rice, *Brassica oleracea*, and a number of other major crops. Genetically, individual plants within a DH population contain two identical gametes and, therefore, each contains only one uniquely informative gamete. Therefore, DH individuals are equivalent to first-generation backcross individuals in terms of their information content about recombination.

In taxa that are not tolerant of inbreeding, fewer options are available. This includes many autopolyploid plants such as potato, sugarcane, alfalfa, and many forage and turf grasses. The most common approach in mapping of these taxa is the production of F_1 hybrids from crosses between highly heterozygous, unrelated individuals. By this approach, one is mapping the segregation and recombination that occurred simultaneously in each of two heterozygous nuclei, one from each parent, and therefore the result is two maps, one of each parent. The use of at least a framework of RFLP or SSR markers permits one to align the different members of a homologous series of chromosomes within each map. Replication, important to the study of stress tolerance, is sometimes possible in that many autopolyploids can be propagated clonally.

CONCLUSION

Molecular maps are now available for most major crops, and are likely to assist improvement of plant stress tolerance in at least two ways. First, basic research that integrates measurement of plant physiological parameters with study of yield and quality promises to shed light on the specific components of stress tolerance that account for genetic variation in modern crop gene pools. Second, the resulting DNA markers promise to accelerate manipulation of stress-tolerance genes in breeding programs toward genotypes that combine these genes with the many other attributes that are necessary to produce a successful cultivar. The detection of QTLs related to stress tolerance also represents a crucial first step toward cloning of stress-tolerance genes, which would be very helpful for use in the analysis of the physiological mechanisms involved.

REFERENCES

Albert, R. and M. Popp (1977). Chemical composition of halophytes from the Neusiedler lake region in Austria. *Oecologia (Berlin)* 27:157-170.

Armiger, H.W., C.D. Foy, A.L. Fleming, and B.E. Caldwell (1968). Differential tolerance of soybean varieties to an acid soil high in exchangeable aluminum. *Agronomy Journal* 59:415-418.

Bernstein, L. (1975). Effect of salinity and sodicity on plant growth. *Annual Review of Phytopathology* 13:295-312.

Blum, A. (1988). *Plant Breeding for Stress Environments.* Boca Raton, FL: CRC Press.

Botstein, D., R.L. White, M. Skolnick, and R.W. Davis (1980). Construction of a genetic linkage map in man using restriction fragment length polymorphisms. *American Journal of Human Genetics* 32:314-333.

Bushamuka, N.V. and R.W. Zobel (1998). Maize and soybean tap, basal, and lateral root responses to a stratified acid, aluminum toxic soil. *Crop Science* 38:416-421.

Campbell, T.A., P.R. Jackson, and Z.L. Xia (1994). Effect of aluminum stress on alfalfa root proteins. *Journal of Plant Nutrition* 17:461-471.

Casa, A.M., C. Brouwer, A. Nagel, L. Wang, Q. Zhang, S. Kresovich, and S.R. Wessler (2000). The MITE family Heartbreaker (Hbr): Molecular markers in maize. *Proceedings of the National Academy of Sciences (USA)* 97:10083-10089.

Castleberry, R.M. (1983). Breeding programs for stress tolerance in corn. In *Crop Reactions to Water and Temperature Stresses in Humid, Temperate Climates,* C.D. Raper Jr. and P.J. Kramer (Eds.). Boulder, CO: Westview Press, pp. 277-287.

Collander, R. (1941). Selective absorption of cations by higher plants. *Plant Physiology* 16:691-721.

Delhaize, E. and P.R. Ryan (1995). Aluminum toxicity and tolerance in plants. *Plant Physiology* 107:315-321.

Deng, X.M., R.J. Joly, and D.T. Hahan (1990). The influence of plant water deficit on distribution of ^{14}C-labeled assimilates in cacao seedlings. *Annual Botany* 66:211-217.

Dudal, R. (1976). Inventory of the major soils of the world with special reference to mineral stress hazards. In *Plant Adaptation to Mineral Stress in Problem Soils,* M.J. Wright (Ed.). Ithaca, NY: Cornell University Agricultural Experiment Station, Cornell University, pp. 3-13.

Dwyer, L.M. and D.W. Stewart (1986). Leaf-area development in field-grown maize. *Agronomy Journal* 78:334-343.

Epron, D. (1997). Effect of drought on photosynthesis and on the thermotolerance of photosystem II in the seedling of cedar *(Cedrus atlantica* and *C. libani). Journal of Experimental Botany* 48:1835-1841.

Epstein, E. (1980). Responses of plants to saline environments. In *Genetic Engineering of Osmoregulation: Impact on Plant Productivity for Food, Chemicals, and Energy,* D.W. Rains, R.C. Valentine, and A. Hollauender (Eds.). New York: Plenum Press, pp. 7-21.

Escobar-Guti, A.J., R.B. Zipperlin, F. Carbonne, A. Moing, and J.P. Gaudill (1998). Photosynthesis, carbon partitioning, and metabolite content during drought stress in peach seedlings. *Australian Journal of Plant Physiology* 25:197-205.

Fageria, N.Y., R.J. Wright, and V.C. Baligar (1988). Rice cultivar response to aluminum in nutrient solution. *Communication in Soil Science and Plant Analysis* 19:1133-1142.

Fischer, R.A. and N.C. Turner (1978). Plant productivity in the arid and semiarid zones. *Annual Review of Plant Physiology* 29:277-317.

Flowers, T.J., P.F. Troke, and A.R. Yeo (1977). The mechanism of salt tolerance in halophytes. *Annual Review of Plant Physiology* 28:89-121.

Foy, C.D. (1974). Effects of aluminum on plant growth. In *The Plant Root and Its Environments,* E.W. Carson (Ed.). Charlottesville, VA: University Press of Virginia, pp. 601-642.

Foy, C.D. (1983). The physiology of plant adaptation to mineral stress. *Iowa State Journal of Research* 57:355-391.

Foy, C.D. and T.A. Campbell (1984). Different tolerances of *Amaranthus* strains to high levels of aluminum and manganese in acid soil. *Journal of Plant Nutrition* 7:1365-1388.

Foy, C.D., R.L. Chaney, and M.C. White (1978). The physiology of metal toxicity in plants. *Annual Review of Plant Physiology* 29:511-566.

Gassmann, W. and J.I. Schroeder (1994). Inward-rectifying K^+ channels in root hairs of wheat. A mechanism for aluminum-sensitive low affinity K^+ uptake. *Plant Physiology* 105:1399-1408.

Gogorcena, Y., I. Iturbe-Ormaetxe, P.R. Escuredo, and M. Becana (1995). Antioxidant defenses against activated oxygen in pea nodules subjected to water stress. *Plant Physiology* 108:753-759.

Han, B. and A.R. Kermode (1996). Dehydrin-like proteins in bean seeds and seedlings are differentially produced in response to ABA and water-deficit-related stresses. *Journal of Experimental Botany* 47:933-939.

Hashem, A., M.N. Amin Majumdar, A. Hamid, and M.M. Hossain (1998). Drought stress on seed yield, yield attributes, growth, cell membrane stability and gas exchange of synthesized *Brassica napus* (L). *Journal of Agronomy and Crop Science* 180:129-136.

Hopkins, M.S., A.M. Casa, T. Wang, S.E. Mitchell, R.E. Dean, G.D. Kochert, and S. Kresovich (1999). Discovery and characterization of polymorphic simple sequence repeats (SSRs) in peanut. *Crop Science* 39:1243-1247.

Howeler, R.H. and L.F. Cadavid (1976). Screening of rice cultivars for tolerance to Al-toxicity in nutrient solutions as compared with a field screening method. *Agronomy Journal* 68:551-555.

Huang, B. and Y. Jiang (2002). Physiological and biochemical responses of plants to drought and heat stress. In *Crop Improvement Challenges in the Twenty-First Century,* M.S. Kang (Ed.). Binghamton, NY: Food Products Press, pp. 287-313.

Jacoby, B. (1965). Sodium retention in excised bean stems. *Physiologia Plantarum* 18:730-739.

Jacoby, B. and A. Ratner (1973). Mechanisms of sodium exclusion in bean and corn plants: A reevaluation. In *Plant Analysis and Fertilizer Problems,* J. Wehrmann (Ed.). Hanover: German Society of Plant Nutrition, pp. 175-184.

Jagtap, V., S. Bhargava, P. Streb, and J. Feierabend (1998). Comparative effect of water, heat, and light stresses on photosynthetic reaction in *Sorghum bicolor* (L). Moench. *Journal of Experimental Botany* 49:1715-1721.

Kang, M.S. (1998). Using genotype-by-environment interaction for crop cultivar development. *Advances in Agronomy* 62:199-252.

Khatiwada, S.P., D. Senadhira, A.L. Carpena, R.S. Zeigler, and P.G. Fernandez (1996). Variability and genetics of tolerance for aluminum toxicity in rice (*Oryza sativa* L.). *Theoretical and Applied Genetics* 93:738-744.

Kochian, L.V. (1995). Cellular mechanisms of aluminum toxicity and resistance in plants. *Annual Review of Plant Physiology and Plant Molecular Biology* 46:237-260.

Konieczny A. and F. Ausubel (1993). A procedure for mapping *Arabidopsis* mutations using co-dominant ecotype-specific PCR-based markers. *Plant Journal* 4:403-410.

Kramer, D. (1984). Cytological aspects of salt tolerance on higher plants. In *Salinity Tolerance in Plants: Strategies for Crop Improvement,* R.C. Staples and G.H. Toenniessen (Eds.). New York: John Wiley and Sons, Inc., pp. 3-15.

Kronfub, G., A. Polle, M. Tausz, W.M. Havranek, and G. Wieser (1998). Effect of ozone and mild drought stress on gas exchange, antioxidants and chloroplast pigments in current-year needles of young Norway spruce [*Picea abies* (L) Karst.]. *Trees* 12:482-489.

Kuo, C.G., H.M. Chen, and H.C. Sun (1993). Membrane thermostability and heat tolerance of vegetable leaves. In *Adaptation of Food Crops to Temperature and Water Stress,* C.G. Kuo (Ed.). Shanhua, Taiwan: Asian Vegetable Research and Development Center, pp. 160-168.

Liu, S.-C., S.D. Kowalski, T.-H. Lan, K. Feldmann, and A.H. Paterson (1996). Global high-resolution mapping of plant genomes, using *Arabidopsis thaliana* as a model. *Genetics* 142:247-258.

Magnavaca, R. and A.F.C. Bahia Filho (1993). Success in maize acid soil tolerance. *Proceedings of the Workshop on Adaptation of Plants to Soil Stresses,* J.W. Maranville, B.V. Baligar, R.R. Duncan, and J.M. Yohe (Eds.). INTSORMIL Publ. No.94-2. Lincoln, NE: University of Nebraska, pp. 209-220.

Massacci, A., A. Battistelli, and F. Loreto (1996). Effect of drought stress on photosynthetic characteristics, growth, and sugar accumulation of field-grown sweet sorghum. *Australian Journal of Plant Physiology* 23:331-340

Massot, N., C. Poschenrieder, and J. Barcelo (1992). Differential response of three bean *(Phaseolus vulgaris)* cultivars to aluminum. *Acta Botanica Neerlandica* 41:293-298.

Matsumoto, H. (1991). Biochemical mechanism of the toxicity of aluminum and the sequestration of aluminium in plant cell. In *Plant-Soil Interaction at Low pH*, R. J. Wright, V.C. Baligar, and R.P. Murrmenn (Eds.). Dordrecht, The Netherlands: Kluwer Academic, pp. 825-838.

Morimura, S., E. Taakahashi, and H. Matsumoto (1978). Association of aluminum with nucleic and inhibition of cell division in onion *(Allium cepa)* roots. *Zeitschrift für Pflanzenphysiologie* 88:395-401.

Naidoo, G., J. Mc.D. Steward, and R.J. Lewis (1978). Accumulation sites of Al in snap-bean and cotton roots. *Agronomy Journal* 70:489-492.

Nguyen, H.T., R.C. Babu, and A. Blum (1997). Breeding for drought resistance in rice: Physiology and molecular genetics considerations. *Crop Science* 37:1426-1434.

Nguyen, H.T. and C.P. Joshi (1993). Molecular strategies for the genetic dissection of water and high-temperature stress adaptation in cereal crops. In *Adaptation of Food Crop to Temperature and Water Stress,* C.G. Kuo (Ed.). Taipei, Taiwan: Asian Vegetable Research and Development Center, pp. 3-19.

Olsson, M., K. Nilsson, C. Liljenberg, and G.A.F. Hendry (1996). Drought stress in seedlings: Lipid metabolism and lipid peroxidation during recovery from drought in *Lotus corniculatus* and *Cerastium fontanum. Plant Physiology* 96:577-584.

Paakkonen, E., J. Vahala, M. Pohjolia, T. Holopainen, and L. Karenlampi (1998). Physiological, stomatal and ultrastructural ozone responses in birch *(Betula pendula)* are modified by water stress. *Plant, Cell and Environment* 21:671-684.

Pan, W.L., A.G. Hopkin, and W.A. Jackson (1989). Aluminum inhibition of shoot lateral branches of *Glycine max* and reversal by exogenous cytokinins. *Plant and Soil* 120:1-9.

Paterson, A.H., J. Bowers, M. Burow, X. Draye, C. Elsik, C. Jiang, C. Katsar, T. Lan, Y. Lin, R. Ming, and R. Wright (2000). Comparative genomic of plant chromosomes. *Plant Cell* 12:1523-1539.

Paterson, A.H., E.S. Lander, J.D. Hewitt, S. Paterson, S. Lincoln, and S.D. Tanksley (1988). Resolution of quantitative traits into Mendelian factors by using a complete linkage map of restriction fragment length polymorphisms. *Nature* 335:721-726.

Perdomo, P., J.A. Murphy, and G.A. Berkowitz (1996). Physiological changes associated with performance of Kentucky bluegrass cultivars during summer stress. *Horticultural Science* 317:1182-1186

Quisenberry, J.S. (1982). Breeding for drought resistance and plant water-use efficiency. In *Breeding Plants for Less Favorable Environments,* M.N. Christiansen and C.F. Lewis (Eds.). New York: John Wiley and Sons, pp. 193-212.

Rains, D.W. (1969). Sodium and potassium absorption by stem tissue of bean and cotton. *Plant Physiology* 44:547-554.

Rains, D.W. (1981). Salt tolerance: New developments. In *Advances in Food Producing Systems for Arid and Semiarid Lands,* J.T. Manassah and E.J. Briskey (Eds). New York: Academic Press, pp. 431-456.

Rengel, Z. (1992). Role of calcium in aluminum toxicity. *New Phytologist* 121:499-513.

Ryan, P., J.E. Shadd, and L.V. Kochian (1992). Aluminum toxicity in root: Correlation among ionic content, ion fluxes, and root elongation in aluminum-sensitive and aluminum-tolerant wheat cultivars. *Plant Physiology* 99:1193-1200.

Ryan, P., D.J. Tomaso, and L.V. Kochian (1993). Aluminum toxicity in roots: An investigation of spatial sensitivity and role of the root cap. *Journal of Experimental Botany* 44:437-446.

Sairam, P.K., P.S. Deshmukh, and D.S. Shukla (1997). Tolerance of drought and temperature stress in relation to increases in antioxidant enzymes in wheat. *Journal of Agronomy and Crop Science* 178:171-178.

Sanchez, P.A. and J.G. Salinas (1981). Low-input technology for managing oxisol and ultisols in tropical America. *Advances in Agronomy* 34:280-298.

Saranga, Y., M. Menz, C. Jiang, R. Wright, D. Yakir, and A.H. Paterson (2001). Genomic dissection of genotype × environment adaptation conferring adaptation of cotton to arid conditions. *Genome Research* 11:1988-1995.

Sivaguru, M., M.R. James, P.R. Anbudurai, and T. Balakumar (1992). Characterization of differential aluminum tolerance among rice genotypes cultivated in South India. *Journal of Plant Nutrition* 15:233-246.

Sivaguru, M. and K. Paliwal (1993). Differential aluminum tolerance in some tropical rice cultivars: I. Growth performance. *Journal of Plant Nutrition* 16:1705-1716.

Sloane, R.J., R.P. Patterson, and T.E. Carter Jr. (1990). Field drought tolerance of a soybean plant introduction. *Crop Science* 30:118-123.

Sousa, C.N.A.D. (1998). Classification of Brazilian wheat cultivars for aluminum toxicity in acid soils. *Plant Breeding* 117:217-221.

Tanji, K.K. (1990). Nature and extent of agricultural salinity. In *Agricultural Salinity Assessment and Management,* Manuals and Reports on Engineering Practice 71, K.K. Tanji (Ed.). New York: American Society of Civil Engineers (ASCE), pp. 1-17.

Tanksley, S., S. Grandillo, T. Fulton, D. Zamir, Y. Eshed, V. Petiard, J. Lopez, and T. Beck-Bunn (1996). Advanced backcross QTL analysis in a cross between an elite processing line of tomato and its wild relative L. *pimpinellifolium. Theoretical and Applied Genetics* 92:213-224.

Taylor, G.J. (1988). The physiology of aluminum phytotoxicity. In *Metal Ions in Biological Systems.* Volume 24, *Aluminum and its role in biology,* H. Sigel (Ed.). New York: Marcel Dekker, pp. 123-163.

Taylor, G.J. (1991). Current views of the aluminum stress response: *The physiological basis of tolerance. Current Topics in Plant Biochemistry and Physiology* 10:57-93.

Thomson, W.W. (1975). The structure of salt glands. In *Plants in Saline Environments,* A. Poljakoff-Mayber and J. Gale (Eds.). Berlin, Germany: Springer-Verlag, pp. 118-145.

Urrea-Gomez, R., H. Ceballos, S. Pandey, F.C. Antonio, B. Filho, and L.A. Leon (1996). A greenhouse screening technique for acid soil tolerance in maize. *Agronomy Journal* 88:806-812.

Van den Broeck, D., T. Maes, M. Sauer, and Z. De Keukeleire (1998). Transposon display identifies individual transposable elements in high copy number lines. *The Plant Journal* 13:121-129.

Volaire, F., H. Thomas, and F. Lelievre (1998). Survival and recovery of perennial forage grasses under prolonged Mediterranean drought. I. Growth, death, water relations and solute content in herbage and stubble. *New Phytologist* 140:439-499.

Vos P., R. Hogers, M. Bleeker, M. Reijans, T. van de Lee, M. Hornes, A. Frijters, J. Pot, J. Peleman, M. Kuiper, and M. Zabeau (1995). AFLP: A new technique for DNA fingerprinting. *Nucleic Acids Research* 23:4407-4414.

Wallace, S.U. and I.C. Anderson (1984). Aluminum toxicity and DNA synthesis in wheat roots. *Agronomy Journal* 76:5-8.

Wang, Z., B. Quebedeaux, and G.W. Stutte (1996). Partitioning of ^{14}C-glucose into sorbitol and other carbohydrates in apple under water stress. *Australian Journal of Plant Physiology* 23:245-251

Welsh, J. and M. McClelland (1990). Fingerprinting genomes using PCR with arbitrary primers. *Nucleic Acids Research* 18:7213-7218.

Williams, J., A. Kubelik, K. Livak, J. Rafalski, and S. Tingey (1990). Oligonucleotide primers of arbitrary sequence amplify DNA polymorphisms which are useful as genetic markers. *Nucleic Acids Research* 18: 6531-6535.

Winter, E. (1982). Salt tolerance of *Trifolium alexandrinum* L. 3. Effects of salt on ultrastructure of phloem and xylem transfer cells in petioles and leaves. *Australian Journal of Plant Physiology* 9:239-250.

Winter, E. and J. Preston (1982). Salt tolerance of *Trifolium alexandrinum* L. 4. Ion measurements by X-ray microanalysis in unfixed, frozen hydrated leaf cells at various stages of salt treatment. *Australian Journal of Plant Physiology* 9:251-259.

Xiao, J., J. Li, S. Grandillo, S. Ahn, S. McCouch, S. Tanksley, and L. Yuan (1996). A wild species contains genes that may significantly increase the yield of rice. *Nature* 384:223-224.

Zhang, J.X. and M.B. Kirkham (1994). Drought-stress-induced changes in activities of superoxide dismutase, catalase, and peroxidase in wheat species. *Plant Cell Physiology* 35:785-791.

Zhang, Q., J. Arbuckle, and S. Wessler (2000). Recent, extensive and preferential insertion of members of the MITE family *Heartbreaker (Hbr)* into genic regions of maize. *Proceedings of the National Academy of Sciences (USA)* 97:1160-1165.

Zhao, X.J., E. Sucoff, and E.J. Stadelmann (1987). Al^{3+} and Ca^{2+} alteration of membrane permeability of *Quercus rubra* root cortex cells. *Plant Physiology* 83:159-162.

Chapter 5

Breeding for Salinity Tolerance

Rajinder S. Malhotra
Thomas Blake

INTRODUCTION

Land degradation is the greatest impediment to agricultural sustainability in developing nations. In 1980, more than 3×10^6 ha of arable land were considered saline (Brinkman, 1980). This area has more than doubled in the past two decades. Salinization through mismanaged irrigation has destroyed civilizations and today contributes to international discord. While appropriate water management is crucial to avoid salinization and restore salinized soils, salt-tolerant crops can provide interim solutions to farmers in degraded landscapes.

THE PROBLEM OF SALINIZATION

Plants Deal with Excess Salt

Growth in hypersaline environments results in plants that suffer hyperosmotic effects. Sodium chloride is the most common causal agent of saline stress in terrestrial environments. Enabling plants to better manage excess sodium is key to improving plant productivity in these environments. The consequences of hyperosmotic stress include membrane disorganization, metabolic toxicity, disruption of photosynthesis, and, in extreme cases, plant death.

Halophytes, including members of the *Atriplex* genus, tolerate impressive levels of salt stress (Glenn et al., 1999). These salt-tolerant plants utilize vacuolar sodium storage and synthesize organic osmotic protectants, including trehalose and fructans; protein protectors, including chaperones and glycine betaine; and a host of compounds to scavenge reactive oxygen species resulting from cellular damage and to act as metabolic sinks. Some halophytes develop specialized salt storage organs and glands, an apparently recent evolutionary adaptation.

Glycophytes utilize similar strategies for coping with saline stress, but fail to cope as rapidly and as completely as do halophytes. This reduced rate of adaptation is responsible for the relative inability of crop plants to withstand rapid changes in soil salinity and rapid desiccation. A host of transcription factors have been uncovered that appear to regulate the rate of adaptive response. One of these, DREB2 (Niu et al., 1993) has been utilized by crop improvement programs worldwide in an attempt to improve tolerance to cellular desiccation.

Our understanding of the genetic underpinnings of stress tolerance has dramatically advanced in the past decade thanks to the development of genome mapping technologies and quantitative trait locus (QTL) analysis. The most dramatic outcome of these experiments is the recognition that not only do similar physiological mechanisms control variation for abiotic stress tolerance across species but also that homeologous genes, or at the least genes in very similar chromosomal locations, control variation for key components of these phenotypes (Gale and Devos, 1998). The major genes regulating the vernalization response reside in, as far as we can tell, identical chromosomal locations in wheat and barley (Hayes et al., 1996), and those controlling the "stay green" phenotype reside on homeologous sites in maize and sorghum (Zhang et al., 2001). Comparative analysis may permit the use of thoroughly characterized genomes such as those of rice and *Arabidopsis* to provide a platform for cloning the transcription factors and structural genes required for dramatic improvement in salt tolerance.

Salinization: The Process

Soil degradation due to improper water management is one of the major threats facing humankind. Salinization not only undermines the productive capacity of an ecosystem but also affects global climate. In 1980, approximately 323 million ha of the world's soils were considered saline or sodic (Brinkman, 1980). These soils occur mainly in arid and semiarid regions, where evaporation considerably exceeds precipitation, leading to salt accumulation in surface soil. Such soils are particularly common in Central and West Asia and in some parts of Australia and the United States. Salinity is also a problem in the coastal regions as a result of the ingress of seawater. Soil degradation due to salinization is a growing problem, and if it continues will impact our ability to increase food production, alleviate rural poverty, and compensate for predicted population increases in the developing world. Apart from global considerations, salinization will specifically impact specific socioeconomic and ecological niches, especially in countries in Central and West Asia.

Growing populations in developing countries have demanded increased access to food and fiber, and one of the easiest (albeit expensive) interventions supporting increased crop production is increased irrigation. Irrigation does not always result in soil salinization. When evapotranspiration exceeds precipitation, careful use of applied water and nutrients is in order. Reservoir-based water management systems have provided electricity for poor populations, mitigated flood potential, and supported recreation-based industries worldwide. Unfortunately, in many canal-irrigated areas rising water tables and inadequate drainage has resulted in salt accumulation. Such problems have increased further where saline groundwater has been used for irrigation. This has become common in many coastal areas, where excessive exploitation of groundwater has caused sea-water intrusion, resulting in ever-increasing salinity problems. This is also a growing problem where coal-bed methane is being tapped and coal-seam sodic water is released for irrigation.

Although technologies for reclamation and rehabilitation of saline soils have been developed, they tend to be expensive and cumbersome. Plant-based salt harvesting, use of best-available soil-water management regimes, and careful use of fertilizer are inexpensive, albeit slow alternatives.

Breeding crop varieties for increased salt tolerance is a promising, energy efficient, and economical complement to thoughtful land-use management. Other methods of reclamation and soil amelioration are often beyond the means of farmers in the marginal areas and developing nations. Soil salinity synergistically interacts with heat and moisture deprivation. These interactions play a great role in limiting high-yielding crop varieties in saline environments. Genotypes intended for moisture-stressed, saline environments must be selected and evaluated at target sites having appropriate stresses to permit identification of dependable sources of stress tolerance.

Before any investment in genetic improvement of salinity tolerance, for example in legumes, is made several points need to be considered. Because of their relative sensitivity to salinity among crop plants (Maas and Hoffman, 1977; Läuchli, 1984), legumes would rank low as an initial crop choice for salt-prone areas. Soil reclamation through the use of salt-harvesting halophytes can be an initial step toward land reclamation. Procedures for doing this, including development of a crop sequence from salt-tolerant species to salt-sensitive ones, have been developed and tested (Meiri and Plaut, 1985; Sharma and Gupta, 1986). Still, increasing the salinity tolerance of forage and food legumes to levels comparable to salt-tolerant crops such as barley would provide poor farmers alternative options for production of crops with improved nutrition and economic returns in the desalinization process and improve the soil fertility, among other benefits.

Economic Impact of Salt-Affected Soils

Soil salinization has a great impact on the economy at the farm as well as national level. At the farm level, soil salinization reduces crop productivity and reduces appropriate crop alternatives, resulting in reduced farm profitability. This often results in farm abandonment and reduced economic productivity. At the national level, the consequences include displacement of labor from agriculture, widening of income disparities, and reduced sustainability of other sectors dependent on the produce from these areas. Reduced crop production also affects the secondary sector as several agro-industries depend on their output. Any decline in crop production leads to unemployment and underemployment of the rural poor.

Salinization and the concomitant reduction in economic productivity contribute to population displacement, urbanization, and eventually emigration. The developed world will likely find it easier to assist with the remediation of degraded environments than suffer the consequences of doing nothing. When high-value crops disappear from a degraded landscape, a clear signal is given that attention must be given to land rehabilitation.

Evidence from Natural Salt-Tolerant Populations

The limits of salinity and alkalinity tolerance differ in different plant species. Algal genera such as *Dunaliella* survive in amazingly sodic environments (Bental et al., 1990). *Atriplex* species provide their most common terrestrial counterparts. Unfortunately, there are few naturally occurring salt-tolerant higher-plant species. *Agrostis stolonifera* L. and *Festuca rubra* L. have provided exceptional tolerance to salinity. Several species, including *Holcus lanatus, Dactylis glomerata, Agrostis castellana,* and *Lolium perenne,* which are not associated with saline habitats, exhibited tremendous variability for salt tolerance (Ashraf et al., 1986). Plainly, broad evaluation of germplasm from a wide array of species will be required to identify those carrying the best component genes for the assembly of salt-tolerant plants.

MECHANISMS OF SALINITY RESISTANCE

Compartmentalization of salt is critical to survival in saline environments. Many glycophytes adapt to moderately saline environments by retaining normal water status in rapidly growing, meristematic regions of the plant while permitting salt to accumulate in lower, senescing leaves. Halophytes actively partition sodium into vacuoles. This is energy dependent and provides a metabolic cost to the plant. Some halophytes actively exclude salt into glands. Osmotic adjustment is obtained in both glycophytes and halophytes through the accumulation of organic compounds, including proline, glycine betaine, trehalose, sorbitol, and mannitol, which all act to maintain homeostasis in salt-stressed plants. The rate at which production of these osmolytes is initiated and their terminal cytosolic concentrations are keys to adaptive salt response (Hasegawa et al., 2000).

The basic mechanisms contributing to osmotic adjustment appear to be common across all plants. However, different plant species contain different levels of genetic variation for the genes responsible for appropriate response to toxic salt concentrations. Understanding the mechanisms at each species's disposal for alleviating toxic effects is fundamental to designing appropriate species-specific selection techniques (Johansen, 1987). As an example, the ability to exclude Na and Cl from shoots distinguishes more resistant chickpea genotypes from sensitive genotypes (Läuchli, 1984; Keating and Fisher, 1985). Among 160 chickpea genotypes screened for salinity resistance in solution culture, the most resistant genotype, L 550, accumulated the least Na in shoots at yield-reducing salt levels (Lauter and Munns, 1986). This suggests that shoot Na concentration would be a good selection criterion for salinity resistance in chickpea.

The physiological approaches used to identify germplasm with relatively greater salt tolerance typically uses static tests, while in nature salinity stress varies with time. This temporal variability due to variation in precipitation, water application, and temperature is a critical and often underestimated component of the environment. Mutant analysis in *Arabidopsis* has provided critical insights into the mechanisms underlying physiological adaptation. The salt-response signaling cascade, while utilizing ABA and various cytokinins (Finkelstein and Zeevaart, 1994), also involves inositols and the products of several enzymes, including invertases and glutathione-S-transferase.

Studies integrating whole transcriptome analysis methods that provide a view of the rate and degree of adaptation to salt stress with carefully constructed field and greenhouse screening techniques will undoubtedly fill the journals in the next several years. It remains useful to review the results of past selection experiments to identify areas where progress was easily made and those that remain recalcitrant.

ARTIFICIAL SELECTION FOR SALINITY TOLERANCE

Legumes

To quantify effects of salinity on plant growth it is necessary to establish critical values, relating salt concentrations in either the rooting medium or plant tissue to reductions in growth or yield. It should be realized that these values are experiment-specific as salinity response and, hence, critical value is much influenced by other growth factors such as soil water status (Saxena, 1987), relative humidity (Salim, 1989), temperature (Ayoub, 1974), and nutrition (Kamel, 1986). Further, there are several alternative methods of calculating critical values (e.g., Bresler et al., 1982; Johansen et al., 1990).

In developing salinity screening techniques, it is also necessary to know the plant-growth stage most sensitive to salinity. As is the case for most legumes, germination of chickpea is relatively less affected by salinity than subsequent seedling growth (Goel and Varshney, 1987; Yadav et al., 1989). Kumar (1985) indicated that later stages of chickpea growth were more sensitive than earlier stages. Salinity effects also increased over time in pea (Siddiqui and Kumar, 1985). However, Maas and Poss (1989) pointed out that, in such studies, apparent sensitivity at later stages may be a result of salt accumulation over a longer period. Experiments in which salinity treatments were imposed at different growth stages showed that cowpea was most sensitive at the early (vegetative) growth stage. In a similar study with pea, Lal (1985) found that earlier growth stages were more sensitive to salinity.

In establishing appropriate salinity screening techniques, it is also important to understand which plant processes are most sensitive to salinity. For legumes, a primary consideration is symbiotic nitrogen fixation. Free-living rhizobia demonstrate much higher salt tolerance than their host plants (Lauter et al., 1981; Sprent, 1984). Hence, there is little point in screening free-living rhizobia for salt tolerance and expecting this to result in improved nitrogen fixation by symbiotic plants under saline conditions. There is, however, evidence that the functioning of the symbiotic process itself is more sensitive to salinity than growth of the host plant per se (Siddiqui et al., 1985). Chickpea and faba bean supplied with mineral N were more salt tolerant than symbiotically dependent plants (Lauter et al., 1981; Yousef and Sprent, 1983).

In lentil grown in saline soil, there were large interactions between *Rhizobium* strain inoculated and lentil genotype (Rai, 1983; Rai et al., 1985). Thus, as Lauter and colleagues (1981) suggest, it may be necessary to screen for salinity response with both N-fed and symbiotic plants. Kumar and Garg (1981) further showed that increasing salinity and alkalinity positively interact in depressing nitrogen fixation of pea, mainly by inhibiting the infection process.

It is important to consider the composition of the medium used for salinity resistance screening because the different ions involved have different uptake and translocation characteristics. For example, Cl^- uptake by plant roots usually greatly exceeds that of SO_4^{2-} and, thus, salt damage is greater when Cl^- rather than SO_4^{2-} is the dominant anion. This has been tested in both chickpea (Lauter and Munns, 1986; Manchanda and Sharma, 1989) and pea (Hasson-Porath et al., 1972). Sheoran and Garg (1983), however, showed that Na_2SO_4 more adversely affected germination and early seedling growth of chickpea than did KCl, K_2SO_4, and NaCl. Kumar and Promila (1983) found that Cl^- and SO_4^{2-} salinity had different effects on different stages and processes of nitrogen fixation in chickpea. Nodule number increased but nodule weight decreased with Cl^- as compared with SO_4^{2-} salinity.

The concentration of Ca in the medium considerably influences salinity response because of its effects on Na absorption, K/Na selectivity, and membrane integrity (Läuchli, 1984; Cramer et al., 1985). Ca levels in screening media need to be carefully selected and monitored.

Kavita and Singh (1999) evaluated lentil genotypes using petri dishes containing 10 ml salt solutions at three salt-stress levels (0.0, 8.0, and 12.0 $dS{\cdot}m^{-1}$). The germination percentage, germination relative index (GRI), seed vigor, and mobilization efficiency (ME), were measured in six-day-old seedlings. The germination percentage, seed vigor, GRI, and ME decreased with increasing stress-level salt-susceptible genotypes.

Moreno and colleagues (2000) studied the genotypic variability in bean cultivars (*Phaseolus vulgaris* L.) for resistance to salinity (sodium chloride) at the seedling stage in pots. Highly significant differences were found among genotypes and among treatments for plant height, seedling fresh and dry weight, and root length. Some of the genotypes tolerant to salinity were selected. Higher root growth and mineral acquisition in the roots in the salinity stress test were related to the mechanism of resistance at the seedling stage.

Nonlegumes

Mano and Takeda (2001) tested about 1300 wheat cultivars for salt tolerance at germination and the seedling stage. Salt tolerance at these two stages showed wide and continuous variation among the wheat varieties tested, and some highly salt-tolerant varieties were selected. Varieties from Korea included a high proportion of tolerant varieties at either stage. Hexaploid (AABBDD) varieties were more tolerant than tetraploid (AABB) and diploid (AA) varieties at both stages, indicating that genetic factor(s) controlling salt tolerance may be located on the D genome. Salt tolerance at germination was considered to be independent of that at the seedling stage.

Yue and colleagues (2000) conducted a study to determine if salt-resistance genes could be transferred into wheat from its phylogenetically remotely related species. Wheat cv. Jian 177 protoplasts were fused with UV-irradiated protoplasts of *Aeluropus littoralis,* a salt-tolerant species intertribal to wheat. The early formed regenerated clones were identified as hybrids through chromosome, isoenzyme, and RAPD (random amplification of polymorphic DNA) analysis. The salt-tolerant ability of the hybrids was compared with their parents in terms of relative growth, proline accumulation, and NA^+/K^+ ratio under salt stress. These parameters were higher in the hybrids, indicating that some corresponding salt-tolerance genes were transferred. Unfortunately, no fertile progeny were recovered.

Velu and Srivastava (2000) studied the status of sodium-potassium as a basis of salinity tolerance in rice using 25-day-old seedlings of a few rice cultivars transplanted in pots containing soil treated with 2.5 liters of 50, 75, 100, or 150 mM NaCl solution. Na, P, and the Na:P ratio of plants were determined 10 and 20 days after sowing and 20 and 25 days after transplanting. Na content and Na:P ratio of all cultivars increased with increasing NaCl concentration, while the P content was reduced. Across all salinity levels, the sensitive cultivar, Pusa-Bas 1, recorded higher mean Na contents and Na:P ratios and lower P contents than the moderately tolerant cultivars, CSR 13 and CSR 10.

Nasir and colleagues (2000) screened sugarcane lines for salt tolerance at the seedling stage using gravel culture and half-strength Hoagland's solution with two salinity levels (50 and 100 mM NaCl). Gravel was flushed with half-strength Hoagland's solution alone as the control. Shoot fresh and dry weights of all lines decreased significantly with increase in salinity. Two highly tolerant lines were identified, with the overall conclusion that mass screening at the seedling stage in solution culture was a reasonably reliable technique.

Asch and colleagues (2000) studied leaf K/Na ratio to predict salinity-induced yield loss in irrigated rice. Twenty-one rice genotypes were grown over seven seasons in a field trial in Ndiaye, Senegal, between 1991 and 1995, and were subjected to irrigation with moderately saline water (3.5 $mS \cdot cm^{-1}$ electrical conductivity) or irrigation with fresh water. Potassium/sodium ratios of the youngest three leaves (K/Na leaves) were determined by flame photometry at the late vegetative stage. Grain yield was determined at maturity. All cultivars showed strong log-linear correlations between K/Na leaves and grain yield, although intercept and slope of those correlations differed between seasons for a given genotype and between genotypes. There was a highly significant correlation ($P< 0.001$) between K/Na ratio and salinity-induced grain yield reduction. Although there were major differences in the effects of salinity on crops in both the hot, dry seasons and the wet seasons, the correlation was equally significant across cropping seasons. The earliest possible time to establish the relationship between K/Na leaves under salinity and grain yield reduction due to salinity was investigated in additional trials. About 60 days after sowing, salinity-induced yield loss could be predicted through foliar K/Na analysis.

Isla and colleagues (1998) assessed relationships between grain yield, carbon isotope discrimination (δ), canopy temperature, stomatal conductance, and grain ash content in a set of barley cultivars grown in a soil-salinity gradient imposed by a triple line-source sprinkler system, during 1992-1994 at Zaragoza, Spain. A linear increase in soil salinity resulted in significant and linear decreases in grain yield and δ, and increases in midday differential canopy temperature (δT). For unit increase in soil-solution electrical conductivity (ECss), δ decreased by 0.20 percent and δT increased by 0.3°C. Increasing soil salinity from the control (nonsaline) to a highly saline level ($Ecss = 22\ dS \cdot m^{-1}$) resulted in decreased leaf stomatal conductance (by 65 percent) and increased grain ash content (by 11 to 29 percent). Grain yield and grain δ of 34 barley cultivars were significantly correlated in nonsaline conditions, and it is suggested that δ may be a useful indicator of yield potential in barley. However, there was no significant correlation between grain yield and grain δ under highly saline conditions. A multiple linear regression of δ, stomatal conductance, and grain ash content on grain yield was highly significant in nonsaline conditions, but not in highly saline conditions. It is concluded that none of the studied characters would be useful in screening for high yield under saline environments, and that grain yield under salt stress remains the only reliable means of identifying high salt tolerance in barley.

Salt-tolerant callus tissues were selected for chickpea (Gosal and Bajaj, 1984; Pandey and Ganapathy, 1984) and pea (Gosal and Bajaj, 1984), but it is yet to be demonstrated that whole plants regenerated from such tissue have significantly superior performance. Sexual transmission of in vitro-selected salt resistance is not always observed, even though apparently stable both in culture and in regenerated plants in the absence of selection pressure (Dix et al., 1986). Nevertheless, success of selection at the cellular level translating into improved salt resistance has been claimed for tobacco (Nabors et al., 1980).

Physiologically oriented genotype screening experiments, while initially promising, have typically failed when expanded to include the numbers of lines that might be utilized by plant breeders. As we move toward "breeding by design," i.e., assembling and "pyramiding" genes from different sources that will contribute to improved salt tolerance, deeper understanding of the relationships between phenotype and genotype is required. QTL analysis and marker-assisted selection (MAS) provided some insight into the potential power of "breeding by design."

SOURCES OF GENETIC VARIATION FOR SALINITY RESISTANCE

A prerequisite to genetic improvement of salinity resistance is the existence of genotypic variability, with at least some genotypes having critical values approaching the desired values. Genotypic variability has been demonstrated in chickpea, but it occurs over a very narrow range of salinity levels (Lauter and Munns, 1986; Johansen et al., 1990). There were differences among four Iranian chickpea genotypes in their germination response to NaCl, but these differences were relatively small (50 percent reduction at 0.5 to 0.7 percent NaCl) (Kheradnam and Ghorashy, 1973). In potted soil, critical ECe values, as assessed from ECe at half-maximal growth, differed by 2 $dS{\cdot}m^{-1}$; whereas mean critical ECe values between different pot experiments under apparently similar growth conditions differed by more than this. Effects of environment, therefore, considerably exceeded those of genotype (Johansen et al., 1990). There was greater genotypic variability in salinity response of wild species related to chickpea, but this was because some species were much more salt sensitive than cultivated chickpea (Johansen et al., 1990). In field plots differing in exchangeable sodium percentage (ESP; 15 to 30 percent) but at a constant ECe (2.0 to 2.4 $dS{\cdot}m^{-1}$), Kumar (1985) found some genotypic differences in response of nine chickpea genotypes but, again, critical values as estimated by half-maximal grain yield occurred over a narrow range of ESP (18 to 23 percent).

For lentil grown in sand culture, Rai (1983) found that half-maximal yield of genotype L-9-12 was at 0.5 percent NaCl and that of Pant 639 at 1.1 percent NaCl. Genotypic differences in salinity response of faba bean have also been reported (Salih, 1983; Knobel, 1987). Poljakoff-Mayber and colleagues (1981) found variation in salinity response, in terms of root extension rate, among three pea cultivars as well as *Pisum elatius* and *P. fulvum*. Cerda and colleagues (1982) found useful variation in salt tolerance of pea cultivars in that half-maximal yield of SP-290 was obtained at 6 $dS{\cdot}m^{-1}$ and that of Durana at 10 $dS{\cdot}m^{-1}$.

Salt-tolerant indica rice cultivars seem to have originated or been selected in the coastal areas of India in the states of Kerala and West Bengal. Examples include Pokkali, Nona Bokra, Cheriviruppu and SR26B. An evaluation at the International Rice Research Institute (IRRI) of 250 traditional cultivars collected from Orissa and Tamil Nadu in India has revealed some salt-tolerant types from the southeastern coastal areas of India. Among the tolerant cultivars identified, NV, Pat, Solla and DH were found to have tolerance similar to the popular Pokkali. Similar tests with coastal rice cultivars of Indonesia, Thailand, and Vietnam identified three highly tolerant types: Ketumbar (Indonesia), Khao Seetha (Thailand), and Soc Nau (Vietnam). Again, tolerance mechanisms appear similar to Pokkali, but cultivars such as Pat, Ketumbar, and Soc Nau have better grain quality, with white fine grains and intermediate amylose, than Pokkali, with red bold grain and high amylose.

Since the start of the Genetic Evaluation and Utilization (GEU) Program at IRRI in 1969, 138,000 rice genotypes have been screened for salt tolerance (DeDatta et al., 1993). Of these, 17 percent had acceptable tolerance at electrical conductivity (EC) of 10 $dS \cdot m^{-1}$ at the seedling stage. Traditional cultivars were the most tolerant to salinity stresses. Cultivars Pokkali, Cheriveruppu, Nona Bokra, SR26B, Damodar, and Getu were tolerant of salinity but exhibited poor agronomic traits, being tall and photosensitive, with poor grain quality and yielding ability. These salt-tolerant cultivars were also tolerant of P deficiency, but exhibited susceptibility to pests and diseases and sensitivity to Zn deficiency (Senadhira and Akbar, 1991).

Attempts have also been made to identify exotic sources of salt tolerance. Wild species of *Oryza* and the African *Oryza glaberrima* were screened at the two-three leaf stage at EC 12 $dS \cdot m^{-1}$ for 14 days. All were highly susceptible with seedling survival range 0 to 16 percent while the tolerant control, Nona Bokra, showed complete survival (Akbar et al., 1987). However, some wild rice accessions *(0. rufipogon)* from Sri Lanka showed salinity tolerance at the seedling stage comparable to that of Pokkali. A distant relative of *Oryza (Porteresia coarctata)*, a halophyte that needs high salinity (EC of 12 to 18 $dS \cdot m^{-1}$) is available, but production of fertile interspecific hybrids remains difficult (Jain and Selvaraj, 1997).

CLASSICAL GENETICS AND MARKER-ASSISTED SELECTION

Foliar potassium content in tolerant varieties tends to be greater than in susceptible varieties grown under similar conditions. Gregorio and Senadhira (1993) studied the inheritance of Na/K·ratio in shoots of rice and showed that both additive and dominance gene effects govern salinity tolerance. The trait exhibited overdominance and is controlled by at least two groups of dominant genes. Environmental effects were large and narrow-sense heritability was low (19 percent). These findings suggest that early generation breeding populations must be large and replicated, and selection for tolerance must be made in a later generation and under controlled conditions to minimize environmental effects. This further suggests the use of modified-bulk and single-seed descent (SSD) as breeding methods to develop salinity-tolerant rice varieties.

Maximizing heritability demands minimizing environmental variance. This can be done either by rigorous environmental control or through increased replication. The goal of the QTL mapper is to identify genetic markers that cosegregate with genes, contributing to significant components of genetic variation. These genetic markers may then be utilized as selection tools in assembly of ideal populations and varieties. Molecular marker technology has progressed rapidly, and dense, informative maps are available for most crops. QTLs controlling components of stress response have been identified in many crops (Hayes et al., 1996; Cattivelli et al., 2002), and these are now guiding efforts to clone the genes responsible for cold and drought tolerance (Dubcovsky, personal communication).

QTLs for tolerance to adverse soils are being mapped, with loci for salinity and iron toxicity being identified. MAS can be used to transfer important QTLs into cultivars with the broadest range of adaptation. As an example, MAS could be used to introduce QTLs for tolerance to salinity into a cultivar that produces high yields in a specific region. As the cost of MAS technology declines, the process can be applied to ever smaller ecogeographic regions. In the future, molecular marker technology will enable the identification of specific genes conferring increased yield as well as resistance to biotic and abiotic stresses. This will give breeders an unprecedented ability to tailor cultivars to the specific environments where they will be grown.

Guo and colleagues (2000) studied different crosses of soybean to screen and identify PCR markers associated with salt-tolerant genes. Three crosses of salt-tolerant and susceptible cultivars (Wenfeng 7 × Union, Jindou 33 × Hark, and Tiefeng 8 × Zaoshu 6) were used as materials in this experiment. Two codominant PCR markers were identified via analysis of the segregation of F2 plants. The markers were closely linked with salt-tolerant/susceptible alleles, and are now being utilized to select salt-tolerant, high-yielding lines.

Prasad and colleagues (2000) mapped QTLs controlling various rice seedling traits conferring salt tolerance using a doubled haploid population derived from a cross between IR64 and Azucena. Seven QTL were identified (threshold LOD [log of odds ratio] = 2.00) for seedling traits under salt stress, i.e., two for seed germination (percent), one for seedling root length (cm), three for seedling dry matter (mg), and one for seedling vigor. Among the seven QTLs, four were located on chromosome 6. A QTL for root length on chromosome 6 flanked by restriction fragment length polymorphism (RFLP) markers RG162-RG653, exhibited a high phenotypic variance of 18.9 percent and a peak LOD score of 2.852. Homozygous genotypes containing quantitative genes tolerant to salt stress were identified. The markers tightly linked to salt tolerance can be used in marker-assisted selection after testing their stability and reliability.

Rapid and reliable techniques have been developed to detect tolerance to salinity. However, these techniques have limited capacity and are expensive. Recent research has, however, focused on developing MAS techniques that are efficient and inexpensive. Most of the important traits have been tagged in some crop plants and are now being closely mapped for use in MAS and in the breeding programs by national-program scientists. In rice, the genes governing tolerance in Pokkali were mapped using F_8 recombinant inbred lines (RILs) derived from the cross between Pokkali and the salinity-sensitive variety IR29. Because of the strong interaction between salinity tolerance and environmental factors, the phenotyping was done in highly controlled environmental conditions in a phytotron at the seedling stage using salinized nutrient solution ($EC = 12\ dS \cdot m^{-1}$). A major gene for salinity tolerance was mapped on chromosome 1 (IRRI, 1997) and may be the same as the *SalT* gene, which codes for a protein of 140 amino acid residues (Causse et al., 1994). This protein accumulates very rapidly in sheaths and roots of mature plants and seedlings when subjected to salt (Claes et al., 1990).

A quantitative measure of the Na and K concentrations in the shoot at the seedling stage in a salinized environment was taken for each line of the population. QTLs governing high K absorption, low Na absorption, and low Na/K ratio were mapped to five chromosomes. For high K absorption, in addition to the major gene on chromosome 1, QTLs were detected on chromosomes 4 and 12. For Na absorption QTLs were found on chromosomes 1, 10, and 3. For Na/K ratio, three QTLs were mapped to chromosomes 1, 10, and 12. A common major QTL was detected on chromosome 1 for the three traits putatively associated with salinity tolerance (IRRI, 1998). In a different population (Agarni × 1R63352-AC202) developed at IRRI, a major QTL was also mapped on the same segment of chromosome 1. Breeding programs for tolerance of salinity should emphasize developing MAS techniques that can increase selection efficiency and permit the simultaneous selection of several traits associated with abiotic stresses.

GENETIC ENGINEERING FOR SALINITY RESISTANCE

Transgenic technology offers an attractive alternative to conventional techniques for the genetic improvement of soil salinity tolerance because it enables the introduction of one or more genes into a leading cultivar without affecting its genetic background. Although faba bean remains recalcitrant, transformation of chickpea, lentil, alfalfa, and many other food and feed legumes is reasonably routine (M. Baum, personal communication). An array of some useful genes and transcription factors is now available for deployment in different crop plants. Many of these are now being tested in rice to determine their impact on abiotic stress resistance (Jain and Selvaraj, 1997). No single gene is likely to offer the ultimate solution to soil salinity, although it appears reasonable to be optimistic about the future of salinity tolerance improvement.

CONCLUSION

Effective utilization of available genetic resources, improved understanding of the relative importance of variation in both transcription factors and the genes they regulate, and efficient deployment of well-designed varieties of appropriate crops for saline environments can halt and will hopefully reverse the debilitating impact of soil salinization. We are nearing far more complete understanding of the interaction between genetic variation and the environment, and will soon be able to tailor genotypes to fit the needs of farmers and the societies that depend upon them.

In addition to providing appropriate cropping choices to farmers in saline-prone environments, we must also encourage legislative bodies to promote safe water use. Economic models that would demonstrate the relative costs and benefits of different water-use strategies should be of great assistance to policy makers as well as to scientists attempting to design the crops to fit their farming communities.

REFERENCES

Akbar, M., K.K.K. Jena, and D.V. Seshu (1987). Salt tolerance in wild rice. *International Rice Research Newsletter* 12:15.

Asch, F., M. Dingkuhn, K. Dorffling, and K. Miezan (2000). Leaf K/Na ratio predicts salinity induced yield loss in irrigated rice. *Euphytica* 113:109-118.

Ashraf, M., T. McNeilly, and A.B. Bradshaw (1986). Heritability of sodium chloride tolerance in seven grass species. *Euphytica* 35:935-940.

Ayoub, A.T. (1974). Effect of calcium on sodium salinization of beans (*Phaseolus vulgaris* L.). *Journal of Experimental Botany* 25:245-252.

Bental, M., U. Pick, M. Avron, and H. Degani (1990). The role of intracellular orthophosphate in triggering osmoregulation in the alga *Dunaliella salina. European Journal of Biochemistry* 188:117-122.

Bresler, E., B.L. McNeal, and D.L. Carter (1982). *Saline and Sodic Soils: Principles—Dynamics—Modeling.* Berlin, Germany: Springer-Verlag.

Brinkman, R. (1980). Saline and sodic soils. In *Problem Soils: Their Reclamation and Management,* K.J. Beck, W.A. Blokhuis, P.M. Driessen, N. van Breemen, R. Brinkman, and L.J. Pans (Eds.). Wageningen, The Netherlands: International Soil Reference and Information Centre, pp. 62-68.

Cattivelli, L., P. Baldi, C. Crosatti, N. DiFonzo, P. Faccioli, M. Grossi, A.M. Mastrangelo, N. Peccioni, and A.M. Stanca (2002). Chromosome regions and stress-related sequences involved in resistance to abiotic stress in *Triticeae. Plant Molecular Biology* 48:649-665.

Causse, M.A., T.M. Fulton, Y.G. Cho, S.N. Ahn, J. Chunwongse, K. Wu, J. Xiao, Z. Vu, P.C. Ronald, S.E. Harrington, et al. (1994). Saturated molecular map of rice genome based on an interspecific backcross population. *Genetics* 138:1251-1274.

Cerda, A., M. Caro, and F.G. Fernandez (1982). Salt tolerance of two pea cultivars. *Agronomy Journal* 74:796-798.

Claes, B., R. Dekeyser, R. Villaroel, M. Van Den Hulcke, and G. Bauwi (1990). Characterization of a rice gene showing organ-specific expression in response to salt stress and drought. *Plant Cell* 2:19-27.

Cramer, G.R., A. Läuchli, and V.S. Polito (1985). Displacement of Ca^{2+} by Na^+ from the plasmalemma of root cells: A primary response to salt stress. *Plant Physiology* 79:207-211.

DeDatta, S.K., D. Senadhira, H.U. Neue, and C. Quijano (1993). Success in rice improvement for poor soils. In *Proceedings of the Workshop on Adaptation of Plants to Soil Stresses,* J.W. Maranville, B.V. Baligar, R.R. Duncan, and J.M. Yohe (Eds.). Lincoln, NE: University of Nebraska, pp. 248-268.

Dix, P.I., U.A. McLysaght, and A. Plunkett (1986). Salt stress: Resistance mechanisms and in vitro selection procedures. In *Plant Tissue Culture and Its Agricultural Applications,* L.A. Withers and P.G. Alderson (Eds.). London, UK: Butterworths, pp. 469-478.

Finklestein R.R. and J.A.D. Zeevaart (1994). Gibberellin and ABA biosynthesis and response. In *Arabidopsis,* E. Meyerowitz and C.R. Somerville (Eds.). New York: Cold Spring Harbor Press, pp. 1172-1179.

Gale, M. and K. Devos (1998). Comparative genetics in the grasses. *Proceedings of the National Academy of Sciences, USA* 95:1971-1974.

Glenn, E.P., J.J. Brown, and E. Blumwald (1999). Salt tolerance and crop potential of halophytes. *Critical Reviews in Plant Sciences* 18:227-255.

Goel, N. and K.A. Varshney (1987). Note on seed germination and early seedling growth of two chickpea varieties under saline conditions. *Legume Research* 10:34-36.

Gosal, S.S. and Y.P. Bajaj (1984). Isolation of sodium chloride resistant cell lines in some grain legumes. *Indian Journal of Experimental Biology* 22:209-214.

Gregorio, G.B. and D. Senadhira (1993). Genetic analysis of salinity tolerance in rice. *Theoretical and Applied Genetics* 86:333-338.

Guo, B., Q. LiJusn, S. GuiHua, C. RuZhen, L. LiHong, Z. ZhanYou, L. XiangHua, S. JianYing, B. Guo, L.J. Qiu, et al. (2000). Tagging salt tolerant gene using PCR markers in soybean. *Scientia Agricultura Sinica* 33:10-16.

Hasegawa, P.M., R.A. Bressan, J.-K. Zhu, and H.J. Bohnert (2000). Plant cellular and molecular responses to high salinity. *Annual Review of Plant Physiology and Plant Molecular Biology* 51:463-499.

Hasson-Porath, B., I. Kahane, and A. Poljakoff-Mayber (1972). The effect of chloride and sulphate types of salinity on growth and osmotic adaptation of pea seedlings. *Plant and Soil* 36:449-359.

Hayes, P.M., F.Q. Chen, A. Corey, A. Pan, T.H.H. Chen, W. Baird, W. Powell, W. Thomas, R. Waugh, I. Karsai, et al. (1996). The Dicktoo-Morex population: A model for dissecting components of winterhardiness in barley. In *Plant Cold Hardiness,* P.H. Li and T.H. Chen (Eds.). New York: Plenum Press, pp. 77-87.

IRRI (1997). *Program Report for 1996.* Los Banos, Philippines: International Rice Research Institute.

IRRI (1998). *Program Report for 1997.* Los Banos, Philippines: International Rice Research Institute.

Isla, R., R. Araguese, and A. Royo (1998). Validity of various physiological traits as screening criteria for salt tolerance in barley. *Field Crops Research* 58:97-107.

Jain, R.K. and G. Selvaraj (1997). Molecular genetic improvement of salt tolerance in plants. *Biotechnology Annual Review* 3:245-267.

Johansen, C. (1987). Mechanisms of salt tolerance and their relevance. In *Adaptation of Chickpea and Pigeonpea to Abiotic Stresses.* Proceedings of the Consultants' Workshop, Patancheru, India, December 1984. Patancheru, India: ICRISAT, pp. 109-113.

Johansen, C., N.P. Saxena, Y.S. Chauhan, G.V. Subbarao, R.P.S. Pundir, J.V.D.K. Kumar Rao, and M.K. Jana (1990). Genotypic variation in salinity response of chickpea and pigeonpea. In *Proceedings of the International Congress of Plant Physiology,* Volume 1, S.K. Sinha, P.V. Sane, S.C. Bhargava, and P.K. Aagrwal (Eds.). New Delhi, India: Society of Plant Physiology and Biochemistry, pp. 977-983.

Kamel, W. (1986). Effect of potassium and nitrogen on germination and primary growth of faba beans irrigated with diluted sea water. *FABIS Newsletter* 14: 27-29.

Kavita, S. and R.A. Singh (1999). Response of lentil (*Lens culinaris* Medik) genotypes to salt stress during germination. *Journal of Applied Biology* 9:172-175.

Keating, B.A. and M.J. Fisher (1985). Comparative tolerance of tropical grain legumes to salinity. *Australian Journal of Agricultural Research* 36:373-383.

Kheradnam, M. and S.R. Ghorashy (1973). Salt tolerance of chickpea varieties during germination. *Agronomy Journal* 65:329.

Knobel, W. (1987). Influence of salt stress on the development of nodules of food legumes. In *ACIAR Proceedings*, Series No.18. Canberra, Australia: ACIAR, p. 262.

Kumar, D. (1985). Emergence, establishment and seed yield of chickpea as affected by sodicity. *Annals of Arid Zone* 24:334-340.

Kumar, S. and O.P. Garg (1981). Effect of a shift to saline-alkaline conditions on nodulation, nitrogen fixation and growth in pea (*Pisum sativum* L.). *Indian Journal of Plant Physiology* 24:212-217.

Kumar, S. and K. Promila (1983). Effects of chloride and sulphate types of salinization and desalinization on nodulation and nitrogen fixation in chickpea. *Indian Journal of Plant Physiology* 26:396-401.

Lal, R.K. (1985). Effect of salinity applied at different stages of growth on seed yield and its constituents in field peas (*Pisum sativum* L. var. arvensis). *Indian Journal of Agronomy* 30:296-299.

Läuchli, A. (1984). Salt exclusion: An adaptation of legumes for crops and pastures under saline conditions. In *Salinity Tolerance in Plants: Strategies for Crop Improvement,* R.C. Staples and G.H. Toenniessen (Eds.). New York: John Wiley & Sons, pp. 171-187.

Lauter, D.J. and D.N. Munns (1986). Salt resistance of chickpea genotypes in solutions salinized with NaCl or Na_2SO_4. *Plant and Soil* 95:271-279.

Lauter, D.J., D.N. Munns, and K.L. Clarkin (1981). Salt response of chickpea as influenced by nitrogen supply. *Agronomy Journal* 73:961-966.

Maas, E.V. and G.J. Hoffman (1977). Crop salt tolerance-current assessment. *Journal of the Irrigation and Drainage Division,* American Society of Civil Engineers 103:115-134.

Maas, E.V. and J.A. Poss (1989). Salt sensitivity of cowpea at various growth stages. *Irrigation Science* 10:313-320.

Manchanda, H.R. and S.K. Sharma (1989). Tolerance of chloride and sulphate salinity in chickpea *(Cicer arietinum). Journal of Agricultural Science (Cambridge)* 113:407-410.

Mano, Y. and K. Takeda (2001). Genetic resources of salt tolerance at germination and the seedling stage in wheat. *Japanese Journal of Crop Science* 70:215-220.

Meiri, A. and Z. Plaut (1985). Crop production and management under saline conditions. *Plant and Soil* 89:253-271.

Moreno, L.S., R.K. Maiti, A.N. Gonzales, J.V. Star, R. Foroughbakhch, and H.G. Gonzales (2000). Genotypic variability in bean cultivars (*Phaseolus vulgaris* L.) for resistance to salinity at the seedling stage. *Indian Agriculturist* 44:1-12.

Nabors, M.W., S.E. Gibbs, C.S. Bernstein, and M.B. Meis (1980). NaC1-tolerant tobacco plants from cultured cells. *Zeitschrift für Pflanzenphysiologie* 97:13-17.

Nasir, N.M., R.H. Qureshi, and M. Aslam (2000). A quick screening technique for salt tolerance in sugarcane (*Saccharum officinarum* L.). *Pakistan Sugar Journal* 15:2-7.

Niu, X., M.L. Narasimhan, R.A. Salzman, R.A. Bressan, and P.M. Hasegawa (1993). NaCl regulation of plasma membrane H^+- ATPase gene expression in a glycophyte and a halophyte. *Plant Physiology* 103:713-718.

Pandey, R. and P.S. Ganapathy (1984). Effects of sodium chloride stress on callus cultures of *Cicer arietinum* L. cv BG-203: Growth and ion accumulation. *Journal of Experimental Botany* 35:1194-1199.

Poljakoff-Mayber, A., N. Bar-Nun, B. Hasson, and O. Heichal (1981). Respiratory carbohydrate metabolism of different pea varieties under saline conditions. *Botanical Gazette* 142:431-437.

Prasad, S.R., P.G. Bagali, S. Hittalmani, and H.E. Shashidhar (2000). Molecular mapping of quantitative trait loci associated with seedling tolerance to salt stress in rice (*Oryza sativa* L.). *Current Science* 78:162-164.

Rai, R. (1983). The salt tolerance of Rhizobium strains and lentil genotypes and the effect of salinity on aspects of symbiotic nitrogen fixation. *Journal of Agricultural Science (Cambridge)* 100:81-86.

Rai, R., S.K.T. Nasar, S.J. Singh, and V. Prasad (1985). Interactions between Rhizobium strains and lentil (*Lens culinaris* L.) genotypes under salt stress. *Journal of Agricultural Science (Cambridge)* 104:199-205.

Salih, F.A. (1983). Effect of soil salinity and alkalinity on faba bean growth in Sudan: A review. *FABIS Newsletter* 7:8-9.

Salim, M. (1989). Effects of salinity and relative humidity on growth and ionic relations of plants. *New Phytologist* 113:13-20.

Saxena, M.C. (1987). Agronomy of chickpea. In *The Chickpea,* M.C. Saxena and K.B. Singh (Eds.). Wallingford, UK: CAB International, pp. 207-232.

Senadhira, D. and M. Akbar (1991). Development of improved rice cultivars for problem soils. In *Rice Production on Acid Soils of the Tropics,* P. Detruck and F.N. Ponnamperuma (Eds.). Kandy, Sri Lanka: Institute of Fundamental Studies, pp. 367-377.

Sharma, S.K. and I.C. Gupta (1986). *Saline Environment and Plant Growth.* Bikaner, India: Agro Botanical Publishers.

Sheoran, I.S. and O.P. Garg (1983). Effect of different types of salinities on gram (*Cicer arietinum* L.) during germination. 1. Seedling growth and water relations. *Indian Journal of Plant Physiology* 26:363-369.

Siddiqui, S. and S. Kumar (1985). Effect of salinisation and desalinisation on growth and development of pea (*Pisum sativum* L.). *Indian Journal of Plant Physiology* 28:151-156.

Siddiqui, S., S. Kumar, and H.R. Sarma (1985). Studies on the effects of salinity on nodulation and nitrogen fixation in pea (*Pisum sativum* L.). *Indian Journal of Plant Physiology* 28:369-375.

Sprent, J.I. (1984). Effects of drought and salinity on heterotrophic nitrogen fixing bacteria and on infection of legumes by rhizobia. In *Advances in Nitrogen Fixation Research,* C. Veeger and W.E. Newton (Eds.). The Hague, The Netherlands: Martinus Nijhoff, pp. 295-302.

Velu, G. and G.C. Srivastava (2000). Status of sodium-potassium as a basis of salinity tolerance in rice. *Indian Journal of Plant Physiology* 5:300-302.

Yadav, H.D., O.P. Yadav, O.P. Dhankar, and M.C. Oswal (1989). Effect of chloride salinity and boron on germination, growth and mineral composition of chickpea (*Cicer arietinum* L.). *Annals of Arid Zone* 28:63-67.

Yousef, A.N. and J.I. Sprent (1983). Effects of NaCl on growth, nitrogen incorporation and chemical composition of inoculated and fertilized *Vicia faba* L. plants. *Journal of Experimental Botany* 34:941-950.

Yue, W., M.X. Guang, D.Y. Zhi, H.M. Chen, W. Yue, G.M. Xia, D.Y. Zhi, and H.M. Chen (2000). Transfer of salt tolerance from *Aeluropus littoralis sinensia* to wheat (*Triticum aestivum* L.) via asymmetric somatic hybridization. *Plant Science* 161:259-266.

Zhang, J., H.G. Zheng, A. Aarti, G. Pantuwan, T.T. Nguyen, J.N. Tripathy, A.K. Sarial, S. Robin, R.C. Babu-Bey, D. Nguyen, et al. (2001). Locating genomic regions associated with components of drought resistance in rice: Comparative mapping within and across species. *Theoretical and Applied Genetics* 103:19-29.

Chapter 6

Breeding for Drought Resistance

D. Kumar

INTRODUCTION

Drought may be defined as a period during which soil or plant water deficiencies affect growth and yield. Such moisture deficiencies may occur through a reduction in the moisture supply or through an increase in demand. Drought is a recurrent problem of rainfed agriculture throughout the world. Marginal and submarginal lands in semiarid and arid regions, characteristically depending on rains for their cultivation, are particularly prone to such water deficit problems. The intensity and duration of droughts faced by these lands may, however, vary considerably over time and between climatic zones. These so-called drought-prone zones, including desert regions, bestowed with low precipitation, high ambient temperature, and drought-induced secondary stresses, are commonly known for dismally poor agricultural productivity. The rainfed zones subject to droughts are vast with global occurrence. Arid and semiarid regions, under the serious threat of drought, occupy almost 6 billion hectares of land in more than 100 countries. For instance, out of 1,474 million hectares (Mha) of cultivated lands worldwide, rainfed lands occupy 1,247 Mha. These rainfed lands are found predominantly in Asia (309 Mha), followed by North America (248 Mha), the former Soviet Union (233 Mha), Africa (174 Mha), South America (133 Mha), Europe (123 Mha), and Australia (47 Mha) (Ghassemi et al., 1995). Thus, 84 percent of the world's cultivated lands are rainfed and this has major implications for the global economy and trade. The occurrence of drought depends on soil, climate, social factors, and cropping intensity; and severity and time of occurrence may not easily be predicted. Throughout the world, however, drought is considered to be the main agricultural menace limiting the successful realization of land potential and consequently reducing crop productivity.

THE PROBLEM OF DROUGHT

Economic Impacts

The severe impacts of drought as an agricultural impediment are numerous, including reduced crop yields, decreased availability of fodder and feed, nonplanting of crops, abandonment of land, and reduced livestock populations. Drought may also bring about changes in land use, migration of humans and animals, and reduced industrial production. Moreover, large expenditure by governments in combating drought and providing compensation for losses are common in arid areas. Drought leads to increases in regional prices of food and other commodities because of reduced production, ultimately affecting world trade. For instance, countries in the Horn of Africa, Ethiopia, Sudan, and Somalia, have experienced severe economic and political difficulties, and their limited natural resources have always suffered from inherent drought.

As a further example, the reduction in world grain production by about 5 percent during 1983-1984 was due to a severe drought problem in the United States and adverse weather in Africa and part of Europe (Wisner and Chase, 1984). The impacts of the 1988 drought in North America on world cereal production and the resultant price increase were considerable. World wheat production in 1988 was reduced by 10 million tons (Mt) compared with 1987, and, consequently, the price of wheat increased from \$140/t in 1987 to \$200/t in 1988. Similarly, the world maize price also shot up to \$160/t in 1988 compared with \$140/t at the beginning of the year.

Similarly, drought effects in Africa also have been experienced extensively. For instance, in 1991-1992, a severe drought reduced maize output by about 60 percent to 3.1 Mt, forcing the importation of almost 5.0 Mt of maize in the 1992-1993 marketing year to meet domestic demands. The continued drought situation in the southern hemisphere autumn and winter reduced South African wheat production in 1992-1993 by almost 1.0 Mt to 1.3 Mt, also causing livestock losses and a reduction in cash crops (Sandene, 1993). As far as maize is concerned, drought causes average annual yield losses in maize of about 17 percent in the tropics (Edmeades et al., 1992), but yield losses in individual seasons have approached 60 percent in regions such as southern Africa (Rosen and Scott, 1992). Droughts in the Indian subcontinent, particularly in northwestern parts, are a regular feature and cause of all kinds of extensive losses.

Need for Breeding for Resistance to Drought

Quantitative estimates of yield losses due to drought in the tropics appear scarce and imprecise. According to an estimate made by Subbarao and colleagues (1995), the yield increase over the present level could be 49, 57, and 29 percent for chickpea, pigeonpea, and groundnut, respectively, whereas maize production could be increased by about 17 percent, if soil moisture was not a limiting factor. If precipitation could be increased it could effectively alleviate drought and enhance crop production. Alternatively, irrigation can bring cultivation to many drought lands. The assured production through steady and timely irrigation can bring stable and reliable crop yields. However, continued and unplanned irrigation, with excessive use and without provision for drainage, may prove uneconomic because of a slow build-up of salinity in the effective root zone. In addition, irrigation is a costly input, particularly in rainfed areas, and may prove uneconomic in certain areas with a very deep water table. For instance, to increase irrigation capacity by about 10 percent in Australia, an investment of $85 to $175 billion would be required, therefore forcing a continued reliance on dryland farming.

During the cropping period in drought-affected areas, there seems to be no reliable way to increase precipitation. Moreover, if, as is common in arid areas, irrigation facilities are not available, the only water management options are to employ cultural practices to increase the availability of stored soil moisture, to improve water-use efficiency, or to exploit other methods by which plants can yield reasonably well under water deficit conditions (Boyer, 1982). Approaches involving the provision of irrigation water or the use of cultural methods for alleviating drought are cumbersome, uneconomic, inconvenient, and require specific knowledge. Biological approaches, on the contrary, are considered efficient in achieving effective water use in crop production. The biological approach includes domestication and cultivation of drought-tolerant crop species and the incorporation of the drought-resistance traits into plants through appropriate breeding tools. The provision of seed of the crop varieties that can effectively avoid, escape, or tolerate drought may prove the most acceptable technology to counter drought. This technology has proved sustainable and has paid rich dividends on drought-hit soils. For instance, in arid areas that are subject to frequent droughts, the crops that have been cultivated for centuries are those that are highly adapted to drought. These crops, including pearl millet, drought-tolerant legumes, barley, and certain oilseeds predominate in dry areas. Thus, rainfed arid and semiarid lands subject to unpredictable drought can best be utilized for crop production cultivation of agricultural crops and their varieties with optimum adaptation to such hostile soil and weather.

Evidence from Populations Naturally Tolerant to Drought

Plants showing improved growth with limited water are said to be drought tolerant regardless of how the improvement occurred. Plants do tolerate drought in natural habitats in many ways, by improving water-use efficiency or by dehydration avoidance or tolerance. Water is a necessity of life and is required for biological activities. Some studies indicate that water-use efficiency remains unchanged even if water availability is scarce throughout the plant's life cycle. Thus, continued water limitations throughout the entire life cycle, possibly for decades in annuals, may lead to acclimation and, therefore, dry matter and harvest index of drought-adapted plants may also remain unchanged. However, in many situations of variable soil-moisture availability during the life cycle, dry matter partitioning is influenced by water availability. This particularly affects the reproductive parts, leading to fluctuations in dry matter accumulation and, therefore, harvest index.

In naturally dry habitats, there are examples of several different adaptations. There are some plant species with an escape mechanism, maturing rapidly before the onset of the dry season or reproducing after rainfall. California poppy (*Eschscholtzia californica* Cham.), for instance, completes its life cycle in a few weeks, but coffee (*Coffea arabica* L.) and cacao (*Theobroma cacao* L.) flower and fruit when rains follow a drought period (Alvim, 1985). Other plant species avoid drought by possessing a deep root system, by preventing water loss from transpiration by rolling or folding leaves, or by covering bud tissues with scales. Still others survive by storing water in fleshy tissues. These are all said to be dehydration postponers and include upland rice (*Oryza sativa* L.), which develops deeper roots than those of paddy rice (Chang et al., 1974), agave *(Agave deserti),* and some cactus species, which are succulents. There are other plant species, including *Selaginella lepidophylla* (Hook and Crev), that tolerate desiccation to an air-dry state.

Northwest India includes 3.2 Mha of land that experiences inherent drought, with an annual rainfall of 150 to 350 mm and a highly erratic distribution. Only drought-tolerant crops are cultivated under rainfed conditions. These include the arid legumes, clusterbean [*Cyamopsis tetragonoloba* (L.) Taub], dew bean [*Vigna aconitifolia* (Jacq.) Marechal], and cowpea (*Vigna unguiculata* L. Walp). Indeed, 80 percent of land devoted to these crops in India is in the western desertic regions. These crops are drought avoiders, characterized by deep taproots and slow growth with generally long maturity. These so-called arid legumes have been growing in these regions for centuries and have developed specific adaptation to drought; they do not respond well to high rainfall. Pearl millet is another example of a natural drought-tolerant cereal being cultivated worldwide in arid and semiarid regions. In the rapeseed/mustard group of crops salad rocket (*Eruca sativa* L.) is grown with conserved soil moisture in the rabi (dry) season and has the potential to yield 500 to 600 kg·ha^{-1} with one presowing irrigation of 60 mm only. This oil crop is noted for its deep-penetrating root system and fleshy leaves, which store water. It is specifically cultivated in naturally drought-prone pockets. Thus, in arid regions, the plants experiencing the most severe droughts represent the crops sustained, those characteristically having an inherent drought avoidance or escape mechanism. The habitat is so specific that other crops with high moisture requirement do not survive, hence, showing poor crop diversification.

Tolerance

Drought limits the productivity of field crops by preventing them from expressing their full genetic potential. Drought effects on plants are complex and variable and are greatly accentuated by a number of primary and secondary factors. The expression of drought-induced symptoms appears extremely unpredictable because they are generally confounded with the effects of heat, drought-induced pests and diseases, and nutritional deficiencies. The impacts of drought are further influenced by soil type, rooting pattern, plant density, and the root pathogens of the crop (Boyer, 1996). Drought-induced impacts are complicated still further by rainfall pattern and distribution, wind velocity, evaporative demands, crop species, variety, and growth stage (Werry et al., 1994). Water stress, induced directly or indirectly, produces both adverse and stimulatory effects on plants. The stimulation in plant growth may sometimes be explained by compensation growth resulting from reduced emergence and early seedlings mortality. This leads to a sparse plant population, thereby lowering the competition for soil moisture and nutrient availability and allowing increased growth of individual plants. Water deficit, more usually, may simply reduce those growth and yield traits associated with plant size and structure. Other characters may vary depending on the crop species and the treatment. In spite of the complexity of impacts and expression of drought, the effective utilization of limited soil moisture by biological means can be achieved basically by two approaches, increasing plant water-use efficiency and enhancing plant drought-tolerance potential. A plant or a group of plants showing better growth and productivity with limited soil moisture than other plants in a given set of similar environments is understood to be tolerant to drought (Hall, 1993).

Plant tolerance to soil-moisture deficit is an adaptive feature involving the cell membrane of individual cells and whole plants. Biochemically, specific proteins and the synthesis and accumulation of organic solutes are involved, thus making plant tolerance to drought highly complex. In morphological terms, tolerance can be achieved by substantial and proportional growth reduction and by diversion of carbohydrates to economic parts (sinks). In tolerance of plants to drought, roots have been observed to play an important role, with growth and biomass increase greater in the root than in the shoot. For instance, in pigeon pea, the taproots were almost 18 cm longer in low rainfall than in wet situation (Onim, 1983). Similarly, in wilting conditions of the Indian arid zone, much longer roots of chickpea (root:shoot ratio >2) were associated with drought situations (Gupta et al., 1995). A deep rooting pattern makes a major contribution to plant tolerance in soil-moisture deficits. Maize and sorghum roots can grow 2 to 3 m deep into the soil, allowing their survival even when surrounding grasses or crops with shallow roots succumb. Similarly, upland rice grows and yields better than paddy rice in upland soils because the former has deeper roots (Chang et al., 1974). The importance of a better-developed root system in *B. juncea* than in *B. carinata, B. napus,* and *B. campestris* has been related to the greater drought tolerance of that species (Liang et al., 1992).

The importance of the cuticle in plants showing tolerance to drought has been demonstrated in many studies (Jordan et al., 1984; Sinclair and Ludlow, 1986; Rawson and Clarke, 1988). In the absence of stomata, as in fruit and stems, or when the stomata are closed due to high temperature or high water stress, water loss is mainly controlled by the cuticle, with some residual vapor diffusion also through stomata. Studies have clearly shown enormous genetic differences in survival under severe dehydration when stomata are closed. Such differences occur between species, for example, between pigeon pea, black gram, soybean, and cowpea (Sinclair and Ludlow, 1986). Differences also occur within species such as wheat, oats, maize, and soybean (Paje et al., 1988; Rawson and Clarke, 1988). These differences could be used to select genotypes with decreased conductance and, thus, reduced water loss when stomata are closed. Drought tolerance is not simply affected and determined by the depth and pattern of root behavior and by cuticular resistance. A number of morphobiochemical factors such as stomatal sensitivity and number, cell turgor, chemical traits, stability of membrane proteins, phospholipids, and metabolic activities are also important and actively involved.

Plants can tolerate or avoid dehydration stress by escape, avoidance, and tolerance mechanisms. Escape is a morphological phenomenon of plants, achieved by early maturity, whereby they can complete their life cycles within the wet season or can reproduce whenever sufficient soil moisture is available. Drought avoidance, however, is a physiological adaptation. The plants maintain high internal water potential, maintain cell turgor, and avoid secondary drought-induced stresses so that, in presence of low external water potential, there is no induced indirect metabolic or direct injury. Tolerance usually refers to specific tolerance of drought-related injuries, but may also be interpreted as a complete phenomenon, comprising both escape and avoidance mechanisms.

Oilseed *Brassicae* tolerate drought by changed and modified leaf morphology and structure. For instance, in Poland, swede rape cv. Gorez anksi tolerated drought with leaf expansion due to formation of central vacuoles in mesophyll cells. Thus, desiccation tolerance appeared to be achieved by avoidance of cytoplasm dehydration (Obloi and Kacperska, 1991). Under progressive drought, differential leaf modification was displayed by rape cv. Darmor in France. The older leaves, developed before the onset of water deficit, wilted gradually, whereas the youngest leaves continued to harden when the shoot water potential was below –3 MPa and the leaf water saturation deficit was about 60 percent. The hardening of leaves was distinguished by leaf turgor and a bluish wax bloom color (Reviron et al., 1992).

Morphologically, plants also adapt to water stress by early flowering; decreased plant height, leaf area, and leaf dry weight; and increased leaf conductance (Yadava and Hari Singh, 1996). Tolerant plants under water stress appear to shift carbohydrate utilization from shoot growth into the production of additional root biomass in the early stages of adaptation. However, at later stages, carbohydrate concentration may shift toward the shoots (Mathur and Wattal, 1996; Wright et al., 1996), accompanied by growth retardation.

Physiological changes in water potential and relative water content of the water-stressed leaves through osmoregulation and osmotic adjustment have been observed in *Brassicae*. In the process of physiological adaptation, maintenance of turgor pressure appears to be the central process. In a study of seven *Brassica* species in India, tolerance was associated with high water use, high stomatal conductance, and osmoregulation (Kumar et al., 1987). Net assimilation rates were closely correlated with stomatal conductance. For instance, common bean cultivars Carioca and Negro had greater tolerance and higher water retention capacity than the susceptible cultivars A 320 and Xodo. Leaf dehydration of the former cultivars was slower because they showed high membrane resistance (electrolyte leakage test) and osmotic adjustment (Franca et al., 2000). Osmotic adjustment involves the accumulation of organic solutes, intracellularly, which is believed to restore turgor pressure. The organic solutes (osmolytes) generally accumulated are polyols and amino acid derivatives (Ashraf and Mehmood, 1990). More than one type of osmolyte may occur in a cell, but specific classes of osmolytes have also been found by many workers.

Solute leakage from the cell membrane of plants exposed to drought stress results from increased membrane permeability due to stress-induced disturbance in its integrity. Accordingly, at the cellular level, integrity and stabilization of cell membrane structure, rather than repair mechanisms have been observed to operate in rape hypocotyls dehydrated to a water-saturation deficit of 60 percent (Scherbakova and Kacperska-Palacz, 1980, 1983). On closure of stomata, plants show various protective mechanisms to reduce photoinhibition, including thermal dissipation of energy, removal of active oxygen species (AOS), and increase in certain enzyme activities. For instance, in a dehydration stress tolerance study on rice in the Philippines, after three days of stress, the catalase activity in all the cultivars tested increased and, after six days of stress, SOD activity also increased in all the cultivars. However, ascorbic peroxidase (APX) activity increased in only the tolerant cv. IR 52. Thus, in the initial stress, rice plants increased photorespiration but, with continued stress, plants activated the AOS system to avoid production of extra energy. Thus, the AOS system and APX enzyme play key roles in alleviating drought (Holt et al., 2000).

An important aspect of severe desiccation is that water content in the cells becomes so scarce that enzyme activities are inhibited (Vertucci and Leopold, 1987a,b). Developing seeds of a range of crops, including rape and mustard, accumulate hydrophilic proteins in the embryo as the desiccation begins (Dure et al., 1989). In certain cases, an alpha helix is present that remains structurally stable during desiccation, and this portion of protein has been observed to act as a membrane-stabilizing factor. The mRNAs for these proteins are specifically induced by severe desiccation in very young rape (Harada et al., 1989) and barley (Close and Chandler, 1990) plants. The expression of mRNA was especially increased in shoots that are directly exposed to dehydration under natural conditions. This cellular response suggests that the dehydrin-Em-LEA protein plays an active role in desiccation tolerance of rape seedlings. Drought-stress treatment was found to induce change in the protein pattern in the root system of rape cv. Darmor. Of 13 new polypeptides of low molecular weight identified in the tap root, 12 were present in short, tuber-sized roots, a specific root type induced in response to soil moisture deficit (SMD) situations. Surprisingly, however, reversal of these induced proteins in just three days of rehydration suggested involvement of specific proteins in drought tolerance (Vartanian et al., 1987). Similarly, a specific drought-induced protein, BnD22, was shown to accumulate in *B. napus* leaves adapted to progressive osmotic stresses induced by salt and water stresses. The protein might be involved in decreased protease activity in the drought-adapted leaves, thus contributing to delayed leaf senescence (Ilami et al., 1997).

METHODOLOGY AND PARAMETERS OF DROUGHT TOLERANCE

Not only is the expression of drought complex, the methodologies being used are also unsystematic and have varied among researchers. The methodologies used to simulate targeted drought include irrigation on the basis of percent depletion of available soil moisture, irrigation following irrigation water:cumulative pan evaporation (IW:CPE) ratios or at different growth stages, withholding water up to wilting of plants and then irrigating (Ashraf and Mehmood, 1990), lysimeter soil drying studies, creation of drought cycles in a glasshouse (Paclik et al., 1996); water-stress studies in pots (Champoliver and Merrien, 1996), experiments under contrasting evaporative demands (Mogensen et al., 1996), screening under rainfed situations, and laboratory studies using polyethylglycol to withhold soil water. In view of this, results may lack uniformity even for the same genotype.

Various criteria for drought tolerance are also employed. For instance, absolute seed yield, relative seed yield, whole-plant fresh and dry weight, plant dry weight at different growth stages, harvest index, regression slope between leaf water potential and osmotic potential, water-use efficiency, and drought susceptibility indices are all utilized to judge the drought tolerance potential of cultivars. Screening for resistance to drought is, thus, still complicated by the lack of fast, reproducible, simple, and convenient parameters, and the inability to routinely create defined and repeatable water-stress conditions where large numbers of genotypes can be screened efficiently (Ramirez-Vallejo and Kelly, 1998). Selection based solely on yield under extreme environments is a poor method because the plants selected may give poor yields in optimum environments. Conversely, the carry-over effect of high-yielding genotypes takes place only up to a given level of stress then disappears under severe stress. Breeding lines selected under optimum conditions may, thus, not perform well in a stress environment (Ceccarelli, 1996). Several reports confirm that selection in the target environment is almost invariably more efficient than indirect selection (Chisi et al., 1996; Shakhatreh et al., 2001).

Difficulties in suggesting a usable and reliable morphophysiological parameter for selecting genotypes to targeted drought environments has compelled breeders to use average yield against a set of environments as the main indicator. Thus, many regression equations providing a drought susceptibility index (Fischer and Maurer, 1978) or superiority index (Lin and Binna, 1988) have been used to estimate genotypic adaptability over a range of environments. The weakness of such regression techniques lies mainly in the quantification of drought environments. These regression techniques and the indices employed, based on yield levels in response to environments, provide only a rough estimate of tolerance because it is difficult to separate genotype difference in resistance and grain yield. Thus, indices based on water-relation parameters and other physiological parameters may be more reliable. A list of possible selection indices reported worldwide on field crops is given in Table 6.1. Our practical experience and consultation of the large body of literature on drought tolerance suggests the following convincing, conceivable, and workable approaches:

1. It is the dry-matter accumulation of a genotype that indicates its water-use efficiency under water-scarce situations. However, more important is the partitioning of dry matter to economic plant parts, resulting in a high harvest index. Thus, high dry-matter accumulation toward reproductive parts at anthesis may be a sensible and convenient parameter for selection under dry conditions.
2. Leaf area expansion initiation under drought conditions may indicate the beginning of the relief of a water-stress situation. Thus, high leaf area/canopy at the active drought period may also prove a sensible selection aid.
3. Cell-membrane integrity is a central feature of almost all the osmotic stress-tolerance mechanisms, requiring structurally and functionally intact proteins and phospholipids. Thus, electrolyte leakage (low EC of the leachate) may serve as a criterion in growth-chamber studies at the seedling stage.
4. In field studies, evaluation of genotypes at 3 to 5 percent soil moisture at the active growth stage may separate tolerant and susceptible lines.
5. A thick cuticular layer with wax coating, controlling leaf water loss when stomata are closed, is a simple criterion and, hence, can form the basis of selection from large numbers of germplasm lines.

Genetics

Information on the genetics of drought tolerance is a prerequisite for plant improvement. However, published results solely for drought tolerance are scant. Efforts have mainly centered on the extent of genetic variance and genetic associations, on genetic studies of yield and yield traits, and, to a

TABLE 6.1. Drought-tolerance selection indices used for field crops.

Index of drought tolerance	Crop	Source
Higher leaf-water potential (LWP) and higher acetylene reduction indicating more N_2-fixation	Cowpea	Walker and Miller, 1982
Higher whole-plant dry matter at 277.5 mm rainfall in Burkina Faso	Cowpea	Aggarwal, 1987
Higher levels of relative water content (RWC) and LWP under drought in Japan	Rice	Fujii and Horie, 2001
Lower values of delta (C_{13}) concomitant with higher values of water-use efficiency (WUE), leaf area (LA), and xylem ABA	Cowpea; Wheat	Ngugi et al., 1994; Merah et al., 2001
Earliness, large seeds, and high dry seed yield in Egypt	Cowpea	Mohammed et al., 2000
Regression analysis of yield on water-potential index	Wheat; Faba bean	Karamanos and Papatheohari, 1999
Larger tap root system under wilting conditions in India	Cowpea	Itani, 1992
Strong stability in the phospholipase D enzyme (PLD) under water stress in India	Cowpea	El Maarouf et al., 1999
Higher activities of NAD- and NADP-dependent enzymes	Horse gram	Reddy et al., 1990
High stomatal conductance leading to high water potential, improved turgor resulting from osmotic adjustment, positive association with grain yield	Chickpea; Indian mustard	Singh et al., 1985; Yadav et al., 1990
Midday transpirational cooling of the canopy observed by infrared thermometer at podding	Indian mustard	Choudhary et al., 1991
Thickness of cuticle, opening and closing of stomata, root depth and extension, osmotic adjustment, and production of antioxidants	Wheat	Szegletes et al., 2000

lesser extent, on physiobiological traits. The information available is largely on traits under rainfed situations only. Deliberate attempts to study the genetics of drought tolerance in simulated situations appear quite rare. The main obstacle in studying this could be that tolerance is conferred by various morphological, physiological, and biochemical traits, and these involve many complex pathways. Drought is not an independent factor and its expression, impacts, and evaluation are influenced by secondary and, sometimes, tertiary factors also. Variation in the targeted environment, including soil, climate, and the period and intensity of drought, is also a barrier in this context. However, the results available do indicate that morphological and physiological characters contributing toward drought tolerance show different inheritance patterns (monogenic and polygenic) and gene actions (additive and nonadditive types) in various agricultural crops (Bouwkamp and Summers, 1982; Sangwan et al., 1998; Mai-Kodomi et al., 1999; Singh et al., 1999; Mitra, 2001).

A simple and convincing study of inheritance of drought tolerance for shoot growth was reported in cowpea following crossing of two tolerant genotypes (Tvu-11986 and Dan IIa) and one susceptible genotype (T Vu-7778) at Kano, Nigeria (Singh et al., 1999; Mai-Kodomi et al., 1999). Tolerance was a dominant character governed by single genes (*Rds 1* or *Rds 2*). In this study, only the stomatal behavior and/or osmotic adjustments were involved. This study, therefore, indicates simple inheritance, focusing on shoot rather than root or other characters. In another study, an 8 × 8 diallale analysis of wheat in irrigated and rainfed conditions in Iran revealed higher estimates of heritability (h^2B) for harvest index (HI), grain yield, and earliness; and that predominantly additive gene effects were involved in inheritance of these traits (Farshadfar et al., 2001). Additive genetic variances were also reported for seed yield and number of capsules per plants in sesame under drought conditions in India (Kandaswami, 1985). Similarly, genetic variances for drought-susceptibility index (DSI) were also shown to be additive in sunflower in Italy (Kamali and Miller, 1982).

Nonadditive gene effects were noted in the expression of leaf-water potential, relative water content (RWC), and number of stomata per unit area in drought conditions in 23 parents, 60 F_1s, and 62 F_2s, following line × tester analysis in Indian mustard (Wani and Srivastava, 1989). Inheritance studies in respect of water-relation characters carried out with Indian mustard under drought conditions at Hisar, India, also revealed nonadditive gene effects for the expression of these traits (Choudhary et al., 1989). In wheat, both additive and nonadditive gene actions were observed for drought tolerance. For instance, overdominance in inheritance of stress yield, relative water loss, HI, and excised leaf water retention was exhibited while RWC and yield potential were controlled by additive genes (Farshadfar et al., 2001).

Utility of a particular trait in breeding for drought tolerance depends primarily on the extent of its heritability estimates in the targeted environment. Estimates under extreme situations of soil-moisture and heat stress have been reported to be low (Marrow and Hall, 1992). Broad-sense heritability for proline content (43 percent) in *Brassica juncea* and for seed germination

(55 percent) in *B. napus* were, for instance, low in a simulated drought situation of –17.0 bar osmotic potential (OP) in pots (Richards and Thurling, 1979), indicating difficulty in their use for improvement in drought tolerance. On the contrary, higher estimates of h^2B for yield have been observed under dry conditions. For instance, estimates ranged from 35 percent for siliquae length to 88.5 percent for seed yield, thus justifying their use in breeding for drought tolerance (Choudhary et al., 1989). Similarly, higher estimates of h^2B and genetic gain were observed for excised leaf water retention, RWC, and HI in wheat under drought conditions (Farshadfar et al., 2001). Evaluation of 32 sunflower hybrids in irrigated, rainfed, and drought conditions in Mexico indicated high estimates of h^2B. Traits with low h^2B estimates were those related to reproductive ability, but traits with high heritability estimates were generally related to biomass production (Gomez-Sanchez et al., 1999).

Thus, it appears that monogenic inheritance of drought tolerance is rare unless a simple and single trait contributes toward drought tolerance. Drought tolerance as reflected in a complex trait such as yield appears to be governed by polygenes. Both additive and nonadditive gene actions are involved in inheritance of this complex trait. Some examples reveal high heritability estimates, thus making improvement for drought tolerance easier.

BREEDING FOR DROUGHT TOLERANCE

Breeding crops for tolerance to drought may not prove helpful under catastrophic conditions: long drought cycles of unpredictable duration; poor water-holding capacity of the soils; and exceptionally high evapotranspiration demands, as commonly observed in Sahelian regions and occasionally in western, arid parts of India. Drought tolerance may, however, be of some use for enhanced and sustained production under rainfed conditions, where precipitation during cropping season is expected to occur, as is customarily experienced in many semiarid parts of the Indian subcontinent, South and Southeast Asia, the Middle East, and other parts of the world. However, in spite of the great surge in the practical needs of these regions, efforts and strategies in this context have remained limited. There are many reasons for this. A lack of interdisciplinary approaches, specific and convenient screening techniques, and know-how on the genetics of drought tolerance; poor association of drought resistance with plant productivity; and unavailability of appropriate genes to obtain transgenic plants are the main constraints to genetic improvement of crop plants for drought resistance (Mitra, 2001).

Improved grain yield in dry conditions is by far regarded as the most desirable aim of breeding. However, the growing environment is known to have strong effects on the response to selection and improvement in grain yield. Several studies have, therefore, been directed to studying the effects of growing environments on the response to selection for yield (Allen et al., 1978; Turner, 1979). Many studies also indicate that genetic variation, heritability estimates, and genetic gains are generally more when selections were made in optimal conditions (Frey, 1964; Johnson and Frey, 1967). The response to selection for yields would be greater in a favorable rather than an unfavorable environment, whereas response to selection for drought tolerance would be greater in the target environment. Thus, breeding approaches for drought tolerance can be directed toward two types of environmental situations.

Adaptation to Specific Environments

This approach postulates the development of varieties that are specifically suited to drought environments. This approach is based on the fact that varieties suited to drought and having drought-related traits may not behave so for these traits in ideal situations (Hurd, 1971, 1975, 1976) and vice versa. This approach must, therefore, be useful for genotypes that can complete their life cycle on the soil moisture stored during the previous season. Some classic reports on wheat and cotton indicate that widely adapted and high-yielding genotypes may not show high efficiency in stress environments. Thus, a superior variety for a moisture-stress environment must be selected and evaluated under such an environment. However, this approach may not be useful for environments having widely differing levels of precipitations over the years, as occurs in some semiarid and arid regions. Genotypes developed through this approach may not necessarily yield well during the years of relatively high precipitation. In addition, heritability and genetic gain may be extremely low in hot, dry weather, thereby reducing selection efficiency.

Adaptation to Variable Environments

This approach appears justified to develop crop varieties adapted to a wide range of environmental conditions and may be suitable in situations when optimal and suboptimal precipitations are expected during the cropping cycle. Average yield per se and the stability in yield may be used as criteria for evaluation of genotypes suited to drought conditions. In this approach it may be possible to breed varieties having high adaptation and high yield. Roy and Murty (1970) have suggested that selection for yield in wheat should initially be performed under a range of optimal and suboptimal conditions. Similarly, Mederski and Jeffers (1973) have reported that optimum conditions may give higher genotypic variations. Therefore, selection for high yield can be done in optimal-condition situations and selected lines may also retain high yields in stress conditions.

Physiogenetic Approach

A third school of thought proposes that resistance/adaptation to drought and yield potential are two separate traits and governed by different genes or gene blocks and, hence, are liable to be selected independently. This approach envisages combining both improved yield and high drought-tolerance potential from two sources. An attempt is made, therefore, to select traits associated with drought resistance in soil-moisture deficit conditions and to select yield and other economic traits with optimal soil moisture. This approach can best be utilized by choosing a suitable morphological trait associated with drought tolerance; for instance a thick cuticular layer, leaf area expansion, detached leaf water loss, or cuticular wax layer on leaf. The latter trait reduces water loss through cuticular transpiration during the night when stomata are closed. It is convenient and can be used with a large number of crosses made between high-yielding genotypes and a variety with a thick cuticular layer.

The following year, the seeds from the selections are raised in progeny rows in moisture-stress conditions. A number of plants with thick cuticular layers are selected and no attention is paid toward high yield. The selections based on high cuticular layer are evaluated in irrigated conditions for two years, irrespective of the yield potential of these selections. Later on, yield testing is done under moisture-stress conditions similar to those of the targeted growing conditions. Additional selections among high-yielding lines can be attempted for wax coating. Thus, a wax coating developed on the leaf of plants in response to soil-moisture stress can be used as selection criterion.

Breeding Priorities

In order to make progress in yield improvement of field crops under dryland conditions, the following three priorities may be set:

1. Reduce the risk of crop failure and maximize yields by manipulating sowing dates and curtailing the maturity period in accordance with the rainfall pattern of the region concerned (drought escape).
2. Maximize water use when soil moisture is available by incorporating important traits such as fast growth and well-developed root systems, rapid leaf-area expansion, and physiological efficiency (drought avoidance).
3. Incorporate plant traits responsible for increasing and stabilizing yields and providing resistance to catastrophic events during water-stress periods (drought resistance in the complete sense).

Our experience with a variety of crops to date indicates that selection for drought escape followed by drought avoidance has been by far the most successful and effective, and this approach is likely to be continued because it favors higher yields vis-à-vis tolerance mechanisms (resistance) under water stress. A number of early maturing varieties in several agricultural crops have been developed and yield potentials have increased under soil-moisture deficit situations in spite of curtailment in the growth period.

Selection of Traits

Efficient selection based on simple traits is logically an important exercise while breeding for stress tolerance. Even simple selection of heritable traits governed by a few genes may revolutionize yield under drought conditions. Thus, selection is an important breeding tool under drought-prone situations. For empirical selection, morphological and phenological traits must precede physiological and biochemical traits because the former are easily identifiable, measurable, and transferable. However, with continued selection, and when the improvement reaches an advanced stage, biological and physiological traits may be required for yield improvement. Metabolic and related biochemical traits are not always universal and appear to differ among the crops. High ABA accumulation for instance, represents drought tolerance in sorghum and maize, but causes drought susceptibility in wheat (Jones, 1983).

Usefulness of the Traits

Examples of the control of drought resistance by single genes are rare. Moreover, it is not wise to invest important resources into breeding for physiological response to water stress. It is also impractical to prove the benefits of such traits in directly improving yield under water stress. Cost-effectiveness of screening for physiological traits is an important consideration. Kanangara and colleagues (1983) observed that the correlation coefficient between ABA levels and sorghum yields was similar to that between yields and the amount of irrigation applied. Therefore, there was no advantage in measuring ABA levels as a drought-tolerance selection criterion.

Earliness in certain crops due to their indeterminant growth habit is more complex, but it has been exploited as a character for improving drought resistance. The selection for drought tolerance is associated with early flowering and partitioning of dry matter to the reproductive parts. High harvest index obtained under nonirrigated conditions can be influenced by differences in proportion of dry-matter partitioning to reproductive parts on initiation of reproductive growth. Differences in early partitioning between cowpea selections and populations were observed to be greater than with irrigation (Hall and Grantz, 1981). We must, therefore, try to improve crops under drought rather than under optimal conditions. The main problem of selection under drought conditions is that anthesis, partitioning, and maturity are greatly influenced by water stress, making it difficult to detect differences due to drought. It may be more effective to select for earliness using wide spacing, as it is then easier to identify individual plants under water-limiting situations. Early partitioning of dry matter in cowpea was observed to be greater under wide than under close spacing (Grantz, 1979). It has been observed that selecting plants that bear more mature pods early in the season can be an effective method for improving drought resistance of crops such as indeterminant legumes. This trait is easier to assess than early flowering, as the latter requires precise, repeated, and careful observations of individual plants.

Drought-Tolerant Varieties

Some known, promising, drought-tolerant varieties of field crops are given in Table 6.2. These varieties originate from many different regions and are the result of different breeding methodologies. The majority of them, however, owe their origin to the evaluation of germplasm lines developed while breeding for optimum conditions. Hardly any have resulted from deliberate attempts to develop drought-tolerant varieties.

GENETIC ENGINEERING FOR RESISTANCE TO DROUGHT

Difficulty in breeding for drought resistance stems from the complexity of the targeted environment, polygenic inheritance of this trait, lower heritability estimates in hot, dry environments, and the lack of sound selec-

TABLE 6.2. Examples of drought-tolerant cultivars of field crops.

Crop	Cultivar	Country	Reference
Cowpea	IGFR-1440, HFG-42-1	India	Yadav and Patil, 1984
Cowpea	Katumani-80 (dual purpose)	Kenya	Shakoor et al., 1984
Cowpea	New era	Nigeria	Babalola, 1980
Cowpea	Rogue, KVX, 30-305, 3.6	Norway	Aune, 1992
Mungbean	VI-281,VC-2755A	Korea	Kang et al., 1986
Chickpea	G-130	Delhi	Singh and Mehra, 1980
Chickpea	Tyson	Australia	Rettke and Keys, 1985
Chickpea	Zimistoni and Ustoichvyi 3/68	Russia	Pirmakhmadov, 1972
Chickpea	Garnet, Tammasy	Ethiopia	Muchilbauer and Kaiser, 1987
Chickpea	Somava-80	Mexico	Duron Noriega, 1986
Pigeonpea	KCS-74-1, KCS-22-4	Kenya	Shakoor et al., 1984
Soybean	416937 (PI)	United States	Chipman et al., 2001
Maize	Soim, Cocor Fundulea 376	Romania	Sarca et al., 1996
Cotton	Pima S6	United States	McDaniel et al., 2000

tion aids. Breeders, therefore, desire alternatives to the conventional approaches being used. Furthermore, improvement in both grain yield and drought resistance is difficult, as the two seem to be negatively correlated. Thus, the genetic basis of these characters must be studied for agronomic improvement.

Plant molecular biology has made spectacular growth in the past two decades and offers several new opportunities for strategic research on the production of improved plant types, exploiting transgenic and nontransgenic approaches. For instance, embryo culture helps breeders attempting wide crosses between ecotypes that are difficult to cross by the conventional approach. Embryo rescue technique has enabled transfer of genes for resistance to white-backed plant hopper, bacterial blight, and blast from wild species into breeding lines of rice. Similarly, anther culture allows faster stabilization of breeding lines.

Genetic engineering, which involves transfer of a gene or a block of genes to a variety from an alien source, has three distinct advantages over the conventional breeding methods: (1) it is useful in reorientation of the genes in the same species after in vitro modification; (2) it is useful in transferring alien genes, derived from bacteria, fish, insects, and so on into the crops; and (3) it allows introduction of one or two well-characterized genes into the background of favorable genotypes.

The methods involved in tissue culture and molecular biology have been standardized for several crops and the protocols for introduction of desired genes to the crop varieties have been developed (Birch, 1997; Hanken and Wright, 1999). Different vector systems have been constructed for the introduction of transgenes into monocot and dicot plants. Transgenic plants overexpressing a number of different genes have been produced with resistance to insect pests, viruses, and fungal pathogens, in addition to herbicide tolerance (Grover and Minhas, 2000).

Recent efforts to produce transgenic plants for improved tolerance to salt stress have been made by Tarczynski and colleagues (1993). Following this initial report, work on development of transgenic lines with tolerance to salt and water stress achieved considerable momentum employing various means (Grover, 1999; Grover and Minhas, 2000; Katiyar-Agarwal et al., 2001). Genes encoding various structural proteins have been used to produce these transformants. For production of stress-tolerant plants, both regulatory and structural genes have been employed; however, regulatory genes have been preferred. The regulatory machinery involved in synthesis and control of transacting factors has emerged as a central point in regulating stress-responsive genes. Specific examples of the use of regulatory genes for abiotic stress resistance are shown in Table 6.3. However, in some cases, increased drought tolerance was found to be associated with negative impacts on the morphology of the transformed plants. A stress-induced promoter DREBIA cDNA, controlled by a promoter rdzgA, showed overexpression (Kasuga et al., 1999). Overexpression of DREBIA cDNA in transgenic plants resulted in minimal adverse effects on growth and greater stress tolerance than the expression controlled by the CaMV promoter. Genetic, biochemical, and physiological approaches have provided a great deal of information on many signal-transduction pathways in plants. The manipulation of specific activities of signal reception and transduction might be helpful in controlling multiple genes through single transfers. In spite of the requirement for basic research to unravel the mechanisms of perception and transduction of stress signals from the outside of the nuclei, where transcriptional activation of the specific genes takes place, good progress has been made. Genetic engineering has been successfully applied to identify and ultimately transfer different genes responsible for the biosynthesis of metabolites such as trehalose, which is responsible for late-embryogenesis-abundant proteins. This has been transferred by a shotgun approach to develop drought-resistant transgenic rice (Mitra, 2001). There are further examples of the production of transgenic lines resistant to drought using structural and regulatory genes (Table 6.3). For instance,

TABLE 6.3. Drought and tolerant transgenic plants produced using structural and regulatory genes.

Gene	Protein	Source	Cellular role	Trans host	Stress type	Reference
Regulatory Genes						
At-dbf 2	Cell-cycle-regulated phospho protein	*Arabidopsis thaliana*	Kinase	*Arabidopsis thalania*	Osmotic, salt, high temperature, cold	Lee et al., 1999
Dreb	DRE binding protein	*Arabidopsis thaliana*	Transcription factor	*Arabidopsis thalania*	Cold, drought	Kasuga et al., 1999
Oscdpk7	Calcium dependent protein kinase	*Oryza sativa*	Protein kinase	*Oryza sativa*	Drought, salinity	Saijo et al., 2000
Structural Genes						
Hva 1	Leaf protein	*Hordeum vulgare*	Unknown	*Arabidopsis thalania*	Drought, salinity	Xu et al., 1996
Msalr	NADPH-dependent aldose/ aldehyde reductase	*Medicago sativa*	Detoxification	*Nicotiana tabacum*	Drought	Oberschall et al., 2000
P5cs	1-pyrroline 5-carboxylate synthase	*Vigna aconitifolia*	Proline biosynthesis	*Nicotiana tabacum*	Drought	Hong et al., 2000
Tps1	Trehalose 6-phosphate synthase	*Arabidopsis thaliana*	Trehalose biosynthesis	*Nicotiana tabacum*	Drought, salinity	Holmstrom et al., 1996

drought-resistant transgenic tobacco (*Nicotiana tabacum* L.) has been produced using alien genes from *Arabidopsis thaliana, Vigna aconitifolia,* and *Medicago sativa.* In rice (*Oryza sativa* L.), transgenic lines affecting calcium-dependent protein kinase have been developed (Saijo et al., 2000). Three genes having potential to improve drought tolerance in cereals, identified from alien sources, were transformed to wheat. These genes are (1) *MtlD/MtlDH* gene from Escherichia; (2) *CodA* gene from Anthrobacter globiliformis; and (3) *TPS1* gene from yeast (Jenes et al., 2000). Thus, genetic engineering may lead to progress in developing drought resistance. Progress will, however, depend on isolation of sources of drought tolerance carrying no negative effects and additional experimentation on the adaptation of transgenic to real habitats.

LIMITATIONS AND FUTURE PROSPECTS

Management of drought is the key issue for sustained agricultural production in areas depending on rains for their cultivation. Providing irrigation is, by and large, considered the simplest way to overcome the drought problem. The transport of water in drought-hit areas is probably the costliest input and, hence, is often not feasible in arid zones. During cultivation of crops, soil and water conservation techniques may improve the water availability to a limited extent. However, biological approaches, through the use of drought-tolerant cultivars, may prove to be the most feasible and sustainable practices for combating drought. Drought tolerance, in itself, is a complex trait both in expression and in methods for its evaluation. Hence, breeding for drought appears to have many practical limitations, as detailed in the following sections.

Inconsistency in the Targeted Growing Environment

The environment for which a drought-resistant variety is required is itself not constant and may vary with soil type, climate, rainfall pattern, and cropping intensity. Broadly, three types of soil environments are important for selecting a suitable drought-tolerant variety: stored soil environment, variable moisture environment, and optimal moisture environment. Thus, for different growing environments, plants with different growth and root characters are required. For instance, for environments with long and unpredictable droughts, plants with indeterminate growth and deep roots may be preferred; whereas, for more assured rains, determinate plants with well-spread roots rather than deep roots may be required.

Lack of Universal, Dependable Screening Procedures

A great variance in screening methods being employed by different workers at various places is a serious concern for the identification of genotypes with proven performance. The level and intensity of drought and the availability of soil moisture may vary, leading to mismatching of critical growth and critical drought. A uniform and sensible screening procedure that can identify critical growth and yield components along with lines tolerant to drought remains an objective of plant breeders.

Lack of Selection Indices

Breeding strategies for drought tolerance are advancing at a relatively slow speed because of the lack of a workable, cheap, and universal selection parameter. More and more parameters are being added to a lengthy list, but still no parameter appears feasible and workable. Physiobiochemical parameters are time-consuming to monitor, costly, and require specific knowledge. Roots have direct relevance to drought resistance, but their study is difficult. Growth and morphological traits are under great environmental influence and their expression changes considerably between environments. Large numbers of germplasm lines, segregating materials, and plants from dry habitats await evaluation.

Complexity of the Traits

The end product of any breeding strategy for different environmental situations is the realization of improved yield potential. Many reports, however, indicate the independent nature of yield and drought tolerance as far as genetic inheritance and pattern are concerned. Thus, it is very difficult to bring together two independent traits governed by several genes. Reports on inheritance of drought tolerance governed by only a few genes are rare, as is true also for grain yield. Trait-based selection for drought tolerance thus still appears rare.

Tolerance for Obligatory Traits

Drought is known to induce direct and indirect effects on plants. In other words, yield reduction in drought situations may also occur due to reduced nutrient availability, drought-induced heat stress, diseases, and insect pests. Thus, realization of complete tolerance to both the primary and secondary adverse effects appears difficult. Tolerance to secondary stresses such as heat, insects, pests, and diseases would result in complete proofing against drought. Such reports are, however, lacking.

Genetic Knowledge

The number of genes involved in drought tolerance is still unconfirmed. Moreover, both additive and nonadditive gene effects are involved in its inheritance. Heritability estimates are generally low under hot, dry conditions, limiting selection in such situations. Thus, incomplete genetic information on drought resistance is a major barrier in breeding crops for tolerance to osmotic stresses.

Promising and Potential Sources of Drought Tolerance

In the majority of the field crops, known promising and discrete sources of drought tolerance are not available. Varieties showing better performance and yield in rainfed/drought conditions are available, but may not be technically termed as tolerant to drought. Such genotypes may be treated as high yielders due to carryover effects but not as drought tolerant.

In view of a number of aforesaid limitations associated with breeding for drought tolerance and the vast information available on the aspect, the following future strategies can be set out:

1. Targeted drought environments are greatly variable. A dry environment represented by 3 to 4 percent available soil moisture in the effective root zone would be effective in separating tolerant and susceptible annual field crops, particularly during the active growth stage.
2. Osmotic stress is hardly independent of transpirational stress; thus, selection of plants from natural habitats representing hot, dry climates may be made. Plants in such hostile situations have adapted and developed acquired characters needed for drought tolerance.
3. Long-duration, deep-rooted, slow-growing crops, acclimatized to osmotic stress, may have inbuilt drought-avoidance phenomena. Induction of earliness in such a background would, therefore, be useful as far as tolerance to drought is concerned. This approach may lead to accumulation of genes for both drought escape and avoidance.
4. Large numbers of germplasm lines, segregating lines, advanced breeding lines, and many varieties of major field crops are available that need to be precisely evaluated for drought tolerance. However, owing to the lack of appropriate selection parameters, breeding for drought tolerance is still at the takeoff stage. Hence, money, workers,

and time resources may be channeled into identification of the most common and simple morphological and growth parameters only, rather than into physiobiochemical parameters. This strategy is being investigated by many workers around the world, but practically nothing has emerged which could be of universal relevance.

5. More emphasis must be given to isolation of contrasting and effective donors for drought-tolerance and high-yield traits in drought-prone situations, so as to use these two types of donors in breeding for drought tolerance and for high productivity separately in such situations. In the present-day scenario, high-yielding varieties having sizeable carryover effects for yield in drought conditions are referred to as drought tolerant, which is, however, a misconception as far as true drought tolerance is concerned.
6. Conventional breeding approaches have a vital role to play, particularly with reference to the ecotypes of natural drought habitats. Crop plants have been growing for centuries in these environments and have survived drought, heat, disease, and insect pest attacks. Careful collection of germplasm from true natural drought habitats providing high intensities of osmotic and transpirational stresses and drought-induced biotic stresses as well, may represent an ecotype resistance to drought. Selection from such habitats and meaningful evaluation under desired, controlled conditions may result in development of a real drought-resistant genotype.
7. Efforts to develop transgenic plants using alien and unrelated parental sources of drought tolerance may be developed and tested. Such lines would certainly be needed after exhaustion of conventional approaches and natural variability of the targeted crop after a decade or so.

REFERENCES

Aggarwal, V.D. (1987). Selecting cowpea varieties for harsh, semiarid, sahelian areas. *IITA Research Briefs, International Institute for Tropical Agriculture* 8(2):2.

Allen, F.L., R.E. Comstock, and D.C. Rasmusson (1978). Optimal environments for yield testing. *Crop Science* 18:747-751.

Alvim, P. de T. (1985). *Theobroma cacao*. In *Handbook of Flowering*, A.H. Halevy (Ed.). Boca Raton, FL: CRC Press, pp. 357-365.

Ashraf, M. and S. Mehmood (1990). Response of four *Brassica* species to drought stress. *Environmental and Experimental Botany* 30:93-100.

Aune, J.B. (1992). Cowpea cultivars for rainfed cultivation in Gourma, Mali. *Nigerian Journal of Agricultural Sciences* 6:145-160.

Babalola, O. (1980). Water relations of three cowpea cultivars. *Plants and Soil* 56:59-69.

Birch, R.G. (1997). Plant transformation: Problems and strategies for practical application. *Annual Review of Plant Physiology and Plant Molecular Biology* 48:297-326.

Bouwkamp, J.C. and W.L. Summers (1982). Inheritance of resistance to temperature–drought stress in snap bean. *Journal of Heredity* 73:385-386.

Boyer, J.S. (1982). Plant productivity and environment. *Science* 218:443-448.

Boyer, J.S. (1996). Advances in drought tolerances in plants. *Advances in Agronomy* 56:187-218.

Ceccarelli, S. (1996). Positive interpretation of genotype by environment extraction in relation to sustainability and bio-diversity. In *Plant Adaptation and Crop Improvement,* M. Cooper and C. Hammers (Eds.). Wallingford, UK: CAB International, pp. 467-486.

Champoliver, L. aad A. Merrien (1996). Effect of water stress applied at different growth stages to *Brassica napus* L. var. oleifera on yield, yield components and seed quality. *European Journal of Agronomy* 5:153-160.

Chang, T.T., G.C. Lorcsto, and O. Tagumpay (1974). Screening rice germplasm for drought resistance. *SABRAO Journal* 6:9-16.

Chipman, R.B., C.D. Raper Jr., and R.P. Patterson (2001). Allocation of nitrogen and dry matter for two soybean genotypes in response to water stress during reproductive growth. *Journal of Plant Nutrition* 24:873-884.

Chisi, M., P.J. Bramel-Cox, M.D. Will, M.M. Claassen, and D.J. Andrews (1996). Breeding for grain yield stability using full sib family recurrent selection in sorghum. *Crop Science* 36:1083-1087.

Choudhary, B.D., D.P. Singh, P. Singh, and A. Kumar (1989). Association among morphological characters in chickpea. *Indian Journal of Plant Physiology* 32:35-42.

Choudhary, B.D., P. Singh, and D.P. Singh (1991). Association and modified path coefficient analysis to partition first and second order morpho-physiological characters in *Brassica* under moisture stress conditions. *Indian Journal of Plant Physiology* 34:92-96.

Close, T.J. and P.M. Chandler (1990). Cereal dehydration: Serology gene mapping and potential functional roles. *Australian Journal of Plant Physiology* 17:333-334.

Dure, L., M. Crouch, J. Harda, T.H.D. Ho, J. Humdy, R. Quatrano, T. Thomas, and Z.R. Sung (1989). Common amino acids sequence domains among the LEA proteins of higher plants. *Plants Molecular Biology* 12:475-486.

Duron Noriega, L.J. (1986). Determination of water regime for chickpeas in a heavy soil on the coast of Hermosillo, Sonora. *Agricultural Tecnica en Mexico* 12:105-120.

Edmeades, G.O., J. Bolanos, and H.R. Lafitte (1992). Progress in breeding for drought tolerance in maize. *Proceedings of Forty-Seventh Annual Corn and Sorghum Industry Research Conference,* D.B. Wilkinson (Ed.). Washington, DC: American Seed Trade Association, pp. 93-111.

El Maarouf, H., Y. Juily-Fodil, M.D. Gareil, A. d'Arcy-Lameta, and Arh-Thu-Phan-Thi (1999). Enzymatic activity and gene expression under water stress of phospholipase D in two cultivars of *V. unguiculata* L. Walp. differing in drought tolerance. *Plant Molecular Biology* 39:1257-1265.

Farshadfar, E., M. Farshadfar, and J. Sutka (2001). Combining ability analysis in drought tolerance in wheat over different water regimes. *Acta Agronomica Hungarica* 48:353-361.

Fischer, R.A. and R. Maurer (1978). Drought resistance in spring wheat cultivars. *Australian Journal of Agricultural Research* 29:897-912.

Franca, M.G.C., Anh-Thu-Pham-Thi, C. Pimentel, R.O.P. Rossiello, Y. Zuily-Fodil, and D. Laffray (2000). Differences in growth and water relations among *Phaseolus vulgaris* cultivars in response to induced drought stress. *Environmental and Experimental Botany* 43:227-237.

Frey, K.J. (1964). Adaptation reaction of oat strains selected under stress and non-stress environmental conditions. *Crop Science* 4:55-58.

Fujii, M. and T. Horie (2001). Relative contributions of tolerance and avoidance to drought resistance in dry matter production of different rice cultivars at different fertilization levels. *Japanese Journal of Crop Science* 70:59-70.

Ghassemi, F., A.J. Jakeman, and H.A. Nix (1995). *Salinization of Lands and Water Resources: Human Causes, Extent, Management and Case Studies*. Sydney, Australia: UNSW Press and Wallingford, UK: CAB International.

Gomez-Sanchez, D., M. Baldini, D. Aguilera-Charles, and G.P. Vannozzi (1999). Genetic variances and heritability of sunflower traits associated with drought tolerance. *Helia* 22:23-32.

Grantz, D.A. (1979). "Earliness and drought adaptation of cowpea." MSc thesis, University of California; Riverside.

Grover, A. (1999). A novel approach for raising salt-tolerant transgenic plants based on altering stress signaling through Ca^{2+}/calmodulin-dependent protein phosphatase calcineurin. *Current Agriculture* 76:136-137.

Grover, A. and D. Minhas (2000). Toward production of abiotic stress tolerant transgenic rice plants: Issues, progress and future research needs. *Proceedings of Indian National Academy of Sciences* B 66:13-32.

Gupta, S.N., B.S. Dahiya, B.P.S. Malik, and N.R. Bishnoi (1995). Response of chickpea to water deficits and drought stress. *Haryana Agricultural University Journal of Research* 1/2:11-19.

Hall, A.E. (1993). Is dehydration tolerance relevant to genotypic differences in leaf sequence and crop adaptation to dry environment? In *Plant Responses to Cellular Dehydration During Environmental Stresses,* T.J. Close and E.A. Bray (Eds.). Rockville, MD: The American Society of Plant Pathologists, pp. 1-10.

Hall, A.E. and D.A. Grantz (1981). Drought resistance of cowpea improved by selecting for early appearance of mature pods. *Crop Science* 21:461-464.

Hanken, G. and M.S. Wright (1999). Recent advances in the transformation of plants. *Trends in Plant Sciences* 4:226-231.

Harada, J.J., A.J. Delisle, C.S. Baden, and M.L. Crouch (1989). Unusual sequence of an abscisic acid inducible mRNA which accumulates late in *Brassica napus* seed development. *Plant Molecular Biology* 12:395-401.

Holmstrom, K.O., E. Mantyl, B. Welin, and E.T. Palva (1996). Drought tolerance in tobacco. *Nature* 379:683-684.

Holt, D.F., N. Kawano, and O. Ito (2000). Involvement of active oxygen-scavenging system in drought tolerance. *Philippines Journal of Crop Science* 25(Supp. 1):50.

Hong, Z., K. Lakkineni, Z. Zhang, and D.P.S. Verma (2000). Removal of feedback inhibition of 1-pyrroline 5-carboxylate synthetase results in increased proline accumulation and protection of plants from osmotic stress. *Plant Physiology* 122:1129-1136.

Hurd, E.A. (1971). Can we breed for drought resistance? In *Drought Injury and Resistance in Crops,* Special Publication No. 2, K.L. Larson and J.D. Eastin (Eds.). Madison, WI: Crop Science Society of America, pp. 77-88.

Hurd, E.A. (1975). Phenotype and drought tolerance in wheat. In *Plant Modification for More Efficient Water Use,* First Edition, J.F. Stone (Ed.). New York: Elsevier, pp. 39-57.

Hurd, E.A. (1976). Plant breeding for drought resistance. In *Water Deficits and Plant Growth,* Volume 4, T.T. Kozlowski (Ed.). New York: Academic Press, pp. 317-353.

Ilami, G., C. Nespoulous, J.C. Huet, N. Vartanian, and J.C. Pernollet (1997). Characterization of BND_{22}, a drought-induced protein expressed in *Brassica napus* leaves. *Phytochemistry* 45:1-8.

Itani, J. (1992). Studies on water absorption ability of cowpea. *Japanese Journal of Tropical Agriculture* 36:37-44.

Jenes, B., J. Pauk, O. Toldi, Z. Banfalvi, R. Doczi, A. Oreifig, G. Dallmann, and G. Hazdina (2000). Improvement of tolerance against dehydration stress in cereals. In *Use of Agriculturally-Important Genes in Biotechnology: Proceedings of the NATO Advanced Research Workshop,* Szeged, Hungary, 17-21 Oct. 1999, G. Hrazdina (Ed.). Amsterdam: IOS Press, pp. 7-13.

Johnson, G.R. and K.J. Frey (1967). Heritabilities of quantitative attributes of oats (*Avena* sp.) at varying levels of environmental stress. *Crop Science* 7:43-46.

Jones, H.G. (1983). *Plant and Microclimate.* Cambridge, UK: Cambridge University Press.

Jordan, W.R., P.J. Shouse, A. Blum, E.R. Miller, and R.L. Monk (1984). Environmental physiology of sorghum. II. Epicuticular wax load and cuticular transpiration. *Crop Science* 24:1168-1173.

Kamali, V. and J. Miller (1982). The inheritance of drought tolerance in sunflower. In *Proceedings of Tenth International Sunflower Conference.* Toowoomba, Australia: Australian Sunflower Association, pp. 228-231.

Kanangara, T., N. Seetharama, R.C. Durley, and G.M. Simpson (1983). Drought resistance of *Sorghum bicolor:* Changes in endogenous growth regulators of plants grown across an irrigation gradient. *Canadian Journal of Plant Sciences* 63:147-155.

Kandaswami, M. (1985). Genetic variation and genotype-environmental interaction in sesame (*Sesamum indicum* L.). *Madras Agricultural Journal* 72:156-161.

Kang, C.W., G.C. Duo, and F.R. Son (1986). Screening mungbean for drought tolerance. 2. Field Trial. *Reports on Rural Development and Administration, Korea Republic Crops* 28:156-161.

Karamanos, A.J. and A.Y. Papatheohari (1999). Assessment of drought resistance of crop genotypes by means of water potential index. *Crop Science* 39:1792-1797.

Kasuga, M., O. Liu, S. Mitra, K. Yamaguchi-Shinozaki, and K. Shinozaki (1999). Improving plant drought, salt and freezing tolerance by genetic transfer of a single stress inducible transcription factor. *Nature Biotechnology* 17:287-291.

Katiyar-Agarwal, S., M. Agarwal, and A. Grover (2001). Emerging trends in agricultural biotechnology research: Use of abiotic stress-induced promoter to drive expression of a stress resistance gene in the transgenic system leads to high level stress tolerance associated with minimal negative effects on growth. *Current Science* 77:1577-1579.

Kumar, A., D.P. Singh, and P. Singh (1987). Genotypic variation in the response of *Brassica* sp. to water deficit. *Journal of Agricultural Science* 109:615-618.

Kumar, D. (2001). Breeding approaches in arid legumes for abiotic stresses. *National Symposium for Sustainable Agriculture and Nutritional Security,* April 17-19, 2001, New Delhi, India. Kanpur, India: Indian Society of Pulses Research and Development, p. 16.

Lee, J.H., M.V. Montagu, and N. Verbruggen (1999). A highly conserved kinase is an essential component of stress tolerance in yeast and in plant cells. *Proceedings of National Academy of Sciences, USA* 96:5873-5877.

Liang, Z.S., Z.R. Ding, and S.T.R. Wang (1992). Study on type of water stress adaptation in both *Brassica napus* and *B. juncea* L. species. *Acta Botanika* 12:38-45.

Lin, C.S. and M.R. Binna (1988). A superiority measure of cultivar performance for cultivar × location data. *Canadian Journal of Plant Sciences* 68:193-198.

Mai-Kodomi, Y., B.B. Singh, T. Terao Jr., O. Myers, J.H. Yopp, and P.J. Gibson (1999). Inheritance of drought tolerance in cowpea. *Indian Journal of Genetics* 53:317-323.

Marrow, K.O. and A.K. Hall (1992). Inheritance of heat tolerance during pod set in cowpea. *Crop Science* 32:912-918.

Mathur, D. and P.N. Wattal (1996). Physiological analysis of growth and development in three species of rapeseed and mustard (*Brassica juncea, B. campestris* and *B. napus*) under irrigated and unirrigated conditions. *Indian Journal of Plant Physiology* 1:171-174.

McDaniel, R.G., P. Dugger, and D. Richter (2000). Genetic manipulation of cotton leaf stomatal density to enhance drought tolerance. *Proceedings of the Beltwide Cotton Conferences,* Volume 1, P. Dugger and D. Richter (Eds.). Memphis: National Cotton Council, pp. 562-564.

Mederski, H.J. and D.L. Jeffers (1973). Yield response of soybean varieties grown at two soil moisture stress levels. *Agronomy Journal* 65:410-412.

Merah, O., E. Deleens, A. Al-Hakimi, and P. Monneveux (2001). Carbon isotope discrimination and grain yield variations among tetraploid wheat species cultivated under contrasting precipitation regimes. *Japan Journal of Agronomy and Crop Science* 186:129-134.

Mitra, J. (2001). Genetics and genetic improvement of drought resistance in crop plants. *Current Science* 80:758-763.

Mogensen, V.O., C.R. Jensen, G. Mortensen, J.H. Thaoe, J. Koribidis, and A. Ahmed (1996). Spectral reflectance index as an indicator of drought of field grown oilseed rape (*Brassica napus* L.). *European Journal of Agronomy* 5:125-135.

Mohammed, M.F., M.I. Ferreira, and H.G. Jomes (2000). Useful tepeary and cowpea lines in rationalizing water use for pulse legumes production in Southern Egypt. *Acta Horticulturae* 537:813-822.

Muchilbauer, F.J. and W.J. Kaiser (1987). Registration of Garnet and Tammany chickpea. *Crop Science* 27:1087-1088.

Ngugi, E.C.K., N.W. Galwey, and R.B. Austin (1994). Genotype × environment interaction in carbon isotope discrimination and seed yield in cowpea. *Euphytica* 73:213-224.

Oberschall, A., M. Deak, K. Torok, L. Sass, I. Vass, I. Kovacs, A. Feher, D. Dudits, and G.V. Horvarth (2000). A novae aldose/aldehyde reductase protects transgenic plants against lipid peroxidation under chemical and drought stresses. *Plant Journal* 24:437-446.

Obloi, H. and A.A. Kacperska (1991). Desiccation tolerance changes in winter rape leaves grown under different environmental conditions. *Biologia Plantarum* 23:209-213.

Onim, J.F.M. (1983). Association between grain yield and drought resistance in pigeon pea in marginal rainfall areas of Kenya. In *More Food from Better Technology*, J.C. Holmes and W.M. Tahir (Eds.). Rome, Italy: FAO, pp. 864-872.

Paclik, R., L. Sakova, and V. Curn (1996). Reaction of different cultivars of *Brassica napus* subsp. *oleifera* to water stress. *Phytochemica* 13:55-62.

Paje, M.C.M., M.M. Ludlow, and R.J. Lawn (1988). Variation among soybean (*Glycine max* L.) accessions in epidermal conductance of leaves. *Australian Journal of Agricultural Research* 39:363-373.

Pirmakhmadov, K. (1972). A study of promising chickpea varieties on rainfed soil. In *Bgoan. Zemledelic Dushanbe, Tajik SSR:* 100-106.

Ramirez-Vallejo, P. and J.D. Kelly (1998). Traits related to drought resistance in common bean. *Euphytica* 99:127-136.

Rawson, H.M. and J.M. Clarke (1988). Nocturnal transpiration in wheat. *Australian Journal of Plant Physiology* 15:397-406.

Reddy, P.S., C. Sudhakar, and K. Veeranjaneyulu (1990). Water stress induced changes in enzymes of N_2 metabolism in horse gram. *Indian Journal of Experimental Biology* 28:273-276.

Rettke, M.L. and P.J. Keys (1985). Chickpea a new crop for Queensland. *Queensland Agricultural Journal* 111:53-55.

Reviron, M.P., N. Vartanian, M. Sallantin, J.C. Huet, T.C. Pernollet, and D.E. de Vienne (1992). Characterization of a novel protein induced by progressive on rapid drought and salinity in *Brassica napus* leaves. *Plant Physiology* 100:1486-1493.

Richards, R.A. and N. Thurling (1979). Genetic analysis of drought stress response in rapeseeds (*Brassica campestris* and *B. napus*). II. Yield improvement and application of selection indices. *Euphytica* 28:169-177.

Rosen, S. and L. Scott (1992). Famine grips Sub-Saharan Africa. *Agricultural Outlook* 191:20-24.

Roy, N.N. and B.R. Murty (1970). A selection procedure in wheat for stress environment. *Euphytica* 19:509-521.

Saijo, Y., S. Hata, J. Kyozuka, K. Shimamoto, and K. Izui (2000). Over expression of a single Ca^{2+} dependent protein kinase confers both cold and salt/drought tolerance on rice plants. *Plant Journal* 23:319-327.

Sandene, P. (1993). South African corn situation. *World Agriculture Production* 2:8-61.

Sangwan, R.S., G.P. Lodhi, and C. Kishore (1998). Inheritance of fodder yield and its traits in cowpea. *Haryana Agriculture University Journal of Research* 28:91-94.

Sarca, T., O. Cosmin, I. Ciocazanu, N. Bica, and C. Bagiu (1996). Maize breeding for drought tolerance. *Romanian Agricultural Research* 5-6:1-11.

Scherbakova, A. and A. Kacperska-Palacz (1980). Modification of stress tolerance by dehydration pretreatment in winter rape hypocotyl. *Journal of Experimental Botany* 48:560-563.

Scherbakova, A. and A. Kacperska-Palacz (1983). Water stress injuries and tolerance as related to potassium efflux from winter rape hypocotyls. *Physiologia Plantarum* 57:296-300.

Shakhatreh, Y., O. Kafawin, S. Ceccarelli, and H. Saoub (2001). Selection of barley lines for drought tolerance in low rainfall areas. *Journal of Agronomy and Crop Science* 186:119-127.

Shakoor, A., E.C.K. Ngugi, H.S. Muthoka, and J.W. Kamau (1984). Improvement of cowpea varieties for dry areas in Kenya. *East African Agricultural and Forestry Journal* 44(Special Issue):306-311.

Sinclair, T.R. and M.M. Ludlow (1986). Influence of soil water supply on the plant water balance of four tropical grain legumes. *Australian Journal of Plant Physiology* 13:329-341.

Singh, B.B., Y. Mai-Kodomi, and Y. Terao (1999). A simple screening method for drought tolerance in cowpea. *Indian Journal of Genetics* 59:211-220.

Singh, G., P.S. Thakur, and V.K. Rai (1985). Free amino acid pattern in stressed leaves of two contrasting resistant and susceptible cultivars of chickpea. *Experimental Agriculture* 41:40-41.

Singh, S.P. and R.B. Mehra (1980). Adaptability studies in Bengal gram (*Cicer arietinum* L.). *Tropical Grain Legumes Newsletter* 19:51-54.

Subbarao, G.V., C. Johansen, A.E. Slinkard, R.C. Nageswar Rao, N.P. Saxena, and Y.S. Chauhan (1995). Strategies for improving drought resistance in grain legumes. *Critical Reviews in Plant Sciences* 14:469-523.

Szegletes, Z., L. Erdei, I. Tari, and L. Cseuz (2000). Accumulation of osmoprotectants in wheat cultivars of different drought tolerance. *Cereal Research Communication* 28:403-410.

Tarczynski, M.C., R.G. Jensen, and H.J. Bohnert (1993). Stress protection of transgenic tobacco by production of the osmolyte mannitol. *Science* 259:508-510.

Turner, N.C. (1979). Drought resistance and adaptation to water deficits in crop plants. In *Stress Physiology in Crop Plants,* H. Mussell and R.C. Staples (Eds.). New York: Wiley-Interscience, pp. 343-372.

Vartanian, N., C. Damernval, and D. de Vienne (1987). Drought induced changes in protein pattern of *Brassica napus* var. oleifera roots. *Plant Physiology* 84:989-992.

Vertucci, C.W. and A.C. Leopold (1987a). The relationship between water binding and desiccation tolerance in tissues. *Plant Physiology* 85:232-238.

Vertucci, C.W. and A.C. Leopold (1987b). Water binding in legume seeds. *Plant Physiology* 85:224-231.

Walker, D.W. and J.C. Miller (1982). The relationship between leaf water potential and biological nitrogen fixation in drought resistant and susceptible genotypes of cowpea (*Vigna unguiculata* (L.) Walp). *Abstract Horticulture Sciences* 17(3, II):503.

Wani, S.A. and A.N. Srivastava (1989). Genetic analysis of characters associated with drought tolerance in Indian mustard. *Journal of Oilseeds Research* 6:97-101.

Werry, J., S.N. Silim, E.J. Knight, R.S. Malhotra, and R. Cousin (1994). Screening techniques and sources of tolerance to extremes of moisture and air temperature in cool season food legumes. *Euphytica* 73:73-83.

Wisner, R.N. and C.A. Chase (1984). *World Food Trade and US Agriculture 1960-1983.* Ames, IA: World Food Institute, Iowa State University.

Wright, P.R., J.M. Morgan, and R.S. Jessop (1996). Comparative adaptation to canola *(Brassica napus)* and Indian mustard *(B. juncea)* to soil water deficit: Plant water relations and growth. *Field Crops Research* 49:51-64.

Xu, D.P., X.L. Duan, B.Y. Wang, B.M. Hong, T.H.D. Ho, and R. Wu (1996). Expression of a late embryogenesis abundant protein gene, *HVA1* from barley confers tolerance to water deficit and salt stress in transgenic rice. *Plant Physiology* 110:249-257.

Yadav, R.B.R. and B.D. Patil (1984). Screening of cowpea (*Vigna unguiculata* L.) varieties for drought tolerance. *Zeitschrift für Pflanzenzuchtung* 93:259-263.

Yadav, S.K., A. Kumar, and D.P. Singh (1990). Micro-environment, water relations and yield of mustard cultivars in irrigated and unirrigated soils. *Beitrage zur Tropischen Landwirtschaft und Veterinarmedizin* 28:399-406.

Yadava, T.P. and Hari Singh (1996). Morpho-physiological determinants of yield under water stress conditions in Indian mustard *(Brassica juncea). Acta Horticulturae* 407:55-160.

Chapter 7

Molecular Genetics and Breeding for Flooding Tolerance

Apichart Vanavichit
Somvong Tragoonrung
Theerayuth Toojinda

INTRODUCTION

Flooding devastates many crops because most plants were not evolved to cope with hypoxic situations. Waterlogging has increasingly become a major problem for rice-based cropping systems where the rhizosphere is partially saturated by irrigation water or monsoon. In Southeast Asia alone, about 15 percent of the total maize-growing area is affected by flooding. In crops such as rice grown on lowlands near riverbanks, flooding can completely submerge the whole plant for a few days to months. In general, flooding is defined as any situation with excess water. The impact of excess water on plant performance is crop-dependent and growth stage-dependent. Under such abrupt changes, their sessile nature forces plants to adapt, acclimate, and tolerate these suboptimum situations.

Flooding in Terrestrial Crops

Flooding stress in terrestial species is referred to as waterlogging. The first visual symptoms of terrestial species such as maize, soybean, and tomato after a few days of waterlogging is wilting or epinasty. In tomato (Jackson, 1956) and maize (Rathore et al., 1998), in addition to an epinastic response, adventitious roots are formed along the submerged stem portion, stem growth ceases, and leaves abscise.

Flooding in Semiaquatic Crops

Flooding is the major natural disaster affecting half of the 40 Mha of rainfed lowland rice in South and Southeast Asia. Rice, the most important semiaquatic plant, can adapt to different regimes of flooding that often confront farmers in the tropics. Flooding can be roughly classified into flash flooding and prolonged flooding. In the tropics, flash flooding is devastating, causing economic loss on any paddy field, including irrigated areas, following continual heavy rains. The rapid inundation causes severe stress at every stage of plant growth and development. The situation is different near some riverbanks and deltas, where water increases more slowly to its maximum during the monsoon and slowly recedes at the end of the season.

Paddy fields can be classified into medium-deep, deepwater, or floating, where the maximum depth of water is <1 m, >1 m, or 2 to 5 m, respectively. In irrigated areas, the situation can also be devastating because most popular semidwarf rice cultivars have not evolved a resistance mechanism to cope properly with even a medium depth of water (75 cm). High priority, therefore, has long been set for breeding submergence-tolerant rice.

TOLERANCE TO FLOODING

Flooding tolerance means different things in different crops and flooding situations. Tolerance mechanisms acquired for terrestrial crops are not the same as for semiaquatic species such as rice. Under different flooding situations, even the same plant species acquires different means to survive. Tolerance to flooding can be generalized as the ability to survive and continue growing after several days in submerged conditions (Adkins et al., 1990). The definition signifies two important aspects of tolerance to stresses; the ability to survive and, consequently, the ability to recover from possible damage caused by submergence stresses and continue growing.

Development of Aerenchyma

In terrestrial species such as maize, soybean, and tomato, waterlogging creates a special apparatus for breathing called aerenchyma. In waterlogged soil, water replaces oxygen in the soil macro- and micropores and depletes oxygen, causing hypoxic stresses to root systems. One mechanism plants use to alleviate this constraint is to develop aerenchymas, special structures developed from cell collapse in response to hypoxic conditions to facilitate air conductance from shoots to roots. Under normal conditions, root porosity is maintained at 7.3 to 7.7 percent during tasseling in maize (Rathore et al., 1998). Waterlogging increases root porosity in maize to 8.3 to 10.5 percent at knee height and to 10.3 to 12.5 percent at tasseling.

Development of Adventitious Roots

In most aquatic trees, aerial roots are well-developed to conduct gas exchange for the submerged root system. Adventitious roots are observed a few days after terrestial plants face waterlogging conditions. In maize, adventitious root development is the most typical trait of waterlogging resistance (Rathore et al., 1998). Normally, adventitious roots are developed above the root collar near the water surface. These adventitious roots developed on four or five nodes provide mechanical support to plants in wet and loose soil conditions. Adventitious roots are also thought to alleviate underground roots from hypoxic conditions. Thus, adventitious roots and the formation of aerenchyma provide indices to waterlogging tolerance.

Energy Conservation

One crucial adaptation for survival under depleted oxygen environment is metabolic adjustment toward alcoholic fermentation. Under hypoxic conditions, alcohol pyruvate decarboxylase (PDC) and alcohol dehydrogenase (ADH) are two key enzymes involved in converting pyruvate to ethanol. The increased rate of alcoholic fermentation is needed for energy production to ensure plant growth and survival under hypoxic and anoxic stresses (Taylor, 1942; Avadhani et al., 1978; Alpi and Beevers, 1983; Waters et al., 1991; Setter and Ella, 1994; Setter et al., 1994). The nicotinamide adenine dinucleotide (NAD) generated in the fermentation pathway can be used for conversion of glucose to pyruvate. However, to ensure a high energy charge for cell division and elongation, rice plants must hydrolyze carbohydrate reserves for rapid growth. In aerobic cells, starch is degraded to glucose-1-P by alpha amylase (Plaxton, 1996). If sucrose is metabolized by sucrose synthase, UDP-glucose-6-P pyrophosphorylase, and PPi-dependent phosphofructokinase, 4 ATP is the net gain (Plaxton, 1996).

Fast Underwater Stem Elongation

In semiaquatic species such as rice, *Rumex palustris, Potamogeton pectinatus, Ranunculus sceleratus,* and *Callitriche platycarpa,* one immediate response to aid survival of flooding is the elongation of the whole plant and upper leaves (Musgrave et al., 1972; Musgrave and Walters, 1973; Metraux and Kende, 1983; Jackson et al., 1987; Voesenek and Blom, 1989). In some rice varieties, the increment of plant height may be as great as 20 to 25 cm per day, enabling rice plants to emerge from 1 m depth of flooding in less than a week (Catling, 1992). This phenomenon is also true for *R.*

palustris (Voesenek et al., 1990), *Ranunculus sceleratus,* and *Numhoides peltata* (Ridge, 1987).

Semiaquatic plants that grow near riverbanks where the water stays high during the monsoon evolve underwater stem elongation as a mechanism to maintain leaves above water. The elongation was thought to be initiated from intercalary meristems located slightly above the youngest internode (Metraux and Kende, 1983). In later studies on internodal elongation, using thymidine incorporation (^3H), a rapid-cell-division zone was revealed, originating at the intercalary meristem, 2 to 3 mm above the node. This generates new cells for elongation and differentiation (Metraux and Kende, 1984). The small, brick-like cells produced in the intercalary meristem increase and expand threefold to fivefold in response to submergence (Kende et al., 1998). Stem elongation in *Ranunculus sceleratus* and *Nymphoides peltata* result from a combination of cell division and cell elongation (Ridge, 1987). On the other hand, stem and leaf elongation of *Rumex palustris* results from cell elongation, but not cell division (Voesenek et al., 1990).

Slow Underwater Stem Elongation

Under flash flooding conditions, however, because water level accelerates so fast, plant elongation is seen as self-destructive and inductive to susceptibility (Setter et al., 1997). As an energy-driven process, the rapid plant elongation competes with cell maintenance for limited energy and hydrocarbon skeletons when a plant faces stressful conditions (Greenway and Setter, 1995). Rapid stem elongation is an energy-demanding process facing plants under anaerobiosis. The acceleration of elongation under flood conditions combines the elongation of both leaf sheath (Mazaredo and Vergara, 1982) and leaf blade (Jackson et al., 1987). Moreover, this accelerated growth in height underwater is agronomically undesirable because after flooding the elongated stems become weak and prone to lodging. Such plants, if they recover, would lose their yield potential.

In the case of slow flooding, the intercalary meristem is the source of plant elongation (Metraux and Kende, 1983). The question is where does elongation occur in flash flooding? When plant elongation was compared under flash flooding (100 cm per day) and slow flooding (10 cm per day) using submergence-tolerant and -susceptible strains, in the short-term flooding, elongation was measurable only on the youngest leaves, but not on the intercalary meristem, in contrast to the response to long-term flooding. The rate of leaf elongation depends mainly on genotype and the speed of floodwater. With flash flooding, the tolerant strains barely elongated and withheld their new leaves. In susceptible strains, stem elongation was rapid under both flash and slow flooding.

Genetic evidence linking fast elongation underwater and susceptibility to flooding was not positive in earlier experiments (Yamada, 1959; Mazaredo and Vergara, 1982). Recently, in a survey of 903 entries in the International Rice Research Institute (IRRI) seed bank, poor tolerance was always associated with fast underwater elongation, but slow elongation was not always linked to tolerance of submergence (Setter and Laureles, 1996). The link between submergence tolerance and resistance to fast elongation underwater was recently confirmed by Toojinda and colleagues (2003) and Siangliw and colleagues (2003). In those experiments, recovery from submergence and plant elongation were explored in recombinant inbred lines (RIL) and doubled haploid (DH) populations. Strong phenotypic correlation among traits responsive to submergence stress was observed in the DH and RIL populations, indicating common mechanisms for submergence tolerance. Intolerant plants must generate more energy for elongation during flooding by greater respiration. On the other hand, short plants suffered less mortality and recovered rapidly after desubmergence. This could be due to greater respirable reserves in short plants (Karin et al., 1982; Emes et al., 1988; Setter, Greenway, and Kupkanchanakul, 1989; Setter, Waters, et al., 1989). The negative relationships between plant survival and plant elongation and tolerance score and plant elongation also support this hypothesis (Toojinda et al., 2003). This evidence supports the hypothesis that hypoxic-induced elongation lends plants to susceptibility to flash flooding.

Stay Green

In waterlogged conditions, leaf senescence and wilting have been observed in tomato and maize. In rice, one of the most important indicators of slow recovery from submergence is leaf senescence promoted while underwater. The phenotypic association between low leaf senescence and high plant survival is highly positive (Toojinda et al., 2003). Leaf senescence and tolerance score were indeed linked genetically to plant elongation when rice plants experience submergence. Rapid plant elongation and leaf senescence, being negatively correlated with plant survival in rice, were parallel to the responses caused by exogenous ethylene (Jackson et al., 1987). All of these data suggest plant elongation and leaf senescence are directly linked physiologically to plant survival after submergence.

Roles of Phytohormones in Submergence Tolerance

The gaseous hormone ethylene plays important roles in triggering disease resistance, ripening, senescence, and, most important, plant elongation. Rice seedlings accumulate up to 12 ppm of ethylene in the shoots after 55 h under submerged conditions (Jackson et al., 1987). This level of ethylene in most terrestrial plants can inhibit growth. Under low partial pressures of oxygen such as in flooded plants, ethylene-accelerated elongation in coleoptiles and seedlings of rice occurs at a faster rate than in air (Jackson, 1985). Although the rate of ethylene synthesis in air and while submerged is apparently no different, more ethylene is detected in submerged plants. Moreover, the accumulation of ethylene in submerged tissues is physically entrapped in submerged tissues. In water, the diffusion of ethylene is 10,000 times less than in air (Jackson, 1985). However, in long-term flooding, additional ethylene synthesized in submerged internodes enhances fast elongation (Raskin and Kende, 1984a,b,c). Rice seedlings mimic the ethylene responses when grown in hypoxic conditions. Ethylene, therefore, plays some part in rapid shoot elongation.

The presence of a low amount of oxygen in hypoxia promotes aerenchyma formation in root cortex and in leaves (Jackson, 1989). The aerenchymatous leaf sheath forms interconnections that facilitate long-distance diffusion of ethanol, methane, and CO_2 from roots to leaves (Vartapetian and Jackson, 1997). Ethylene plays important roles in promoting aerenchyma formation in adventitious roots of rice (Justin and Armstrong, 1991). In anoxic conditions, roots developed fewer aerenchyma than in hypoxic conditions because oxygen is required for ethylene production. The formation of aerenchyma is thought to be a programmed cell death (Jackson and Armstrong, 1999; Drew et al., 2000). The cell collapse and death of large cells start in the mid cortex and involve programmed cell death (Kawai et al., 1998; Samarajeewa et al., 1999).

In *Potamogeton pectinatus,* a semiaquatic monocot, ethylene was totally absent under submerged conditions (Summers and Jackson, 1998). Therefore, other hormones may promote internode elongation. Gibberellic acid was the prime cause of internode elongation in deepwater rice (Kende et al., 1998). Inhibition of plant elongation by a gibberellin biosynthesis inhibitor increased tolerance of 14-day-old intolerant seedlings by 98 percent (Setter and Laureles, 1996; Setter et al., 1997).

GENETICS OF TOLERANCE TO WATERLOGGING AND FLOODING

Flooding Tolerance in Flash-Flooded Rice

Submergence tolerance is defined as the ability of a rice plant to survive and continue growing after being completely submerged in water for ten to fourteen days (Catling, 1992). Rice plants are less tolerant of submergence at an early growth stage (Richharia and Parasuram, 1963; Adkins et al., 1990). Varietal differences in the degree of tolerance to submergence have been reported (Palada and Vergara, 1972; Supapoj et al., 1982). However, submergence tolerance is rare in the collected rice germplasm. It is a genetically determined trait with relatively high heritability, controlled by one or a few genes with major effects and minor modifiers (Suprihatno and Coffman, 1981; Mohanty and Khush, 1985; Sinha and Saran, 1988; Haque et al., 1989). Phenotypic screening of rice germplasm at the Huntra Rice Experiment Research Station in Thailand and IRRI revealed that only 2 to 6 percent of 21,271 rice cultivars/lines survived ten days of submergence (Setter et al., 1987; Setter and Laureles, 1996). A selected rice cultivar, known as Dhullaputia or FR13A, grown in Orissa, India, is the most impressive submergence-tolerant cultivar known (Mackill, 1986). This cultivar was released in the 1940s by the Central Rice Research Institute, Cuttack, India.

Most genetic and physiological studies of submergence tolerance have utilized FR13A, Kurkaruppan, Thavalu 15325, and Thavalu 15314, landrace varieties from India that are adapted to flood-prone areas (Mazerado and Vergara, 1982). To date, the genetic and physiological bases of submergence tolerance have not been well-understood. The genetic and physiological basis of tolerance shown by the unrelated Sri Lankan cultivar, Goda Heerati (Setter et al., 1997), and by Veidehi, a tolerant cultivar from Bihar in eastern India (Sarkarung and Prayongsap, 2002), has not been explored.

Segregation analysis of progenies from crosses between tolerant and intolerant cultivars revealed a genetic model for submergence tolerance, in which a small number of genes are partially or completely dominant, but with their expression being strongly affected by environmental factors (Suprihatno and Coffman, 1981; Mishra et al., 1996). This model is supported by the success of conventional breeding in transferring submergence tolerance to desirable rice cultivars (Mackill et al., 1993; Siangliw et al., 2003).

Flooding Tolerance in Floating Rice

Tolerance to flooding in floating rice and semiaquatic species such as *Rumex palustris* is defined as the ability to maintain leaves above water at all times and to continue to produce grain yield at a satisfactory level after the water receded. Fast elongation ability is a complex trait and also donor-specific. Inheritance of elongation ability is controlled by (1) partially dominant genes with additive effects (Hamamura and Kupkanchanakul, 1979); (2) two dominant complementary genes, one for initial plan height and the other for elongation ability (Suge, 1988; Thakur and Hillerislambers, 1988); (3) a digenic inheritance with dominant epistatis (Saha et al., 1994); and (4) an overdominant effect of elongation ability and a positive transgressive segregation for plant height (Mishra et al., 1996).

Screening Methods

Submergence tolerance is a complex trait; it cannot be attributed just to one or a few physiological and morphological characteristics (Xu and Mackill, 1996). Five traits have been practically considered by breeders as indicators of tolerance. However, the results of screening vary with location of experiment, age of seedlings, water condition, climate, and duration of flooding. A simple and convenient screening method for rice involving direct seeding in a plot, on a puddle seed bed constructed in the base of a 60 m^2 pond, was reported by Toojinda and colleagues (2003) (Figures 7.1 and 7.2). Each plot included two rows, 75 cm in length with 25 cm between rows. Alternatively, tests have been carried out on plants directly seeded in greenhouse tanks (Vergara and Mazaredo, 1975). Under field conditions, three- or four-week-old seedlings were subjected to complete submergence in 30 cm of water above the tallest plants to prevent their leaf tips emerging into the atmosphere for seven to ten days, depending on the average light intensity and light transmissions. Under greenhouse conditions, ten-day-old seedlings were normally subjected to submergence in 30 cm of water for seven days (Vergara and Mazaredo, 1975). One to five days after water is drained, the seedlings were measured as follows:

1. Percent plant survival (PPS) was calculated as total number of surviving seedlings counted ten days after the water was drained from the submerging ponds, divided by total number of seedlings counted before submergence. This value was then multiplied by 100. A tolerance score (TS) was derived that summarizes the ability of plants to survive flash flooding. The Standard Evaluation System (SES) associating

single-digit scores for submergence tolerance with PPS was used to compute the TS (IRRI, 1988). Each line was scored for tolerance immediately after desubmergence using a scale of 1 to 9 (1 = all plants survive and 9 = all plants completely dead).

2. Total plant submergence elongation under water (TSE) was used as a measure of the increment in shoot height during submergence and calculated as the average difference in shoot height (in practice, shoot length) before and after desubmergence. Relative plant submergence elongation under submergence (RSE) was used as a measure of the shoot expansion during submergence. To compare the impact of submergence on shoot elongation, the extension in height in each individual line was set to be 100 percent under nonsubmerged conditions and elongation growth underwater compared with this value. The resulting value is termed RSE. Values of RSE smaller than 100 indicate suppression of growth under submergence, whereas values of RSE above 100 indicate faster growth under water. RSE = (elongation growth under submergence × 100)/elongation growth under nonsubmerged conditions.
3. Leaf senescence (LS) is characterized by dramatic yellowing resulting from chloroplast degeneration. Submergence-tolerant plants are able to retain green leaves for longer than intolerant lines when under water. A chlorophyll meter was used to measure leaf senescence based on the amount of chlorophyll in the leaves. Ten randomly selected youngest leaves for each individual line are normally used as samples. The measurements were made at the base, middle, and tip of each leaf and average LS values of 30 readings were used.

GENOME MAPPING AND QTL ANALYSIS

Efforts to isolate genes governing flooding tolerance have been reported in soybean (Van Toai et al., 2001) and rice (Xu and Mackill, 1996; Nandi et al., 1997; Xu et al., 2000; Siangliw et al., 2003; Toojinda et al., 2003). In the case of waterlogging, two RIL populations have been used, Archer × Minsoy and Archer (Siangliw et al., 2003; Toojinda et al., 2003). In soybean, improved plant growth in waterlogging conditions has great impact on final grain yield and was used as a criterion for mapping (Van Toai et al., 2001). In flash flooding conditions, recovery ten days after desubmergence has been the most important trait for lowland rice (Xu and Mackill, 1996; Nandi et al., 1997; Xu et al., 2000; Siangliw et al., 2003; Toojinda et al.,

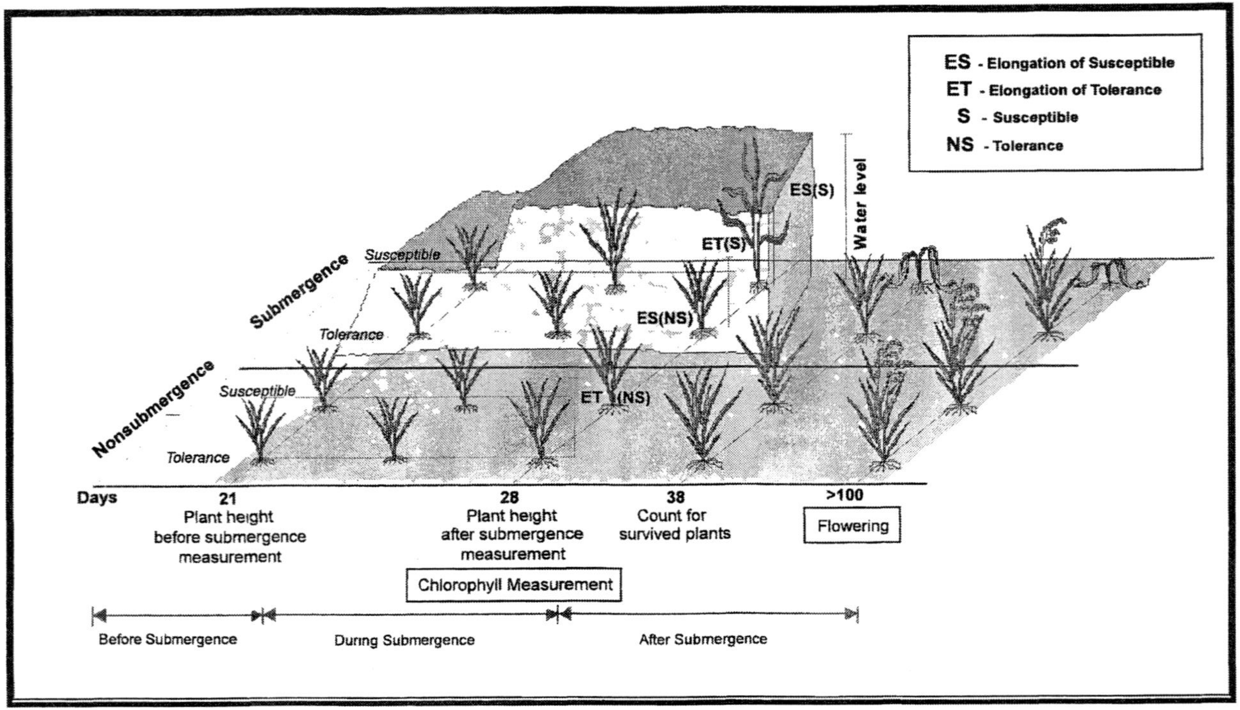

FIGURE 7.1. Design of experiment for flooding tolerance taking rice as example (see color figures section).

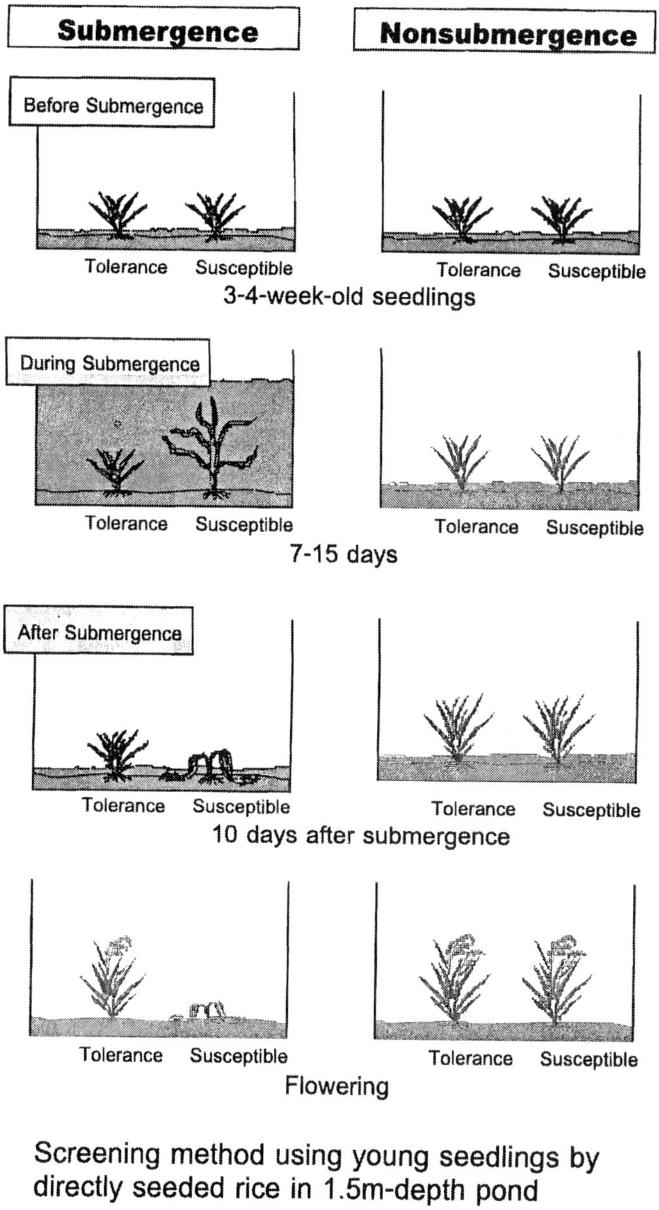

FIGURE 7.2. Differential responses between tolerance and susceptible genotypes during seedling and flowering in rice (see color figures section).

2003). Nonetheless, QTL analysis has also focused deeply on correlation traits such as fast underwater elongation and stay green in order to shed light on the mechanism underlying submergence tolerance in rice (Siangliw et al., 2003; Toojinda et al., 2003). In stagnant flooding, fast elongation was reported as the most important survival mechanism (Sripongpangkul et al., 2000).

To link physiological to genetic studies, genetic control of physiological traits contributing to the final plant performance must be considered. QTL analysis is a powerful technique to dissect a quantitative trait into several discrete Mendelian factors and associate them with specific genomic locations. Strong phenotypic correlation among traits associated with tolerance to submergence stress such as plant survival, total plant elongation, and leaf senescence were observed in various mapping populations used to study submergence-tolerant QTLs (Xu and Mackill, 1996; Nandi et al., 1997; Xu et al., 2000; Toojinda et al., 2003). It is apparent that there is a common mechanism for submergence tolerance in rice. To date, most of the data suggest that fast plant elongation when submerged and the associated leaf senescence are direct physiological causes of plant injury resulting from submergence. However, the molecular basis of submergence tolerance has not yet been completely resolved.

Understanding the responses and state of the plants before, during, and after submergence is a prerequisite for understanding the physiological basis and determining the genomic location of genes associated with submergence tolerance. With advances in molecular biology and computational analysis, visual observation of symptoms caused by submergence can be used to identify genes associated with tolerance via the QTL analysis approach. Using this approach, genomic segments called QTLs, conferring characteristics associated with submergence tolerance, have been reported on chromosomes 1, 2, 5, 6, 7, 9, 11, and 12 (Xu and Mackill, 1996; Nandi et al., 1997; Xu et al., 2000; Siangliw et al., 2003; Toojinda et al., 2003) (Figures 7.3a and b). Of these, the major QTL explained more than 80 percent of explainable genetic variation and has been shown to reside on chromosome 9. It was designated a *SUB1* locus by Xu and Mackill (1996). In recent years, many thousands of DNA markers have been mapped onto the rice genome based mainly on restriction fragment length polymorphisms (RFLPs), simple sequence length polymorphisms (SSLPs), amplified fragment length polymorphisms (AFLPs), and expressed sequence tags (ESTs) (McCouch et al., 1988; Kurata et al., 1994). These have played an essential part in highly refining the location of the tolerant genes (Xu et al., 2000; Kamolsukyunyong et al., 2001). The molecular markers (i.e., short lengths of genomic DNA with sequence homology to labelled probes of known sequence) that lie very close to the *SUB1* locus have been identified.

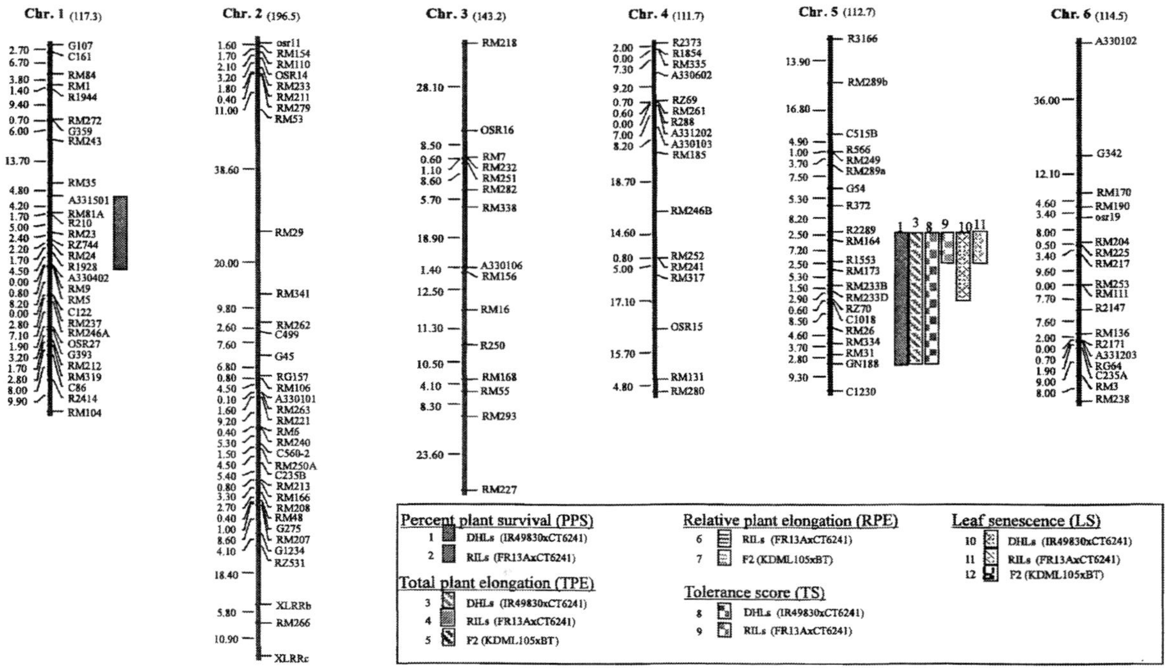

FIGURE 7.3a. QTL mapping for flash flooding tolerance in rice illustrates few minor QTLs located on chromosomes 1 and 5 (see color figures section).

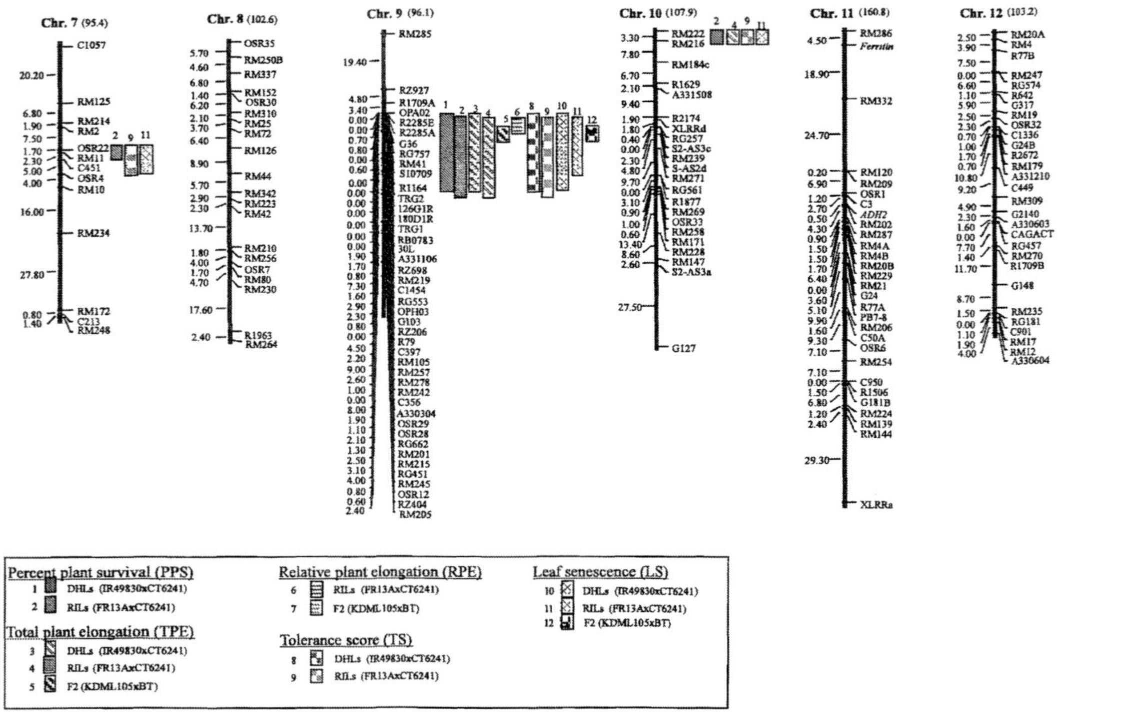

FIGURE 7.3b. QTL mapping for flash flooding tolerance in rice illustrates the dominant major QTL located on chromosome 9 and a few minor QTLs located on chromosomes 7 and 10 (see color figures section).

The role of secondary QTLs in submergence tolerance had been reported in a few cases. The contributions to submergence tolerance of secondary QTLs have been reported on chromosomes 1, 2, 5, 6, 7, 10, 11, and 12 (Xu and Mackill, 1996; Nandi et al., 1997; Toojinda et al., 2003; Xu et al., 2000). These QTLs are detected in particular environments and have high genotypic × environment interaction. To understand the role of secondary QTLs, development of near-isogenic lines will be necessary. Such populations would also provide useful information about epistatic interactions and clarify the true nature of epistatic interaction among the various QTLs.

A physical map (map-based positional cloning) covering the *SUB1* locus has now been constructed using large insert libraries based on bacterial artificial chromosomes (BACs) and yeast artificial chromosomes (YACs) (Kamolsukyunyong et al., 2001). Ultimately, this approach should enable the base sequence of the locus to be deduced and allow putative functions of the genes in the region to be assigned on the basis of sequence homology (Figure 7.4).

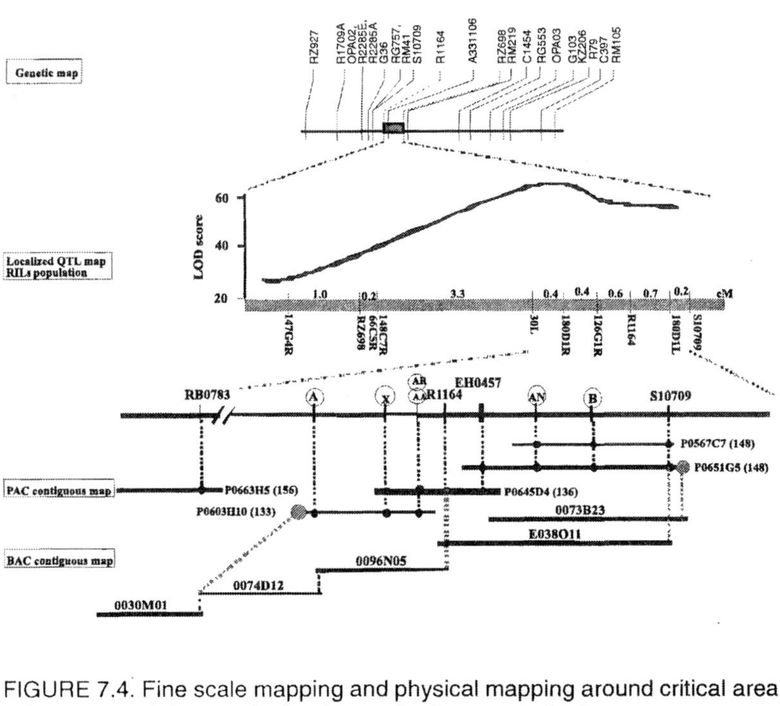

FIGURE 7.4. Fine scale mapping and physical mapping around critical area for submergence tolerance in rice (see color figures section).

BREEDING STRATEGIES FOR SUBMERGENCE TOLERANCE

Although many physiological studies on the mechanisms of flooding tolerance have been reported, there are few breeding lines for flooding tolerance that have been successful. The highly flooding-tolerant varieties reported to date from conventional breeding are usually not as productive as susceptible ones under normal circumstances because of genetic drag from landrace donors. Based on extensive breeding at IRRI, a few successes in breeding for submergence tolerance have been reported, but advanced lines with strong submergence tolerance have yet to be developed (Mackill et al., 1996). These improved lines should have desirable agronomic features and show no negative correlation between yield and submergence tolerance. Reasons for the slow rate of improvement of flooding-tolerant lines using conventional breeding approaches include: (1) highly tolerant cultivars normally possess undesirable agronomic characteristics such as awns, shattering, low yield, and poor cooking quality, (2) although heritability of the plant survival trait is high, secondary traits such as plant elongation and leaf senescence have a high genotypic × environment interaction, and (3) rice breeders are not only selecting for high submergence tolerance but also for other abiotic stress tolerance and biotic stress resistance. Selection for multiple traits using conventional breeding can be accomplished, but is laborious and time consuming. To date, much molecular data of the rice genome is available. Therefore, molecular marker technology should be a method of choice to overcome these constraints of conventional breeding.

DNA Marker-Assisted Selection (DMAS)

Recently, the genomic locations of QTLs conferring submergence tolerance have been reported (Xu and Mackill, 1996; Nandi et al., 1997; Xu et al., 2000; Siangliw et al., 2003; Toojinda et al., 2003). This information coupled with an advance in molecular biology will enhance and increase the efficiency of development of submergence-tolerant cultivars through breeding and selection processes. The objective is to select submergence-tolerant offspring based on the markers closely linked to the submergence-tolerant genes or the candidate gene markers. Therefore, new submergence-tolerant lines can be efficiently developed (Mackill et al., 1999; S. Sarkarung and J. Prayongsap, personal communication). The success of transferring the submergence tolerance trait to commercial rice cultivars via the marker-assisted selection (MAS) approach has been demonstrated in new submergence-tolerant lines derived from the well-known Thai jasmine rice 'Khao Dawk Mali 105 (KDML 105)' (Siangliw et al., 2003).

Adding submergence tolerance into elite rice cultivars can be done by conventional backcross breeding. When markers tightly linked to the submergence tolerance gene/QTL from the donor parent are available, the heterozygous genotype can be visualized from the homozygous dominant/recessive genotype at the marker locus (Figure 7.5). By using marker-assisted backcrossing, rice plants possessing submergence tolerance can be selected and undesirable regions surrounding the linkage drag to be transferred to the next generation can be minimized. The ability to recover the rest of the recurrent genome in a few backcrosses is the most important target for backcross breeding. In conventional backcrossing, the rate of donor genome replacement is approximately half for each round of backcrossing. Different locations may be replaced at different rates. At the end of six rounds backcrossing, 99 percent of the donor genome is replaced. In marker-assisted backcrossing, breeders can recover 99 percent of the recurrent genome within two or three rounds of marker-assisted backcrossing when three or four untargeted marker loci on each chromosome are checked. The success of a marker-assisted selection program depends mainly on the precision of genome location and how tightly linked DNA markers are with the gene conferring submergence tolerance. This limitation can be overcome by positional cloning of the *Sub1* gene and using the candidate gene marker to select the rice plants.

FLOODING TOLERANCE AT THE MOLECULAR LEVEL

Anaerobic Induction of Protein Synthesis

When oxygen is depleted from a growing environment, each cell depresses the synthesis of most normal proteins and stimulates a new set of polypeptides (Sachs et al., 1996). In maize primary roots, a set of 20 new polypeptides is induced at the expense of the normal protein synthesis when the root environment is shifted from aerobic to hypoxic conditions (Sachs et al., 1980, 1996; Bailey-Serres et al., 1988). Most of the anaerobic proteins are enzymes involved in glycolysis and alcoholic fermentation (Walker, Llewellyn, et al., 1987; Walker, Howard, et al. 1987; Kelly, 1989; Mahajan et al., 1989; Ricard et al., 1989; Umeda and Uchimiya, 1994). The rest of the induced proteins were involved in specific functions, such as aerenchyma formation, senescence, stem elongation, and cell division.

Hypoxia induces two enzymes of alcoholic fermentation, alcohol dehydrogenase (ADH) and pyruvate decarboxylase (PDC). Several enzymes involved in starch degradation and glycolysis are detected in maize roots during hypoxia, such as sucrose synthase (SS), alpha amylase, aldolase, enolase, glucose phosphate isomerase (GPI), glyceraldehyde phosphate dehydrogenase (GAPDH), pyrophosphate-dependent phosphofructokinase

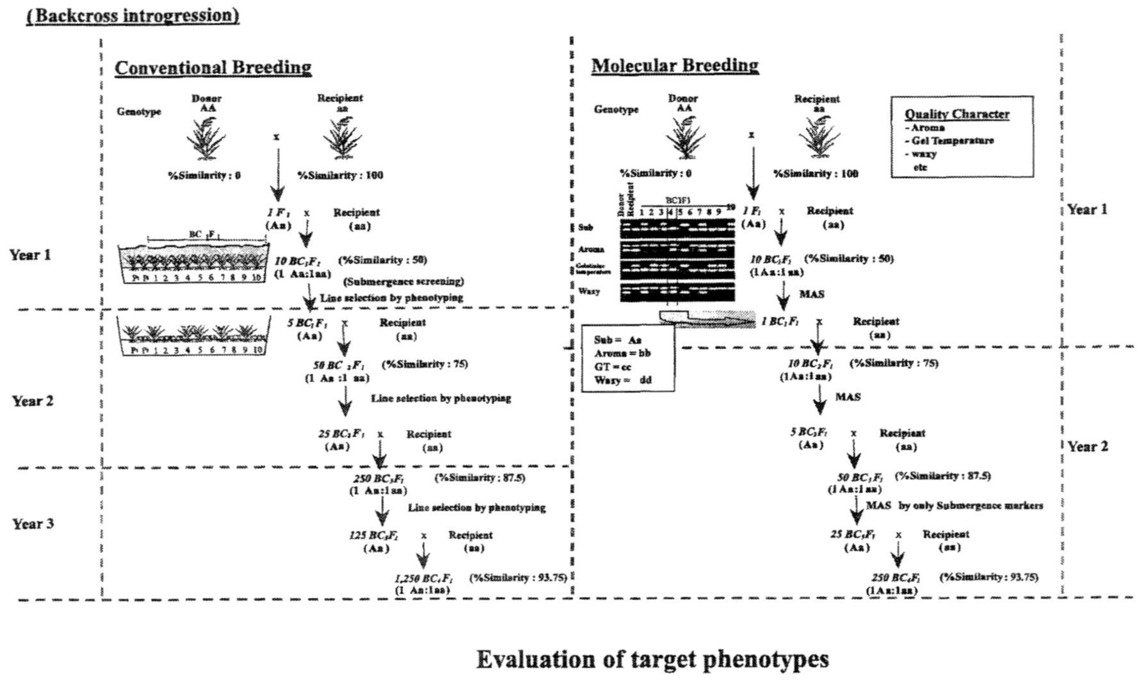

If hybridization can produce 10 F1 seeds in each backcrossing and A is completely dominant.

FIGURE 7.5. Molecular-assisted backcrossing (MAB) for submergence tolerance in rice (see color figures section).

(PPFK), and pyruvate orthophosphate dikinase (PPDK) (Walker, Llewellyn, et al., 1987; Kelly, 1989; Ricard et al., 1989; Mertens et al., 1993; Sachs et al., 1996; Moons et al., 1998). A similar shift in normal protein synthesis was also observed in tomato (Tanksley and Jones, 1981), rice (Reggiani et al., 1990), barley (Hoffman et al., 1986), pea (Llewellyn et al., 1987), and soybean (Tihanyi et al., 1989).

Enzymes of ethylene biosynthesis are also reported to be induced by the depletion of oxygen. The two enzymes, 1-amonocyclopropane-1-carboxylate synthase (ACCS), which converts the S-adenosyl methionine (SAM) to ACC, and ACC oxidase, which mediates the last step to ethylene, were stimulated in hypoxic conditions in rice (Mekhedov and Kende, 1996) and in maize root tips (Sachs et al., 1996). Anoxic conditions, however, inhibited ACC oxidase but not ACC synthase (Sachs et al., 1996).

During rapid elongation, deepwater rice acquires a new set of proteins presented in the epidermal layer of internodes and differentiating vascular tissues called expansins (Cho and Kende, 1997a,b, 1998). Expansins are thought to disrupt the hydrogen bonds linking cellulose and other wall polymers (McQueen-Mason and Cosgrove, 1995; Cosgrove, 1996). These wall-loosening proteins promote long-term extension of isolated cell walls (McQueen-Mason et al., 1993; Cosgrove and Durachko, 1994). The causal link between expression of expansins and flooding could be the cytosolic pH. Expansins are normally active at low pH; a similar change is detected when cells face hypoxia. In *Rumex* species, the elongating *R. palustris* contains at least six expansins, while the nonelongating *R. acetosa* contains five copies of the wall-loosening genes (Cho and Kende, 1997c; Vriezen et al., 2000). *Rp-EXP1*, one of the expansins from *R. palustris*, was abundantly expressed in both petiole and leaves, in particular the youngest leaves, upon submergence (Vriezen et al., 2000). The acid-induced wall loosening responds not only to low pH but also to elevated ethylene trapped in leaf tissues in *Rumex* species (Voesenek et al., 1997). However, unlike the cases of elongating *R. palustris* and rice, neither ethylene nor submergence induced expansin in petioles and leaves of *R. acetosa*, the flood-susceptible *Rumex* (Vriezen et al., 2000). In addition to ethylene, auxin seems to play the dominant role in acid-induced wall loosening because growth induction was measurable within 10 min upon treatment with auxin alone (Rayle and Cleland, 1992). Moreover, expansin plays important roles in photomorphogenesis and hypocotyl growth, the two developmental changes regulated by auxin in tomato (Caderas et al., 2000). In internodal cells in deepwater rice, evidence supports gibberellin as the ultimate factor mediating flood-induced elongation (Hoffman-Benning and Kende, 1992; Raskin and Kende, 1984a). Therefore, expression of expansins is up-regulated in response to flooding via increased ethylene, gibberellin, and auxin, the hormones that play important roles in acid-induced wall loosening.

Identification of Genes Underlying Flooding Tolerance

New rice varieties that are tolerant of stresses, responsive to production inputs, and highly nutritious are the most desirable for the next century. In the case of rice, conventional breeding still plays important roles in generating new cultivars. Therefore, technology that can improve breeding efficiency will have a great impact on developing new strains of rice. In the past few years, molecular markers and copies of expressed genes have been identified and mapped to segregating populations. With those technologies, genes responsible for biotic and abiotic stress resistance, yield components, and cooking quality have been mapped in several chromosomes. Innovative breeding programs, collectively called marker-assisted selection, that utilize tightly linked markers to genes of interest have not been implemented successfully in every case. One of the constraints has been the use of molecular markers that are not tightly linked to the responsible genes. In addition, when the QTLs are governed by many loci, marker-assisted selection becomes less effective. Also, cost-effectiveness for marker-assisted selection is still not attractive if genes underlying the QTL are not directly used. In addition, the reproductive barrier between cultivated rice and wild relatives is an important constraint to exploit germplasm for selective breeding and map-based cloning. A map-based approach has allowed scientists to discover a few genes at a time.

Most genetic traits important to agriculture or human diseases are manifested as observable, quantitative phenotypes called quantitative trait loci (QTL). In many instances, the complexity of the phenotype/genotype interaction and the general lack of clearly identifiable gene products render the direct molecular cloning approach ineffective, thus additional strategies such as genome mapping are required to identify the QTL in question. Genome mapping requires no prior knowledge of the gene function but utilizes statistical methods to identify the most likely gene location. To completely characterize genes of interest, the initially mapped region of a gene location will be narrowed down to a size that is suitable for cloning and sequencing. Strategies for gene identification within the critical region have to be applied after the sequencing of a potentially large clone or set of clones that contain this gene (Vanavichit et al., 2001). Tremendous successes of positional cloning have been shown for cloning genes responsible for human diseases, including cystic fibrosis and muscular dystrophy, as well as plant disease resistance genes (Martin et al., 1993; Bent et al., 1994; Grant et al., 1995; Song et al., 1995).

Rice has become an attractive model for map-based cloning for several reasons. Rice has not only been used extensively for QTL analysis but also for physical mapping. Physical mapping based on YAC, BAC, P1-derived artificial chromosomes (PAC), and fingerprinted contigs (FPC) are publicly available. Recently, rice has been used for whole genome sequencing. Rice also has established available transformation systems that allow functional complementation by transformation. The availability of the major QTL for flooding tolerance mapped onto a 6 cM subcentromeric region on chromosome 9 in rice, which has been one of the largest QTLs ever reported (Toojinda et al., 2003), is ideal for understanding how genes function. With the available physical mapping tools such as FPC, YAC, and BAC clones, the major QTL for submergence tolerance (located in the 6.8 cM interval flanked by two RFLP markers, S10709 and RZ 698) was narrowed down to as small an interval as 0.1 cM (Kamolsukyunyong et al., 2001). The average physical to genetic distance of 170 kb per cM was surprisingly high considering that the sequencing area is proximal to the centromere.

To learn how the genes controlling the QTL regulate submergence tolerance, it is necessary to sequence the whole genomic region and look for genes of interest. The 1.5 Mb finished sequences from four PAC and five BAC clones encompassing the flooding tolerance on rice chromosome 9 were recently reported (http://www.ncbi.nih.gov/GenBank/index.html). Within the first megabase region, more than 80 genes were predicted and confirmed by homology search. High gene density with small physical to genetic distance makes this subcentromeric region particularly gene-rich and a hotspot for recombination. One feature that might contribute to this is the high number of retrotransposon flanking genes. In total, ten genes are hypothetical proteins, nine genes have no match, five genes are transcription factors, and seven genes play roles in signal transduction (Tables 7.1 and 7.2). Neither genes encoding enzymes involved in fermentation pathways such as ADH and PDC, nor genes of the glycolytic pathway were present in the QTL region. Regulatory genes are clustered into two distinct subgenomic regions surrounded by a large number of retrotransposons forming two gene-dense islands. These two gene islands are essential for maintaining submergence tolerance. Genes located on the first island are involved in G-protein signal transduction, such as a hypothetical protein containing G-protein coupled factor, a Gamma response I protein, and a small Ras guanosine triphosphate (GTP)-binding protein. On the second island, two EREBP, ethylene response element binding proteins, a GTPase regulator, and transcription factors containing zinc-finger motif and WRKY are located.

The possible roles of these regulatory proteins in submergence tolerance are elucidated in the model (Figure 7.6). At least in the case of submergence tolerance, genes underlining the QTL play important roles in signal transduction pathways that involve the plant hormones, gibberellin, ethylene, ABA, and auxin.

TABLE 7.1. Gene products with similarity to those identified in the submergence region.

Functional group	Predicted genes	BLASTP match	Functional domain	Organism	Genebank accession no.
Hormonal signal transduction	1. D04.1	Gamma response I	G-path	*Arabidopsis*	T49091
	2. D04.2	Ras-related GTP-binding protein, *rgp1*	P-loop, Ras family ARF, CAAX box	Rice	X59276
	3. D04.4	Hypothetical protein	GPCR, TM protein	*Arabidopsis*	AC011437
	4. H05.1	Putative to unknown protein (GTPase regulator)	Zinc finger, C3HC4	Rice	AC079029
Transcription factor	5. G05.1	bZIP transcription factor	BZIP	*Arabidopsis*	AB010697
	6. D04.5	Zinc finger protein	Zf-C2H2, zf-MYND	*Arabidopsis*	T06753
	7. H05.2	Hypothetical protein	WRKY protein, TM protein, Na_H_Exchanger	*Arabidopsis*	AC005287
	8. H05.3	AP2/EREBP	AP2, NLS_BP	Rice	AF190770
	9. H05.4	AP2/EREBP	AP2	Rice	AU071242
Resistance gene	10. H10.1	Putative gene resistance gene analog	LRR, NB_ARC	Rice	AP002540
Others	11. H10.3	Hypothetical protein	ARG_RICH, GLY_RICH, NLS_BP	*Arabidopsis*	AP002482
	12. G05.2	Anthranilate phosphoribosyl transferase	—	*Arabidopsis*	T46010
	13. G05.3	Hypothetical protein	TIM17	None	T00930
	14. G05.4	Hypothetical protein	RHODANESE	*Arabidopsis*	T00931

15. G05.5	Plastid-specific ribosomal protein2 precursor	ARG_RICH, mm, NLS_BP	*Spinacia oleracea*	AF240462
16. D04.3	Hypothetical protein	—	*Arabidopsis*	AY044323
17. H10.2	Hypothetical protein	SER_RICH	*Arabidopsis*	AP002484
18. H05.5	Unknown protein	—	Rice	—
19. H05.6	Unknown protein	—	Rice	—
20. H05.7	Unknown protein	—	Rice	—

TABLE 7.2. Predicted genes with their functional domains from genomic sequence where the major QTL was identified on rice chromosome 9.

Domain type	Gene family function	Promoter element	Genetically characterized factor
MYB	Secondary metabolism, cellular morphogenesis, signal transduction in plant growth, abiotic and biotic stress responses, circadian rhythm, and dorsoventrality	MRE-like sequences	*AtMYB2, ATR1, CCA1, CPC, GL1, LHY, WER*
AP2/EREBP	Flower development, cell proliferation, secondary metabolism, abiotic and biotic stress responses, ABA response, and ethylene response	GCC box	*ABI4, ANT, AP2, CBF1-3/DREB1A-C, DREB2A, ERF1*
BZIP	Seed-storage gene expression, photomorphogenesis, leaf development, flower development, defense response, ABA response, and GA biosynthesis	G-box, Dof, OCS	*ABI5, HY5, PAN*
Z-C_2H_2	Flower development, flowering time, seed development, and root nodule development	GARE	*FIS2, SUP, HRT*
WRKY	Defense response	W Boxes	*TDBA12*

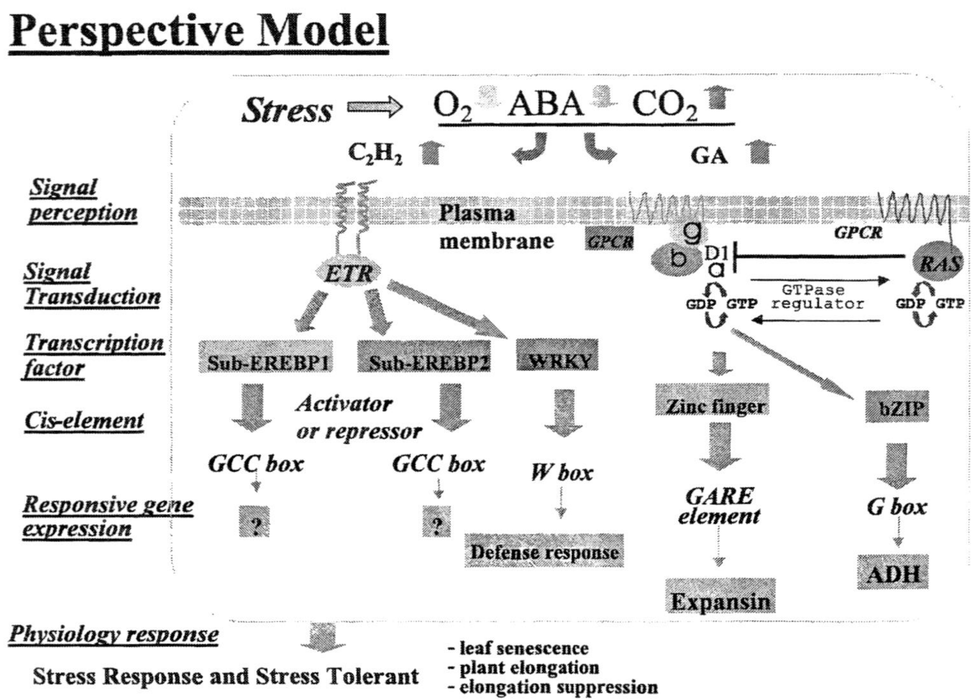

FIGURE 7.6. Perspective model depicting how GPCR, RAS, GTPase regulator, EREBPs, and two transcription factors, candidate genes that mediate submergence-induced elongation and leaf senescence interact with responsive genes such as ADH, PDC, expansin, etc. (see color figures section).

REFERENCES

Adkins, S.W., T. Shiraishi, and J.A. McComb (1990). Submergence tolerance of rice: A new glasshouse method for experimental submergence of plant. *Physiologia Plantarum* 80:642-646.

Alpi, A. and H. Beevers (1983). Effect of O_2 concentration on rice seedlings. *Plant Physiology* 71:30-34.

Avadhani, P.N., H. Greenway, R. Lefroy, and L. Prior (1978). Alcoholic fermentation and malate metabolism in rice germinating seeds at low oxygen concentrations. *Australian Journal of Plant Physiology* 5:15-25.

Bailey-Serres, J., B. Kloeckener-Gruissem, and M. Freeling (1988). Genetic and molecular approaches to the study of the anaerobic response and tissue-specific gene expression in maize. *Plant Cell and Environment* 11:351-357.

Bent, A.F., B.N. Kunkel, D. Dahlbeck, K.L. Brown, and R. Schinidt (1994). *RPS2* of *Arabidopsis thaliana:* A leucine-rich repeat class of plant disease resistance gene. *Science* 265:1856-1860.

Caderas, D., M. Muster, H. Vogler, T. Mendel, J.K.C. Rose, S.J. McQueen-Mason, and C. Kuhlemeier (2000). Limited correlation between expansin gene expression and elongation growth rate. *Plant Physiology* 123:1399-1413.

Catling, D. (1992). *Rice in Deep Water.* London, England: Macmillan.

Cho, H. and H. Kende (1997a). Expansins and internodal growth of deepwater rice. *Plant Physiology* 113:1145-1151.

Cho, H. and H. Kende (1997b). Expansins in deepwater rice internodes. *Plant Physiology* 113:1137-1143.

Cho, H. and H. Kende (1997c). Expression of expansin genes is correlated with growth in deepwater. *Plant Cell* 9:1661-1671.

Cho, H. and H. Kende (1998). Tissue localization of expansins in deepwater rice. *Plant Journal* 15:805-812.

Cosgrove, D.J. (1996). Plant cell enlargement and the action of expansins. *BioEssays* 18:533-540.

Cosgrove, D.J. and D.M. Durachko (1994). Autolysis and extension of isolated walls from growing cucumber hypocotyls. *Journal of Experimental Botany* 45:1711-1719.

Drew, M.C., C.J. He, and P.W. Morgan (2000). Programmed cell death and aerenchyma formation in roots. *Trends in Plant Science* 5:123-127.

Emes, M.J., C.P. Wilkins, P.A. Smith, K. Kupkanchanakul, K. Hawker, W.A. Charlton, and E.G. Cutter (1988). Starch utilization by deepwater rices during submergence. In *International Deepwater Rice Workshop.* Manila, The Philippines: International Rice Research Institute, pp. 319-326.

Grant, M.R., L. Godiard, E. Straube, T. Ashfield, and J. Lewald (1995). Structure of the *Arabidopsis RPM1* gene enabling dual specificity disease resistance. *Science* 296:843-846.

Greenway, H. and T.L. Setter (1995). *Text for Experimental Workshop on Adaptation to Waterlogging and Submergence.* Prachinburi, Thailand: Prachinburi Rice Research Center.

Hamamura, K. and T. Kupkanchanakul (1979). Inheritance of floating ability in rice. *Japan Journal Breeding* 29:211-216.

Haque, Q.A., D. Hille Ris Lambers, N.M. Tepora, and N.M. de la Cruz (1989). Inheritance of submergence tolerance in rice. *Euphytica* 41:247-251.

Hoffman, N.E., A.F. Bent, and A.D. Hanson (1986). Induction of lactate dehydrogenase isozymes by oxygen deficits in barley root tissue. *Plant Physiology* 82:658-663.

Hoffmann-Benning, S. and H. Kende (1992). On the role of abscisic acid and gibberellin in the regulation of growth in rice. *Plant Physiology* 99:1156-1161.

IRRI (1988). *Standard Evaluation System for Rice,* Third Edition. Manila, The Philippines: International Rice Research Institute, pp.1-54.

Jackson, M.B. (1956). Flooding injury studied by approach graft and split root system techniques. *American Journal of Botany* 43:496-502.

Jackson, M.B. (1985). Ethylene and responses of plant to soil waterlogging and submergence. *Annual Review of Plant Physiology* 36:145-174.

Jackson, M.B. (1989). Regulation of aerenchyma formation in roots and shoots by oxygen and ethylene. In *Cell Separation in Plants: Physiology, Biochemistry, Molecular Biology,* D.J. Osborn and M.B. Jackson (Eds.). Berlin, Germany: Springer, pp. 263-274.

Jackson, M.B. and W. Armstrong (1999). Formation of aerenchyma and the processes of plant ventilation in relation to soil flooding and submergence. *Plant Biology* 1:274-287.

Jackson, M.B., I. Waters, T.L. Setter, and H. Greenway (1987). Injury to rice plants caused by complete submergence: A contribution by ethylene. *Journal of Experimental Botany* 38:1826-1838.

Justin, S.H.F.W. and W. Armstrong (1991). Evidence for the involvement of ethylene in aerenchyma formation in adventitious roots of rice (*Oryza sativa* L.). *New Phytologist* 118:49-62.

Kamolsukyunyong, W., V. Ruanjaichon, M. Siangliw, S. Kawasaki, T. Sasaki, A. Vanavichit, and S. Tragoonrung (2001). Mapping of quantitative trait locus related to submergence tolerance in rice with aid of chromosome walking. *DNA Research* 8:63-171.

Karin, S., B.S. Vergara, and A.M. Mazaredo (1982). Anatomical and morphological studies of rice varieties tolerant of and intolerant to complete submergence at the seedling stage. In *Proceedings of the 1981 International Deepwater Rice Workshop.* Los Banos, The Philippines: International Rice Research Institute, pp. 285-291.

Kawai, M., P.K. Samarajeewa, R.A. Barrero, M. Nishiguchi, and H. Uchimiya (1998). Cellular dissection of the degradation pattern of cortical cell death during aerenchyma formation of rice roots. *Planta* 204:277-287.

Kelly, P.M. (1989). Maize pyruvate decarboxylase mRNA is induced anaerobically. *Plant Molecular Biology* 13:213-222.

Kende, H., E. van der Knaap, and H.T. Cho (1998). Deepwater rice: A model plant to study stem elongation. *Plant Physiology* 118:1105-1110.

Kurata, N., Y. Nagamura, K. Yamamoto, Y. Harushima, N. Sue, J. Wu, B.A. Antonio, A. Shomura, T. Shimizu, S.-Y. Lin, et al. (1994). A 300 kilobase interval genetic map of rice including 883 expressed sequences. *Nature Genetics* 8:365-372.

Llewellyn, D.J., E.J. Finnegan, J.G. Ellis, E.S. Dennis, and W.J. Peacock (1987). Structure and expression of an alcohol dehydrogenase I gene from *Pisum sativum* (cv. Greenfast). *Journal of Molecular Biology* 195:115-123.

Mackill, D.J. (1986). Varietal improvement for rainfed lowland rice in South and Southeast Asia: Result of a survey. In *Progress in Rainfed Lowland Rice.* Manila, The Philippines: International Rice Reseach Institute, pp. 115-144.

Mackill, D.J., M.M. Amante, B.S. Vergara, and S. Sarkarung (1993). Improved semi-dwarf rice lines with tolerance to submergence of seedlings. *Crop Science* 33:749-753.

Mackill, D.J., W.R. Coffman, and D.P. Garrity (1996). *Rainfed Lowland Rice Improvement.* Manila, The Philippines: International Rice Reseach Institute.

Mackill, D.J., H.T. Nguyen, and J. Zhang (1999). Use of molecular markers in plant improvement programs for rainfed lowland rice. *Field Crops Reseach* 64:177-185.

Mahajan, R., R. Randir-Singh, and R. Singh (1989). Properties of pyrophosphate-dependent phosphofructokinase from endosperm of developing wheat (*Triticum aestivum* L.). *Plant Physiology* 91:424-426.

Martin, G.B., S.H. Brommonschenkel, J. Chunwongse, A. Frary, and M.W. Ganal (1993). Map-based cloning of a protein kinase gene conferring disease resistance in tomato. *Science* 262:1432-1436.

Mazaredo, A.M. and B.S. Vergara (1982). Physiological differences in rice varieties tolerant and susceptible to complete submergence. In *Proceedings of the 1981 International Deepwater Rice Workshop.* Manila, The Philippines: International Rice Research Institute, pp. 327-341.

McCouch, S.R., G. Kochert, Z.H. Yu, Z.Y. Wang, G.S. Khush, W.R. Coffman, and S.D. Tanksley (1988). Molecular mapping of rice chromosome. *Theoretical and Applied Genetics* 76:815-829.

McQueen-Mason, S.J. and D.J. Cosgrove (1995). Expansin mode of action on cell walls. *Plant Physiology* 107:87-100.

McQueen-Mason, S.J., S.C. Fry, D.M. Durachko, and D.J. Cosgrove (1993). The relationship between xyloglucan endotransglycosylase and in-vitro cell wall extension in cucumber hypocotyls. *Planta* 190:327-331.

Mekhedov, S.L. and H. Kende (1996). Submergence enhances expression of a gene encoding 1-aminocyclopropane-1-carboxylate oxidase in deepwater rice. *Cell Physiology* 37:531-537.

Mertens, E., J. De Jonckheere, and F. Van Schaftingen (1993). Pyrophosphate-dependent phosphofructokinase from the amoebae *Naegleria fowler,* and AMP-sensitive enzyme. *Biochemistry Journal* 292:797-803.

Metraux, J.P. and H. Kende (1983). The role of ethylene in the growth response of submerged deep water rice. *Plant Physiology* 72:441-446.

Metraux, J.P. and H. Kende (1984). The cellular basis of the elongation response in submerged deep-water rice. *Planta* 160:73-77.

Mishra, S.B., D. Senadhira, and N.L. Manigbas (1996). Genetics of submergence tolerance in rice (*Oryza sativa* L.). *Field Crops Research* 46:177-181.

Mohanty, H.K. and G.S. Khush (1985). Diallel analysis of submergence tolerance in rice (*Oryza sativa* L.). *Theoretical and Applied Genetics* 70:467-473.

Moons, A., R. Valcke, and M. van-Montagu (1998). Low-oxygen stress and water deficit induce cyclosolic pyruvate orthophosphodikinase (PPDK) expression in roots of rice, a C_3 plant. *Plant Journal* 15:89-98.

Musgrave, A., M.B. Jackson, and E. Ling (1972). *Callitriche* stem elongation is controlled by ethylene and gibberellin. *Nature New Biology* 238:93-96.

Musgrave, A. and J. Walters (1973). Ethylene-stimulated growth and auxin transport in *Ranunculus scelerates* petioles. *New Phytologist* 72:783-789.

Nandi, S.P., K. Subudhi, D. Senadhira, N.L. Manigbas, S. Sen-Mand, and N. Huang (1997). Mapping QTLs for submergence tolerance in rice by AFLP analysis and selective genotyping. *Molecular and General Genetics* 255:1-8.

Palada, M.C. and B.S. Vergara (1972). Environmental effects on the resistance of rice seedlings to complete submergence. *Crop Science* 12:209-212.

Plaxton, W.C. (1996). The organization and regulation of plant glycolysis. *Annual Review of Plant Physiology and Plant Molecular Biology* 47:185-214.

Raskin, I. and H. Kende (1984a). Effect of submergence on translocation, starch content and amylolytic activity in deep-water rice. *Planta* 162:556-559.

Raskin, I. and H. Kende (1984b). Regulation of growth in stem sections of deep-water rice. *Planta* 160:66-72.

Raskin, I. and H. Kende (1984c). Role of gibberellin in the growth response of submergence deep water rice. *Plant Physiology* 76:947-950.

Rathore, T.R., M.Z.K. Warsi, N.N. Singh, and S.K. Vasal (1998). Production of maize under excess soil moisture (waterlogging) condition. In *Proceedings of the Seventh Asian Regional Maize Workshop,* S.K. Vasal, F.G. Ceniceros, and F. Xingming (Eds.). Los Banos, The Philippines: PCARRD, pp. 232-239.

Rayle, D.J. and R.E. Cleland (1992). The acid growth theory of auxin-induced cell elongation is alive and well. *Plant Physiology* 99:1271-1274.

Reggiani, R., I. Brambilla, and A. Bertani (1990). Anaerobic protein synthesis and glycosylation in rice seedlings. *Plant Physiology* 137:116-119.

Ricard, B., J. Rivoal, and A. Pradet (1989). Rice cytosolic glyceraldehyde 3-phosphate dehydrogenase contains two subunits differentially regulated by anaerobiosis. *Plant Molecular Biology* 12:131-139.

Richharia, R.H. and N.A. Parasuram (1963). Resistance to submergence in rice. *Science and Culture* 29:149-150.

Ridge, I. (1987). Ethylene and growth control in amphibious plants. In *Plant Life in Aquatic and Amphibious Habitats,* R.M.M. Crawford (Ed.). Oxford, England: Blackwell, pp. 53-76.

Sachs, M.M., M. Freeling, and R. Okimoto (1980). The anaerobic protein of maize. *Cell* 20:761-767.

Sachs, M.M., C.C. Subbainah, and I.N. Saab (1996). Anaerobic gene expression and flooding tolerance in maize. *Journal of Experimental Botany* 47:1-15.

Saha, R.P.K., D. Hillerislambers, and N.M. Tepora (1994). Genetics of stem elongation ability in rice (*Oryza sativa* L.) *Euphytica* 74:137-141.

Samarajeewa, P.K., R.A. Barrero, C. Umeda-Hara, M. Kwai, and H. Uchimiya (1999). Cortical cell death, cell proliferation, macromolecular movements and rTip1 expression pattern in roots of rice (*Oryza sativa* L.) under NaCl stress. *Planta* 207:354-361.

Setter, T.L. and E.S. Ella (1994). Relationship between coleoptile elongation and alcoholic fermentation in rice exposed to anoxia. I. Importance of treatment conditions and different tissues. *Annals of Botany* 74:265-271.

Setter, T.L., E.S. Ella, and A.P. Valdez (1994). Relationship between coleoptile elongation and alcoholic fermentation in rice exposed to anoxia. II. Cultivar differences. *Annals of Botany* 74:273-279.

Setter, T.L., M. Ellis, E.V. Laureles, E.S. Ella, D. Senadhira, S.B. Mishra, S. Sarkarung, and S. Datta (1997). Physiology and genetics of submergence tolerance in rice. *Annals of Botany* 79 (supp. A):67-77.

Setter, T.L., H. Greenway, and T. Kupkanchanakul (1989). Submergence in rice. II. Adverse effect of low CO_2 concentration. *Australian Journal Plant Physiology* 16:265-278.

Setter, T.L., M.B. Jackson, I. Waters, I. Wallace, and H. Greenway (1987). Floodwater carbon dioxide and ethylene concentrations as factors in chlorosis development and reduced growth of completely submerged rice. In *Proceedings of the 1987 International Deepwater Rice Workshop.* Los Banos, The Philippines: International Rice Research Institute, pp. 301-310.

Setter, T.L. and E.V. Laureles (1996). The beneficial effect of reduced elongation growth on submergence tolerance of rice. *Journal of Experimental Botany* 47:1551-1559.

Setter, T.L., I. Waters, I. Wallace, P. Bhhhekasut, and H. Greenway (1989). Submergence of rice. I. Growth and photosynthetic response to CO_2 enrichment of floodwater. *Australian Journal of Plant Physiology* 16:251-263.

Siangliw, M., T. Toojinda, S. Tragoonrung, and A. Vanavichit (2003). Molecular genetics of submergence tolerance in rice: Introgression of QTLs conditioning submergence tolerance. *Annals of Botany* 91:255-261.

Sinha, M.M. and S. Saran (1988). Inheritance of submergence tolerance in lowland rice. *Oryza* 25:351-354.

Song, W.Y., G.L. Wang, L.L. Chen, H.S. Kim, and L.Y. Pi (1995). A receptor kinase-like protein encoded by the rice disease resistance gene, *Xa-21*. *Science* 270:1804-1806.

Sripongpangkul, K., G.B.T. Posa, D.W. Senadhira, D. Brar, N. Huang, G.S. Khus, and Z.K. Li (2000). Genes/QTLs affecting flood tolerance in rice. *Theoretical and Applied Genetics* 101:1074-1081.

Suge, H. (1988). Physiological genetics of internodal elongation in submerged deepwater rice. In *Proceedings of the International Deepwater Rice Workshop.* Manila, The Philippines: International Rice Research Institute, pp. 29-36.

Summers, J.E. and M.B. Jackson (1998). Light- and dark-grown *Potamogeton pectinatus*, an aquatic microphyte, make no ethylene (ethene) but retain responsiveness to the gas. *Australian Journal of Plant Physiology* 25:599-608.

Supapoj, N., C. Boonwite, J.C. O'Toole, S.K. De Datta, and B.P. Jackson (1982). Using check varieties to adjust visual scores for nonuniform soil drying in drought resistance field screening. *International Rice Research Newsletter* 7(4):12.

Suprihatno, B. and W.R. Coffman (1981). Inheritance of submergence tolerance of rice (*Oryza sativa* L.). *Sabrao Journal* 13:98-108.

Tanksley, S.D. and R.A. Jones (1981). Effect of O_2 stress on tomato alcohol dehydrogenase activity: Description of a second ADH coding gene. *Biochemical Genetics* 19:397-409.

Taylor, D.L. (1942). Influence of oxygen tension on respiration, fermentation, and growth of wheat and rice. *American Journal of Botany* 29:721-738.

Thakur, R. and D. Hillerislambers (1988). Inheritance of elongation ability in rice. In *Proceedings of the International Deepwater Rice Workshop*. Manila, The Philippines: International Rice Research Institute, pp. 269-274.

Tihanyi, K., A. Fontanell, and J.P. Thirion (1989). Gene regulation during anaerobiosis in soya roots. *Biochemical Genetics* 27:719-730.

Toojinda, T., M. Siangliw, S. Tragoonrung, and A. Vanavichit (2003). Molecular genetics of submergence tolerance in rice: Quantitative traits loci (QTLs) analysis of traits associated with submergence tolerance. *Annals of Botany* 91:243-253.

Umeda, M. and H. Ushimiya (1994). Differential transcript levels of genes associated with glycolysis and alcoholic fermentation in rice plants (*Oryza sativa* L.) under submergence stress. *Plant Physiology* 106:1015-1022.

Vanavichit, A., S. Tragoonrung, and T. Toojinda (2001). Genomic mapping and positional cloning, with emphasis on plant science. In *Biotechnology*, Volume 5B: *Genomics and Bioinformatics*, C.W. Sensen (Ed.). Weinheim, Germany: Wiley-VCH, pp. 166-178.

Van Toai, T.T., S.K. St. Martin, K. Chase, G. Boru, V. Schnipke, A. Fred Schmitthenner, and K.G. Lark (2001). Identification of a QTL associated with tolerance of soybean to soil waterlogging. *Crop Science* 41:1247-1252.

Vartapetian, B.B. and M.B. Jackson (1997). Plant adaptations to anaerobic stress. *Annals of Botany* 79(supp. A):3-20.

Vergara, B.S. and A.M. Mazaredo (1975). Screening for resistance to submergence under greenhouse conditions. In *Proceedings of the International Seminar On Deepwater Rice*. Joydebpur, Dacca, Bangladesh: Bangladesh Rice Research Institute, pp. 67-70.

Voesenek, L.A.C.J. and C.W.P.M. Blom (1989). Growth responses of *Rumex* species in relation to submergence and ethylene. *Plant Cell and Environment* 12:433-439.

Voesenek, L.A.C.J., P.J.M. Perik, C.W.P.M. Blom, and M.M.A. Sessen (1990). Petiole elongation in *Rumex* species during submergence and ethylene exposure: The relative contribution of cell division and cell elongation. *Journal of Plant Growth Regulation* 9:13-17.

Voesenek, L.A.C.J., W.H. Vriezen, M.J.E. Smekens, F.H.M. Huitink, G.M. Bogemann, and C.W.P.M. Blom (1997). Ethylene sensitivity and response sensor expression in petioles of *Rumex* spicies at low O_2 and high CO_2 concentrations. *Plant Physiology* 114:1501-1509.

Vriezen, W.H., B.D. Graaf, C. Mariani, and L.A.C.J. Voesenek (2000). Submergence induces expansin gene expression in flooding-tolerant *Rumex palustris* and not in flooding-intolerant *R. acetosa*. *Planta* 210:956-963.

Walker, J.C., E.A. Howard, E.A. Dennis, and W.J. Peacock (1987). DNA sequence required for anaerobic expression of the maize alcohol dehydrogenase 1 gene. *Proceedings of the National Academy of Science, USA* 84:6624-6628.

Walker, J.C., D.J. Llewellyn, L.E. Mitchell, and E.S. Dennis (1987). Anaerobically regulated gene expression molecular adaptation of plants for survival under flooding conditions. *Plant Molecular and Cellular Biology* 4:71-93.

Waters, I., S. Morrell, H. Greenway, and T.D. Colmer (1991). Effect of anoxia on wheat seedlings. II. Influence of O_2 supply prior to anoxia on tolerance for anoxia, alcoholic fermentation, and sugar level. *Journal of Experimental Botany* 42:1437-1447.

Xu, K.N. and D.J. Mackill (1996). A major locus for submergence tolerance mapped on rice chromosome 9. *Molecular Breeding* 2:219-224.

Xu, K.N., X. Xu, P.C. Ronald, and D.J. Mackill (2000). A high-resolution linkage map of the vicinity of rice submergence tolerance locus *Sub1*. *Molecular General Genetics* 263:681-689.

Yamada, N. (1959). Physiological basis of resistance of rice plant against overhead flooding. *Bulletin of the National Institute of Agricultural Sciences, Series D (Plant Physiology, Genetics and Crops in General)* 8:1-112.

Chapter 8

Breeding for Metal Tolerance

Wilfried H. O. Ernst

INTRODUCTION

Demographic pressure by the growing human population is demanding increased food production, which is expanding into less favorable soils with naturally high concentration of aluminum (Al) and manganese (Mn) at pH <5.0. The emission of sulfur dioxide, ammonium, and other soil-acidifying compounds has enhanced the availability of chemical elements to wild and cultivated plants. Therefore, these soils are a major source of metal toxicity (Foy, 1988), which demands a solution. The scope for breeding crops for such soils is to reduce the transfer of aluminum and manganese to consumable plant parts. Industrial activities have created another toxicity problem: increased levels of heavy metals in soils and consequently in plants and food chains. If the metal level of these soils is to be decreased, the scope for breeding plants for such soils demands the opposite approach, i.e., enhancing the translocation of heavy metals into harvestable plant parts (Ernst, 1998b; Schat and Verkleij, 1998).

BREEDING FOR ALUMINUM TOLERANCE

Expansion of food production into marginal soils and increasing deposition of protons in acidic soils both mean that the plants have to be adapted to enhanced concentrations of aluminum and manganese (Foy, 1988). In contrast to manganese, which is an essential minor nutrient for all organisms, including humans (Garrow and James, 1993; Marschner, 1995), aluminum has no essential function in any organism. Aluminum toxicity is a severe problem for crop production (Clark, 1982); therefore, a considerable emphasis is placed on the selection of Al-tolerant cultivars of crop and fodder plant species (Table 8.1). In all these studies, the aluminum concentration in the target plant part was usually not analyzed, and the Al-tolerant cultivar was often not grown up to the relevant development stage. Breeding of Al-

TABLE 8.1. Examples of crops and fodder plants with selected cultivars tolerant to high levels of aluminum.

Plant species	Consumable part	References
Anthoxanthum odoratum	Shoot	Davies and Snaydon, 1973
Cucurbita pepo	Fruit	Ahn et al., 2001
Glycine max	Seed	Yang et al., 2000
Hordeum vulgare	Grain	Reid et al., 1980
Ipomoea batatas	Root	Ritchey, 1991
Lotus corniculatus	Shoot	Pedersen et al., 1986
Oryza sativa	Grain	Howeler, 1991
Phaseolus vulgaris	Fruit, seed	Foy et al., 1972
Sorghum bicolor	Leaf, grain	Flores, Gourley, et al., 1991; Flores, Clark, et al., 1991
Trifolium repens	Shoot	Caradus et al, 1991
Triticum aestivum	Grain	Slaski et al., 1996
Zea mays	Leaf, grain	Llugany et al., 1995

tolerant plants should entail the following three aspects: (1) the survival of plants in Al-enriched soil, (2) the completion of a full life cycle of those plants when the final harvestable product is grains or seeds, and (3) a strong reduction of Al accumulation in the edible plant parts.

Most research on Al tolerance has concentrated on and is still emphasizing the first aspect, the response of roots of seedlings to Al exposure (for recent reviews see Matsumoto, 2000; Barceló and Poschenrieder, 2002). It is quite obvious that avoidance mechanisms, diminishing the bioavailability by excretion of organic acids, and real tolerance after the uptake of aluminum into the cell are both involved in Al tolerance at the root cell level. Less information is available on the compartmentation and sequestration of aluminum in leaves, fruits, grains, and seeds. The second aspect, survival from seedling up to harvest of the grain or seed and the amount of harvestable biomass, are reported only rarely, for example in *Oryza sativa* (Mahadevappa et al., 1991) and *Sorghum bicolor* (Flores, Gourley, et al., 1991). Unfortunately, the third most relevant parameter, the aluminum concentration of the consumable plant part, is almost never considered. It may be expected that the aluminum concentration of the tolerant cultivars is as high as that of the sensitive cultivars, as shown for crops such as *Sorghum bicolor* (Flores, Clark, et al., 1991), wild plants such as *Chamaenerion angustifolium* (De Neeling and Ernst, 1986), and for manganese-tolerant cultivars of rice (Mahadevappa et al., 1991). Leaves contain more aluminum than grains and seeds; therefore, consumption of leafy food grown on acidic soils will bear the greatest health risk. The relationship between high Al intake and Alzheimer's disease is a warning. Thus, the final scope of the selection of Al-tolerant crops has overemphasized food productivity and neglected food security for animals and humans. It is disappointing that the breeding effort for Al-tolerant crops has failed due to insufficient consideration of the food web consequences.

HEAVY METAL CONTAMINATION AND THE DEMAND FOR PHYTOREMEDIATION

Without human impact, heavy metal–enriched soils in the vicinity of undisturbed outcrops of ore veins can be vegetated only by those plant species that have been able to evolve metal tolerance (Ernst, 1974). As a result of mining and processing of ores near the site of ore outcrops, the metal-enriched soils were disturbed and the remnants of the mining activity were piled up into spoil heaps, causing the first metal contamination in the Bronze Age (Ernst et al., 2000). With progress in metal mining, tailings were established on formerly metal-poor soils. As long as a remnant heavy metal vegetation was present in the neighborhood, highly adapted plants could colonize these sites. Historically, uncontrolled and wasteful smelting procedures have polluted all ecosystems in the northern hemisphere, as shown by metal deposits in fens, lake sediments, and the Arctic ice (Lee and Thallis, 1973; Hong et al., 1994, 1996; Becker et al., 2001). Therefore, there is no environment in the northern hemisphere with a metal background for copper, lead, and zinc that has not been increased by human activities. From the late nineteenth century onward, sites of metal smelting and processing were established far away from ore outcrops and, due to imperfect techniques, resulted in a huge soil contamination around the smelters, refineries, and other metal-processing industries up to the present day. The metal operation determines the degree of contamination and the purity of the metal being processed: high contamination of cadmium, lead, and zinc in the vicinity of a lead smelter (Gzyl, 1995), cadmium and zinc near a zinc smelter (Ernst, 1972), or copper in the immediate environment of a copper refinery (Wu and Bradshaw, 1972).

The development of metal-based pesticides was a new source of metal contamination, having its highest impact on soils of vineyards and hop agriculture (Schramel et al., 2000; Besnard et al., 2001). Electricity pylons and copper high-tension lines have contaminated soils with zinc (Al-Hiyaly et al., 1988) and copper (Kraal and Ernst, 1976), respectively. The use of lead-based petrol additives was, and is still in many countries, another source of metal contamination (Daniel et al., 1997), with an adverse effect on human health. The substitution of lead by platinum-group elements in three-way catalytic converters is not an ecologically neutral process and results in enhanced concentration of Pd, Pt, and Rh in soils and plants (Ballach, 1997). The application of Cd-contaminated phosphate fertilizer (Mordvedt, 1987) and of metal-contaminated sewage sludge onto agricultural soils (Juste and Mench, 1992; Krebs et al., 1998) has more directly affected plant, animal, and human nutrition. All these human activities, taken together, have resulted in a high diversity of metal-contaminated soils varying in element composition, metal concentration, and metal speciation influencing biological availability.

In the 1970s, an awareness of the revegetation of metal-contaminated sites (Gemmell, 1977) and the potential of metal extraction by vegetation (Ernst, 1974) did not receive public and political interest. In the early 1990s, a new consciousness of environmental organizations, governments, and finally the public has demanded a cleanup of metal-contaminated sites, often in a political overreaction after a long period of denial of the problem. It was concluded that cleaning metal-contaminated soils by physical techniques (moving vast amounts of contaminated earth and containing it within impermeable plastic liners, or electrolytic extraction) and chemical procedures (chemical leaching) was incomplete and very costly (Fox et al., 2001). Therefore, the interest in biologically based cleanup procedures has grown and is described by the term "phytoremediation."

In contrast to organic soil contaminants, which can be broken down by bacteria and fungi to CO_2 and H_2O (Alexander, 1994) with only a small amount of some remaining mineral elements such as B, Cl, I, N, P, or S, heavy metals and metalloids remain stable, owing to their character as basic elements. As soon as these metals can be extracted from the soil with the help of microorganisms or plants, the process is defined as "phytoextraction." Another aspect of managing metal-contaminated soils is revegetation of bare soils with metal-resistant plants to prevent soil erosion by water and wind and leaching of the metal to the groundwater, sometimes stimulated by soil additives (Mench et al., 1994). For both approaches of phytoremediation, the principal question to be answered is: Have we species that can do the work or have we to improve plant species prior to starting soil cleaning?

METAL RESISTANCE IN PLANTS AND DECONTAMINATION POTENTIAL

During plant evolution, several species or ecotypes of widespread plant species have evolved metal-resistant populations on soils in the vicinity of ore outcrops (Ernst, 1974). Considering only higher plants (Magnoliophyta, angiosperms), the capacity for the evolution of metal resistance and high metal accumulation (hyperaccumulation) is not present in all higher taxonomic ranks (Ernst, 1996). At the family level not all taxa, and even within a species not all populations, bear or express the genes for metal tolerance (Ernst, 2001). Plant species that are able to grow on soils enriched with heavy metals and/or metalloids are defined as metallophytes. A real metal-resistant population has to be able to complete its life cycle by seeds and to cope with the often-high amount of metals in the various plant tissues. Genetic analysis of naturally metal-tolerant plant species has shown that the metal tolerance is (1) metal specific and (2) steered by one or two genes (Macnair, 1993; Schat et al., 1996; Schat and Vooijs, 1997). Several metabolic processes are involved in the regulation of uptake and cellular compartmentalization of a metal and its translocation from the roots into the shoot (Ernst, 1998b). Consequently, it is likely that the number of genes involved in metal tolerances is underestimated. The capacity for metal uptake, transfer, and accumulation in the above-ground biomass has been known for more than a century, since the first analysis of heavy metals in plants such as *Armeria maritima* ssp. *halleri*, *Silene vulgaris* (syn. *S. cucubalus*, *S. inflata*), *Thlaspi caerulescens* (syn. *T. calaminare*, *T. alpestre* ssp. *calaminare*), and *Viola calaminaria* (Sachs, 1865). More than 100 years later, a high accumulation capacity of heavy metals by plants has been defined as hyperaccumulation by Brooks and colleagues (Brooks et al., 1977) if the maximum value found in a plant species is above 1,000 mg metal/kg. Jaffré (1980) has refined the values to metal-specific limits, now being 100 $mg \cdot kg^{-1}$ for Cd, 1,000 $mg \cdot kg^{-1}$ for Cu, and above 10,000 $mg \cdot kg^{-1}$ dry mass for Mn, Ni, and Zn.

Plants for Decontamination of Slightly to Moderately Metal-Contaminated Soils

To achieve the goal of a biological decontamination of soils, it is necessary that the plant species used in the phytoextraction approach have a high capacity (1) to take up the metals and transfer them to the harvestable above-ground plant parts, (2) to accumulate them to a high degree, and (3) to produce a high biomass within one growing season. For effective metal phytoextraction, the morphology of the root system is decisive, as already known for more than fifty years in biogeochemical ore prospecting (Hawkes and Webb, 1962). Plants can take up only those metals that are in the vicinity of their roots. Thus, a shallow root system can help to decontaminate only the upper soil layer, whereas plants with deep root systems can extract metals from more than 30 m depth. The density and activity of the root system, together with the symbiotic mycorrhizal fungi, will determine the efficiency of the metal extraction.

As mentioned previously, the levels of heavy metals in many contaminated soils are well above the advised acceptable values but far below concentrations that do serious harm to plant growth. The reduction of the metal load of these slightly to moderately contaminated soils demands only the selection of a high-yielding agricultural crop with a high metal accumulation in the shoot and a relatively high heavy metal toxicity threshold (Paschke et al., 2000). Many cultivars of maize, a highly zinc-demanding crop, and soybean are able to take up sufficient amounts of cadmium and zinc and translocate them to the shoot (Bogess et al., 1978; Hinsley et al., 1978; Florijn et al., 1993). Differentiation of plants with regard to iron uptake and translocation is well-known and offers many possibilities for tackling iron-enriched soils. All of the grasses belong to the strategy II plants and release Fe^{3+}-binding phytosiderophores into the rhizosphere and mobilize iron so that it can reach the plasma membrane. During transfer across the plasma membrane, Fe^{3+} is reduced to Fe^{2+} by a ferric-chelate reductase. Wild grasses from acidic soils scarcely release phytosiderophores, in contrast to plants from alkaline soils, thus avoiding iron toxicity (Gries and Runge, 1992). Classical breeding and selection have resulted in many Fe-efficient genotypes and ecotypes of grasses (Römheld and Marschner, 1990; Von Wirén et al., 1997) and dicots (Romera et al., 1992) that will happily grow on Fe-deficient soils but will not survive on iron-enriched soils.

Further improvement of a plant's ability to decontaminate heavy metal–enriched soils can be achieved by considering the community of organisms of the plant rhizosphere. Plant roots have an intense interaction with the root environment by excreting compounds that are useful for the metabolism of heterotrophic organisms such as bacteria and fungi. These root communities may be helpful in stimulating plant growth in metal-contaminated soils, as shown for some bacteria (Whiting et al., 2001) and many fungi. The majority of higher plant species have a symbiotic relationship with mycorrhizal fungi, which occur in most heavy metal–contaminated soils and may have evolved metal tolerances (Ietswart et al., 1992; Gucwa-Przepiora and Turnau, 2001). The evolution of metal tolerances is independent of the taxonomic class of the mycorrhizal fungi: Ascomycetes (Martino et al., 2000), Basidiomycetes (Colpaert et al., 2000), or Zygomycetes (Del Val et al., 1999). Herbs and grasses can develop an intimate symbiotic relationship with these so-called arbuscular mycorrhizal (AM) fungi belonging mostly to the class of Zygomycetes. Without any breeding, the addition of AM fungi, depending on the inoculum source, can protect the metal-sensitive plants (Dueck et al., 1986). In moderately metal-contaminated soils the addition of spores of an AM fungus, *Glomus* sp., isolated from the roots of the zinc violet, *Viola calaminaria,* and spores of a commercially available strain of *Glomus intraradices* has enhanced growth of maize nearly tenfold (Hildebrandt et al., 1999). The combination of the metal-resistant AM fungus with maize could restrict the metal translocation into the shoots. Therefore, a well-designed combination of a metal-resistant inoculum of the AM fungus with a highly productive crop species may be a helpful alternative in phytoextraction of weakly to moderately metal-contaminated soils and in phytostabilization of highly metal-contaminated soils without any breeding effort.

Another procedure is followed for diminishing a nonmetal element, i.e., selenium from Se-contaminated soils. Several plant species, especially the nonmycorrhizal members belonging to the mustard family (Brassicaceae), have the ability to volatilize selenium. In contrast to mercury, the volatilization of selenium is a natural process with a strong interdependence of plants and microorganisms (Zieve and Peterson, 1981). Due to selenium deficiency in many soils and its requirement by cattle and humans, the transport and deposition of selenium on a regional and global scale can be beneficial to animal and human health. In the past decade, many plant species from the mustard family were tested for their capability to volatilize selenium, with *Brassica juncea* as the most promising species (Zayed and Terry, 1994). The metabolic steps of the Se volatilization are well-known because of their similarity to the metabolic process for sulfur (Hanson et al., 1994), maintaining the global sulfur cycle (Steudler and Peterson, 1984). Introduction of the *Arabidopsis APS1* gene, encoding a plastidic ATP sulfurylase, leads to an overexpression, and this moderately increased Se tolerance in *B. juncea* (Pilon-Smits et al., 1999) but could not improve Se volatilization. Other transgenic lines of *B. juncea,* overexpressing several enzymes of this volatilization pathway, also were not successful. Up to now it has not been possible to construct a transgenic plant that can enhance Se volatilization (Zayed et al., 2000), obviously demonstrating the limits of transgenic plants if the selection is started from nonadapted plant species. It is very surprising that the molecular-genetic approach has not considered the high biodiversity of selenium-hyperaccumulating plant species on naturally selenium-enriched soils (Rosenfeld and Beath, 1964) and does not exploit the high ecotypic variation in Se-accumulation in *Stanleya pinnata,* a member of the Brassicaceae (Feist and Parker, 2001).

Metal Hyperaccumulators Can Decontaminate Highly Metal-Contaminated Soils

Completely different problems must be solved in the case of highly metal-contaminated soils. These occur mostly in the vicinity of metal processing industries. Here, metal contamination has affected not only the upper soil horizons (Ernst, 1972; Wu and Bradshaw, 1972; Dueck et al., 1984) but often also the groundwater (Johnson, 2000) at soil depths that cannot be explored by plant roots. In the vicinity of mining industries, giant tailings of highly metal-enriched materials with low soil fertility have often piled up over the past 500 years (Ernst, 1974). Metal-enriched ponds (O'Sullivan et al., 1999) and metal-contaminated river beds (Förstner and Wittmann, 1981) are the result of inadequate metal management. If these soils and raw materials have to be phytoextracted, then highly adapted plant species must be selected.

With the exception of the high-biomass producing nickel accumulator, *Berkheya coddii,* producing up to 20 t biomass/ha (Robinson et al., 1997), all other hyperaccumulators of one or more heavy metals have a low biomass, often combined with a rosette growth form, which is difficult to mow and thus to harvest (Ernst, 1974). Without any improvement of these plant species, it will take decades to centuries to decontaminate a highly metal-contaminated soil to background values of 0.007 mmol Cd/kg soil and 3.06 mmol Zn/kg soil (Felix, 1997), depending on the uptake capacity of the ecotypes (Table 8.2).

BREEDING FOR METAL TOLERANCE

Improving the growth form of a hyperaccumulator to a more erect habit will be more difficult than the isolation of relevant genes and their transfer into nonhyperaccumulating, high-productive plants. Therefore, genetic engineering for enhanced metal accumulation and, at the same time, maintaining metal tolerance may be a very interesting option. Several approaches have been tried by growing the genetically modified plants in agar and nutrient solution, but only a very few have been tested in field situations (see also review by Krämer and Chardonnens, 2001).

Somatic Hybridization

To overcome the negative relationship between metal (hyper)accumulation and biomass production in wild plants, attempts have been made to hybridize cells from a Zn- and Cd-hyperaccumulating 'Prayon' population of

TABLE 8.2. Phytoextraction of heavy metals from highly metal-contaminated soils by annual harvest of shoots of the Zn hyperaccumulator, *Thlaspi caerulescens*.

Site	Annual harvest (t dry mass/ha)	Soil metal concentration ($mmol \cdot kg^{-1}$)		Years for decontamination[d]		Reference
		Cd	Zn	Cd	Zn	
Dornach	13.4[a,b]	0.036	12.4	—	54	Felix, 1997
Tiefenbach	14.0[a,b]	0.059	1.5	186	—	Felix, 1997
Sala	18.2[a,c]	0.125	33.2	310	281	Knight et al., 1997
Arras	18.2[a,c]	0.211	26.3	545	243	Knight et al., 1997
Felton	18.2[a,c]	0.516	49.9	230	368	Knight et al., 1997
Les Malines	5.2[d]	1.450	581.4	15	650	Robinson et al., 1998

[a]Harvested biomass extrapolated from small-scale experiments and therefore likely to be overestimated
[b]Ecotype 'Le Prayon'
[c]Ecotype 'St. Laurent-Le-Minier'
[d]The calculation of the numbers of years for decontamination assumes a constant yearly extraction rate.

Thlaspi caerulescens with those of a nonaccumulating but highly productive cultivar, Westar, of *Brassica napus* (Brewer et al., 1999), both members of the family Brassicaceae. Owing to differences in the basic chromosome numbers ($n = 7$ in *T. caerulescens*, $n = 19$ in *B. napus*), it is not surprising that the regeneration of normal plants was quite low: 64 plants from 4024 calli, and only seven of these 64 plants were able to flower. Only two plants had a zinc tolerance level comparable to that of *T. caerulescens*, 0.05 percent of the callus population. No data have been published on the improvement of biomass production of this metal-resistant interspecies hybrid. In conclusion, this method will demand an enormous improvement of the hybridization technique if it is to become commercially interesting.

Modifying Genes for Volatilization of Metal(loid)s

Another approach is the modification of the genome of a nonaccumulating, metal-sensitive plant by the introduction of foreign genes to help the plant not only remove the metal from the soil but also avoid toxic levels in the plant itself. This genetic engineering was applied for the removal of mercury. A modified bacterial gene (mercuric ion-reductase) was introduced into *Arabidopsis thaliana* (Rugh et al., 1996), *Liriodendron tulipifera* and *Nicotiana tabacum* (Rugh et al., 1998), and *Brassica napus* (Meagher et al., 2000). The transgenic yellow poplar and tobacco plants were grown in a contaminated soil and could diminish the Hg concentration of the soil by Hg uptake and volatilization. Meagher and colleagues (2000), however, confessed that volatile mercury has a "few risks to the local environment." In contrast to selenium, phytovolatilization of a nonessential heavy metal such as mercury is *not* an environmentally sound phytoremediation approach. It is part of the old-fashioned dilution policy, which involves transferring the pollutant from the soil into the atmosphere. Therefore, it is partly comparable with the enhancement of the height of chimneys for transporting the local sulfur dioxide pollution from Central and Western Europe to Northern Europe. This technical solution was also propagated as environmentally sound, but, as it turned out later, it caused an enhanced acidification of weakly buffered ecosystems with great ecological disasters in the European forests. The argument of a low environmental risk is not a valid one because the dilution of mercury is spread on the entire Earth's surface, thus spreading pollution, not solving pollution.

Enhanced Metal Uptake and Detoxification by Genetic Engineering

An important improvement of high-biomass producing plants will be the introduction of those characters that combine an enhanced expression of high affinity metal transporters in the roots with maintenance of a high metal resistance, a strong transport into the shoot, and high metal accumulation in the leaves. Such a construction of a hyperaccumulating transgenic plant demands changes at the cellular, tissue, and whole-plant levels. The metal specificity of metal uptake and transfer by plants and yeasts is regulated by specific metal transporters. For the heavy metals essential for plant growth (Cu, Fe, Mn, Ni, Zn), only a few metal transporters have been identified so far. For the genetic engineering of plants with enhanced metal metabolism, three different steps at the cellular level have to be considered and realized:

1. Each metal is taken up from the soil solution across the plasma membrane by one or a few specific transporters that are regulated by different affinities for the external metal concentration.
2. A surplus of free metal ions in the cytosol has to be avoided by either a direct rapid removal of the metal-ion and/or after a transient, often metal-specific, complexation.

3. The removal of the metal from the cytosol requires transport across the tonoplast into the vacuole, based on the general concept of compartmentalization of heavy metals (Ernst, 1975, 1998a; Vögeli-Lange and Wagner, 1990; Hall, 2002), or exudation into the apoplast.

GENETIC ENGINEERING AT THE CELLULAR LEVEL

Transporters Associated with the Plasma Membrane

Most investigations of metal transporters have focused on plant species that are nonmycorrhizal, i.e., species belonging to the families of the Brassicaceae and Caryophyllaceae. As to be expected from the classical physiological and genetic experiments, metal transport across the plasma membrane is mostly metal specific. The activity of the transporters of all metabolically essential heavy metals seems to be dependent on the external bio-available metal concentration and seems to have high and low affinity to the metal in all organisms investigated. Most of the research has focused on high-affinity metal transporters of Cu, Fe, and Zn in metal-deficient crops (Hacisalihoglu et al., 2001) and is just exploring the transporters in Ni and Zn hyperaccumulators. Metal tolerance has to involve the protection of plasma membrane integrity, especially against oxygen radical-producing metal species (De Vos et al., 1991) and of metal-sensitive cell organelles, especially the chloroplasts (Lolkema and Vooijs, 1986). In *A. thaliana,* Cu uptake may be mediated by at least one transporter coded by the *COPT* gene. It is a homologue to the *CTR* transporter in yeast; if overexpressed it increases the sensitivity to copper toxicity (Kampfenkel et al., 1995). Because of this reaction pattern, it is very likely that *COPT* is a high-affinity Cu transporter (Table 8.3). During the transport across the plasma membrane, copper is reduced from Cu^{2+} to Cu^{+}, which is a very aggressive copper species causing oxidative stress if not immediately bound to organic complexes (De Vos et al., 1993). Once copper is inside the cell, it may efflux back into the apoplast by a metal-transporting P-type ATPase (Van Hoof, Koevoets, et al. 2001) which is likely encoded by the *RAN1* gene, isolated from *A. thaliana* (Tabata et al., 1997) or transferred into the vacuole.

On alkaline soils, iron mobility is low and maintaining the supply to plants demands specific adaptation. *IRT1,* a gene which encodes an iron metal transporter, is expressed only in roots under iron-limiting growth conditions of *A. thaliana* (Eide et al., 1996). The *IRT2* promoter is mainly active in the external cell layers of roots (Vert et al., 2001). All dicots, including *A. thaliana,* have to reduce Fe^{3+} to Fe^{2+} during its transport across root plasma membranes (strategy I plants; Bienfait et al., 1983). Fe^{3+} chelate

TABLE 8.3. Genes and proteins involved in metal transport across the plasma membrane.

Element	Affinity	Gene/ Protein	Plant species	Reference
Copper	High	COPT	*Arabidopsis thaliana*	Kampfenkel et al., 1995
Iron	High	IRT1	*Arabidopsis thaliana*	Eide et al., 1996
Iron		IRT2	*Arabidopsis thaliana*	Vert et al., 2001
Iron		FRO2	*Arabidopsis thaliana*	Robinson et al., 1999
Iron		ZMRPP	*Zea mays*	Von Wirén et al., 1997
Molybdenum		AMA1	*Arabidopsis thaliana*	Palmgren and Harper, 1999
Zinc	High	ZIP1	*Arabidopsis thaliana*	Grotz et al., 1998
Zinc		ZIP3	*Arabidopsis thaliana*	Grotz et al., 1998
Zinc		ZNT1	*Thlaspi caerulescens*	Pence et al., 2000
Zinc		ZNT1	*Thlaspi caerulescens*	Assuncao et al., 2001
Zinc		ZNT2	*Thlaspi caerulescens*	Assuncao et al., 2001
Zinc	Low	ZIP2	*Arabidopsis thaliana*	Grotz et al., 1998

reductase is encoded by the *FRO2* gene (Robinson et al., 1999). In grasses (strategy II plants), the transport of iron bound to phytosiderophores is enabled by root plasma polypeptides (ZMRPP) highly expressed in *Zea mays* under Fe deficiency (Von Wirén et al., 1997). Because of the high colonization of grass roots with arbuscular mycorrhizal fungi, the ecological importance of IRT and ZMRPP has still to be demonstrated.

A metal that occurs in the anionic form is molybdenum. Its uptake is regulated by a molybdenum-specific pump in *A. thaliana,* which belongs to the P-type heavy metal ATPase and is encoded by the *AMA1* gene (Palmgren and Harper, 1999). More emphasis in research is given to zinc transporters that belong to different families in the plasma membrane and the tonoplast. All zinc transporters in the plasma membranes belong to the *ZIP* family and are strongly expressed under zinc deficiency. Zn uptake is regulated by transporters that are encoded by the *ZIP1, ZIP2,* and *ZIP3* genes, being isolated from *A. thaliana* by mutant complementation (Grotz et al., 1998). Zn uptake is governed by the expression of a similar transporter gene *(ZNT1)* in the metal-resistant ecotype 'Prayon' of *Thlaspi caerulescens* (Pence et al., 2000), which was introduced in 1903 by ores to the smelter site at Prayon from an unknown origin in Belgium (Duvigneaud and Jortay, 1987) as well as by two transporter genes *(ZNT1, ZNT2)* in the Cd- and Zn-resistant ecotype 'La Calamine' and two metal-sensitive ecotypes, originally collected from a serpentine and a normal soil (Assuncao et al., 2001). *ZIP1, ZIP3, ZNT1,* and *ZNT2* are encoding a high-affinity, and *ZIP2* a low-affinity, Zn transporter, comparable to similar Zn transporters *ZRT1, IRT1,* and *ZRT2* in *Saccharomyces cerevisiae* (Zhao and Eide, 1996a, 1996b). *ZRT1* and *ZRT2* activity is regulated in response to Zn availability at both the transcriptional and posttranscriptional levels, which may also be expected for the Zn transporter in higher plants. This differentiation into low and high affinity is well correlated with the physiological biphasic uptake pattern of copper and zinc by higher plants (Marquenie-van der Werff and Ernst, 1979; Reid et al., 1996). It has been suggested that hyperaccumulation of zinc in *T. caerulescens* relies on a decreased, Zn-imposed down-regulation of the genes for the high-affinity Zn transporter and/or on a high density of the Zn transporter(s) in the plasma membrane (Pence et al., 2000; Assuncao et al., 2001). Other micronutrients and nonessential metals may be taken up by specific transporters or may use existing systems, as shown for arsenate. This element is taken up by the phosphate transporter (Meharg and Macnair, 1992). A down-regulation of the high-affinity phosphate transporter is part of the mechanism governing arsenate resistance in the grass *Holcus lanatus* and is dependent on the phosphorus status of the soil.

Metal Binding in the Cytosol and Chloroplast

As soon as heavy metals and metalloids, especially those that do not have specific functions in a plant's metabolism, such as As, Cd, Cr, Hg, and Pb, are accumulating in the cytosol and cell organelles, they cause a deregulation of the metabolism (Ernst, 1998a). Several metal-binding compounds in the cytosol may contribute to a transient metal complexation (Table 8.4); other compounds are known for their role in copper trafficking to metal-demanding enzymes and proteins (chaperones). This latter group will not be discussed because it was extensively considered in recent reviews (Himmelblau and Amasino, 2000; Huffman and O'Halloran, 2001). The most investigated metal-complexing peptides are the phytochelatins (PCs). They are synthesized nontranslationally using glutathione as a substrate (Grill et al., 1987) by PC synthase. Genes for PC synthase, *AtPCS1,* have been isolated from *A. thaliana* (Clemens et al., 1999; Ha et al., 1999; Vatamaniuk et al., 1999), formerly named *CAD1* (Howden and Cobbett, 1992). As soon as a metal surplus is accumulating in the cytosol as a symptom of insufficient decontamination, synthesis of phytochelatins is stimulated by a variety of metal ions, but with a different affinity in the sequence at equimolar concentration: Hg > Cd, As, Te > Ag >> Cu, Ni > Sb, Au >> Sn, Se > Bi > Pb, W, Zn >> Cr, Fe, Mn, Mo. The role of PCs in metal resis-

TABLE 8.4. Genes and proteins involved in metal complexation in cytosol and chloroplasts.

Element	Gene/ Protein	Plant species	Reference
Cytosol			
Cadmium	CAD1	*Arabidopsis thaliana*	Howden and Cobbett, 1992
Cadmium	AtPCS1	*Arabidopsis thaliana*	Vatamaniuk et al., 1999
Cadmium	MT-1	*Brassica napus* *Nicotiana tabacum*	Misra and Gedamu, 1989
Cadmium	MT2	*Arabidopsis thaliana*	Murphy and Taiz, 1995
Cadmium	hMTIA	*Nicotiana tabacum*	Pan et al., 1993
Copper	CUP1	*Brassica oleracea*	Hasegawa et al., 1997
Copper	MT2	*Arabidopsis thaliana*	Zhou and Goldsbrough, 1994
Chloroplast			
Cadmium	cpGR	*Brassica juncea*	Pilon-Smits et al., 2000
Copper	cpGR	*Brassica juncea*	Pilon-Smits et al., 2000
Zinc	ZIP4	*Arabidopsis thaliana*	Grotz et al., 1998

tance still remains to be answered. Phytochelatins do not remain in the cytosol but have to be removed into the vacuole.

A lack of the synthesis of precursors of phytochelatins and of phytochelatins themselves resulted in Cd sensitivity of *A. thaliana* where the *CAD2* gene encodes for the gamma-glutamylcystein synthetase (Howden et al., 1995; Cobbett et al., 1998). Overexpressing a bacterial gamma-glutamylcysteine synthetase in the cytosol of a transgenic poplar could enhance Cd accumulation but not improve its Cd tolerance (Arisi et al., 2000). In the metabolic steps from activated sulfate to phytochelatins, overexpressing the *Escherichia coli gshII* gene encoding glutathione synthetase in the cytosol of transgenic *Brassica juncea* resulted in a slightly enhanced Cd tolerance (Zhu et al., 1999). But overexpression of the plastidic glutathione reductase (cpGR) in the same plant species *(B. juncea)* resulted in the same level of chlorophyll with Cd exposure as in the wild type without Cd and did not enhance biomass production or Cd resistance at the level of the whole plant (Pilon-Smits et al., 2000). This result was interpreted by the authors as enhanced Cd tolerance at the chloroplast level. But the impact of the *cpGR* gene was not at the chloroplast level alone; it strongly diminished Cd uptake and translocation to the shoot so that it may be also interpreted as a prevention of premature senescence through increased activity of galactolipase (Krupa and Baszynski, 1995). In every case, this latter example indicates the problems in shaping transgenics for phytoextraction within the same plant species, because all steps from the precursor to the final product have to be transformed.

Do phytochelatins play an important ecological role in metal resistance? There are contrasting opinions and results. The hypothesis of a low impact of PCs per se in Cd, Cu, and Zn resistance is substantiated by wild plants. In a comparative analysis of metal-resistant and metal-sensitive ecotypes of *Silene vulgaris,* PC synthesis starts at lower external Cd and Cu concentration in the sensitive ecotypes than in resistant ecotypes, establishing the enhanced synthesis of PCs as a stress reaction, but not involved in the Cu and Zn tolerance mechanisms (De Vos et al., 1992; De Knecht et al., 1995). A Cd- and Zn-resistant, but Cu-sensitive ecotype of the same plant species, growing on a wide diversity of polymetallic soils, synthesized only PCs on Cu-enriched soils (Ernst et al., 2000). Recently, Ebbs and colleagues (2002) have confirmed for the Zn/Cd hyperaccumulator *Thlaspi caerulescens* that PCs are not involved in Cd tolerance, thus confirming earlier findings in *S. vulgaris* (Schat and Kalff, 1992). On arsenate-enriched soils, PCs may play a role in the very moderate arsenate resistance of *H. lanatus* (Hartley-Whitaker et al., 2001) and of *S. vulgaris* (Paliouris and Hutchinson, 1991; Sneller et al., 1999).

Another option is the binding of metals by metallothioneins. In ten ecotypes of *A. thaliana,* a correlation was found between the Cu sensitivity and the expression of MT2 RNA (Murphy and Taiz, 1995). Because of progress in the knowledge of Cu trafficking in the cell (Huffman and O'Halloran, 2001), most of the MTs are thought to operate as chaperones for the normal supply of enzymes and organelles with copper. In Cu-resistant plants, MTs may also contribute to a rapid transfer of the metal from the cytosol into the vacuole (Van Hoof, Hassinen, et al., 2001). MT genes have been transferred from human, animal, and yeast origins to establish genetically modified higher plants. Expression of the *MT-1* gene from mouse in *Nicotiana tabacum* resulted in a 20-fold increase of Cd resistance on agar (Misra and Gedamu, 1989); a moderate increase of Cd resistance was achieved by the expression of human metallothionein-1 gene in transgenic tobacco (Pan et al., 1993). Enhanced expression of the *CUP1* metallothionein gene from *Saccharomyces cerevisiae* in *Brassica oleracea* increased its Cd resistance (Hasegawa et al., 1997). However, in the case of Cd, it may be that the expression of MT genes demands more time than the activation of PC synthase, as shown for a wild-type *B. juncea* during Cd exposure (Haag-Kerwer et al., 1999).

The regulation of free metal ions in the cytosol by complexation without removal from the cytosol can only be a short-term, transient solution (Schat and Kalff, 1992). Real metal resistance is based on a transfer and deposition of the metal in vacuoles, at least in the case of As, Cd, Ni, and Zn.

Metal Compartmentalization in the Vacuole

Starting with the first analysis of Zn compartmentation, it was shown that Zn is accumulated primarily in the vacuole (Ernst, 1969), later supported by experiments using ^{65}Zn (Brune et al., 1994). Nickel has a similar preference for accumulation in the vacuoles (Küpper et al., 2001). Second, Zn may be precipitated in the cell walls (Ernst and Weinert, 1972) as Zn silicates (Bringezu et al., 1999). A Zn-resistant ecotype of *Silene vulgaris* has a 2.5 times enhanced efficiency to transport Zn across the tonoplast into the vacuole compared with a Zn-sensitive ecotype (Verkleij et al., 1998; Chardonnens, Koevoets, et al., 1999). In the meantime, several metal-specific transporters in the tonoplast have been characterized in yeast (Li and Kaplan, 1998) and higher plants (Table 8.5). They belong to the family of cation diffusion facilitators (CDF), later renamed the cation efflux (CE) family (Mäser et al., 2001), or to the ZIP family (Grotz et al. 1998; MacDiarmid et al., 2000; Assuncao et al., 2001).

Overexpression of the tonoplast zinc transporter (ZAT) in transgenic *A. thaliana* resulted in a two times higher Zn resistance, although this improvement may not be sufficient to survive on zinc-contaminated soils. Similar transporter genes have been found in *Thlaspi goesingense* (Persans et al., 2001) and *T. montanum* var. *fendleri* (Mäser et al., 2001). A specific situation for the transport into the vacuole has been established for Cd, which is either transported as Cd-PC complex across the tonoplast by an ATP-dependent ABC-cassette transporter (Vögeli-Lange and Wagner, 1990; Salt and Rauser, 1995) or via a Cd^{2+}/H^{+} antiporter (Salt and Wagner, 1993).

TABLE 8.5. Proteins and genes involved in the transport of metals across the tonoplast into the vacuole.

Element	Gene/ protein	Plant species	References
Zinc	ZAT	*Arabidopsis thaliana*	Van der Zaal et al., 1999
Zinc	ZIP4	*Arabidopsis thaliana*	Grotz et al., 1998
Zinc	ZTP1	*Thlaspi caerulescens*	Assuncao et al., 2001
Nickel	TgMTP	*Thlaspi goesingense*	Persans et al., 2001
Nickel	TmMTP	*Thlaspi montanum*	Mäser et al., 2001
Cadmium	Cd^{2+}/H^{+}-antiporter	*Avena sativa*	Salt and Wagner, 1993
Cadmium	ABC-like cassette	*Nicotiana tabacum*	Vögeli-Lange and Wagner, 1990

Metal Resistance at the Whole-Plant Level

Real metal resistance demands a phased transport of the metal from the root into the shoot, and within the shoot a strongly diminished loading of the seed (Ernst, 1974). It is conceivable that metal-resistant and metal-sensitive plants differ in their rate of transfer of the metal from root to shoot, as shown for cadmium, copper, and zinc in *S. vulgaris* (Lolkema et al., 1984; Verkleij and Prast, 1989; Harmens et al., 1993). Crosses between a zinc-hyperaccumulating ecotype of *Arabidopsis halleri* and a nonhyperaccumulating ecotype of *A. petraea* have indicated that the characters of zinc tolerance and zinc hyperaccumulation are genetically independent (Macnair et al., 1999). Obviously, the transfer of a metal from the root cortex via the endodermis into the xylem parenchyma and further into the xylem for transport to the shoot may be regulated by other transporters that have still to be identified. Once in the leaves, heavy metals are differentially distributed across the various leaf tissues. There is a preference for accumulation of cadmium, copper, nickel, and zinc in the photosynthetically inactive epidermal cells of the leaves (Heath et al., 1997; Mesjasz-Przybylowicz et al., 1997; Chardonnens, Ten Bookum, et al., 1999), with low concentrations in the stomata. This pattern fits well with the high sensitivity of chloroplasts to a surplus of heavy metals (Krupa and Baszynski, 1995). How this pattern is directed by metabolism is not known and will demand a thorough investigation prior to breeding of metal-accumulating target species. As a general rule, in contrast to grains, seeds of metal-resistant plants hardly have an enhanced metal concentration (Ernst, 1974; Mesjasz-Przybylowicz et al., 2001). One reason may be the very restricted xylem-transported compounds so that selection of a crop for phytoremediation may have economical profit. The elucidation of genes involved in the different aspects of metal resistance opens possibilities for an introduction of gene(s) for hyperaccumulation in nonhyperaccumulating target plant species as soon as the gene(s) are isolated. But it seems that there is a long way to go.

Constraints of Metal-Enriched Mining Sites

Breeding for metal resistance of plant species to colonize highly metal-contaminated sites has to consider more aspects of those environments. Remnants of the mining activities are raw materials, not really soils, which are poor in nitrogen and phosphorus (Ernst, 1974). Application of fertilizers may improve the nutrient supply for the plant, but at the same time phosphorus will diminish metal availability and thus prolong the time necessary for phytoextraction. Another aspect of mining sites is the poor water-holding capacity (Ernst, 1974). Copper-resistant ecotypes of *Silene vulgaris* growing on copper shales have a constitutive expression of an increased proline level (Schat et al., 1997), one of the osmotic compounds facilitating drought tolerance.

CONCLUSION

Breeding of high-productive, metal-accumulating target plants will demand the isolation and introduction of a large number of genes, which have to operate in concert in the target plants. Once brought into the field, the dispersal of the propagules of the transgenic plants has to be restricted and strongly controlled. Therefore, plant species with wind-dispersed propagules are not desirable as this restricts the application of metal-resistant *Berkheya, Senecio, Salix,* and *Populus* species for this process. Independent of the plant species used and its genetical improvement for phytoextraction, an environmentally relevant solution of soil decontamination is possible only if the extracted metals are processed in such a way that they can be stored or recycled in industrial processes (Ernst, 2000). Unfortunately, the postharvest treatment, as generally agreed upon, is only conceptual up to now (Mench et al., 2000).

REFERENCES

Ahn, S.J., M. Sivaguru, H. Osawa, G.C. Chung, and H. Matsumoto (2001). Aluminium inhibits the H^+-ATPase activity by permanently altering the plasma membrane surface potentials in squash roots. *Plant Physiology* 126:1381-1390.

Alexander, M. (1994). *Biodegradation and Bioremediation*. San Diego: Academic Press.

Al-Hiyaly, S.A., T. McNeilly, and A.D. Bradshaw (1988). The effects of zinc contamination from electricity pylons: Evolution in a replicated situation. *New Phytologist* 110:571-580.

Arisi, A.C.M., B. Mocquot, A. Lagriffoul, M. Mench, C.H. Foyer, and L. Jouanin (2000). Responses to cadmium in leaves of transformed poplars overexpressing gamma-glutamylcysteine synthetase. *Physiologia Plantarum* 109:143-149.

Assuncao, A.G.L., F. Da Costa Martins, S. De Folter, R. Vooijs, H. Schat, and M.G.M. Aarts (2001). Elevated expression of metal transporter genes in three accessions of the metal hyperaccumulator *Thlaspi caerulescens*. *Plant, Cell and Environment* 24:217-226.

Ballach, H.J. (1997). Ozone and heavy metals from automobile catalytic converters. *Environmental Science and Pollution Research* 4:131-139.

Barceló, J. and C. Poschenrieder (2002). Fast root growth responses, root exudates, and internal detoxification as clues to the mechanisms of aluminium toxicity and resistance: A review. *Environmental and Experimental Botany* 48:75-92.

Becker, A., W. Klöck, K. Friese, P. Schreck, H.C. Treutler, B. Spettel, and M.C. Duff (2001). Lake Süsser See as a natural sink for heavy metals from copper mining. *Journal of Geochemical Exploration* 74:205-217.

Besnard, E., C. Chenu, and M. Robert (2001). Influence of organic amendments on copper distribution among particle-size and density fractions in Champagne vineyard soils. *Environmental Pollution* 112:329-337.

Bienfait, H.F., R.J. Bino, A.M. van der Bliek, J.F. Duivenvoorden, and J.M. Fontaine (1983). Characterization of ferric reducing activity in roots of Fe-deficient *Phaseolus vulgaris*. *Physiologia Plantarum* 59:196-202.

Bogess, S.F., S. Willavize, and D.E. Koeppe (1978). Differential response of soybean varieties to soil cadmium. *Agronomy Journal* 70:756-760.

Brewer, E.P., J.A. Saunders, J.S. Angle, R.L. Chaney, and M.S. McIntosh (1999). Somatic hybridization between the zinc accumulator *Thlaspi caerulescens* and *Brassica napus*. *Theoretical and Applied Genetics* 99:761-771.

Bringezu, K., O. Lichtenberger, I. Leopold, and D. Neumann (1999). Heavy metal tolerance of *Silene vulgaris*. *Journal of Plant Physiology* 154:536-546.

Brooks, R.R., J. Lee, R.D. Reeves, and T. Jaffré (1977). Detection of nickeliferous rocks by analysis of herbarium specimens of indicator plants. *Journal of Geochemical Exploration* 7:49-57.

Brune, A., W. Urbach, and K.J. Dietz (1994). Compartmentation and transport of zinc in barley primary leaves as basic mechanisms involved in zinc tolerance. *Plant, Cell and Environment* 17:153-162.

Caradus, J.R., A.D. Mackay, and S. Wewala (1991). Selection for tolerance and susceptibility to aluminium within *Trifolium repens* L. In *Plant-Soil Interactions at Low pH*, R.J. Wright, V.C. Baligar, and R.P. Murrmann (Eds.). Dordrecht, The Netherlands: Kluwer Academic Publishers, pp. 1029-1036.

Chardonnens, A.N., P.L.M. Koevoets, A. van Zanten, H. Schat, and J.A.C. Verkleij (1999). Properties of enhanced tonoplast zinc transport in naturally selected zinc-tolerant *Silene vulgaris*. *Plant Physiology* 120:779-785.

Chardonnens, A.N., W.M. Ten Bookum, S. Vellinga, H. Schat, J.A.C. Verkleij, and W.H.O. Ernst (1999). Allocation patterns of zinc and cadmium in heavy metal tolerant and sensitive *Silene vulgaris*. *Journal of Plant Physiology* 155:778-787.

Clark, D.T. (1982). Plant response to mineral element toxicity and deficiency. In *Breeding Plants for Less Favorable Environments*, M.N. Christiansen and C.F. Lewis (Eds.). New York: John Wiley and Sons, pp. 71-142.

Clemens, S., E.J. Kim, D. Neumann, and J.I. Schroeder (1999). Tolerance to toxic metals by a gene family of phytochelatin synthases from plants and yeast. *EMBO Journal* 18:3325-3333.

Cobbett, C.S., M.J. May, R. Howden, and B. Rolls (1998). The glutathione-deficient, cadmium-sensitive mutant, cad2-1, of *Arabidopsis thaliana* is deficient in gamma-glutamylcysteine synthetase. *Plant Journal* 16:73-78.

Colpaert, J.V., P. Vandenkoornhuyse, K. Adriaensen, and J. Vangronsveld (2000). Genetic variation and heavy metal tolerance in the ectomycorrhizal basidiomycete *Suillus luteus*. *New Phytologist* 147:367-379.

Daniel, P., B. Kovacs, J. Prokisch, and Z. Gyori (1997). Heavy metal dispersion detected in soils and plants alongside roads in Hungary. *Chemical Speciation and Bioavailability* 9:83-93.

Davies, M.S. and R.W. Snaydon (1973). Physiological differences among populations of *Anthoxanthum odoratum* L. collected from the Park Grass Experiment, Rothamsted. II. Response to aluminium. *Journal of Applied Ecology* 10:47-55.

De Knecht, J.A., N. van Baren, W.M. Ten Bookum, H.W. Wong Fong Sang, P.L.M. Koevoets, H. Schat, and J.A.C. Verkleij (1995). Synthesis and degradation of phytochelatins in cadmium-sensitive and cadmium-tolerant *Silene vulgaris*. *Plant Science* 106:9-18.

Del Val, C., J.M. Barea, and C. Azcón-Aguilar (1999). Assessing the tolerance of heavy metals of arbuscular mycorrhizal fungi isolated from sewage sludge-contaminated soils. *Applied Soil Ecology* 11:261-269.

De Neeling, A.J. and W.H.O. Ernst (1986). Response of an acidic and a calcareous population of *Chamaenerion angustifolium* (L.) Scop. to iron, manganese and aluminium. *Flora* 178:85-92.

De Vos, C.H.R., H. Schat, M.A.M. De Waal, R. Vooijs, and W.H.O. Ernst (1991). Increased resistance to copper-induced damage of the root cell plasmalemma in copper tolerant *Silene cucubalus*. *Physiologia Plantarum* 82:523-528.

De Vos, C.H.R., W.M. Ten Bookum, R. Vooijs, H. Schat, and L.J. De Kok (1993). Effect of copper on fatty acid composition and peroxidation of lipids in the roots of copper tolerant and sensitive *Silene cucubalus*. *Plant Physiology* 31:151-158.

De Vos, C.H.R., M.J. Vonk, R. Vooijs, and H. Schat (1992). Glutathione depletion due to copper-induced phytochelatin synthesis causes oxidative stress in *Silene cucubalus*. *Plant Physiology* 98:853-858.

Dueck, T.A., W.H.O. Ernst, J. Faber, and F. Pasman (1984). Heavy metal emission and genetic constitution of plant populations in the vicinity of two metal emission sources. *Angewandte Botanik* 58:47-59.

Dueck, T.A., P. Visser, W.H.O. Ernst, and H. Schat (1986).Vesicular-arbuscular mycorrhizae decrease zinc toxicity to grasses growing in zinc-polluted soils. *Soil Biology and Biochemistry* 18:331-333.

Duvigneaud, J. and A. Jortay (1987). Un site intéressant de la region liègoise: La partie méridionale du vallon des Fonds de Foret (Foret et Magnée, province de Liège). *Les Naturalistes Belges* 68:33-48.

Ebbs, S., I. Lau, B. Ahner, and L. Kochian (2002). Phytochelatin synthesis is not responsible for Cd tolerance in the Zn/Cd hyperaccumulator *Thlaspi caerulescens* (J.and C. Presl). *Planta* 214:635-640.

Eide, D., M. Broderius, J. Fett, and M.L. Guerinot (1996). A novel iron-regulated metal transporter from plants identified by functional expression in yeast. *Proceedings of the National Academy of Sciences USA* 93:5624-5628.

Ernst, W.H.O. (1969). Physiology of heavy metal plants: Subcellular sites of zinc accumulation (in German). *Berichte der Deutschen Botanischen Gesellschaft* 82:161-164.

Ernst, W.H.O. (1972). Impact of zinc and cadmium emissions on soils and plants in the vicinity of a zinc smelter (in German). *Berichte der Deutschen Botanischen Gesellschaft* 85:295-300.

Ernst, W.H.O. (1974). *Heavy Metal Vegetation of the World* (in German). Stuttgart, Germany: G. Fischer.

Ernst, W.H.O. (1975). Physiology of heavy metal resistance in plants. In *Proceedings of the International Conference on Heavy Metals in the Environment,* Volume 2, T.C. Hutchinson (Ed.). Toronto, Canada: University of Toronto, pp. 121-136.

Ernst, W.H.O. (1996). Phytotoxicity of heavy metals. In *Fertilizers and Environment,* C. Rodriguez-Barrueco (Ed.). Dordrecht, The Netherlands: Kluwer Academic Publishers, pp. 423-430.

Ernst, W.H.O. (1998a). Effects of heavy metals in plants at the cellular and organismic level. In *Ecotoxicology: Ecological Fundamentals, Chemical Exposure, and Biological Effects,* G. Schüürmann and B. Markert (Eds.). New York, Heidelberg: J. Wiley and Sons, and Spektrum Akademischer Verlag, pp. 587-620.

Ernst, W.H.O. (1998b). The origin and ecology of contaminated, stabilized and non-pristine soils. In *Metal Contaminated Soils: In Situ Inactivation and Phytorestoration,* J. Vangronsveld and S.D. Cunningham (Eds.). Berlin, Germany: Springer, pp. 17-29.

Ernst, W.H.O. (2000). Evolution of metal hyperaccumulation and phytoremediation hype. *New Phytologist* 146:357-358.

Ernst, W.H.O. (2001). Evolutionary mechanisms of adaptation to soils enriched with heavy metals (in German). In *Ökophysiologie der Pflanzen,* W. Larcher (Ed.). Stuttgart, Germany: E. Ulmer, pp. 353-355.

Ernst, W.H.O., H.J.M. Nelissen, and W.M. Ten Bookum (2000). Combination toxicology of metal-enriched soils: Physiological responses of a Zn- and Cd-resistant ecotype of *Silene vulgaris* on polymetallic soils. *Environmental and Experimental Botany* 43:55-71.

Ernst, W.H.O. and H. Weinert (1972). Localization of zinc in leaves of *Silene cucubalus* (in German). *Zeitschrift für Pflanzenphysiologie* 66:258-264.

Feist, L.J. and D.R. Parker (2001). Ecotypic variation in selenium accumulation among populations of *Stanleya pinnata. New Phytologist* 149:61-69.

Felix, H. (1997). Field trials for in situ decontamination of heavy metal polluted soils using crops of metal-accumulating plants. *Zeitschrift für Pflanzenernährung und Bodenkunde* 160:525-529.

Flores, C.L., R.B. Clark, J.F. Pedersen, and L.M. Gourley (1991). Leaf mineral element concentrations in sorghum *(Sorghum bicolor)* hybrids and their parents grown at varied aluminium saturations on an Ultisol. In *Plant-Soil Interactions at Low pH,* R.J. Wright, V.C. Baligar, and R.P. Murrmann (Eds.). Dordrecht, The Netherlands: Kluwer Academic Publishers, pp. 1095-1104.

Flores, C.L., L.M. Gourley, J.F. Pedersen, and R.B. Clark (1991). Inheritance of acid-soil tolerance in sorghum *(Sorghum bicolor)* grown on an Ultisol. In *Plant-Soil Interactions at Low pH,* R.J. Wright, V.C. Baligar, and R.P. Murrmann (Eds.). Dordrecht, The Netherlands: Kluwer Academic Publishers, pp. 1081-1093.

Florijn, P.J., J.A. De Knecht, and M.L. Van Beusichem (1993). Phytochelatin concentrations and binding state of Cd in roots of maize genotypes differing in shoot/root Cd partitioning. *Journal of Plant Physiology* 142:537-542.

Förstner, U. and G.T.W. Wittmann (1981). *Metal Pollution in the Aquatic Environment.* Berlin, Heidelberg, New York: Springer-Verlag.

Fox, C.A., L.J. Circeo, and R.C. Martin (2001). In-situ plasma remediation of contaminated soils. *Remediation Journal* 11(4):3-13.

Foy, C.D. (1988). Plant adaptation to acid, aluminium-toxic soils. *Communication in Soil Science and Plant Analysis* 19:959-987.

Foy, C.D., A.L. Fleming, and G.C. Gerloff (1972). Differential aluminium tolerance in two snapbean varieties. *Agronomy Journal* 64:815-818.

Garrow, J.S. and W.P.T. James (1993). *Human Nutrition and Dietetics,* Ninth Edition. Edinburgh, Scotland: Churchill Livingstone.

Gemmell, R.P. (1977). *The Colonization of Industrial Waste Land.* London, England: Edward Arnold.

Gries, D. and M. Runge (1992). The ecological significance of iron mobilization in wild grasses. *Journal of Plant Nutrition* 15:1727-1737.

Grill, E., E.L. Winnacker, and M.H. Zenk (1987). Phytochelatins, a class of heavy-metal-binding peptides from plants, are functionally analogues to metallothionein. *Proceedings of the National Academy of Sciences USA* 84:439-443.

Grotz, N., T.C. Fox, E. Conolly, W. Park, M.L. Guerinot, and D. Eide (1998). Identification of a family of zinc transporter genes from *Arabidopsis* that respond to zinc deficiency. *Proceedings of the National Academy of Sciences USA* 95:7220-7724.

Gucwa-Przepiora, E. and K. Turnau (2001). Arbuscular mycorrhiza and plant succession on zinc smelter spoil heap in Katowice-Welnowice. *Acta Societatis Botanicorum Poloniae* 70:153-158.

Gzyl, J. (1995). Ecological impact and remediation of contaminated sites around lead smelters in Poland. *Journal of Geochemical Exploration* 52:251-258.

Ha, S.B., A.P. Smith, R. Howden, W.M. Dietrich, S. Bugg, M.J. O'Connell, P.B. Goldsbrough, and C.S. Cobbett (1999). Phytochelatin synthase genes from *Arabidopsis* and the yeast *Schizosaccharomyces pombe. Plant Cell* 11:1153-1163.

Haag-Kerwer, A., H.J. Schäfer, S. Heiss, C. Walter, and T. Rausch (1999). Cadmium exposure in *Brassica juncea* causes a decline in transpiration rate and leaf expansion without effect on photosynthesis. *Journal of Experimental Botany* 50:1827-1835.

Hacisalihoglu, G., J.J. Hart, and L.V. Kochian (2001). High- and low affinity zinc transport systems and their possible role in zinc efficiency in bread wheat. *Plant Physiology* 125:456-463.

Hall, J.L. (2002). Cellular mechanisms for heavy metal detoxification and tolerance. *Journal of Experimental Botany* 53:1-11.

Hanson, A.D., J. Rivoal, L. Paquet, and D.A. Gage (1994). Biosynthesis of 3-dimethylsulfoniopropionate in *Wollastonia biflora* (L.) DC: Evidence that S-methylmethionine is an intermediate. *Plant Physiology* 105:103-110.

Harmens, H., N.G.C.P.B. Gusmao, P.R. Den Hartog, J.A.C. Verkleij, and W.H.O. Ernst (1993). Uptake and transport of zinc in zinc-sensitive and zinc-tolerant *Silene vulgaris. Journal of Plant Physiology* 141:309-315.

Hartley-Whitaker, J., G. Ainsworth, R. Vooijs, W. Ten Bookum, H. Schat, and A.A. Meharg (2001). Phytochelatins are involved in differential arsenate tolerance in *Holcus lanatus. Plant Physiology* 126:299-306.

Hasegawa, I., E. Terada, M. Sunairi, H. Wakita, F. Shinmachi, A. Noguchi, M. Nakajima, and J. Yakazi (1997). Genetic improvement of heavy metal tolerance in plants by transfer of the yeast metallothionein gene (CUP1). *Plant and Soil* 196:277-281.

Hawkes, H.E. and J.S. Webb (1962). *Geochemistry in Mineral Exploration.* New York: Harper and Row.

Heath, S.M., D. Southworth, and J.A. D'Allura (1997). Localization of nickel in epidermal subsidiary cells of leaves of *Thlaspi montanum* var. *sikiyouense* (Brassicaceae) using energy-dispersive X-ray microanalysis. *International Journal of Plant Science* 158:184-188.

Hildebrandt, U., M. Kaldorf, and H. Bothe (1999). The zinc violet and its colonization by arbuscular mycorrhizal fungi. *Journal of Plant Physiology* 154:209-217.

Himmelblau, E. and R.M. Amasino (2000). Delivering copper within plant cells. *Current Opinion in Plant Biology* 3:205-210.

Hinsley, T.D., D.E. Alexander, E.L. Ziegler, and G.L. Barett (1978). Zinc and cadmium accumulation by corn inbreds grown on sludge-amended soil. *Agronomy Journal* 70:425-428.

Hong, S.M., J.P. Candelone, C.C. Patterson, and C.F. Boutron (1994). Greenland ice evidence of hemispheric pollution for lead two millennia ago by Greek and Roman civilizations. *Science* 265:1841-1843.

Hong, S.M., J.P. Candelone, M. Soutif, and C.F. Boutron (1996). A reconstruction of changes in copper production and copper emissions to the atmosphere during the past 7000 years. *Science of the Total Environment* 188:183-193.

Howden, R. and C.S. Cobbett (1992). Cadmium-sensitive mutants of *Arabidopsis thaliana. Plant Physiology* 100:100-107.

Howden, R., P.B. Goldsbrough, C.R. Andersen, and C.S. Cobbett (1995). Cadmium-sensitive, cad1, mutants of *Arabidopsis thaliana* are phytochelatin deficient. *Plant Physiology* 107:1059-1066.

Howeler, R.H. (1991). Identifying plants adaptable to low pH conditions. In *Plant-Soil Interactions at Low pH,* R.J. Wright, V.C. Baligar, and R.P. Murrmann (Eds.). Dordrecht, The Netherlands: Kluwer Academic Publishers, pp. 885-904.

Huffman, D.L., and T.V. O'Halloran (2001). Function, structure, and mechanism of intracellular copper trafficking proteins. *Annual Review of Biochemistry* 70:677-701.

Ietswart, J.H., W.A.J. Griffioen, and W.H.O. Ernst (1992). Seasonality of VAM infection in three populations of *Agrosits capillaris* (Gramineae) on soil with and without heavy metal enrichment. *Plant and Soil* 139:67-73.

Jaffré, T. (1980). Étude écologique du peuplement végétal des sols derives de roches ultrabasiques en Nouvelle-Calédonie. *Travaux et Documents de l'O.R.S.T.O.M.* 124.

Johnson, B. (2000). Biological removal of sulfurous compounds from inorganic wastewaters. In *Environmental Technologies to Treat Sulfur Pollution: Principles and Engineering,* P. Lens and L. Hulshoff Pol (Eds.). London, England: IWA Publishing, pp. 175-205.

Juste, C. and M. Mench (1992). Long-term application of sewage sludge and its effects on metal uptake by crops. In *Biogeochemistry of Trace Metals,* D.C. Adriano (Ed.). Boca Raton: Lewis Publishers, pp. 159-193.

Kampfenkel, K., S. Kushnir, E. Babiychuk, D. Inzé, and M. Van Montagu (1995). Molecular characterization of a putative *Arabidopsis thaliana* copper transporter and its yeast homologue. *Journal of Biological Chemistry* 270:28479-28486.

Knight, B., F.J. Zhao, S.P. McGrath, and Z.G. Shen (1997). Zinc and cadmium uptake by the hyperaccumulator *Thlaspi caerulescens* in contaminated soils and its effect on the concentration and chemical speciation of metals in soil solution. *Plant and Soil* 197:71-78.

Kraal, H. and W. Ernst (1976). Influence of copper high tension lines on plants and soils. *Environmental Pollution* 11:131-136.

Krämer, U. and A.N. Chardonnens (2001). The use of transgenic plants in the bioremediation of soils contaminated with trace elements. *Applied Microbial Biotechnology* 55:661-672.

Krebs, R., S.K. Gupta, G. Furrer, and R. Schulin (1998). Solubility and plant uptake of metals with and without liming of sludge-amended soils. *Journal of Environmental Quality* 27:18-23.

Krupa, Z. and T. Baszynski (1995). Some aspects of heavy metal toxicity toward photosynthetic apparatus: Direct and indirect effects on light and dark respiration. *Acta Physiologiae Plantarum* 17:177-190.

Küpper, H., E. Lombi, F.J. Zhao, G. Wieshammer, and S.P. McGrath (2001). Cellular compartmentation of nickel in the hyperaccumulators *Alyssum lesbiacum, Alyssum bertolonii* and *Thlaspi goesingense. Journal of Experimental Botany* 52:2291-2300.

Lee, J.A. and J.H. Thallis (1973). Regional and historical aspects of lead pollution in Britain. *Nature* 245:216-218.

Li, L. and J. Kaplan (1998). Defects in the yeast high affinity iron transport system result in increased metal sensitivity because of the increased expression of transporters with a broad transition metal specificity. *Journal of Biological Chemistry* 273:22181-22187.

Llugany, M., C. Poschenrieder, and J. Barceló (1995). Monitoring of aluminium-induced inhibition of root elongation in four maize cultivars differing in tolerance to aluminium and proton toxicity. *Physiologia Plantarum* 93:265-271.

Lolkema, P.C., M.H. Donker, A.J. Schouten, and W.H.O. Ernst (1984). The possible role of metallothioneins in copper tolerance of *Silene cucubalus. Planta* 162:174-179.

Lolkema, P.C. and R. Vooijs (1986). Copper tolerance in *Silene cucubalus*. Subcellular distribution of copper and its effects on chloroplasts and plastocyanin synthesis. *Planta* 167:30-36.

MacDiarmid, C.W., L.A. Gaither, and D. Eide (2000). Zinc transporters that regulate vacuolar zinc storage in *Saccharomyces cerevisiae. EMBO Journal* 19:2845-2855.

Macnair, M.R. (1993). The genetics of metal tolerance in vascular plants. *New Phytologist* 124:541-559.

Macnair, M.R., V. Bert, S.B. Huitson, P. Saumitou-Laprade, and D. Petit (1999). Zinc tolerance and hyperaccumulation are genetically independent characters. *Proceedings of the Royal Society of London* B 266:2175-2179.

Mahadevappa, M., A.V. Bulbule, R. Siddaramappa, and R.S. Kulkarni (1991). Field screening of rice cultivars for growth, yield and nutrient uptake in acidic soil. In *Plant-Soil Interactions at Low pH,* R.J. Wright, V.C. Baligar, and R.P. Murrmann (Eds.). Dordrecht, The Netherlands: Kluwer Academic Publishers, pp. 959-965.

Marquenie-van der Werff, M. and W.H.O. Ernst (1979). Kinetics of copper and zinc uptake by leaves and roots of an aquatic plant, *Elodea nuttallii. Zeitschrift für Pflanzenphysiologie* 92:1-10.

Marschner, H. (1995). *Mineral Nutrition of Higher Plants,* Second Edition. London: Academic Press.

Martino, E., K. Turnau, M. Girlanda, P. Bonfante, and S. Perotto (2000). Ericoid mycorrhizal fungi from heavy metal polluted soils: Their identification and growth in the presence of zinc ions. *Mycorrhizal Research* 104:338-344.

Mäser, P., S. Thomine, J.I. Schroeder, J.M. Ward, K. Hirschi, H. Sze, I.N. Talke, A. Amtmann, F.J.M. Maathuis, D. Sanders, et al. (2001). Phylogenetic relationships with cation transporter families of *Arabidopsis. Plant Physiology* 126:1646-1667.

Matsumoto, H. (2000). Cell biology of aluminium toxicity and tolerance in higher plants. *International Review in Cytology* 200:1-46.

Meagher, R.B., C.L. Rugh, M.K. Kandasamy, G. Gragson, and N.J. Wang (2000). Engineered phytoremediation of mercury pollution in soil and water using bacterial genes. In *Phytoremediation of Contaminated Soil and Water,* N. Terry and G. Banuelos (Eds.). Boca Raton: Lewis Publishers, pp. 201-219.

Meharg, A.A. and M.R. Macnair (1992). Suppression of the high-affinity phosphate-uptake system: A mechanism of arsenate tolerance in *Holcus lanatus* L. *Journal of Experimental Botany* 43:519-543.

Mench, M., J. Vangronsveld, H. Clijsters, N.W. Lepp, and R. Edwards (2000). In situ metal immobilization and phytostabilization of contaminated soils. In *Phytoremediation of Contaminated Soil and Groundwater,* N. Terry and G. Banuelos (Eds.). Boca Raton: Lewis Publishers, pp. 323-358.

Mench, M., J. Vangronsveld, V. Didier, and H. Clijsters (1994). Evaluation of metal mobility, plant availability and immobilization by chemical agents in a limed-silty soil. *Environmental Pollution* 86:279-286.

Mesjasz-Przybylowicz, J., K. Grodzinska, W.J. Przybylowicz, B. Godzik, and G. Szarek-Lukaszewska (2001). Nuclear microprobe studies of elemental distribution in seeds of *Biscutella laevigata* L. from zinc wastes in Olkusz, Poland. *Nuclear Instruments and Methods in Physics Research Section B: Beam Interactions with Materials and Atoms* 181:634-639.

Mesjasz-Przybylowicz, J., W.J. Przybylowicz, V.M. Prozesky, and C.A. Pineda (1997). Quantitative micro-PIXE comparison of elemental distribution in Ni-hyperaccumulating and non-accumulating genotypes of *Senecio coronatus*. *Nuclear Instruments and Methods in Physics Research Section B: Beam Interactions with Materials and Atoms* 130:368-373.

Misra, S. and L. Gedamu (1989). Heavy metal tolerant transgenic *Brassica napus* L. and *Nicotiana tabacum* L. plants. *Theoretical and Applied Genetics* 78:161-178.

Mordvedt, J.J. (1987) Cadmium levels in soils and plants from some long-term soil fertility experiments in the United States of America. *Journal of Environmental Quality* 16:137-142.

Murphy, A. and L. Taiz (1995). Comparison of metallothionein gene expression and nonprotein thiols in ten *Arabidopsis* ecotypes. *Plant Physiology* 109:945-954.

O'Sullivan, A.D., O.M. McCabe, D.A. Murray, and M.L. Otte (1999). Wetlands for rehabilitation of metal mine wastes. *Proceedings of the Royal Irish Academy* 99(B):11-17.

Paliouris, G. and T.C. Hutchinson (1991). Arsenic, cobalt and nickel tolerances in two populations of *Silene vulgaris* (Moench) Garcke from Ontario, Canada. *New Phytologist* 117:449-459.

Palmgren, M.G. and J.F. Harper (1999). Pumping with plant P-type ATPases. *Journal of Experimental Botany* 50:883-893.

Pan, A., F. Tie, Z. Duau, M. Yang, Z. Wang, L. Li, Z. Chen, and B. Ru (1993). Alpha-domain of human metallothionein IA can bind metals in transgenic tobacco plants. *Molecular and General Genetics* 242:666-674.

Paschke, M.W., E.F. Redente, and D.B. Levy (2000). Zinc toxicity threshold for important reclamation grass species of the western United States. *Environmental Toxicology and Chemistry* 19:2751-2756.

Pedersen, J.F., R.L. Haaland, and C.S. Hoveland (1986). Registration of 'AU Dewey' birdsfoot trefoil. *Crop Science* 26:1081.

Pence, N.S., P.B. Larsen, S.D. Ebbs, D.L.D. Letham, M.M. Lasat, D.F. Garvin, D. Eide, and L.V. Kochian (2000). The molecular physiology of heavy metal transport in the Zn/Cd hyperaccumulator *Thlaspi caerulescens*. *Proceedings of the National Academy of Sciences USA* 97:4956-4960.

Persans, M.W., K. Nieman, and D.E. Salt (2001). Functional activity and role of cation-efflux family members in Ni hyperaccumulation in *Thlaspi goesingense*. *Proceedings of the National Academy of Sciences USA* 98:9995-10000.

Pilon-Smits, E.A.H., S.B. Hwang, C.M. Lytle, Y.L. Zhu, J.C. Tai, R.C. Bravo, Y.C. Chen, T. Leistek, and N. Terry (1999). Overexpression of ATP sulfurylase in Indian mustard leads to increased selenate uptake, reduction, and tolerance. *Plant Physiology* 119:123-132.

Pilon-Smits, E.A.H., Y.L. Zhu, T. Sears, and N. Terry (2000). Overexpression of glutathione reductase in *Brassica juncea:* Effects on cadmium accumulation and tolerance. *Physiologia Plantarum* 110:455-460.

Reid, D.A., L.A.J. Slootmaker, O. Stolen, and J.C. Craddock (1980). Registration of barley composite cross XXXIV (Reg. No. G.P. 33). *Crop Science* 20:416-417.

Reid, R.J., J.D. Brookes, M.A. Tester, and F.A. Smith (1996). The mechanism of zinc uptake in plants. Characterization of the low-affinity system. *Planta* 198:39-45.

Ritchey, K.D. (1991). Evaluating sweet potato tolerance to aluminium toxicity: Comparison of rapid test method and field results. In *Plant-Soil Interactions at Low pH,* R.J. Wright, V.C. Baligar, and R.P. Murrmann (Eds.). Dordrecht, The Netherlands: Kluwer Academic Publishers, pp. 939-945.

Robinson, B.H., R.R. Brooks, A.W. Howes, J.H. Kirkman, and P.E.H. Gregg (1997). The potential of the high-biomass nickel accumulator *Berkheya coddii* for phytoremediation and phytomining. *Journal of Geochemical Exploration* 60:215-226.

Robinson, B.H., M. LeBlanc, D. Petit, R.R. Brooks, J.H. Kirkman, and P.E.H. Gregg (1998). The potential of *Thlaspi caerulescens* for phytoremediation of contaminated soils. *Plant and Soil* 203:47-56.

Robinson, N.J., C.M. Proctor, E.L. Connolly, and M.L. Guerinot (1999). A ferric-chelate reductase for iron uptake from soils. *Nature* 397:694-697.

Romera, F.J., E. Alcantara, and M.D. de la Gardia (1992). Role of roots and shoots in the regulation of the Fe deficiency responses in sunflower and cucumber. *Physiologia Plantarum* 85:141-146.

Römheld, V. and H. Marschner (1990). Genotypical differences among graminaceous species in release of phytosiderophores. *Plant and Soil* 123:147-153.

Rosenfeld, J. and O.A. Beath (1964). *Selenium: Geobotany, Biochemistry, Toxicity and Nutrition.* New York: Academic Press.

Rugh, C.L., J.F. Senecoff, R.B. Meagher, and S.A. Merkle (1998). Development of transgenic yellow poplar for mercury phytoremediation. *Nature Biotechnology* 16:925-928.

Rugh, C.L., H.D. Wilde, N.M. Stack, D.M. Thompson, A.O. Summers, and R.B. Meagher (1996). Mercuric ion reduction and resistance in transgenic *Arabidopsis* thaliana plants expressing a modified bacterial *merA* gene. *Proceedings of the National Academy of Sciences USA* 93:3182-3187.

Sachs, J. (1865). *Handbuch der Experimental-Physiologie der Pflanzen.* Leipzig, Germany: Verlag Wilhelm Engelmann.

Salt, D.E. and W.E. Rauser (1995). MgATP-dependent transport of phytochelatins across the tonoplast of oat roots. *Plant Physiology* 107:1293-1301.

Salt, D.E. and G.J. Wagner (1993). Cadmium transport across tonoplast of vesicles from oat roots: Evidence for a Cd^{2+}/H^{+} antiport activity. *Journal of Biological Chemistry* 268:12297-12302.

Schat, H. and M.M.A. Kalff (1992). Are phytochelatins involved in differential metal tolerance or do they merely reflect metal-imposed strain? *Plant Physiology* 99:1475-1480.

Schat, H., S.S. Sharma, and R. Vooijs (1997). Heavy metal-induced accumulation of free proline in a metal-tolerant and a nontolerant ecotype of *Silene vulgaris. Physiologia Plantarum* 101:477-482.

Schat, H. and J.A.C. Verkleij (1998). Biological interactions: The role for non-woody plants in phytorestoration: Possibilities to exploit adaptive heavy metal tolerance. In *Metal-Contaminated Soils: In Situ Inactivation and Phytorestor-*

ation, J. Vangronsveld and S.D. Cunningham (Eds.). Berlin, Germany: Springer, pp. 51-65.

Schat, H. and R. Vooijs (1997). Multiple tolerance and co-tolerance to heavy metals in *Silene vulgaris*: A co-segregation analysis. *New Phytologist* 136:489-496.

Schat, H., R. Vooijs, and E. Kuiper (1996). Identical major gene loci for heavy metal tolerances that have independently evolved in different local populations and subspecies of *Silene vulgaris*. *Evolution* 50:1888-1895.

Schramel, O., B. Michalke, and A. Kettrup (2000). Study of copper distribution in contaminated soils of hop fields by single and sequential extraction procedures. *Science of the Total Environment* 263:11-22.

Slaski, J.J., G.C. Zhang, U. Basu, J.L. Stephens, and G.J. Taylor (1996). Aluminium resistance in wheat *(Triticum aestivum)* is associated with rapid, Al-induced changes in activities of glucose-6-phosphate dehydrogenase and 6-phosphogluconate dehydrogenase in root apices. *Physiologia Plantarum* 98:477-484.

Sneller, F.E.C., L.M. van Heerwaarden, F.J.L. Kraaijeveld-Smit, W.M. Ten Bookum, P.L.M. Koevoets, H. Schat, and J.A.C. Verkleij (1999). Toxicity of arsenate in *Silene vulgaris*, accumulation and degradation of arsenate-induced phytochelatins. *New Phytologist* 144:223-232.

Steudler, P.A. and B.J. Peterson (1984). Contribution of sulfur from salt marshes to the global sulfur cycle. *Nature* 311:455-457.

Tabata, K., S. Kashiwagi, H. Mori, and T. Mizuno (1997). Cloning a cDNA encoding a putative metal-tranporting P-type ATPase from *Arabidopsis thaliana*. *Biochimica et Biophysica Acta* 1326:1-6.

Van der Zaal, B.J., L.W. Neuteboom, J.E. Pinas, A.N. Chardonnens, H. Schat, J.A.C. Verkleij, and P.J.J. Hooykaas (1999). Overexpression of a novel *Arabidopsis* gene related to putative zinc-tranporter genes from animals can lead to enhanced zinc resistance and accumulation. *Plant Physiology* 119:1047-1055.

Van Hoof, N.A.L.M., V.H. Hassinen, H.W.J. Hakvoort, K.F. Ballentijn, H. Schat, J.A.C. Verkleij, W.H.O. Ernst, S.O. Kärenlampi, and A.I. Tervahauta (2001). Enhanced copper tolerance in *Silene vulgaris* (Moench) Garcke populations from copper mines is associated with increased transcript levels of a 2b-type metallothionein gene. *Plant Physiology* 126:1519-1526.

Van Hoof, N.A.L.M., P.L.M. Koevoets, H.W.J. Hakvoort, W.M. Ten Bookum, H. Schat, J.A.C. Verkleij, and W.H.O. Ernst (2001). Enhanced ATP-dependent copper efflux across the root cell plasma membrane in copper-tolerant *Silene vulgaris*. *Physiologia Plantarum* 113:225-232.

Vatamaniuk, O.K., S. Mari, Y.P. Lu, and P.A. Rea (1999). AtPCS1, a phytochelatin synthase from *Arabidopsis*: Isolation and in vitro reconstitution. *Proceedings of the National Academy of Sciences USA* 96:7110-7115.

Verkleij, J.A.C., P.L.M. Koevoets, M.M.A. Blake-Kalff, and A.N. Chardonnens (1998). Evidence for an important role of the tonoplast in the mechanism of naturally selected zinc tolerance in *Silene vulgaris*. *Journal of Plant Physiology* 153:188-191.

Verkleij, J.A.C. and J.E. Prast (1989). Cadmium tolerance and co-tolerance in *Silene vulgaris* (Moench) Garcke (= *S. cucubalus* L.) Wib. *New Phytologist* 111:637-645.

Vert, G., J.F. Briat, and C. Curie (2001). *Arabidopsis* IRT2 gene encodes a root-periphery iron transporter. *Plant Journal* 26:181-189.

Vögeli-Lange, R. and G.J. Wagner (1990). Subcellular localization of cadmium and cadmium-binding peptides in tobacco leaves. *Plant Physiology* 92:1086-1093.

Von Wirén, N., J.B. Peltier, D. Rouquié, and J.F. Briat (1997). Four root plasmalemma polypeptides under-represented in the maize mutant ys1 accumulate in a Fe-efficient genotype in response to iron-deficiency. *Plant Physiology and Biochemistry* 35:945-950.

Whiting, S.N., M.P. De Souza, and N. Terry (2001). Rhizosphere bacteria mobilize Zn for hyperaccumulation by *Thlaspi caerulescens. Environmental Science and Technology* 35:3144-3150.

Wu, L. and A.D. Bradshaw (1972). Aerial pollution and the rapid evolution of copper tolerance. *Nature* 238:167-169.

Yang, Z.M., M. Sivaguru, W.J. Horst, and H. Matsumoto (2000). Aluminium tolerance is achieved by exudation of citric acid from roots of soybean *(Glycine max). Physiologia Plantarum* 110:72-77.

Zayed, A., E. Pilon-Smits, M. De Souza, Z.Q. Lin, and N. Terry (2000). Remediation of selenium-polluted soils and waters by phytovolatilization. In *Phytoremediation of Contaminated Soil And Water*, N. Terry and G. Banuelos (Eds.). Boca Raton: Lewis Publishers, pp. 61-83.

Zayed, A. and N. Terry (1994). Selenium volatilization in roots and shoots: Effects of shoot removal and sulfate level. *Journal of Plant Physiology* 143:8-14.

Zhao, H. and D. Eide (1996a). The yeast ZRT1 gene encodes the zinc transporter protein of a high-affinity uptake system induced by zinc limitation. *Proceedings of the National Academy of Sciences USA* 93:2454-2458.

Zhao, H. and D. Eide (1996b). The yeast ZRT2 gene encodes the low-affinity zinc transporter in *Saccharomyces cerevisiae. Journal of Biological Chemistry* 271:23203-23210.

Zhou, J. and P.B. Goldsbrough (1994). Functional homologs of fungal metallothionein genes from *Arabidopsis. The Plant Cell* 6:875-884.

Zhu, Y.L., E.A.H. Pilon-Smits, L. Jouanin, and N. Terry (1999). Overexpression of glutathione synthetase in Indian mustard enhances cadmium accumulation and tolerance. *Plant Physiology* 119:73-79.

Zieve, R. and P.J. Peterson (1981). Factors influencing volatilization of selenium from soil. *Science of the Total Environment* 19:277-284.

Chapter 9

Breeding Crops for Adaptation to Environments with Low Nutrient Availability

Zed Rengel

INTRODUCTION

Nutrient deficiencies in plants are widespread throughout much of the agricultural regions of the world and can result in reduced growth or a decline in yields of many crop species (e.g., Graham, 1984; Sillanpaa and Vlek, 1985; Welch et al., 1991). Losses from environmental stresses, such as water and nutrient deficiency stresses, in agricultural production are enormous and have been estimated at 75 percent of the potential yield in U.S. agriculture (Boyer, 1982). Plant responses to a range of nutrient deficiencies have been discussed in a number of recent reviews (Graham and Rengel, 1993; Rengel et al., 1994; Randall, 1995; Welch, 1995; Jolley et al., 1996; Jungk and Claassen, 1997; Pearson and Rengel, 1997b; Cianzio, 1999; Crowley and Rengel, 1999; Rengel, 1999a, 2000, 2001; Engels et al., 2000; Jungk, 2001).

Plants suffer nutrient deficiency stress when the availability of soil nutrients and/or the amount of nutrients taken up is below that required for sustaining metabolic processes in a particular growth stage. This may be the result of an inherently low nutrient status in the soil, low mobility of nutrients within the soil, or poor solubility of the given chemical form of the nutrient. The mobility of nutrients within the soil is governed by a number of factors, including mass flow of water, adsorption capacity of the soil, and soil pH (Jungk, 1991; Shuman, 1991; Jungk and Claassen, 1997; Jungk, 2001). The chemical form of nutrients within the soil also determines the extent of availability; for instance, Fe^{2+} is available for plant uptake, but Fe^{3+} is not (Crowley and Rengel, 1999).

Availability of nutrients in the rhizosphere, the soil around the root that is influenced by the root, is controlled by the combined effects of soil properties, plant characteristics, and the interactions of plant roots with microorganisms and the surrounding soil (Bowen and Rovira, 1999). The rhizosphere extends up to a few millimeters from the root surface into the surrounding soil and is characterized by a relatively high concentration of easily degradable compounds exuded from roots (Lynch and Whipps, 1990), especially at the root tip (Marschner et al., 1987). A better knowledge of the rhizosphere processes that contribute to nutrient uptake efficiency is essential for breeding genotypes that perform well in nutrient-limiting soils (Crowley and Rengel, 1999).

Modern agriculture has mostly been developed on the premise of supplying N, P, and K in sufficient, if not abundant, quantities as synthetic fertilizers. As a consequence, the selection and breeding of new genotypes for agriculture has resulted in the development of cultivars that are highly responsive to fertilizers but without traits that are necessary for growth under nutrient-limiting or adverse soil conditions (Lynch, 1998). However, this tendency has been challenged by the current emphasis on developing better cultivars for sustainable, low-input agriculture, fueled by the increasing cost of fertilizers, government restrictions on fertilizer applications to minimize environmental damage, and the increased use of poor-quality land (Sattelmacher et al., 1994). These trends have led to renewed interest in the physiology of nutrient stress and breeding genotypes adapted to low-nutrient environments (see Lynch, 1998).

Among macronutrients, breeding for adaptation to low-nutrient environments has advanced to a considerable extent only for P, and to a much smaller extent for N and K (Isfan, 1993; Caradus, 1994; Gorny, 1996; Stelling et al., 1996; Kessel and Becker, 1998; Wu et al., 1998; Agrama et al., 1999; Cianzio, 1999; Harrison et al., 2000; Sanginga et al., 2000; Manske et al., 2001; Wissuwa and Ae, 2001). For P and K, deficiency is caused in most cases by poor availability of otherwise substantial quantities of these nutrients present in soil (Jungk and Claassen, 1989; Schilling et al., 1998; Jungk, 2001). In contrast, for N the deficiency problem is mostly due to insufficient total amounts existing in soil rather than due to a small fraction of the total N present being available to plants (Jungk, 1991; Jungk and Claassen, 1997).

Among micronutrients, some breeding efforts have been reported in the case of Fe, Mn, Zn, Cu, and B (Stevens et al., 1993; Owouoche et al., 1996; Rerkasem and Jamjod, 1997; Cianzio, 1999; Rengel, 1999a, 2001; Lin, Baumer, et al., 2000; Graham et al., 2001). Deficiency of micronutrients is generally caused by poor plant availability of pools present in soil. However, crop species and genotypes capable of increasing the fraction of total micronutrient soil pool that is plant-available have been identified (e.g., Graham and Rengel, 1993; Cakmak, Öztürk, et al., 1996; Cakmak, Sari, Marschner, Ekiz, et al., 1996; Cakmak, Sari, Marschner, Kalayci, et al., 1996; Cakmak, Yilmaz, et al., 1996; Cakmak, Ekiz, et al., 1997; Grewal et al., 1997; Pearson and Rengel, 1997b; Xue et al., 1998; Grewal and Williams, 1999; Rengel, 1999a; Stangoulis et al., 2000).

Growing nutrient-efficient genotypes of crop plants on soils of low nutrient availability represents an environmentally friendly approach that would reduce land degradation by reducing the use of machinery (e.g., Thongbai et al., 1993) and by minimizing application of chemicals, i.e., fertilizers, on agricultural land. A danger of exhaustion of soil-nutrient resources ("land mining") would be negligible, at least for micronutrients and P (Graham, 1984), because the total supply of micronutrients and P in soils is sufficient for hundreds of years of sustainable cropping by new, efficient genotypes that are able to gain access to the soil-nutrient pools not available to inefficient genotypes.

NUTRIENT EFFICIENCY

Nutrients with limited mobility in soils (P, K, Fe, Zn, Mn, and Cu) are usually present in relatively large total amounts, but the plant-available fraction and the concentration in the soil solution may be insufficient to satisfy plant requirements. In such a case, a desirable characteristic of efficient genotypes would be a capacity to increase a nutrient fraction that is available to them and hence to take up relatively larger amounts of that nutrient (uptake efficiency), even though nutrient utilization efficiency, e.g., increased amounts of dry matter produced per unit of nutrient taken up, may also be important (Rengel, 1999a, 2001).

Uptake efficiency of soil-grown plants may consist of increased capacity to solubilize nonavailable nutrient forms into plant-available ones, and/or increased capacity to transport nutrients across the plasma membrane (Rengel, 2001). However, it appears likely that the increased capacity to convert nonavailable into available nutrient forms is of greater importance for efficient uptake, especially for nutrients that are transported to roots by diffusion, a relatively slow process that frequently limits nutrient uptake. For diffusion-supplied nutrients, larger amounts are transported toward roots if a larger concentration gradient between the root surface and the bulk soil can be maintained by vigorous nutrient uptake at the root surface. However, when a capacity of root cells to take up nutrients exceeds the rate of nutrient replenishment at the root surface, the uptake rate is governed by the nutrient supply rather than by the capacity of plants to take up nutrients (Rengel, 1993). Therefore, an increased capacity of root cells to take up nutrients due to an increased expression of high-affinity nutrient uptake systems in the plasma membranes under nutrient-deficiency stress because of a greater number of transporters, a greater affinity of these transporters for the nutrient, or a greater rate of turnover of transporters is expected to be of secondary importance as an efficiency mechanism for diffusion-supplied nutrients, e.g., P, K, Zn, Mn, and Cu, where the supply to the root surface is the rate-limiting step. It is important to acknowledge a theoretical possibility that plants with an enhanced capacity to take up nutrients at the root surface, e.g., as a consequence of genetic manipulation, will maximize the nutrient gradients between the rhizosphere and the bulk soil and will therefore keep diffusion operating at its maximum. However, even at the maximum level, diffusion is still inherently slow. Arguably, a much greater effect will be achieved by modifications obtained by transformation of plants to increase the capacity to exude chelating and other agents into the rhizosphere (e.g., Herrera-Estrella, 2000; Koyama et al., 2000; Lopez-Bucio et al., 2000), resulting in increased solubilization and conversion of nutrients into plant-available forms that can be transported across the root-cell plasma membrane (but see also conflicting opinions, e.g., Delhaize et al., 2001).

A number of possible mechanisms of plant adaptation to nutrient deficiencies have been suggested (Marschner, 1995; Welch, 1995; Rengel, 1997, 1999a, 2001). These mechanisms can be separated into two main categories: uptake efficiency and utilization efficiency. By identifying efficient traits for particular nutrients and locating the gene(s) responsible, it may be possible to introduce these genes to cultivars or species with other desirable attributes that are inefficient at acquiring or utilizing that particular nutrient.

In terms of uptake efficiency, plants may differ in the method used to acquire nutrients from environments with low nutrient availability, thus avoiding nutrient-deficiency stress. Plants may change the morphology of their root systems, which may be inherently different among species or nutrient deficiencies (e.g., Römheld, 1990; Dong et al., 1995; Halsted and Lynch, 1996; Johnson, Allan, et al., 1996; Rengel and Wheal, 1997a; Caradus and Woodfield, 1998), so that the plants are better able to take up nutrients (Römheld, 1990; Rengel and Graham, 1995c; Cakmak, Öztürk, et al., 1996; Rengel and Wheal, 1997a). Plants may also form symbiotic associations that aid the uptake of nutrients from the atmosphere or soil, such as with *Rhizobia* or mycorrhizae (Rengel, 2002a). There may also be differences in the physiology and biochemistry of the plant, such as increases in the affinity of specific nutrient uptake systems on the plasma membrane (Hawkesford and Belcher, 1991; Von Wiren et al., 1995; Rengel and Graham, 1996; Rengel and Wheal, 1997b; Erenoglu et al., 1999; Hawkesford, 2000). Plants may also modify the rhizosphere by secretion of chelating agents, reducing compounds, and so on to increase the availability of nutrients (Zhang et al., 1991; Hoffland et al., 1992; Cakmak et al., 1994; Cakmak, Sari, Marschener, Ekiz, et al., 1996; Zhang et al., 1997; Rengel et al., 1998; Neumann et al., 1999; Neumann and Römheld, 1999, 2001;

Rengel, 1999a,b,c; Erenoglu et al., 2000; Rengel and Römheld, 2000b; Gaume et al., 2001).

Increased efficiency at which the nutrient is used within the plant (utilization efficiency) may be due to increasing the rate at which the nutrient is transported (both long- and short-distance transport) or compartmented within cells (Graham and Rengel, 1993; Rengel, 1995a,b; Cakmak, Öztürk, et al., 1997; Rengel and Hawkesford, 1997; Tong et al., 1997; Caradus et al., 1998; Dong et al., 1998, 1999; Yu et al., 1998, 1999). Retranslocation of nutrients under nutrient-deficiency stress within the shoot may also differ between species and cultivars (Cakmak, Sari, Marschner, Ekiz, et al., 1996, Cakmak, Sari, Marschner, Kalayci, et al., 1996; Cakmak, Ekiz, et al., 1997; Pearson and Rengel, 1997a; Rengel et al., 1998; Cakmak, Tolay, Ozdemir, et al., 1999).

A mechanistic explanation of differential efficiency among genotypes of crop plants is still lacking for most nutrients studied (Rengel, 1999a, 2001). It may, however, be assumed that (1) efficiency mechanisms vary among crop species and genotypes, (2) more than one mechanism is often responsible for the level of nutrient efficiency in a particular genotype, (3) increased efficiency of one genotype in comparison to the other is due to involvement of additional mechanisms not present (expressed) in the less-efficient genotype, and (4) expression of more than one efficiency mechanism is likely to result in an additive effect. The mechanisms of nutrient efficiency may operate at various levels of plant organization (molecular, physiological, structural, or developmental).

Elucidation of mechanisms of nutrient efficiency is hampered by an inability to distinguish between causes and effects of nutrient deficiency (Rengel, 1999a). The time between withdrawal of a nutrient and the time when measurable changes can be observed is relatively long: hours for macronutrients and frequently days for micronutrients (Rengel, 1999a, 2001). Strictly speaking, this should be the time of imposition of deficiency rather than withdrawal of a nutrient because, in reality, plants do not become deficient immediately but only when internal nutrient stores required for normal metabolic activities are exhausted. As a consequence, a large number of secondary (indirect) effects may govern the observed physiological and morphological responses.

ADAPTATION TO ENVIRONMENTS LOW IN P, FE, ZN, OR MN

Plants exude a variety of organic compounds, including carboxylate anions, phenolics, carbohydrates, amino acids, enzymes and other proteins, and inorganic ions, including protons, phosphate, and other nutrients, into the rhizosphere to change the chemistry and biology of the rhizosphere and enhance adaptation to a particular environment (Crowley and Rengel, 1999). Common conditions under which plants will increase exudation of organic compounds into the rhizosphere encompass deficiency of various nutrients, e.g., P, Fe, Zn, or Mn (Marschner and Römheld, 1994; Azaizeh et al., 1995; Rengel, 1997, 1999a, 2001; Subbarao et al., 1997; Rengel et al., 1998).

Understanding the role that root exudates play in increasing plant adaptation to the given set of environmental conditions is sketchy at present (for a recent review see Rengel, 2002b). Similarly, regulation of the complete exudation sequence as well as underlying genetics is in only the initial stages of being unraveled. The two most popular strategies in increasing root exudation, overexpressing genes coding for critical enzymes in the biosynthesis of carboxylate anions and engineering plants to increase exudation of phosphatase and phytase (Delhaize, 1995), are promising approaches currently being tested in several laboratories. In addition, we should also utilize genotypic differences among existing crop germplasm, including wild relatives (Cakmak, Öztürk, et al., 1996) and, particularly for cereals, rye as the cereal most tolerant to a wide range of nutrient-deficiency and ion-toxicity stresses (see Cakmak, Ekiz, et al., 1997). Using genotypes from the local crop germplasm in the breeding program would be advantageous because development of cultivars with tolerance to rhizosphere-related stresses may be expedited without severely disrupting the broad adaptation already selected and proved to work effectively, including regulation of relevant biochemical reactions involved in root exudation at appropriately enhanced levels.

Phosphorus Deficiency

Plant species as well as genotypes within the species differ in P efficiency when grown in soils with low P availability (e.g., Caradus, 1994, 1995; Randall, 1995; Mamo et al., 1996; Trolove et al., 1996). Generally, three broad categories of P-efficiency mechanisms exist in plants to increase availability and uptake of P under P deficiency (Randall, 1995; Pearson and Rengel, 1997b; Rengel, 1999a, 2001): (1) secretion or exudation of chemical compounds into the rhizosphere, (2) alteration of the geometry or architecture of the root system, and (3) association with microorganisms.

Differences between genotypes in utilization efficiency are sometimes found when the amount of shoot dry matter produced per unit of P acquired is considered (Hedley et al., 1994; Stelling et al., 1996; Rao et al., 1997; Subbarao et al., 1997), but there are numerous contrary examples (e.g., Buso and Bliss, 1988; Jones et al., 1989; Gourley et al., 1994; Kemp and Blair, 1994). Attempts to screen genotypes for P efficiency based on an index of utilization efficiency, while sometimes apparently successful with a greater utilization index associated with P-efficient genotypes (e.g., Trolove et al., 1996), should be avoided (Jones et al., 1989; Blair, 1993) because such indices may not reflect efficiency mechanisms but just a generally better physiological status of the P-efficient in comparison to P-inefficient genotypes when grown under P deficiency. However, both P uptake and P utilization efficiency are important in selecting wheat genotypes for capacity to produce good grain yield, even though uptake efficiency was more important on low-P soil and utilization efficiency on high-P soil (Manske et al., 2001). In addition, by combining high P-uptake efficiency of the donor variety with a high harvest index characteristic of modern varieties, it was possible to more than triple the grain yield of rice under P deficiency (Wissuwa and Ae, 2001).

Under P deficiency, plants exude a wide range of organic and inorganic compounds to increase mobilization of P from sparingly soluble sources (for references see Rengel, 1999a, 2001). The genetic basis of regulation of carboxylate anion exudation, now in the process of being deciphered (Rengel, 2002b), will provide important knowledge for breeding and transforming crop species for enhanced root exudation.

In white lupin cluster roots subjected to P deficiency, carbon required for relatively large amounts of citrate exuded comes primarily from nonphotosynthetic CO_2 fixation (Hoffland et al., 1992; Johnson, Allan, et al., 1996) brought about by increased activity of phosphoenolpyruvate carboxylase (PEPC) (Johnson, Vance, and Allan, 1996; Neumann et al., 1999). Malate dehydrogenase and citrate synthase may also be increased (Allan and Vance, 1999), even though Neumann and colleagues (1999) did not find an increase in citrate synthase activity under P deficiency. The PEPC gene has been cloned from white lupin cluster roots subjected to P deficiency; in situ hybridization showed increased expression of that gene in the cortex of cluster roots (Allan and Vance, 1999), thus confirming its involvement in organic acid anion accumulation and exudation. Intensive work is underway in several laboratories to decipher the sequence of events leading to increased exudation of organic acid anions and cloning of the genes involved.

In addition to cloning genes involved in production of organic acid anions in white lupin, genes involved in biosynthesis of acid phosphatase are being cloned at an increasing pace (e.g., Ozawa et al., 1995; Wasaki et al., 1997; Allan and Vance, 1999; Haran et al., 2000; Baldwin et al., 2001; Hegeman and Grabau, 2001; Miller et al., 2001; Nishikoori et al., 2001; Yan et al., 2001). Cloned genes will allow transformation of various plant species in order to increase exudation of acid phosphatase and thus enhance plant capacity to utilize organic forms of P.

Many crop plants may form symbiotic associations with arbuscular mycorrhizae (AM) and thus gain access to soil pools of P that are unavailable to nonmycorrhizal plants (e.g., Smith et al., 1994; Rengel, 2002a). However, considerable genetic variability in the AM colonization capacity exists among various genotypes of host species, e.g., bell pepper and tomato (Nemec and Datnoff, 1993), barley (Boyetchko and Tewari, 1995), wheat (Al-Karaki and Al-Raddad, 1997), and grapevine (Karagiannidis et al., 1995). In addition, the genotype of the fungus is important because some degree of host specificity exists in AM fungi (Boyetchko and Tewari, 1995). Substantial variation also exists among crop genotypes in responsiveness to mycorrhizae as well as in P efficiency in the absence of mycorrhizae, e.g., in maize (Kaeppler et al., 2000).

A number of genes involved in AM symbiosis have been cloned and characterized (see Gianinazzi-Pearson, 1996; Murphy et al., 1997; Krajinski et al., 1998, 2000; Ruiz-Lozano et al., 1999; Salzer et al., 2000). However, it would be fair to say that genes known so far represent just the tip of the iceberg of what is involved in the effective symbioses. With the expanding number of genes cloned and with insights into transcription regulation and product function, we should see the benefits from translating that knowledge into breeding for better symbioses (Rengel, 2002a).

Clearly, it would be too dangerous to try to enhance the symbiosis, mycorrhizae as well as nodulation, by down-regulating plant defense responses because susceptibility to pathogen attack may increase. This is the case even though Myc$^-$ (e.g., David-Schwartz et al., 2001; Roussel et al., 2001) and Nod$^-$ (e.g., Bhatia et al., 2001; Oldroyd, 2001) mutants do not appear to suffer from increased susceptibility to pathogen attack (Rengel, 2002a), indicating considerable specificity in the infection pattern and localization of changes in plant defense responses of AM fungi and rhizobia as opposed to pathogens. In contrast, the fact that Myc$^-$ mutants are also Nod$^-$ mutants, at least in pea, indicate that there are common mechanisms regulating plant–microbe interactions in the two symbioses (Gianinazzi-Pearson, 1996), thus raising the possibility that breeding efforts to improve one symbiosis may fortuitously result in improvements to the other.

Iron Deficiency

Plants exhibit two distinctly different strategies of coping with Fe deficiency (for reviews see Marschner and Römheld, 1994; Mori, 1994; Welch,

1995; Jolley et al., 1996; Pearson and Rengel, 1997b; Crowley and Rengel, 1999; Rengel, 1999a, 2001). Strategy I, characteristic for dicots and nongraminaceous monocots, involves enhanced reduction of Fe^{3+} to Fe^{2+} at the root-cell plasma membrane; chemical changes in the rhizosphere, namely acidification and exudation of reducing and/or chelating compounds; and root histology and morphology changes, including formation of rhizodermal transfer cells, induced root branching, and more root hairs. Strategy II, employed by grasses, involves increased exudation of phytosiderophores into the rhizosphere. Within both Strategy I and Strategy II plants, various species and genotypes within species differ in Fe efficiency (Römheld and Marschner, 1990; Jolley and Brown, 1994; Pearson and Rengel, 1997b; Rengel, 1999a, 2001).

In Strategy II plants, the release of phytosiderophores and the subsequent uptake of the Fe-phytosiderophore complex are under different genetic controls (Römheld and Marschner, 1990). The rate of release of phytosiderophores is positively related to the Fe efficiency of various crop species (Römheld and Marschner, 1990; Mori, 1994). Under Fe-deficiency conditions, the Fe-efficient genotypes of wheat (Hansen et al., 1995, 1996) and oats (Jolley and Brown, 1989; Hansen and Jolley, 1995) exuded more phytosiderophores than Fe-inefficient genotypes, making a phytosiderophore-exudation test suitable for screening genotypes in the breeding program.

A number of genes involved in biosynthesis of phytosiderophores and their precursor nicotianamine have been cloned (Mori, 1997; Nakanishi et al., 2000; Yamaguchi et al., 2000a,b), including a gene family coding for nicotianamine synthase (Higuchi et al., 1999). Regulation of gene expression and the mosaic of biosynthetic pathways involved in phytosiderophore production are being deciphered at an increasing pace (Itai et al., 2000; Higuchi et al., 2001; Kobayashi et al., 2001). The availability of genes as well as an understanding of their induction and regulation will be crucial in future attempts to genetically engineer Strategy II crops for increased capacity to synthesize and exude phytosiderophores and thus improve Fe nutrition in environments low in available Fe.

Zinc Deficiency

In comparison with Zn-inefficient genotypes, Zn-efficient ones have a number of efficiency mechanisms that allow them to better withstand the Zn-deficiency stress (Rengel, 1999a, 2001). In wheat, there is usually a good relationship between root exudation of phytosiderophores and differential Zn efficiency when durum (inefficient) and bread (efficient) wheats are compared (Cakmak et al., 1994; Cakmak, Sari, Marschner, Ekiz, et al., 1996; Cakmak, Sari, Marschner, Kalayci, et al., 1996; Rengel et al., 1998; Rengel, 1999a, 2001). However, the positive correlation between phytosiderophore exudation and tolerance of bread wheat genotypes to Zn deficiency may not always hold, as shown for six (Erenoglu et al., 1996) and seven bread wheat genotypes (Cakmak et al., 1998).

Wheat genotypes considered Zn efficient (Cakmak et al., 1994; Walter et al., 1994; Cakmak, Sari, Marschner, Ekiz, et al., 1996; Cakmak, Sari, Marschner, Kalayci, et al., 1996; Cakmak, Ekiz, et al., 1997; Cakmak et al., 1998; Hopkins et al., 1998; Rengel et al., 1998) as well as those that are Fe efficient (Hansen and Jolley, 1995; Hansen et al., 1996) exude considerably larger amounts of phytosiderophores into the rooting medium than genotypes sensitive to Zn and Fe deficiencies (Zn and Fe inefficient) in order to increase mobilization of Zn and Fe from sparingly soluble sources. However, the possibility that Fe and Zn efficiencies are genetically similar and coded for by the same genes appears somewhat remote (Rengel and Römheld, 2000a,b), even though some genotypes posseses both efficiencies.

When grown in soils of low Zn availability, Zn-efficient wheat genotypes had greater fertilizer efficiency (grain yield produced per unit of fertilizer applied) and a greater harvest index than did the Zn-inefficient genotypes (Rengel and Graham, 1995b). The greater fertilizer efficiency of the Zn-efficient wheat genotypes compared to the Zn-inefficient ones offers some promise in managing soils low in available Zn by a combination of (1) growing Zn-efficient genotypes and (2) applying Zn fertilizers at low rates. Given that Zn deficiency is typically patchy in a single field (Kubota and Allaway, 1972), growing Zn-efficient genotypes can overcome spatial variability in the horizontal and vertical direction (see Nable and Webb, 1993) as well as temporal variation in Zn availability to plants.

Different seed Zn contents may differentially influence performance of different genotypes under Zn-deficient conditions. Higher seed Zn is more beneficial to Zn-inefficient than to Zn-efficient genotypes (Rengel and Graham, 1995a,b; Grewal and Graham, 1997; Rengel et al., 1999), the latter being able to make up for low seed Zn by extracting more Zn from the soil.

Grain is a poor sink for Zn loading in comparison with other tissues under Zn deficiency (Rengel and Graham, 1995b), but it remains to be established whether various crop genotypes have different potential for loading of nutrients into various seed tissues and/or remobilization of these nutrients to a young seedling (Pearson and Rengel, 1994; Pearson et al., 1995, 1998, 1999). This is of particular interest for genotypes that differ in nutrient efficiency, because greater nutrient efficiency may start as a faster and more complete remobilization of nutrients from various seed tissues. Preliminary results indeed show that wheat genotypes have different abilities to load the nutrients to, and remobilize them from, different parts of seeds (Moussavi-Nik et al., 1997). However, no correlation was found between the level of Zn efficiency of the genotypes studied and the early growth while dependent on seed-nutrient reserves.

Zn-efficient genotypes of bean *(Phaseolus vulgaris)* loaded more Zn into seeds than Zn-inefficient genotypes in field experiments (Moraghan and Grafton, 1999). There was also a significant difference in the capacity of 285 rice genotypes grown in a single trial to load Zn into seeds, with *indica* varieties having two times greater Zn content in polished grain than *japonica* types (Yang et al., 1998). Greater Zn concentration in grain destined for human consumption will improve the nutrition of populations reliant mostly, or even exclusively, on grain for a supply of micronutrients.

Manganese Deficiency

Yield of cereals grown on calcareous soils is frequently limited by Mn deficiency resulting from low Mn availability, rather than from low Mn content in the soil (Welch et al., 1991; Rengel, 2000). On such soils, Mn fertilizers are unsuitable for correction of crop deficiency because of a really low residual value and conversion to plant-unavailable forms. An alternative strategy of correcting problems associated with soils containing low amounts of available Mn is to breed crop cultivars with greater Mn efficiency. Mn-efficient genotypes are usually able to extract more Mn from soils with limited Mn availability (Graham, 1988; Rengel et al., 1994; Rengel, 2000). The physiological mechanisms underlying Mn efficiency are poorly understood (Graham, 1988; Huang et al., 1993, 1994; Huang, Webb, and Graham, 1996; Rengel et al., 1994; Rengel, 2000).

Some attempts at breeding for increased Mn efficiency have been done in South Australia because Mn deficiency is the most intractable trace element problem in crops grown on many soils in that region (Rengel et al., 1994). The interest in breeding for Mn efficiency was also stimulated by a string of Mn-inefficient cultivars, released unwittingly by cereal breeders in South Australia, e.g., 'Galleon' and 'Skiff' barley, 'Coorong' triticale, and 'Condor,' 'Bayonet,' and 'Millewa' wheat. Field variety trials on Mn-deficient soils using paired subplots of +Mn or –Mn treatments for each genotype showed a wide range of Mn efficiencies at each site, evidence that considerable improvement in Mn efficiency is possible in modern cultivars with the germplasm already available. Further improvement would be possible through the transfer of Mn efficiency from rye that is outstandingly efficient by comparison with wheat. From studies on rye addition lines, it appears that efficiency is carried on the 2R chromosome (Graham, 1988;

Cakmak, Derici, et al., 1997; Cakmak, Ekiz, et al., 1997; Schlegel et al., 1997).

Until recently, little was known about the genetics of Mn efficiency in higher plants. Recent experiments with barley suggest the involvement of simple mechanisms of Mn efficiency controlled by a single, major, dominant gene (McCarthy et al., 1988; Longnecker et al., 1990). In the study of 72 barley genotypes from a world collection, the pedigree relationships within the most efficient and the most inefficient groups were consistent with single, major-gene inheritance (see Rengel et al., 1994). In the study in which progeny from the cross between barley genotypes Weeah (Mn efficient) and WI-2585 (Mn inefficient) was visually assessed for the leaf-chlorosis score (McCarthy et al., 1988), the distribution of F_2 individuals followed a 3:1 ratio characteristic of a single dominant gene for Mn efficiency without overdominance. In further work with F_3 individuals, this conclusion was substantiated, and a narrow-sense heritability of 71 percent for Mn efficiency was found (see Rengel et al., 1994).

In durum wheat (*Triticum turgidum* L. var. *durum*), the existence of considerable genetic variation in the current germplasm of durum wheat and the development of a consistent selection criterion, Mn content of 35-day-old seedlings, have made breeding for Mn efficiency feasible (Khabaz-Saberi, Graham, and Rathjen, 1999). The F_1 progeny of the cross between Mn-efficient and Mn-inefficient genotypes was intermediate to parents, i.e., medium efficient and inefficient, indicating incomplete dominance and dependence on external Mn concentration, while segregating ratios for F_2 and F_3 families indicated inheritance by two genes (Khabaz-Saberi, Graham, and Rathjen, 1999). However, breeding for increased Mn efficiency is hampered by the importance of seed Mn content in screening durum wheat germplasm (Khabaz-Saberi et al., 2000), making screening via molecular markers (Khabaz-Saberi, Barker, et al., 1999) an attractive possibility.

MECHANISMS GOVERNING NITROGEN EFFICIENCY

Nitrogen-Use Efficiency

Nitrogen-use efficiency, units of dry matter produced per units of N taken up, measures the capacity of a given genotype to take up, and convert to yield, nitrogen applied as fertilizer. Worldwide, nitrogen-use efficiency (NUE) in production of wheat, maize, rice, barley, sorghum, *Pennisetum glaucum*, oats, and rye is approximately 33 percent. The unaccounted for 67 percent represents a US$15.9 billion annual loss of N fertilizer, assuming fertilizer-soil equilibrium (Raun and Johnson, 1999). In Germany, NUE in crop production decreased from around 100 percent in 1951 to 81 percent in 1990, with an increase in N fertilization from 27 to 108 $kg \cdot ha^{-1}$, and a corresponding N surplus increasing from 7 to 95 $kg \cdot ha^{-1}$ in the same period (van der Ploeg et al., 1997). Clearly, one of the basic principles of sustainable agriculture is to optimize NUE in agriculture.

Slow-release N fertilizers (Wiesler, 1998) generally result in higher NUE by a range of crops and crop genotypes, including rice (Ohnishi et al., 1999; Wani et al., 1999) and sunflower *(Helianthus annuus)* (Reddy and Sharma, 2000). Various other agronomic improvements in placements of fertilizer and planting geometry may improve N-use efficiency, e.g., for onion *(Allium cepa),* lettuce *(Lactuca sativa)* (Stone, 2000), and sugarcane (Yadav et al., 1997). NUE decreases with an increase in N supply (e.g., Buah et al., 1998; Maman et al., 1999). However, for wild species, NUE increases with an increase in nutrient availability in their preferred native habitat up to a certain nutrient availability and then decreases (Vazquez de Aldana and Berendse, 1997). NUE is related to N-uptake efficiency of roots and metabolic efficiency of absorbed N in roots and shoots (Jiang et al., 2000). Among wheat genotypes, some lines had high efficiency in taking up fertilizer N, while others preferred nonfertilizer soil N, probably due to differences in the root system (Anderson and Hoyle, 1999). Similarly, in lupin, characteristics of the root-growth response as well as the capacity to take up N are determinants of the nitrate-capturing capacity of different lupin species (Dunbabin et al., 2001a,b).

Genotypic variation in NUE has been reported for a number of crops, e.g., rice (Singh et al., 1998; Ohnishi et al., 1999), maize (Bertin and Gallais, 2000), Kentucky bluegrass *(Poa pratensis)* (Jiang et al., 2000), timothy *(Phleum pratense)* (Michaud et al., 1998), sorghum (Buah et al., 1998), bread and durum wheats (Ehdaie et al., 2001), bread wheat (Anderson and Hoyle, 1999), canola *(Brassica napus)* (Kessel and Becker, 1998), and sweet potato *(Ipomoea batatas)* (Villagarcia et al., 1998), thus setting the stage for selection and breeding for increased NUE.

Various strategies for selecting and breeding for increased NUE have been suggested. In canola, a breeding strategy for improved N efficiency in canola can be based on selection for low N content in dropped leaves and for high N harvest index by reducing N yield of straw (Kessel and Becker, 1998). In wheat, selection of genotypes for N-uptake efficiency under low N input should improve NUE for both yield and grain protein content (Ehdaie et al., 2001). In contrast, in the field experiment with a range of Australian wheat genotypes, no lines were both yield and protein efficient, suggesting that lines either use fertilizer nitrogen preferentially in yield production or in production of protein (Anderson and Hoyle, 1999).

An improved understanding of nitrogen assimilation will be vital if improvements in crop NUE are to be made to reduce the need for excessive input of fertilizers (Harrison et al., 2000). It appears that leaf nitrate accumulation and the reactions catalyzed by nitrate reductase and glutamine synthetase (glutamate-ammonia ligase) are coregulated and represent key elements controlling NUE in maize (Hirel et al., 2001). However, it should be borne in mind that the limiting steps in N metabolism appeared to be different at low and high levels of N fertilizer (Bertin and Gallais, 2000). Using tobacco and *Lotus* as model plants, glutamine synthetase (glutamate-ammonia ligase) and glutamate synthase (GOGAT) activities have been altered by stimulating or inhibiting the expression of the corresponding genes in an organ- or tissue-specific manner (Harrison et al., 2000).

N_2 Fixation

Dinitrogen fixation provides more N to the agricultural ecosystems worldwide than the total amount of fertilizer N applied. However, a significant potential exists to further improve symbiotic and associative N_2 fixation by breeding genotypes with a greater capacity to sustain these interactions with bacteria (Rengel, 2002a). Better N_2-fixing symbiosis may be brought about by manipulating both rhizobia and plant hosts and by eventually creating an artificial rhizosphere. An important aim also is to improve the symbiotic relationship in suboptimal environmental situations related to soil-borne or environmental stress. However, it is still unclear whether attempts to increase the symbiotic capacity where symbiosis is already fully functional (overexpression, supernodulation, etc.) will be beneficial in terms of increasing crop yields.

The Nodulation Process in the Rhizobia-Legume Symbiosis

Inheritance of the process of nodulation may be relatively simple (Rengel, 2002a). Recent discovery of nonfixing mutants with altered patterns of nodulation has revealed that genes involved in the control of the nodulation process are recessive and may not be too numerous (Tsyganov et al., 1998). Faba bean mutants unable to nodulate as well as those with ineffective nodulation and supernodulation had mutations of single recessive genes (Duc, 1995). However, grafting experiments showed that a supernodulating mutant was influenced by the shoot genotype, while the nonnodulating and ineffective nodulation mutants were influenced by the root genotype, indicating that the regulation of gene expression may be more complicated.

Knowledge of the genetics behind exudation of specific *nod*-gene inducers in host root exudates is sketchy at present (Rengel, 2002a). Advancements may come from studying legume host genotypes that differ in the components of the signaling pathway, e.g., in rice (Soejima et al., 1992), bean *(Phaseolus vulgaris)* (Hungria and Phillips, 1993), and lucerne *(Medicago sativa)* (Hernandez et al., 1995), but considerably more genotypic testing needs to be done, including the development of isogenic lines (see Hungria and Phillips, 1993; Shelp et al., 1998), to provide an excellent genetic material for studies aimed at discerning genetics of *nod*-gene inducer exudation.

Transgenic crops will provide excellent material for studying various aspects of the symbiotic processes and will thus prove valuable in breeding for more effective symbioses. For example, transgenic clover *Trifolium repens* (Rolfe et al., 1997) has been used to study genetic programs activated during nodule initiation and development.

Genotypic Differences in Symbiotic N_2 Fixation

A continuous and coordinated selection of the most effective combinations of host and microbial symbionts is a prerequisite for profitable and sustainable agricultural systems. Different legume genotypes differ in the efficacy of symbiotic N_2 fixation they can support (for references see Rengel, 2002a). Genotypic differences in activities of glutamine synthetase (glutamate-ammonia ligase) and GOGAT in nodules were related to genotypic differences in symbiotic N_2-fixation efficiency, at least in faba bean (Caba et al., 1993). Therefore, screening for increased activities of these enzymes might be used in a breeding program to increase symbiotic efficiency of faba bean genotypes. A number of other symbiotic characteristics can be considered in breeding for better symbiosis. For example, chickpea genotypes differed in the total number and weight of nodules per plant as well as the weight of individual nodules (Dangaria et al., 1994). Other criteria related to the functioning of nodules may also be considered. In pea cultivars, N accumulation in shoots is a more suitable criterion than the shoot mass for selection of either highly effective or highly competitive *R. leguminosarum* bv. *viciae* strains in the pot experiments (Fesenko et al., 1995).

Various rhizobia strains differ in N_2-fixation capacity; it is therefore possible to select for more efficient strains (Hartmann et al., 1998; Hungria et al., 1998). As the effectiveness of the nodulation and the N_2-fixation process varies widely in different rhizobia-host combinations, it would be worthwhile to identify the highly effective rhizobial strains, followed by identifying responsible genes that regulate high effectiveness of the symbiosis. These genes could then be transferred into commercial strains of rhizobia to increase efficiency of N_2 fixation or their host range (Rengel, 2002a).

It would be particularly important to develop genotypes that maintain high efficiency of N_2 fixation in the presence of small or moderate levels of nitrate (Rengel, 2002a). Otherwise, nitrate produced from mineralization of the organic matter may not be taken up by legumes (Dunbabin et al., 2001a,b), but instead would leach into the lower horizons or ground water (see Anderson et al., 1998) and/or may hinder the symbiotic process. Genotypic variability exists among host genotypes in the degree of tolerance of the N_2-fixation process to nitrate (Raffin and Roumet, 1994). Additional genotypic variability and tolerance to nitrate can be introduced by mutagenesis, e.g., in soybean (Neo et al., 1996). Also, a genetic variability within nodule bacteria germplasm in tolerance of the N_2-fixation process to nitrate (e.g., Leung et al., 1994; Nour et al., 1994) needs to be taken into account when creating a breeding program for better symbioses.

Screening genotypes for nodulation and N_2 fixation in hydroponics under controlled environmental conditions needs to be calibrated against field performance of standard cultivars because nodulation of, for example, bean genotypes, may be different between controlled conditions and the field, even when growth in the two environments appears similar (Hernandez et al., 1993). Early screening of soybean genotypes for nodulation and N_2 fixation, 20 to 60 days after planting in both greenhouse and field, may generate useful parameters for breeding programs aimed at improving efficiency of the legume-rhizobia symbiosis (Pazdernik et al., 1997).

Wild genotypes of crop plants may serve as a source of genes for improved and more effective N_2 fixation (Andriolo et al., 1994). Similarly, host mutants, e.g., white sweetclover *(Melilotus alba)* (Utrup et al., 1993), unable to achieve complete and functional symbiosis with rhizobia, should be studied more to increase our understanding of the symbiotic process.

BREEDING GENOTYPES FOR GREATER NUTRIENT EFFICIENCY

Nutrient-efficient genotypes are important in modern agriculture because they can produce greater yields on soils where the effectiveness of fertilizers may be limited by chemical and biological reactions, topsoil drying, subsoil constraints, and/or disease interactions (e.g., Graham and Rengel, 1993; Rengel et al., 1994; Lynch, 1998; Rengel, 1999a, 2001). Some of these nutrient-efficient genotypes can be used in conventional breeding as a source of valuable genes conferring increased capacity for exudation of organic compounds into the rhizosphere. However, the relatively slow progress in deciphering the genetics, physiology, and biochemistry behind the mechanisms of nutrient efficiency has hampered development of genotypes of superior efficiency (tolerance to nutrient deficiency) through conscious breeding efforts geared specifically toward that purpose.

Breeding for greater Zn efficiency, at least in wheat, should be possible by (1) using local landraces that show significant Zn efficiency (Kalayci et al., 1999), (2) transferring rye genes controlling the Zn efficiency trait (Cakmak, Derici, et al., 1997; Schlegel et al., 1997), (3) using wild relatives of wheat (Schlegel et al., 1998; Cakmak, Cakmak, et al., 1999; Cakmak, Tolay, Ozkan, et al., 1999), and (4) transferring alien chromosomal material into wheat (Schlegel et al., 1998).

Breeding wheat and possibly other cereals for greater tolerance to micronutrient deficiency is possible by transferring rye genes controlling these tolerance traits. Individual genes conferring nutrient efficiencies have not been deciphered yet, but transferring rye chromosomes 1R and 7R into wheat has increased its Zn efficiency (Cakmak, Derici, et al., 1997), and transferring chromosomes 2R and 7R increased Mn and Fe efficiency in wheat (Schlegel et al., 1997). In contrast, there is no report linking Cu efficiency with root exudation. However, transferring the rye chromosome 5R (Schlegel et al., 1997) or transferring just a small piece of chromatin from the long arm of the chromosome 5R to wheat increases its Cu efficiency (Graham et al., 1987; Schlegel et al., 1993). It is known that genes for mugineic acid synthase and 3-hydroxymugineic acid synthase, the enzymes involved in biosynthesis of common phytosiderophores, are on the rye chromosome 5R (Mori et al., 1990), but they map outside the region conferring Cu efficiency to the wheat-rye addition line (Schlegel et al., 1993). These findings are in accordance with the fact that phytosiderophores do not appear to be involved in the uptake and transport of Cu (see Schlegel et al., 1993), but the possibility cannot be excluded that the function of the gene products coded for by the genes conferring Cu efficiency may in some way be connected with root exudation.

Studies of diploid, tetraploid, and hexaploid wheat showed that both A and D genomes appear to contribute to increased Zn efficiency (Cakmak, Tolay, Ozdemir, et al., 1999). In addition, wild relatives of wheat, particularly *Aegilops speltoides* var. *ligustica*/SS genome/ and *Ae. triuncialis* /UUCC genome/, may be donors of Zn-efficiency-related genes (Cakmak, Cakmak, et al., 1999; Cakmak, Tolay, Ozkan, et al., 1999). Some more distant relatives, and even alien plant species such as *Agropyron* and *Haynaldia*, have chromosomes carrying major Zn-efficiency-related genes that are clearly expressed upon transfer into wheat (Schlegel et al., 1998).

It is yet unclear to what extent possible overexpression of Zn transporter genes (Eide, 1997; Van der Zaal et al., 1999; Guerinot, 2000; Lasat et al., 2000) in crop and pasture plants may alter their capacity to take up Zn from soils low in available Zn, but it is likely that the effect will not be large (see discussion in the section on Nutrient Efficiency). Nevertheless, with rapid advances in cloning membrane transporters of other nutrients, we may see in the near future a wave of attempts to genetically engineer crop and pasture plants via overexpression of genes coding for relevant membrane transporters in order to increase plant capacity to grow and yield well on soils with low nutrient availability (Rengel, 1999a, 2001). For example, increased exudation after transformation of plants with a single gene may result in new genotypes with improved adaptation to environmental stresses due to enhanced exudation of specific compounds (e.g., Herrera-Estrella, 2000; Koyama et al., 2000; Lopez-Bucio et al., 2000). Nevertheless, availability of promoters whose regulation is cell or tissue specific in roots (e.g., De Pater and Schilperoort, 1992; Nitz et al., 2001; Swoboda et al., 2001) is important in that respect. In addition, genetic engineering of border cells that are released from roots into the rhizosphere (O'Connell et al., 1996; Miyasaka and Hawes, 2001) offers a potential to influence the rhizosphere processes in addition to any exudation of specific organic and inorganic compounds from roots.

The various mechanisms of efficiency are likely to be additive, as shown for Zn efficiency in rice (Majumder et al., 1990), putting an emphasis on stepwise compounding of genetic information for a number of nutrient-efficiency mechanisms into one locally adapted crop cultivar. In such a breeding program, genotypes with genes that control a particular mechanism of efficiency may be important, even though they themselves may not show phenotypically high nutrient efficiency. Using genotypes from local crop germplasm would be advantageous because the development of cultivars with improved nutrient efficiency may be achieved without disrupting previously selected adaptation to the local environmental conditions.

QTLs have been recently identified for macronutrient (N, P, and K) efficiency in maize and rice (Ni et al., 1998; Wissuwa et al., 1998; Wu et al., 1998; Agrama et al., 1999; Kaeppler et al., 2000; Bertin and Gallais, 2001; Hirel et al., 2001; Hu et al., 2001) and for Fe deficiency in soybean (Lin et al., 1998; Cianzio, 1999), but similar results are yet to be reported for the efficiency of other nutrients. Availability of QTLs is expected to speed up breeding for increased nutrient efficiency.

It is encouraging that there was good correspondence between QTLs identified in soybean screening for Fe efficiency in nutrient solutions and those identified in field screening (Lin, Baumer, et al., 2000), indicating that both screening systems identified similar genetic mechanisms of Fe uptake and/or utilization. Similarly, there was good correspondence between screening canola genotypes in a greenhouse pot trial and in the field (Grewal et al., 1997).

Screening for Nutrient Efficiency

Simple, fast, and inexpensive techniques are needed in breeding programs for assessment of a large number of genotypes in segregating populations without relying on expensive field testing (Graham, 1984; Graham and Rengel, 1993; Rengel, 1999a, 2001). The better nutrient efficiency mechanisms are understood, the more likely it is that a successful screening technique will be developed.

One of the major problems with nutrient deficiency is that the level of stress is frequently assessed based on plant growth. However, growth, as a very complex character influenced by a myriad of interrelated processes, is clearly not a suitable parameter for distinguishing the causes and effects of nutrient-deficiency stress (Rengel, 1999a, 2001). Better screening techniques need to be developed by targeting specific processes that cause or prevent nutrient deficiency, rather than those that appear as a consequence of it.

It is expected that in the near future molecular methods will dominate selection for nutrient efficiency. The development of molecular markers for regions of the genome segregating with the trait of a particular nutrient efficiency will be crucial in that respect. However, molecular markers may be population dependent, e.g., markers for Fe efficiency in soybeans (Cianzio, 1999; Lin, Grant, et al., 2000), making them unsuitable for use in breeding programs.

Even in screening based on molecular methods, seed nutrient content may need to be taken into consideration, as differential seed nutrient content frequently presents a confounding factor when screening methods are based on growth, e.g., Zn (Rengel and Graham, 1995a,b; Grewal and Graham, 1997; Genc et al., 1999, 2000), because the expression of two genes associated with Fe efficiency in barley (*Ids1* and *Ids2*) was greater after three than after two weeks of growth, possibly indicating exhaustion of seed Fe reserves after prolonged growth (Huang, Graham, et al., 1996).

Screening for Increased Root Exudation

Increasing the root exudation capacity may adversely affect crop yield because of increased partitioning of assimilates toward roots (Rengel, 2002b). Indeed, it appears that the selection of genotypes in breeding programs since the 1980s resulted in decreased capacity to exude organic compounds into the rhizosphere, e.g., in rice (Waschutza et al., 1992). As a result, advocating selection toward increased exudation may appear to be a step backward. However, instead of generally increased rates of exudation, a better and more economic solution obviously would be to increase exudation of only targeted compounds that perform a specific function in the rhizosphere, a function that will result in improved tolerance to nutrient deficiencies or in a more effective nodulation process (Rengel, 2002b). For such a selection pressure to be included in the breeding program, breeders would need to have well-defined screening procedures that rely heavily on understanding the physiological processes involved. Such knowledge at present appears to be inadequate.

Screening techniques based on characterizing root exudation are not in routine use at present, even though HPLC determination of phytosiderophores exuded under Fe deficiency has potential (Brown and Jolley, 1989; Hansen and Jolley, 1995; Hansen et al., 1995, 1996). The better the exudation sequence is understood, the more likely it is that a successful screening technique will be developed. However, plants are often grown in nutrient solution and root exudates are also collected in the nutrient solution or, even worse, in distilled water. However, mechanical impedance increases root exudation significantly (Boeuf Tremblay et al., 1995), suggesting that the rates of exudation calculated for soil-grown plants from the data on exudation by solution-grown plants have been underestimated (Rengel, 2002b).

CONCLUSION

Genotypes adapted to environments with low nutrient availability show increased nutrient uptake efficiency and/or utilization efficiency. Mechanisms underpinning nutrient efficiency are diverse, complex, and still poorly understood, except for Fe efficiency. Absence of effective screening techniques as well as a lack of specific molecular markers for nutrient efficiency hamper the efforts to breed genotypes with increased efficiency. However, traditional breeding techniques will be combined with transformation of plants with genes coding for nutrient transporters or specific enzymes in the biosynthesis of organic compounds and/or those involved in regulation of efflux of organic compounds across the membrane are gaining momentum.

The importance of genotypes adapted to environments with low nutrient availability is emphasized by the fact that the area of good agricultural land is shrinking, thus forcing agriculture onto more marginal land where nutrient availability may be jeopardized by soil chemical and physical conditions. Nutrient-efficient genotypes can be effectively used in modern agriculture in conjunction with fertilization, especially for nutrients with low fertilizer effectiveness and in areas where spatial and temporal variability hinders nutrient availability. These nutrient-efficient genotypes will allow more sustainable and profitable farming in the future to feed the rapidly increasing world population.

REFERENCES

Agrama, H.A.S., A.G. Zakaria, F.B. Said, and M. Tuinstra (1999). Identification of quantitative trait loci for nitrogen use efficiency in maize. *Molecular Breeding* 5:187-195.

Al-Karaki, G.N. and A. Al-Raddad (1997). Drought stress and VA mycorrhizal fungi effects on growth and nutrient uptake of two wheat genotypes differing in drought resistance. *Crop Research (Hisar)* 13:245-257.

Allan, D.L. and C.P. Vance (1999). Altered metabolism and gene expression in cluster roots of phosphorus-deficient white lupin. Presented at the 16th International Botanical Congress, August 1-7, 1999, St. Louis, MO.

Anderson, G.C., I.R.P. Fillery, F.X. Dunin, P. Dolling, and S. Asseng (1998). Nitrogen and water flows under pasture-wheat and lupin-wheat rotations in deep sands in Western Australia: 2. Drainage and nitrate leaching. *Australian Journal of Agricultural Research* 49:345-361.

Anderson, W.K. and F.C. Hoyle (1999). Nitrogen efficiency of wheat cultivars in a Mediterranean environment. *Australian Journal of Experimental Agriculture* 39:957-965.

Andriolo, J., P.A.A. Pereira, and R.A. Henson (1994). Variability among wild *Phaseolus vulgaris* lines for traits related to biological N_2 fixation. *Pesquisa Agropecuaria Brasileira* 29:831-837.

Azaizeh, H.A., H. Marschner, V. Römheld, and L. Wittenmayer (1995). Effects of a vesicular-arbuscular mycorrhizal fungus and other soil microorganisms on growth, mineral nutrient acquisition and root exudation of soil-grown maize plants. *Mycorrhiza* 5:321-327.

Baldwin, J.C., A.S. Karthikeyan, and K.G. Raghothama (2001). LEPS2, a phosphorus starvation-induced novel acid phosphatase from tomato. *Plant Physiology* 125:728-737.

Bertin, P. and A. Gallais (2000). Genetic variation for nitrogen use efficiency in a set of recombinant maize inbred lines. I. Agrophysiological results. *Maydica* 45:53-66.

Bertin, P. and A. Gallais (2001). Genetic variation for nitrogen use efficiency in a set of recombinant inbred lines. II. QTL detection and coincidences. *Maydica* 46:53-68.

Bhatia, C.R., K. Nichterlein, and M. Maluszynski (2001). Mutations affecting nodulation in grain legumes and their potential in sustainable cropping systems. *Euphytica* 120:415-432.

Blair, G. (1993). Nutrient efficiency: What do we really mean? In *Genetic Aspects of Plant Mineral Nutrition,* P.J. Randall, E. Delhaize, R.A. Richards, and R. Munns (Eds.). Dordrecht, The Netherlands: Kluwer Academic Publishers, pp. 205-213.

Boeuf Tremblay, V., S. Plantureux, and A. Guckert (1995). Influence of mechanical impedance on root exudation of maize seedlings at two development stages. *Plant and Soil* 172:279-287.

Bowen, G.D. and A.D. Rovira (1999). The rhizosphere and its management to improve plant growth. *Advances in Agronomy* 66:1-102.

Boyer, J.S. (1982). Plant productivity and environment. *Science* 218:443-448.

Boyetchko, S.M. and J.P. Tewari (1995). Susceptibility of barley cultivars to vesicular-arbuscular mycorrhizal fungi. *Canadian Journal of Plant Science* 75:269-275.

Brown, J.C. and V.D. Jolley (1989). Plant metabolic responses to iron deficiency stress. *Bioscience* 39:546-551.

Buah, S.S.J., J.W. Maranville, A. Traore, and P.J. Bramel-Cox (1998). Response of nitrogen use efficient sorghums to nitrogen fertilizer. *Journal of Plant Nutrition* 21:2303-2318.

Buso, G.S.C. and F.A. Bliss (1988). Variability among lettuce cultivars grown at two levels of available phosphorus. *Plant and Soil* 111:67-74.

Caba, J.M., C. Lluch, and F. Ligero (1993). Genotypic differences in nitrogen assimilation in *Vicia faba*: Effect of nitrate. *Plant and Soil* 151:167-174.

Cakmak, I., O. Cakmak, S. Eker, A. Ozdemir, N. Watanabe, and H.J. Braun (1999). Expression of high zinc efficiency of *Aegilops tauschii* and *Triticum monococcum* in synthetic hexaploid wheats. *Plant and Soil* 215:203-209.

Cakmak, I., R. Derici, B. Torun, I. Tolay, H.J. Braun, and R. Schlegel (1997). Role of rye chromosomes in improvement of zinc efficiency in wheat and triticale. *Plant and Soil* 196:249-253.

Cakmak, I., H. Ekiz, A. Yilmaz, B. Torun, N. Köleli, I. Gültekin, A. Alkan, and S. Eker (1997). Differential response of rye, triticale, bread and durum wheats to zinc deficiency in calcareous soils. *Plant and Soil* 188:1-10.

Cakmak, I., K.Y. Gülüt, H. Marschner, and R.D. Graham (1994). Effect of zinc and iron deficiency on phytosiderophore release in wheat genotypes differing in zinc efficiency. *Journal of Plant Nutrition* 17:1-17.

Cakmak, I., L. Öztürk, S. Eker, B. Torun, H.I. Kalfa, and A. Yilmaz (1997). Concentration of zinc and activity of copper/zinc superoxide dismutase in leaves of rye and wheat cultivars differing in sensitivity to zinc deficiency. *Journal of Plant Physiology* 151:91-95.

Cakmak, I., L. Öztürk, S. Karanlik, H. Marschner, and H. Ekiz (1996). Zinc-efficient wild grasses enhance release of phytosiderophores under zinc deficiency. *Journal of Plant Nutrition* 19:551-563.

Cakmak, I., N. Sari, H. Marschner, H. Ekiz, M. Kalayci, A. Yilmaz, and H.J. Braun (1996). Phytosiderophore release in bread and durum wheat genotypes differing in zinc efficiency. *Plant and Soil* 180:183-189.

Cakmak, I., N. Sari, H. Marschner, M. Kalayci, A. Yilmaz, S. Eker, and K.Y. Gülüt (1996). Dry matter production and distribution of zinc in bread and durum wheat genotypes differing in zinc efficiency. *Plant and Soil* 180:173-181.

Cakmak, I., I. Tolay, A. Ozdemir, H. Ozkan, L. Ozturk, and C.I. Kling (1999). Differences in zinc efficiency among and within diploid, tetraploid and hexaploid wheats. *Annals of Botany* 84:163-171.

Cakmak, I., I. Tolay, H. Ozkan, A. Ozdemir, and H.J. Braun (1999). Variation in zinc efficiency among and within *Aegilops* species. *Journal of Plant Nutrition and Soil Science* 162:257-262.

Cakmak, I., B. Torun, B. Erenoglu, L. Ozturk, H. Marschner, M. Kalayci, H. Ekiz, and A. Yilmaz (1998). Morphological and physiological differences in the response of cereals to zinc deficiency. *Euphytica* 100:349-357.

Cakmak, I., A. Yilmaz, M. Kalayci, H. Ekiz, B. Torun, B. Erenoglu, and H.J. Braun (1996). Zinc deficiency as a critical problem in wheat production in Central Anatolia. *Plant and Soil* 180:165-172.

Caradus, J.R. (1994). Selection for improved adaptation of white clover to low phosphorus and acid soils. *Euphytica* 77:243-250.

Caradus, J.R. (1995). Genetic control of phosphorus uptake and phosphorus status in plants. In *Genetic Manipulation of Crop Plants to Enhance Integrated Nutrient Management in Cropping Systems. 1. Phosphorus,* C. Johansen, K.K. Lee, K.K. Sharma, G.V. Subbarao, and E.A. Kueneman (Eds.). Patancheru, Andhra Pradesh, India: International Crops Research Institute for the Semi-Arid Tropics, pp. 55-74.

Caradus, J.R., L.D. Kennedy, and A. Dunn (1998). Genetic variation for the ratio of inorganic to total phosphorus in white clover leaves. *Journal of Plant Nutrition* 21:2265-2272.

Caradus, J.R. and D.R. Woodfield (1998). Genetic control of adaptive root characteristics in white clover. *Plant and Soil* 200:63-69.

Cianzio, S.R. (1999). Breeding crops for improved nutrient efficiency: soybean and wheat as case studies. In *Mineral Nutrition of Crops: Fundamental Mechanisms and Implications,* Z. Rengel (Ed.). Binghamton, NY: Food Products Press, pp. 267-287.

Crowley, D.E. and Z. Rengel (1999). Biology and chemistry of nutrient availability in the rhizosphere. In *Mineral Nutrition of Crops: Fundamental Mechanisms and Implications,* Z. Rengel (Ed.). Binghamton, NY: Food Products Press, pp. 1-40.

Dangaria, C.J., R. Parameshwarappa, P.M. Salimath, and B.S. Annigeri (1994). Genetic divergence for nodulating characters in chickpea. *Legume Research* 17:32-36.

David-Schwartz, R., H. Badani, W. Smadar, A.A. Levy, G. Galili, and Y. Kapulnik (2001). Identification of a novel genetically controlled step in mycorrhizal colonization: Plant resistance to infection by fungal spores but not extra-radical hyphae. *Plant Journal* 27:561-569.

Delhaize, E. (1995). Genetic control and manipulation of root exudates. In *Genetic Manipulation of Crop Plants to Enhance Integrated Nutrient Management in Cropping Systems. 1. Phosphorus,* C. Johansen, K.K. Lee, K.K. Sharma, G.V. Subbarao, and E.A. Kueneman (Eds.). Patancheru, Andhra Pradesh, India: International Crops Research Institute for the Semi-Arid Tropics, pp. 145-152.

Delhaize, E., D.M. Hebb, and P.R. Ryan (2001). Expression of a *Pseudomonas aeruginosa* citrate synthase gene in tobacco is not associated with either enhanced citrate accumulation or efflux. *Plant Physiology* 125:2059-2067.

De Pater, B.S. and R.A. Schilperoort (1992). Structure and expression of a root-specific rice gene. *Plant Molecular Biology* 18:161-164.

Dong, B., Z. Rengel, and E. Delhaize (1998). Uptake and translocation of phosphate by *pho2* mutant and wild-type seedlings of *Arabidopsis thaliana*. *Planta* 205:251-256.

Dong, B., Z. Rengel, and R.D. Graham (1995). Root morphology of wheat genotypes differing in zinc efficiency. *Journal of Plant Nutrition* 18:2761-2773.

Dong, B., P.R. Ryan, Z. Rengel, and E. Delhaize (1999). Phosphate uptake in *Arabidopsis thaliana*: Dependence of uptake on the expression of transporter genes and internal phosphate concentrations. *Plant, Cell and Environment* 22:1455-1461.

Duc, G. (1995). Mutagenesis of faba bean (*Vicia faba* L.) and the identification of five different genes controlling no nodulation, ineffective nodulation or supernodulation. *Euphytica* 83:147-152.

Dunbabin, V., Z. Rengel, and A. Diggle (2001a). *Lupinus angustifolius* has a plastic uptake response to heterogeneously supplied nitrate while *Lupinus pilosus* does not. *Australian Journal of Agricultural Research* 52:505-512.

Dunbabin, V., Z. Rengel, and A. Diggle (2001b). The root growth response to heterogeneous nitrate supply differs for *Lupinus angustifolius* and *Lupinus pilosus*. *Australian Journal of Agricultural Research* 52:495-503.

Ehdaie, B., M.R. Shakiba, and J.G. Waines (2001). Sowing date and nitrogen input influence nitrogen-use efficiency in spring bread and durum wheat genotypes. *Journal of Plant Nutrition* 24:899-919.

Eide, D. (1997). Molecular biology of iron and zinc uptake in eukaryotes. *Current Opinions in Cell Biology* 9:573-577.

Engels, C., G. Neumann, T.S. Gahoonia, E. George, and M. Schenk (2000). Assessing the ability of roots for nutrient acquisition. In *Root Methods: A Handbook*, A.L. Smit, A.G. Bengough, C. Engels, M. Van Noordwijk, S. Pellerin, and S.C. Van de Geijn (Eds.). Berlin/Heidelberg/New York: Springer-Verlag, pp. 403-459.

Erenoglu, B., I. Cakmak, H. Marschner, V. Römheld, S. Eker, H. Daghan, M. Kalayci, and H. Ekiz (1996). Phytosiderophore release does not relate well with Zn efficiency in different bread wheat genotypes. *Journal of Plant Nutrition* 19:1569-1580.

Erenoglu, B., I. Cakmak, V. Römheld, R. Derici, and Z. Rengel (1999). Uptake of zinc by rye, bread wheat and durum wheat cultivars differing in zinc efficiency. *Plant and Soil* 209:245-252.

Erenoglu, B., S. Eker, I. Cakmak, R. Derici, and V. Römheld (2000). Effect of iron and zinc deficiency on release of phytosiderophores in barley cultivars differing in zinc efficiency. *Journal of Plant Nutrition* 23:1645-1656.

Fesenko, A.N., N.A. Provorov, I.F. Orlova, V.P. Orlov, and B.V. Simarov (1995). Selection of *Rhizobium leguminosarum* bv. *viceae* strains for inoculation of

Pisum sativum L. cultivars: Analysis of symbiotic efficiency and nodulation competitiveness. *Plant and Soil* 172:189-198.

Gaume, A., F. Machler, C. De Leon, L. Narro, and E. Frossard (2001). Low-P tolerance by maize (*Zea mays* L.) genotypes: Significance of root growth, and organic acids and acid phosphatase root exudation. *Plant and Soil* 228:253-264.

Genc, Y., G.K. McDonald, and R.D. Graham (2000). Effect of seed zinc content on early growth of barley (*Hordeum vulgare*) under low and adequate soil zinc supply. *Australian Journal of Agricultural Research* 51:37-45.

Genc, Y., G.K. McDonald, Z. Rengel, and R.D. Graham (1999). Genotypic variation in the response of barley to Zn deficiency. In *Plant Nutrition: Molecular Biology and Genetics*, G. Gissel-Nielsen and A. Jensen (Eds.). Dordrecht, The Netherlands: Kluwer Academic Publishers, pp. 205-221.

Gianinazzi-Pearson, V. (1996). Plant cell responses to arbuscular mycorrhizal fungi: Getting to the roots of the symbiosis. *The Plant Cell* 8:1871-1883.

Gorny, A.G. (1996). Genetic variation in the response of winter wheat to low level of nitrogen and phosphorus supply. *Plant Breeding and Seed Science* 40:125-130.

Gourley, C.J.P., D.L. Allan, and M.P. Russelle (1994). Plant nutrient efficiency: A comparison of definitions and suggested improvement. *Plant and Soil* 158:29-37.

Graham, R.D. (1984). Breeding for nutritional characteristics in cereals. *Advances in Plant Nutrition* 1:57-102.

Graham, R.D. (1988). Genotypic differences in tolerance to manganese deficiency. In *Manganese in Soils and Plants*, R.D. Graham, R.J. Hannam, and N.C. Uren (Eds.). Dordrecht: Kluwer Academic Publishers, pp. 261-276.

Graham, R.D., J.S. Ascher, P.A.E. Ellis, and K.W. Shepherd (1987). Transfer to wheat of the copper efficiency factor carried on rye chromosome arm 5RL. *Plant and Soil* 99:107-114.

Graham, R.D. and Z. Rengel (1993). Genotypic variation in zinc uptake and utilization. In *Zinc in Soils and Plants*, A.D. Robson (Ed.). Dordrecht, The Netherlands: Kluwer Academic Publishers, pp. 107-118.

Graham, R.D., R.M. Welch, and H.E. Bouis (2001). Addressing micronutrient malnutrition through enhancing the nutritional quality of staple foods: Principles, perspectives and knowledge gaps. *Advances in Agronomy* 70:77-142.

Grewal, H.S. and R.D. Graham (1997). Seed zinc content influences early vegetative growth and zinc uptake in oilseed rape (*Brassica napus* and *Brassica juncea*) genotypes on zinc-deficient soil. *Plant and Soil* 192:191-197.

Grewal, H.S., J.C.R. Stangoulis, T.D. Potter, and R.D. Graham (1997). Zinc efficiency of oilseed rape (*Brassica napus* and *B. juncea*) genotypes. *Plant and Soil* 191:123-132.

Grewal, H.S. and R. Williams (1999). Alfalfa genotypes differ in their ability to tolerate zinc deficiency. *Plant and Soil* 214:39-48.

Guerinot, M.L. (2000). The ZIP family of metal transporters. *Biochimica et Biophysica Acta* 1465:190-198.

Halsted, M. and J. Lynch (1996). Phosphorus responses of C_3 and C_4 species. *Journal of Experimental Botany* 47:497-505.

Hansen, N.C. and V.D. Jolley (1995). Phytosiderophore release as a criterion for genotypic evaluation of iron efficiency in oat. *Journal of Plant Nutrition* 18:455-465.

Hansen, N.C., V.D. Jolley, W.A. Berg, M.E. Hodges, and E.G., Krenzer (1996). Phytosiderophore release related to susceptibility of wheat to iron deficiency. *Crop Science* 36:1473-1476.

Hansen, N.C., V.D. Jolley, and J.C. Brown (1995). Clipping foliage differentially affects phytosiderophore release by two wheat cultivars. *Agronomy Journal* 87:1060-1063.

Haran, S., S. Logendra, M. Seskar, M. Bratanova, and I. Raskin (2000). Characterization of Arabidopsis acid phosphatase promoter and regulation of acid phosphatase expression. *Plant Physiology* 124:615-626.

Harrison, J., N. Brugiere, B. Phillipson, S. Ferrario-Mery, T. Becker, A. Limami, and B. Hirel (2000). Manipulating the pathway of ammonia assimilation through genetic engineering and breeding: Consequences to plant physiology and plant development. *Plant and Soil* 221:81-93.

Hartmann, A., J.J. Giraud, and G. Catroux (1998). Genotypic diversity of *Sinorhizobium* (formerly Rhizobium) *meliloti* strains isolated directly from a soil and from nodules of alfalfa (*Medicago sativa*) grown in the same soil. *FEMS Microbiology Ecology* 25:107-116.

Hawkesford, M.J. (2000). Plant responses to sulphur deficiency and the genetic manipulation of sulphate transporters to improve S-utilization efficiency. *Journal of Experimental Botany* 51:131-138.

Hawkesford, M.J. and A.R. Belcher (1991). Differential protein synthesis in response to sulphate and phosphate deprivation: Identification of possible components of plasma-membrane transport systems in cultured tomato roots. *Planta* 185:323-329.

Hedley, M.J., G.J.D. Kirk, and M.B. Santos (1994). Phosphorus efficiency and the forms of soil phosphorus utilized by upland rice cultivars. *Plant and Soil* 158:53-62.

Hegeman, C.E. and E.A. Grabau (2001). A novel phytase with sequence similarity to purple acid phosphatases is expressed in cotyledons of germinating soybean seedlings. *Plant Physiology* 126:1598-1608.

Hernandez, G., M. Ramirez, R. Suarez, and S.I. Fuentes (1995). Root exuded *nod*-gene inducing signals limit the nodulation capacity of different alfalfa varieties with *Rhizobium meliloti*. *Plant Cell Reports* 14:626-629.

Hernandez, G., H. Vasquez, V. Toscano, M. Sanchez, T. Penate, A. Franchi-Alfaro, N. Mendez, and J.J. Drevon (1993). Nodulation and growth of common bean (*Phaseolus vulgaris* L.) cultivars in hydroponic culture and in the field. *Tropical Agriculture* 70:230-234.

Herrera-Estrella, L.R. (2000). Genetically modified crops and developing countries. *Plant Physiology* 124:923-925.

Higuchi, K., K. Suzuki, H. Nakanishi, H. Yamaguchi, N.-K. Nishizawa, and S. Mori (1999). Cloning of nicotianamine synthase genes, novel genes involved in the biosynthesis of phytosiderophores. *Plant Physiology* 119:471-479.

Higuchi, K., S. Watanabe, M. Takahashi, S. Kawasaki, H. Nakanishi, N.-K. Nishizawa, and S. Mori (2001). Nicotianamine synthase gene expression differs in barley and rice under Fe-deficient conditions. *Plant Journal* 25:159-167.

Hirel, B., P. Bertin, I. Quillere, W. Bourdoncle, C. Attagnant, C. Dellay, A. Gouy, S. Cadiou, C. Retailliau, M. Falque, and A. Gallais (2001). Toward a better understanding of the genetic and physiological basis for nitrogen use efficiency in maize. *Plant Physiology* 125:1258-1270.

Hoffland, E., R. Van Den Boogaard, J. Nelemans, and G. Findenegg (1992). Biosynthesis and root exudation of citric and malic acids in phosphate-starved rape plants. *The New Phytologist* 122:675-680.

Hopkins, B.G., D.A. Whitney, R.E. Lamond, and V.D. Jolley (1998). Phytosiderophore release by sorghum, wheat, and corn under zinc deficiency. *Journal of Plant Nutrition* 21:2623-2637.

Hu, B., P. Wu, C.Y. Liao, W.P. Zhang, and J.J. Ni (2001). QTLs and epistasis underlying activity of acid phosphatase under phosphorus sufficient and deficient condition in rice (*Oryza sativa* L.). *Plant and Soil* 230:99-105.

Huang, C., R.D. Graham, S.J. Barker, and S. Mori (1996). Differential expression of iron deficiency-induced genes in barley genotypes with differing manganese efficiency. *Journal of Plant Nutrition* 19:407-420.

Huang, C., M.J. Webb, and R.D. Graham (1993). Effect of pH on Mn absorption by barley genotypes in a chelate-buffered nutrient solution. *Plant and Soil* 155/156: 437-440.

Huang, C., M.J. Webb, and R.D. Graham (1994). Manganese efficiency is expressed in barley growing in soil system but not in a solution culture. *Journal of Plant Nutrition* 17:83-95.

Huang, C., M.J. Webb, and R.D. Graham (1996). Pot size affects expression of Mn efficiency in barley. *Plant and Soil* 178:205-208.

Hungria, M., L.H. Boddey, M.A. Santos, and M.A.T. Vargas (1998). Nitrogen fixation capacity and nodule occupancy by *Bradyrhizobium japonicum* and *B. elkanii* strains. *Biology and Fertility of Soils* 27:393-399.

Hungria, M. and D.A. Phillips (1993). Effects of a seed color mutation on rhizobial *nod*-gene-inducing flavonoids and nodulation in common bean. *Molecular Plant-Microbe Interactions* 6:418-422.

Isfan, D. (1993). Genotypic variability for physiological efficiency index of nitrogen in oats. In *Optimization of Plant Nutrition*, M.A.C. Fragosos and M.L. Van Beusichem (Eds.). Dordrecht, The Netherlands: Kluwer Academic Publishers, pp. 189-195.

Itai, R., K. Suzuki, H. Yamaguchi, H. Nakanishi, N.-K. Nishizawa, E. Yoshimura, and S. Mori (2000). Induced activity of adenine phosphoribosyltransferase (APRT) in iron-deficient barley roots: A possible role for phytosiderophore production. *Journal of Experimental Botany* 51:1179-1188.

Jiang, Z., W.M. Sullivan, and R.J. Hull (2000). Nitrate uptake and nitrogen use efficiency by Kentucky bluegrass cultivars. *HortScience* 35:1350-1354.

Johnson, J.F., D.L. Allan, C.P. Vance, and G. Weiblen (1996). Root carbon dioxide fixation by phosphorus-deficient *Lupinus albus*. Contribution to organic acid exudation by proteoid roots. *Plant Physiology* 112:19-30.

Johnson, J.F., C.P. Vance, and D.L. Allan (1996). Phosphorus deficiency in *Lupinus albus*. Altered lateral root development and enhanced expression of phosphoenolpyruvate carboxylase. *Plant Physiology* 112:31-41.

Jolley, V.D. and J.C. Brown (1989). Iron inefficient and efficient oat cultivars. II. Characterization of phytosiderophore released in response to iron deficiency stress. *Journal of Plant Nutrition* 12:923-937.

Jolley, V.D. and J.C. Brown (1994). Genetically controlled uptake and use of iron by plants. In *Biochemistry of Metal Micronutrients in the Rhizosphere*, J.A. Manthey, D.E. Crowley, and D.G. Luster (Eds.). Boca Raton, FL: Lewis Publishers, pp. 251-266.

Jolley, V.D., K.A. Cook, N.C. Hansen, and W.B. Stevens (1996). Plant physiological responses for genotypic evaluation of iron efficiency in strategy I and strategy II plants. A review. *Journal of Plant Nutrition* 19:1241-1255.

Jones, G.P.D., G.J. Blair, and R.S. Jessop (1989). Phosphorus efficiency in wheat: A useful selection criterion. *Field Crops Research* 21:3-4.

Jungk, A. (1991). Dynamics of nutrient movement at the soil-root interface. In *Plant roots: The Hidden Half*, J. Waisel, A. Eshel, and U. Kafkafi (Eds.). New York: Marcel Dekker, pp. 455-481.

Jungk, A. (2001). Root hairs and the acquisition of plant nutrients from soil. *Journal of Plant Nutrition and Soil Science* 164:121-129.

Jungk, A. and N. Claassen (1989). Availability in soil and acquisition by plants as the basis for phosphorus and potassium supply to plants. *Zeitschrift für Pflanzenernährung und Bodenkunde* 152:151-157.

Jungk, A. and N. Claassen (1997). Ion diffusion in the soil-root system. *Advances in Agronomy* 61:53-110.

Kaeppler, S.M., J.L. Parke, S.M. Mueller, L. Senior, C. Stuber, and W.F. Tracy (2000). Variation among maize inbred lines and detection of quantitative trait loci for growth at low phosphorus and responsiveness to arbuscular mycorrhizal fungi. *Crop Science* 40:358-364.

Kalayci M. B. Torun, S. Eker, M. Aydin, L. Ozturk, and I. Cakmak (1999). Grain yield, zinc efficiency and zinc concentration of wheat cultivars grown in a zinc-deficient calcareous soil in field and greenhouse. *Field Crops Research* 63:87-98.

Karagiannidis, N., N. Nikolaou, and A. Mattheou (1995). Influence of three VA-mycorrhiza species on the growth and nutrient uptake of three grapevine rootstocks and one table grape cultivar. *Vitis* 34:85-89.

Kemp, P.D. and G.J. Blair (1994). Phosphorus efficiency in pasture species: VIII. Ontogeny, growth, P acquisition and P utilization of Italian ryegrass and phalaris under P deficient and P sufficient conditions. *Australian Journal of Agricultural Research* 45:669-688.

Kessel, B. and H.C. Becker (1998). Genetic variation of nitrogen-efficiency in field experiments with oilseed rape (*Brassica napus* L.). In *Plant Nutrition: Molecular Biology and Genetics*, G. Gissel-Nielsen and Jensen A. (Eds.). Dordrecht, The Netherlands: Kluwer Academic Publishers, pp. 391-395.

Khabaz-Saberi, H., S.J. Barker, R.D. Graham, and A.J. Rathjen (1999). The application of amplified fragment length polymorphism (AFLP) for breeding Mn efficiency in durum wheat (*Triticum turgidum* L. var. *durum*). In *Plant Nutrition:*

Molecular Biology and Genetics, G. Gissel-Nielsen and Jensen A. (Eds.). Dordrecht, The Netherlands: Kluwer Academic Publishers, pp. 409-416.

Khabaz-Saberi, H., R.D. Graham, J.S. Ascher, and A.J. Rathjen (2000). Quantification of the confounding effect of seed manganese content in screening for manganese efficiency in durum wheat (*Triticum turgidum* L. var. *durum*). *Journal of Plant Nutrition* 23:855-866.

Khabaz-Saberi, H., R.D. Graham, and A.J. Rathjen (1999). Inheritance of manganese efficiency in durum wheat. *Journal of Plant Nutrition* 22:11-21.

Kobayashi, T., H. Nakanishi, M. Takahashi, S. Kawasaki, N.-K. Nishizawa, and S. Mori (2001). In vivo evidence that *Ids3* from *Hordeum vulgare* encodes a dioxygenase that converts 2'-deoxymugineic acid to mugineic acid in transgenic rice. *Planta* 212:864-871.

Koyama, H., A. Kawamura, T. Kihara, T. Hara, E. Takita, and D. Shibata (2000). Overexpression of mitochondrial citrate synthase in *Arabidopsis thaliana* improved growth on a phosphorus-limited soil. *Plant and Cell Physiology* 41:1030-1037.

Krajinski, F., A. Biela, D. Schubert, V. Gianinazzi-Pearson, R. Kaldenhoff, and P. Franken (2000). Arbuscular mycorrhiza development regulates the mRNA abundance of *Mtaqp1* encoding a mercury-insensitive aquaporin of *Medicago truncatula*. *Planta* 211:85-90.

Krajinski, F., F. Martin-Laurent, S. Gianinazzi, V. Gianinazzi-Pearson, and P. Franken (1998). Cloning and analysis of *psam2*, a gene from *Pisum sativum* L. regulated in symbiotic arbuscular mycorrhiza and pathogenic root-fungus interactions. *Physiological and Molecular Plant Pathology* 52:297-307.

Kubota, J. and W.H. Allaway (1972). Geographic distribution of trace element problems. In *Micronutrients in Agriculture*, J.J. Mortvedt, P.M. Giordano, and W.L. Lindsay (Eds.). Madison, WI: Soil Science Society of America, pp. 525-554.

Lasat, M.M., N.S. Pence, D.F. Garvin, S.D. Ebbs, and L.V. Kochian (2000). Molecular physiology of zinc transport in the Zn hyperaccumulator *Thlaspi caerulescens*. *Journal of Experimental Botany* 51:71-79.

Leung, K., S.R. Strain, F.J. De Bruijn, and P.J. Bottomley (1994). Genotypic and phenotypic comparisons of chromosomal types within an indigenous soil population of *Rhizobium leguminosarum* bv. *trifolii*. *Applied and Environmental Microbiology* 60:416-426.

Lin, S.F., J.S. Baumer, D. Ivers, S.R. Cianzio, and R.C. Shoemaker (1998). Field and nutrient solution tests measure similar mechanisms controlling iron deficiency chlorosis in soybean. *Crop Science* 38:254-259.

Lin, S.F., J.S. Baumer, D. Ivers, S. Cianzio, and R. Shoemaker (2000). Nutrient solution screening of Fe chlorosis resistance in soybean evaluated by molecular characterization. *Journal of Plant Nutrition* 23:1915-1928.

Lin, S.F., D. Grant, S. Cianzio, and R. Shoemaker (2000). Molecular characterization of iron deficiency chlorosis in soybean. *Journal of Plant Nutrition* 23:1929-1939.

Longnecker, N.E., R.D. Graham, K.W. McCarthy, D.H.B. Sparrow, and J.P. Egan (1990). Screening for manganese efficiency in barley (*Hordeum vulgare* L.). In

Genetic Aspects of Plant Mineral Nutrition, N. El Bassam (Ed.). Dordrecht, The Netherlands: Kluwer Academic Publishers, pp. 273-280.

Lopez-Bucio, J., O.M. De la Vega, A. Guevara-Garcia, and L. Herrera-Estrella (2000). Enhanced phosphorus uptake in transgenic tobacco plants that overproduce citrate. *Nature Biotechnology* 18:450-453.

Lynch, J.M. and J.M. Whipps (1990). Substrate flow in the rhizosphere. *Plant and Soil* 129:1-10.

Lynch, J.P. (1998). The role of nutrient-efficient crops in modern agriculture. In *Nutrient Use in Crop Production*, Z. Rengel (Ed.). Binghamton, NY: The Haworth Press, pp. 241-264.

Majumder, N.D., S.C. Rakshit, and D.N. Borthakur (1990). Genetic effect on uptake of selected nutrients in some rice (*Oryza sativa* L.) varieties in phosphorus deficient soils. *Plant and Soil* 123:117-120.

Maman, N., S.C. Mason, T. Galusha, and M.D. Clegg (1999). Hybrid and nitrogen influence on pearl millet production in Nebraska: Yield, growth, and nitrogen uptake, and nitrogen use efficiency. *Agronomy Journal* 91:737-743.

Mamo, T., C. Richter, and A. Hoppenstedt (1996). Phosphorus response studies on some varieties of durum wheat (*Triticum durum* Desf.) and tef (*Eragrostis tef* [Zucc.] Trotter) grown in sand culture. *Journal of Agronomy and Crop Science* 176:189-197.

Manske, G.G.B., J.I. Ortiz-Monasterio, M. van Ginkel, R.M. Gonzalez, R.A. Fischer, S. Rajaram, and P.L.G. Vlek (2001). Importance of P uptake efficiency versus P utilization for wheat yield in acid and calcareous soils in Mexico. *European Journal of Agronomy* 14:261-274.

Marschner, H. (1995). *Mineral Nutrition of Higher Plants*, Second Edition. London: Academic Press.

Marschner, H. and V. Römheld (1994). Strategies of plants for acquisition of iron. *Plant and Soil* 165:261-274.

Marschner, H., V. Römheld, and M. Kissel (1987). Localization of phytosiderophore release and of iron uptake along intact barley roots. *Physiologia Plantarum* 71:157-162.

McCarthy, K.W., N.E. Longnecker, D.H.B. Sparrow, and R.D. Graham (1988). Inheritance of manganese efficiency in barley (*Hordeum vulgare* L.). In *International Symposium on Manganese in Soils and Plants*, M.J. Webb, R.O. Nable, R.D. Graham, and R.J. Hannam (Eds.). Adelaide: Manganese Symposium, Inc., pp. 121-122.

Michaud, R., G. Belanger, A. Bregard, and J. Surprenant (1998). Selection for nitrogen use efficiency and N concentration in timothy. *Canadian Journal of Plant Science* 78:611-613.

Miller, S.S., J. Liu, D.L. Allan, C.J. Menzhuber, M. Fedorova, and C.P. Vance (2001). Molecular control of acid phosphatase secretion into the rhizosphere of proteoid roots from phosphorus-stressed white lupin. *Plant Physiology* 127:594-606.

Miyasaka, S.C. and M.C. Hawes (2001). Possible role of root border cells in detection and avoidance of aluminum toxicity. *Plant Physiology* 125:1978-1987.

Moraghan, J.T. and K. Grafton (1999). Seed-zinc concentration and the zinc-efficiency trait in navy bean. *Soil Science Society of America Journal* 63:918-922.

Mori, S. (1994). Mechanisms of iron acquisition by graminaceous (strategy II) plants. In *Biochemistry of Metal Micronutrients in the Rhizosphere*, J.A. Manthey, D.E. Crowley, and D.G. Luster (Eds.). Boca Raton, FL: Lewis Publishers, pp. 225-249.

Mori, S. (1997). Reevaluation of the genes induced by iron deficiency in barley roots. *Soil Science and Plant Nurition* 43:975-980.

Mori, S., N.K. Nishizawa, and J. Fujigaki (1990). Identification of rye chromosome 5R as a carrier of the genes for mugineic acid synthetase and 3-hydroxymugineic acid synthetase using wheat-rye addition lines. *Japanese Journal of Genetics* 65:343-352.

Moussavi-Nik, M., Z. Rengel, J.N. Pearson, and G. Hollamby (1997). Dynamics of nutrient remobilisation from seed of wheat genotypes during imbibition, germination and early seedling growth. *Plant and Soil* 197:271-280.

Murphy, P.J., P. Langridge, and S.E. Smith (1997). Cloning plant genes differentially expressed during colonization of roots of *Hordeum vulgare* by the vesicular-arbuscular mycorrhizal fungus *Glomus intraradices*. *The New Phytologist* 135:291-301.

Nable, R.O. and M.J. Webb (1993). Further evidence that zinc is required throughout the root zone for optimal plant growth and development. *Plant and Soil* 150:247-253.

Nakanishi, H., H. Yamaguchi, T. Sasakuma, N.-K. Nishizawa, and S. Mori (2000). Two dioxygenase genes, *Ids3* and *Ids2*, from *Hordeum vulgare* are involved in the biosynthesis of mugineic acid family phytosiderophores. *Plant Molecular Biology* 44:199-207.

Nemec, S. and L. Datnoff (1993). Pepper and tomato cultivar responses to inoculation with *Glomus intraradices*. *Advances in Horticultural Science* 7:161-164.

Neo, H.H., S. Hunt, and D.B. Layzell (1996). Can genotypes of soybean *(Glycine max)* selected for nitrate tolerance provide good "models" for studying the mechanism of nitrate inhibition of nitrogenase activity. *Physiologia Plantarum* 98:653-660.

Neumann, G., A. Massonneau, E. Martinoia, and V. Römheld (1999). Physiological adaptations to phosphorus deficiency during proteoid root development in white lupin. *Planta* 208:373-382.

Neumann, G. and V. Römheld (1999). Root excretion of carboxylic acids and protons in phosphorus-deficient plants. *Plant and Soil* 211:121-130.

Neumann, G. and V. Römheld (2001). The release of root exudates as affected by the plant's physiological status. In *Rhizosphere: Biochemistry and Organic Substances at the Soil-Plant Interface*, R. Pinton, Z. Varanini, and P. Nannipieri (Eds.). New York: Marcel Dekker Inc., pp. 41-93.

Ni, J.J., P. Wu, D. Senadhira, and N. Huang (1998). Mapping QTLs for phosphorus deficiency tolerance in rice (*Oryza sativa* L.). *Theoretical and Applied Genetics* 97:1361-1369.

Nishikoori, M., K. Washio, A. Hase, N. Morita, and H. Okuyama (2001). Cloning and characterization of cDNA of the GPI-anchored purple acid phosphatase and its root tissue distribution in *Spirodela oligorrhiza. Physiologia Plantarum* 113:241-248.

Nitz, I., H. Berkefeld, P.S. Puzio, and F.M.W. Grundler (2001). Pyk10, a seedling and root specific gene and promoter from *Arabidopsis thaliana. Plant Science* 161:337-346.

Nour, S.M., J.C. Cleyet-Marel, D. Beck, A. Effosse, and M.P. Fernandes (1994). Genotypic and phenotypic diversity of rhizobium isolated from chickpea (*Cicer arietinum* L.). *Canadian Journal of Microbiology* 40:345-354.

O'Connell, K.P., R.M. Goodman, and J. Handelsman (1996). Engineering the rhizosphere: Expressing a bias. *Trends in Biotechnology* 14:83-88.

Ohnishi, M., T. Horie, K. Homma, N. Supapoj, H. Takano, and S. Yamamoto (1999). Nitrogen management and cultivar effects on rice yield and nitrogen use efficiency in Northeast Thailand. *Field Crops Research* 64:109-120.

Oldroyd, G.E.D. (2001). Dissecting symbiosis: Developments in Nod factor signal transduction. *Annals of Botany* 87:709-718.

Owouoche, J.O., K.G. Briggs, and G.J. Taylor (1996). The efficiency of copper use by 5A/5RL wheat-rye translocation lines and wheat (*Triticum aestivum* L.) cultivars. *Plant and Soil* 180:113-120.

Ozawa, K., M. Osaki, H. Matsui, M. Honma, and T. Tadano (1995). Purification and properties of acid phosphatase secreted from lupin roots under phosphorus-deficiency conditions. *Soil Science and Plant Nutrition* 41:461-469.

Pazdernik, D.L., P.H. Graham, and J.H. Orf (1997). Variation in the pattern of nitrogen accumulation and distribution in soybean. *Crop Science* 37:1482-1486.

Pearson, J.N. and Z. Rengel (1994). Distribution and remobilization of Zn and Mn during grain development in wheat. *Journal of Experimental Botany* 45:1829-1835.

Pearson, J.N. and Z. Rengel (1997a). Genotypic differences in the production and partitioning of carbohydrates between roots and shoots of wheat grown under zinc or manganese deficiency. *Annals of Botany* 80:803-808.

Pearson, J.N. and Z. Rengel (1997b). Mechanisms of plant resistance to nutrient deficiency stress. In *Mechanisms of Environmental Stress Resistance in Plants*, A.S. Basra and R.K. Basra (Eds.). Amsterdam: Harwood Academic Publishers, pp. 213-240.

Pearson, J.N., Z. Rengel, and R.D. Graham (1999). Regulation of zinc and manganese transport into developing wheat grains having different zinc and manganese concentrations. *Journal of Plant Nutrition* 22:1141-1152.

Pearson, J.N., Z. Rengel, C.F. Jenner, and R.D. Graham (1995). Transport of zinc and manganese to developing wheat grains. *Physiologia Plantarum* 95:449-455.

Pearson, J.N., Z. Rengel, C.F. Jenner, and R.D. Graham (1998). Dynamics of zinc and manganese movement in developing wheat grains. *Australian Journal of Plant Physiology* 25:139-144.

Raffin, A. and P. Roumet (1994). Shoot-root control of nitrate tolerance of N_2 fixation in spontaneously tolerant soybean lines: Reciprocal grafting experiments. *Agronomie* 14:473-480.

Randall, P.J. (1995). Genotypic differences in phosphate uptake. In *Genetic Manipulation of Crop Plants to Enhance Integrated Nutrient Management in Cropping*

Systems. 1. Phosphorus, C. Johansen, K.K. Lee, K.K. Sharma, G.V. Subbarao, and E.A. Kueneman (Eds.). Patancheru, Andhra Pradesh, India: International Crops Research Institute for the Semi-Arid Tropics, pp. 31-47.

Rao, I.M., V. Borrero, J. Ricaurte, R. Garcia, and M.A. Ayarza (1997). Adaptive attributes of tropical forage species to acid soils. III. Differences in phosphorus acquisition and utilization as influenced by varying phosphorus supply and soil type. *Journal of Plant Nutrition* 20:155-180.

Raun, W.R. and G.V. Johnson (1999). Improving nitrogen use efficiency for cereal production. *Agronomy Journal* 91:357-363.

Reddy, D.D. and K.L. Sharma (2000). Effect of amending urea fertilizer with chemical additives on ammonia volatilization loss and nitrogen-use efficiency. *Biology and Fertility of Soils* 32:24-27.

Rengel, Z. (1993). Mechanistic simulation models of nutrient uptake: A review. *Plant and Soil* 152:161-173.

Rengel, Z. (1995a). Carbonic anhydrase activity in leaves of wheat genotypes differing in Zn efficiency. *Journal of Plant Physiology* 147:251-256.

Rengel, Z. (1995b). Sulfhydryl groups in root-cell plasma membranes of wheat genotypes differing in Zn efficiency. *Physiologia Plantarum* 95:604-612.

Rengel, Z. (1997). Root exudation and microflora populations in rhizosphere of crop genotypes differing in tolerance to micronutrient deficiency. *Plant and Soil* 196:255-260.

Rengel, Z. (1999a). Physiological mechanisms underlying differential nutrient efficiency of crop genotypes. In *Mineral Nutrition of Crops: Fundamental Mechanisms and Implications*, Z. Rengel (Ed.). Binghamton, NY: The Haworth Press, pp. 227-265.

Rengel, Z. (1999b). Physiological responses of wheat genotypes grown in chelator-buffered nutrient solutions with increasing concentrations of excess HEDTA. *Plant and Soil* 215:193-202.

Rengel, Z. (1999c). Zinc deficiency in wheat genotypes grown in conventional and chelator-buffered nutrient solutions. *Plant Science* 143:221-230.

Rengel, Z. (2000). Uptake and transport of manganese in plants. In *Metal Ions in Biological Systems*, A. Sigel and H. Sigel (Eds.). New York: Marcel Dekker, pp. 57-87.

Rengel, Z. (2001). Genotypic differences in micronutrient use efficiency in crops. *Communications in Soil Science and Plant Analysis* 32:1163-1186.

Rengel, Z. (2002a). Breeding for better symbiosis. *Plant and Soil* 245:147-162.

Rengel, Z. (2002b). Genetic control of root exudation. *Plant and Soil* 245:59-70.

Rengel, Z., G.D. Batten, and D.E. Crowley (1999). Agronomic approaches for improving the micronutrient density in edible portions of field crops. *Field Crops Research* 60:27-40.

Rengel, Z. and R.D. Graham (1995a). Importance of seed Zn content for wheat growth on Zn-deficient soil: I. Vegetative growth. *Plant and Soil* 173:259-266.

Rengel, Z. and R.D. Graham (1995b). Importance of seed Zn content for wheat growth on Zn-deficient soil: II. Grain yield. *Plant and Soil* 173:267-274.

Rengel, Z. and R.D. Graham (1995c). Wheat genotypes differ in Zn efficiency when grown in chelate-buffered nutrient solution. II. Nutrient uptake. *Plant and Soil* 176:317-324.

Rengel, Z. and R.D. Graham (1996). Uptake of zinc from chelate-buffered nutrient solutions by wheat genotypes differing in zinc efficiency. *Journal of Experimental Botany* 47:217-226.

Rengel, Z. and M.J. Hawkesford (1997). Biosynthesis of a 34-kDa polypeptide in the root-cell plasma membrane of a Zn-efficient wheat genotype increases upon Zn deficiency. *Australian Journal of Plant Physiology* 24:307-315.

Rengel, Z., J.F. Pedler, and R.D. Graham (1994). Control of Mn status in plants and rhizosphere: Genetic aspects of host and pathogen effects in the wheat take-all interaction. In *Biochemistry of Metal Micronutrients in the Rhizosphere*, J.A. Manthey, D.E. Crowley, and D.G. Luster (Eds.). Boca Raton, FL: Lewis Publishers/CRC Press, pp. 125-145.

Rengel, Z. and V. Römheld (2000a). Genotypic differences in tolerance to Fe and Zn deficiencies in wheat. *Euphytica* 113:219-225.

Rengel, Z. and V. Römheld (2000b). Root exudation and Fe uptake and transport in wheat genotypes differing in tolerance to Zn deficiency. *Plant and Soil* 222:25-34.

Rengel, Z., V. Römheld, and H. Marschner (1998). Uptake of zinc and iron by wheat genotypes differing in tolerance to zinc deficiency. *Journal of Plant Physiology* 152:433-438.

Rengel, Z. and M.S. Wheal (1997a). Herbicide chlorsulfuron decreases growth of fine roots and micronutrient uptake in wheat genotypes. *Journal of Experimental Botany* 48:927-934.

Rengel, Z. and M.S. Wheal (1997b). Kinetic parameters of Zn uptake by wheat are affected by the herbicide chlorsulfuron. *Journal of Experimental Botany* 48:935-941.

Rerkasem, B. and S. Jamjod (1997). Genotypic variation in plant response to low boron and implications for plant breeding. *Plant and Soil* 193:169-180.

Rolfe, B.G., M.A. Djordjevic, J.J. Weinman, U. Mathesius, C. Pittock, E. Gartner, K.M. Ride, Z. Dong, M. McCully, and J. McIver (1997). Root morphogenesis in legumes and cereals and the effect of bacterial inoculation on root development. *Plant and Soil* 194:131-144.

Römheld, V. (1990). The soil-root interface in relation to mineral nutrition. *Symbiosis* 9:19-27.

Römheld, V. and H. Marschner (1990). Genotypical differences among graminaceous species in release of phytosiderophores and uptake of iron phytosiderophores. *Plant and Soil* 123:147-153.

Roussel, H., D. Van Tuinen, P. Franken, S. Gianinazzi, and V. Gianinazzi-Pearson (2001). Signalling between arbuscular mycorrhizal fungi and plants: Identification of a gene expressed during early interactions by differential RNA display analysis. *Plant and Soil* 232:13-19.

Ruiz-Lozano, J.M., H. Roussel, S. Gianinazzi, and V. Gianinazzi-Pearson (1999). Defence genes are differentially induced by a mycorrhizal fungus and *Rhizobium* sp. in wild-type and symbiosis-defective pea genotypes. *Molecular Plant-Microbe Interactions* 12:976-984.

Salzer, P., A. Bonanomi, K. Beyer, R. Vogeli-Lange, R.A. Aeschbacher, J. Lange, A. Wiemken, D. Kim, D.R. Cook, and T. Boller (2000). Differential expression of eight chitinase genes in *Medicago truncatula* roots during mycorrhiza formation, nodulation, and pathogen infection. *Molecular Plant-Microbe Interactions* 13:763-777.

Sanginga, N., O. Lyasse, and B.B. Singh (2000). Phosphorus use efficiency and nitrogen balance of cowpea breeding lines in a low P soil of the derived savanna zone in West Africa. *Plant and Soil* 220:119-128.

Sattelmacher, B., W.J. Horst, and H.C. Becker (1994). Factors that contribute to genetic variation for nutrient efficiency of crop plants. *Zeitschrift für Pflanzenernährung und Bodenkunde* 157:215-224.

Schilling, G., A. Gransee, A. Deubel, G. Lezovic, and S. Ruppel (1998). Phosphorus availability, root exudates, and microbial activity in the rhizosphere. *Zeitschrift für Pflanzenernährung und Bodenkunde* 161:465-478.

Schlegel, R., I. Cakmak, B. Torun, S. Eker, and N. Koleli (1997). The effect of rye genetic information on zinc, copper, manganese and iron concentration of wheat shoots in zinc deficient soil. *Cereal Research Communications* 25:177-184.

Schlegel, R., I. Cakmak, B. Torun, S. Eker, I. Tolay, H. Ekiz, M. Kalayci, and H.J. Braun (1998). Screening for zinc efficiency among wheat relatives and their utilisation for alien gene transfer. *Euphytica* 100:281-285.

Schlegel, R., R. Kynast, T. Schwarzacher, V. Römheld, and A. Walter (1993). Mapping genes for copper efficiency in rye and the relationship between copper and iron efficiency. *Plant and Soil* 154:61-65.

Shelp, B.J., B.N. Kaiser, and A.M. Deschesne (1998). Registration of five near-isogenic genetic stocks of "Juneau" pea with altered nodulation and nitrate reductase deficiency: A317I, nod3I, A317nod3I, E135I, and R25I. *Crop Science* 38:554.

Shuman, L.M. (1991). Chemical forms of micronutrients in soils. In *Micronutrients in agriculture,* Second Edition, J.J Mortvedt, F.R. Cox, L.M. Shuman, and R.M. Welch (Eds.). Madison, WI: Soil Science Society of America, pp. 113-144.

Sillanpaa, M. and P.L.G. Vlek (1985). Micronutrients and the agroecology of tropical and Mediterranean regions. *Fertilizer Research* 7:1-3.

Singh, U., J.K. Ladha, E.G., Castillo, G. Punzalan, A. Tirol-Padre, and M. Duqueza (1998). Genotypic variation in nitrogen use efficiency in medium- and long-duration rice. *Field Crops Research* 58:35-53.

Smith, S.E., V. Gianinazzi-Pearson, R. Koide, and J.W.G. Cairney (1994). Nutrient transport in mycorrhizas: Structure, physiology and consequences for efficiency of the symbiosis. *Plant and Soil* 159:103-113.

Soejima, H., T. Sugiyama, and K. Ishihara (1992). Changes in cytokinin activities and mass spectrometric analysis of cytokinins in root exudates of rice plants (*Oryza sativa* L.). Comparison between cultivars Nipponbare and Akenohoshi. *Plant Physiology* 100:1724-1729.

Stangoulis, J.C.R., H.S. Grewal, R.W. Bell, and R.D. Graham (2000). Boron efficiency in oilseed rape: I. Genotypic variation demonstrated in field and pot grown *Brassica napus* L. and *Brassica juncea* L. *Plant and Soil* 225:243-251.

Stelling, D., S.H. Wang, and W. Römer (1996). Efficiency in the use of phosphorus, nitrogen and potassium in topless faba beans (*Vicia faba* L.): Variability and inheritance. *Plant Breeding* 115:361-366.

Stevens, W.B., V.D. Jolley, N.C. Hansen, and D.J. Fairbanks (1993). Modified procedures for commercial adaptation of root iron-reducing capacity as a screening technique. *Journal of Plant Nutrition* 16:2507-2519.

Stone, D.A. (2000). The effects of starter fertilizers on the growth and nitrogen use efficiency of onion and lettuce. *Soil Use and Management* 16:42-48.

Subbarao, G.V., N. Ae, and T. Otani (1997). Genetic variation in acquisition and utilisation of phosphorus from iron-bound phosphorus in pigeonpea. *Soil Science and Plant Nutrition* 43:511-519.

Swoboda, I., P.L. Bhalla, H. Xu, Y. Zhang, I. Mittermann, R. Valenta, and M.B. Singh (2001). Identification of *pronp1*, a tobacco profilin gene activated in tip-growing cells. *Plant Molecular Biology* 46:531-538.

Thongbai, P., R.J. Hannam, R.D. Graham, and M.J. Webb (1993). Interaction between zinc nutritional status of cereals and Rhizoctonia root rot severity. I. Field observation. *Plant and Soil* 153:207-214.

Tong, Y., Z. Rengel, and R.D. Graham (1997). Interactions between nitrogen and manganese nutrition of barley genotypes differing in manganese efficiency. *Annals of Botany* 79:53-58.

Trolove, S.N., M.J. Hedley, J.R. Caradus, and A.D. Mackay (1996). Uptake of phosphorus from different sources by *Lotus pedunculatus* and three genotypes of *Trifolium repens*. 1. Plant yield and phosphate efficiency. *Australian Journal of Soil Research* 34:1015-1026.

Tsyganov, V.E., E.V. Morzhina, S.Y. Stefanov, A.Y. Borisov, V.K. Lebsky, and I.A. Tikhonovich (1998). The pea (*Pisum sativum* L.) genes *sym33* and *sym40* control infection thread formation and root nodule function. *Molecular and General Genetics* 295:491-503.

Utrup, L.J., A.J. Cary, and J.H. Norris (1993). Five nodulation mutants of white sweetclover (*Melilotus alba* Desr.) exhibit distinct phenotypes blocked at root hair curling, infection thread development, and nodule organogenesis. *Plant Physiology* 103:925-932.

van der Ploeg, R.R., H. Ringe, G. Machulla, and D. Hermsmeyer (1997). Postwar nitrogen use efficiency in West German agriculture and groundwater quality. *Journal of Environmental Quality* 26:1203-1212.

Van der Zaal, B.J., L.W. Neuteboom, J.E. Pinas, A.N. Chardonnens, H. Schat, J.A.C. Verkleij, and P.J.J. Hooykaas (1999). Overexpression of a novel *Arabidopsis* gene related to putative zinc-transporter genes from animals can lead to enhanced zinc resistance and accumulation. *Plant Physiology* 119:1047-1055.

Vazquez de Aldana, B.R. and F. Berendse (1997). Nitrogen-use efficiency in six perennial grasses from contrasting habitats. *Functional Ecology* 11:619-626.

Villagarcia, M.R., W.W. Collins, and C.D. Raper Jr. (1998). Nitrate uptake and nitrogen use efficiency of two sweetpotato genotypes during early stages of storage root formation. *Journal of the American Society for Horticultural Science* 123:814-820.

Von Wiren, N., H. Marschner, and V. Römheld (1995). Uptake kinetics of iron-phytosiderophores in two maize genotypes differing in iron efficiency. *Physiologia Plantarum* 93:611-616.

Walter, A., V. Römheld, H. Marschner, and S. Mori (1994). Is the release of phytosiderophores in zinc-deficient wheat plants a response to impaired iron utilization? *Physiologia Plantarum* 92:493-500.

Wani, A.R., B. Hasan, and A.S. Bali (1999). Relative efficiency of some modified urea materials in transplanted rice under temperate conditions of Kashmir. *Oryza* 36:382-383.

Wasaki, J., M. Ando, K. Ozawa, M. Omura, M. Osaki, H. Ito, H. Matsui, and T. Tadano (1997). Properties of secretory acid phosphatase from lupin roots under phosphorus-deficient conditions. *Soil Science and Plant Nutrition* 43:981-986.

Waschutza, S., N. Hofmann, E.G., Niemann, and I. Fendrik (1992). Investigations on root exudates of Korean rice. *Symbiosis* 13:181-189.

Welch, R.M. (1995). Micronutrient nutrition of plants. *Critical Reviews in Plant Sciences* 14:49-82.

Welch, R.M., W.H. Allaway, W.A. House, and J. Kubota (1991). Geographic distribution of trace element problems. In *Micronutrients in Agriculture*, Second Edition, J.J. Mortvedt, F.R. Cox, L.M. Shuman, and R.M. Welch (Eds.). Madison, WI: Soil Science Society of America, pp. 31-57.

Wiesler, F. (1998). Comparative assessment of the efficacy of various nitrogen fertilizers. In *Nutrient Use in Crop Production*, Z. Rengel (Ed.). Binghamton, NY: The Haworth Press, pp. 81-114.

Wissuwa, M. and N. Ae (2001). Genotypic variation for tolerance to phosphorus deficiency in rice and the potential for its exploitation in rice improvement. *Plant Breeding* 120:43-48.

Wissuwa, M., M. Yano, and N. Ae (1998). Mapping of QTLs for phosphorus-deficiency tolerance in rice (*Oryza sativa* L.). *Theoretical and Applied Genetics* 97:777-783.

Wu, P., J.J. Ni, and A.C. Luo (1998). QTLs underlying rice tolerance to low-potassium stress in rice seedlings. *Crop Science* 38:1458-1462.

Xue, J., M. Lin, R.W. Bell, R.D. Graham, X. Yang, and Y. Yang (1998). Differential response of oilseed rape (*Brassica napus* L.) cultivars to low boron supply. *Plant and Soil* 204:155-163.

Yadav, R.L., R.V. Singh, R. Singh, and V.K. Srivastava (1997). Effect of planting geometry and fertilizer N rates on nitrate leaching, nitrogen use efficiency, and sugarcane yield. *Tropical Agriculture* 74:115-120.

Yamaguchi, H., H. Nakanishi, N.-K. Nishizawa, and S. Mori (2000a). Induction of the *IDI1* gene in Fe-deficient barley roots: A gene encoding a putative enzyme that catalyses the methionine salvage pathway for phytosiderophore production. *Soil Science and Plant Nutrition* 46:1-9.

Yamaguchi, H., H. Nakanishi, N.-K. Nishizawa, and S. Mori (2000b). Isolation and characterization of *IDI2*, a new Fe-deficiency-induced cDNA from barley roots, which encodes a protein related to the alpha subunit of eukaryotic initiation factor 2B (elF2B alpha). *Journal of Experimental Botany* 51:2001-2007.

Yan, X., H. Liao, M. Trull, S.E. Beebe, and J.P. Lynch (2001). Induction of a major leaf acid phosphatase does not confer adaptation to low phosphorus availability in common bean. *Plant Physiology* 125:1901-1911.

Yang, X., Z.Q. Ye, C.H. Shi, M.L. Zhu, and R.D. Graham (1998). Genotypic differences in concentrations of iron, manganese, copper, and zinc in polished rice grains. *Journal of Plant Nutrition* 21:1453-1462.

Yu, Q., L. Osborne, and Z. Rengel (1998). Micronutrient deficiency changes activities of superoxide dismutase and ascorbate peroxidase in tobacco plants. *Journal of Plant Nutrition* 21:1427-1437.

Yu, Q., C. Worth, and Z. Rengel (1999). Using capillary electrophoresis to measure Cu/Zn superoxide dismutase concentration in leaves of wheat genotypes differing in tolerance to zinc deficiency. *Plant Science* 143:231-239.

Zhang, F.S., J. Ma, and Y.P. Cao (1997). Phosphorus deficiency enhances root exudation of low-molecular weight organic acids and utilization of sparingly soluble inorganic phosphates by radish (*Raphanus sativus* L.) and rape (*Brassica napus* L.) plants. *Plant and Soil* 196:261-264.

Zhang, F.S., V. Römheld, and H. Marschner (1991). Release of zinc mobilizing root exudates in different plant species as affected by zinc nutritional status. *Journal of Plant Nutrition* 14:675-686.

Chapter 10

Genetic Improvements of Tolerance to High Temperature

Catherine J. Howarth

INTRODUCTION

Temperature is a critical factor in plant growth. Consequently, the distribution of species throughout the world, both as natural populations and as crop plants, is influenced by temperature. It is a plant's ability to respond to environmental change that helps determine its ecological niche and often affects its usefulness to humans. In the absence of other biotic and abiotic constraints, plant growth and subsequent yield is maximal when a crop is cultivated as close to its temperature optimum as possible. However, the physical environment is neither constant nor optimal for crop production; fluctuations, often extreme, occur both in time and space. Such changes occur over the different seasons of the year and often more rapidly over the course of an individual day. Longer-term environmental changes can also occur, such as appears to be occurring at present due to increased greenhouse gas emissions. Current climate models project that the mean annual global surface temperature will increase by 1.7 to 3.8°C by 2100 (Wigley and Raper, 1992). Such changes in temperature could have radical effects on agriculture. Relatively small changes in the mean temperature could have a marked effect on the frequency of extreme temperature events occurring. For example, the number of very hot days, i.e., above a particular threshold, which can cause damaging heat injury to plants, could increase markedly as a result of a small increase in mean temperature. The effect of global warming shows considerable regional variation and it is likely that developing countries will be affected to a much greater degree (Rosenzweig and Parry, 1994). As many agricultural environments in such countries are already marginal and vulnerable to interannual climate fluctuations, this will result in increased food insecurity in such regions. Developing countries are also highly vulnerable to climate change because many are located in arid and semiarid regions. In the tropics and subtropics, where some crops are near their maximum temperature tolerance and where dryland, nonirrigated agriculture predominates, yields are likely to decrease. The livelihoods of subsistence farmers and pastoral peoples, who make up a large portion of rural populations in some regions, also could be negatively affected. This indicates the enhanced requirement to improve the adaptation of plants to their temperature environment and to understand the mechanisms by which such adaptation occurs.

Unlike animals, plants are unable to escape from adverse environmental conditions. Plants have evolved to survive such conditions, and they show adaptation to the environments from which they originate. They may be able to avoid the consequences of stress by restricting growth to favorable environmental conditions. For example, an annual plant can avoid the most extreme seasons, although its seed has to survive if further generations are to result. Similarly, a deciduous tree may be dormant during winter. Perennial crops, such as temperate grasses, need to both maximize growth under optimum environmental conditions and possess the ability to survive extreme conditions. However, plants are able to acclimate. Periods of moderate stress often induce tolerance, which results in the ability to survive more extreme conditions. There is a limit beyond which acclimation cannot permit survival; this can be termed the "adaptive range." The acclimation of plants to changes in the environment involves both short-term chemical, molecular, and physiological responses and longer-term physiological, structural, and morphological modifications. Successful acclimation results in a plant with some form of change in its metabolism or structure so that it survives if the stress persists or increases in severity, and also so that it recovers from the consequences of that stress. Thus, the previous environmental history of a plant affects its response to a subsequent change in the environment. Such acclimation is freely reversible when conditions return to a more favorable range, assuming that the adaptive range has not been exceeded. Stress resistance in agronomic terms is a balance between (1) high yield and growth rate during stress, and (2) dormancy and a conservative growth strategy during stress. For agricultural systems based on annual crops, maximizing yield under the prevalent environmental conditions is a high priority, whereas with perennial crops and natural vegetation the ability to survive periods of environmental constraint is an essential characteristic for success. A survival strategy may be more important than a growth strategy, particularly in more extreme or variable environments. The balance of these two strategies may need to be shifted toward survival in crop improvement programs as crops are grown increasingly at the limit of their adaptation. At the same time, the more frequent occurrence of extreme events predicted with global climate change may mean that the ability to tolerate environmental constraints is essential for a cultivar to give reliable production.

Stress can be defined as a condition that limits a plant in realizing its potential for growth, development, and reproduction; extremes of stress result in irreversible damage and ultimately in plant death. Environmental factors may be stressful because they are in limited supply (suboptimal) or because they are in excess supply (superoptimal). Within this, there are degrees of stress; i.e., the farther from the optimum, the more stressful the environment. As plants rarely grow in optimal environments, they can be said to be under some degree of stress at all times. The environmental sensitivity of a plant varies throughout its development (Levitt, 1980). Seedling establishment and floral development are two particularly vulnerable stages. Thus, a given level of an environmental variable may be a greater stress during certain parts of a plant's life cycle than during others. For annual crops, temperature extremes are often greatest near the start and end of the season, when the plant is at its most sensitive developmental stages. There is also an interaction between the degree of stress, its duration, and the rate of increase of temperature. A long period at a moderately high temperature, may be as injurious as a brief exposure to an extreme temperature. The time scale of fluctuations in the environment can be a matter of minutes or seasons. The rate of temperature change will affect the intensity of the stress as well; a plant may well survive a given temperature if the rate of increase of temperature is relatively slow. In terms of crop production, an environment is stressful if harvestable yield is reduced, although this may simply be because the rate of growth is accelerated sufficiently by high temperature, resulting in a shorter duration between growth stages. For example, when the duration of the grain-filling period is shortened by high temperature, grain weight is generally reduced in wheat (Wiegand and Cuellar, 1981) as the shorter duration for grain filling is not compensated by an increased rate of grain filling, probably due to the shorter period available for photosynthetic fixation of carbon dioxide and subsequent translocation to the grain.

The changes in temperature that a plant is subjected to are more rapid than many other stress factors in the natural environment, e.g., drought and nutrient stress. This has meant both that the plant has developed mechanisms for minimizing the damage caused by such changes and that it is a very amenable experimental system for investigation. Although in the laboratory it is possible to control one environmental variable such as temperature very precisely, in the field it must be appreciated that there are often complex interactions between stresses and that heat stress may not occur in isolation. For example, an increase in temperature results in increased evaporative demand, both of the soil and plant, resulting in high transpiration rates. Although the latter can cause considerable cooling to the plant, it also can result in plant water deficit. Water shortage in the soil can limit this heat dissipating process, thus exacerbating the effect of high temperature.

In order to understand the effect of environmental variables on plant growth and survival in the field, it is important that the climate is characterized in detail. Mean daily air temperatures do not indicate either the diurnal variation in temperature that occurs or the extremes reached. For example, on a day at the start of the growing season in Rajasthan, India, with a mean air temperature of 35°C, the actual maximum air temperature was 46°C. If soil temperatures are considered, then more extreme temperatures are reached, with the soil surface reaching up to 60°C at midday and a predawn minimum of less than 20°C (Howarth, 1991; Peacock et al., 1993). These conditions may persist for many days (Figure 10.1). Even if a plant is able to withstand the high air temperatures, many consequences of the extremely high soil-surface temperatures are encountered. In graminaceous species, leaf and shoot growth occurs from meristems situated near the soil surface and it is the temperature of this region that determines growth rate (Peacock, 1975). Heat injury of these meristems would affect further shoot growth. However, a number of species, and particular genotypes of those

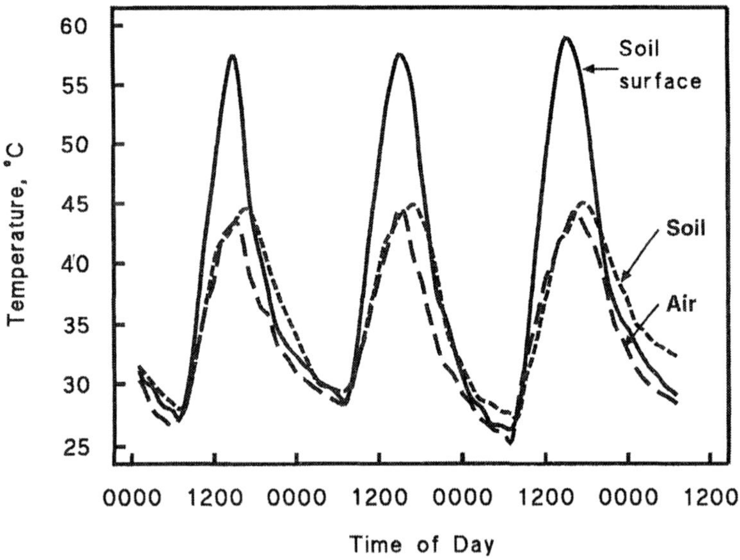

FIGURE 10.1. Diurnal temperature data recorded at Fatehpur, Rajasthan, India (latitude 27° 37' N), in June 1990. Each measurement is the mean value from three thermocouples placed at 5 cm depth of soil (- - -); 0.5 cm depth of soil (——); or 150 cm above the soil surface(— — —).

species, are capable of not only surviving these conditions but also producing a harvestable crop. Understanding the mechanisms involved could be invaluable in improving crop production under high-temperature stress conditions.

One approach to understanding the ability of plants to acclimate to, and tolerate, environmental stress is to identify stress-induced changes in gene expression (Howarth and Ougham, 1993). It has been hypothesized that particular proteins, whose synthesis is induced by stress conditions, are critical in the survival of that stress. Changes in gene activation, transcription, and translation often occur during the acclimation process and thus are thought to be involved in the induction of tolerance. Investigations into the molecular mechanisms of survival under extreme environmental conditions provide valuable information both on how plants survive and on what genes are involved. At the molecular level, the response of plants to temperature extremes is the best-characterized response to the environment, and this will be discussed in detail here. Advancements in the identification of specific genes associated with survival of high temperature bring with them the possibility of directly improving plant tolerance to high-temperature stress. In addition, developments in molecular genetics now have the potential to enhance this process by identifying specific markers linked to regions of the genome associated with high-temperature tolerance. Plant response to temperature stress is a complex process conditioned by a number of component responses, each controlled by one or more genes. In recent years, developments in molecular marker technologies and their use in quantitative trait loci (QTL) analysis have provided effective new opportunities for the study of plant responses to the environment (Lebreton et al., 1995; Prioul et al., 1997; Yadav et al., 2002). Molecular marker technology provides opportunities to not only identify such QTL that determine complex phenotypes such as heat and drought tolerance but also to improve greatly the efficiency of genetic improvement by facilitating introgression of desirable traits through the use of linked markers (Tanksley, 1993; Mohan et al., 1997).

HIGH-TEMPERATURE-INDUCED CHANGES IN GENE EXPRESSION

The daytime temperatures that plants are exposed to in many tropical areas are often above their optimum for growth; an increase in temperature can have dramatic effects on growth and survival. As can be seen from Figure 10.2, a small increase in temperature above the optimum has a large effect on growth rate, and a 10°C rise in temperature can be sufficient to kill the plant. This contrasts with the effect of a decrease in temperature from the optimum, where growth rates may be much reduced, but plant death does not occur until much lower temperatures are reached. Temperature affects both the rate of development and the duration of various developmental stages. In addition to speeding up phenological processes, high temperatures have deleterious effects on photosynthesis, respiration, reproduction and, in particular, seedling survival.

Only short-term exposure to high temperatures is required for shoot growth to cease (Figure 10.2). In this case, pearl millet seedlings were exposed to a given temperature for 2 h and the subsequent growth rate after return to optimal temperatures was measured. Similarly, if the growth temperature of a plant is increased beyond a particular point, a sudden decline in the ability to synthesize protein occurs. However, tolerance of both protein synthesis and seedling growth to a previously lethal high temperature can be induced by prior exposure to a sublethal temperature (Figure 10.2). This induced thermotolerance can result from either a short-term heat shock or a relatively gradual increase in temperature, as might be experienced by a plant in the field. Such exposure of living tissue to high-temperature stress results in a dramatic change in gene expression. Above a threshold temperature, the normal pattern of protein synthesis is repressed and a new set of proteins is synthesized from newly transcribed mRNA. These are the heat-shock proteins (HSPs). The heat-shock (HS) response is not unique to plants but is found in every organism so far examined (Schlesinger et al., 1982; Vierling, 1991). There are a number of major families of HSPs, classified on the basis of molecular weight: high-molecular-weight HSPs (>80 kDa), the HSP70 family (i.e., HSPs of molecular weight approximately 70 kDa), and low-molecular-weight HSPs (14 to 30 kDa). Most of these proteins have chaperonin function; that is, they either prevent irreversible protein denaturation due to high temperature or promote refolding of proteins to their native structure (Lee et al., 1997; Glover and Lindquist, 1998). The temperature at which maximal HSP synthesis occurs is positively correlated with the optimal growth temperature for that organism. The HS response is finely tuned to its thermal adaptation (Howarth and Ougham, 1993). For example, the temperature at which HSPs accumulate maximally is 45°C in one-day-old pearl millet and sorghum seedlings (Howarth, 1989), 40°C in soybean and maize (Key et al., 1981; Cooper and Ho, 1983), 35°C in the temperate grass *Lolium temulentum* (Ougham, 1987); and 25°C in the snow fungus *Monographella nivale* (Cairns et al. 1995). In general, the HS response is induced by a 10°C rise from the optimal growth temperature.

The response of gene expression to an increase in temperature is very finely modulated and is dependent on the severity of the heat treatment. A small increase in temperature (5°C) results in only selected HSP synthesis

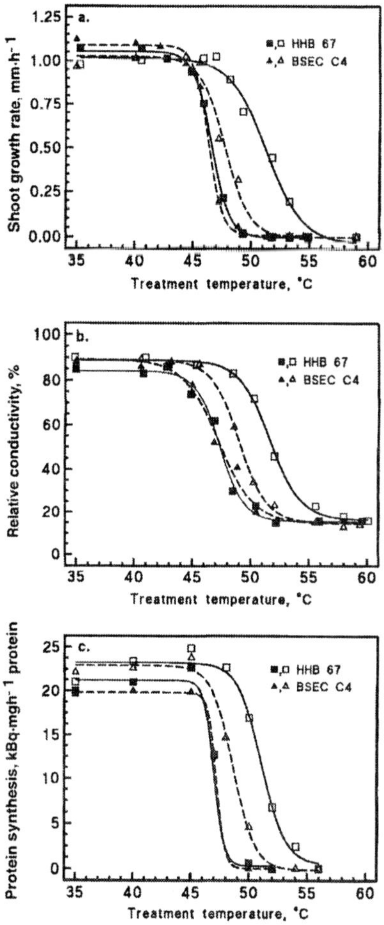

FIGURE 10.2. Effect of treatment temperature on (a) growth; (b) electrolyte leakage; and (c) protein synthesis of two-day-old pearl millet seedlings. Two-day-old seedlings were exposed to the temperatures indicated using a thermal gradient bar for 2 h, either with no pretreatment (closed symbols) or immediately following a 2 h pretreatment at 43°C (open symbols). For growth studies, seedlings were returned to 35°C following treatment and the subsequent growth measured after 24 h. Electrolyte leakage was determined as described by Howarth et al. (1997). Protein synthesis was determined by the use of [^{35}S]-methionine during the 2 h treatment on the thermal gradient bar as described by Howarth (1989). Each experiment was repeated three times for the two pearl millet accessions tested (HHB 67; —, ■, □, BSEC C4; ---, ▲, Δ), the mean values determined for each temperature point and sigmoid logistic curves fitted.

as well as the continued synthesis of normal cellular proteins (Howarth and Skøt, 1994). As the temperature is increased, the synthesis of non-heat shock proteins is preferentially inhibited and HSPs predominate. A difference in treatment temperature of as little as 1°C can have a considerable effect on gene expression. At even higher temperatures, HSP synthesis is also inhibited, with the synthesis of HSP70 being detectable at temperatures where all other protein synthesis has ceased. The extent to which normal protein synthesis is repressed during HS is reported to vary from species to species. However, rather than these being conflicting reports concerning the degree of translational control of normal protein synthesis, they may simply be due to a different relative severity of HS. The heat-shock response is very rapidly induced and is detectable within minutes of the onset of high-temperature conditions. It is a transient phenomenon; a prolonged high-temperature treatment does not result in a continuation of the HS response, and on return to optimal temperatures HSP synthesis is rapidly repressed (Howarth and Skøt, 1994).

THERMOTOLERANCE

The HS response is not only universal in all species examined, including bacteria, yeast, plants, invertebrates, and vertebrates, but it has also been found in virtually all tissue and cell types. Many of the individual HSPs show remarkable homology; antibodies against HSPs show cross-reactivity between species (Kelley and Schlesinger, 1982; Vierling, 1991). HSP cDNA probes also indicate homology between different species. So, if they are among the most highly conserved proteins in nature, do they also have a universal function?

When organisms are exposed to extremely high temperatures, they are killed. If, prior to such exposure, they are given a mild heat treatment, this enables survival of an otherwise lethal temperature (Figure 10.2). During this mild heat treatment, HSPs are synthesized. Since thermotolerance appears to correlate closely with HSP synthesis, it has been hypothesized that the function of at least one heat-shock gene is implicated in causing thermotolerance. The kinetics of synthesis of HSPs closely mirrors that of the development of thermotolerance (Howarth and Skøt, 1994). HSP synthesis can be induced by chemical treatments such as arsenite in the absence of HS. As these chemicals also induce thermotolerance, this provides further evidence for the involvement of HSPs in high-temperature survival (Howarth, 1990). More direct evidence is provided by the existence of mutants defective in HSP synthesis that are also unable to acquire thermotolerance (Sanchez and Lindquist, 1990). A number of studies question the involvement of HSPs, or individual HSPs, in thermotolerance. Other HSPs may compensate for the absence of an individual HSP. Many studies investigating the link between thermotolerance and HSPs may be confounded by a number of problems. First, at least some HSPs, including members of the HSP 70 family, are found synthesized in the cell under normal growth conditions. Second, the presence of HSPs in the cell does not imply that the cell is in a thermotolerant state. HSPs persist for a number of days following high-temperature conditions (Chen et al., 1990; Howarth, 1991). Thermotolerance, however, does not persist (Howarth, 1991; Howarth and Skøt, 1994). Reheat shock, during which HSPs are again synthesized from newly transcribed mRNAs, returns the tissue to its thermotolerant state. This suggests either that de novo synthesis of HSPs is required for thermotolerance or that the acquisition of thermotolerance was not related to the presence of HSPs. It is also possible that the role of HSPs is not in the development of thermotolerance but in the repair of heat-induced damage. Transgenic studies are required to ascertain the function of individual HSPs.

The correlation between the synthesis of HSPs and the development of induced or acquired thermotolerance has long enticed researchers with the prospect that plants could be engineered to better tolerate high-temperature stress. Until recently, the evidence for the direct involvement of HSPs in thermotolerance has been equivocal, but recent results suggest that they do have a direct role. Some of the best-characterized HSPs, in both prokaryotes and eukaryotes, are proteins of the HSP100 family; these proteins are a class of molecular chaperones found in bacteria, yeast, and higher plants where they are known as ClpB, Hsp104, and Hsp101, respectively, but not in other eukaryotes (Sanchez et al., 1992; Schirmer et al., 1996). In yeast, deletion of the gene encoding the HSP100 protein, *Hsp104*, reduces the ability of cells to survive high-temperature exposure by three to four orders of magnitude (Sanchez and Lindquist, 1990). The HSP101 encoded by a number of plant species can complement this thermotolerance defect in yeast caused by the deletion of the *Hsp104* gene (Lee et al., 1994; Schirmer et al., 1994; Wells et al., 1998; Young et al., 2001), suggesting conservation of the molecular chaperone activity required for the acquisition of thermotolerance. Queitsch and colleagues (2000) have recently provided more direct evidence that HSP101 from *Arabidopsis* is essential for survival under severe heat stress and for the acquisition of thermotolerance. The requirement of HSP101 for acquired thermotolerance is demonstrated by exposure of seedlings to the lethal temperature of 45°C for 2 h with and without preconditioning at 38°C. All plants are killed by direct exposure to the severe heat stress. Pretreatment at 38°C, however, enables wild-type controls to survive the subsequent severe heat, whereas plants with reduced HSP101 levels are drastically impaired or killed despite pretreatment. This result provides clear evidence that HSP101 expression is essential for induced thermotolerance. Furthermore, Hong and Vierling (2000) identified a mutant of *Arabidopsis* defective in the acquisition of tolerance to high-temperature stress. These mutant plants not only possessed a defective *Hsp101* gene but also were unable to acquire thermotolerance. Complementation of the mutant plants by transformation with wild-type *Hsp101* genomic DNA restored the mutant plants to the wild-type phenotype. This provides direct evidence that *Hsp101,* which is required for thermotolerance in bacteria and yeast, is also essential for thermotolerance in a complex eukaryote. Thermotolerance also can be improved by manipulating the heat-shock transcription factors that control the synthesis of HSPs. However, this approach has had mixed results and potentially may lead to unwanted consequences because many of the HSPs induced have roles in normal growth and development (Lee and Schöffl, 1996; Prändl et al., 1998).

Survival of high-temperature conditions also requires that the tissue recovers its normal cellular functioning once a heat shock has ceased. The recovery of normal protein synthesis following heat shock is dependent on the severity of that heat shock (Howarth and Skøt, 1994). Three stages are apparent during recovery; first, the synthesis of HSPs continues; second, all protein synthesis is apparently repressed; and third, a gradual increase in the synthesis of normal cellular proteins occurs, with a decline in the synthesis of HSPs. The higher the heat-shock temperature, the longer the time period in which HSPs are synthesized on return to normal growth temperatures. Recovery from a mild heat shock results in a rapid resumption in normal protein synthesis. If HSPs are involved in coping with the consequences of heat stress then it might be expected that the more extreme the stress, the greater the requirement for HSPs.

Most of the experiments described here involve treating tissue with a sudden increase in temperature. Although temperature changes in the field can be very rapid, particularly in tropical areas, plants are not exposed to a temperature shock as such but to a relatively gradual increase in temperature. Field conditions can be mimicked in the laboratory, and it is found that the heat-shock response is induced not only by a sudden increase in temperature (DeRocher et al., 1991; Howarth, 1991). Moreover, thermotolerance is also induced by a gradual increase in temperature. The upper limit for both protein synthesis and survival when the temperature is increased gradually is much higher than for a sudden increase, as the acclimation process actually occurs during the change in temperature. Plants in the field are often exposed to high-temperature conditions on a daily basis, occurring for a few hours at midday and recurring each day, and this can have a detrimental effect. Experiments with pearl millet seedlings have shown that although seedlings may be able to synthesize HSPs on their first exposure to high temperature, on subsequent days this ability is lost (Howarth, 1991). It is possible that the seedlings have not recovered from the consequences of a preceding heat shock by the time they are exposed to further high-temperature conditions. The ability to survive repeated heat shocks is of prime importance in parts of the world with high midday temperatures, and genotypic differences in this ability have been shown (Howarth, 1991). In the field, there can also be a spatial distribution of stress. A seedling emerging from the soil is likely to be exposed to a wide range of temperatures. This is particularly evident in tropical areas, where at midday a seedling may be exposed to a root-zone temperature of 40°C, a soil-surface temperature of 60°C, and an air temperature of 45°C. HSPs may be important in the survival of such wide fluctuations in temperature by stabilizing this meristematic tissue.

ASSESSMENT OF HIGH-TEMPERATURE TOLERANCE

To enable improvements in thermotolerance, it is necessary to determine the genetic variation available for this trait and to identify sources of tolerance. Screening techniques are required to characterize this variation, to identify appropriate breeding material, and ultimately for selection for improved adaptation. For routine screening, the technique used must be applicable for large numbers and be economical, rapid, and straightforward to assess. Screening techniques in general can be based in either the field or the laboratory. Screening in the target environment has the advantage of using relevant stress levels. However, there are many difficulties with respect to field screening, particularly in relation to abiotic stress: the natural climate is not always reliable and is certainly variable both day to day and year to year; screening is often limited to small portions of the year; and the target environment is not always conveniently situated. Moreover, in the field, the interactions of water stress, high temperature, and other edaphic constraints on plant responses complicate the assessment process. Laboratory screening techniques are not affected by these problems, in that they can be conducted under controlled conditions without recourse to the variability of the natural environment. They are often based on one aspect of response to stress and not the integrated effects of the environment on many physiological and developmental processes. Dissecting a complex process such as acclimation to, and survival of, high-temperature stress into component parts that are under more simple genetic control should permit rapid and precise improvement. An understanding of the physiology of the response to stress is required to enable the development of such screening techniques. It is important, however, that a laboratory-based technique has a significant relationship to field performance. Most laboratory tests are, therefore, based on analysis of whole plants or plant segments in controlled environments. Recently, molecular techniques have been used to detect cultivar differences in levels of thermotolerance. The challenge is to identify a simple screening test that can be used quickly and easily on large numbers of plants, that is quantitative, that has a heritable response, and that displays a close correlation to performance in the field. Such screening techniques can then be used either directly in a breeding program for improvement of seedling thermotolerance or combined with genetic analysis to identify molecular markers linked to the trait of interest.

Considerable genetic variation exists for the ability to both survive and grow after exposure to high-temperature conditions. A field technique to screen pearl millet entries for their emergence and survival at high soil-surface temperatures has been developed and indicates that considerable genetic variation exists for this trait (Peacock et al., 1993). As shown in Figure 10.2, there were significant differences in the ability of two contrasting genotypes of pearl millet to grow after exposure to high temperature once they had been acclimated. However, little difference was found between nonacclimated genotypes with any method used, suggesting that their basal thermotolerance before acclimation was similar. A method of assessing heat tolerance in cowpea during reproductive development has also been developed both in the field and greenhouse (Warrag and Hall, 1983; Ismail and Hall, 1998). The greenhouse method is extremely effective for screening for reproductive-stage heat tolerance but has a limitation on the number of plants that can be screened as compared to a field nursery. One advantage of using a hot greenhouse compared with most hot field conditions is that air temperatures are relatively stable from day to day and over weeks. There is also evidence for differences in the intrinsic temperature response of different genotypes of the same species. For example, studies with maize indicate that the optimum temperature for growth and development in Mexican highland cultivars of maize may be 19 to 20°C, which is some 10°C lower than for lowland cultivars (Ellis et al.,1992).

A number of physiological processes have been examined in the past for their use in the assessment of tolerance to high temperatures. Cellular membrane dysfunction results from high-temperature stress (Levitt, 1980). The leakage of electrolytes into a bathing medium after exposure to stress provides an indication of the cellular membrane damage caused by that stress; the more damage caused by the stress, the more solutes will leak into the bathing medium and the higher the conductivity of that medium will be. Electrolyte leakage has been used to assess stress-induced damage in many species (Sullivan, 1972; Martineau et al., 1979; Saadalla, Shanahan, and Quick, 1990; Saadalla, Quick, and Shanahan, 1990; Shanahan et al., 1990).

A cell membrane system that remains functional during heat stress appears central to adaptation of plants to high temperatures (Raison et al., 1980). Thus, conductometric measurements of solute leakage have been used as a measure of the thermal stability of cellular membranes to assess the thermotolerance of genotypes (Sullivan and Ross, 1979; Blum, 1988; Howarth et al., 1997), but care must be taken since injury to membrane stability is influenced by age of the leaf and the part of the leaf sampled (Wardlaw et al., 1989), and the expression of heat tolerance is dependent on acclimation (Martin and Wehner, 1987; Blum, 1988) and its previous thermal history. It is important that leaves of an appropriate age are assessed for membrane thermostability, as it is a trait that varies throughout the life cycle of the plant. Positive associations between membrane thermostability, as measured by electrolyte leakage, and grain yield under heat stress have been reported for two spring wheat populations (Blum et al., 2001). For cowpea, membrane thermostability of leaf disks was associated with reproductive-stage heat tolerance (Ismail and Hall, 1999). In seedlings of pearl millet and sorghum, a close correlation was found between membrane thermostability, growth after exposure to high temperature, and field performance under heat stress (Howarth et al., 1997). However, Wehner and Watschke (1984) found no significant differences between various turfgrasses in efflux of cell solutes from tissue sections heated to 43 to 49°C. They concluded that disruption of some other physiological process might account for the heat-tolerance rankings of the grasses. Wallner and colleagues (1982) showed that quantitative differences in heat tolerance, measured by electrolyte leakage of leaf segments, were consistent with qualitative descriptions of drought resistance for most species and that the best criterion to illustrate differences of heat tolerance among species was the killing time. Howarth and colleagues (1997) showed a close correlation between membrane thermostability of pearl millet seedlings following acclimation and their survival in the field in Rajasthan; unacclimated seedlings, however, showed little difference in membrane thermostability.

Seedling vigor measurements have also been used to assess their significance on seedling survival of heat stress. Fast seedling growth and consequent early seedling establishment could be one strategy to escape a stressful environment, particularly as conditions for seedling establishment often become increasingly less optimal with time after sowing. On the other hand, rapid development of a greater leaf area might result in faster depletion of limited soil-water resources. In fact, measurements of leaf growth in the field indicate that a conservative growth strategy may favor survival in extreme environments.

During the vegetative stage, high day temperatures can cause damage to components of leaf photosynthesis, reducing carbon dioxide assimilation rates, compared with environments having more optimal temperatures. Sensitivity of photosynthesis to heat may be due mainly to damage to components of photosystem II located in the thylakoid membranes of the chloroplast and membrane properties (Al-Khatib and Paulsen, 1999). Chlorophyll fluorescence can be used to measure the temperature acclimation of photosynthesis (Yamasaki et al., 2002).

One particularly sensitive metabolic process is protein synthesis and, as indicated in the requirement for the synthesis of specific proteins for the acquisition of thermotolerance, is critical in the survival process. In Figure 10.2, the results from examining the effect of temperature on protein synthesis of two genotypes of pearl millet are shown. Treatment temperatures of up to 45°C resulted in a slight increase in protein synthesis compared with that found at the standard growth temperature of 35°C, but treatment temperatures above this resulted in a rapid cessation of protein synthesis, with barely detectable synthesis at a temperature of 48°C. The effect of a pretreatment of 2 h at 43°C immediately prior to the addition of the [^{35}S]-methionine is also shown. This resulted not only in significant acclimation in all entries but also in genotypic differences. The genotype HHB 67, which also displays enhanced thermotolerance by other screening techniques and in the field, exhibited an enhanced ability to acclimate the process of protein synthesis to high temperature, whereas the thermosensitive genotype BSEC C4 showed little ability to acclimate. This method employs the use of radioactivity in its determination, which limits its utility as a screening technique, so the use of specific proteins might have greater value. As discussed earlier, the heat-shock response shows considerable correlation with acclimation to high-temperature conditions, but can it be used to measure high-temperature tolerance? The development of both specific antibodies to individual HSPs and of complementary (c) DNA probes to HSPs mRNAs can be exploited in the development of quantitative screening systems. These could measure rapidly and precisely the abundance of a given gene or set of genes, or their products, in specific tissue under defined physiological conditions. Even if a given gene is not directly involved in the mechanism of acquired thermotolerance, this does not mean that that gene or its product cannot be used in a screening technique if a close correlation exists between its presence and the ability to survive high temperatures. Such a screening technique could then be used to assess the thermosensitivity of genotypes. It is also necessary to determine if HSPs contribute to genetically determined differences in heat tolerance. Differences between cultivars in the actual HSPs synthesized and the temperature thresholds for their synthesis clearly exist (Howarth, 1989; Jorgensen, et al. 1993; DiMascio et al., 1994; Park et al.,1996). However, there is considerable polymorphism in the actual HSPs synthesized in different species, and correlating both qualitative and quantitative differences in HSPs and thermotolerance can be misleading unless appropriate genetic tests are also conducted (Vierling, 1991). Detailed analysis of the HSPs synthesized is also required; unless high-resolution, two-dimensional electrophoresis is conducted, interpretation of HSP patterns obtained can be misleading (Jorgensen and Nguyen, 1995). Another approach is to look for polymorphism within HSP genes from genotypes of differing thermotolerance and determine whether such polymorphism is related to genetic differences in the ability to survive high temperature.

A divergent selection program for improved seedling thermotolerance of pearl millet comparing two laboratory-based screening techniques and field screening is currently being conducted (Weltzien et al.,1994). There are few studies reported where both the genetic variation identified and screening systems developed have been exploited in crop improvement. To do this it is necessary to both select for improved adaptation and then to evaluate the response to selection for improved thermotolerance in the field. Many screening techniques tend to be developed following the detailed analysis of the response of genotypes with proven resistance or susceptibility to high temperature. The problems here are that genotypes differ in many traits as well as the target ones, and it is difficult to know which ones are relevant. The results obtained are often specific to the actual cultivars and growing conditions used. It is more difficult to distinguish genotypes showing an intermediate response to a given stress.

FUTURE PROSPECTS FOR BREEDING FOR TOLERANCE TO HIGH-TEMPERATURE STRESS

In this review, the effects of high temperature on plants have been considered, particularly in relation to resultant changes in gene expression. Acclimatory changes resulting in an increase in tolerance to these stresses often occur, and it is possible to investigate changes in gene expression associated with this. It is much more difficult to demonstrate unambiguously that a particular observed change in gene expression contributes to the survival process. Changes in the expression of large numbers of genes occur due to stress, yet it is probable that only some of these genes are directly involved in stress tolerance. Indeed, it is possible that in some cases the synthesis of a protein indicates sensitivity to a stress rather than being part of a tolerance mechanism. Transgenic plants provide a direct way of testing the role of a given gene in the acclimation process. However, acclimation involves changes in a large number of genes, and it is not known how many of these are actually involved in improving survival. The transfer or manipulation of a single gene may not be sufficient to convert a stress-susceptible genotype into one tolerant of the same stress. One goal of understanding the mechanism of plant acclimation to environmental stress is to be able to manipulate the response of plants so that areas inhospitable to plants, due to natural or man-made causes, can be exploited. Transgenic plants can be used not only for investigating gene function but also as a method of transferring genes across the species barrier as the ultimate way of improving tolerance. Considerable knowledge of the function of the proteins involved, their control pathways, and associated biochemistry and physiology is needed before such improvements can be contemplated. It may be that stress-sensitive plants actually possess the genes necessary for tolerance but that they are not expressed in the same developmental stage or tissue as in tolerant plants. In that case, it would be relatively easy to modify the stress response of the tolerant plant. An understanding of gene expression under stress also has potential for the exploitation of these gene products in quantitative screening systems. The big question remains: Having established that at least some HSPs are directly responsible for thermotolerance, is it possible to use the techniques of genetic engineering to improve temperature tolerance? Although deletion of single HSP genes has been shown to confer thermosensitivity, many other studies indicate the involvement of a number of different genes and interaction between their gene products (e.g., Lee and Vierling, 2000). One approach is to examine the variation in thermotolerance between genotypes of a species and determine variation in HSPs between them. However, it is likely that multiple factors are involved in survival of high temperature, particularly as high-temperature stress in the field is often a complex and interactive phenomenon.

Biotechnology can aid breeding by accessing or creating novel sources of specific plant variation by the introduction of cloned genes into the genome using DNA-manipulative techniques (genetic manipulation). One of the major differences between conventional plant breeding and direct genetic manipulation is that while the former is mainly concerned with the transfer of large numbers of genes, the characteristics of which are known only through phenotypic measurements, the latter aims to transfer small numbers of genes with high precision. Genetic manipulation does, however, require the precise definition of the physiological, morphological, and biochemical bases of action of individual genes as well as their identification and cloning. In only a few cases do we currently have the required knowledge to do this. However, as shown previously, genetic manipulation can provide an excellent tool for understanding basic plant biology and the processes underlying traits of agronomic importance.

As Boyer (1982) points out, there is often a dramatic difference between maximal and average yields for any given crop species. The actual yield achieved is dependent on the environmental conditions the crop encounters. In addition, variation exists in the ability to survive and grow under stress conditions, and this needs to be combined with increased ability for yield, so that not only is the discrepancy between yield under optimal conditions and actual yield reduced but also an increased stability of yield obtained. This indicates that the capacity for yield improvement can be found within a species and is amenable to conventional breeding techniques. Over the years, considerable effort has been made by plant breeding to extend the range of distribution of many crop plants, with a number of notable successes; for example, improved low-temperature tolerance of maize has resulted in extensive production in Europe (Mechin et al., 2001). Wheat, which originated in the Middle East, is now grown in many diverse environments, from tropical to temperate, coping with long, cold winter periods and hot summers. However, the breeding process used for domesticated crops has resulted in a very narrow genetic base. Landraces and related wild species may need to be used to extend the range of variation for tolerance to high temperature that can be used in crop-improvement programs. These considerations are important for the development of new stress-tolerant cultivars of agricultural species. Improvements in temperature tolerance have at present all been made using conventional plant breeding. Developments in molecular genetics, improved screening techniques, and knowledge of appropriate germplasm can now exploit this natural variation to a greater effect. It is possible, however, to improve only the degree of tolerance; there will always be a level beyond which it is not possible to improve. Plant physiologists have extensively explored the relationship of many traits to plant adaptation and yield potential, but the evaluation of such traits has generally proved too costly to permit their use by plant breeders as selection criteria in segregating populations. However, the advent of marker-assisted selection means that such traits can be reconsidered. Once the value of a trait and the presence of genetic variation have been established, routine selection can be based on a linked marker rather than the trait per se.

Of the emerging genetic technologies, marker-assisted selection (MAS) is the most likely to have an immediate impact on plant breeding for complex traits such as stress tolerance. The first stage involves the detection of DNA variation among individuals using a variety of techniques, such as restriction fragment length polymorphisms (RFLPs), simple sequence repeats (SSRs or microsatellites), expressed sequence tags (ESTs), and amplified fragment length polymorphisms (AFLPs) (Henry, 2001). Such genetic mapping combined with physiological screening permits the mapping, identification, manipulation, and combination of specific genes involved in tolerance. The challenge is to identify specific physiological or biochemical processes and to develop rapid, high-throughput screening techniques based on them. Molecular marker maps exist for many crop species and have been used to identify QTLs associated with stress tolerance in a wide range of species (e.g., Lebreton et al., 1995; Mohan et al., 1997; Yadav et al., 2002). A similar approach is being used to detect QTLs associated with seedling thermotolerance (Howarth et al., 1994). Both qualitative and quantitative variation in heat-shock protein synthesis has been found in maize (Frova and Gorla, 1993) and in sorghum (Howarth, 1989). Subsequent genetic analysis confirmed the multigenic control of HSP expression and has revealed both qualitative and quantitative polymorphisms (Jorgensen and Nguyen, 1995) with evidence for at least four unlinked gene loci. QTL associated with adaptation to temperature in a population produced between highland- and lowland-adapted maize have also been identified (Jiang et al., 1999).

MAS can be used to precisely improve the required character by following closely the movement of desired and undesirable gene segments in the breeding process. The potential of these modern, precision-breeding methods is considerable. Mapping potential physiological and biochemical components of adaptation also provides information on their involvement in adaptation and is a new way of elucidating the mechanism of plant responses to the environment. Genetic mapping shows in a much clearer fashion not only how traits are genetically correlated but also how they are linked on the chromosomes. Active collaboration between geneticists, molecular biologists, physiologists, breeders, germplasm collectors, and other relevant disciplines are required to ensure success. Once regions of chromosomes are identified that carry genes to confer stress tolerance, it is then possible to clone those genes and ultimately lead to a more complete understanding of the genetic components of plant responses to temperature stress.

Modern conventional plant breeding is using both these new marker technologies and the increased understanding of the mechanisms and inheritance of complex traits. Although research associated with the identification of QTLs is expanding rapidly, examples of the application of marker-assisted selection methods in breeding programs are still relatively limited. Despite the promising potential of MAS, precision in locating QTLs can often be poor (Kearsey and Farquhar, 1998). Moreover, when quantitative traits are evaluated in different environments, QTL × environment (QTL × E) interaction is a commonly reported phenomenon (Paterson et al., 1991; Lee et al., 1996; Yadav et al., 2002). Such interaction raises a serious concern as to which QTL to use in a marker-assisted selection program. QTLs that are more commonly detected across environments can be used for improving those traits across environments, while those more specific to a particular environment will have their application in specific environments only (Romagosa et al., 1996). These QTL × E interactions result from the differential genotype response to environment, resulting in varying performance of individual genotypes across environments. More work is required on the practical use of MAS selection in breeding programs and the value of specific markers. MAS should, in particular, be most efficient when traits are difficult or expensive to measure.

Genetic improvement of temperature tolerance is thus possible by both conventional plant breeding, either aided by specific screening techniques or the use of molecular markers linked to the trait of interest, and also by genetic manipulation via transformation. Plant breeding, however, is not the only way forward, and agronomic solutions must also be considered. Timing of planting, seedbed preparation, mulching, and irrigation can all affect a plant's exposure to temperature extremes. Increasing the tolerance to temperature extremes through a combination of well-adapted, improved cultivars and management practices remain a challenge. It must also be appreciated that before crop improvement is conducted, it is necessary first to understand both what the major reasons for crop failure or yield reduction are in the farmer's field and what the prevalence of a given stress is. Each environment poses a different set of problems and, in marginal environments where the climate is highly variable, it is often difficult to define precisely what the causes of crop failure are—but this must be at least approximated if successful crop improvement is to take place. For a farmer to adopt a change in these high-risk environments, demonstration of improvements actually on farmers' fields is necessary. It is particularly in these areas that genetic improvement of temperature tolerance is most needed, especially if increases in population pressure result in the expansion of crop production into more and more marginal areas. Climatic change could exacerbate this need. Locally adapted germplasm is often capable of surviving the environmental conditions and should be widely used in breeding for improved temperature tolerance. Targeting of crop improvement research specifically for these environments is thus required.

REFERENCES

Al-Khatib, K. and G.M. Paulsen (1999). High-temperature effects on photosynthetic processes in temperate and tropical cereals. *Crop Science* 39:119-125.

Blum, A. (1988). *Plant Breeding for Stress Environments*. Boca Raton, FL: CRC Press.

Blum, A., N. Klueva, and H.T. Nguyen (2001). Wheat cellular thermotolerance is related to yield under heat stress. *Euphytica* 117:117-123.

Boyer, J.S. (1982). Plant productivity and environment. *Science* 218:443-448.

Cairns, A.J., C.J. Howarth, and C.J. Pollock (1995). Submerged batch culture of the psychrophile *Monographella nivalis* in a defined medium: Growth, carbohydrate utilization and responses to temperature. *New Phytologist* 129:299-308.

Chen, Q., L.M. Lauzon, A.E. DeRocher, and E. Vierling (1990). Accumulation, stability and localization of a major chloroplast heat-shock protein. *Journal of Cell Biology* 110:1873-1883.

Cooper, P. and T.H.D. Ho (1983). Heat shock proteins in maize. *Plant Physiology* 95:215-222.

DeRocher, A.E., K.W. Helm, L.M. Lauzon, and E. Vierling (1991). Expression of a conserved family of cytoplasmic low molecular weight that shock proteins during heat stress and recovery. *Plant Physiology* 96:1038-1047.

DiMascio, J.A., P.M. Sweeney, T.K. Danneberger, and J.C. Kamalay (1994). Analysis of heat shock response in perennial ryegrass using maize heat shock protein clones. *Crop Science* 34:798-804.

Ellis, R.H., R.J. Summerfield, G.O. Edmeades, and E.H. Roberts (1992). Photoperiod, temperature, and the interval from sowing to tassel initiation in diverse cultivars of maize. *Crop Science* 32:1225-1232.

Frova, C. and M.S. Gorla (1993). Quantitative expression of maize HSPs: Genetic dissection and association with thermotolerance. *Theoretical and Applied Genetics* 86:213-220.

Glover, J.R. and S. Lindquist (1998). Hsp104, Hsp70, and Hsp40: A novel chaperone system that rescues previously aggregated proteins. *Cell* 94:73-82.

Henry, R.J. (2001). *Plant Genotyping: The DNA Fingerprinting of Plants*. Wallingord, UK: CABI Publishing.

Hong, S.-W. and Vierling, E. (2000). Mutants of *Arabidopsis thaliana* defective in the acquisition of tolerance to high temperature stress. *Proceedings of the National Academy of Science USA* 97:4392-4397.

Howarth, C.J. (1989). Heat shock proteins in *Sorghum bicolor* and *Pennisetum americanum*. I. Genotypic and developmental variation during seed germination. *Plant, Cell and Environment* 12:471-477.

Howarth, C.J. (1990). Heat shock proteins in sorghum and pearl millet: Ethanol, sodium arsenite, sodium malonate and the development of thermotolerance. *Journal of Experimental Botany* 41:877-883.

Howarth, C.J. (1991). Molecular responses of plants to an increased incidence of heat shock. *Plant, Cell and Environment* 14:831-841.

Howarth, C.J., G.P. Cavan, K.P. Skøt, R.W.H. Layton, C.T. Hash, and J.R. Witcombe (1994). Mapping QTLs for heat tolerance in pearl millet. In *Use of Molecular Markers in Sorghum and Pearl Millet Breeding for Developing Countries*, J.R. Witcombe and R.R. Duncan (Eds.). London: Overseas Development Administration, pp. 80-85.

Howarth, C.J. and H.J. Ougham (1993). Gene expression under stress. Tansley Review. *New Phytologist* 125:1-26.

Howarth C.J., C.J. Pollock, and J.P. Peacock (1997). Development of laboratory-based methods for assessing seedling thermotolerance in pearl millet. *New Phytologist* 137:129-139.

Howarth, C.J. and K.P. Skøt (1994). Detailed characterisation of heat shock protein synthesis and induced thermotolerance in seedlings of *Sorghum bicolor* L. *Journal of Experimental Botany* 45:1353-1363.

Ismail, A.M. and A.E. Hall (1998). Positive and negative effects of heat-tolerance genes in cowpea. *Crop Science* 38:381-390.

Ismail, A.M. and A.E. Hall (1999). Reproductive-stage heat tolerance, leaf membrane thermostability and plant morphology in cowpea. *Crop Science* 39:1762-1768.

Jiang, C., G.O. Edmeades, I. Armstead, H.R. Lafitte, M.D. Hayward, and D. Hoisington (1999). Genetic analysis of adaptation differences between highland and lowland tropical maize using molecular markers. *Theoretical and Applied Genetics* 99:1106-1119.

Jorgensen, J.A. and H.T. Nguyen (1995). Genetic analysis of heat shock proteins in maize. *Theoretical and Applied Genetics* 91:38-46.

Jorgensen, J.A., D.T. Rosenow, and H.T. Nguyen (1993). Genotypic comparison of heat shock protein synthesis in sorghum. *Crop Science* 33:638-641.

Kearsey, M.J. and A.G.L. Farquhar (1998). QTL analysis in plants: Where are we now? *Heredity* 80:137-142.

Kelley, P.M. and M.J. Schlesinger (1982). Antibodies to two major chicken heat shock proteins cross react with similar proteins in widely divergent species. *Molecular Cell Biology* 2:267-274.

Key, J.L., C.Y. Lin, and Y.M. Chen (1981). Heat shock proteins of higher plants. *Proceedings of the National Academy of Science USA* 78:3526-3530.

Lebreton, C., V. Lazicjancic, A. Stead, S. Pekic, and S.A. Quarrie (1995). Identification of QTL for drought responses in maize and their use in testing causal relationships between traits. *Journal of Experimental Botany* 46:853-865.

Lee, G.J., A.M. Roseman, H.R. Saibil, and E. Vierling (1997). A small heat shock protein stably binds heat-denatured model substrates and can maintain a substrate in a folding-competent state. *EMBO Journal* 16:659-671.

Lee, G.J. and E. Vierling (2000). A small heat shock protein cooperates with heat shock protein 70 systems to reactivate a heat-denatured protein. *Plant Physiology* 122:189-198.

Lee, J.H. and F. Schöffl (1996). An Hsp70 antisense gene affects the expression of HSP70/HSC70, the regulation of HSF, and the acquisition of thermotolerance in transgenic *Arabidopsis thaliana*. *Molecular and General Genetics* 252:11-19.

Lee S.H., M.A. Bailey, M.A.R. Mian, T.E. Carter, D.A. Ashley, R.S. Hussey, W.A. Parrott, and H.R. Boerma (1996). Molecular markers associated with soybean plant height, lodging, and maturity across locations. *Crop Science* 36:728-735.

Lee, Y.R., R.T. Nagao, and J.L. Key (1994). A soybean 101-kD heat shock protein compliments a yeast HSP104 deletion mutant in acquiring thermotolerance. *Plant Cell* 6:1889-1897.

Levitt, J. (1980). *Responses of Plants to Environmental Stress,* Volume 1, *Chilling, Freezing and High Temperature Stress.* New York: Academic Press.

Martin, D.L. and D.J. Wehner (1987). Influence of pre-stress environment on annual bluegrass heat tolerance. *Crop Science* 27: 579-585.

Martineau J.R., J.E. Specht, J.H. Williams, and C.Y. Sullivan (1979). Temperature tolerance in soybeans. I. Evaluation of a technique for assessing cellular membrane thermostability. *Crop Science* 19:75-78.

Mechin, V., O. Argillier, Y. Hébert, E. Guingoa, L. Moreau, A. Charcosset, and Y. Barrière (2001). Genetic analysis and QTL mapping of cell wall digestibility and lignification in silage maize. *Crop Science* 41:690-697.

Mohan, M., S. Nair, A. Bhagwat, T.G. Krishna, M. Yano, C.R. Bhatia, and T. Sasaki (1997). Genome mapping, molecular markers and marker-assisted selection in crop plants. *Molecular Breeding* 3:87-103.

Ougham, H.J. (1987). Gene expression during leaf development in *Lolium temulentum*: Patterns of protein synthesis in response to heat-shock and cold-shock. *Physiologia Plantarum* 70:479-484.

Park, S.Y., R. Shivaji, J.V. Krans, and D.S. Luthe (1996). Heat-shock responses in heat-tolerant and non-tolerant variants of *Agrostis palustris* Huds. *Plant Physiology* 111:515-524.

Paterson, A., S. Damon, J.D. Hewitt, D. Zamir, H.D. Robinowich, S. Lincoln, E.S. Lander, and S.D. Tanksley (1991). Mendelian factors underlying quantitative traits in tomato: Comparison across species, generations, and environments. *Genetics* 127:181-197.

Peacock, J.M. (1975). Temperature and leaf growth in *Lolium perenne*. II. The site of temperature perception. *Journal of Applied Ecology* 12:115-123.

Peacock, J.M., P. Soman, R. Jayachandran, A.U. Rani, C.J. Howarth, and A. Thomas (1993). High soil surface temperature effects on seedling survival in pearl millet. *Experimental Agriculture* 29:215-225.

Prändl, R., K. Hinderhofer, G. Eggers-Schumacher, and F. Schöffl (1998). HSF3, a new heat shock factor from *Arabidopsis thaliana*, derepresses the heat shock response and confers thermotolerance when overexpressed in transgenic plants. *Molecular and General Genetics* 258:269-278.

Prioul, J.-L., S. Quarrie, M. Causse, and D. de Vienne (1997). Dissecting complex physiological functions through the use of molecular quantitative genetics. *Journal of Experimental Botany* 48:1151-1163.

Queitsch, C., S.-W. Hong, E. Vierling, and S. Lindquist (2000). Heat shock protein 101 plays a crucial role in thermotolerance in *Arabidopsis*. *Plant Cell* 12:479-492.

Raison, J.K., J.A. Berry, P.A. Armond, and C.S. Pike (1980). Membrane properties in relation to the adaptation of plants to temperature stress. In *Adaptation of Plants to Water and High Temperature Stress*, N.C. Turner and P.J. Kramer (Eds.). New York: John Wiley and Sons, pp. 261-273.

Romagosa, I., S.E. Ullrich, F. Hann, and M.H. Hayes (1996). Use of the additive main effects and multiplicative interaction model in QTL mapping for adaptation in barley. *Theoretical and Applied Genetics* 93:30-37.

Rosenzweig, C. and M.L. Parry (1994). Potential impact of climate-change on world food supply. *Nature* 367:133-138.

Saadalla, M.M., J.S. Quick, and J.F. Shanahan (1990). Heat tolerance in winter wheat: II. Membrane thermostability and field performance. *Crop Science* 30:1248-1251.

Saadalla, M.M., J.F. Shanahan, and J.S. Quick (1990). Heat tolerance in winter wheat: I. Hardening and genetic effects on membrane thermostability. *Crop Science* 30:1243-1247.

Sanchez, Y. and S.L. Lindquist (1990). HSP104 is required for induced thermotolerance. *Science* 248:1112-1115.

Sanchez, Y., J. Taulien, K.A. Borkovich, and S. Lindquist (1992). HSP104 is required for thermotolerance to many forms of stress. *EMBO Journal* 11:2357-2364.

Schirmer, E.C., J.R. Glover, M.A. Singer, and S. Lindquist (1996). HSP100/Clp proteins: A common mechanism explains diverse functions. *Trends in Biochemical Sciences* 21:289-296.

Schirmer, E.C., S. Lindquist, and E. Vierling (1994). An *Arabidopsis* heat shock protein complements a thermotolerance defect in yeast. *Plant Cell* 6:1899-1909.

Schlesinger, M.J., M. Ashburner, and A. Tissières (1982). *Heat Shock from Bacteria to Man.* Cold Spring Harbor, NY: Cold Spring Harbor Laboratory Press.

Shanahan, J.F., I.B. Edwards, J.S. Quick, and J.R. Fenwick (1990). Membrane thermostability and heat tolerance of spring wheat. *Crop Science* 30:247-251.

Sullivan, C.Y. (1972). Mechanism of heat and drought resistance in grain sorghum and methods of measurement. In *Sorghum in the Seventies,* N.G.P. Rao and L.R. House (Eds.). New Delhi, India: Oxford and India Book House, pp. 247-264.

Sullivan, C.Y. and W.M. Ross (1979). Selecting for drought and heat resistance in grain sorghum. In *Stress Physiology in Crop Plants*, H. Mussel and R.C. Staples (Eds.). New York: Wiley and Sons, pp. 263-281.

Tanksley, S.D. (1993). Mapping polygenes. *Annual Review of Genetics* 27:205-234.

Vierling, E. (1991). The roles of heat shock proteins in plants. *Annual Review of Plant Physiology and Plant Molecular Biology* 42:579-620.

Wallner, S.J., M.R. Beewar, and J.D. Butler (1982). Measurement of turfgrass heat tolerance in vitro. *Journal of the American Society of Horticultural Science* 107:608-613.

Wardlaw, I.F., I.A. Dawson, P. Munibi, and R. Fewster (1989). The tolerance of wheat to high temperatures during reproductive growth. I. Survey procedures and general responses pattern. *Australian Journal of Agricultural Research* 40:1-13.

Warrag, M.O.A. and A.E. Hall (1983). Reproductive responses of cowpea to heat stress: Genotypic differences in tolerance to heat at flowering. *Crop Science* 23:1088-1092.

Wehner, D.K. and T.L. Watschke (1984). Heat stress effects on protein synthesis and exosmosis of cell solutes in three turfgrass species. *Agronomy Journal* 76:16-19.

Wells, D.R., R.L. Tanguay, H. Le, and D.R. Gallie (1998). HSP101 functions as a specific translational regulatory protein whose activity is regulated by nutrient status. *Genes and Development* 12:3236-3251.

Weltzien, R.E., C.J. Howarth, R. Jayachandran, and F.R. Bidinger (1994). Seedling thermotolerance: A comparison of screening methods. *1994 Agronomy Abstracts,* p. 113.

Wiegand, C.L. and J.A. Cuellar (1981). Duration of grain filling and kernel weight of wheat as affected by temperature. *Crop Science* 21:95-101.

Wigley, T.M.L. and S.C.B. Raper (1992). Implications for climate and sea level of revised IPCC emission scenarios. *Nature* 357:293-300.

Yadav, R.S., C.T. Hash, F.R. Bidinger, G.P. Cavan, and C.J. Howarth (2002). Quantitative trait loci associated with traits determining grain and stover yield in pearl millet under terminal drought stress conditions. *Theoretical and Applied Genetics* 104:67-83.

Yamasaki, T., T. Yamakawa, Y. Yamane, H. Koike, K. Satoh, and S. Katoh (2002). Temperature acclimation of photosynthesis and related changes in photosystem II electron transport in winter wheat. *Plant Physiology* 128:1087-1097.

Young, T.E., J. Ling, C.J. Geisler-Lee, R.L. Tanguay, C. Caldwell, and D.R. Gallie (2001). Developmental and thermal regulation of the maize heat shock protein, HSP101. *Plant Physiology* 127:777-791.

Chapter 11

Breeding for Cold Tolerance

Pedro Revilla
Ana Butrón
M. Elena Cartea
Rosa Ana Malvar
Amando Ordás

INTRODUCTION

Plant breeding deals with materials that have already undergone evolutionary change. These changes have sometimes been profound and have occurred over long periods of time (Simmonds and Smartt, 1999). Among these, a prominent position should be assigned to cold adaptation. Strictly speaking, a domesticated plant is not the same as a cultivated plant (Harlan, 1992), although the terms are usually regarded as synonymous. According to Harlan (1992), a domesticated plant or animal is one that has been brought into the household to serve those who live there. Domestication is the process through which this plant or animal is so altered from its wild state that it is no longer able to survive if not cultivated by humans. The information available on domestication is mainly historical, and many conclusions must remain speculative.

The domestication of plants began when humans observed that sowing seeds, undoubtedly of plants used for food, led to growth in a desired place. Consequently, people no longer had to look for food. Now, they could settle on the banks of a river with year-round water and make a palisade to defend themselves against their enemies. The Neolithic Age had begun, and villages began to flourish.

But how did domestication come about? Consider, for instance, the way wheat, one of the first species domesticated (Feldman et al., 1995), became a crop. Wheat was a good candidate for domestication for several reasons: it has large seeds, it is an annual plant, and is mostly self-pollinating. Its domestication was somewhat hindered, however, by its method of seed dispersal. Wild wheats have several mechanisms that assure seed dispersal, but the most important is the possession of a fragile rachis. Ancient farmers must have chosen, in a more or less conscious way, those plants with nonshattering spikes, and they must have picked only the grains of plants that were mature at a certain date (Cubero, 1999). Given the tremendous variability usually found in wild populations, this would have brought about an enormous reduction in the variability of the incipient crop; repeated over many generations, only a small portion of the total gene pool of the original species would be left. However, and in spite of this reduction in the number of alleles, wheat is very successful. The same can be said of many other crops that almost certainly followed a similar pattern of domestication.

The earliest evidence of the use of wheat comes from Israel, where a tetraploid form has been dated c. 17,000 B.C. (Kislev, 1992, cited by Feldman et al., 1995). Most of the evidence for wheat domestication comes from the Middle and Near East (Feldman et al., 1995). This area is one of Vavilov's centers of diversity. Although his theories on the origin of crops have been very influential, they have suffered considerable modification. Most of the places where archaeological data have been found are arid, such as the eastern Mediterranean, Mexico, and Peru. Such places favor the preservation of human artifacts and even plant remains. The latter would have little chance of prolonged preservation in warm, moist climates, so our sample is somewhat biased (Simmonds and Smartt, 1999). Nevertheless, the archaeological and historical evidence leads us to believe with some certainty that most if not all plant domestication took place in ancient times in these parts of the world.

If the domestication of plants began in warm areas, how is it possible today to find successful crops in cold places such as Canada or Northern Europe? The answer is simple; there has been intensive selection, either conscious or unconscious, for cold hardiness. In the Old World, as civilizations moved northward, farmers had to select those plants that better survived in the cooler areas being settled. Those alleles not well adapted to the new situation were undoubtedly lost, while mutations that conferred a selective advantage against the cold were favored and eventually fixed. Not only resistance/tolerance to cooler conditions was selected but also adaptations to changes in the photoperiod. This process lasted many generations, eventually leading to maize, for instance, being grown in such northern latitudes as those of Minnesota, where more than 2.5 million hectares were planted in 1997. If one compares teosinte, the wild ancestor of maize, which is adapted to warm climates and a 12 h photoperiod, with modern maize grown in temperate areas, the differences are enormous, and not just from the standpoint of productivity but also with respect to photoperiod response and sensitivity to the cold. The most primitive races of maize grown today are the six ancestral race parents of Mangelsdorf's lineages (Mangelsdorf, 1974). Three of these races come from Peru, two from Colombia, and one from Mexico, places with very different photoperiods and climates to those of Minnesota, Ontario, or Northern France, where maize now flourishes. Although crop domestication is not a common activity nowadays, it still goes on. For instance, the U.S. Department of Agriculture runs a new crops research program.

Selection for cold resistance in established crops is a way of moving them to areas where they would not normally succeed. Further, obtaining more cold-resistant genotypes is a good way of improving the stability of a variety by diminishing susceptibility to the cold during the first stages of life. Environmental stresses are the most limiting factors of agricultural productivity in temperate regions. They are also significant barriers to the introduction of crops into new areas (Galiba, 1994). Generally, more than one stress occurs at a time, and there is wide temporal and spatial variation in their incidence and intensity. Mechanisms of stress tolerance may be related in some cases, such as cold and desiccation tolerance, but often specific adaptations are required that may affect plant development and production.

Greaves (1996) defines suboptimal temperature stress as

> any reduction in growth or induced metabolic, cellular or tissue injury that results in limitations to the genetically determined yield potential, caused as a direct result of exposure to temperatures below the thermal thresholds for optimal biochemical and physiological activity or morphological development.

Each crop has an optimal thermal threshold. Tropical and subtropical crops can be seriously injured at low, but above 0°C, temperatures, while temperate plants have the ability to cold acclimatize and overcome temperatures from –5° to –30°C. The perennials of the northern forests can even survive temperatures below –30°C (Thomashow, 1998).

In temperate areas or highland regions, suboptimal temperatures during spring decrease productivity and cause poor yield stability (Stamp, 1984). Reductions in grain yield are caused by direct damage to cold-season crops, the limitation of the area where crops can be sown, and the use of warm-season crops or early varieties that are generally less productive. For example, low temperatures limit the production of about 7 million hectares of modern rice cultivars in South and Southeast Asia. It is also an important constraint in temperate areas (Khan et al., 1986); high-yielding cultivars cannot be cultivated in many temperate areas because of their susceptibility to cold stress (Majumder et al., 1989). In some temperate areas, where spring is cool and damp, it is useful to have cold-tolerant cultivars, with high emergence and rapid growth under low spring temperatures, for early planting. Cold-tolerant, warm-season cultivars could also be planted early to lengthen the growing cycle, promote early pollination and harvest, and avoid other stresses (Pendleton, 1965; Mock and McNeill, 1979). Cultivation in higher latitudes, as well as early planting, requires high percent emergence and vigorous seedling growth under cool temperatures.

In higher latitudes, low temperature is the most important limiting factor, especially around the arctic circle where forage occupies the major part of agricultural land (Rognli, 1988; Helgadóttir and Björnsson, 1994). Perennial grasses are the basis for dairy and meat production in countries such as Norway. In marginal northern areas, the winters are long and conditions severe, and the growing season is short and characterized by long photoperiods and low temperatures (Larsen, 1994). In less-marginal areas, most agricultural land is used for winter-crop varieties. Breeding and seed production has to be performed in situ, even though the market for such a crop is quite limited, resulting in problems of profitability for seed companies. Cold temperatures limit the perennation of forage crops in northern regions (Steponkus, 1978), as well as the distribution of trees in the polar circle (Savolainen, 1996). Transferring pines from latitude 66° N to just 60° N would probably increase wood production by as much as 100 percent (Beuker, 1994).

Food demand rises as the human population increases. Consequently, many crops, such as maize and rice, have expanded toward higher latitudes and altitudes, where temperatures are lower. Most crops are susceptible to freezing at some point, and many of tropical or subtropical origin are damaged by temperatures below 15°C or even higher, depending on the growth stage. Protection against cold stress can be achieved by agronomic methods or by plant breeding. Among the former are late sowings of warm-season crops to avoid cold temperatures at germination, and the use of early varieties to avoid cold at maturation. Germplasm collected in cold regions is therefore not necessarily cold tolerant, since crops can be managed to avoid low temperatures (Caradus and Christie, 1998; Revilla et al., 1998). Agricultural methods to defend plants against stress and escape mechanisms such as late planting or early maturity lay outside this chapter's scope.

Most plants do not suffer chilling injury when temperatures are above 10°C, although some important crops, such as rice, suffer injury even at 15°C. The term used to describe tolerance of temperatures below freezing point is "frost tolerance." Cold stress refers to temperatures between 0°C and 10 to 15°C, while frost stress occurs below 0°C. Cold and frost tolerance are genetically independent (Stone et al., 1993). Usually, injury increases with the degree of chilling and time of exposure to suboptimal temperatures.

Cold and freezing tolerance are different subjects from the genetic and physiological point of view. However, from a breeder's perspective, the sources of germplasm and methodologies for attaining such characteristics are similar. Therefore, in this chapter, freezing tolerance is mentioned within sections on cold tolerance. Usually, cold tolerance is required for annual warm-season crops, while freezing tolerance is necessary for annual cold-season or perennial crops.

Winterhardiness is a complex characteristic that includes the ability to survive the interacting factors of a winter environment, including soil heaviness, low or rapidly fluctuating temperatures, low light intensity, desiccation, wind, snow or ice cover, disease, insects, and management practices (Humphreys, 1989). Only some of these aspects concern cold tolerance, although freezing tolerance is a common selection criterion for improving winterhardiness (Waldron, Ehlke, Vellekson, and White, 1998; Waldron, Ehlke, Wyse, and Vellekson, 1998). Breeding for winterhardiness is breeding for adaptation and is critical in breeding cereal grains and perennial forage crops. The acclimation of plants to the cold has recently been reviewed by Thomashow (1998, 1999, 2001), who focused mainly on molecular aspects. The present review therefore concerns acclimation only collaterally. Changes produced during acclimation cannot be considered detrimental effects of stress but adaptive alterations of the phenotype according to changing environmental conditions.

Scientific breeding for cold tolerance has been undertaken since the beginning of the twentieth century, and several reviews have been published during its latter half, though most have focused on the physiological explanation of cold injury (Neal, 1950; Rinke, 1953; Herczegh, 1970). Steponkus (1978), who summarized cold hardiness and freezing injury of agronomic crops, concluded there to be a wide diversity of freezing tolerance, and that relatively small increases in the hardiness of a particular crop could have a significant impact on world food production. Li (1984) documented major metabolic changes during cold acclimation in herbaceous plants. Bochicchio (1985) reviewed chilling conditions at emergence in maize, though he focused on chilling effects rather than breeding for tolerance. Layne (1992) reported the state of breeding for cold-hardy peaches and nectarines. Monti (1993) reviewed cold breeding in potato and tomato and other horticultural crops. Sethi and colleagues (1993) reviewed the breeding of bread wheat for cold stress. Seyedbagheri and Fallahi (1994) reviewed cold hardiness in grapevines. Galiba (1994) reviewed the physiological, metabolic, and genetic aspects of adaptation to stresses, particularly osmoregulation and specific metabolic changes associated with osmotic adjustment, and cold hardiness in wheat. Gusta and colleagues (1996) summarized the induction and regulation of freezing tolerance in plants, the use of plant growth regulators related to acclimation, and the prospects of molecular markers and genetic engineering in plant breeding for cold tolerance. Finally, Palonen and Buszard (1997) reviewed cold hardiness in tree fruit and berry crops, explaining that although cold hardiness is a complex phenomenon, several efficient methods are available for producing more cold-tolerant varieties.

Blum (1988) reviewed breeding for cold tolerance, focusing mainly on physiology. He also reported several former reviews of cold stress with the same physiological background. This author concluded that selection for improved seedling vigor or seedling dry weight under chilling stress can be an efficient method, depending on the available genetic material and the selection pressure applied.

The purpose of this chapter is to review breeding for cold tolerance from a breeder's perspective. The main issues covered are cold tolerance, frost tolerance, and the cold-temperature aspects of winterhardiness.

Effects of Cold Temperature

This section examines the responses of plants exposed to low temperatures as well as the effects of cold on growth and development. Freezing stress is outside the context of cold stress discussed here but is the subject of several extensive reviews (Levitt, 1980; Blum, 1988; Andrews, 1996). Low-temperature-induced responses will be discussed in several important agricultural plants, but particularly in maize.

Plant cold resistance takes the form of physiological or morphological adaptations that allow plants to tolerate or avoid cold stress. Tolerance and avoidance mechanisms may arise at the cellular, tissue, or whole-plant levels. Mechanisms of cold avoidance such as seed or vegetative dormancy may be present in both annual and perennial plants. Some species from temperate regions may avoid low temperatures by becoming dormant when temperatures are freezing. Some herbaceous annual species survive both cold and freezing conditions, changing from a dormant to a reproductive stage in the spring. These plants are very cold resistant and will continue to grow and develop, although more slowly, under nonfreezing cold conditions.

For crop plants cultivated outside their original areas, exposure to low temperatures may cause reductions in growth or induce damage at the metabolic, cellular, or tissue levels. Maize is moderately sensitive to low temperature, and cellular and tissue injuries occur as a result of exposure to temperatures below 5°C, temperatures that are very common in northern Europe, the United States, and Canada during early spring. Below 15°C, growth processes are slow. The minimum temperature for chilling injury to seeds and seedlings is about 0 to 5°C, whereas diseases in the root and seedling start at about 8 to 12°C (Miedema, 1982). The term "chilling injury" describes the modified metabolism and reductions in growth rate and development following exposure to temperatures between 0 and 12°C. Chilling stress can be measured at different developmental stages, such as germination and growth, as well as in terms of photosynthetic activity, development of reproductive organs, and fruit ripening (Blum, 1988). Therefore, at the higher levels of plant organization, cold damage may have several effects, such as reduced growth, poor seedling establishment, wilting, chlorosis, necrosis, or poor fruiting. Chilling injury at different developmental stages, as well as changes in photosynthetic activity, will be discussed in this section.

Reduced Growth and Development

Cold tolerance may change during the life cycle, particularly in rice that shows differences in its vegetative and reproductive growth stages (Bertin, Busogoro, et al., 1996). In maize, susceptibility to low temperatures during germination, emergence, and early growth can be explained by chilling injuries to the seed, embryo, and seedling. Minimum temperatures depend on the phenological stage of the plant. For maize and millet, most authors consider that the temperature for germination should be at least 10°C (Herczegh, 1970; Blacklow, 1972; Zauralov, 1998); normal leaf development in maize requires temperatures above 15°C (Nie et al., 1992); the leaves of wild *Solanum* species can resist frost (–10°C for *S. acaule*, –5°C for *S. bukasovii*, and –11°C for *S. rionegrium*) (Mastenbroek, 1956); temperatures of around –3°C kill potato plants (Nyman and Waara, 1997); spikelet sterility in rice occurs at 20°C (Khan et al., 1986); storage of potatoes at 3°C may cause accumulation of reducing sugars (Sowokinos, 2001); some apple trees do not freeze even at temperatures as low as –55°C (Saveljev and Michurin, 1988); and Fowler and Carles (1979) found that the maximum cold hardiness of plants fully acclimated to winter was –13°C in oats, –15°C in barley, –21°C in wheat, and –30°C in rye.

The germinating seed and developing seedling pass through several stages with different sensitivities to low temperatures. The first, seed imbibition, and the last, beginning of photosynthesis, phases are the most sensitive, and the intermediate stages are less sensitive. As temperature decreases below the optimum, growth is retarded. In recent years, several reviews have focused on the effects of temperature on growth and development. Bochicchio (1985) reviewed the effects of cold conditions at sowing on maize, with special focus on chilling effects, and concluded that low temperatures cause reductions in percent emergence, rate of emergence, dry matter accumulation, seedling vigor, color, and sterility. Several studies have discussed the effect of low temperatures on other important agricultural species such as tomato (Nieuwhof et al., 1997), which can suffer reduced growth, scarce fruit formation, and rapid death. Kuwabara and colleagues (1997) showed that in wheat, shoot weight, crown weight, and the ratio of dry weight/fresh weight are the main growth parameters affected by cold. Under cold stress, spikelet fertility, plant height, and 1000-grain weight are the main influences of grain yield in rice (Gupta et al., 1998).

The biochemical and physiological requirements of germination and emergence may be different (Simon et al., 1976). Germination begins with the first signs of metabolic activity during imbibition and ends with the emergence of the radicle from the seed. Development from this point onward to the emergence of the shoot above the soil is called early seedling growth. Cold is probably the main environmental stress suffered during germination, particularly for warm-season crops grown in temperate regions. The reduction of growth in these conditions depends on several factors: processes of imbibition, enzymatic breakdown of stored kernel reserves, cellular division, and cellular extension.

Seed imbibition is very sensitive to chilling. In maize, as in other cold-susceptible plants, imbibition under cold stress results in exudation of amino acids and carbohydrates from the kernels (Miedema, 1982) as a consequence of membrane disorder. Embryo growth and differentiation is supported by seed reserves, and Blum (1988) concluded that the growing embryo and seedling remain cold tolerant as long as the endosperm has not been completely used. When this occurs, however, a considerable increase in sensitivity can be noted. Another effect of cold stress during imbibition of low-moisture seeds is the formation of lesions in imbibed embryonic tissues. Among these, the mesocotyl plays a decisive role in determining the ability of seedlings to emerge under adverse conditions (Stewart et al., 1990) such as deep planting or cold soils.

The effect of cold on germination can be described by its effect on emergence since its rate is negatively affected by low temperatures. Low temperature at germination not only reduces the germination rate but also has an effect on subsequent seedling growth in terms of dry matter. It has been suggested that failure to germinate at low temperatures may result from the denaturation of proteins (Orr et al., 1983). Several studies have shown that differences between maize inbred lines for improved seedling emergence rate under chilling conditions are mainly determined by nuclear genes. These affect the rate of conversion of seed reserves into new root and shoot tissues rather than the capacity of the conversion process (Eagles, 1982).

The spring temperatures of more northerly latitudes are usually too low for the favorable germination and emergence of temperate crops such as maize. Most maize genotypes germinate very slowly at temperatures below 6 to 8°C, the critical temperatures for maize germination and emergence being about 5 and 9°C, respectively (Blacklow, 1972; Watts, 1972; Eagles, 1982; Eagles and Hardacre, 1979; Miedema, 1982; Stewart et al., 1990; Kiel and Stamp, 1992; Ellis et al., 1992; Greaves, 1996). Low temperatures in the range of 10 to 15°C limit growth performance, particularly when such conditions are maintained over a long period. The minimum temperature for shoot growth before and after emergence in maize is between 8 and 15°C, which is above the critical temperature for physical damage (Blacklow, 1972; Miedema, 1982).

Soil temperature (Bochicchio, 1985; Barlow et al., 1977; Richner et al., 1996), depth of sowing, and seed moisture (Ismail et al., 1997) are three important parameters that influence percent emergence, emergence rate, and seedling vigor. Any decrease in soil temperature may affect root growth, nutrient uptake, water absorption, and hormone production in the roots, as well as leaf elongation. Relatively small reductions (2 to 3°C) in soil temperatures near 15°C may reduce root development in cold-sensitive maize seedlings (Richner et al., 1996). Under chilling conditions in the field, reducing the seed moisture content results in decreased percentage emergence and rapid electrolyte leakage in the cowpea, while deep sowing results in slow and low emergence (Ismail et al., 1997).

Early seedling growth can be expressed as the elongation rate of either the primary root or the shoot (Greaves, 1996). During early seedling life, one can distinguish between effects on heterotrophic development, such as the use of kernel reserves and the growth of roots and leaves, and effects on autotrophic development such as photosynthesis and nutrient uptake. Cold damage occurring at the seedling stage often reduces vigor for the entire life cycle, despite what any later temperatures may be. Important alterations occur in seedling metabolism at chilling temperatures, but recovery is quick if exposure is no longer than 48 h. However, in some instances, plants cannot recover the leaf area or dry matter lost due to chilling.

The seedling may continue to grow although cellular and tissue injuries occur. However, a reduction in growth rate and water stress is observed because of the loss of root tissue. Failure of root growth at low temperatures implies a reduced capacity for water and mineral uptake, and subsequent nutritional effects on plant growth. Root growth is reduced at temperatures much higher than the critical chilling temperature. The effect of root temperature on the growth of maize is complex (Ellis et al., 1992; Greaves, 1996). During the early stages, before tassel initiation, root zone temperatures may affect shoot growth directly. Ellis and colleagues concluded that it is impossible to characterize shoot performance at low temperatures based on anatomical traits. After tassel initiation and stem elongation, when the meristematic regions are above the soil surface, shoot growth becomes less influenced by root temperature and more sensitive to air temperature. The effect of temperature on root development has been studied in forage legumes (Brar et al., 1990) and maize (Kiel and Stamp, 1992).

Maize seedlings exposed to temperatures below 16°C are sensitive to several types of physiological damage (Blacklow, 1972; Watts, 1972; Miedema, 1982; Bochicchio, 1985; Stewart et al., 1990; Kiel and Stamp, 1992). They lose their ability to expend dry matter on the development of the root in favor of the shoot. The shoot-weight/root-weight ratio therefore increases. A similar situation has been reported for cereals from temperate regions when grown at 5°C (Abbas Al-Ani and Hay, 1983). Seedling growth responses to chilling may be related to maturity adaptation in maize genotypes. For example, Castleberry and colleagues (1978) found that early-maturing germplasm led to higher dry weights and greater rates of leaf area expansion than late-maturing germplasm. Miedema (1979) studied the effect of low temperature on seedling growth in several populations of maize by comparing early and late sowing. He found that early sowing increased shoot dry weight. The difference can be attributed to shoot length since early-sown plants are shorter than those sown later. Seedlings appear to be more susceptible to cold than adult plants because chilling events may cause considerable reduction in their growth. However, chickpea is more susceptible in late vegetative stages than at the seedling stage (Singh et al., 1989).

Reliable germination and rapid growth in cold soils should raise the stability of performance across environments and should allow early sowing. The benefits of the latter have already been reported in the literature (Bochicchio, 1985; Blum, 1988). Leaf weight and area are important attributes from the perspectives of providing photosynthetic area and nutritive quality to plants. Leaf extension is more controlled by the temperature of the shoot apex than by the temperature of the root (Watts, 1972). The lower thermal limit for leaf expansion in maize is close to 14°C, although it depends on the developmental stage of the leaf. For example, leaves at 7°C are able to elongate, but continuous exposure to 10°C causes a decrease in the extension rate, leaf damage, and finally the death of the shoot apical meristem (Miedema, 1982). Giauffret and colleagues (1995) found that the temperature threshold varied from 7.1 to 16°C for leaf-tip appearance rate, from 8.3 to 12.8°C for the elongation rate of leaf number 6, and from 7.1 to 13.5°C for the elongation rate of leaf number 9.

The influence of low temperature on leaf-area development and foliar damage has been described in maize (Miedema, 1982; Stamp, 1988; Giauffret et al., 1995) and durum wheat (Elings and Nachit, 1993). The main effects of low temperature on maize development at the vegetative stage are reduced photosynthesis and growth, loss of leaf area, and leaf chlorosis (Miedema, 1982).

After emergence, leaves appear successively and grow until their final size is reached. Young leaf tissue is more cold sensitive than that of mature leaves (Greaves, 1996), and leaves are more susceptible to the cold than other organs, owing to the sensitivity of their chloroplasts. Under normal growth conditions, leaf-area development closely correlates with dry-matter accumulation. This relationship was found to hold after exposure to low temperatures (Hanson, 1971). The leaf-elongation response of maize seedlings to low temperature is due to cold-water stress and a reduction in the temperature of the shoot apical meristem (Barlow et al., 1977). These authors found that leaf elongation is more sensitive to low soil temperatures or water stress than photosynthesis. When cold-susceptible plants are subjected to low temperatures, a rapid fall in leaf-water potential occurs. Two processes contribute to this effect, the slow closure of stomata at low temperatures and the reduced permeability of root membranes. As a result, there is a loss of leaf turgidity (Stamp, 1988).

The role of temperature in the final shape of the leaf was determined in tobacco plants by Raper and Thomas (1972). These authors concluded that continuous low temperatures induce an abnormal elongation of the upper leaves, probably by affecting cell division. The final leaf shape is an important parameter, since the morphological characteristics of leaves may affect photosynthetic efficiency. The relationship between foliar damage and grain yield has also been evaluated by measuring cold tolerance in durum wheat landraces (Elings and Nachit, 1993). Foliar damage reduces straw yield but not grain yield. Maize leaf number is also sensitive to temperature immediately prior to tassel initiation. Temperatures of about 18°C result in a reduction in the number of leaves (Warrington and Kanemasu, 1983).

Another growth stage significantly affected by cold stress is flowering. Interest in cold tolerance at flowering is largely associated with either the abnormal development of flowers or the failure to set seed or fruit at chilling temperatures (Blum, 1988). Reproduction is particularly susceptible to low temperatures, and the influence of cold stress on flower induction or sex expression has been reported in many species, including maize (Ellis et al., 1992; Bonhomme et al., 1994), wheat (Elings and Nachit, 1993; Subedi et al., 1998), sorghum (Lansac et al., 1996), common bean (Wallace et al., 1991), pepper (Mercado et al., 1997), tomato (Lozano et al., 1998), and *Picea abies* (Johnsen et al., 1996).

Some stages of reproduction, such as meiosis, pollen tube growth, early embryogenesis, and embryo competition, may be susceptible to chilling.

Low temperatures may cause the deformation of flowers, or parts thereof, and affect pollen viability. At low temperatures, cold-sensitive plants produce less pollen, resulting in a reduction in fruit number and size, particularly in the case of rice (Chiba et al., 1998). Reduced pollen viability has been associated with anther morphology and pollen proline content (Lansac et al., 1996).

The exact cause of damage to flowers is unclear. Elings and Nachit (1993) found that floret damage caused in durum wheat by cold stress decreased the grain yield yet increased the straw yield. In soybean, the evaluation of cold tolerance during flowering is based on the number of flowers produced, the capacity to retain them, and the ability to replace those lost (Gass et al., 1996). The role of flower morphology and position within the inflorescence may also affect fertility and the rate of fruit production (Blum, 1988). Nevertheless, the biological significance of flower position with respect to fertility under cold stress is unclear. The relationships between cold temperatures and grain setting in spring wheat have been shown by Subedi and colleagues (1998). These authors found that temperatures below 10°C prolonged the time to anthesis and led to greater sterility in cold-susceptible cultivars. Rylski (1979) concluded that fruit deformation may result as the effect of low temperatures on ovary development and that ovaries may abort or produce parthenocarpic fruits.

Physiological and Biochemical Responses

Results with cold-susceptible plants support the possibility of cold-induced toxicity, since low temperatures induce oxidative stress. The ability of plants to survive under cold conditions depends on their capacity to increase the activity of their antioxidant enzymes. When maize grows at 14°C, the four enzymes of the active oxygen cycle (superoxide dismutase, glutathione reductase, ascorbate peroxidase, and catalase) show total leaf activity that differs from that of plants grown at 25°C. The role of these enzymes is to protect maize during germination and to defend seedlings against chilling injury. Several authors have examined this response by analyzing isozyme profiles and enzyme activities in different maize organs (Massacci et al., 1995; Kocsy et al., 1996; Pinhero et al., 1997; Prasad, 1997; Kocsy et al., 2001). When maize is exposed to low temperatures, the electron transport rate may be in excess of carbon metabolism, and electrons may thus be used to reduce oxygen (Massacci et al., 1995). It has been suggested that an increase in the activity of oxygen-detoxifying enzymes may protect maize leaves from the accumulation of oxygen radicals at low temperature.

The role of oxygen enzymes in germination capacity at low temperatures has been determined by Tanida (1996) in seeds of rice cultivars. This author found that the specific activity of catalase increased during germination and could be detected even in the embryo of dry seeds, before imbibition. He also showed a positive correlation between the initial level of catalase activity and germination rate at low temperature. However, this type of correlation was not observed between the superoxide dismutase activity of dry seed embryos and low-temperature germination rate.

In the field, low temperatures cause chlorotic leaves. When the light energy captured by the plant is not used in the process of carbon assimilation, electron acceptors such as oxygen become chemically activated to form free radicals, which cause cell damage. This results in leaf chlorosis and finally necrosis. Several studies indicate that cold temperatures cause chlorosis in maize at temperatures between 10 and 15°C (Miedema, 1982; Hodgins and van Huystee, 1985; Stamp, 1988; Baker, 1994). It is often observed in the field immediately after emergence, but cold temperatures can also cause chlorotic growth in later stages of development. Severe chlorosis results in irreversible leaf damage. As a result, young leaves become pale green; yellow, because of carotenoids; or even white. The first signs of cold damage are the wilting and dying off of the leaf tips. The loss of green area during cold stress has been compared in several populations of acclimated and nonacclimated maize (Stamp, 1988). Acclimated maize loses less green area.

Susceptibility to chlorosis may be caused by retarded chloroplast development, decreased chlorophyll synthesis, or increased rates of chlorophyll photooxidation. The thermal threshold for chlorosis in maize is generally some 3 to 5°C above that for cell division and extension. This phenomenon mainly occurs in developing leaf tissue, especially in the youngest. Genotypes with fast development are more susceptible than those with slow development (Greaves, 1996).

In the vegetative stage, resistance to cold is related to the capacity to photosynthesize under conditions that partially or completely suppress other metabolic processes. The biochemical responses of plants exposed to low temperatures have been widely documented (Burke et al., 1976; Graham and Patterson, 1982; Steponkus, 1984; Guy et al., 1985; Guy, 1990; Bartels and Nelson, 1994; Nishida and Murata, 1996; Thomashow, 1999, 2001), but the mechanisms by which low-temperature signals are perceived and translated into specific biochemical responses are largely unknown (Bowler and Chua, 1994).

Although cold stress has several effects on the physiology of plants, photosynthesis is among the first to be adversely affected when those susceptible experience low temperatures. The stability of the photosynthetic apparatus under cold conditions depends partly on the stage of development. It is well documented that photosynthesis in maize and other tropical grasses can be altered greatly by temperatures slightly below 15°C, although the chlorophyll content of fully green leaves is very stable under the same conditions (Stamp, 1988). When the temperature of leaves, roots, and shoots is below about 10°C, photosynthesis in leaves drops immediately (Long et al., 1994). The most likely site for cold damages is photosystem II. Photosynthetic activity decreases as temperatures fall below the optimum and visible lesions on the leaves may arise. This effect has been demonstrated in agriculturally important crops such as maize (Long et al., 1983; Nie and Baker, 1991; Long et al., 1994), sugarcane (Du et al., 1999), common bean (Powles et al., 1983), sorghum and soybean (Taylor and Craig, 1971), and rice (Bertin, Bouharmont, and Kinet, 1997; JinHong et al., 1997). In these reports, cold led to a reduction in the photosynthetic activity of photosystem II and to alterations in thylakoid membranes.

As chlorophyll production is inhibited by low temperature in growing leaves, cold tolerance is reflected in the pigment ratios that develop under stress. These relationships have been widely examined in maize (Hetherington, Smillie, Hardacre, and Eagles, 1983; Nie and Baker, 1991; Andrews et al., 1995; Massacci et al., 1995; Verheul et al., 1995; Haldimann, 1997, 1998, 1999; Fracheboud et al., 1999), sweet sorghum, and Sudan grass (Havaux, 1989). The ability of the cold-tolerant genotypes to retain high levels of pigments and to form a competent photosynthetic apparatus at low temperature is the basis of their more vigorous growth in cool climates.

The chlorophyll content and activities of both RuBP carboxylase and NADP malate dehydrogenase can also be severely affected by chilling temperatures in different ways (Stamp, 1988). In young maize plants, effects on CO_2 assimilation and the photochemical processes of the thylakoids have been described by Long and colleagues (1983) and Nie and Baker (1991). It appears that desaturation of lipids in thylakoid membranes protects the photosynthetic machinery from photoinhibition at low temperatures (Murata, 1983).

In sensitive plants, the major physiological and biochemical changes observed in response to cold temperatures are the disruption of cell membranes, alterations in membrane-lipid composition, increased levels of soluble proteins, antioxidant enzymes, amino acids (proline), abscisic acid (ABA), and carbohydrates. It is generally agreed that the primary physical manifestation of chilling injury occurs in cell membranes (Steponkus, 1984; Blum, 1988). The most important cause of damage by severe chilling seems to be phase changes of membrane lipids. In chilling-sensitive crops, membrane lipids undergo a transition from the normal liquid-crystal to an abnormal solid gel phase, a process that is complete below 8 to 12°C. Many of the biochemical manifestations of chilling injury are believed to originate from this. Severe cold may cause greater permeability of the membrane and kill the cell due to an efflux of ions and metabolites.

This change in the physical state of the membrane can also lead to several organellar disorders such as vacuolization, disruption of the cytoplasmic reticulum, and vesiculation, as well as a loss of turgor. Walker and colleagues (1991) studied chilling injury in thylakoid membranes as well as the alteration of biochemical and functional characteristics in two species of tomato, resistant and sensitive. Cold stress decreased stomatal conductance, dry-matter accumulation, and chlorophyll fluorescence ratio more in sensitive than in resistant genotypes.

Changes in membrane physiology are accompanied by a number of alterations in biochemical processes responsible for maintaining membrane stability in different plant organs during growth and development. Chemical compartmentalization changes the composition of cell membranes, increasing unsaturated/saturated fatty acids, phospholipids/sterines, and lipids/proteins ratios. Physical compartmentalization leads to the accumulation of cryoprotective substances, together with others that increase the osmotic pressure of the cytosol and its water-retaining capacity. These changes allow the lipid bilayer to remain liquid at low temperatures, thus preserving the complex functions of cell membranes.

A survey of the fatty acid composition of 74 plant species (Murata, 1983) confirmed the relationship between cold tolerance and membrane fatty acid composition. The temperature that produces phase transition in membranes depends on the ratio of saturated to unsaturated fatty acids (Murata, 1983; Nishida and Murata, 1996). Cold-resistant plants tend to have membrane lipids with higher ratios of unsaturated/saturated fatty acids. Polyunsaturated membrane fatty acids such as linolenic acid are considered essential for maintaining cellular function and plant viability at low temperatures. Changes in the fatty acid content of maize seedlings because of low temperatures have been reported in the literature (Grubbels, 1974; Duhalde et al., 1991; De Santis et al., 1999). The relative levels of oleic and linoleic fatty acids in the phospholipids of maize kernels are reported to be adaptive responses to low temperatures (Zemetra and Cuany, 1991). Modifications of membrane lipid composition have also been related to increases in cold tolerance in other crops. For example, Palta and colleagues (1993) reported the lipid changes associated with increased cold tolerance in two potato species, while Samala and colleagues (1998) reported changes in fatty acid composition in different organs (leaves, crowns, and roots) after Bermuda grass accessions were grown at low temperature. The latter authors concluded that different tissues respond differently to low temperatures, with crowns showing the most dramatic changes in total fatty acid content and composition.

Cold sensitivity can be manipulated by modulating the levels of desaturation of fatty acids in membrane lipids. Specific desaturase enzymes are essential in controlling membrane lipid/fatty acid composition in response to low temperatures and therefore in avoiding winter damage in sensitive plants. The role of membrane lipid desaturation in tolerance and the response to low temperatures has been examined using transgenic and mutant plants (Somerville and Browse, 1990; Nishida and Murata, 1996).

The transition of the membrane phase, and consequently the alteration of permeability levels, leads to the disruption of membrane-linked enzymatic systems. Plants respond to environmental stresses such as cold by inducing the synthesis of specific proteins. The increased expression of at least 20 proteins associated with low temperatures supports their essential role in the development of cold tolerance. Several studies have focused on the role of cold-defense proteins (Graham and Patterson, 1982; Guy, 1990). The effect of low temperature (around 6°C) on the protein patterns associated with membranes was studied in cold-tolerant and cold-sensitive inbred maize lines in order to identify specific polypeptides (Gasparíková et al., 1996). During the winter, the protein concentration remained high, but it declined rapidly during the breaking of dormancy and spring growth. This response has been found in seedlings of other important agricultural crops such as spinach (Guy et al., 1985), rapeseed (Meza-Basso et al., 1986), barley (Marmiroli et al., 1986), tomato (Cooper and Ort, 1988), rice (Hahn and Walbot, 1989), wheat (Perras and Sarhan, 1989), and common bean (Zaiter et al., 1994).

Meza-Basso and colleagues (1986) demonstrated an increase in the relative synthesis of specific polypeptides at 0°C in rapeseed seedlings, as well as in the relative amounts of the corresponding mRNAs. A class of proteins known as dehydrins has been found to accumulate in several species of plants during cold stress (Borovskii et al., 2000). However, the synthesis of other proteins, such as the small subunit of ribulose 1,6-bisphosphate carboxylase, is apparently repressed. Many others might be insensitive to cold stress.

The effect of low temperatures on gene expression was not analyzed in detail until the 1980s, and it was not clear whether the changes in protein content and enzyme activity resulted from changes in mRNA abundance or translation efficiency, inactivation or activation of preexisting enzymes, or from some other mechanism. The results of Meza-Basso and colleagues (1986) demonstrated that the relative concentration of different specific mRNAs could either increase or decrease at low temperatures. Other studies on spinach leaves provided evidence for an increase in specific mRNAs correlated with improved freezing tolerance (Guy et al., 1985). The accumulation of rRNAs in different cultivars of wheat and barley growing at low temperatures has been studied by Paldi and colleagues (1997).

Plants respond to cold stress by generally increasing their content of free proline and other nitrogenous compounds such as putrescine, agmantine, or glycine-betaine. Several authors have reported the effect of low temperature on the synthesis of these compounds, which are regarded as stress markers. For example, Dory and colleagues (1990) found an increase in plant proline content after exposing the leaves of two maize inbred lines with different degrees of cold tolerance to a temperature of 5°C. Svenning and colleagues (1997) concluded that the contents of total soluble sugars, sucrose, and the amino acids proline and arginine were significantly higher in cold-resistant clover than in susceptible plants. Murelli and colleagues (1995) evaluated the cold acclimation of two barley genotypes, frost resistant and frost sensitive, and observed an increase in proline in both cultivars. It was found that the high concentration of proline in grapefruit peel could improve tolerance to low temperatures in storage environments (Purvis, 1981). The effect of proline-enhancing cold tolerance has also been observed in in vitro culture. For example, in maize cells, proline and ABA increase cold tolerance by different mechanisms (Xin and Li, 1993). Proline was also found to protect cell suspensions of potato against low temperatures. Thus, proline applied exogenously can result in increased tolerance to low temperatures (Chen et al., 1983).

When plants experience cold stress, there is a reduction in the water uptake and transport capacity of the roots known as physiological dryness. Proline, however, can help maintain full hydration of enzymes and membrane proteins during such times. Increases in shoot proline levels have been attributed to its function in the preservation of the osmotic potential of tissues.

In pollen, the accumulation of proline during acclimation to cold has been proposed in various crops such as *Lolium perenne*, wheat, barley, and alfalfa. Thomas and James (1993) concluded that cold acclimation induced osmotic adjustment and increased concentrations of proline and amino acids in 12 genotypes of *Lolium perenne* with different chilling sensitivity. The accumulation of proline and other amino acids may be the consequence of reduced protein synthesis or increased protein degradation during extreme conditions. Proline probably lessens cold damage via the direct protection of cellular components from cold stress (Xin and Li, 1993).

ABA-induced cold tolerance appears to be associated with an increase in the net synthesis of some proteins. It also regulates embryo maturation and seed dormancy, as well as inhibiting mitotic activity in root meristems. During stress conditions such as water deficit, which becomes important in tissues affected by low temperatures, it also induces stomatal closure. The effectiveness of ABA application in improving the water status of cold-sensitive plants suggests a possible role for this hormone in cold tolerance (Chen et al., 1983; Smolenska-Sym et al., 1995). This effect has been shown at different growth stages. For example, Wilen and colleagues (1994) showed improvement in germination and flowering after ABA or ABA-analogue treatment of *Brassica rapa* and *Brassica napus* seedlings. Smolenska-Sym and colleagues (1995) observed an increase in ABA concentration in the leaves of winter rapeseed plants grown in cold conditions. This increase took place during the first three days of cold treatment. After six days growth in the cold, the ABA level started to descend or remained constant. Welbaum and colleagues (1997) found that ABA and cold treatment increased the freezing tolerance of pea epicotyl tissues, shoot tips, and root tips. The relationship between cold tolerance and ABA accumulation has also been studied in inbred lines of maize with different degrees of cold tolerance (Capell and Dörffling, 1993; Janowiak and Dörffling, 1996). Changes in ABA accumulation and water status were measured in the third leaves before, during, and after a cold period. It was concluded that cold tolerance in maize is related to the capacity for greater and faster ABA accumulation and better stabilization of water status.

Low-molecular-weight carbohydrates are considered important in the improvement of cold tolerance owing to their capacity to regulate osmotic pressure and their involvement in the preservation of cell membranes during cold stress. Goldstein and Nobel (1991) suggest that an increase in concentration of low-molecular-weight solutes may act as cryoprotectants as well as helping to retain liquid water inside cells. Many studies support the idea that changes in carbohydrate metabolism can enhance cold tolerance in cotton (Conner et al., 1972), rye (Olien and Lester, 1985; Olien and Clark, 1995), muskmelon (Mitchell and Madore, 1992), alfalfa (Castonguay et al., 1995), *Lolium perenne* (Bredmeijer and Esselink, 1995), maize (Hodges et al., 1997), and clover (Svenning et al., 1997; Turner and Pollock, 1998). Leaves of young plants subjected to low temperatures accumulate carbohydrates, mainly glucose and fructose (Olien and Lester, 1985; Olien and Clark, 1995). The contents of total soluble sugars in leaf tissues of cold-resistant and tolerant genotypes are significantly and positively correlated with plant viability. During hardening conditions, cold tolerance and soluble sugar levels increase, with a concomitant decrease in starch in cultivars of alfalfa (Castonguay et al., 1995) or buffelgrass (Hussey and Bashaw, 1996). In both these studies, the most winterhardy entries had the highest carbohydrate concentration; the least winterhardy had the lowest.

An increase in carbohydrate composition and in the concentration of sugar-metabolism enzymes has been found during cold acclimation in several species (Blum, 1988; Goldstein and Nobel, 1991; Castonguay et al., 1995). During the winter, annual species maximize the production of sugars with a possible cryoprotective function and accumulate sufficient carbohydrate reserves to support basal metabolism and regrowth in the spring. For example, sucrose is known to increase in response to cold stress, desiccation stress, and cold hardening (Chatterton et al., 1988; Hoekstra and van Roekel, 1988). Its cleavage enzyme, sucrose synthetase, has been the subject of several studies. Newsted and colleagues (1990) described the effect of cold acclimation on its expression in winter and spring wheat plants and found it correlated with low temperature in both. Soluble carbohydrates concentration can show a complex pattern of changes throughout the season. For example, Turner and Pollock (1998) determined starch concentration variation in stolon tissues of white clover and pointed out the role of starch reserves in winter survival. Yoshida and colleagues (1998) determined carbohydrate levels in the crowns and leaves of wheat plants in autumn and winter. During the first stage of cold hardening, the pattern of sugar accumulation was similar in all cultivars, but during the second stage, the greatest accumulation was observed among freezing-tolerant cultivars. Fry and colleagues (1991) found an increase in sucrose and a decrease in starch in Augustine grass stolons between November and December. However, on all other sampling dates, the changes in sucrose and starch were variable. The increase in carbohydrates in leaves under cold conditions owed more to a reduction in the rate of assimilate use than to a fall in photosynthetic rate.

Carbohydrate levels have been referred to in the early vigor of maize seedling leaves by Sowínski and colleagues (1999). Sucrose accumulation induced by cold stress was not correlated with early vigor. On the other hand, the accumulation of starch at low temperature was negatively correlated with early vigor in cold conditions.

SOURCES OF RESISTANCE

The difficulty of transferring genes for cold tolerance from a source to a target genotype increases with the genetic distance between them. Most reports deal with transference of cold tolerance, with variable success, between cultivars of the same species. For example, Bardhan Roy and colleagues (1982) and Khan and colleagues (1986) obtained a significant response to selection for cold tolerance in rice from crosses of susceptible by resistant parents. Other breeders have moved favorable genes from different types of crops within the same species, for example, from field corn to sweet corn (Revilla et al., 1998) or from winter to spring wheat (Braun,

1997). Others still have moved cold-tolerance genes between compatible species, e.g., rye and wheat (Rizza et al., 1997). Finally, when genes have to be taken from incompatible species, several methods exist, ranging from protoplast fusion (Heath et al., 1994) to genetic engineering (Sarhan and Danyluk, 1998). Most variability comes from local varieties, but it can be also generated in the laboratory. The appendix shows sources of cold tolerance identified by breeders during recent years.

Somaclonal Variation

Variability has been artificially generated through somaclonal variation. This often appears in plants after tissue culture involving a callus stage (Larkin and Scowcroft, 1981). It is common in *Brassica* crops and potato and has produced wide phenotypic variation in somatic potato hybrids (Preiszner et al., 1991). Bertin and colleagues (1995) submitted embryo-derived callus of rice varieties to chilling stress for three generations while practicing mass selection. They observed in vitro-induced, preexisting, epigenetic, and heritable variation. Culture-derived lines are characterized by a highly skewed distribution with lower means than control lines. Thus, Bertin, Busogoro, and Kinet (1997) postulate that a great number of lines would have to be produced and screened to identify any improvement.

Distinct mechanisms of chilling tolerance improvement may be obtained through somaclonal variation. Mechanisms may differ from one cultivar to another (Bertin, Bouharmont, and Kinet, 1997). Somaclonal-tolerant cell lines have been obtained and regenerated from suspensions of nucellar callus cells of *Citrus sinensis* (DingBo et al., 1999). Ezura and colleagues (1995) regenerated plants with low-temperature germinability from somatic embryos of melon, and these selections showed greater fruit growth rate and greater fruit volume than the original cultivar. Freezing tolerance has been improved through somaclonal variation in winter wheat (Lazar et al., 1988; Galiba and Sutka, 1989).

Although cold-tolerant lines have been obtained from somaclonal variation (Bertin, Bouharmont, and Kinet, 1996; Bertin et al., 1998), the usefulness of somaclonal variation in plant breeding for cold tolerance has not yet been clearly demonstrated. Most of the presumably resistant varieties obtained so far have not been adequately tested (Bertin and Bouharmont, 1997). Genetic improvements of cold tolerance based on somaclonal variation of cell lines can be maintained, as in rice, or lost, as in *Nicotiana sylvestris,* and failures can be attributed to the low correlation between in vitro and in vivo resistance to low temperatures (Bertin et al., 1995).

INHERITANCE

Studies addressing the inheritance of cold tolerance in cold-sensitive and cold-tolerant species are reviewed here, although the main focus is on maize. This crop is now cultivated well outside its center of origin, and it is a clear example of the need for selection of more efficient growth habits under chilling stress in order to optimize yield in northern latitudes.

Cold tolerance is a complex trait whose inheritance is under genetic control (Tokuhisa and Browse, 1999; Thomashow, 2001). Several factors contribute to this complexity. A number of studies have demonstrated that cold tolerance at different growth stages is inherited independently in many crops. In maize, the genetics involved in cold tolerance at germination, emergence, and early growth are certainly independent (Bocsi and Kovács, 1990; Hodges et al., 1997; Revilla et al., 2000). Dickson and Petzoldt (1987) have pointed out that cold tolerance during germination, seedling, and flowering is inherited independently in beans. The same conclusions were drawn from a study on rice by Sthapit and Witcombe (1998); cold tolerance levels measured as the rate of germination or as plumule greening were not correlated. There is also seasonal variation of cold tolerance. For example, factors controlling frost tolerance in the autumn in a woody plant, the coastal Douglas fir, are, in general, different from those controlling frost tolerance in spring (Jermstad et al., 2001). Most genetic studies on cold tolerance have been performed at early stages of plant development (germination and emergence). Fewer studies have focused on the more advanced stages. A factor contributing to the complexity of cold tolerance inheritance is the strong interaction between genotypes and environments (GE interaction). Revilla and colleagues (2000) detected significant GE interactions with respect to maize vigor at the two-leaf stage under cool field conditions. GE interactions were also significant for yellowing, purpling, and drying of leaves in a diallel among eight maize inbred lines evaluated under cold stress (Mahajan et al., 1993). Ali and Johnson (2000) found that the heritability of freezing tolerance of lentil seedlings, measured on a five-point visual scale, was severely reduced in field conditions compared with controlled environments. GE interactions for cold tolerance have also been found in woody species such as poplar (Howe et al., 2000). They are even significant for traits measured under more controlled conditions because small fluctuations at temperatures below the optimum can cause significant GE interactions (McConnell and Gardner, 1979a).

Further evidence of the great influence of environmental conditions on cold tolerance lies in the fact that the inheritance of nonacclimated freezing tolerance is independent of acclimation capacity and acclimation tolerance in different species (Stone et al., 1993; Palta, 1994; Teutonico et al., 1995). In agreement, Gullord and colleagues (1975) suggested that mechanisms for wheat frost tolerance under high-intensity freezing were different from those under low-intensity freezing. Sutka and Veisz (1988) showed that, even when the genes involved were the same, gene action could be changed depending on the temperature at which freezing tolerance was measured. Hence, controlled conditions for evaluating cold tolerance have been widely used to limit GE interactions as much as possible. Revilla and colleagues (2000), however, warned about the generalized use of cold chambers to estimate cold tolerance in the field, since cold stress under field conditions is unpredictable and performance in a cold chamber is poorly correlated with field performance.

It has generally been assumed that the inheritance of cold tolerance is polygenic (Guy, 1990; Hughes and Dunn, 1996; Fowler et al., 1999; Thomashow, 1999). The segregation distribution of populations obtained when crossing genotypes differing in cold tolerance agrees with this assumption. Cold tolerance among the hybrid clones *Hibiscus rosa-sinensis* × *Lavatera thuringiaca* ranges from the chilling sensitivity of *Lavatera* to the chilling tolerance of *Hibiscus,* suggesting a multigenic nature for this characteristic (Vázquez-Thello et al., 1996). *Brassica rapa* and *Brassica napus* populations display transgressive segregation for freezing tolerance, suggesting it to be a quantitative trait controlled by many genes (Teutonico et al., 1995). Monosomic and substitution analyses have also revealed the multigenic nature of cold tolerance, since the quantitative trait loci (QTLs) for wheat cold tolerance are located on 10 of the 21 pairs of chromosomes (Sutka, 1994).

However, some studies suggest that in some cases cold tolerance may be controlled by only a few genes. For example, winterhardiness in roses is controlled by very few or closely linked genetic factors (Svejda, 1979). Stone and colleagues (1993) stated that nonacclimated freezing tolerance and acclimation capacity in populations derived from interspecific crosses between *Solanum commersonii* and *Solanum cardiophyllum* are controlled by relatively few genes. One or two major genes are involved in low-temperature-induced browning in soybean seed coats (Takahashi and Abe, 1994). Frost tolerance among recombinant inbred lines (RILs) of lentil (*Lens culinaris* Medik.) is actually under monogenic control (Eujayl et al., 1999). Further, the approximately bimodal fertility segregation of three populations derived from crosses between *Oryza sativa* ssp. *indica* and *O. sativa* ssp. *japonica* suggested that the low-temperature-sensitive sterility is controlled by only one or a few genes (Li et al., 1996). Finally, the analyses of segregation for cold tolerance at the rice booting stage shows that differences among cultivars are due to the action of only a few genes (Nishimura, 1995).

Across different crops, heritability estimates for cold tolerance under controlled conditions range from moderate to high, suggesting the contribution of genes with strong additive effects (Eagles and Hardacre, 1979; Volker et al., 1994; Foolad and Lin, 1998; Sthapit and Witcombe, 1998; Waldron, Ehlke, Wyse, and Vellekson, 1998; Foolad et al., 1999; Ali and Johnson, 2000). Additive effects have been reported as the most important gene effects involved in cold tolerance in maize (Bocsi and Kovács, 1990; Revilla et al., 2000), sorghum (Singh, 1985), wheat (Gullord et al., 1975; Sutka, 1981; Veisz et al., 1997), barley (Rhode and Pulham, 1960), oats (Amirshashi and Patterson, 1956; Jenkins, 1969), rice (Sthapit and Witcombe, 1998), tomato (Foolad and Lin, 1998), common bean (Otubo et al., 1996), potato (Chen, Bamberg, and Palta, 1999), chickpea (Malhotra and Singh, 1991), blueberry (Arora et al., 2000), St. Augustine grass (Philley et al., 1998), perennial ryegrass (Waldron, Ehlke, Wyse, and Vellekson, 1998), *Rhododendron* (Lim et al., 1998), and Douglas fir (Jermstad et al., 2001). However, it is not always possible to reliably predict hybrid cold tolerance from knowledge of inbred responses (Aidun et al., 1991; Giauffret et al., 1997; Hodges et al., 1997) since nonadditive gene effects may be relevant.

Some studies on the inheritance of cold tolerance show most genetic variability to be nonadditive (McConnell and Gardner, 1979a; Parodi et al., 1983), and significant nonadditive gene effects have been found in the majority of investigations. Among nonadditive gene effects, dominant effects are the most important and are involved in cold tolerance during early and late growth. They are involved in cold tolerance in many species, such as maize (McConnell and Gardner, 1979a; Eagles, 1982; Maryam and Jones, 1985; Jha and Sinha, 1989; Mahajan et al., 1993; Giauffret et al., 1997; Revilla et al., 2000), potato (Stone et al., 1993), chickpeas (Malhotra and Singh, 1991), rice (Yea, 1995, 1996; Anand et al., 1999; Han and Koh, 2000), wheat (Parodi et al., 1983; Sutka, 1994; Sutka et al., 1997; Veisz et al., 1997), peas (Auld, Adams, et al., 1983), cucumber (Cai et al., 1995), and blueberry (Rowland et al., 1999; Arora et al., 2000). The direction of dominance toward sensitivity or tolerance depends on the material evaluated. Chen, Bamberg, and Palta (1999) suggested that contradictory observations concerning whether freezing tolerance is dominant or recessive in potato could be explained, in part, by the relative level of ploidy of the tolerant parent compared to that of the sensitive parent. Previously, Ternovskaya and Zhirov (1993) found that frost resistance in bread wheat was controlled by genes of genomes D and AB. Those of the former had dominant and additive epistatic effects when combined with genes of the other two genomes. Genes of the A and B genomes interacted additively at the interallelic and intergenic level.

The existence of interlocus gene interactions for cold tolerance depends on the genotypes involved (Revilla et al., 2000) rather than on plant life stage. Additive-dominance models are adequate for cold tolerance in some cases (McConnell and Gardner, 1979a; Maryam and Jones, 1985; Stone et al., 1993; Foolad and Lin, 1998; Revilla et al., 2000), while significant epistatic effects are involved in others (McConnell and Gardner, 1979a; Bocsi and Kovács, 1990; Malhotra and Singh, 1991; Kaw, 1999; Arora et al., 2000; Han and Koh, 2000). Maternal gene effects appear to restrict their contribution to cold tolerance to the beginning of plant development. These have been detected in maize at the germination and emergence stages (Pesev, 1970; Eagles and Hardacre, 1979; Maryam and Jones, 1983; Bochicchio, 1985; Bocsi and Kovács, 1990) and are important for the extent of electrolyte leakage from cowpea seeds at chilling temperatures (Ismail et al., 1997).

Intrapopulational recurrent selection or even mass selection is therefore recommended for improving cold tolerance since additive effects are the most important gene effects. As the final expression of the genetic potential conditioned by environmental factors, yield under cold conditions is the best trait to evaluate cold tolerance. However, variation for yield under stress is largely environmental, not genetic (Johnson and Frey, 1967), and selection based exclusively on yield performance in such conditions would be inadvisable. Greaves (1996) proposes the use of genetic variation for several cold-tolerance traits in order to improve performance under cold stress. These traits should measure cold tolerance during early plant establishment as well as autotrophic growth. However, it is necessary to use a selection index that takes cold tolerance into account at all stages threatened by suboptimal temperatures. For instance, to improve maize performance in northern latitudes, it would be advisable to increase emergence as well as early growth under low temperatures. In addition, selection for cold tolerance should be performed under controlled thermal conditions to avoid GE interactions. To be able to predict field performance, such conditions should be as close as possible to the average field temperatures at the stages when the plant is exposed to cold stress.

Molecular markers such as RFLP (restriction fragment length polymorphism) and SSR (simple sequence repeats), among others, make it possible to dissect complex quantitative traits in QTLs that can be located in the plant genome and whose gene effects and interactions can be measured. The following examples illustrate some achievements in locating the QTLs for cold tolerance using molecular techniques. At least three QTLs are responsible for growth at low temperatures in a population developed from the cross *Lycopersicon esculentum* × *L. hirsutum* (Vallejos and Tanksley, 1983). Two chromosomal regions have been identified as being involved in cold tolerance during germination in a population derived from the interspecific cross between *L. esculentum* and *L. pimpinellifolium;* QTLs accounted for the 43.8 percent of phenotypic variance (Houde et al., 1992). Oyanedel and colleagues (2000) detected three QTLs on chromosomes 1, 5, and 6 for photoinhibition under chilling stress in tomato.

A single QTL for winter survival was detected on chromosome 7 in a cross between winter and spring barley. This QTL accounted for 31 or 79 percent of the variability, depending on environmental conditions (Pan et al., 1994; Chen et al., 1996). This QTL appears to be linked to the locus *Sh2,* a vernalization gene involved in the determination of the growth habit (spring versus winter), since the former region accounts for 47 percent of variation in this respect. In later work, Tuberosa and colleagues (1997) detected four markers for barley freezing tolerance at −12°C, and six for −14°C.

Molecular markers are now being used to confirm the location of QTLs for frost tolerance in wheat that were originally detected by substitution and monosomic analysis. Galiba and colleagues (1995) mapped the frost-tolerance gene *Fr1,* closely linked to the vernalization gene *Vrn1* on the long arm of chromosome 5A of wheat. The interval *Vrn1-Fr1* explains between 71 and 91 percent of the variation for lethal temperature in 50 percent of experimental plants (Storlie et al., 1998). Molecular markers have been used more recently to localize QTLs for frost tolerance on chromosomes 5A and 5D of wheat (Snape et al., 1997, 2001). The gene *Fr2,* homologous to *Fr1,* and the loci on chromosomes 7 of barley and 5R of rye, were located on the long arm of chromosome 5D. Gene *Fr2* was linked to *Vrn3* (a vernalization response gene). Vernalization genes in cereals have been identified as the key factor in determining the duration and expression of cold-induced genes (Fowler, Limin, et al., 1996; Sarhan et al., 1997). As a consequence, the less-tolerant spring cultivars have less chance of achieving the levels of tolerance reached by hardy winter cultivars.

Genes for cold tolerance at the booting stage of rice have been mapped on chromosomes 3 and 4 (Saito et al., 1995; Kato et al., 1996). Thirteen QTLs associated with cold tolerance in rice pollen were detected on chromosomes 1, 3, 9, and 11 (Misawa et al., 2000). The QTL-holding chromosomes 3, 9, and 11 showed synteny with chromosome 5 of wheat and 7 of barley. This agrees with the conclusions derived from the characterization of several genes involved in cold tolerance (Bartels and Nelson, 1994). These authors stated that some cold-tolerance genes seem to occur in a wide variety of plants.

Teutonico and colleagues (1995) detected four chromosomal regions in *Brassica rapa* associated with acclimated freezing tolerance and acclimation ability, and two unlinked regions associated with nonacclimated freezing tolerance. The variability explained by each individual QTL and gene effect were reported. A single QTL was located for radiation frost tolerance among recombinant inbred lines of lentil when plants were 50 days old (Eujayl et al., 1999). In general, in spite of the complexity of cold tolerance, few QTLs have been found. Karsai and colleagues (1997) stated that differential QTL expression might mask gene complexity for cold tolerance. Fewer attempts have been made to locate QTLs for cold tolerance in woody perennials. Two QTLs for cold tolerance in *Eucalyptus nitens* were identified in the same linkage group (Byrne et al., 1997) and QTLs for spring and autumn cold hardiness in coastal Douglas fir have been recognized (Jermstad et al., 2001). Eleven QTLs controlled autumn cold hardiness and 15 controlled spring cold hardiness; only one linkage group contained QTLs for spring and autumn cold hardiness.

Rowland and colleagues (1999) proposed the genetic mapping of blueberry as a model for other woody perennials. Such genetic mapping could allow the establishment of the relationship between chilling-responsive genes, chilling requirements, and cold hardiness. However, previous studies involving different genotypes of overwinter, freezing-tolerant crops, such as winter rape *(Brassica napus oleifera)* and barley, have revealed no relationship between cold tolerance and vernalization requirements (Doll et al., 1989; Markowski and Rapacz, 1994). Nevertheless, in wheat and barley, a close relationship was observed between the time to vernalization saturation and the start of the decline of low-temperature tolerance (Fowler, Limin, et al., 1996).

Besides QTL analysis, mutational approaches have been postulated as useful tools for identifying the genes required for low-temperature growth (Tokuhisa et al., 1997). Once gene locations are identified, chromosomal walking techniques can be employed to isolate the gene(s) of interest. It is then necessary to prove the role of the isolated gene under cold stress. Extensive information on genes involved in the response to cold stress can be found in several reviews (Hughes and Dunn, 1996; Thomashow, 1990, 1999, 2001). This knowledge provides the potential for new strategies since traditional breeding approaches can achieve a limited improvement in the cold tolerance of important crops (Sarhan and Danyluk, 1998).

BREEDING PROGRAMS

Cold tolerance becomes a breeding objective for plant breeders when a species expands into marginal areas to which it is poorly adapted or when other stress conditions require the modification of growth habits. The general objective of breeding for cold tolerance is to obtain genotypes with the ability to emerge from the soil and grow vigorously in low soil and air temperatures. Breeding programs may have several objectives with respect to low-temperature tolerance, depending on the crop. Among annual crops, cold- and warm-season types can be distinguished which condition these objectives. A common objective in the breeding of perennial and cold-season crops in temperate and cold areas is to achieve winter hardening and frost tolerance. Cold tolerance is of paramount importance for some warm-season crops cultivated in temperate regions.

Selection Criteria

The definition of selection criteria is an essential step in any breeding program, and particularly so when trying to improve cold tolerance. Chen, Bamberg, and Palta (1999) attribute the slow progress in breeding programs for cold tolerance in potato to the lack of effective selection criteria. Generally, breeders focus on one of several traits affecting one stage of development, but cold tolerance is a complex characteristic, particularly if the whole plant is considered. Breeding for several traits simultaneously, however, may lead to failure, as reported by Holubowicz and Legutko (1995).

The objectives of breeding for cold tolerance vary from germination to postharvest conservation. For example, maize cultivated in Canada requires rapid growth at low temperatures if it is to reach normal physiological maturity in short seasons. Thus, Canadian maize breeders are trying to adapt maize to early planting in order to obtain earlier canopy development to increase competition against weeds, pollination prior to the hot days of summer, and earlier maturity to avoid late-season frosts (Aidun et al., 1991). The situation is quite similar in non-Mediterranean Europe (Ordás et al., 1994).

Several approaches to selecting for cold tolerance can be adopted. The most common is to choose those genotypes with higher germination or growth at low temperatures. Alternatively, it is possible to estimate cold tolerance as the ratio of performance in cold conditions/performance in optimum conditions. In tomato, Foolad and Lin (2000) used a germination tolerance index [the inverse of the ratio (germination time under cold stress)/(germination time under control conditions)], and a vegetative growth tolerance index [the ratio of (shoot dry weight under cold stress)/(shoot dry weight under control conditions)]. Some traits commonly used in breeding for cold tolerance during the first stages are percent germination or emergence, rate of emergence, dry matter accumulation, vigor, and color. Rate of emergence has been used in several crops and the following equation proposed for evaluations in tomato (Smith and Millet, 1964):

$$RE = \sum p \cdot d / Tp$$

where p = number of plants emerged on a day, d = number of days after planting, and Tp = total number of plants emerged at 30 days after planting.

Selection for early vigor is not necessarily included in breeding programs for cold tolerance, but it can lead to improvements (Martiniello, 1985). Selection for early vigor should be possible by most breeding methods, as shown by Ordás and colleagues (1994). Root length is associated with cold tolerance in pigeon pea, while shoot length is not. However, the recommended selection criterion for early vigor is the shoot-length/root-length ratio (Kumar et al., 1995). Ratinam and colleagues (1994b) found that this ratio showed a consistent relationship to cold tolerance irrespective of the growing conditions and season in several forage legumes. Other selection criteria recommended by Ratinam and colleagues (1994b) for *Lathyrus ochrus* were the number of primary branches per plant, root weight, and shoot weight. Adak and Eser (1993) reported positive and significant correlations in barley between cold hardiness and crown depth, root length, dry crown weight, and number of tillers, and negative correlations with seedling height and crown water content.

With regard to vegetative growth, the main temperature stress is frost and the main selection criterion for freezing resistance is survival or overwintering. In wheat, Fowler and Gusta (1979) constructed a field survival index that included a freezing test, which has been subsequently modified. The agreement between freezing tests and winter survival depends on the crop, e.g., the results of freezing tests with winter wheat and barley correlate better with overall survival than the same tests carried out on winter rye (Larsen, 1994). Winter survival is an essential characteristic for winter cereals such as oats, barley, wheat, and rye (Fowler and Carles, 1979; Hensleigh et al., 1992). In white lupin, frost tolerance requires large roots; it is affected by the physiological state of the apex and by the hardening ability of the leaves (Huyghe, 1997) and is required for perennial plants such as white clover (Rhodes et al., 1994). In perennial crops, the selection criterion for cold tolerance is survival of the plant, or part of it, over the winter. For example, for white clover, the trait is stolon survival (Rhodes et al., 1994), while for peaches and nectarines the selection criterion is cold hardiness (Layne, 1992).

At flowering, the main constraint is sterility. An important cold-related trait is spikelet sterility in rice, which causes severe reductions in yield (Khan et al., 1986). Low temperatures can damage rice at any stage of development, causing shortening of plants, lengthening of the growth period, poor panicle exertion, and yellowing of leaves (Bardhan Roy et al., 1982). However, it is the reproductive stage that is most susceptible, the male part of the flower being particularly sensitive (Bardhan Roy et al. 1982; Khan et al., 1986). In soybean, cold-tolerance evaluation during flowering is based on the number of flowers produced, the capacity to retain them, and the capacity to replace those lost (Gass et al., 1996).

Winterhardiness is related to all vegetative stages of plant growth. The main selection criteria are hardening, freezing, and recovery (Waldron, Ehlke, Vellekson, et al. 1998). Fuller and Eagles (1978) predicted winterhardiness by using controlled freezing of perennial ryegrass seedlings. Humphreys and Eagles (1988) used a glycol freezing tank to evaluate tiller mortality in perennial ryegrass as a selection criterion for winterhardiness. Testing for freezing tolerance seems to give the best indication of the overall wintering ability of grasses (Larsen, 1994).

Cold storage is somewhat related to cold tolerance in that there may be injury to stored fruit and vegetables due to cold temperatures (Wang, 1982; Muñoz et al., 2001; Sowokinos, 2001). The tolerance of crops to cold temperatures is variable during storage (Wang, 1982; Alique et al., 1994; Griffiths et al., 1998).

Usually, the genetic regulation of cold tolerance of each plant stage is independent of the others, as shown by Revilla and colleagues (2000) in maize and by Foolad and Lin (2001) in tomato. The biochemical and physiological requirements of germination and emergence may not be the same (Hegarty, 1973; Simon et al., 1976). Chilling sensitivity varies during the life cycle. For example, low-temperature tolerance in rice is different at the vegetative and reproductive growth stages (Bertin, Busogoro, et al., 1996). Some stages are not independent; for instance, chilling stress at successive leaf-stages is similar in rice (Bertin et al., 1995) and in maize (Revilla et al., 2000).

Improving the cold tolerance at different stages in a combined fashion might be possible by using selection protocols that include all critical stages. In temperate areas, most cold-tolerance breeding programs focus on germination and early development. Several authors (Hodges et al., 1995, 1997; Revilla et al., 1998) suggest that the ability to germinate and survive under cold conditions may be necessary, but these characteristics by themselves do not ensure early vigor. Hodges and colleagues (1997) and Revilla and colleagues (2000) found that germination and seedling growth are under the control of different genetic factors and that it is possible to combine both high germination and vigorous early growth under cold conditions.

Selection for several traits can be accomplished by a variety of methods. Layne (1992) explains that it is possible to select simultaneously for bud and wood hardiness in peaches and nectarines if controlled freezing of fully acclimated shoots is undertaken. When the traits selected behave independently, the most obvious procedure is sequential selection. Seed survival or germination is the first trait observed, and this limits the evaluation of subsequent traits. Seedling vigor can then be estimated, but if this is done by measuring dry-matter accumulation, no further traits can be measured. Crosbie and colleagues (1980) compared several indices for the simultaneous selection of several traits and concluded that the best predicted results for all traits were given by a rank summation index, a multiplicative, and a base index. They also found that selection for dry weight per plot identified lines with excellent percentage emergence and seedling dry weight. Hoard and Crosbie (1985) concluded that selection for cold tolerance would be effective in maize if a selection index was used that included percent emergence and seedling dry weight based on data from more than one year. Layne (1992) developed a multiple-trait selection index based on 15 traits. A model combining several traits for screening sugarcane for low-temperature resistance was developed by Singh and Singh (2000). Humphreys (1989) and Larsen (1994) recommended that freezing tolerance be selected either as a truncation or as an index selection.

Selection Conditions

A major influence on the effectiveness of breeding for cold tolerance is the environment used for evaluations. Field breeding programs suffer environmental variations that limit or impede progress (McConnell and Gardner, 1979a), while laboratory programs are limited by the correlation between field and laboratory performance. Hodges and colleagues (1994) and Hodges and colleagues (1997) state that most studies evaluate emergence either in the laboratory or in the field, and that few studies have evaluated seedling growth under laboratory conditions. Field variation can be somewhat controlled. For example, Fletcher (1983, cited by Blum, 1988) suggested that when stress occurs, plants going through the appropriate phenological stage should be marked for subsequent measurement. Alternatively, synchronization between phenological stage and stress could be tried if stress can be predicted and plant development manipulated. Such field control, however, is often impossible since even if some plants are at the correct stage when stress occurs, and even if synchronization is possible, the intensity of stress can vary greatly, perhaps causing the death or escape of some genotypes (Martin, 1927). Some authors (Martin, 1927; Blum, 1988) conclude that spatial or temporal variation can be dealt with by replicating field tests in several locations and over several years, and by complementary laboratory tests under controlled environments. Hensleigh and colleagues (1992) indicate that field survival is the ultimate measurement of winterhardiness, but field survival tests are often characterized by incomplete or irregular winter kill. An accurate, consistent technique would, therefore, be very helpful.

There can be some diversity in laboratory evaluations, depending on the species and equipment used. Schwab and colleagues (1996) compared several laboratory evaluation conditions and proposed a test in which field-hardened alfalfa entries of a similar size are packed in dry vermiculite and exposed to temperatures of –12 to –16°C. Often, cold tolerance is measured as germination at an established low temperature compared with the optimum temperature (Landi et al., 1992). Herczegh (1970) reviewed some of the early techniques used for the evaluation of maize under cold tolerance. Each developmental stage of each crop has different temperature requirements. The temperature used for evaluation for cold tolerance in maize is generally around 10°C. For example, cold tolerance was evaluated at 10°C (Eagles and Hardacre, 1979), 9.5°C (Landi et al., 1992), cycles of 14 and 8°C (Revilla et al., 2000), and cycles of 14 and 10°C (Hotchkiss et al., 1997). Also, the substrate might differ: rolled paper towels (Bacon, 1986), soaked filter paper (Landi et al., 1992), sterilized peat (Revilla et al., 1998), sterilized sand (Hotchkiss et al., 1997), hydroponic solutions (Hodges et al., 1995), and so on; and evaluations can be made without light (Landi et al., 1992) or using cycles of dark/light, as in Revilla and colleagues (2000) and Hotchkiss and colleagues (1997). Cold tolerance of sorghum during germination has been measured at 14°C (Bacon, 1986), while in cucumber this has been performed at 15°C (Nienhuis et al., 1983) and in maize at 10°C (Herczegh, 1970).

Evaluation facilities described in the literature are also quite variable, even though standard growing chambers often allow a precise environment for most objectives, especially for improving germination. Cold chambers are commonly used for selecting for cold tolerance and for complementing field evaluations (Haskell and Singleton, 1949). It has long been known that the correlation between field performance and cold chamber evaluations are generally poor (Schultz, 1941), though controlled freezing tests can provide a good estimate of field performance (Chen, Bamberg, and Palta, 1999). In addition to this major shortcoming, cold chambers have some limitations related to heterogeneous humidity and temperature conditions, depending on the position within the chamber. Therefore, ventilation is essential if the homogeneity of temperature and humidity is to be attained. Zimmerman and Buck (1977) designed a modified chamber with a turntable to expose plants to homogeneous conditions, an apparatus successfully used in evaluations of safflower. When advanced plant stages are to be evaluated under cold conditions, such as fruit ripening or seed production, plants should be grown in an appropriate environment to avoid confounding factors such as the effect of nutrients or light. For example, for rice, Khan and colleagues (1986) used a phytotron for evaluations of spikelet sterility while Bardhan Roy and colleagues (1982) used a method called rapid generation advance for cold-tolerance selection.

In addition to errors due to variable stress environments and the methodologies used for evaluation, stressful conditions can cause large experimental errors. Another source of error is the origin of the seed itself. Seeds of the same genotype produced under different environments can yield significantly different results, thus confounding any interpretation. Under stress conditions, it is important to evaluate seeds produced in the same environment. Ideally, several production environments should be used in order to estimate the GE interactions on seed production, particularly if the production environment is stressful.

Breeding for stress can be performed in nonstress environments when varieties that perform well in the absence of stress also perform well under stress (Baker, 1994). Rather than increasing production under stress as a proportion of the production under nonstress, the best strategy is to breed for higher production under stress (Baker, 1994). Van Oosterom and Ceccarelli (1993) demonstrated that selection in high-yielding environments for winter-plant ideotype and heading date can be an efficient method of selecting for barley yield under stress.

Direct Selection

Traditionally, farmers have made some kind of mass selection for cold tolerance as they have expanded crops into cooler regions. This selection has been reinforced by inevitable natural selection. The response to this historical mass selection is clearly manifested in the wide range of cultivation of several crops. Hensleigh and colleagues (1992) demonstrated that natural selection of winter barley improved winter survival and affected several agronomic traits. The expansion of some crops, such as maize, toward higher latitudes continues indirect adaptation for cold tolerance as a correlated response to breeding for yield under cooler conditions. Frei (2000) states that the grain/stover yield antagonism is most severe in cooler climates, where assimilates are diverted from grain filling to plant maintenance when temperatures drop. The cold can cause partial sterility of pollen or silks, yet great improvements in chilling tolerance in rice have been associated with lower fertility and reduced panicle length and plant height. Moderate improvements have had nonsignificant effects on other agronomic traits (Bertin et al., 1995). Breeding designed to adapt maize to northern Europe has resulted in the release of compact, low-stover hybrids (Frei, 2000). Similarly, in Canada, most of the genetic yield improvement of maize is attributable to increased stress tolerance (Tollenaar and Wu, 1999). The genetic improvement of maize has been accompanied by a decrease in plant-to-plant variability, which in turn results in more stress tolerance. In the future, global warming may introduce new goals for adaptation breeding. As pointed out by Veisz and colleagues (1996), predictions suggest that in Central Europe there will be a rise in winter temperatures, which will improve the overwintering of winter cereals.

The relationship between cold tolerance and yield is inconsistent. Mastenbroek (1956) found no link between tuber yield and frost tolerance in potato. Mahajan and colleagues (1992) indicated that yield is not directly related to cold tolerance during germination, and Hides (1979) found no correlated response in growth potential after selection for cold tolerance in Italian ryegrass. Chen, Palta, and Bamberg (1999) concluded that there were no significant correlations between undesirable tuber traits and freezing tolerance in potato. These authors even found that some cold-tolerant progenies had increased yield. On the other hand, Hoard and Crosbie (1986a) found that selection for cold tolerance may result in correlated improvements in yield, though they did not obtain such improvements due to an increase of inbreeding depression caused by the small population size used in their program. A significant increase in yield was found as an indirect response of a breeding program for cold tolerance in sorghum (Bacon, 1986).

Many breeding programs in Europe have routinely included cold tolerance in their selection of new varieties in species as diverse as maize (Koshelyaev and Serkov, 2000), pea (Ranalli et al., 1999), and grapevine (Korbuly, 2000). Maize grown in Europe needs some cold tolerance during early development to be able to stand the normally cool and wet spring. Revilla and colleagues (1998) showed that European maize is a promising source of cold tolerance. The adaptation of maize to cooler regions can be accomplished gradually and may not require major efforts, except for some special maize such as sweet corn, which is an endosperm mutant with physiological limitations, and which has poorer germination, early vigor, and growth rates than field corn (Hunsperger and Davis, 1987; Treat and Tracy, 1994).

Sweet corn does not grow well during the early stages of development in cold and humid environments such as the northwest of Spain (Ordás et al., 1994). Some European sweet corn breeders are currently working on improving its cold tolerance (Revilla et al., 1998). According to Hensleigh and colleagues (1992), conventional breeding methods have not been successful in developing cultivars with adequate winterhardiness for many northern regions. However, the improvement of winterhardiness in oats allowed the expansion of this crop from the southern United States to northern regions (Livingston and Elwinger, 1995). According to these authors, a long-term program for improving winterhardiness in this crop produced significant genetic gains from 1935 to 1970, with progress decreasing since the mid-1970s.

Mass selection has been used by some maize breeders with limited success (Mosely et al., 1984). Three cycles of mass selection for ear performance were performed in two high-density maize populations. Some cold-tolerance response was obtained in one of the populations, along with improved agronomic performance and yield due to early planting. In tomato, Foolad and Lin (2001) obtained significant improvements in germination under cold conditions, with a realized heritability of 0.75 that encourages the use of mass selection. Marshall (1966) found improvements in cold tolerance from natural mass selection in oats. Mass selection in peaches and nectarines for most desired traits, including cold hardiness, is the most effective breeding strategy (Layne, 1992). Begum and colleagues (1994) concluded that exposure of a segregating population to cold stress in the field is an effective selection procedure for identifying resistant genotypes. GE interactions limit genetic gains through mass selection in breeding programs for cold tolerance (Layne, 1992).

Pedigree selection is another, generally successful, practical method used in many breeding programs for cold tolerance, as reported by Malhotra and Kalloo (1995) for tomato and by Caradus and Christie (1998) for white clover. Waldron, Ehlke, Vellekson, and White (1998) conducted successful selection in perennial ryegrass using half-sib families. Breeding methods have used several types of relatives, including half-sib families (Staub et al., 1988), full-sib families (Landi et al., 1992), and S_1-families (Mock and Bakri, 1976; Martiniello, 1985; Hoard and Crosbie, 1985). S_1-family selection is commonly used in maize breeding when the base population has not been previously selected, since selfing allows the rapid detection and elimination of undesirable alleles. However, the sample size and intensity of selection must be appropriate to avoid inbreeding depression (Hoard and Crosbie, 1985). Helgadóttir and Björnsson (1994) successfully carried out a breeding program using half-sib families of Timothy grass for northern Europe. However, few reports on the efficiency of the most common methods have been published, probably because these applied methods are designed for improving varieties and not for publishing scientific articles, owing to the lack of systematic design or the absence of novelty.

Most genetic studies have shown that cold tolerance is controlled by polygenic systems. It has therefore been suggested that cold tolerance may be improved through recurrent selection methods (Grogan, 1970). Due to the complexity of cold tolerance, GE interactions, and differences in the inheritance across genetic backgrounds, recurrent selection for cold tolerance has been unsuccessful on occasion. Recurrent selection would be expected to change allelic frequencies and, hence, population means, which would directly affect the selected traits. Recurrent selection for cold tolerance has also been made in sorghum (Bacon, 1986) using a phenotypic recurrent selection method in the field. A significant linear response in emergence in cold temperatures was obtained. Bacon (1986) found significant response in earliness and increase in seed color and yield. Correlation with earliness can be negative, as in the faba bean, where frost tolerance is associated with lateness of flowering (Duc, 1997).

Mock and Eberhart (1972) found that the predicted response on cold-tolerance breeding was substantial for two maize populations. Subsequently, Mock and Bakri (1976) undertook an S_1-recurrent selection program with these two populations, obtaining progress in percent emergence and seedling dry weight but not in emergence index. One of the populations responded significantly to selection, while the other did so less consistently. Some correlated response toward earlier plants was also observed. McConnell and Gardner (1979b) undertook an S_1-recurrent selection program combining field and laboratory evaluations of cold tolerance in maize. Laboratory-produced, significant gains in selection were seen, with no detrimental effects on other agronomic traits, but there was poor correlation between laboratory and field evaluations. The authors justify the poor results of the field-based program in that the evaluation years were mild. This illustrates the common limitations of field trials for cold tolerance.

Nienhuis and colleagues (1983) reported that half-sib recurrent selection for cold tolerance was effective for improving germination in cucumber. Hoard and Crosbie (1986b) also obtained significant genetic gains in two maize populations from an S_1-recurrent selection program for percent emergence, seedling dry weight, and vigor. They concluded, however, that the rate of emergence was not a dependable selection trait. They also found that recurrent selection for cold tolerance in two maize populations altered GE interactions for traits under direct selection but not necessarily for correlated traits. Half-sib recurrent selection for improved germinability at 15°C did not change horticulturally important characteristics in cucumber (Staub et al., 1988). Cousin (1997) reported a recurrent selection program for improving cold tolerance in peas. He found a high combining ability between spring and winter cultivars, and then actually combined them to bring together the main genes conferring cold resistance.

Selection can be effective both for high and low germination proportions. Landi and colleagues (1992) found significant responses in divergent full-sib recurrent selection for germination at low temperatures; and response to selection was significant when the selected populations were compared at low temperature, but not when the comparison was made at 25°C. De Santis and colleagues (1999) evaluated the populations selected by Landi and colleagues (1992), focusing on associated changes in mitochondrial properties. They found a correlated response in the interaction between membrane lipids and cytochrome C oxidase content. Further, selection for cold tolerance apparently affected the allelic frequencies of the genes controlling this interaction.

Breeding for freezing tolerance is more complex and difficult than breeding for cold tolerance and has had less-practical outcomes. Traditional plant breeding methods for improving freezing tolerance by using field selection have achieved limited success due to a lack of genetic diversity, lack of effective selection criteria, and limited and inconsistent knowledge regarding genetic control (Stone et al., 1993). The difficulty grows as the target temperature changes farther from the temperature to which the crop is adapted. Published breeding programs for low-temperature tolerance report a great variety of techniques, from mass selection to recently developed genetic engineering. These reports mainly concern maize, but *Brassica* crops, tomato, and others are also included. Breeding for low-temperature tolerance uses the designs developed for plant breeding in general, with few methodological modifications.

Indirect Selection

As previously explained, selection for cold tolerance entails some problems of evaluation because of the unpredictable climatic variation of field trials and the inconsistent correlation between controlled environment and field performance. Further, reproducing field conditions in the laboratory is difficult and expensive. Moreover, heritability measured under stressful conditions generally decreases because stressful environments raise the GE variance. This can lead to indirect selection or selection under nonstress conditions, given that the correlation of the selected traits compensates for the reduced heritability. Baker (1994) suggested the use of an index combining data from stress and nonstress environments when heritability under nonstress is much higher than under stress.

As shown by Tollenaar and Wu (1999) and Frei (2000), breeding for yield in cool regions results in indirect responses on cold tolerance. Also, some stresses are genetically or physiologically related, so that breeding for one produces improvements in the others. This is the case of breeding for salt and cold tolerance in tomato (Foolad et al., 1999). These authors found that progenies selected for high germination in salty or cold conditions performed well under either stress, as well as in nonstress conditions. Associations between different stresses allow the indirect improvement of cold tolerance when the breeding program focuses on another stress.

Many indirect criteria have been identified as selection criteria for cold tolerance. Indirect selection criteria can be anything from a component of stress tolerance to a molecular marker associated with stress-related loci. A component of productivity under stress can be useful if it is highly correlated to productivity under stress, amenable to selection, and economically assessable (Baker, 1994). An indirect criterion must have high heritability and high genetic correlation with productivity under stress.

Of the many traits used as indirect selection criteria for improving cold tolerance, electrolyte leakage, an indication of freezing tolerance that can be easily estimated by conductance measurements, deserves particular notice. It has been successfully used for evaluations of freezing tolerance in many species, such as maize (Mahajan et al., 1992), potato (Chen, Bamberg, and Palta, 1999), *Pennisetum* species (Stair et al., 1998), and other crops (Blum, 1988). Bertin, Bouharmont, and Kinet (1996) and Bertin, Kinet, and Bouharmont (1996) concluded that low-temperature electrolyte leakage and low-temperature fluorescence are good indicators of chilling tolerance during the vegetative growth of rice.

Another important method for indirect selection for cold tolerance is the measurement of chlorophyll fluorescence in leaves. This method allows the rapid assessment of photosynthetic activity in early growth stages of several crops. The usefulness of chlorophyll fluorescence in screening plants for cold tolerance has been widely demonstrated (Hetherington, Smillie, Hardacre, and Eagles, 1983; Hetherington, Smillie, Malagamba, and Huaman, 1983; Blum, 1988; Nie and Baker, 1991; Schapendonk et al., 1992; Andrews et al., 1995; Verheul et al., 1995; Bertin, Bouharmont, and Kinet, 1997; JinHong et al., 1997; Haldimann, 1998; Janda et al., 1998; Fracheboud et al., 1999). Fracheboud and colleagues (1999) explain that this method is based on the ability of tolerant plants to maintain a higher efficiency of excitation energy captured by open photosystem II (PSII) reaction centers. This trait is low-temperature specific. Quenching analysis using the saturation pulse technique (Schreiber et al., 1986) has been used to investigate responses to stress in the laboratory and in the field (Fracheboud et al., 1999). When a leaf is exposed to light after a dark period, fluorescence increases to a level called F0—initial fluorescence when all PSII reaction centers are open. Soon, fluorescence increases to the level Fm—the maximum fluorescence when all reaction centers are closed. The increase in fluorescence is called variable fluorescence (Fv). The Fv/Fm ratio is a measure of the intrinsic efficiency of PSII photochemistry and appears to be of primary importance in studies of stress effects on plants (Bertin, Bouharmont, and Kinet, 1997). The maximum rate of the induced rise in chlorophyll fluorescence at cold temperatures, Fr, has also been used in maize breeding (Hetherington, Smillie, Hardacre, and Eagles, 1983).

MacRae and colleagues (1986) compared several techniques for evaluating cold tolerance in crops and found that none was able to distinguish cold tolerance in all species. This suggests that the mechanisms of chilling tolerance are diverse. Ratinam and colleagues (1994a) detected a strong correlation between chlorophyll fluorescence analysis and frost sensitivity in *Lathyrus ochrus*. Sthapit and colleagues (1995) found that the Fv/Fm ratio was strongly correlated to the adaptation of rice genotypes to the cold highlands of Nepal. Bertin and Bouharmont (1997) used the Fv/Fm ratio for evaluating cold-tolerant rice genotypes obtained from somaclonal variation. Chlorophyll fluorescence allowed detectable changes in chilling tolerance in several, but not all, rice genotypes, and they concluded that different mechanisms were implicated in chilling-tolerance improvement (Bertin, Bouharmont, and Kinet, 1997). Finally, JongUn and colleagues (2000) found that chlorophyll fluorescence measured by Fv/Fm was very useful in evaluating the relative levels of freezing resistance in tea.

Several morphological features have been used as markers of cold tolerance. In peas, susceptible genotypes are generally early lines or lines with long internodes, large leaf area, or wrinkled seeds (Cousin, 1997). Floral morphology has been proposed as an indirect criterion for selecting for cold tolerance in rice (Suzuki, 1982) and eggplant (Nothman et al., 1983). Dwarfing genes in rice are associated with cold tolerance. Some seem to produce cold tolerance while others could cause sensitivity at the seedling (Murai et al., 2000) and booting stages (Bahadur et al., 2000). In peaches and nectarines, several traits have been used for indirect selection for cold hardiness, particularly early defoliation (Hansche, 1990), flower bud density (Byrne, 1986), and disease resistance (Layne, 1992; Cousin, 1997), although their inheritance and genetic linkages are still being investigated. Zauralov (1998) proposed the growth of the primary root of sprouted seeds as an indirect selection criterion for improving frost resistance of millet varieties. In potatoes, paraquat tolerance was related to cold acclimation but not to frost tolerance (Seppanen et al., 1998). Martin (1927) found an association between dark narrow leaves and prostrate growth habit of winter cereals and winterhardiness. Hensleigh and colleagues (1992) found an association between short seedling-leaf length and winter survival in barley. In peas, Cousin (1997) reports significant associations between winterhardiness and pigmented hilum, pigmented seed coats, and yellow cotyledons.

Chemical measurements have been assayed as indirect selection criteria for cold tolerance. However, their ability to predict cold tolerance is insufficient, and they have rarely been used for selection (Blum, 1988). Isozymes, proteins, and DNA markers have been extensively studied, and many have been suggested as indirect criteria for selection for cold tolerance. Isozyme patterns of UDP-glucose pyrophosphorylase have been suggested as markers for cold-sweetening resistance in potatoes (Sowokinos, 2001). Kang and colleagues (1998) found isozymes associated with cold tolerance in *Cucurbita* spp. The isozyme markers *Pgi-1* and *Adh-2* appear to be linked to the better low-temperature tolerance of *Lycopersicon hirsutum* pollen than that of *L. esculentum* (Zamir et al., 1982). The enzyme locus *Pgi-1* has also been proposed as a useful marker for growth at low temperature in a population derived from the cross *Lycopersicon esculentum* × *L. hirsutum* (Vallejos and Tanksley, 1983). In a meadow fescue population, *dd*-genotypes for the *Gpi-2* alloenzyme locus were found more cold tolerant than all other genotypes (Rognli and Myhre, 1996). In another study, 11 isozyme markers were examined in an attempt to modify maize cold tolerance, but the results were not promising (Guse et al., 1988). Zabrodina and colleagues (1996) observed that after three winters under field conditions, a population of annual rye and another of perennial rye increased their frequency of specific alleles of the *Ast3* and *Est1* isozyme loci.

Other molecules are also associated with cold tolerance. Dehydrins are proteins that may play a critical role in stabilizing cell functions during freezing and other dehydrative stresses (Lim et al., 1999). These authors concluded that the relative level of a dehydrin can serve as a physiological indicator of freezing tolerance under a range of development conditions. In winter wheat, frost tolerance correlates with the accumulation of free proline during cold acclimation (Dörffling et al., 1990).

The cross-reaction between WCS120-antibodies, with the *Wcs120* gene strongly induced during cold acclimation of wheat, and proteins associated with the development of freezing tolerance has been suggested as a potential tool for selecting freezing-tolerant plants in the *Gramineae* (Houde et al., 1992). Crosatti and colleagues (1996) suggested that the threshold induction temperature of COR14a, a cold-regulated protein, could be a biochemical marker for the identification of frost-resistant barley genotypes since these authors found a relationship between that threshold and the loci controlling winterhardiness.

Lee and colleagues (1995) suggested the use of changes in ABA and putrescine levels as physiological markers for the chilling tolerance of rice seedlings. Fatty acid composition is also related to cold tolerance and has been proposed for use in the evaluation of cold tolerance in rice (Majumder et al., 1989; Bertin and Bouharmont, 1997; Bertin et al., 1998).

Several reports suggest that soluble sugars and oligosaccharides combined with proteins are involved in frost tolerance. There is a relationship between carbohydrates and cold hardiness in fruit crops (Palonen and Buszard, 1997) and annual bluegrass (Dionne et al., 2001). Humphreys (1989) concluded that the amount of water-soluble carbohydrates maintained over the winter appeared to be important for winter survival in perennial ryegrass. Significant correlations between cold damage and the concentration of sugars were found in the roots of various forage legumes grown in the field and subjected to natural frost (Ratinam et al., 1994b). Fowler and colleagues (1981) also found a strong association between cold hardiness in winter wheat and sucrose content. On the other hand, frost resistance appears not to be significantly associated with Brix degree or sucrose or reducing sugar contents in sugarcane, although it is correlated with fiber content (Tai and Miller, 1996). Gusta and colleagues (1996) summarize several investigations showing that the sugars associated with winterhardiness are fructosans in grasses, raffinose in *Ajuga reptans,* and polyhydric alcohols in *Gardenia* and *Sorbus*. The relationship between sugars and cold tolerance has been known for many years (see Dexter, 1933). Further, sugar solutions in association with heat-stable proteins function in concert to protect plants against stress (Gusta et al., 1996).

In addition to providing a more detailed insight into the genetics of quantitative traits, molecular markers could be used in marker-assisted selection or in screening possible sources for cold tolerance. The gene *T/t,* which controls pubescence color in soybean seeds, could be used as a marker for chilling tolerance since genotypes with the *T* allele show no browning around the hilum region, show partially suppressed seedcoat cracking, and also produce a greater yield after chilling stress during flowering (Takahashi and Asanuma, 1996). However, when genotypes with yellow hilum and gray pubescence *(tI)* are preferred, alleles for the four maturity genes, E_1, E_5, e_3, and e_4, could be useful in reducing seedcoat pigmentation and cracking (Takahashi and Abe, 1999). In a previous study, selection for delayed maturity was pointed out as an indirect method for improving soybean chilling tolerance (Takahashi and Abe, 1994).

Grossnickle and colleagues (1996) proposed the relative abundance of interior spruce nuclear ribosomal RNA in seedlings from hybrids between Sitka and interior spruce as a marker for freezing tolerance. Storlie and colleagues (1998) suggested the use of the probe *Xwg644* as a marker for vernalization and cold hardiness in wheat, and since the level of *Fcor1* transcript accumulation correlated with freezing tolerance in strawberry cultivars, *Fcor1* was suggested as a useful molecular marker for freezing tolerance in related species of the *Rosaceae* family by Ndong and colleagues (1997). A RAPD marker has also been suggested for screening germplasm resources of lentil for radiation frost injury (Eujayl et al., 1999).

In Vitro Selection

In vitro selection has been used for obtaining cold-tolerant cells for several crops (Dix and Street, 1976; van Swaay et al., 1985), although most reports fall mainly in the field of physiology. It has been used for speeding up screening when using in vitro media with specific constituents and for overcoming the limitations of seed production. This method of selection may work if the mechanisms of cold tolerance in both plant and tissue cultures are the same, as found by Bertin and Bouharmont (1997). Selection at the cell level is based on the hypothesis that mitosis of cold-tolerant mutants is less inhibited by suboptimal temperatures than in the wild type (Preil et al., 1991). The efficiency of regenerating cold-tolerant mutants of *Chrysanthemum molinifolium* Ramat. with and without in vitro cold tolerance was assessed by Huitema and colleagues (1989). These authors used in vitro selection for detecting cold tolerance in mutants developed by irradiation of cell suspension cultures and found that the method of harvest had a great effect on the number of plants obtained. While investigating the effect of in vitro cold treatment on cold-tolerant mutant selections of *Chrysanthemum molinifolium,* Preil and colleagues (1991) found that low temperature significantly reduced the regeneration of callus. Huitema and colleagues (1991) tested the efficiency of using preselection in vitro at low temperature. They concluded that this preselection step had no clear advantage and recommended the slow regeneration of callus.

Proline and ABA induce chilling tolerance in many species (Rikin et al., 1979). The addition of proline to culture media increases chilling tolerance of callus (Duncan and Widholm, 1987) and cultured maize cells (Songstad et al., 1990). ABA and proline were found to have induced cold tolerance by different mechanisms in maize (Xin and Li, 1993). Tantau and Dörffling (1991) used hydroxyproline as a constituent of in vitro media for selecting a spring wheat variety more frost tolerant than the wild type. By the same procedure, Dörffling and colleagues (1993, 1997) obtained frost-tolerant inbreds of wheat.

Interspecific hybridization has been possible with advances in protoplast fusion and can be used to overcome the obstacles of interspecific transmission of cold tolerance. Protoplast fusion between unrelated species overcomes sexual barriers to gene transfer and induces genetic variability in plastid and mitochondrial genomes. Protoplast fusion is currently being used with *Brassica* species in interspecific crosses, but most of the literature on this is not related to cold tolerance. The relationship between protoplast fusion and cold tolerance appeared in *Brassica* species because protoplast fusion has been used to transmit cold tolerance from tolerant parents to susceptible cultivars with male-sterile cytoplasm. The male-sterile Ogura cytoplasm obtained from a variety of *Raphanus sativus* L. becomes chlorotic below 15°C (Walters et al., 1992). This is due to an unfavorable interaction between genes of the *Raphanus* chloroplast and *Brassica* nucleus. This cytoplasm, following its conversion to a cold-tolerant form, has been transferred to several *Brassica* species via protoplast fusion and used in *Brassica* breeding (Heath et al., 1994; Heath and Earle, 1996). Other male-sterile cytoplasms obtained from wild *Brassica* have been found to be cold tolerant, for example, Anand cytoplasm (Cardi and Earle, 1997).

Some researchers have tried to fuse the protoplasts of different genera. Kisaka and Kameya (1998) attempted to make somatic hybrids between barley and carrot, and they had some success at the callus level. The perspectives of protoplast fusion are restricted to a few crops in which it is possible, and it has been practiced in *Brassica* and potato. The material produced can suffer somatic incompatibility that limits growth and agronomic performance, and can also produce sterile hybrids (Bastia et al., 2000). Besides technical limitations, however, this methodology faces other problems such as genetic instability and somatic incompatibility. Male sterility of hybrids often limits the use of wild *Solanum* species in breeding for cold tolerance (Tucci et al., 1996). The usefulness of protoplast fusion is related to the effect of the cytoplasm on a particular trait. In the previous examples, the cytoplasm seems to play a significant role in the expression of the trait, but it has no significant effect on freezing tolerance in wheat (Gullord et al., 1975), rye, or triticale (Limin and Fowler, 1984).

Somatic fusion has been used in potato as a way of transferring freezing tolerance from wild germplasm to cultivated species (Chen, Bamberg, and Palta, 1999). The main problems in potato breeding programs have been the polyploid nature of the species and the crossing barriers. The cultivated potato, *Solanum tuberosum,* is a vegetatively propagated species with tetrasomic inheritance and a high level of heterozygosity. In contrast, most of the wild *Solanum* species are diploid and are a significantly greater source of allelic diversity than standard cultivars (Carputo et al., 1997). Interspecific gene transfer in potato has been made possible by the recent development of somatic hybridization (Cardi et al., 1993) and the combination of doubling of the ploidy level and producing 2*n* gametes (Carputo et al., 1997). In potato, somatic hybrids between diploid and tetraploid potatoes can be selfed or crossed with tetraploid cultivated potatoes for transferring cold tolerance from wild to cultivated potatoes (Chen, Palta, and Bamberg, 1999). Several authors have obtained somatic hybrids in potato with cold tolerance. For example, Preiszner and colleagues (1991) obtained frost-resistant cultivars by interspecific somatic hybridization between potato (*Solanum tuberosum* L.) and *S. brevidens*. Nyman and Waara (1997) and

Bastia and colleagues (2000) produced somatic hybrids between *Solanum commersonii* and *S. tuberosum* and found some interspecific hybrids with cold tolerance coming from the wild parent while the appearance was similar to the cultivated *S. tuberosum*. Similarly, Chen, Palta, Bamberg, and colleagues (1999) obtained freezing-tolerant somatic hybrids between *S. commersonii* or *S. brevidens* and *S. tuberosum*. In tomato, Brüggemann and colleagues (1995) obtained chilling-tolerant varieties by using somatic hybridization between *Lycopersicon esculentum* and *L. peruvianum*. Galiba (1994) concluded that somaclonal variation and cryoselection can be useful tools for increasing the genetic variability of frost tolerance in wheat.

In vitro selection has not been widely adopted by plant breeders working on cold tolerance, probably because it provides no advantage over traditional methods. Requirements of equipment, complexity, and price can be cited among the limitations of in vitro selection. Several species undergo in vitro selection very poorly. In species in which this technique can be used, such as rice, its success depends on the original genotype (Bertin, Busogoro, et al., 1997). Callus cold stress must be adapted in each variety to its most efficient; the higher the varietal tolerance, the harder the callus chilling treatment (Bertin and Bouharmont, 1997). According to Galiba (1994), the hardiness reached after cold acclimation, at both the intact plant and callus levels, is comparable, but in vitro selection in its present state is not recommendable in wheat breeding programs designed to obtain frost-resistant germplasm.

Engineering Cold Tolerance

Once the major genes controlling cold tolerance are identified and cloned, they can be transferred to other organisms. Nevertheless, the multigenic nature of cold tolerance is a deterrent to attempts to improve cold tolerance through genetic engineering, owing to the difficulty of transferring large numbers of genes. This complex characteristic, however, appears to be under the regulation of a few major genes (Fowler, Chauvin, et al., 1996) as demonstrated in *Arabidopsis thaliana* (Jaglo-Ottosen et al., 1998). It is known that in this species CRT/DRE (the C-repeat/dehydration responsive element) stimulates *COR* (cold-regulated) genes in response to low temperatures, drought, and salinity (Yamaguchi-Shinozaki and Shinozaki, 1994). Jaglo-Ottosen and colleagues (1998) have demonstrated that constitutive overexpression of the locus *CBF1* (CBF1 is CRT/DRE binding factor 1) in transgenic *Arabidopsis* triggers the expression of CRT/DRE-controlled *COR* genes without exposure to low temperatures. Moreover, other studies have revealed that all cereals, cold tolerant as well as cold sensitive, have homologues of low-temperature-regulated wheat genes, but that these are expressed only in tolerant cereals. In wheat, differences in gene expression are closely correlated with differences in freezing tolerance (Danyluk et al., 1994; Sarhan and Danyluk, 1998). Therefore, Fowler and colleagues (1999) suggested that the key to improving cold tolerance in chilling-sensitive species depends on the primary regulation of endemic genes rather than on the introgression of exotic structural genes.

A preliminary study on the overexpression of the *Arabidopsis CBF* genes in canola *(Brassica napus)* is mentioned by Thomashow (2001). Overexpression increased freezing tolerance in nonacclimated and cold-acclimated canola plants. The suitability of the CBF cold-acclimation pathway in other plants should be determined and *CBF* transgenes used to increase low-temperature tolerance in other crops. But there could be other, independent cold-acclimation signaling pathways, as suggested by Xin and Browse (1998). This is deduced from the behavior of *esk1* mutants. These plants do not overexpress CRT/DRE-regulated cold-responsive genes but do increase freezing tolerance in nonacclimated and cold-acclimated plants compared to the wild-type.

The first attempts at improving cold tolerance through genetic engineering tried to introduce flounder (*Pseudopleuronectes americanus* L.) antifreeze protein genes into plant cells (Georges et al., 1990; Hightower et al., 1991). The resulting transgenic plants marginally increased freezing tolerance. Most genetic engineering for cold tolerance has been focused on genes whose role in cold tolerance is not well-known. Comparative biochemical approaches have suggested a possible role for membrane fatty acid composition in chilling tolerance, as well as for metabolized reactive oxygen intermediates, increased sugar levels, the accumulation of solutes that keep the cytoplasm in osmotic equilibrium, and heat-stable proteins (Bartels and Nelson, 1994; Gusta et al., 1996; Thomashow, 1999). The genes involved in these biochemical processes have been isolated and cloned. Papers concerning the suitability and usefulness of transferring some of these genes to other plant species are now being published.

Several investigations have evaluated the potential of some genes involved in lipid desaturation for engineering cold-tolerant plants. Murata and colleagues (1992) transformed tobacco with glycerol-3-phosphate acyltransferase from either a chilling-sensitive plant (squash) or a chilling-tolerant species *(Arabidopsis)*. This gene is responsible for the level of fatty acid desaturation of phosphatidylglycerol. The degree of chilling sensitivity of tobacco plants transformed with squash and *Arabidopsis* enzyme transformants were similar to those of squash and spinach, respectively. The cDNA for oleate-selective glycerol-3-phosphate acyltransferase from *Arabidopsis thaliana* was also introduced into rice, and the net photosynthetic rate increased by 20 percent compared with the wild-type (Yokoi et al., 1998). Kodama and colleagues (1994) obtained tobacco transgenic plants containing the *Arabidopsis FAD7* gene. These plants showed an increase in the conversion of dienoic to trienoic fatty acids in membrane lipids by 10 percent and reduction in chlorosis after a seven-day exposure to 1°C. From these results, it appears that normal leaf development at low temperatures could be a consequence, in part, of increased trienoic fatty acid synthesis in leaf tissues during chilling acclimation (Kodama et al., 1995). However, transgenic tobacco plants with increased 18/3 content obtained through the introduction of a construct with a tobacco microsomal ω-3 fatty acid desaturase gene did not improve chilling tolerance (Hamada et al., 1998). The disagreement between them could be the consequence of the different target-membrane systems considered. Ishizaki-Nishizawa and colleagues (1996) transformed tobacco plants with the *des9* gene from *Anacystis nudulans,* which reduced the level of saturated fatty acids and 16/0 phosphatidylglycerol in the membrane. The *des9*-expressing tobacco showed no chlorosis when treated for 11 days at 1°C.

Another way to engineer cold-tolerant plants is to introduce genes that increase leaf-sugar concentration (von Schaewen et al., 1990; Sonnewald, 1992; Hincha et al., 1996). Although different transgenes produce tobacco transgenic plants with similar phenotype, when these are exposed to cold stress they differ significantly in their response (Hincha et al., 1996). This could mean that the successful transgene, in addition to increasing the concentration of soluble carbohydrates, leads to the induction of other genes and that their products contribute to cold tolerance. The transgenes for increased sugar concentration do not, therefore, appear to be directly involved in cold tolerance.

Another successful way to obtain transformants with increased levels of cold tolerance has been the introduction of genes that have a protective role under oxidative stress. Sen Gupta and colleagues (1993) and McKersie and colleagues (2000) have had promising results transforming tobacco and alfalfa with superoxide dismutase (SOD) cDNAs. These authors demonstrated that neither subcellular targeting nor SOD type had any effect on field survival. Two transgenic maize lines overexpressing Mn-SOD in chloroplasts showed greater cold tolerance than did nontransgenic lines (Stassart et al., 1995). These results show that SOD is a critical component of the active-oxygen-scavenging system and that modification of its expression can improve plant cold tolerance. It is suggested that overexpression reduces secondary injury symptoms and thereby enhances recovery from cold stress (McKersie et al., 2000).

Several studies have addressed the introduction of the gene *CodA* from *Arthrobacter globiformis* in an attempt to improve plant performance under cold stress (Hayashi et al., 1997; Alia et al., 1998; Hayashi et al., 1998; Sakamoto et al., 2000). This gene is responsible for the synthesis of glycine-betaine, one of the compatible solutes that certain halotolerant plants accumulate in the chloroplasts when exposed to salt or cold stress. Genetic modifications of *Arabidopsis thaliana* to encourage the accumulation of glycine-betaine enhanced the ability to tolerate cold and salt stress during imbibition, germination, and early growth (Alia et al., 1998). In rice transformed with *Arabidopsis* carrying *CodA,* the recovery of photosystem II from the photoinhibited state was accelerated (Hayashi et al., 1998; Sakamoto et al., 1998). When the transgene product was targeted to the chloroplast, cellular damage by cold stress was less than when targeted to the cytosol. This stresses the importance of engineering for the subcellular compartmentalization of the biosynthesis of glycine-betaine (Sakamoto et al., 1998).

Genetically modified organisms (GMOs) could also be developed to obtain better cold acclimation. As this depends partly on ABA-mediated processes, evaluations have been made of the ectopic expression of *ABI3*. This influences some ABA-mediated vegetative responses (Tamminen et al., 2001). Transgenic plants of *Arabidopsis* acclimated faster than wild-type plants, and the maximum tolerance obtained was significantly better. In another study, *Arabidopsis* plants overexpressing COR15 protein (*Cor15* is a cold-regulated gene) showed improved PSII photochemical efficiencies compared to wild-type plants after freezing at –4 and –5 °C (Artus et al., 1994).

The enhancement of cold tolerance achieved by the genetic engineering of plants with cold- or dehydration-responsive genes has been limited. The complexity of cold tolerance means that the overexpression of a single gene results in only small or even undetectable increases in cold tolerance. However, as already pointed out, a deeper knowledge on regulator genes could provide the tools necessary for successfully transferring freezing tolerance to chilling-sensitive plants. According to Pearce (2001), a promising strategy may be to confer tolerance to cellular dehydration.

Gametophytic Selection

Natural selection occurs during pollen germination and pollen tube growth (Mulcahy, 1975). Pollen competitive ability (mainly pollen tube growth rate) correlates with sporophytic traits in maize (Mulcahy, 1971). Gametophytic selection can be efficient for improving cold tolerance since most of the genes expressed in the sporophyte are also expressed in the gametophyte, and pollen selection allows the screening of large numbers of individuals without the confounding effects of dominance and epistasis (Mulcahy, 1979). Pollen can be selected for pollen germination or pollen tube growth since cold reactions are controlled by different sets of genes in the two phases.

Several studies show that pollen selection is useful in breeding for adaptation to specific environments. Zamir and colleagues (1981) demonstrated that pollen from cold-tolerant genotypes pollinated more efficiently than pollen from sensitive plants. Zamir and colleagues (1982) also demonstrated that gametophytic selection for low-temperature tolerance of tomato pollen is determined, at least in part, by genes expressed in the haploid pollen. The usefulness of pollen selection for cold tolerance using intraspecific crosses among tomato varieties with different growth capacities was tested, and it was found that pollen selection was no more efficient than sporophyte breeding (Maisonneuve et al., 1986). In another study, Kovacs and Barnabas (1992) found that pollen germination characteristics correlated strongly with the germination of kernels at low temperature. However, they obtained inconsistent results in successive selection programs involving selection of pollen germination under cold conditions, since female gametes cannot be selected in the same way. To overcome this limitation, they suggested combining sporophytic and gametophytic selection or repeated pollen selection during the inbreeding process. Selection programs for pollen cold tolerance have been successful in the breeding of potatoes (Kristjansdottir, 1990), maize (Frascaroli, 1995), *Brassica* species (Kalinova et al., 1998; Lyakh et al., 1998), chickpea (Lawlor et al., 1998), and rice (Pattanayak et al., 2000).

Kovacs and Barnabas (1997) used anther culture to obtain haploid wheat embryos screened for frost tolerance, and released some double haploid lines resistant to frost. Nyman and Waara (1997) proposed the use of repeated cycles of anther culture and somatic hybridization. Tantau and colleagues (1996) obtained frost-tolerant winter barley cultivars using in vitro selection of hydroxyproline-resistant lines derived from anther culture. Gupta and colleagues (1996) incorporated cold tolerance in rice through anther culture. Gametophytic selection began during the last quarter of the twentieth century, but the number of reports concerning its use has been small, perhaps because it does not simplify or improve the usual evaluation of seed germination.

CONCLUSION

Enough variability seems to be present within most species to allow effective breeding programs without the use of genetically distinct germplasm. The genetic regulation of cold tolerance allows successful breeding programs using simple designs, such as mass selection, although recurrent selection programs are recommended. Selection criteria should be clearly defined at each growth stage, considered as independent traits, and improved using a multiple-trait selection procedure. A combination of laboratory and field evaluations should increase the possibilities of success and the rate of improvement in breeding programs for cold tolerance. The most efficient indirect selection criteria are electrolyte leakage and chlorophyll fluorescence. Alternative methods related to in vitro culture have been successful in a few crops. Genetic transformation is a promising methodology for improving cold tolerance beyond the variability available within the species. Further improvements require a better knowledge of regulatory genes.

APPENDIX: SOURCES OF GERMPLASM FOR COLD TOLERANCE

Crop	Germplasm	Tolerance	Evaluation traits	Test	Germplasm conclusions	Reference
Actinidia spp.	Two-year-old vines from several *Actinidia* species	Freezing	Plant survival, growth, and stem necrosis	Artificial freezing	Three species were more tolerant than *A. deliciosa* indicating their potential for improvement of cold hardiness through interspecific hybridization.	Chat, 1995
Alysicarpus vaginalis, Desmodium heterocarpon (pasture legumes)	Seven accessions of *Desmodium* and 40 of *Alysicarpus*	Winterhardiness	Visual rating	Field evaluations	Some *Desmonium* plants survived and provide a potential source of winterhardiness for development of a perennial warm-season pasture legume.	Pitman et al., 1997
Avena sativa (oat)	Four winter and three spring cultivars	Frost			The winter oat cultivars performed better than spring ones. Significant differences were detected among the winter-type group.	Cattivelli et al., 1995
Brassica napus (rape)	Cultivars of European winter oilseed rape	Winterhardiness Frost resistance			In the past 20 years winterhardiness of double low cultivar has been improved.	Rapacz and Markowski, 1999

Crop	Germplasm	Tolerance	Evaluation traits	Test	Germplasm conclusions	Reference
Brassica napus, Brassica rapa (rapeseed)	Ten cultivars	Winter survival	Ion leakage from leaves exposed to freezing temperatures, winter survival	Laboratory and field	Freezing tolerance increased after acclimation. Cultivars differed for freezing tolerance before and after acclimation. In vitro freezing tolerance after acclimation was correlated with winter survival.	Teutonico et al., 1993
Brassica napus, Brassica rapa	Winter cultivars of two *Brassica* species	Freezing and ice in leaves	Leaf dead after freezing and ice	Controlled environment and field evaluations	Some cultivars of *B. napus* and *B. rapa* showed cold-stress tolerance.	Andrews and Morrison, 1992
Brassica rapa, Brassica oleracea, Brassica napus	One rape, three turnip, and four kale cultivars	Low temperature in seedling emergence	Total emergence	Growth chamber	Differences in total emergence were observed among cultivars even though seed germination potentials of all cultivars were similar.	Rao and Dao, 1987
Cajanus cajan (pigeonpea)	31 genotypes	Low temperatures at germination	Percentage germination and seedling emergence	Laboratory and field	Six cold-tolerant genotypes were identified.	Kumar and Shukla, 1993
Cajanus cajan (pigeonpea)	453 genotypes	Cold	Bud and flower drop	Field evaluations	Fourteen accessions exhibited tolerance and 23 accessions were moderately tolerant.	Singh et al., 1997

Casuarina cunninghamiana	130 open-pollinated from 13 provenances	Frost tolerance	Frost injury rating	Field evaluations	The frost tolerance of *C. cunninghamiana* can be improved by proper choice of provenance and selection within provenance.	Merwin et al., 1995
Cenchrus ciliaris (buffelgrass)	Seedling of five genotypes	Winterhardiness	Field winter survival and yield potential	Field evaluations	Two cultivars had greater field survival than the others but were less winter hardy than the resistant check.	Hussey and Bashaw, 1996
Cicer arietinum (chickpea)	3,276 germplasm accessions and breeding lines	Cold tolerance for winter sowing	Plant damage	Field evaluations	Twenty-one lines were identified as tolerant.	Singh et al., 1989
Cicer arietinum (chickpea)	20 chickpea, under three levels of nitrogen	Cold tolerance for winter sowing	Plant damage	Field evaluations	The difference in the cold susceptibility of genotypes was not as large without N. Three genotypes were tolerant at all levels of N, while three were tolerant without N and susceptible with N.	Malhotra et al., 1995
Cicer arietinum *Cicer* spp. (chickpea)	4,284 kabuli lines, 2,137 desi lines, 857 breeding lines, 59 lines of seven wild *Cicer* species	Cold tolerance for winter sowing	Plant damage	Field evaluations	Eleven kabuli but no desi germplasm lines, and 14 breeding lines were tolerant. Among wild species some lines showed cold tolerance.	Singh et al., 1995

Crop	Germplasm	Tolerance	Evaluation traits	Test	Germplasm conclusions	Reference
Cicer arietinum (chickpea)		Cold tolerance			Registration of three kabuli chickpea germplasm lines	Singh et al., 1997
Cicer spp. (chickpea)	137 lines of eight wild species	Cold tolerance for winter sowing	Plant damage	Field evaluations	Most of the lines of three wild *Cicer* species were tolerant to cold and had significantly higher levels of cold tolerance than cultivated species.	Singh et al., 1990
Citrus sinensis (sweet orange)	Sweet orange trees on cold-hardy trifoliate orange and cold-sensitive citron	Freeze	Freeze survival	Temperature-programmed walk-in freezer room	Nonhardened trees generally did not survive the freeze, whereas cold-hardened trees survived with no wood kill on either rootstock.	Yelenosky and Vu, 1992
Cornus florida	Trees from 15 states	Cold hardiness	Autumn color, flowering, presence of stem cankers, and cold hardiness	Field evaluations	Use of northern seed sources should provide hardier trees with better color displays in northern landscapes.	Heatley et al., 1994
Cucumis sativus (cucumber)	Nine cultivars and breeding lines	Chilling at the first state	Chilling resistance of seedling at cotyledon and first true leaf-growth stages	Controlled-environment chambers	Variation was larger at the first leaf stage than at the cotyledon stage. Differences among genotypes in chilling damage were largest five days after chilling.	Smeets and Wehner, 1997
Cynodon dactylon (Bermudagrass)	Clonally propagated plants of six varieties	Freeze	Regrowth, lowest survival temperatures	Greenhouse, controlled-environment chamber, and low-temperature freezer	Varieties varied in freeze tolerance.	Anderson et al., 1993

Cynodon spp. (Bermudagrass)	Five cultivars	Freeze	Lowest survival temperatures	Laboratory	The differences in freeze tolerance among the cultivars tested indicates substantial variation in the species.	Anderson and Taliaferro, 1995
Chamaecyparis nootkatensis (yellow cedar)	Seedlings of three populations	Frost hardiness	Freeze-induced electrolyte leakage on upper primary branches	Samples from the field	Seedlings of parents from certain families and populations developed greater hardiness than stock of parents from other area.	Hawkins et al., 1994
Fragaria (strawberry)	11 cultivars	Flowers' freeze resistance	Freezing-point depression of expressed cell sap and flower size	Laboratory freeze test	Flowers of all cultivars are susceptible to freeze damage and survive spring frost by freeze avoidance.	Hummel and Moore, 1997
Glycine max (soybean)	20 genotypes of different origins	Low night temperatures	Seed yield, above-ground dry matter, apparent harvest index, seed weight, and days to maturity	Field evaluation	Genetic variations for adaptation to cool nights were identified for all traits.	Seddigh et al., 1988
Gossypium hirsutum (cotton)	Four genotypes	Cool temperature during seedling growth	Root length	Controlled environment and field evaluations	Measuring the amount of seedling root-length inhibition caused by suboptimal temperature may be useful for determining cold sensitivity in cotton.	Bauer and Bradow, 1996

Crop	Germplasm	Tolerance	Evaluation traits	Test	Germplasm conclusions	Reference
Hebe spp.	35 *Hebe* species, cultivars, and hybrids	Frost hardiness	Frost hardiness temperature and lethal temperature	Field evaluations	Varieties showed frost-hardiness differences.	Warrington and Southward, 1995
Hordeum sp.(barley)	109 hull-less barley entries	Cold	Plant survival	Field evaluations	Four highly cold-tolerant lines were identified.	Tahir and Shevtsov, 1994
Hordeum sp.(barley)	ICARDA's winter barley	Cold	Plant survival	Field evaluations	Some cold-tolerant lines were identified.	Tahir et al., 1994
Hordeum sp.(barley)	Exotic barley genotypes	Cold	Frost/cold mortality	Field evaluations	Exotic germplasm survived under minimum air temperatures of –6 to –11°C.	Khan et al., 1997
Lens culinaris (lentil)	23 large seed exotic lentil	Winterhardiness	Seed yield, harvest index, and 100 seed weight	Field evaluations	The local landrace was more winter hardy than the exotic genotypes.	Asghar et al., 1994
Lens culinaris (lentil)	245 accessions of wild lentil, 10 of cultivated lentil, and three accessions of *Vicis montbretti*	Late autumn sown	Percentages of survival plants and visual damage rating	Field evaluations	Accessions of L. *culinaris* ssp. *orientalisk* exhibited the highest level of winterhardiness, on average. Correlation revealed that winterhardiness was concentrated among accessions originating from high-elevation areas.	Hamdi et al., 1996
Linum usitatissimum (flax)	Several varieties	Winter tolerance of varieties sown in the fall	Branched habit of growth, plant height, and plant and tissue age	Field evaluations	Four varieties were the most hardy.	Dillman, 1941

Lolium multiflorum (Italian ryegrass)	Nine varieties of varying origin	Cold and freezing in tiller	Tiller mortality	Cold room	Varieties varied in tolerance from 31.1 to 71.4 percent of tiller mortality.	Hides, 1979
Lolium perenne (English ryegrass)	Seven varieties	Freezing in leaves	Electrolyte leakage	Laboratory	Relatively small differences were observed in freezing tolerance among six varieties. Only one was susceptible.	Bredmeijer and Esselink, 1995
Lolium perenne (English ryegrass)	12 cultivars	Freezing	Percentage of mortality and impedance spectrum	Laboratory	Two cultivars had the least mortality.	Repo and Pulli, 1996
Lolium perenne (perennial ryegrass)	86 accessions	Winter hardiness	14 winter-related characters	Field evaluations	None of the accessions exhibited complete winterhardiness.	Humphreys, 1989
Lolium perenne (ryegrass)	Six diploid and five autotetraploid perennial ryegrass cultivars	Cold	Competitive ability and cold tolerance	Field evaluations	The tetraploid cultivars had significantly lower cold tolerance. Chromosome doubling has a larger effect on competitive ability and cold tolerance.	Sugiyama, 1998
Lycopersicon esculentum (tomato)	Two cultivars	Chilling during seedlings grown	Several growth parameters and amount of leaf pigment	In vivo and in vitro	The relative resistance shown by the two cultivars during in vivo growth was maintained in vitro.	Sebastiani et al., 1996

Crop	Germplasm	Tolerance	Evaluation traits	Test	Germplasm conclusions	Reference
Lycopersicon spp. (tomato)	30 accessions representing six *Lycopersicon* species	Low temperature at germination	Rate of seed germination	Laboratory	Significant differences were found among accessions for cold tolerance. Correlation between germination under control and cold stress was high.	Foolad and Lin, 1999
Lycopersicon spp. (tomato)	31 accession-representing six *Lycopersicon* species	Low temperature at germination and during vegetative growth	Germination tolerance index and growth tolerance index	Laboratory and growth chambers	Genotypic variation for cold tolerance during germination and vegetative growth was found.	Foolad and Lin, 2000
Lycopersicon spp.(tomato)	170 accessions of *S. esculentum* and related species	Pollen tolerance to low temperature	Pollen fertility and pollen shedding		*L. esculentum* cultivars showed poor pollen release. There were wild species with fertile pollen at low temperatures.	Fernández-Muñóz et al., 1995
Malus domestica (apple)	90 scab-resistant cultivars	Low-temperature injury	Injury symptoms	Field evaluations	Twenty-seven cultivars exhibited no terminal die-back injury.	Warner and Nickerson, 1996
Olea europaea (olive)	Four clones belonging to cv. Leccino	Frost hardiness	Ionic leakage		Variability was found within the clones of 'Leccino.' Only one clone showed difference between hardiness at −15 and −20°C.	Porta et al., 1994
Onobrychis sativa (sainfoin)	34 local varieties from France and 17 accessions from other countries	Winterhardiness	Seven agronomic traits	Field evaluations	The local varieties showed better winter growth, a quick start in the spring, and an earlier flowering time while keeping a good level of frost resistance.	Prosperi et al., 1994

Oryza sativa (rice)	16 varieties	Chilling during germination and vegetative growth	Germination, plantlet chilling survival, electrolyte leakage, and low-temperature fluorescence	Controlled environment	Weak correlations were observed between low-temperature germination and plantlet chilling survival.	Bertin, Busogoro, et al., 1996
Oryza sativa (rice)	50 *japonica* rice cultivars	Germination at low temperature	Germination ability of seed	Laboratory	Five cultivars showed cold tolerance.	Tanida, 1996
Oryza sativa (rice)	117 cultivars of *O. japonica* and *O. indica*	Cold	Chlorosis	Laboratory	Considerable diversity existed within groups. *Japonica* cultivars had better seedling-stage cold tolerance.	Mackill and XiaoMao, 1997
Oryza sativa (rice)	18 hybrids	Cold	Cold tolerance yield	Cold conditions in laboratory, yield in normal field conditions	Two hybrids showed good performance.	Kennedy and Rangasamy, 1998
Paspalum vaginatum (paspalum)	Three ecotypes of seashore paspalum	Low temperature	Electrolyte leakage and freeze-shock recovery	Controlled freezing	Bermudagrass, used as control, demonstrated superior cold hardiness than seashore paspalum. The paspalum ecotypes showed differences.	Cardona et al., 1997
Persea americana (avocado)	Trees of three varieties	Freezing in leaves	Electrolyte leakage and visual appearance of browning in leaves	Field evaluations	The plants growing at lower temperature had lower lethal freeze temperature. Varieties varied in freeze tolerance.	McKellar et al., 1992

Crop	Germplasm	Tolerance	Evaluation traits	Test	Germplasm conclusions	Reference
Phaseolus vulgaris (common bean)	14 cultivars/lines	Low temperatures at germination	Germination percentages and germination rates	Germination test	Differences in germination were significant among cultivars at low temperature.	Zaiter et al., 1994
Phaseolus vulgaris (French beans)	24 varieties	Low temperature	Plant morphology, vigor, and health	Growth chamber	Four varieties showed cold tolerance.	Holubowicz and Legutko, 1995
Phleum pratenses (timothy)	Clones from six populations	Winterhardiness	General performance, winterhardiness	Field evaluations	Significant genetic variations were found for winterhardiness within all six populations.	Schjelderup et al., 1994
Picea abies (Norway spruce)	Three clones	Freezing	Chlorophyll fluorescence	Artificial freezing	Differences were observed among clones, indicating clonal differences in cold hardiness and hardiness development.	Westin et al., 1995
Picea glauca (white spruce)	Trees of different origin	Cold hardiness of dormant vegetative buds	Bud hardiness, foliage, and stem hardiness using electrolyte leakage	Differential thermal analysis in an arboretum	Buds of trees from northerly origins were more hardy in early autumn than buds of trees from a more southerly origin.	Simpson, 1994
Pinus sylvestris (pine)	18 plustree clones	Cold	Visual scoring of the proportion of discolored needle tissue	Artificial freezing	Clonal variation in needle cold tolerance was observed. Fluctuation in winter cold hardiness were considerable.	Nilsson and Walfridsson, 1995

Pisum sativum (pea)	14 genotypes of winter-hardy peas and six genotypes of spring peas	Winterhardiness	Percentage seedling survival	Field evaluations and evaluation in a programmable freezer	The winter-hardy lines showed differential survival. All spring pea cultivars failed to survive the winter.	Auld, Ditterline et al., 1983
Populus trichocarpa (poplar)	Clones from populations	Freezing in leaves	Leaf injury	Laboratory and field	Freezing tolerance varied among different populations and was dependent on growing environment.	McCamant and Black, 2000
Prunus	20 *Prunus* species	Flower-bud hardiness	Number of buds killed	Field evaluations	The two hardiest species belonged to the subgenus *Padus*.	Kadir and Proebsting, 1994
Prunus armeniaca (apricot)	27 genotypes	Freeze in shoot and flower buds	Temperature required to kill 50 percent flower bud and 50 percent shoot xylem	Controlled laboratory freezing test	Genotypes differed significantly for freeze tolerance and were placed in three hardiness classes.	Layne and Gadsby, 1995
Prunus persica (peach)	Review	Cold hardiness			A useful range of cold hardiness is present within the species that can and have been exploited to improve cold hardiness of peach and nectarine cultivars.	Layne, 1992
Prunus persica (peach)	One nectarine cultivar and five peach cultivars	Cold hardiness of stem tissue	Electrical conductivity	Samples from the field	Two peach cultivars were the hardiest.	Werner et al., 1993
Quercus petraea (oak)	16 European sessile oak from six European countries	Frost hardiness	Relative conductivity	Field evaluations	German, Polish, and Danish provenances were more frost hardy than French, Austrian, and English provenances.	Deans and Harvey, 1996

Crop	Germplasm	Tolerance	Evaluation traits	Test	Germplasm conclusions	Reference
Rhododendron sp. (rhododendron)	100 cultivars	Flower-bud hardiness	Flower-bud damage	Field evaluations	Cultivars were ranked for cold hardiness.	Gilkey, 1996
Rhododendron sp. (rhododendron)	Diploid cultivars and their tetraploid derivatives	Freezing in leaves and stem; flower-bud hardiness	Leaf and stem injury, and flower-bud injury	Controlled freezing test. Differential thermal analysis	Diploid cultivars showed more leaf-freezing tolerance. There was no difference in the stem hardiness between ploidy levels.	Väinölä and Repo, 1999
Rubus idaeus (red raspberry)	Cultivars and selections	Shoot cold hardiness	Tissue-cultured shoots injury	In vitro	Shoots subjected to stepwise acclimation at low temperatures showed significant differences among the four types of cultivars.	Zatylny et al., 1993
Rubus idaeus (red raspberry)	Four cultivars and six selections	Shoot cold hardiness	Tissue survival	Controlled freezing test and field evaluations	Genotypes differed in their relative hardiness throughout the winter.	Zatylny et al., 1996
Rubus idaeus, *R. neglectus* (raspberry), *R. allegheniensis* (blackberry)	80 raspberry and 42 blackberry genotypes and other *Rubus* species	Cold hardiness	Temperature when 50 percent of the buds, bud base, and cane were killed	Controlled freezing	Red, purple, and black raspberries had a lower T_{50} value that did blackberries. Several purple blackberries were quite cane hardy. The canes of most cultivars were about 2 to 15°C hardier than buds.	Hummer et al., 1995
Saccharum officinarum (sugarcane)	60 genotypes	Low temperature	Bud mortality index, plant kill index, and foliar tissue injury	Laboratory and field evaluations	Seven genotypes were resistant to freezing.	Singh and Kaur, 1993

Saccharum officinarum (sugarcane)	389 clones derived from 'US-87'	Frost	Percentages of green leaf tissue	Field evaluations	The gene pool of this population exhibited wide variation in frost resistance. Resistance to injury was positively correlated with fiber content.	Tai and Miller, 1996
Saccharum officinarum (sugarcane)	Large number of genotypes	Low temperature	Bud mortality index, plant kill index, and natural field reaction		Ten genotypes were determined to be low-temperature-stress resistant genotypes.	Singh and Singh, 2000
Salix viminalis (willow)	Six clones and two hybrid clones	Fall frost resistance	Stem and leaves composition	Laboratory	There were differences in fall frost resistance among the eight clones.	Ogren, 1999
Sorghum bicolor (sorghum)	380 accessions obtained from China, Ethiopia, and Uganda	Ability to germinate, grow, and produce satisfactory grain yields under cold air and soil temperatures	Stand and early vigor rating and percentage of seed set	Early sowing in several fields	Thirty-nine were cold-tolerant and 157 were partially tolerant.	Singh, 1985
Sorghum bicolor (sweet sorghum) *Sorghum sudanense* (Sudan grass)	28 genotypes	Chilling in leaves	Chlorophyll fluorescence	Laboratory	Some sweet sorghum varieties and most of the Sudan grass varieties appeared to possess a certain degree of resistance toward low temperature.	Havaux, 1989

Crop	Germplasm	Tolerance	Evaluation traits	Test	Germplasm conclusions	Reference
Solanum brevidens S. tuberosum	Three *S. brevidens* and one *S. tuberosum* cultivars	Frost	50 percent tissue leaching	Growth chamber	Two *S. brevidens* showed the greatest frost resistance.	Barrientos et al., 1994
Solanum spp.	2,600 accessions representing 101 species	Frost hardiness	Visual rating	Field evaluations	Five species were the most hardy species. An unreported extremely frost-hardy species *(S. paucissectum)* was discovered.	Vega and Bamberg, 1995
Solanum spp., *S. tuberosum* (potato)	Four wild *Solanum* species and eight potato cultivars	Freezing	Conductivity, 50 percent tissue leaching	Growth chamber	The material analyzed was classified into three groups: freezing resistant and able to acclimate to cold after chilling, freezing resistant but unable to cold acclimate, and freezing sensitive.	Irzykowski et al., 1996
Solanum tuberosum Solanum × juzepczupii (potato)	Potato and Andean hybrid cultivars	Low temperatures in leaves	The response of photochemical apparatus of photosynthesis	Laboratory	The photochemical activities of both species were similarly impaired by chilling treatments, indicating that frost resistance does not preclude susceptibility to photoinhibition damage at chilling temperature.	Havaux, 1995

Stenotaphrum secundatum (St. Augustinegrass)	14 genotypes	Freezing	Predicted lethal freeze temperatures	Field evaluations	Predicted lethal temperatures ranged from −7.7 to −4.7 °C.	Philley et al., 1995
Trifolium repens (white clover)	Four varieties	Winter survival and low temperature growth	Damage to stolon over winter. Rates of leaf expansion and leaf dimension	Field and temperature gradient tunnel	Two varieties showed two different mechanisms for cold tolerance.	Rhodes et al., 1994
Trifolium repens (white clover)	450 ecotypes from five landraces and two natural populations	Cold	Dry matter yield, stolon density, production of seed heads, and cold tolerance	Field evaluations	Landraces tended to somewhat greater variation for cold tolerance.	Annichiarico and Piano, 1995
Trifolium repens (white clover)	Several populations	Stolon winter hardiness	Lethal dose 50	Artificial freezing test	There is a considerable amount of genetic variation in cold tolerance within white clover.	Collins and Rhodes, 1995
Triticum aestivum (wheat) *Hordeum vulgare* (barley)	Wheat and barley genotypes	Winter survival	Field survival index Crown freezing tolerance	Laboratory and field evaluations	The winter barley genotypes were clearly more hardy than the spring genotypes but less hardy than the winter wheat genotypes.	Bridge et al., 1996
Triticum aestivum x *Secale cereale* (triticale)	Ten hexaploid triticale strains, six winter *durum* wheat, and four rye genotypes	Freeze	Percentage of survival plants	Phytotrons	The hardiness of the triticale strains was not always correlated with the hardiness of the wheat parents, but it depended to a great extent on hardiness of the rye parental genotype.	Maximov and Kostandi, 1995

Crop	Germplasm	Tolerance	Evaluation traits	Test	Germplasm conclusions	Reference
Triticum aestivum × Secale cereale (triticale)	Ten modern secondary triticale cultivars	Frost	–	Field evaluations and laboratory test	'Aubrac' and 'Torpededo' were identified as the most frost-resistant cultivars.	Rizza et al., 1997
Triticum aestivum, Secale cereale, Hordeum vulgare, Secale × Triticosecale	13 and 23 winter wheat, 10 and 13 rye, three and five triticale, and six and 11 winter barley cultivars	Leaf hardening ability during autumn, and winter-hardiness	Ion leakage from frost-treated leaf segments, field winter survival	Laboratory and field evaluations	Winter survival correlated well with hardening ability determined by electrolyte leakage method.	Hömmö, 1994
Vaccinium corymbosum (blueberry)	Seven southern highbush blueberry cultivars	Freeze in shoot and flower buds	Shoot and flower bud damage	Field evaluations and programmable freezer	Hardiness for several southern highbush clones was similar to that for northern highbush at several dates and temperatures.	Clarck et al., 1996
Vicia sativa (common vetch)	16 cultivars or experimental lines from different origins	Freezing	Survival rate	Field evaluations after freezing	Four varieties exhibited freezing resistance.	Acikgoz, 1982
Vigna unguiculata (cowpea)	Six pairs of closely related lines and eight genotypes	Chilling at germination	Leakage of electrolytes, rate, and percentage of germination	Chamber at low temperature	With chilling genotypes differed in rate but not in maximal percentage of germination.	El-Kholy et al., 1997

Vitis venifera (grape)	47 grape cultivars	Frost damage	Survival of buds and bark damage	Field evaluations	Significant differences in frost tolerance were detected within cultivars according to their nutrient supply and age. Some cultivars have excellent winter tolerance.	Hajdu and Gábor, 1997
Vitis vinifera (grape)	Three varieties	Frost	Yield/vine and yield/ha	Field evaluations	Frost damage did not cause yield loss in one variety.	Dobreva et al., 1995
Vitis vinifera (wine grapes)	Eight table and 18 wine grape cultivars	Winter-hardiness	Vine vigor and fruit set	Field evaluations	Four cultivars showed large number of grape clusters and little injury.	Kiyomoto, 1995
Zea mays (maize)	Strain of sweet corn and field corn	Low temperature at germination	Final germination	Cold room, warm greenhouse. Early sowing in the field	Resistance to cold depends primarily on the embryo genetical constitution rather than endosperm type.	Haskell and Singleton, 1949
Zea mays (maize)	Eight inbred lines, 11 hybrids, and fertile and male sterile lines	Low temperature at germination	Emergence	Cold room and field evaluations	Significant negative correlation was ascertained between the FAO number and emergence. Different cold tolerance was observed in fertile and sterile analogues of inbred lines.	Herczegh, 1970
Zea mays (maize)	34 inbred lines adapted to various latitudes in North America	Cold tolerance	Percentage emergence, emergence index, and seedling dry weight	Field evaluations	Cold tolerance was not related to geographical location of adaptation. Large amount of variability was observed for each of the three tolerance traits.	Mock and McNeill, 1979

Crop	Germplasm	Tolerance	Evaluation traits	Test	Germplasm conclusions	Reference
Zea mays (maize)	Ten cultivars	Low temperature at germination and early growing	Germination and shoot growth	Laboratory	Differences among cultivars were found for all analyzed traits.	Furter and van de Venter, 1990
Zea mays (maize)	Inbred lines	Low temperature at imbibition and germination	Electrolyte leakage after imbibition. Radicle emergence index on temperature gradient	Laboratory	Differences among inbreds were found for leakage. Two cold-tolerance mechanisms were postulated.	Zemetra and Cuany, 1991
Zea mays (maize)	28 hybrids	Cold injury	Yellowing, purpling, and drying of leaves 80 days after sowing	Field evaluations	Three hybrids performed well and suffered less from cold injury under poor environmental conditions.	Mahajan et al., 1993
Zea mays (maize)	Seven inbred lines	Chilling	Percentage of germination, percentage viability, and time to germination	Laboratory germination test and field evaluations in the early spring	Resistant inbred lines were detected in laboratory test and in field evaluations.	Hodges et al., 1994
Zea mays (maize)	Five populations	Low temperature	Percent emergence	Field evaluations	Populations with a high number of mesocotyl roots and root branching had higher percent emergence.	Rahman, et al., 1994

Zea mays (maize)	Eight inbred lines	Chilling responses at the early growth phase	Total, shoot, and root dry mass. Shoot and root water content	Growing plants hydroponically at a constant low temperature. Early sowing in the field	Differences among inbreds were found for chilling tolerance in the laboratory, and the ratio was correlated with field results.	Hodges et al., 1995
Zea mays (maize)	Two inbred lines and two maize populations, who differ in their cold tolerance at germination	Pollen cold reaction	Pollen grain germination and tube elongation	Low temperature applied to tassels	There was consistency between plant and pollen cold tolerance in maize.	Frascaroli, 1995
Zea mays (maize)	Polish genotypes	Chilling periods at seedling	Shoot and root growth	Growth chamber, first chilling period following a higher temperature period	The ability of maize cultivars of different chilling tolerance to regrow after a cold period was strongly correlated to the rate of increase in root length in the cool.	Sowínski and Królikowski, 1995
Zea mays (maize)	Inbred lines of different origins	Chilling	Rates of relative growth, relative area expansion, and assimilation rate	Field evaluations	Genotype variability existed for most of the growth parameters.	Verheul et al., 1996
Zea mays (maize)	Ten populations of field corn and 40 populations × sweet corn inbreds	Low temperature at germination and early state of development	Emergence proportion and score, plant growth, dry-shoot weight, survival, and plant vigor	Cold chamber	Two populations of field corn were the best potential donors to improve the sweet corn cold tolerance.	Revilla et al., 1998

Crop	Germplasm	Tolerance	Evaluation traits	Test	Germplasm conclusions	Reference
Zea mays (maize)	Three cultivars	Low temperature at germination	Electrical conductivity of electrolytic leakage	Laboratory	Differences among cultivars were found for leakage.	Singha et al., 1999
Zea mays (maize)	18 populations from Andean accessions	Low temperatures during heterotrophic and early autotrophic growth	Germination percentage, germination index, and plant growth rate	Controlled environment and in the field with early sowing	None of accessions was superior for all traits, but several Andean accessions outperformed the U.S. corn belt checks for one or more traits.	Brandolini et al., 2000
Zea mays (sweet corn)	24 genotypes with endosperm mutations sugary1 *(su1)*, sugary enhancer *(se1)* or shrunken2 *(sh2)*	Seedling emergence in cold spring soil	Percentage of emergence, vigor, and uniformity	Field evaluations	Data showed a significant positive correlation between kernel starch content and percentage field emergence. Variability among *se1* genotypes indicated that cold emergence could be improved to equal some of the best *su1* varieties.	Douglass et al., 1993
Zea mays (sweet corn)	35 open-pollinated sweet corn populations and controls	Cold in germination	Percentage and time to emergence, seedling color, seedling root, and shoot dry mass	Evaluations in growth chamber	Differences in cold tolerance existed among the entries. Some entries that performed well under low temperatures also performed well under warm conditions.	Hotchkiss et al., 1997

REFERENCES

Abbas Al-Ani, M.K. and R.K.M. Hay (1983). The influence of growing temperature on the growth and morphology of cereal seedling root systems. *Journal of Experimental Botany* 34:1720-1730.

Acikgoz, E. (1982). Parameters of cold tolerance in common vetch *Vicia sativa*, correlation between morphological and physiological characteristics of cultivars. *Euphytica* 31:997-1001.

Adak, M.S. and D. Eser (1993). Effect of fall growth and development on winter hardiness in barley. *Rachis* 12:11-14.

Aidun, V.L., W.N. Migus, and R.I. Hamilton (1991). Use of inbred seedling cold tolerance to predict hybrid cold tolerance in maize (*Zea mays* L.). *Canadian Journal of Plant Science* 71:663-667.

Ali, A. and D.L. Johnson (2000). Heritability estimates for winter hardiness in lentil under natural and controlled conditions. *Plant Breeding* 119:283-285.

Alia, H., T.H.H. Chen, and N. Murata (1998). Transformation with a gene for choline oxidase enhances the cold tolerance of *Arabidopsis* during germination and early growth. *Plant Cell and Environment* 21:232-239.

Alique, R., J.P. Zamorano, M.L. Calvo, C. Merodio, and J.L. De la Plaza (1994). Tolerance of cherimoya (*Annona cherimoya* Mill.) to cold storage. *Journal of the American Society for Horticultural Science* 119:524-528.

Amirshashi, M.C. and F.L. Patterson (1956). Cold resistance of parent varieties, F_2 populations and F_3 lines of 20 oat crosses. *Agronomy Journal* 48:184-188.

Anand, G., A. Amirthadevarathinam, and J.E. Rogbell (1999). Combining ability and heterosis for cold tolerance in rice. *Oryza* 36:114-117.

Anderson, J.A. and C.M. Taliaferro (1995). Laboratory freeze tolerance of field-grown forage Bermudagrass cultivars. *Agronomy Journal* 87:1017-1020.

Anderson, J.A., C.M. Taliaferro, and D.L. Martin (1993). Evaluating freeze tolerance of Bermudagrass in a controlled environment. *HortScience* 28:955.

Andrews, C.J. (1996). How do plants survive ice? *Annals of Botany* 78:529-536.

Andrews, C.J. and M.J. Morrison (1992). Freezing and ice tolerance test for winter *Brassica*. *Agronomy Journal* 84:960-962.

Andrews, J.R., M.J. Fryer, and N.R. Baker (1995). Characterization of chilling effects on photosynthetic performance of maize crops during early season growth using chlorophyll fluorescence. *Journal of Experimental Botany* 46:1195-1203.

Annichiarico, P. and E. Piano (1995). Variation within and among Ladino white clover ecotypes for agronomic traits. *Euphytica* 86:135-142.

Arora, R., L.J. Rowland, J.S. Lehman, C.C. Lim, G.R. Panta, and N. Vorsa (2000). Genetic analysis of freezing tolerance in blueberry (*Vaccinium* section *Cyanococcus*). *Theoretical and Applied Genetics* 100:690-696.

Artus, N.N., S.J. Gilmour, C. Lin, and M.F. Thomashow (1994). Use of transgenic *Arabidopsis* plants to evaluate the roles of cold-regulated genes in freezing tolerance. *Plant Physiology* 105:169.

Asghar, A., A. Sarfra, M.A. Khan, B.R. Khan, and S.A. Jallil (1994). Evaluation and selection of exotic lentil genotypes for continental Mediterranean environments. *Lens Newsletter* 21:10-15.

Auld, D.L., K.J. Adams, J.B. Swensen, and G.A. Murray (1983). Diallel analyses of winterhardiness in peas. *Crop Science* 23:763-766.

Auld D.L., R.L. Ditterline, G.A. Murray, and J.B. Swensen (1983). Screening peas for winterhardiness under field laboratory conditions. *Crop Science* 23:85-88.

Bacon, R.K. (1986). Selection for seedling cold tolerance in grain sorghum. *Crop Science* 26:900-903.

Bahadur, H.K.C., M. Murai, and T. Yoshida (2000). Effect of *d18-k*, a dwarfing gene of rice, on cool tolerance at the booting stage under the genetic background of Norin-PL8, and its relation with water depth. *Environment Control in Biology* 38:149-156.

Baker, R.J. (1994). Breeding methods and selection indices for improved tolerance to biotic and abiotic stresses in cool season food legumes. *Euphytica* 73:67-72.

Bardhan Roy, S.K., G.F. Pateña, and B.S. Vergara (1982). Feasibility of selection for traits associated with cold tolerance in rice under rapid generation advance method. *Euphytica* 31:25-31.

Barlow, E.W.R., L. Boersma, and J.L. Young (1977). Photosynthesis, transpiration, and leaf elongation in corn seedlings at suboptimal soil temperatures. *Agronomy Journal* 69:95-100.

Barrientos, M., E. Mol, G. Peruzzo, A. Contreras, and M. Alberdi (1994). Responses to cold of Chilean wild *Solanum* species. *Environmental and Experimental Botany* 34:47-54.

Bartels, D. and D. Nelson (1994). Approaches to improve stress tolerance using molecular genetics. *Plant Cell and Environment* 17:659-667.

Bastia, T., N. Carotenuto, B. Basile, A. Zoina, and T. Cardi (2000). Induction of novel organelle DNA variation and transfer of resistance to frost and Verticillium wilt in *Solanum tuberosum* trough somatic hybridization with 1EBN *S. commersonii. Euphytica* 116:1-10.

Bauer, P.J. and J.M. Bradow (1996). Cotton genotypes response to early-season cold temperatures. *Crop Science* 36:1602-1606.

Begum, I., S. Ahmad, and B.R. Khan (1994). Bread wheat selection for tolerance to abiotic and biotic stress in highland Balochistan. *Rachis* 13:11-15.

Bertin, P. and J. Bouharmont (1997). Use of somaclonal variation and in vitro selection for chilling tolerance improvement in rice. *Euphytica* 96:135-142.

Bertin, P., J. Bouharmont, and J.M. Kinet (1996). Somaclonal variation and improvement in chilling tolerance in rice: Changes in chilling induced electrolyte leakage. *Plant Breeding* 115:268-272.

Bertin, P., J. Bouharmont, and J.M. Kinet (1997). Somaclonal variation and improvement of chilling tolerance in rice: Changes in chilling induced chlorophyll fluorescence. *Crop Science* 37:1727-1735.

Bertin, P., P. Bullens, J. Bouharmont, and J.M. Kinet (1998). Somaclonal variation and chilling tolerance improvement in rice: Changes in fatty acid composition. *Plant Growth Regulation* 24:31-41.

Bertin, P., J.P. Busogoro, J.P. Tilquin, J.M. Kinet, and J. Bouharmont (1996). Field evaluation and selection of rice somaclonal variants at different altitudes. *Plant Breeding* 115:183-188.

Bertin, P., J.P. Busogoro, J.P. Tilquin, J.M. Kinet, and J. Bouharmont (1997). Field performance of rice somaclones at low temperature: Effects of callus chilling treatment. *Crop Science* 37:1943-1950.

Bertin, P., J.M. Kinet, and J. Bouharmont (1995). Heritable chilling tolerance improvement in rice through somaclonal variation and cell line selection. *Australian Journal of Botany* 44:91-105.

Bertin, P., J.M. Kinet, and J. Bouharmont (1996). Evaluation of chilling sensitivity in different rice cultivars: Relationship between screening procedures applied during germination and vegetative growth. *Euphytica* 89:201-210.

Beuker, E. (1994). Long-term effects of temperature on the wood production of *Pinus sylvestris* L. and of *Picea abies* (L.) Karst in old provenance experiments. *Scandinavian Journal of Forestry Research* 9:34-45.

Blacklow, W.M. (1972). Influence of temperature on germination and elongation of the radicle and shoot of corn (*Zea mays* L.). *Crop Science* 12:647-650.

Blum, A. (1988). *Plant Breeding for Stress Environments*. Boca Ratón, FL: CRC Press.

Bochicchio, A. (1985). *Zea mays* L. and chilling conditions at sowing time: A review. *Maydica* 30:241-256.

Bocsi, J. and G. Kovács (1990). Inheritance of the rate of germination and emergence at low temperatures in maize (*Zea mays* L.). *Acta Agronomica Hungarica* 39:127-135.

Bonhomme, R., M. Derieux, and G.O. Edmeades (1994). Flowering of diverse maize cultivars in relation to temperature and photoperiod in multilocation field traits. *Crop Science* 34:156-164.

Borovskii, G.B., I.V. Stupnikova, A.I. Antipina, C.A. Downs, and V.K. Voinikov (2000). Accumulation of dehydrin-like-proteins in the mitochondria of cold-treated plants. *Journal of Plant Physiology* 156:797-800.

Bowler, C. and N.H. Chua (1994). Emerging themes of plant signal transduction. *Plant Cell* 6:1529-1541.

Brandolini, A., P. Landi, G. Monfredini, and F. Tano (2000). Variation among Andean races of maize for cold tolerance during heterotophic and early autotrophic growth. *Maydica* 111:33-41.

Brar, G.S., J.F. Gómez, B.L. McMichael, A.G. Matches, and H.M. Taylor (1990). Root development of 12 forage legumes as affected by temperature. *Agronomy Journal* 82:1024-1026.

Braun, H.J. (1997). Winter hardiness of bread wheats derived from spring × winter crosses. *Acta Agronomica Hungarica* 45:317-327.

Bredmeijer, G.M.M. and G. Esselink (1995). Sugar metabolism in cold-hardened *Lolium perenne* varieties. *Plant Varieties and Seeds* 8:187-195.

Bridger, G.M., D.E. Falk, B.D. McKersie, and D.L. Smith (1996). Crown freezing tolerance and field winter survival of winter cereals in eastern Canada. *Crop Science* 36:150-157.

Brüggemann, W., A. Wernner, and Y. Sakata (1995). Long-term chilling of young tomato plants under low light. VII. Increasing chilling tolerance of photosyntesis in *Lycopersicon esculentum* by somatic hybridization with *L. peruvianum*. *Plant Science* 108:23-30.

Burke, M.J., L.V. Gusta, H.A. Quamme, C.J. Weiser, and P.H. Li (1976). Freezing and injury in plants. *Annual Review of Plant Physiology* 27:507-528.

Byrne, D.H. (1986). Mechanisms of spring freeze injury avoidance in peach. *HortScience* 21:1235-1236.

Byrne, M., J.C. Murrell, J.V. Owen, E.R. Williams, and G.F. Moran (1997). Mapping of quantitative trait loci influencing frost tolerance in *Eucalyptus nitens*. *Theoretical and Applied Genetics* 95:975-979.

Cai, ZhuHu, Zhu QiJie, and Xu Yong (1995). Studies on inheritance of chilling tolerance in cucumber at seeding stage. *Acta Horticulturae* 402:206-213.

Capell, B. and K. Dörffling (1993). Genotype-specific differences in chilling tolerance of maize in relation to chilling-induced changes in water status and abscisic acid accumulation. *Physiologia Plantarum* 88:638-646.

Caradus, J.R. and B.R. Christie (1998). Winterhardiness and artificial frost tolerance of white clover ecotypes and selected breeding lines. *Canadian Journal of Plant Science* 78:251-255.

Cardi, T. and E.D. Earle (1997). Production of new CMS *Brassica oleracea* by transfer of "Anand" cytoplasm from *B. rapa* through protoplast fusion. *Theoretical and Applied Genetics* 94:204-212.

Cardi, T., K.J. Puite, K.S. Ramulu, F. D'Ambrosio, and L. Frusciante (1993). Production of somatic hybrids between frost tolerant *Solanum commersonii* and *S. tuberosum*: Protoplast fusion, regeneration and isozyme analysis. *American Potato Journal* 70:753-764.

Cardona, C.A., R.R. Duncan, and O. Lindstrom (1997). Low temperature tolerance assessment in paspalum. *Crop Science* 37:1283-1291.

Carputo, D., A. Barone, T. Cardi, A. Sebastiano, L. Frusciante, and S.J. Peloquin (1997). Endosperm balance number manipulation for direct in vivo germplasm introgression to potato from a sexually isolated relative (*Solanum commersonii* Dun.). *Proceedings of the National Academy of Science* 94:12013-12017.

Castleberry, R.M., J.A. Teeri, and J.F. Buriel (1978). Vegetative growth responses of maize genotypes to simulated natural chilling events. *Crop Science* 18:633-637.

Castonguay, Y., P. Nadeau, P. Lechasseur, and L. Chouinard (1995). Differential accumulation of carbohydrates in alfalfa cultivars of contrasting winterhardiness. *Crop Science* 35:509-516.

Cattivelli, L., C. Crossatti, and F. Rizza (1995). Increasing in membrane stability and COR14 accumulation associated with cold hardening in oats. *Journal of Genetics and Breeding* 49:333-337.

Chat, J. (1995). Cold hardiness within the genus *Actinidia*. *Hortscience* 30:329-332.

Chatterton, N.J., W.R. Thornley, P.A. Harrison, and J.H. Bennet (1988). Dynamics of fructan and sucrose biosynthesis in crested wheatgrass. *Plant Cell Physiology* 29:1103-1108.

Chen, T.H.H., P.H. Li, and M.N. Brenner (1983). Involvement of abscisic acid in potato cold acclimation. *Plant Physiology* 71:362-365.

Chen, T.H.H., P.M. Hayes, AiHong Pan, FuQiang Chen, K. van Zee, T.K. Blake, T.J. Close, and I. Karsai (1996). Marker-assisted genetic analysis of winter hardiness in barley. In *Physical Stresses in Plants: Genes and Their Products for Tolerance*, K. Grillo and A. Leone (Eds.). Berlin, Germany: Springer-Verlag, pp. 65-70.

Chen, Y.K.H., J.B. Bamberg, and J.P. Palta (1999). Expression of freezing tolerance in the interspecific F_1 and somatic hybrids of potatoes. *Theoretical and Applied Genetics* 98:995-1004.

Chen, Y.K.H., J.P. Palta, and J.P. Bamberg (1999). Freezing tolerance and tuber production in selfed and backcross progenies derived from somatic hybrids between *Solanum tuberosum* L. and *S. commersonii* Dun. *Theoretical and Applied Genetics* 99:100-107.

Chen, Y.K.H., J.P. Palta, J.P. Bamberg, H. Kim, G.T. Haberlach, and P. Helgeson (1999). Expression of nonacclimated freezing tolerance and cold acclimation capacity in somatic hybrids between hardy wild *Solanum* species and cultivated potatoes. *Euphytica* 107:1-8.

Chiba, B., S. Usuki, H. Hayasaka, K. Nagano, and K. Matsunaga (1998). Selection of extremely cold-tolerant rice lines developed by crossing cold parents from diverse origins. Tohoku Journal of *Crop Science* 41:17-18.

Clarck, J., R. Bourne, and E. Gbur (1996). Flower buds and shoot hardiness of southern highbush blueberry cultivars. *Fruit Varieties Journal* 50:98-104.

Collins, R.P. and I. Rhodes (1995). Stolon characteristics related to winter survival in white clover. *Journal of Agricultural Science* 124:11-16.

Conner, J.W., D.R. Krieg, and J.R. Gipson (1972). Accumulation of simple sugars in developing cotton bolls as influenced by night temperatures. *Crop Science* 12:752-754.

Cooper, P. and D.R. Ort (1988). Changes in protein synthesis induced in tomato by chilling. *Plant Physiology* 88:454-461.

Cousin, R. (1997). Peas (*Pisum sativum* L.). *Field Crops Research* 53:111-130.

Crosatti, C., E. Nevo, A.M. Stanca, and L. Cattivelli (1996). Genetic analysis of the accumulation of COR14 proteins in wild *(Hordeum spontaneum)* and cultivated *(Hordeum vulgare)* barley. *Theoretical and Applied Genetics* 93:975-981.

Crosbie, T.M., J.J. Mock, and O.S. Smith (1980). Comparison of gains predicted by several methods for cold tolerance traits of two maize populations. *Crop Science* 20:649-655.

Cubero J.I. (1999). *Introducción a la Mejora Genética Vegetal* [Introduction to plant genetic improvement]. Madrid, Spain: Editorial Mundi-Prensa.

Danyluk, J., M. Houde, E. Rassart, and F. Sarhan (1994). Differential expression of a gene encoding an acidic dehydrin in chilling sensitive and freezing tolerant *Graminae* species. *FEBS Letters* 344:20-24.

Deans, J.D. and F.J. Harvey (1996). Frost hardiness of 16 European provenances of sessile oak growing in Scotland. *Forestry* 69:5-11.

De Santis, A., P. Landi, and G. Genchi (1999). Changes of mitochondrial properties in maize seedlings associated with selection for germination at low temperature.

Fatty acid composition, cytochrome C oxidase, and adenine nucleotide translocase activities. *Plant Physiologist* 119:743-754.

Dexter, S.T. (1933). Effect of several environmental factors on the hardening of plants. *Plant Physiology* 8:123-139.

Dickson, M.H. and R. Petzoldt (1987). Inheritance of low temperature tolerance in beans at several growth stages. *HortScience* 22:481-483.

Dillman, A.C. (1941). Cold tolerance in flax. *Journal of the American Society of Agronomy* 33:787-799.

DingBo, L., Y. QiuSheng, and S. DeXu (1999). Selection of cold tolerant somaclonal variant from *Citrus sinensis* cv. Jincheng and genetic stability evaluation of its cold tolerance. *Acta Botanica Sinica* 41:136-141.

Dionne, J., Y. Castonguay, P. Nadeau, and Y. Desjardins (2001). Freezing tolerance and carbohydrate changes during cold acclimation of green-type annual bluegrass (*Poa annua* L.) *Crop Science* 41:443-451.

Dix, P.J. and H.E. Street (1976). Selection of plant cell lines with enhanced chilling resistance. *Annals of Botany* 40:903-910.

Dobreva, S., A. Donchev, and T. Slavcheva (1995). Vine resistance to low winter temperatures and its relation to the main agrobiological and physiological characteristics. I. Yield components. *Bulgarian Journal of Agricultural Science* 3:269-274.

Doll, H., V. Haahr, and B. Sogaard (1989). Relationship between vernalization requirement and winter hardiness in doubled haploids of barley. *Euphytica* 42:209-213.

Dörffling, K., H. Dörffling, and G. Lesselich (1993). In vitro-selection and regeneration of hydroxyproline-resistant lines of winter wheat with increased proline content and increased frost tolerance. *Journal of Plant Physiology* 142:222-225.

Dörffling, K., H. Dörffling, G. Lesselich, E. Luck, C. Zimmermann, G. Melz, and H.U. Jürgens (1997). Heritable improvement of frost tolerance in winter wheat by in vitro-selection of hydroxyproline-resistant proline overproducing mutants. *Euphytica* 93:1-10.

Dörffling, K., S. Schulenburg, G. Lesselich, and H. Dörffling (1990). Abscisic acid and proline levels in cold hardened winter shear leaves in relation to variety-specific differences in freezing resistance. *Journal of Agronomy and Crop Science* 165:230-239.

Dory, I., B. Böddi, J. Kissimon, and E. Paldi (1990). Cold stress responses of inbred maize lines with various degrees of cold tolerance. *Acta Agronomica Hungarica* 39:309-318.

Douglass, S.K., J.A. Juvick, and W.E. Splittstoesser (1993). Sweet corn seedling emergence and variation in kernel carbohydrate reserves. *Seed Science and Technology* 21:433-445.

Du, Y.C., A. Nose, and K. Wasano (1999). Effects of chilling temperature on photosynthetic rates, photosynthetic activities and metabolic levels in leaves of three sugarcane species. *Plant Cell and Environment* 22:317-324.

Duc, G. (1997). Faba bean (*Vicia faba* L.) *Field Crops Research* 53:99-109.

Duhalde, M.C., M.I. Aveldaño, and M. Carceller (1991). Major fatty acids and cold tolerance during germination in *Zea mays*. *Maydica* 36:251-256.

Duncan, D.R. and J.M. Widholm (1987). Proline accumulation and its implication in cold tolerance of regenerable maize callus. *Plant Physiology* 83:703-708.

Eagles, H.A. (1982). Inheritance of emergence time and seedling growth at low temperatures in four lines of maize. *Theoretical and Applied Genetics* 62:81-87.

Eagles, H.A. and A.K. Hardacre (1979). Genetic variation in maize (*Zea mays* L.) for germination and emergence at 10°C. *Euphytica* 28:287-295.

Elings, A. and M.M. Nachit (1993). Field evaluation and simulation of frost tolerance in Syrian durum wheat landraces. *Netherlands Journal of Agricultural Science* 41:1-11.

El-Kholy, A.S., A.E. Hall, and A.A. Mohsen (1997). Heat and chilling tolerance during germination and heat tolerance during flowering are not associated in cowpea. *Crop Science* 37:456-463.

Ellis, R.H., R.J. Summerfield, G.O. Edmeades, and E.H. Roberts (1992). Photoperiod, temperature and the interval from sowing to tassel initiation in diverse cultivars of maize. *Crop Science* 32:125-132.

Eujayl, I., W. Erskine, M. Baum, and E. Pehu (1999). Inheritance and linkage analysis of frost injury in lentil. *Crop Science* 39:639-642.

Ezura, H., H. Amagai, I. Kikuta, M. Kubota, and K. Oosawa (1995). Selection of somaclonal variants with low-temperature germinability in melon (*Cucumis melo* L.). *Plant Cell Reports* 14:684-688.

Feldman M., F.G.H. Lupton, and T.E. Miller (1995). Wheats. In *Evolution of Crop Plants*, Second Edition, J. Smartt and N.W. Simmonds (Eds.). London, UK: Longman, pp. 184-192.

Fernández-Muñóz, R., J.J. González-Fernández, and J. Cuartero (1995). Variability of pollen tolerance to low temperatures in tomato and related wild species. *Journal of Horticultural Science* 70:41-49.

Foolad, M.R., J.R. Hyman, and G.Y. Lin (1999). Relationships between cold- and salt-tolerance during seed germination in tomato: Analysis of response and correlated response to selection. *Plant Breeding* 118:49-52.

Foolad, M.R. and G.Y. Lin (1998). Genetic analysis of low-temperature tolerance during germination in tomato, *Lycopersicon esculentum* Mill. *Plant Breeding* 117:171-176.

Foolad, M.R. and G.Y. Lin (1999). Relationships between cold- and salt-tolerance during seed germination in tomato: Germplasm evaluation. *Plant Breeding* 118:45-48.

Foolad, M.R. and G.Y. Lin (2000). Relationships between cold tolerance during seed germination and vegetative growth in tomato: Germplasm evaluation. *Journal of the American Society for Horticultural Science* 125:679-683.

Foolad, M.R. and G.Y. Lin (2001). Relationship between cold tolerance during seed germination and vegetative growth in tomato: Analysis of response and correlated response to selection. *Journal of the American Society for Horticultural Science* 126:216-220.

Fowler, D.B. and R.J. Carles (1979). Growth, development, and cold tolerance of fall-acclimated cereal grains. *Crop Science* 19:915-922.

Fowler, D.B., L.P. Chauvin, A.E. Limin, and F. Sarhan (1996). The regulatory role of vernalization in the expression of low-temperature-induced genes in wheat and rye. *Theoretical and Applied Genetics* 93:554-559.

Fowler, D.B. and L.V. Gusta (1979). Selection for winter hardiness in wheat. I. Identification of genotypic variability. *Crop Science* 19:769-772.

Fowler, D.B., L.V. Gusta, and N.J. Tyler (1981). Selection for winter hardiness in wheat and screening methods. *Crop Science* 21:896-901.

Fowler, D.B., A.E. Limin, and J.T. Ritchie (1999). Low-temperature tolerance in cereals: Model and genetic interpretation. *Crop Science* 39:626-633.

Fowler, D.B., A.E. Limin, S.-Y. Wang, and R.W. Ward (1996). Relationship between low-temperature tolerance and vernalization response in wheat and rye. *Canadian Journal of Plant Science* 76:37-42.

Fracheboud, Y., P. Haldimann, J. Leipner, and P. Stamp (1999). Chlorophyll fluorescence as a selection tool for cold tolerance of photosynthesis in maize (*Zea mays* L.). *Journal of Experimental Botany* 50:1533-1540.

Frascaroli, E. (1995). Pollen cold tolerance of maize genotypes with different ability to germinate at low temperature. *Maydica* 40:229-232.

Frei, O.M. (2000). Changes in yield physiology of corn as a result of breeding in northern Europe. *Maydica* 45:173-183.

Fry, J.D., N.S. Lang, and R.G.P. Clifton (1991). Freezing resistance and carbohydrate composition of "Floratam" St. *HortScience* 26:1537-1539.

Fuller, M.P. and C.F. Eagles (1978). A seedling test for cold hardiness in *Lolium perenne* L. *Journal of Agricultural Science* 91:217-222.

Furter, M. and H.A. van de Venter (1990). A comparison of parameters for the assessment of relative cold tolerance of germinating kernels of maize (*Zea mays* L.). *South African Journal of Plant and Soil* 7:207-211.

Galiba, G. (1994). In vitro adaptation for drought and cold hardiness in wheat. *Plant Breeding Reviews* 12:115-162.

Galiba, G., S.A. Quarrie, J. Sutka, A. Morgounov, and J.W. Snape (1995). RFLP mapping of the vernalization *(Vrn1)* and frost-tolerance *(Fr1)* genes on chromosomes 5A of wheat. *Theoretical and Applied Genetics* 90:1174-1179.

Galiba, G. and J. Sutka (1989). Frost resistance of somaclones derived from *Triticum aestivum* L. winter calli. *Plant Breeding* 102:101-104.

Gasparíková, O., M. Ciamporová, L. Tamás, I. Trginová, and M. Luxová (1996). Cold-induced changes in protein patterns and ultrastructure of root cells of maize seedlings. *Biológia* 51:449-456.

Gass, T., A. Schori, A. Fossati, A. Soldati, and P. Stamp (1996). Cold tolerance of soybean [*Glycine max* L. Merr.] during the reproductive phase. *European Journal of Agronomy* 5:71-80.

Georges, F., M. Saleen, and A.J. Cutler (1990). Design and cloning of synthetic gene for the flounder antifreeze protein and its expression in plant cells. *Gene* 91:159-165.

Giauffret, C., R. Bonhomme, and M. Derieux (1995). Genotypic differences for temperature response and leaf appearance rate and leaf elongation rate in field-grown maize. *Agronomie* 15:123-137.

Giauffret, C., R. Bonhomme, and M. Derieux (1997). Heterosis in maize for biomass production, leaf area establishment, and radiation use efficiency under cool spring conditions. *Maydica* 42:13-19.

Gilkey, R. (1996). Cold hardiness ranking of rhododendrons by means of flower bud damage. *American Rhododendron Society Journal* 50:100-102.

Goldstein, G. and P.S. Nobel (1991). Changes in osmotic pressure and mucilage during low-temperature acclimation of *Opuntia ficus-indica*. *Plant Physiology* 97:954-961.

Graham, D. and B.D. Patterson (1982). Response of plants to low, non-freezing temperature: proteins, metabolism, and acclimation. *Annual Review of Plant Physiology* 33:347-372.

Greaves, J.A. (1996). Improving suboptimal temperature tolerance in maize: The search for variation. *Journal of Experimental Botany* 47:307-323.

Griffiths, D.W., H. Bain, and M.F.B. Dale (1998). Effect of storage temperature on potato (*Solanum tuberosum* L.) tuber glycoalkaloid content and the subsequent accumulation of glycoalkaloids and chlorophyll in response to light exposure. *Journal of Agricultural and Food Chemistry* 46:5262-5268.

Grogan, C.O. (1970). Genetic variation in maize (*Zea mays* L.) for germination and seedling vigor at low temperatures. *Proceedings of the Annual Corn and Sorghum Research Conference* 25:90-98.

Grossnickle, S.C., B.C.S. Sutton, R.S. Folk, and R.J. Gawley (1996). Relationship between nuclear DNA markers and physiological parameters in Sitka × interior spruce populations. *Tree Physiology* 16:547-555.

Grubbels, G.H. (1974). Growth of corn seedlings under low temperature as affected by genotype, seed size, total oil and fatty acid content of the seed. *Canadian Journal of Plant Science* 54:425-426.

Gullord, M., C.R. Olien, and E.H. Everson (1975). Evaluation of freezing hardiness in winter wheat. *Crop Science* 15:153-157.

Gupta, H.S., R.N. Bhuyan, A. Pattanayak, and D.K. Pandey (1996). Development of cold-tolerant rice through anther culture. *International Rice Research Notes* 21:20.

Gupta, J.C., P.C. Katoch, R.P. Kaushik, and S.L. Sharma (1998). Cause-and-effect relationship of yield and its components under cold stress condition in rice (*Oryza sativa*). *Indian Journal of Agricultural Sciences* 68:13-15.

Guse, R.A., J.G. Coors, P.N. Drolsom, and W.F. Tracy (1988). Isozyme marker loci associated with cold tolerance and maturity in maize. *Theoretical and Applied Genetics* 76:398-404.

Gusta, L.V., R.W. Wilen, and P. Fu (1996). Low-temperature stress tolerance: The role of abscisic acid, sugars, and heat-stable proteins. *HortScience* 31:39-46.

Guy, C.L. (1990). Cold acclimation and freezing stress tolerance: Role of protein metabolism. *Annual Review of Plant Physiology and Plant Molecular Biology* 41:187-223.

Guy, C.L., K.J. Niemi, and R. Brambl (1985). Altered gene expression during cold acclimation of spinach. *Proceedings of the National Academy of Sciences USA* 82:3673-3677.

Hahn, M. and V. Walbot (1989). Effects of cold-treatment on protein synthesis and mRNA levels in rice leaves. *Plant Physiology* 91:930-938.

Hajdu, E. and G. Gábor (1997). Winter tolerance of grape varieties in the winter of 1995/96. *Horticultural Science* 29:43-47.

Haldimann, P. (1997). Chilling-induced changes to carotenoid composition, photosynthesis, and the maximum quantum yield of photosystem II photochemistry in two maize genotypes differing in tolerance to low temperature. *Journal of Plant Physiology* 151:610-619.

Haldimann, P. (1998). Low growth temperature-induced changes to pigment composition and photosynthesis in *Zea mays* genotypes differing in chilling sensitivity. *Plant Cell and Environment* 21:200-208.

Haldimann, P. (1999). How do changes in temperature during growth affect leaf pigment composition and photosynthesis in *Zea mays* genotypes differing in sensitivity to low temperature? *Journal of Experimental Botany* 50:543-550.

Hamada, T., H. Kodama, K. Takeshita, H. Utsumi, and K. Iba (1998). Characterization of transgenic tobacco with an increased α-linolenic acid level. *Plant Physiology* 118:591-598.

Hamdi, A., I. Küsmenoglü, and W. Erskine (1996). Sources of winter hardiness in wild lentil. *Genetic Resources and Crop Evolution* 43:63-67.

Han, L.Z. and H.J. Koh (2000). Genetic analysis of growth response to cold water irrigation in rice. *Korean Journal of Crop Science* 45:26-31.

Hansche, P.E. (1990). Heritability of spring bloom and fall leaf abscission dates in *Prunus persica*. *HortScience* 25:1639-1641.

Hanson, W.D. (1971). Selection for differential productivity among juvenile maize plants: Associated net photosynthetic rate and leaf area changes. *Crop Science* 11:334-339.

Harlan, J.R. (1992). *Crops and Man*, Second Edition. Madison, WI: American Society of Agronomy, Crop Science Society of America.

Haskell, G, and W.R. Singleton (1949). Use of controlled low temperature in evaluating the cold hardiness of inbred and hybrid maize. *Agronomy Journal* 41:34-40.

Havaux, M. (1989). Fluorimetric determination of the genetic variability existing for chilling tolerance in sweet sorghum and Sudan grass. *Plant Breeding* 102: 327-332.

Havaux, M. (1995). Temperature sensitive of the photochemical function of photosynthesis in potato *(Solanum tuberosum)* and a cultivate Andean hybrid *(Solanum × juzepczupii)*. *Journal of Plant Physiology* 146:47-56.

Hawkins, B.J., J. Russell, and R. Shortt (1994). Effect of population, environment, and maturation on the frost hardiness of yellow-cedar *(Chamaecyparis nootkatensis)*. *Canadian Journal of Forestry Research* 24:945-953.

Hayashi, H., H. Alia, L. Mustardy, P. Deshnium, M. Ida, and N. Murata (1997). Transformation of *Arabidopsis thaliana* with the codA gene for choline oxidase: Accumulation of glycinebetaine and enhanced tolerance to salt and cold stress. *Plant Journal* 12:133-142.

Hayashi, H., A. Sakamoto, H. Alia, and N. Murata (1998). Enhancement of stress tolerance by gene-engineering of betaine accumulation in plants. In *Photosyn-*

thesis: Mechanisms and Effects, Volume IV, G. Garab (Ed.). Dordrecht, The Netherlands: Kluwer Academic Publishers, pp. 2419-2424.

Heath, D.W. and E.D. Earle (1996). Synthesis of Ogura male sterile rapeseed (*Brassica napus* L.) with cold tolerance by protoplast fusion and effects of atrazine resistance on seed yield. *Plant Cell Reports* 15:939-942.

Heath, D.W., E.D. Earle, and M.H. Dickson (1994). Introgressing cold-tolerant Ogura cytoplasms from rapeseed into Pak choi and Chinese cabbage. *HortScience* 29:202-203.

Heatley, R.C., J.J. Kielbaso, and G.S. Howell (1994). Environmental adaptation and ornamental display of *Cornus florida* ecotypes. *Journal of Arboriculture* 20:305-309.

Hegarty, T.W. (1973). Temperature sensitivity of germination in carrots: Its frequency of occurrence and response to seed advancement. *Journal of Horticultural Science* 48:43-50.

Helgadóttir, Á. and H. Björnsson (1994). Cooperative breeding for the northern marginal areas. *Euphytica* 77:221-230.

Hensleigh, P.F., T.K. Blake, and L.E. Welty (1992). Natural selection on winter barley composite cross XXVI affects winter survival and associated traits. *Crop Science* 32:57-62.

Herczegh, M. (1970). Some problems of cold tolerance. In *Some Methodological Achievements of the Hungarian Hybrid Maize Breeding,* I. Kovacs (Ed.). Budapest, Hungary: Kiado, pp. 271-281.

Hetherington, S.E., R.M. Smillie, A.K. Hardacre, and H.A. Eagles (1983). Using chlorophyll fluorescence in vivo to measure the chilling tolerances of different populations of maize. *Australian Journal of Plant Physiology* 10:247-256.

Hetherington, S.E., R.M. Smillie, P. Malagamba, and Z. Huaman (1983). Heat tolerance and cold tolerance of cultivated potatoes measured by the chlorophyll-fluorescence method. *Planta* 159:119-124.

Hides, D.H. (1979). Winter hardiness in *Lolium multiflorum* Lam. III. Selection for improved cold tolerance and its effect on agronomic performance. *Grass and Forage Science* 34:119-124.

Hightower, R., C. Raden, E. Penses, P. Lund, and P. Dunsmuir (1991). Expression of antifreeze proteins in transgenic plants. *Plant Molecular Biology* 17:1013-1021.

Hincha, D.K., U. Sonnewald, L. Willmitzer, and J.M. Schmitt (1996). The role of sugar accumulation in the leaf frost hardiness: Investigations with transgenic tobacco expressing a bacterial pyrophosphatase or a yeast invertase gene. *Plant Physiology* 147:604-610.

Hoard, K.G. and T.M. Crosbie (1985). S_1-lie recurrent selection for cold tolerance in two maize populations. *Crop Science* 25:1041-1045.

Hoard, K.G. and T.M. Crosbie (1986a). Correlated changes in agronomic traits from S_1-line recurrent selection for cold tolerance in two maize populations. *Crop Science* 26:519-522.

Hoard, K.G. and T.M. Crosbie (1986b). Effects of recurrent selection for cold tolerance on genotype-environment interactions for cold tolerance and agronomic traits in two maize populations. *Crop Science* 26:238-242.

Hodges, D.M., C.J. Andrews, D.A. Johnson, and R.I. Hamilton (1997). Sensitivity of maize hybrids to chilling and their combining abilities at two developmental stages. *Crop Science* 37:850-856.

Hodges, D.M., R.I. Hamilton, and C. Charest (1994). A chilling resistance test for inbred maize lines. *Canadian Journal of Plant Science* 74:687-691.

Hodges, D.M., R.I. Hamilton, and C. Charest (1995). A chilling response test for early phase maize. *Agronomy Journal* 87:970-974.

Hodgins, R. and R.B. van Huystee (1985). Chilling-induced chlorosis in maize (*Zea mays*). *Canadian Journal of Botany* 63:711-715.

Hoekstra, F.A. and I. van Roekel (1988). Desiccation tolerance of *Papaper dubium* L. Pollen during its development in the anther. Possible role of phospholipid composition and sucrose content. *Plant Physiology* 88:626-632.

Holubowicz, R. and W. Legutko (1995). Low temperature tolerance of French beans (*Phaseolus vulgaris* L.). *Folia Horticulturae* 7:27-36.

Hömmö, L.M. (1994). Hardening of some winter wheat (*Triticum aestivum* L.), rye (*Secale cereale* L.), triticale (× *Triticosecale* Wittmack) and winter barley (*Hordeum vulgare* L.) cultivars during autumn and the final winter survival in Finland. *Plant Breeding* 112:285-293.

Hotchkiss, J.R., P. Revilla, and W. F. Tracy (1997). Variation of cold tolerance among open-pollinated sweet corn cultivars. *HortScience* 32:719-723.

Houde, M., R.S. Dhindsa, and F. Sarhan (1992). A molecular marker to select for freezing tolerance in Gramineae. *Molecular and General Genetics* 234:43-48.

Howe, G.T., P. Saruul, J. Davis, and T.H.H. Chen (2000). Quantitative genetics of bud phenology, frost damage, and winter survival in an F_2 family of hybrid poplars. *Theoretical and Applied Genetics* 101:632-642.

Hughes, M.A. and M.A. Dunn (1996). The molecular biology of plant acclimation to low temperature. *Journal of Experimental Botany* 47:291-305.

Huitema, J.B.M., W. Preil, and J. de Jong (1991). Methods for selection of low-temperature tolerant mutants of *Chrysanthemum morifolium* Ramat. using irradiated cell suspension cultures. III. Comparison of mutants selected with or without preselection in vitro at low-temperature stress. *Plant Breeding* 107:135-140.

Huitema, J.B.M., W. Preil, G.C. Gussenhoven, and M. Schneidereit (1989). Methods for selection of low-temperature tolerant mutants of *Chrysanthemum morifolium* Ramat. using irradiated cell suspension cultures. I. Selection of regenerants in vivo under suboptimal temperature conditions. *Plant Breeding* 102:140-147.

Hummel, R.L. and P.P. Moore (1997). Freeze resistance of Pacific Northwest strawberry flowers. *Journal of the American Society for Horticultural Science* 122:179-182.

Hummer, K., H.L. Fuchigami, V. Peters, and N. Bell (1995). Cold hardiness in *Rubus*. *Fruit Varieties Journal* 49:52-58.

Humphreys, M.O. (1989). Assessment of perennial ryegrass (*Lolium perenne* L.) for breeding: II. Components of winter hardiness. *Euphytica* 41:99-106.

Humphreys, M.O. and C.F. Eagles (1988). Assessment of perennial ryegrass (*Lolium perenne* L.) for breeding: I. Freezing tolerance. *Euphytica* 38:75-84.

Hunsperger, M.H. and D.W. Davis (1987). Effects of the sugary-1 locus on plant and ear traits in corn. *Crop Science* 27:1173-1176.

Hussey, M.A. and E.C. Bashaw (1996). Performance of buffelgrass germplasm with improved winter survival. *Agronomy Journal* 88:944-946.

Huyghe, C. (1997). White lupin (*Lupinus albus* L.). *Field Crops Research* 53:147-160.

Irzykowsky, W., T. Rorat, and J. Kakubiec (1996). Analysis of cold resistance in wild and cultivated potatoes species. *Acta Physiologiae Plantarum* 18:217-221.

Ishizaki-Nishizawa, O., T. Fujii, M. Azuma, K. Sekiguchi, N. Murata, T. Ohtani, and T. Toguri (1996). Low-temperature resistance of higher plants is significantly enhanced by a nonspecific cyanobacterial desaturase. *Nature Biotechnology* 14:1003-1006.

Ismail, A.M., A.E. Hall, and T.J. Close (1997). Chilling tolerance during emergence of cowpea associated with a dehydrin and slow electrolyte leakage. *Crop Science* 37:1270-1277.

Jaglo-Ottosen, K.R., S.J. Gilmour, D.G. Zarka, O. Schabenberger, and M.F. Thomashow (1998). *Arabidopsis CBF1* overexpression induces *COR* genes and enhances freezing tolerance. *Science* 280:104-106.

Janda, T., G. Szalai, J.M. Ducruet, and E. Paldi (1998). Changes in photosynthesis in inbred maize with different degrees of chilling tolerance grown at optimum and suboptimum temperatures. *Photosynthetica* 35:205-212.

Janowiak, F. and K. Dörffling (1996). Chilling of maize seedlings: Changes in water status and abscisic acid content in ten genotypes differing in chilling tolerance. *Journal of Plant Physiology* 147:582-588.

Jenkins, G. (1969). Transgressive segregation for frost resistance in hexaploid oats (*Avena* spp.). *Journal of Agricultural Science* 73:477-482.

Jermstad, K.D., D.L. Bassoni, N.C. Wheeler, T.S. Anekonda, S.N. Aitken, W.T. Adams, and D.B. Neale (2001). Mapping of quantitative trait loci controlling adaptive traits in coastal Douglas-fir. II. Spring and fall cold-hardiness. *Theoretical and Applied Genetics* 102:1152-1158.

Jha, P.B. and A.K. Sinha (1989). Combining ability for yield and cold tolerance in early maturing maize. *Crop Improvement* 16:137-140.

JinHong, K., H. HongJin, P. HyunSook, L. ChinBum, M. KwonYoung, and L. ChoonHwan (1997). Differences in the rate of dephosphorylation of the thylakoid proteins during dark incubation after chilling in the light between two rice (*Oryza sativa* L.) varieties. *Plant Science* 128:159-168.

Johnsen, O., T. Skroppa, O. Juntilla, and O.G. Daehlen (1996). Influence of the female flowering environment on autumn frost-hardiness of *Picea abies* progenies. *Theoretical and Applied Genetics* 92:797-802.

Johnson, G.R. and K.J. Frey (1967). Heritabilities of quantitative attributes of oat (*Avena* sp.) at varying levels of environmental stress. *Crop Science* 7:43-46.

JongUn, C., J. InHo, and C. HyoungKog (2000). A simple method for testing freezing resistance based on chlorophyll fluorescence in tea (*Camellia sinensis* L.). *Korean Journal of Crop Science* 45:322-327.

Kadir, S.A. and E.L. Proebsting (1994). Various freezing strategies of flower-bud hardiness in *Prunus*. *Journal of the American Society for Horticultural Science* 119:584-588.

Kalinova, M.G., A.I. Soroka, and V.A. Lyakh (1998). Effect of selection during the period of pollen germination and pollen tube growth in F_1 hybrids on cold resistance in spring rape progeny. *Tsitologiya i Genetika* 32:41-46.

Kang, N., K. HoeTae, and K. KwangYoon (1998). Development of early screening method for cold tolerance in *Cucurbita* spp. by analysis of protein and isozymes. *Journal of the Korean Society for Horticultural Science* 39:697-701.

Karsai, I., K. Mészáros, Z. Bedo, P.M. Hayes, A. Pan, and F. Chen (1997). Genetic analysis of the components of winterhardiness in barley (*Hordeum vulgare* L.). Acta Biologica Hungarica 48:67-76.

Kato, A., K. Saito, K. Nagano, K. Miura, and H. Araki (1996). Mapping of genes responsible for cold tolerance at the booting stage of rice. *International Rice Research Notes* 21:35-36.

Kaw, R.N. (1999). Inheritance of reproductive stage cold tolerance and adaptability in two *japonica* × *indica* rice crosses. *Indian Journal of Genetics and Plant Breeding* 59:29-35.

Kennedy, V.J.F. and P. Rangasamy (1998). Synthesis and evaluation of cold tolerant rice hybrids. *Madras Agricultural Journal* 85:198-200.

Khan, D.R., D.J. Mackill, and B.S. Vergara (1986). Selection for tolerance to low temperature-induced spikelet sterility at anthesis in rice. *Crop Science* 26:694-698.

Khan, M.A., S. Ahmed, I. Begum, S.A. Jalil, and A.S. Alvi (1997). Evaluation of exotic barley germplasm for biotic and abiotic stresses in Mediterranean highland Balochistan. *Rachis* 16:32-40.

Kiel, C. and P. Stamp (1992). Internal root anatomy of maize seedlings (*Zea mays* L.) as influenced by temperature and genotype. *Annals of Botany* 70:125-128.

Kisaka, H. and T. Kameya (1998). Cold and salt tolerance of somatic hybrid calli between barley (*Hordeum vulgare* L.) and carrot (*Daucus carota* L.). *Breeding Science* 48:11-15.

Kiyomoto, R.K. (1995). Testing wine grapes for hardiness and yield helps Connecticut growers select varieties. *Frontiers of Plant Science* 47:7-8.

Kocsy, G., M. Brunner, A. Rüegsegger, P. Stamp, and C. Brunold (1996). Gluthatione synthesis in maize phenotype with different sensitivities to chilling. *Planta* 198:365-370.

Kocsy, G., B. Tóth, T. Berzy, G. Szalai, A. Jednákovits, and G. Galiba (2001). Gluthatione reductase activity and chilling tolerance are induced by a hydroxilamine derivative BRX-156 in maize and soybean. *Plant Science* 160:943-950.

Kodama, H., T. Hamada, G. Horiguchi, M. Nishimura, and K. Iba (1994). Genetic enhancement of cold tolerance by expression of a gene for chloroplast ω-3 fatty acid desaturase in transgenic tobacco. *Plant Physiology* 105:601-605.

Kodama, H., G. Horiguchi, T. Nishiuchi, M. Nishimura, and K. Iba (1995). Fatty acid desaturation during chilling acclimation is one of the factors involved in conferring low-temperature tolerance to young tobacco leaves. *Plant Physiology* 107:1177-1185.

Korbuly, J. (2000). Results of breeding for resistance to winter frost and different pathogens using *Vitis amurensis*. *Acta Horticulturae* 528:551-557.

Koshelyaev, V.V. and V.A. Serkov (2000). Breeding early maize lines in Penza province. *Kukuruza i Sorgo* 5:11-15.

Kovacs, G. and B. Barnabas (1992). Production of highly cold tolerant maize inbred lines by repeated gametophytic selection. In *Angiosperm Pollen and Ovules*, E. Ottaviano, D.L. Mulcay, M. Sari Gorla, and G.B. Mulcahy (Eds.). New York: Springer-Verlag, pp. 359-363.

Kovacs, G. and B. Barnabas (1997). Selection for frost tolerance in anther culture-derived embryos and regeneration of frost-tolerant fertile DH plants in winter wheat. *Acta Agronomica Hungarica* 45:285-293.

Kristjansdottir, I.S. (1990). Pollen germination in vitro at low temperature in European and Andean tetraploid potatoes. *Theoretical and Applied Genetics* 80:139-142.

Kumar, U. and A. Shukla (1993). Cold tolerance in pigeonpea: Germination studies. *Haryana Journal of Agronomy* 9:59-67.

Kumar, U., A. Shukla, and K.C. Pant (1995). Cold tolerance in pigeon pea: Seedling characteristics. *Crop Research* 9:105-113.

Kuwabara, T., J. Abe, M. Moriyama, N. Iriki, M. Yoshida, and K. Takata (1997). Relationship between physiological and ecological traits and snow mold damage in winter wheat varieties (*Triticum aestivum* L.). *Breeding Science* 47:271-277.

Landi, P., E. Frascaroli, and A. Lovato (1992). Divergent full-sib recurrent selection for germination at low temperature in a maize population. *Euphytica* 64:21-29.

Lansac, A.R., C.Y. Sullivan, and B.E. Johnson (1996). Accumulation of free proline in sorghum *(Sorghum bicolor)* pollen. *Canadian Journal of Botany* 74:40-45.

Larkin, P.J. and W.R. Scowcroft (1981). Somaclonal variation: A novel source of variability from cell culture for plant improvement. *Theoretical and Applied Genetics* 60:197-214.

Larsen, A. (1994). Breeding winter hardy grasses. *Euphytica* 77:231-237.

Lawlor, H.J., K.H.M. Siddique, R.H. Sedgley, and N. Thurling (1998). Improving cold tolerance and insect resistance in chickpea and the use of AFLPs for the identification of molecular markers for these traits. *Acta Horticulturae* 461:185-192.

Layne, R.E.C. (1992). Breeding cold hardy peaches and nectarines. *Plant Breeding Review* 10:271-308.

Layne, R.E.C. and M.F. Gadsby (1995). Determination of cold hardiness and estimation of potential breeding value of apricot germplasm. *Fruit Varieties Journal* 49:242-248.

Lazar, M.D., T.H.H. Chen, L.V. Gusta, and K.K. Karthe (1988). Somaclonal variation for freezing tolerance in a population derived from Norstar winter wheat. *Theoretical and Applied Genetics* 75:480-484.

Lee, T.M., H.S. Lur, and C. Chu (1995). Abscisic acid and putrescine accumulation in chilling-tolerant rice cultivars. *Crop Science* 35:502-508.

Levitt, J. (1980). *Responses of Plants to Environmental Stresses*, Volume I, *Chilling, Freezing, and High Temperature Stresses*. New York: Academic Press.

Li, H.B., Q. Zhang, A.M. Liu, J.S. Zou, and Z.M. Chen (1996). A genetic analysis of low- temperature-sensitive sterility in indica-japonica rice hybrids. *Plant Breeding* 115:305-309.

Li, P.H. (1984). Subzero temperature stress physiology of herbaceous plants. *Horticultural Reviews* 6:373-416.

Lim, C.C., R. Arora, and S.L. Krebs (1998). Genetic study of freezing tolerance in *Rhododendron* populations: Implications for cold hardiness breeding. *American Rhododendron Society Journal* 52:143-148.

Lim, C.C., S.L. Krebs, and R. Arora (1999). A 25-kDa dehydrin associated with genotype- and age-dependent leaf freezing-tolerance in *Rhododendron*: A genetic marker for cold hardiness? *Theoretical and Applied Genetics* 99:912-920.

Limin, A.E. and D.B. Fowler (1984). The effect of cytoplasm on cold hardiness in alloplasmic rye (*Secale cereale* L.) and triticale. *Canadian Journal of Genetics and Cytology* 26:405-408.

Livingston, D.P. and G.F. Elwinger (1995). Improvement of winter hardiness in oat from 1935 to 1992. *Crop Science* 35:749-755.

Long, S.P., T. East, and R. Baker (1983). Chilling damage to photosyntesis in young *Zea mays*. I. Effects of light and temperature variation on photosyntetic CO_2 assimilation. *Journal of Experimental Botany* 34:177-188.

Long, S.P., S. Humphries, and P.G. Falkowski (1994). Photoinhibition of photosynthesis in nature. *Annual Review of Plant Physiology and Plant Molecular Biology* 45:633-662.

Lozano, R., T. Angosto, P. Gómez, C. Payan, J. Capel, P. Huijser, J. Salinas, and J.M. Martínez-Zapater (1998). Tomato flower abnormalities induced by low temperature are associated with changes of expression MADS-box genes. *Plant Physiology* 117:91-100.

Lyakh, V.A., A.I. Soroka, and M.G. Kalinova (1998). Pollen storage at low temperature as a procedure for the improvement of cold tolerance in spring rape, *Brassica napus* L. *Plant Breeding* 117:389-391.

Mackill, D.J. and L. XiaoMao (1997). Genetic variation for traits related to temperate adaptation of rice cultivars. *Crop Science* 37:1340-1346.

MacRae, E.A., A.K. Hardacre, and I.B. Ferguson (1986). Comparison of chlorophyll fluorescence with several other techniques used to assess chilling sensitivity in plants. *Physiologia Plantarum* 67:659-665.

Mahajan, V., B.S. Dhillon, A.S. Khehra, and O.S. Singh (1993). Combining ability analysis of response to cold stress in maize. *Field Crops Research* 34:71-81.

Mahajan, V., A.S. Khehra, and O.S. Singh (1992). Cold tolerance in relation to yield and other characters in maize. *International Journal of Tropical Agriculture* 10:75-78.

Maisonneuve, B., H.G. Hogenboom, and A.P.M. Dennijs (1986). Pollen selection in breeding tomato (*Lycopersicon esculentum* Mill.) for adaptation to low temperature. *Euphytica* 35:983-992.

Majumder, M.K., D.V. Seshu, and V.V. Shenoy (1989). Implications of fatty acids and seed dormancy is a new screening procedure for cold tolerance in rice. *Crop Science* 29:1298-1304.

Malhotra, R.S. and K.B. Singh (1991). Gene action for cold tolerance in chickpea. *Theoretical and Applied Genetics* 82:598-601.

Malhotra, R.S., K.B. Singh, and M.C. Saxena (1995). Effect of nitrogen fertilizer application on cold tolerance in chickpea. *International Chickpea and Pigeonpea Newsletter* 2:24-25.

Malhotra, S.K. and G. Kalloo (1995). Breeding tomato for low temperature fruit-set. A review. *Agricultural Reviews* 16:63-72.

Mangelsdorf, P.C. (1974). *Corn: Its Origin, Evolution and Improvement.* Cambridge, MA: Harvard University Press.

Markowski, A. and M. Rapacz (1994). Comparison of vernalization requirements and frost resistance of winter rape lines derived from double haploids. *Journal of Agronomy and Crop Science* 173:184-192.

Marmiroli, N., V. Terzi, M. Odoardi Stanca, C. Lorenzoni, and A.M. Stanca (1986). Protein synthesis during cold shock in barley tissues. *Theoretical and Applied Genetics* 73:190-196.

Marshall, H.G. (1966). Natural selection for cold resistance in winter oat bulk populations. *Crop Science* 6:173-176.

Martin, J.H. (1927). Comparative studies of winter hardiness in wheat. *Journal of the American Society of Agronomy* 35:493-535.

Martiniello, P. (1985). Recurrent visual selection in S_1 progenies for early vigor in maize (*Zea mays* L.) *Maydica* 30:301-308.

Maryam, B. and D.A. Jones (1983). The genetics of maize (*Zea mays* L.) growing at low temperatures. I. Germination of inbred lines and their F_1s. *Euphytica* 32:535-542.

Maryam, B. and D.A. Jones (1985). The genetics of maize (*Zea mays* L.) growing at low temperatures. II. Harvesting time, number of kernels and plant height at maturity. *Euphytica* 34:475-482.

Massacci, A., M.A. Iannelli, F. Pietrini, and F. Loreto (1995). The effect of growth at LT on PS characteristics and mechanisms of photoprotection of maize leaves. *Journal of Experimental Botany* 46:119-127.

Mastenbroek, C. (1956). Some experiences in breeding frost-tolerant potatoes. *Euphytica* 5:289-297.

Maximov, N.G. and G.V. Kostandi (1995). Frost hardiness of primary hexaploid triticale strains and their parental species. *Cereal Research Communications* 23: 391-395.

McCamant, T. and R.A. Black (2000). Cold hardiness in coastal, montane and inland populations of *Populus trichocarpa. Canadian Journal of Forest Research* 30:91-99.

McConnell, R.L. and C.O. Gardner (1979a). Inheritance of several cold tolerance traits in corn. *Crop Science* 19:847-852.

McConnell, R.L. and C.O. Gardner (1979b). Selection for cold tolerance in two corn populations. *Crop Science* 19:765-768.

McKellar, M.A., D.W. Buchanan, D.L. Ingram, and C.W. Campbell (1992). Freezing tolerance of avocado leaves. *HortScience* 27:341-343.

McKersie, B.D., J. Murnaghan, K.S. Jones, and S.R. Bowley (2000). Iron-superoxide dismutase expression in transgenic alfalfa increases winter survival without a detectable increase in photosynthetic oxidative stress tolerance. *Plant Physiology* 122:1427-1437.

Mercado, J.A., B. Viñegla, and M.A. Quesada (1997). Effects of hand-pollination, paclobutrazol treatments, root temperatures and genotype on pollen viability and seed fruit content on winter-grown pepper. *Journal of Horticultural Science* 72:893-900.

Merwin, M.L., J.A. Martin, and R.D. Westfall (1995). Provenance and progeny variation in growth and frost tolerance of *Casuarina cunninghamiana* in California, USA. *Forest Ecology and Management* 79:161-171.

Meza-Basso, L., M. Alberdi, M. Raynal, M.L. Ferrero-Cadinanos, and M. Delseny (1986). Changes in protein synthesis in rapeseed (*Brassica napus* L.) seedlings during a low temperature treatment. *Plant Physiology* 82:733-738.

Miedema, P. (1979). Potential use of CIMMYT gene pool material for improving low-temperature adaptation in *Zea mays*. *Euphytica* 28:661-664.

Miedema, P. (1982). The effects of low temperature on *Zea mays*. *Advances in Agronomy* 35:93-128.

Misawa, S., N. Mori, S. Takumi, S. Yoshida, and C. Nakamura (2000). Mapping of QTLs for low temperature response in seedlings of rice (*Oryza sativa* L.). *Cereal Research Communications* 28:33-40.

Mitchell, D.E. and M.A. Madore (1992). Patterns of assimilate production and translocation in muskmelon (*Cucumis melo* L.) II. Low temperature effects. *Plant Physiology* 99:966-971.

Mitra, J. and F. Steward (1961). Growth induction in cultures of *Haplopappus gracilis*. II. The behavior of the nucleus. *American Journal of Botany* 48:358-368.

Mock, J.J. and A.A. Bakri (1976). Recurrent selection for cold tolerance in maize. *Crop Science* 16:230-233.

Mock, J.J. and S.A. Eberhart (1972). Cold tolerance in adapted maize populations. *Crop Science* 12:466-469.

Mock, J.J. and M.J. McNeill (1979). Cold tolerance of maize inbred lines adapted to various latitudes in North America. *Crop Science* 19:239-242.

Monti, L.M. (1993). Breeding for cold tolerance to abiotic stress in garden vegetables. *Informatore Agrario* 49:49-51.

Mosely, P.R., T.M. Crosbie, and J.J. Mock (1984). Mass selection for improved cold and density tolerance of two maize populations. *Euphytica* 33:263-269.

Mulcahy, D.L. (1971). A correlation between gametophytic and sporophytic characteristics in *Zea mays* L. *Science* 171:1155-1156.

Mulcahy, D.L. (1975). The biological significance of gamete competition. In *Gamete Competition in Plants and Animals*, D.L. Mulcahy (Ed.). Amsterdam, The Netherlands: North-Holland, pp. 1-4.

Mulcahy, D.L. (1979). The rise of angiosperms: A genecological factor. *Science* 26:20-23.

Muñoz, T., J. Ruiz-Cabello, A.D. Molina-García, M.I. Escribano, and C. Merodio (2001). Chilling temperature storage changes the inorganic phosphate pool distribution in cherimoya *(Annona cherimola)* fruit. *Journal of the American Society for Horticultural Science* 126:122-127.

Murai, M., H.K.C. Bahadur, T. Yoshida, and K. Ise (2000). Relation between dwarfing genes and chilling injury at seedling stage in rice. *Environmental Control in Biology* 38:135-147.

Murata, N. (1983). Molecular species composition on phosphatidyl-glycerols from chilling-sensitive and chilling-resistant plants. *Plant Cell Physiology* 24:81-86.

Murata, N, O. Ishizaki-Nishizawa, S. Higashi, H. Hayashi, Y. Tasaka, and I. Nishida (1992). Genetically engineered alteration in the chilling sensitivity of plants. *Nature* 356:710-713.

Murelli, C., F. Rizza, F.M. Albini, A. Dulio, V. Terzi, and L. Cativelli (1995). Metabolic changes associated with cold-acclimation in contrasting cultivars of barley. *Physiologia Plantarum* 94:87-93.

Ndong, C., F. Ouellet, M. Houde, and F. Sarhan (1997). Gene expression during cold acclimation in strawberry. *Plant and Cell Physiology* 38:863-870.

Neal, N.P. (1950). Breeding corn for tolerance to cold. In *Progress in Corn Production*. Chicago, IL: American Seed Trade Association, pp. 68-78.

Newsted, W.J., N.R. Chibbar, and F. Georges (1990). Effect of low temperature stress on the expression of sucrose synthetase in spring and winter wheat plants. Development of a monoclonal antibody against wheat germ sucrose synthetase. *Biochemical Cell Biology* 69:36-41.

Nie, G.Y. and N.R. Baker (1991). Modifications to thylakoid composition during development of maize leaves at low growth temperatures. *Plant Physiology* 95:184-191.

Nie, G.Y., S.P. Long, and N.R. Baker (1992). The effects of development at suboptimal growth temperatures on photosynthetic capacity and susceptibility to chilling-dependent photoinhibition in *Zea mays*. *Physiologia Plantarum* 85:554-560.

Nienhuis, J., R.L. Lower, and J.E. Staub (1983). Selection for improved low temperature germination in cucumber. *Journal of the American Society for Horticultural Science* 108:1040-1043.

Nieuwhof, M., L.C. Keizer, and J.C. van Oeveren (1997). Effect of temperature growth and development of adult plants of genotypes of tomato (*Lycopersicum esculentum* Mill.). *Journal of Genetics and Breeding* 51:185-193.

Nilsson, J.E. and E.A. Walfridsson (1995). Phenological variation among plus-tree clones of *Pinus sylvestris* (L.) in northern Sweden. *Silvae Genetica* 44:20-28.

Nishida, I. and N. Murata (1996). Chilling sensitivity in plants and cyanobacteria: The crucial contribution of membrane lipids. *Annual Review of Plant Physiology and Plant Molecular Biology* 47:541-568.

Nishimura, M. (1995). Inheritance of cool tolerance booting stage of rice cultivars in Hokkaido. *Breeding Science* 45:479-485.

Nothman, J., I. Rylski, and M. Spiegelman (1983). Floral morphology and position, cluster size and seasonal set in different eggplant cultivars. *Journal of Horticultural Science* 58:403.

Nyman, M. and S. Waara (1997). Characterization of somatic hybrids between *Solanum tuberosum* and its frost tolerant relative *Solanum commersonii. Theoretical and Applied Genetics* 95:1127-1132.

Ogren, E. (1999). Fall frost resistance in willows used for biomass production. II. Predictive relationships with sugar concentration and dry matter content. *Tree Physiology* 19:755-760.

Olien, C.R. and J.L. Clark (1995). Freeze-induced changes in carbohydrates associated with hardiness of barley and rye. *Crop Science* 35:496-502.

Olien, C.R. and G.E. Lester (1985). Freeze-induced changes in soluble carbohydrates in rye. *Crop Science* 25:285-290.

Ordás, A., P. Revilla, R.A. Malvar, and M.E. Cartea (1994). Development of sweet corn hybrids adapted to the environmental conditions of the northwest of Spain. *Maydica* 39:171-175.

Orr, W., A.I. de la Roche, J. Singh, and H. Voldeng (1983). Imbibitional chilling injury in cultivars of soybean differing in temperature sensitivity to pod formation and maturation period. *Canadian Journal of Botany* 61:2996-2998.

Otubo, S.T., M.A.P. Ramalho, A. de B. Abreu, and J.B. dos Santos (1996). Genetic control of low temperature tolerance in germination of the common bean (*Phaseolus vulgaris* L.). *Euphytica* 89:313-317.

Oyanedel, E., D.W. Wolfe, T.G. Owens, A.J. Monforte, and S.D. Tanksley (2000). Quantitative trait loci analysis of photoinhibition under chilling stress in tomato. *Acta Horticulturae* 521:227-231.

Paldi, E., G. Szalai, and T. Janda (1997). Effect of low temperature on nucleic acid synthesis in wheat and barley. *Acta Agronomica Hungarica* 45:215-221.

Palonen, P. and D. Buszard (1997). Current state of cold hardiness research on fruit crops. *Canadian Journal of Plant Science* 77:399-420.

Palta, J.P. (1994). Sorting genes controlling freezing stress resistance. Strategy for moving desired traits by merging physiological and genetic approaches. In *Biochemical and Cellular Mechanisms of Stress Tolerance in Plants. Proceedings of the NATO Advanced Research Workshop,* J.H. Cherry (Ed.). Berlin, Germany, p. 569.

Palta, J.P., B.D. Whitaker, and L.S. Weiss (1993). Plasma membrane lipids associated with genetic variability in freezing tolerance and cold acclimation of *Solanum* species. *Plant Physiology* 103:793-803.

Pan, A., P.M. Hayes, F. Chen, T.H.H. Chen, T. Blake, S. Wright, I. Karsai, and Z. Bedö (1994). Genetic analysis of the components of winterhardiness in barley (*Hordeum vulgare* L.). *Theoretical and Applied Genetics* 89:900-910.

Parodi, P.C., W.E. Nyquist, F.L. Patterson, and H.F. Hodges (1983). Traditional combining-ability and Gardner-Eberhart analyses of a diallel for cold resistance in winter wheat. *Crop Science* 23:314-318.

Pattanayak, A., R.N. Bhuyan, H.S. Gupta, M. Sreedhar, and M.S. Prasad (2000). Performance of cold-tolerant rice lines developed through anther culture for mid-altitude areas of Meghalaya, India. *International Rice Research Notes* 25:11-12.

Pearce, R.S. (2001). Plant freezing and damage. *Annals of Botany* 87:417-424.

Pendleton, J.W. (1965). Cultural practises. *Annual Corn Sorghum Research Conference* 20:51-58.

Perras, M. and F. Sarhan (1989). Synthesis of freezing tolerance proteins in leaves, crown, and roots during cold acclimation of wheat. *Plant Physiology* 89:577-585.

Pesev, N.V. (1970). Genetic factors affecting maize tolerance to low temperatures at emergence and germination. *Theoretical and Applied Genetics* 40:351-356.

Philley, H.W., C.E. Watson Jr., J.V. Krans, J.M. Goatley Jr., V.L. Maddox, and M. Tomaso-Peterson (1998). Inheritance of cold tolerance in St. Augustinegrass. *Crop Science* 38:451-454.

Philley, H.W., C.E. Watson Jr., J.V. Krans, J.M. Goatley, and F.B. Matta (1995). Differential thermal analysis of St. Augustinegrass. *HortScience* 30:1388-1389.

Pinhero, R.G., M.V. Rao, G. Paliyath, D.P. Murr, and R.A. Fletcher (1997). Changes in activities of antioxidant enzymes and their relationships to genetic and paclobutrazol-induced chilling tolerance of maize seedlings. *Plant Physiology* 114:695-704.

Pitman, W.D., J.L. Hafley, G.J. Cuomo, and A.E. Kretschmer Jr. (1997). Stand regeneration of *Alysicarpus* and *Desmodium* germplasm in Lousiana. *Crop Science* 37:1373-1376.

Porta, N. La, M. Zacchini, S. Bartolini, R. Viti, and G. Roselli (1994). The frost hardiness of some clones of olive cv. Leccino. *Journal of Horticultural Science* 69:433-435.

Powles, S.B., J.A. Berry, and O. Björkman (1983). Interaction between light and chilling temperature on the inhibition of photosynthesis in chilling-sensitive plants. *Plant Cell and Environment* 6:117-123.

Prasad, T.K. (1997). Role of catalase in inducing chilling tolerance in pre-emergent maize seedlings. *Plant Physiology* 114:1369-1376.

Preil, E., J.B.M. Huitema, and J. de Jong (1991). Methods for selection of low-temperature tolerant mutants of *Chrysanthemum morifolium* Ramat. using irradiated cell suspension cultures. II. Preselection in vivo under low-temperature stress. *Plant Breeding* 107:131-134.

Preiszner, J., A. Fehér, O. Veisz, J. Sutka, and D. Dudits (1991). Characterization of morphological variation and cold resistance in interspecific somatic hybrids between potato (*Solanum tuberosum* L.) and *S. brevidens* Phil. *Euphytica* 57:37-49.

Prosperi, J.M., F. Demarquet, M. Angevain, and P. Mansant (1994). Évaluation agronomique de variétés de pays de sainfoin (*Onobrychis sativa* L.) originaires du sud-est de la France [Agronomic evaluation of local varieties of sainfoin (*Onobrychis sativa* L.) collected in southeast France]. *Agronomie* 14:285-298.

Purvis, A.C. (1981). Free proline in peels of grapefruit and resistance to chilling injury during cold storage. *HortScience* 16:160-161.

Rahman, H., Z.W. Wicks, T.E. Schumacher, and Z.A. Swati (1994). Synthesis of maize populations based on seedling root indices. II. Field evaluations for yield and related traits. *Journal of Genetics and Breeding* 48:245-251.

Ranalli, P., M. di Candilo, B. Parisi, I. Poma, A.S. Frenda, I. Giordano, and V. Pirani (1999). Dry grain pea: Assessment of new breeding lines in various environments. *Rivista di Agronomia* 33:109-117.

Rao, S.C. and T.H. Dao (1987). Soil water effects on low-temperature seedling emergence of five *Brassica* cultivars. *Agronomy Journal* 19:517-519.

Rapacz, M. and A. Markowski (1999). Winter hardiness, frost resistance and vernalization requirement of European winter oilseed rape (*Bassica napus* var. oleifera) cultivars within the last 20 years. *Journal of Agronomy and Crop Science* 183:243-253.

Raper, C.D. Jr and J.F. Thomas (1972). Temperatures in early post-transplant growth: Effect on shape of mature *Nicotiana tabacum* L. leaves. *Crop Science* 12:540-542.

Ratinam M., A.M. Abd-El-Monem, and M.C. Saxena (1994a). Assessment of sensitivity to frost in *ochrus* chickling (*Lathyrus ochrus* [L.] D.C.) by chlorophyll fluorescence analyses. *Journal of Agronomy and Crop Science* 173:338-344.

Ratinam, M., A.M. Abd-El-Monem, and M.C. Saxena (1994b). Variations in sugar content and dry matter distribution in roots and their associations with frost tolerance in certain forage legume species. *Journal of Agronomy and Crop Science* 173:345-353.

Repo, T. and S. Pulli (1996). Application of impedance spectroscopy for selecting frost hardy varieties of English ryegrass. *Annals of Botany* 78:605-609.

Revilla, P., R.A. Malvar, M.E. Cartea, A. Butrón, and A. Ordás (2000). Inheritance of cold tolerance at emergence and during early season growth in maize. *Crop Science* 40:1579-1585.

Revilla, P., R.A. Malvar, M.E. Cartea, and A. Ordás (1998). Identifying open-pollinated populations of field corn as sources of cold tolèrance for improving sweet corn. *Euphytica* 101:239-247.

Rhode, D.R. and C.F. Pulham (1960). Heritability estimates of winter hardiness in winter barley determined by the standard unit method of regression analysis. *Agronomy Journal* 52:584-586.

Rhodes, I., R.P. Collins, and D.R. Evans (1994). Breeding white clover for tolerance to low temperature and grazing stress. *Euphytica* 77:239-242.

Richner, W., A. Soldati, and P. Stamp (1996). Shoot-to-root relations in field-grown maize seedlings. *Agronomy Journal* 88:56-61.

Rikin, A., D. Atsmon, and D. Gitler (1979). Chilling injury in cotton (*Gossypium hirsutum* L.): Prevention by abscisic acid. *Plant Cell Physiology* 20:1537-1546.

Rinke, E.H. (1953). Cold test germination. In *Eighth Hybrid Corn Industry Research Conference*, Chicago, IL. Washington, DC: American Seed Trade Association, pp. 54-58.

Rizza, F., P. Baldi, L. Cattivelli, and G. Delogu (1997). Cold hardening in triticale in comparison with rye and wheat. *Cereal Research Communications* 25:947-954.

Rognli, O.A. (1988). Species aspects of breeding herbage varieties for northern marginal regions. *Journal of Agricultural Science in Finland* 60:181-189.

Rognli, O.A. and A. Myhre (1996). Variation in frost-tolerance and agronomic characters among genotypes of meadow fescue, selected for different allozyme combinations at the glucosephosphate isomerase-2 *(Gpi-2)* locus. *Nordisk Jordbruksforskning* 78:85.

Rowland, L.J., E.L. Ogden, R. Arora, C.C. Lim, J.S. Lehman, A. Levi, and G.R. Panta (1999). Use of blueberry to study genetic control of chilling requirement and cold hardiness in woody perennials. *HortScience* 34:1185-1191.

Rylski, I. (1979). Fruit set and development of seeded and seedless tomato fruits under diverse regimes of temperature and pollination. *Journal of the American Society for Horticultural Science* 104:835-838.

Saito, K., K. Miura, K. Nagano, Y. Hayano-Saito, A. Saito, H. Araki, and A. Kato (1995). Chromosomal location of quantitative trait loci for cool tolerance at the booting stage in rice variety "Norin-PL8." *Breeding Science* 45:337-340.

Sakamoto, A., H. Alia, and N. Murata (1998). Metabolic engineering of rice leading to biosynthesis of glycinebetaine and tolerance to salt and cold. *Plant Molecular Biology* 38:1011-1019.

Sakamoto, A., R. Valverde, H. Alia, T.H.H. Chen, and R. Murata (2000). Transformation of *Arabidopsis* with the *codA* gene choline oxidase enhances freezing tolerance of plants. *Plant Journal* 22:449-453.

Samala, S., Yan JiYu, and W.V. Baird (1998). Changes in polar lipid fatty acid composition during cold acclimation in "Midiron" and "U3" Bermudagrass. *Crop Science* 38:188-195.

Sarhan, F. and J. Danyluk (1998). Engineering cold-tolerant crops: Throwing the master switch. *Trends in Plant Science* 3:289-290.

Sarhan, F., F. Ouellet, and A. Vázquez-Tello (1997). The wheat *wcs120* gene family. A useful model to understand the molecular genetics of freezing tolerance in cereals. *Physiologia Plantarum* 101:439-445.

Saveljev, N.I. and I.V. Michurin (1988). Breeding material and genetic principles of breeding apple trees for frost and cold tolerance. *Acta Horticulturae* 224:165-168.

Savolainen, O. (1996). Pines beyond the polar circle: Adaptation to stress conditions. *Euphytica* 92:139-145.

Schapendonk, A.H.C.M., P.E.L. Van der Putten, O. Dolstra, S.R. Haalstra, and W.J.M. Tonk (1992). Chlorophyll fluorescence: A non-destructive method for detecting damage in the photosynthetic apparatus in plants. *Acta Horticulturae* 304:61-70.

Schjelderup, I., A.H. Aastveit, and K. Aastveit (1994). Winter hardiness in marginal populations of timothy. *Euphytica* 77:193-198.

Schreiber, U., U. Schliwa, and W. Bilger (1986). Continuous recording of photochemical and non-photochemical chlorophyll fluorescence quenching with a new type of modulated fluorometer. *Photosynthesis Research* 10:51-62.

Schultz, H.K. (1941). A study of methods of breeding orchard grass, *Dactylis glomerata* L. *Journal of the American Society of Agronomy* 33:546-558.

Schwab, P.M., D.K. Barnes, C.C. Sheaffer, and P.H. Li (1996). Factors affecting a laboratory evaluation of alfalfa cold tolerance. *Crop Science* 36:318-324.

Sebastiani, L., M. Lucchessini, and C. Vitagliano (1996). In vivo and in vitro evaluating chilling resistance in *Lycopersicon esculentum* Mill. seedlings. *Advances in Horticultural Science* 10:115-119.

Seddigh, M., G.D. Jolliff, and J.H. Orf (1988). Field evaluations of early maturing soybean genotypes for differential adaptation to low night temperatures. *Crop Science* 28:639-643.

Sen Gupta, A.S., J.L. Heinen, A.S. Holaday, J.J. Burke, and R.D. Allen (1993). Increased resistance to oxidative stress in transgenic plants that overexpress chloroplastic Cu/Zn superoxide dismutase. *Proceedings of the National Academy of Science USA* 90:1629-1633.

Seppanen, M.M., M. Majaharju, S. Somersalo, and E. Pehu (1998). Freezing tolerance, cold acclimation and oxidative stress in potato: Paraquat tolerance is related to acclimation but is a poor indicator of freezing tolerance. *Physiologia Plantarum* 102:454-460.

Sethi, G.S., P. Plahe, and K.S. Thakur (1993). Breeding bread wheat for cold stress. *Annals of Biology* 9:165-173.

Seyedbagheri, M.M. and E. Fallahi (1994). Physiological and environmental factors and horticultural practices influencing cold hardiness of grapevines. *Journal of Small Fruit and Viticulture* 2:3-38.

Simmonds N.W. and J. Smartt (1999). *Principles of Crop Improvement*, Second Edition. London, UK: Blackwell Science.

Simon, E.W., A. Minchin, M.M. McMenanin, and J.M. Smith (1976). The low temperature limit for seed germination. *New Phytologist* 77:301-311.

Simpson, D.G. (1994). Seasonal and geographic origin effects on cold hardiness of white spruce buds, foliage, and stems. *Canadian Journal of Forest Research* 24:917-927.

Singh, B.B, D.P. Singh, and N.P. Singh (1997). Genetic variation in pigeonpea germplasm for cold tolerance. *Indian Journal of Genetics and Plant Breeding* 57:425-430.

Singh, K.B., R.S. Malhotra, and M.C. Saxena (1989). Chickpea evaluation for cold tolerance under field conditions. *Crop Science* 29:282-285.

Singh, K.B., R.S. Malhotra, and M.C. Saxena (1990). Sources for tolerance to cold in *Cicer* species. *Crop Science* 30:1136-1138.

Singh, K.B., R.S. Malhotra, and M.C. Saxena (1995). Additional sources of tolerance to cold in cultivated and wild *Cicer* species. *Crop Science* 35:1491-1497.

Singh, K.B., R.S. Malhotra, and M.C. Saxena (1997). Registration of FLIP 91-178C, FLIP 93-53C, and FLIP 93-98C chickpea germplasm lines resistant to *Ascochyta* blight, *Fusarium* wilt, and cold. *Crop Science* 37:633.

Singh, O. and S. Kaur (1993). New promising selections with superior freeze resistance. *Indian Sugar* 43:491-494.

Singh, O. and K. Singh (2000). Screening model for low temperature stress resistance potential in sugarcane. *Indian Sugar* 50:87-96.

Singh, S.P. (1985). Sources of cold tolerance in grain sorghum. *Canadian Journal of Plant Science* 65:251-257.

Singha, N.K., J.K. Handoo, and A.K. Srivastava (1999). Screening technique for low temperature resistant maize varieties. *Journal of Applied Biology* 9:167-168.

Smeets, L. and T.C. Wehner (1997). Environmental effects on genetic variation of chilling resistance in cucumber. *Euphytica* 97:217-225.

Smith, P.G. and A.H. Millet (1964). Germinating and sprouting responses of the tomato at low temperatures. *Journal of the American Society for Horticultural Science* 84:480-484.

Smolenska-Sym, G., H. Gawronska, and A. Kacperska (1995). Modifications of abscisic acid level in winter oilseed rape leaves during acclimation of plants to freezing temperatures. *Plant Growth Regulation* 17:61-65.

Snape, J.W., R. Sarma, S.A. Quarrie, L. Fish, G. Galiba, and J. Sutka (2001). Mapping genes for flowering time and frost tolerance in cereals using precise genetic stocks. *Euphytica* 120:309-315.

Snape, J.W., A. Semikhodskii, L. Fish, R.N. Sarma, S.A. Quarrie, G. Galiba, and J. Sutka (1997). Mapping frost tolerance loci in wheat and comparative mapping with other cereals. *Acta Agronomica Hungarica* 45:265-270.

Somerville, C. and J. Browse (1990). Plant lipids: Metabolism, mutants, and membranes. *Science* 252:80-87.

Songstad, D.D., D.R. Duncan, and J.M. Widholm (1990). Proline and polyamine involvement in chilling tolerance of maize suspension cultures. *Journal of Experimental Botany* 41:289-294.

Sonnewald, U. (1992). Expression of *E. coli* inorganic pyrophosphatase in transgenic plants alters photoassimilase portioning. *Plant Journal* 2:571-581.

Sowínski, P., A. Dalbiak, P. Ochodzki, J. Adamczyk, and Z. Królikowski (1999). Carbohydrate accumulation in maize leaves at moderate low temperature under controlled conditions and early field characteristics of flint and dent genotypes adapted to Polish climate. *Plant Breeding and Seed Science* 43:39-48.

Sowínski, P. and Z. Królikowski (1995). Chilling sensitivity in maize (*Zea mays* L.) seedlings. III. Relations between growth and functioning at low temperatures and during post-stress recovery. *Acta Physiologiae Plantarum* 17:219-224.

Sowokinos, J.R. (2001). Allele and isozyme patterns of UDP-glucose pyrophosphorylase as a marker for cold-sweetening resistance in potatoes. *American Journal of Potato Research* 78:57-64.

Stair, D.W., M.L. Dahmer, E.C. Bashaw, and M.A. Hussey (1998). Freezing tolerance of selected *Pennisetum* species. *International Journal of Plant Science* 159:599-605.

Stamp, P. (1984). Chilling tolerance of young plants demonstrated on the example of maize (*Zea mays* L.) *Journal of Agronomy and Crop Science* 7:1-83.

Stamp, P. (1988). The influence of severe cold stress on the characteristics of young plants of adapted and exotic maize genotypes. *Plant Research and Development* 28:63-77.

Stassart, J.M., L. Slooten, J. Botterman, F. van Breuseghem, N. Meray, and D. Inzé (1995). Cold-tolerance, oxidative stress tolerance and anti-oxidant enzyme levels in non-transgenic and Mn-superoxide dismutase overexpressing maize. In *Photosynthesis: From Light to Biosphere,* Volume IV, P. Mathis (Ed.). Dordrecht, The Netherlands: Kluwer Academic Publishers, pp. 873-876.

Staub, J.E., R.L. Lower, and J. Nienhuis (1988). Correlated response to selection for low temperature germination in cucumber. *HortScience* 23:745-746.

Steponkus, P.L. (1978). Cold hardiness and freezing injury of agronomic crops. *Advances in Agronomy* 30:51-98.

Steponkus, P.L. (1984). Role of the plasma membrane in freezing injury and cold acclimation. *Annual Review of Plant Physiology* 35:543-581.

Stewart, C.R., B.A. Martin, L. Reding, and S. Cerwick (1990). Respiration and alternative oxidase in corn seedling tissues during germination at different temperatures. *Plant Physiology* 92:755-760.

Sthapit, B.R. and J.R. Witcombe (1998). Inheritance of tolerance to chilling stress in rice during germination and plumule greening. *Crop Science* 38:660-665.

Sthapit, B.R., J.R. Witcombe, and J.M. Wilson (1995). Methods of selection for chilling tolerance in Nepalese rice by chlorophyll fluorescence analysis. *Crop Science* 35:90-94.

Stone, J.M., J.P. Plata, J.B. Blamberg, L.S. Weis, and J.F. Harbage (1993). Inheritance of freezing resistance in tuber-bearing *Solanum* species: Evidence for independent genetic control of non-acclimated freezing tolerance and cold acclimation capacity. *Proceedings of the National Academy of Science USA* 90:7869-7873.

Storlie, E.W., R.E. Allan, and M.K. Walker-Simmons (1998). Effect of the *Vrn1-Fr1* interval on cold hardiness level in near-isogenic wheat lines. *Crop Science* 38:483-488.

Subedi, K.D., C.N. Floyd, and C.B. Budhathoki (1998). Cool temperature-induced sterility in spring wheat (*Triticum aestivum* L.) at high altitudes in Nepal: Variation among cultivars in response to sowing date. *Field Crops Research* 55:141-151.

Sugiyama, S. (1998). Differentiation in competitive ability and cold tolerance between diploid and tretaploid cultivars in *Lolium perenne*. *Euphytica* 103:55-59.

Sutka, J. (1981). Genetic studies of frost resistance in wheat. *Theoretical and Applied Genetics* 59:145-152.

Sutka, J. (1994). Genetic control of frost tolerance in wheat (*Triticum aestivum* L.). *Euphytica* 77:277-282.

Sutka, J., G. Galiba, and J.W. Snape (1997). Inheritance of frost resistance in wheat (*Triticum aestivum* L.). *Acta Agronomica Hungarica* 45:257-263.

Sutka, J. and O. Veisz (1988). Reversal of dominance in a gene on chromosome 5A controlling frost resistance in wheat. *Genome* 30:313-317.

Suzuki, S. (1982). Cold tolerance in rice plants with special reference to floral characters. II. Relations between floral characters and the degree of cold tolerance in segregating generations. *Japanese Journal of Breeding* 32:9.

Svejda, F. (1979). Inheritance of winterhardiness in roses. *Euphytica* 28:309-314.

Svenning, M.M., K. Rosnes, and O. Juntilla (1997). Frost tolerance and biochemical changes during hardening and dehardening in contrasting white clover populations. *Physiologia Plantarum* 101:31-37.

Tahir, M. and V. Shevtsov (1994). New hull-less barley germplasm for high-altitude areas. *Rachis* 13:43-46.

Tahir, M., V. Shevtsov, H. Pashayani, A. Ottekin, H. Tosun, and T. Akar (1994). Stress tolerance in winter and facultative barley. *Rachis* 13:5-11.

Tai, P.Y.P. and J.D. Miller (1996). Selection for frost resistance in sugarcane. *International Sugar Journal (Cane Sugar Edition)* 98:631-636.

Takahashi, R. and J. Abe (1994). Genetic and linkage analysis of low temperature-induced browning in soybean seed coats. *Journal of Heredity* 85:447-450.

Takahashi, R. and J. Abe (1999). Soybean maturity genes associated with seed coat pigmentation and cracking in response to low temperatures. *Crop Science* 39:1657-1662.

Takahashi, R. and S. Asanuma (1996). Association of *T* gene with chilling tolerance in soybean. *Crop Science* 36:559-562.

Tamminen, I., P. Makela, P. Heino, and E.T. Palva (2001). Ectopic expression of *ABI3* gene enhances freezing tolerance in response to abscisic acid and low temperature in *Arabipodsis thaliana*. *Plant Journal* 25:1-8.

Tanida, M. (1996). Catalase activity of rice seed embryo and its relation to germination rate at low temperature. *Breeding Science* 46:23-27.

Tantau, H., B. Brettschneider, C. Balko, H.U. Jürgens, G. Melz, and K. Dörffling (1996). Improvement of frost tolerance in winter barley by in vitro-selection of hydroxyproline-resistant lines derived from anther cultures. *Nordisk Jordbruksforskning* 78:106.

Tantau, H. and K. Dörffling (1991). In vitro-selection of hydroxyproline-resistant cell lines of wheat *(Triticum aestivum):* Accumulation of proline, decrease in osmotic potential, and increase in frost tolerance. *Physiologia Plantarum* 82:243-248.

Taylor, A.O. and A.S. Craig (1971). Plants under climatic stress. Low temperature, and high light effects on chloroplast ultrastructure. *Plant Physiology* 47:719-725.

Ternovskaya, T.K. and E.G., Zhirov (1993). Genetics and breeding. *Russian Agricultural Sciences* 10:1-5.

Teutonico, R.A., J.P. Palta, and T.C. Osborn (1993). In vitro freezing tolerance in relation to winter survival of rapeseed cultivars. *Crop Science* 33:103-107.

Teutonico, R.A., B. Yandell, J.M. Satagopan, M.E. Ferreira, J.P. Palta, and T.C. Osborn (1995). Genetic analysis and mapping of genes controlling freezing tolerance in oilseed *Brassica*. *Molecular Breeding* 1:329-339.

Thomas, H. and A.R. James (1993). Freezing tolerance and solute changes in contrasting genotypes of *Lolium perenne* L. acclimated to cold and drought. *Annals of Botany* 72:249-254.

Thomashow, M.F. (1990). Molecular genetics of cold acclimation in higher plants. *Advances in Genetics* 28:99-131.

Thomashow, M.F. (1998). Role of cold-responsive genes in plant freezing tolerance. *Plant Physiology* 118:1-7.

Thomashow, M.F. (1999). Plant cold acclimation: Freezing tolerance genes and regulatory mechanisms. *Annual Review of Plant Physiology and Plant Molecular Biology* 50:571-599.

Thomashow, M.F. (2001). So what's new in the field of plant cold acclimation? Lots! *Plant Physiology* 125:89-93.

Tokuhisa, J. and J. Browse (1999). Genetic engineering of plant chilling tolerance. In *Genetic Engineering 21*, J.K. Setlow (Ed.). New York: Kluwer Academic/Plenum Publishers, pp. 79-93.

Tokuhisa, J.G., K.A. Feldmann, S.T. LaBrie, and J. Browse (1997). Mutational analysis of chilling tolerance in plants. *Plant Cell and Environment* 20:1391-1400.

Tollenaar, M. and J. Wu (1999). Yield improvement in temperate maize is attributable to greater stress tolerance. *Crop Science* 39:1597-1604.

Treat, C.L. and W.F. Tracy (1994). Endosperm type effects on biomass production and on stalk and root quality in sweet corn. *Crop Science* 34:396-399.

Tuberosa, R., G. Galiba, M. Sanguineti, E. Noli, and J. Sutka (1997). Identification of QTL influencing freezing tolerance in barley. *Acta Agronomica Hungarica* 45:413-417.

Tucci, M., D. Carputo, G. Bile, and L. Frusciante (1996). Male fertility and freezing tolerance of hybrids involving *Solanum tuberosum* haploids and diploid *Solanum* species. *Potato Research* 39:345-353.

Turner, L.B. and C.J. Pollock (1998). Changes in stolon carbohydrates during the winter in four varieties of white clover (*Trifolium repens* L.) with contrasting hardiness. *Annals of Botany* 81:97-107.

Väinölä, A. and T. Repo (1999). Cold hardiness of diploid and corresponding autotretaploid rhododendrons. *Journal of Horticultural Science and Biotechnology* 74:541-546.

Vallejos, C.E. and S.D. Tanksley (1983). Segregation of isozyme markers and cold tolerance in an interspecific backcross of tomato. *Theoretical and Applied Genetics* 66:241-247.

Van Oosterom, E.J. and S. Ceccarelli (1993). Indirect selection for grain yield of barley in harsh Mediterranean environments. *Crop Science* 33:1127-1131.

Van Swaay A.C., E. Jacobsen, and W.J. Feenstra (1985). Effect of coldhardening wilting and exogenously applied proline on leaf proline content and frost tolerance of several genotypes of *Solanum. Physiologia Plantarum* 64:230-236.

Vázquez-Thello, A., Li Jun Yang, M. Hidaka, and T. Uozumi (1996). Inherited chilling tolerance in somatic hybrids of transgenic *Hibiscus rosa-sinensis* × transgenic *Lavatera thuringiaca* selected by double-antibiotic resistance. *Plant Cell Reports* 15:506-511.

Vega, S.E. and J.B. Bamberg (1995). Screening the U.S. potato collection for frost hardiness. *American Potato Journal* 72:13-21.

Veisz, O., Z. Bedo, L. Láng, and T. Tischner (1997). Genetic analysis of the frost resistance of wheat through diallel analysis. *Acta Agronomica Hungarica* 45:117-125.

Veisz, O., N. Harnos, L. Szunics, and T. Tischner (1996). Overwintering of winter cereals in Hungary in the case of global warming. *Euphytica* 92:249-253.

Verheul, M.J., P.R. Hassel, and P. van Stamp (1995). Comparison of maize inbred lines differing in low temperature tolerance: Effect of acclimation at suboptimal temperature on chloroplast functioning. *Annals of Botany* 76:7-14.

Verheul, M.J., C. Picatto, and P. Stamp (1996). Growth and development of maize (*Zea mays* L.) seedlings under chilling conditions in the field. *European Journal of Agronomy* 5:31-43.

Volker, P.W., J.V. Owen, and N.M.G. Borralho (1994). Genetic variances and covariances for frost tolerance in *Eucaliptus globulus* and *E. nitens. Silvae Genetica* 43:366-372.

Von Schaewen, A., M. Stitt, R. Schmidt, U. Sonnewald, and L. Willmitzer (1990). Expression of a yeast-derived invertase in the cell wall of tobacco and *Arabidopsis* plants leads to accumulation of carbohydrate and inhibition of photosynthesis and strongly influences growth and phenotype of transgenic tobacco plants. *European Molecular Biology Organization Journal* 9:3033-3044.

Waldron, B.L., N.J. Ehlke, D.J. Vellekson, and D.B. White (1998). Controlled freezing as an indirect selection method for field winterhardiness in turf-type perennial ryegrass. *Crop Science* 38:811-816.

Waldron, B.L., N.J. Ehlke, D.L. Wyse, and D.J. Vellekson (1998). Genetic variation and predicted gain from selection for winterhardiness and turf quality in a perennial ryegrass top-cross population. *Crop Science* 38:817-822.

Walker, M.A., B.D. McKersie, and K.P. Pauls (1991). Effects of chilling on the biochemical and functional properties of thylakoid membranes. *Plant Physiology* 97:663-669.

Wallace, D.H., P.A. Gniffke, P.N. Masaya, and R.W. Zobel (1991). Photoperiod, temperature, and genotype interaction effects on days and nodes required for flowering of bean. *Journal of the American Society for Horticultural Science* 116:534-543.

Walters, T.W., M.A. Mutschler, and E.D. Earle (1992). Protoplast fusion-derived Oruga male sterile cauliflower with cold tolerance. *Plant Cell Reports* 10:624-628.

Wang, C.Y. (1982). Physiological and biochemical responses of plants to chilling stress. *HortScience* 17:173-186.

Warner, J. and C. Nickerson (1996). Winter injury to apple trees, 1993-1997. *Fruit Varieties Journal* 50:114-118.

Warrington, I.J. and E.T. Kanemasu (1983). Corn growth response to low temperature and photoperiod. III. Leaf number. *Agronomy Journal* 75:762-766.

Warrington, I.J. and R.C. Southward (1995). Seasonal frost tolerance of *Hebe* species and cultivars. *New Zealand Journal of Crop and Horticultural Science* 23:437-445.

Watts, W.R. (1972). Role of temperature in regulation of leaf extension in *Zea mays*. *Nature* 229:46-47.

Welbaum, G.E., D. Bian, and D.R. Hill (1997). Freezing tolerance, protein composition, and abscisic acid localization and content of pea epicotyl, shoot, and root tissue in response to temperature and water stress. *Journal of Experimental Botany* 48:643-654.

Werner, C.M., R.M. Crassweller, and T.E. Clark (1993). Cold hardiness of peach stem tissue over two dormant seasons. *Fruit Varieties Journal* 47:72-79.

Westin, J., L.G. Sundblad, and J.E. Hällgren (1995). Seasonal variation in photochemical activity and hardiness in clones of Norway spruce *(Picea abies)*. *Tree Physiology* 15:685-689.

Wilen, R.W., L.V. Gusta, B. Lei, S.R. Abrams, and B.E. Ewan (1994). Effects of abscisic acid (ABA) and ABA analogs on freezing tolerance, low-temperature growth, and flowering in rapeseed. *Journal of Plant Growth Regulation* 13:235-241.

Xin, Z. and J. Browse (1998). *Eskimo1* mutants of *Arabidopsis* are constitutively freezing-tolerant. *Proceedings of the National Academy of Science USA* 95:7799-7804.

Xin, Z. and P.H. Li (1993). Relationship between proline and abscisic acid in the induction of chilling tolerance in maize suspension-cultured cells. *Plant Physiology* 103:607-613.

Yamaguchi-Shinozaki, K. and K. Shinozaki (1994). A novel *cis*-acting element in an *Arabidopsis* gene is involved in responsiveness to drought, low temperature, or high-salt stress. *Plant Cell* 6:251-264.

Yea, JongDoo (1995). Inheritance of some agronomic characters related to cold tolerance under cold water treatment in rice. *Korean Journal of Crop Science* 40:473-480.

Yea, JongDoo (1996). Inheritance of days to heading, spike-let number and fertility under cold water treatment in rice. *Korean Journal of Crop Science* 41:45-52.

Yelenosky, G. and J.C.V. Vu (1992). Ability of "Valencia" sweet orange to cold-acclimate on cold-sensitive citron rootstock. *HortScience* 27:1201-1203.

Yokoi, S., S.I. Higashi, S. Kishitani, N. Murata, and K. Toriyama (1998). Introduction of the cDNA for *Arabidopsis* glycerol-3-phosphate acyltransferase (GPAT) confers unsaturation of fatty acids and chilling tolerance of photosynthesis on rice. *Molecular Breeding* 4:269-275.

Yoshida, M., J. Abe, M. Moriyama, and T. Kuwabara (1998). Carbohydrate levels among winter wheat cultivars varying in freezing tolerance and snow mold resistance during autumn and winter. *Physiologia Plantarum* 103:8-16.

Zabrodina, M.V., D. Ya Silis, and E.E. Khavkin (1996). Isoenzyme markers of perennial habit as related to rye winter hardiness. *Russian Journal of Plant Physiology* 43:200-204.

Zaiter, H., E. Baydoun, and M. Sayyed-Hallak (1994). Genotypic variation in the germination of common bean in response to cold temperature stress. *Planta and Soil* 163:95-101.

Zamir, D., S.D. Tanksley, and R.A. Jones (1981). Low temperature effects on selective fertilization by pollen mixtures of wild and cultivated tomato species. *Theoretical and Applied Genetics* 59:235-238.

Zamir, D., S.D. Tanksley, and R.A. Jones (1982). Haploid selection for low temperature tolerance of tomato pollen. *Genetics* 101:129-137.

Zatylny A.M., J.T.A. Proctor, and J.A. Sullivan (1993). Screening red raspberry for cold hardiness in vitro. *HortScience* 28:740-741.

Zatylny A.M., J.T.A. Proctor, and J.A. Sullivan (1996). Assessing cold hardiness of red raspberry genotypes in the laboratory and field. *Journal of the American Society for Horticultural Science* 121:495-500.

Zauralov, O.A. (1998). Growth of the primary root: A criterion of the frost resistance of millet varieties. *Russian Agricultural Science* 3:10-13.

Zemetra, R.S. and R.L. Cuany (1991). Variation among inbreds for seed response to low temperatures in maize (*Zea mays* L.). *Maydica* 36:17-23.

Zimmerman, L.H. and B.B. Buck (1977). Selection for seedling cold tolerance in safflower with modified controlled environment chamber. *Crop Science* 5:679-682.

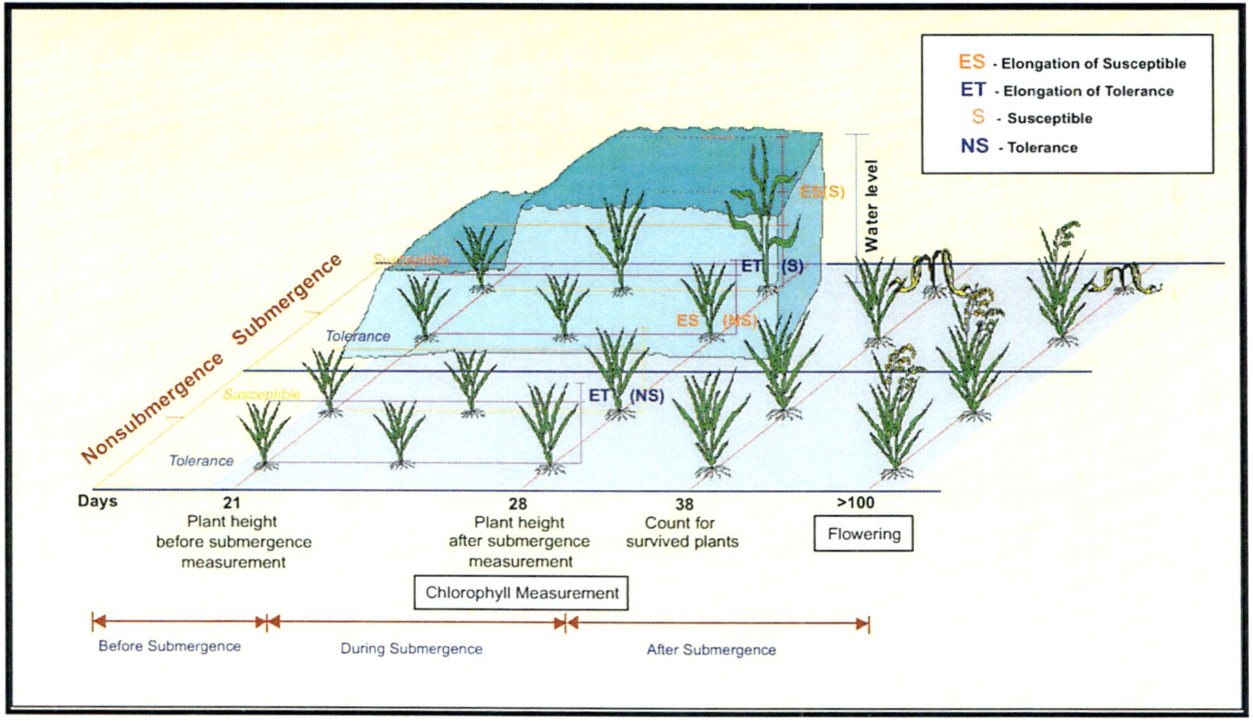

FIGURE 7.1. Design of experiment for flooding tolerance taking rice as example.

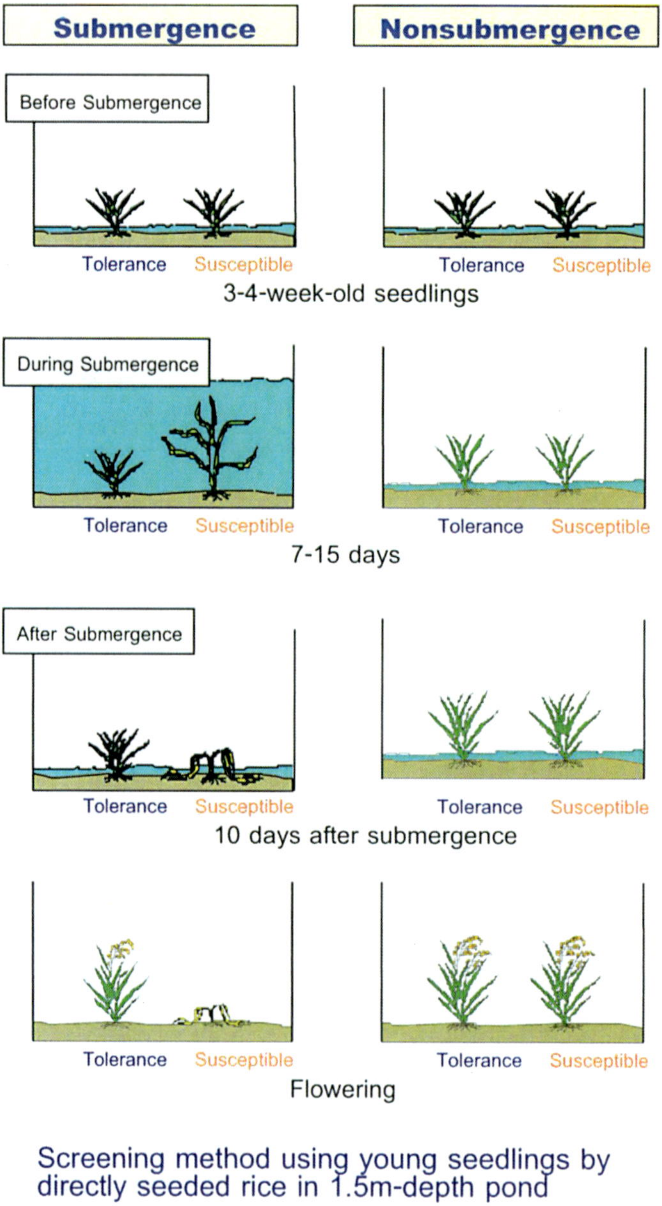

FIGURE 7.2. Differential responses between tolerance and susceptible genotypes during seedling and flowering in rice.

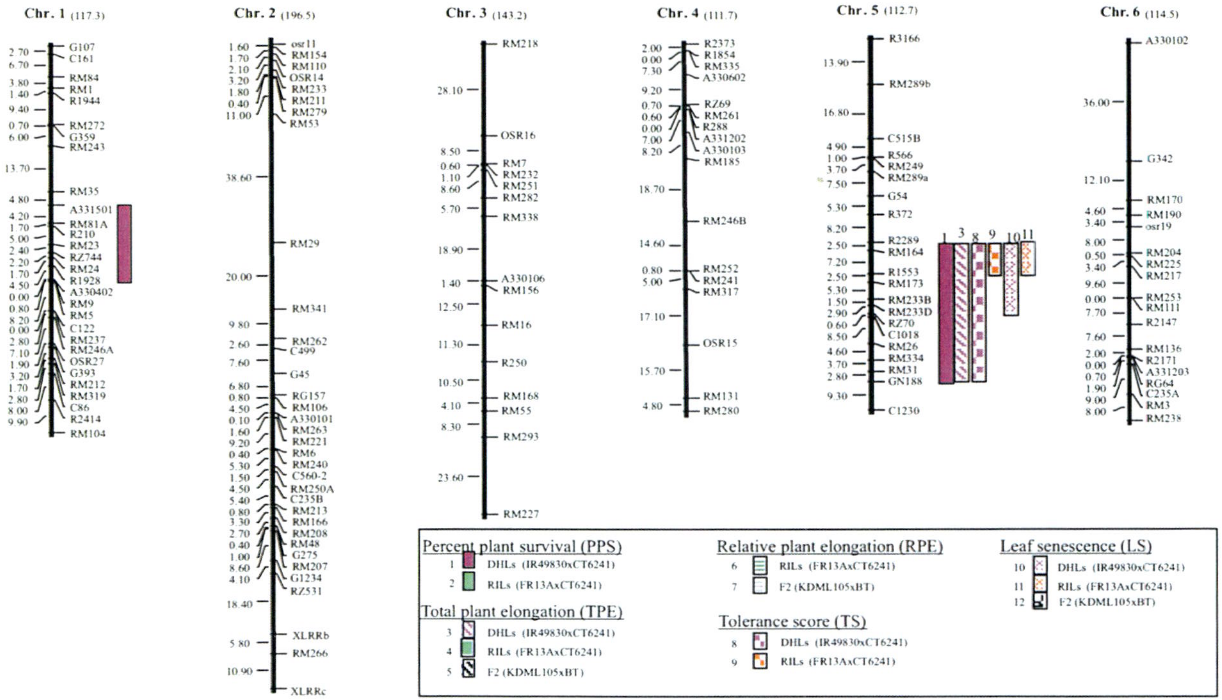

FIGURE 7.3a. QTL mapping for flash flooding tolerance in rice illustrates few minor QTLs located on chromosomes 1and 5.

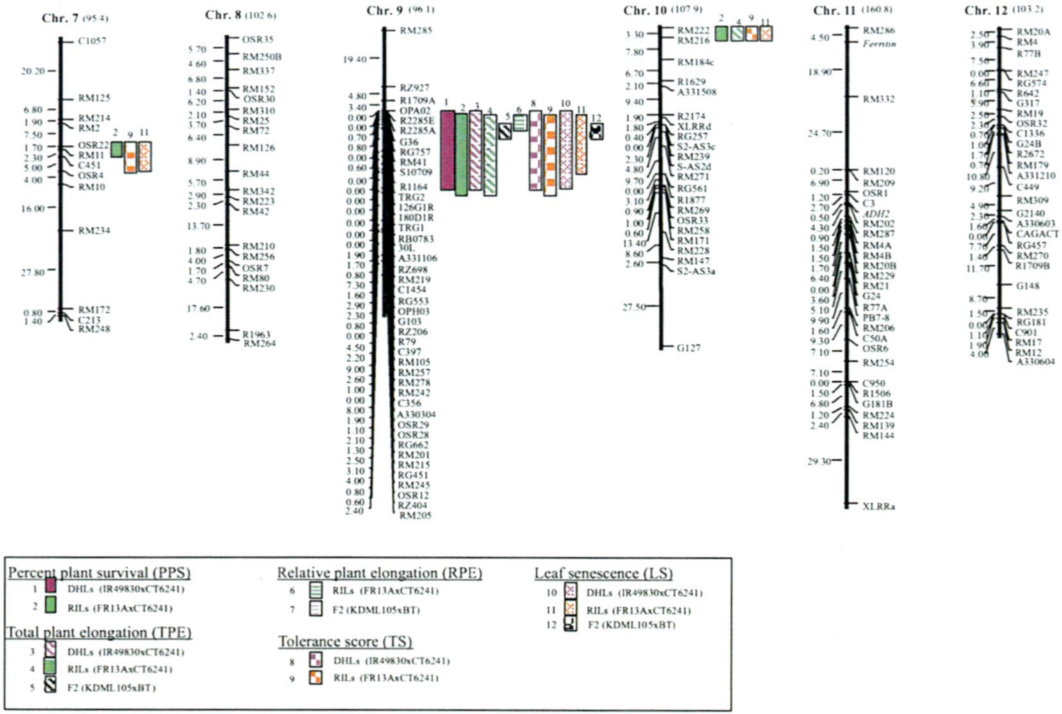

FIGURE 7.3b. QTL mapping for flash flooding tolerance in rice illustrates the dominant major QTL located on chromosome 9 and a few minor QTLs located on chromosomes 7 and 10.

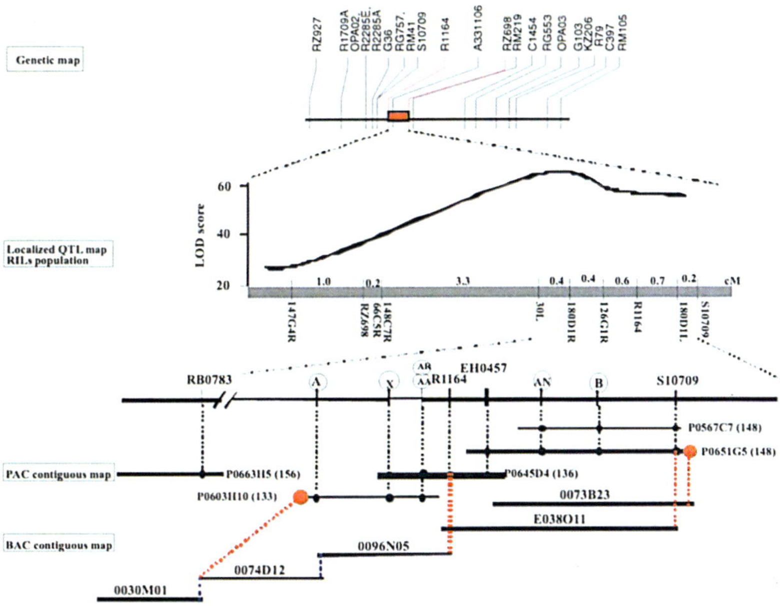

FIGURE 7.4. Fine scale mapping and physical mapping around critical area for submergence tolerance in rice.

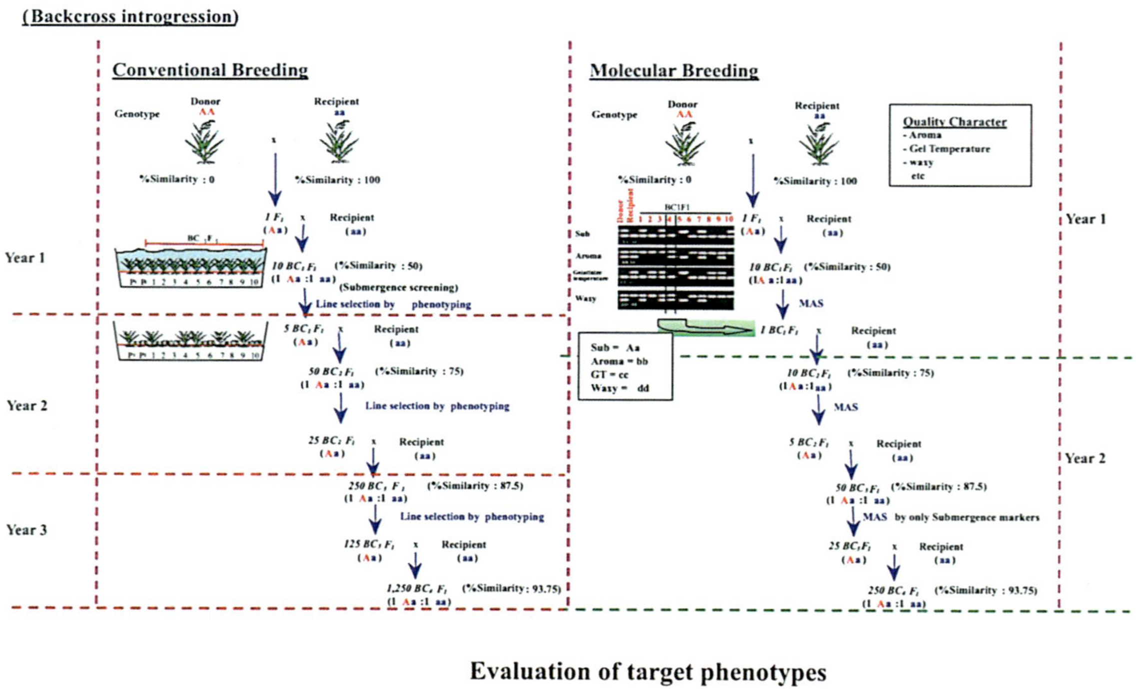

If hybridization can produce 10 F1 seeds in each backcrossing and A is completely dominant.

FIGURE 7.5. Molecular-assisted backcrossing (MAB) for submergence tolerance in rice.

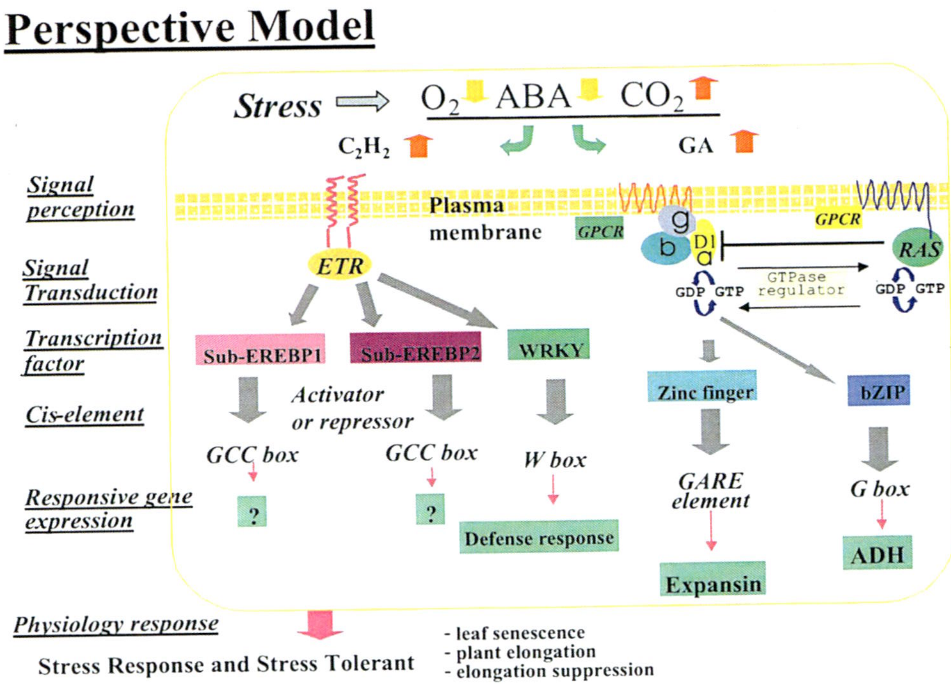

FIGURE 7.6. Perspective model depicting how GPCR, RAS, GTPase regulator, EREBPs, and two transcription factors, candidate genes that mediate submergence-induced elongation and leaf senescence interact with responsive genes such as ADH, PDC, expansin, etc.

FIGURE 12.2. Hypothetical combination of traits conferring good performance under drought in two different genotypes for traits identified by researchers as contributing to drought tolerance.

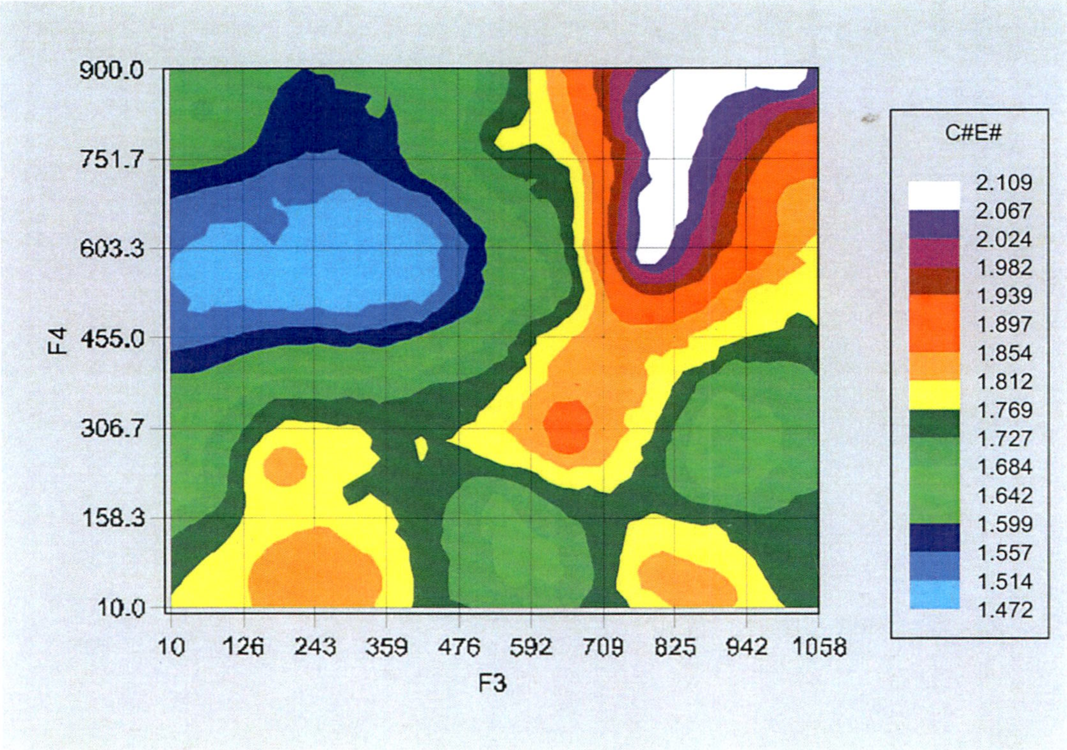

FIGURE 12.3. A representative soil map indicating variability in soil salinity at 0 to 15 cm depth at the CIANO field station located in northwestern Mexico.

Screening for tolerance to moisture stress

Generation of drought stress using gravity-fed irrigation

Generation of drought stress using drip irrigation

FIGURE 12.4. Examples of drought stress generated at CIANO in northwestern Mexico using(a) gravity-fed irrigation and (b) drip irrigation through plastic piping.

PART II: BREEDING FOR ABIOTIC STRESS TOLERANCE IN INDIVIDUAL CROPS

Chapter 12

Breeding for Abiotic Stress Tolerance in Wheat

Wolfgang H. Pfeiffer
Richard M. Trethowan
Maarten van Ginkel
Ivan Ortiz-Monasterio
Sanjaya Rajaram

INTRODUCTION

By 2020, 40 percent more grain will be needed throughout the world, which will require relying heavily on yield increases (Pinstrup-Andersen et al., 1999). Based on projections that land and water resources per capita will diminish during the coming decades, recent studies predict production must increase by 1.6 percent per annum over the next 20 years to meet the increasing global demand for wheat. About half of the required production increases are expected to come from crop management research (CMR), which implies that crop improvement must contribute close to 1 percent per annum. This poses an immense challenge to wheat improvement research, given that in recent years genetic gains of such magnitude have been realized infrequently (Byerlee and Traxler, 1999; Calderini et al., 1999). As Borlaug and Dowswell (1997) observed:

> The only way for agriculture to keep pace with population and alleviate world hunger is to increase the intensity of production in those ecosystems that lend themselves to sustainable intensification, while decreasing intensity of production in the more fragile ecosystems.

Hence, production efficiency in the different agroecological cropping systems must be maximized and at the same time the natural resource base must be protected. Within this context, environmental, cultural, and political sustainability will define the focus of research agendas.

The majority of the world's wheat areas are affected by an abiotic stress and, often, combinations of several stresses. Increasing and stabilizing production under these conditions poses one of the greatest challenges for agricultural research in the twenty-first century, given the fragile and highly variable nature of these areas and the continual deterioration of natural resources. This challenge can be illustrated via the input response of a genotype under nonstress and abiotic stress (Figure 12.1). Constraints are intrinsic; marginal gains for agronomic inputs decrease with increasing moisture stress. Trends in crop improvement are similar to the magnitude of genetic gains falling as environmental stress increases. These realities imply that overproportional genetic gains and production increases are required to attain economic production levels and address low and unstable on-farm productivity in stress environments.

Production increases can originate from various sources: (1) genetic gains in yield potential; (2) genetic gains in tolerance to abiotic and biotic stresses; (3) productivity gains through improved, sustainable CMR production techniques; and (4) synergistic effects among all these factors. In general, productivity in stressed environments is determined by a combination of tolerances to numerous abiotic and biotic stresses, crop management practices, and production economics. Hence, crop improvement cannot concentrate solely on improving abiotic stress tolerance per se but must

	Stress	Nonstress
2	0.30	0.69
	0.48	1.10
4	0.60	1.39
	0.70	1.61
6	0.78	1.79
	0.85	1.95
8	0.90	2.08
	0.95	2.20
10	1.00	2.30
	1.04	2.40
12	1.08	2.48
	1.11	2.56
14	1.15	2.64
	1.18	2.71
16	1.20	2.77
	1.23	2.83
18	1.26	2.89
	1.28	2.94
20	1.30	3.00

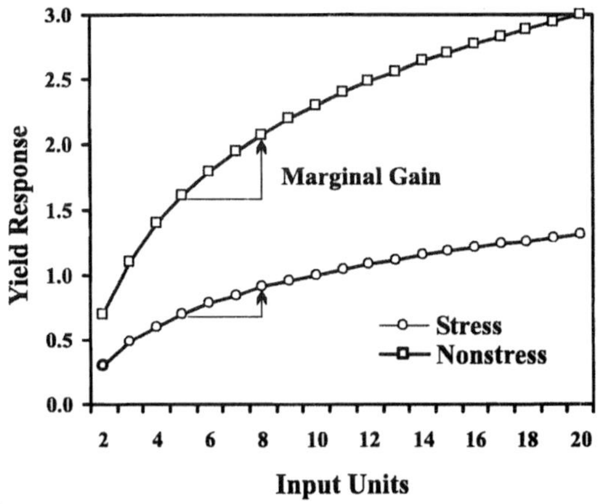

FIGURE 12.1. Hypothetical representation of marginal gains in nitrogen response under abiotic stress and nonstress situations

consider constellations of adaptive traits under defined environmental conditions, including biotic stress tolerance and end-use quality requirements. Most breeding programs use methodologies designed to improve several traits simultaneously. The target traits are selected based on their relevance to local crop production and their economic value. Existing genetic variation, gene action, trait heritability, associations among traits, available and potential (conventional and molecular) screening techniques, and diagnostic tools are commonly used as criteria to characterize traits and project expected genetic progress. In practice, indirect selection for key constraints is frequently more efficient in raising production than selecting for yield or the target abiotic stress. Because many biotic and abiotic factors mask or confound the expression of stress tolerance, many earlier attempts to screen for stress tolerance, in particular drought tolerance, have failed. For example, plants tolerant to nematodes or micronutrient imbalances may be selected as drought tolerant by the plant breeder simply because they have healthier root systems. Thus, breeders often raise productivity by concentrating on improving tolerance to those factors, particularly diseases for which they have known, repeatable variation. Progress in crop improvement for abiotically stressed environments is hampered by a number of factors (Table 12.1).

One complication limiting progress in stressed environments is the high degree of variability from year to year. Singh and Byerlee (1990) analyzed wheat yield variability in 57 countries over 35 years. Yield variability was measured by calculating coefficients of variation of yields around linear trends. Amount and distribution of rainfall was the predominant factor influencing yield variability; countries in which half the wheat was sown in dryland conditions experienced twice as much variability as countries in which wheat is grown mostly under well-watered conditions. Further, yield

TABLE 12.1. Abiotic stress categories.

Moisture stress	Temperature extremes	Micronutrient stress—pH extremes
Terminal	Heat stress, high humidity	Deficiency (e.g., zinc)
Preanthesis	Heat stress, low humidity	Toxicity (e.g., boron)
Residual moisture	Cold stress	Acid soils mineral
Reduced irrigation	Cold stress—late frost	Acid soils volcanic/organic
General low rainfall		Alkaline soils
Shallow, marginal, infertile, eroded lands		Salinity
Waterlogging		

variability also tended to be higher in warmer subtropical countries due to heat stress. A yield simulation model over 50 years for seven rainfed marginal environments displayed dramatic variation in grain yields. Singh and Byerlee (1990) pointed out that actual grain yields are lower since potential yields are defined by moisture and temperature conditions and are not constrained by other abiotic and biotic stresses such as soil fertility. Plant breeders are confronted with selection environments that vary from year to year. Genotypes selected in one year under severe stress, for example, frequently perform poorly in the following year when moderate stress occurs. Consequently, selection gains tend to cancel each other out because genotypes are unlikely to be superior over the wide range of production levels. These changing environmental indices and low trait heritabilities mask genetic potential, while adaptive traits and trait combinations are complex and difficult to identify, and the genetics of abiotic stress tolerance is often poorly understood.

To enhance genetic gains and realize production increases in abiotically stressed environments, research should focus on the following activities:

1. improved characterization and understanding of target environments through the use of physical parameters and probability ranges for climatic variables to identify relevant traits and weighted selection indices;
2. identification of morphophysiological stress adaptive traits and their associated molecular markers with higher heritabilities than yield alone;
3. development of more efficient screening and selection methodologies; and
4. development of sustainable crop management practices to complement crop improvement.

Among the agronomic solutions requiring investigation are the selection and testing of germplasm under moisture conservation practices such as reduced or zero tillage and stubble retention. Tillage regime × genotype interactions are critically important for realizing genetic gains in stress tolerance in farmers' fields. Other practices such as shifting cultivation or periods of fallow and water harvesting (collecting runoff after rainfall for irrigation use) also allow better utilization of available moisture.

BREEDING WHEAT FOR TOLERANCE TO MOISTURE STRESS

Insufficient water is the primary limitation to wheat production worldwide. More than half of the world's estimated 237 million ha of annual wheat production is periodically affected by drought (Table 12.2). In developing countries an estimated 45 percent of the 120 million hectares sown to wheat are prone to drought (Rajaram, 2001). An area is considered to be marginal when average yield levels fall to 70 percent of that expected if crops are fully irrigated (Rajaram, 2001). The percentage of each wheat-growing region currently classified as marginal may in the future change significantly. Increasing urbanization has become a feature of most developing countries, and the available water for irrigation is expected to become increasingly limited (Pingali and Rajaram, 1999). In these traditionally fully irrigated environments, crops are likely to suffer periodic drought stress as the amount and timing of water application become less optimal. Recent projections indicate that wheat production must be increased by 40 percent in order to meet global demand by 2020 (Rosegrant et al., 1997). Once this scenario is placed in the context of diminishing productivity in the future in these high-yielding areas, the fragile nature of the world's food supply is revealed.

It is vital that productivity in the world's wheat-growing areas is not just maintained but substantially increased in time. Attainment of a 40 percent

TABLE 12.2. Approximate area sown to wheat globally and the proportion of this area classified as drought affected.

Region	Approximate wheat area (in thousand ha)	Percent marginal
West Asia/North Africa	28,300	65
Central Asia and Caucasus	15,000	80
South Asia	34,500	35
East Asia	30,100	13
Eastern Africa	1,500	27
Southern Africa	1,300	91
Southern Cone of South America	7,400	60
Andean Region of South America	300	18
Mexico/Central America	900	20
United States and Canada	41,000	75
Europe	24,000	15
Oceania	11,500	90
Russia	42,000	80
Total	237,800	55

Source: Adapted from Rajaram (2001).

increase in production by 2020 as suggested by Rosegrant and colleagues (1997) will depend upon the successful integration of water-use-efficient conservation tillage practices with improved, higher yielding, water-use-efficient cultivars. This chapter will not attempt to cover in detail all aspects of productivity improvement in drought-stressed environments. Instead, it will focus on genetic solutions and the interaction of new, improved water-use-efficient varieties with improved agronomic practices and the wider cropping environment.

Rates of Progress in Productivity in Drought-Affected Environments Over Time

Traditionally, it has been difficult to make significant progress in drought-affected environments by selecting wheat germplasm for improved water-use efficiency (WUE) or drought tolerance. The primary reason for this lack of progress is the extremely variable nature of rainfall from year to year in drought-affected environments where most plant breeders select their elite germplasm. This large, apparent genotype × environment (G × E) interaction masks genetic progress for improved WUE. Nevertheless, progress in improved grain yield has been achieved in many drought-affected environments. Byerlee and Traxler (1995) summarized annual rates of progress in various wheat-growing environments; progress ranged from 0 percent in central India to 1.3 percent in Paraguay (Table 12.3). Progress over time in these regions may not necessarily have been due to improved WUE. More likely, selection for inherited traits influencing root development, such as resistance to crown fungal diseases or nematodes, in combination with improved management practices influenced rates of progress. Trethowan and colleagues (2002) measured progress in yield for more than 1000 yield trials grown in a range of dry environments globally over a 20-year period and found rates of progress between 0.34 and 4.38 percent (Table 12.3). The highest rate of progress was observed in the International Maize and Wheat Improvement Center's (CIMMYT) Semi-Arid Wheat Yield Trial (SAWYT), which contained germplasm bred specifically for adaptation to moisture-limited conditions. The germplasm included in the SAWYT was developed by screening for improved WUE in an arid environment in northern Mexico. Here, water application could be strictly controlled in combination with selection under high rainfall conditions in the central Mexican highlands, where water-responsiveness and disease resistance could be selected for (Trethowan et al., 2001).

TABLE 12.3. Reported rates of progress in mean grain yield with time in dry environments.

Location	Period	Rate of gain (percent/yr)	Data source
Paraguay	1972-1990	1.3[a]	M. Kohli et al. (1991)
Australia (Victoria)	1940-1981	0.8[a]	O'Brien (1982)
Australia (New South Wales)	1956-1984	0.9[a]	Byerlee and Traxler (1995)
Australia (Western Australia)	1884-1992	0.4[a]	Perry and D'Antuono (1989)
Central India	1965-1990	0.0[a]	Byerlee and Traxler (1995)
Global SAWYT[b] data	1992-1998	4.38	Trethowan et al. (2002)
Global ESWYT[c] data	1979-1998	0.34	Trethowan et al. (2002)

[a]Adapted from Byerlee and Traxler (1995).
[b] SAWYT, the Semi-Arid Wheat Yield Trial, distributed by CIMMYT.
[c]ESWYT, the Elite Spring Wheat Yield Trial, distributed by CIMMYT.

Traits Conferring Adaptation to Drought

Genetic control of tolerance to drought is poorly understood. Over the years many researchers have identified traits or characteristics that confer drought tolerance in specific environments or under specified environmental conditions. However, such is the extent of G×E interaction for performance under drought that traits identified in one year or location may not confer tolerance in another year or location. Figure 12.2 lists a range of traits identified by researchers as contributing to drought tolerance and displays the hypothetical combination of different traits conferring performance under drought for two different genotypes. Clearly, control of drought tolerance is likely to be complex and relevant to very specific environmental conditions.

Early research into mechanisms of drought tolerance concentrated on those traits important to drought survival; however, later work indicated that grain yield was the best determinant of drought tolerance. Passioura (1977, cited in Richards et al., 2000) suggested the following relationship, where the improvement in any component of the relationship would translate into an improvement in yield:

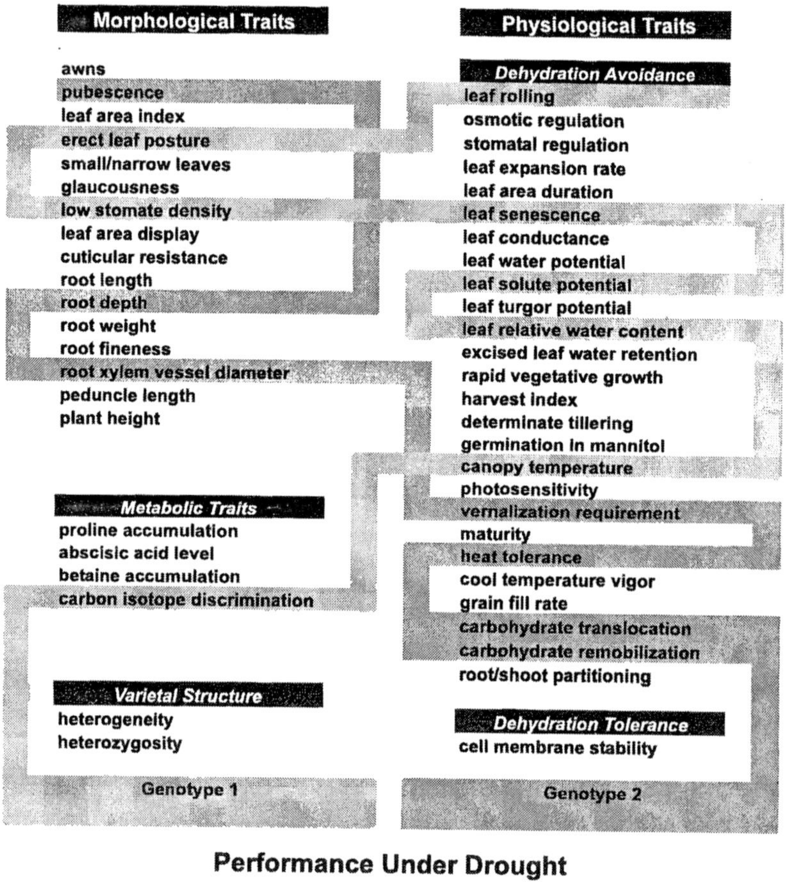

FIGURE 12.2. Hypothetical combination of traits conferring good performance under drought in two different genotypes for traits identified by researchers as contributing to drought tolerance (see color figures section).

$$\text{Grain Yield} = (\text{Crop water use}) \times (\text{Water use efficiency}) \times (\text{Harvest index})$$

Richards and colleagues (2000) conducted an extensive review of morphological and physiological characters that plant breeders can select to improve drought tolerance. Table 12.4 lists some of these traits, their heritability, and expected G×E. Several traits, such as osmotic adjustment and stomatal conductance, will take more time to determine than plant height or flowering date and are much more influenced by environment. The influence of other characters, such as seedling vigor and root depth,

TABLE 12.4. Traits conferring adaptation in drought-affected environments, their heritability, and potential size of genotype × environment (G×E) interaction

Trait	Heritability	Expected G×E interaction
Flowering time	High	Low
Glaucousness	High	Low
Awns	High	Low
Foliar disease resistance	Disease dependent	Low
Plant height	High	Low
Floret fertility	Low	Intermediate
Stay green (maintenance of green leaf area)	Low	High
Large seed size	High	Intermediate
Deeper roots	Low	High
Seedling vigor	High	Low
Tiller inhibition	High	Low
Osmotic adjustment	Low	High
Cooler canopy temperature	Intermediate	High
Stomatal conductance	Intermediate	High
Leaf rolling	Intermediate	High
Long coleoptile length	High	Low
Broad seedling leaves	High	Low
Carbon isotype discrimination	High	Low
Peduncle length	High	Low
Assimilate translocation	Intermediate	High
Pubescence	High	Low
High preanthesis biomass	Intermediate	Intermediate

Source: Adapted from Richards et al. (2001).

will depend upon soil type and moisture availability at sowing. The plant breeder must balance the likely impact of the trait with expected heritability and ease of measurement. For example, early maturity may be a useful mechanism of drought escape in some environments (Ceccarelli et al., 1987), whereas greater plant height may be an important adaptive mechanism in environments characterized by predominately preanthesis moisture stress (Singh et al., 2001). Both of these traits are easily measured and have a high heritability and potentially large impact on adaptation within specific environments. On the other hand, deeper roots may enhance adaptation to environments where farmers sow into residual moisture. However, the difficulty of measuring this characteristic in segregating populations and the cost and low heritability of the trait (Richards et al., 2001) are likely to outweigh the potential benefits of measuring root depth.

Screening Methods Used to Select Germplasm Tolerant to Drought

To successfully develop drought-tolerant cultivars, it is fundamental to develop efficient, repeatable drought screening methods. These methods may address only a particular type of drought stress, such as stress in the vegetative phase prior to flowering or stress during the grain-filling phase. The timing and intensity of moisture stress must be strictly monitored or controlled if genotypes are to be adequately evaluated. Other potentially confounding stresses in the test environment must also be controlled, measured, or eliminated. An example of this may occur in field-based drought screening when nematodes, root rots, and micronutrient imbalances are present. Germplasm may differentiate for these soil-borne constraints rather than for differences in WUE. Some researchers have tried to simulate moisture stress under controlled conditions in the laboratory and greenhouse (Pellegrineschi et al., 2002) or using hydroponics (Mian et al., 1993). Although germplasm will differentiate under these conditions, very little relationship has been noted with drought conditions in the field. Should all confounding biotic and abiotic factors be removed, leaving soil-available moisture as the only differentiating factor, then interaction with soil type will influence transferability of these results across environments with different soil types and textures (Richards et al., 2001).

When developing a repeatable, relevant, drought screening procedure, it is also important to regulate the onset and intensity of moisture stress. For example, a short, fast-drying cycle does not allow the full adaptive potential of plants to express (Babu et al., 1999). Plant dehydration rate and response to dehydration may also vary with development stage, tissue age, and plant size (Blum, 2000). Blum also states that measurement of plant function and plant water status must be clearly distinguished. He cites photosynthesis as an example. Photosynthesis is often used to measure plant response under drought stress; however, photosynthesis changes with leaf water status. If photosynthesis is to be tested as a possible mechanism of drought tolerance, then it must be measured in all genotypes at constant plant water status.

To demonstrate how applied breeding has approached the difficulties of screening for tolerance to drought, reference will be made to the CIMMYT Wheat Program's experience in cultivar development for dry environments. The CIMMYT program uses a field site located near Ciudad Obregón in northwestern Mexico to select for tolerance to moisture stress. This location is arid and uses irrigation. Wheat is sown in mid November and harvested in late April. With negligible precipitation during the cropping cycle, it is possible to manipulate the timing and intensity of drought stress using irrigation. All test sites used by CIMMYT in this area have been characterized for the presence of nematodes and root rots, and soil maps (such as that in Figure 12.3) have been constructed for all potential nutrient stresses. This information enables researchers to choose test locations that will discriminate for moisture stress only. Gravity-fed and drip irrigation systems are used to manipulate timing and intensity of stress (Figure 12.4). This screening approach has been effective in developing cultivars adapted to moisture-stress conditions in many different wheat-growing areas around the globe (Trethowan et al., 2001, 2002).

Sources of Variability or Key Drought-Adaptive Traits

To make progress in the development of cultivars with adaptation to drought stress, it is important that significant and measurable genetic variation exists among parental stocks. All of the traits identified in Table 12.4 as contributing to drought tolerance are found in the existing wheat hexaploid gene pool. However, it is possible to find extreme expression of a number of these traits from wild ancestors such as *Aegilops tauschii,* commonly referred to as goat grass. New synthetic hexaploid wheat can be developed from *A. tauschii* by crossing this diploid species with modern tetraploid durum wheat following embryo rescue and chromosome doubling with colchicine (Figure 12.5). These synthetic hexaploid wheats contain a complete D-genome from *A. tauschii* and offer wheat breeders significant new variation for both biotic and abiotic stress-adaptive traits (Villareal et al., 1994; Valkoun, 2001). The abiotic stress tolerance evident in *A. tauschii* has accumulated over thousands of years of natural selection in the stress-prone regions of North Africa and West Asia. The first spontaneous crosses that gave rise to hexaploid wheat likely occurred 10,000 years ago (Feldman, 2001) and in all probability involved a limited number of diploid goat grasses. More than 1,000 different accessions of goat grass have been collected and evaluated, and new hexaploid combinations have been made with many of these. Significant new variation for a host of different characters ranging from disease resistance (Gill et al., 1985) to grain quality (Peña et al., 1995) and abiotic stresses (Valkoun, 2001) has been found. Other researchers have identified useful variation for moisture-stress tolerance in *Triticum urartu, Triticum boeticum, Triticum dicoccoides* (Valkoun, 2001), and *Aegilops geniculata* (Zaharieva et al., 2001). However, the impact of

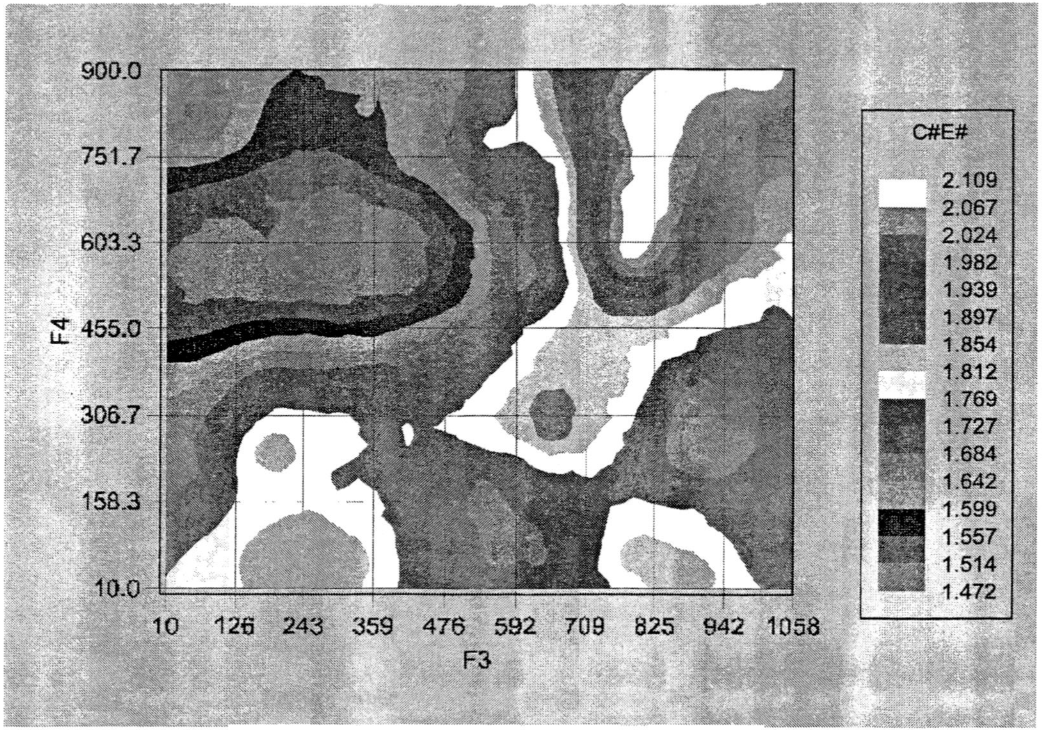

FIGURE 12.3. A representative soil map indicating variability in soil salinity at 0 to 15 cm depth at the CIANO field station located in northwestern Mexico (see color figures section).

FIGURE 12.4. Examples of drought stress generated at CIANO in northwestern Mexico using (a) gravity-fed irrigation and (b) drip irrigation through plastic piping (see color figures section).

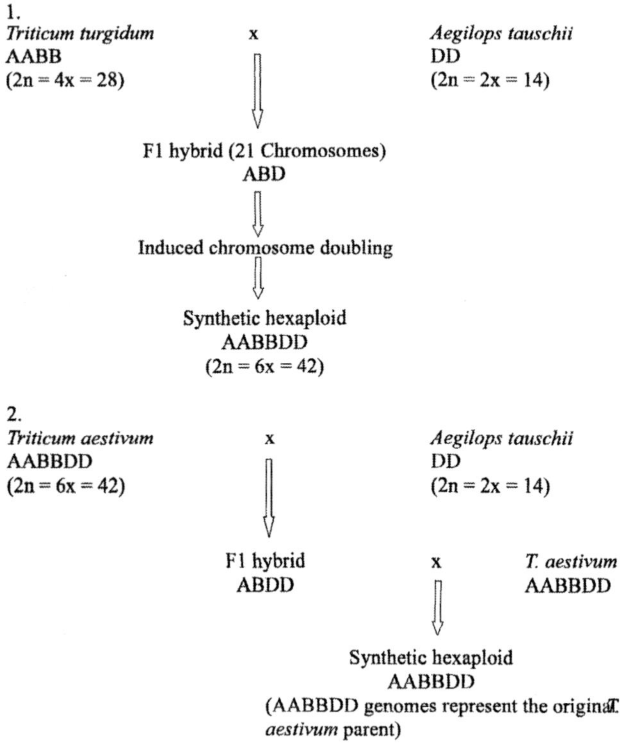

FIGURE 12.5. Schematic representation of the process of synthetic hexaploid production

sources from wild species other than *Aegilops tauchii* in conferring drought tolerance in marginal environments has been small (Skovmand et al., 2001).

Methods for Developing Cultivars for Drought-Affected Environments

In most dry rainfed environments, farmers make the majority of their income in the more productive years, which tend to occur at a much lower frequency than the poorer years. Income received during the better times helps to stabilize farmer income in the longer term. Wheat breeders must be aware of these realities when choosing an appropriate breeding strategy. Any strategy must first identify genotypes with high yield potential and related input responsiveness and an appropriate suite of disease resistance genes for adaptation in the wetter years (Rajaram et al., 1996). Second, the strategy must ensure that stress-adaptive traits such as drought and heat tolerance are identified, selected, and maintained in the germplasm. Third, those abiotic and biotic soil-borne constraints unrelated to drought tolerance or improved WUE must be addressed. When the plant is suffering from moisture deficit, it is less likely to cope with major secondary stresses such as nematodes (Rathjen, 1994). In fully irrigated systems where plants are rapidly growing, many of these problems have little impact on yield.

Two primary schools of thought have influenced plant breeders who target their germplasm to drought-affected areas. The first of these philosophies states that high input responsiveness and inherently high yield potential, combined with stress-adaptive traits including water-use efficiency, will improve performance in drought-affected environments (Rajaram and van Ginkel, 2001). The second is the belief that progress in yield and adaptation in drought-affected environments can be achieved only by selecting under the prevailing conditions found in those environments (Ceccarelli and Grando, 1991).

Breeding for input responsiveness will require an environment where moisture and fertilizer inputs are optimally managed. If the target region in which the breeder is working frequently experiences drought, then these conditions will need to be generated. Assuming that genetic variability exists among parental lines for yield potential, appropriate disease resistance, and adaptation to drought stress, the following breeding scheme adapted from Rajaram and van Ginkel (2001) has been proposed.

F1 (generation): Grown under well-watered conditions, optimally fertilized, and inoculated with disease.

F2: Grown under well-watered conditions, optimally fertilized, and inoculated with disease.

F3 and F4: Selected plants are evaluated in either a modified pedigree/bulk or selected bulk system (Singh et al., 1998) under moisture-stress conditions.

F5 and F6: Grown under well-watered conditions, optimally fertilized with high disease pressure.

Advanced lines are tested for yield potential and drought tolerance under contrasting evaluation conditions. Those lines exhibiting superior performance under both optimally irrigated and drought conditions are promoted. This germplasm shuttle between environments contrasting for moisture status was validated by Kirigwi and colleagues (2000). The authors found that alternating the moisture level between optimal irrigation (applied in F2, F4, and F6) and severe stress (applied in F3 and F5) produced wheat lines better adapted to a range of moisture regimes, including severe drought, than methods involving selection always under optimal conditions or drought stress. The CIMMYT wheat-breeding program in Mexico currently employs a modified version of this scheme.

Breeding for adaptation to drought stress by selection in the target region presumes that all key elements of adaptation can be found in the target environment and are most efficiently selected for under those conditions. Several researchers advocate this strategy on the premise that the selected germplasm will perform well in farmers' fields if selected under the prevailing conditions in these regions (Ceccarelli, 1987; Ceccarelli and Grando, 1991; Rathjen, 1994). If factors such as nematodes, root rots, or nutrient deficiencies are important constraints, then selection within the region should enable breeders to accumulate genes for adaptation. However, the primary weakness of this approach is that input responsiveness, so important in the wetter, admittedly less frequent but much more productive years cannot be easily maintained in the germplasm. The method also assumes yield crossover will occur below a certain yield threshold. Evidence for the existence of crossover (Ceccarelli, 1989) and nonexistence of crossover (Rajaram and van Ginkel, 2001) in environments ranging from moisture stressed to nonstressed has been reported.

Role of Biotechnology in Developing Drought-Tolerant Wheat Cultivars

It appears unlikely that traditional marker-assisted selection will be used by wheat breeders to improve tolerance to drought in the near future. These marker systems aim to identify key genes that control certain physiological processes in the plants. However, low levels of polymorphism in wheat and the large G×E observed for drought tolerance greatly limit the use of marker-assisted selection. Drought tolerance under any given set of environmental conditions is likely to be controlled by quantitatively inherited genes. For this reason, some researchers have explored the use of QTLs (quantitative trait loci) as a tool for detecting drought tolerance in wheat (Cooper et al., 2000), rice (Li et al., 2000), and barley (This and Teulat-Merah, 2000). In the case of wheat, Cooper and colleagues (2000) found major QTLs for traits such as height, maturity, and tolerance to root lesion nematode *(Pratylenchus thornei)*. These traits are related to performance in the drier areas of the Queensland wheat belt but not to drought tolerance.

However, in barley, QTLs for osmotic adjustment have been identified (This and Teulat-Merah, 2000).

Comparative genomics, a technique that identifies genes through the rapid generation of large amounts of DNA sequence data, may shed some light on the genetic control of drought tolerance. The degree of conserved linearity of gene order among the genomes of the different crop species is relatively high (Bennetzen and Freeling, 1997; Devos and Gale, 1997), and the different crop species tend to differ less in the genes they carry than in the alleles of these genes (Bennetzen, 2000). Evidence suggests that similar biochemical pathways control tolerance to drought among different species. Therefore, it should be possible to enhance our understanding of the mechanisms of drought tolerance by exchanging information among the different crop species. By testing the plant for differences in RNA levels before, during, and after drought stress, it is possible to detect which genes are expressed and potentially important in conferring drought tolerance (Ruan et al., 1998; Bennetzen, 2000). This technique provides information on a host of different genes, and variation in the RNA levels of these genes may also indicate their relative importance in the control of drought tolerance under a particular set of environmental conditions. As the intensity and timing of drought stress varies among environments, it is necessary to examine gene expression across multiple test locations or stress patterns. It may then be possible to find commonality in expression for a subset of genes potentially important in all moisture-stress scenarios. To date, the complete genome of *Arabidopsis* and rice have been sequenced, and a large number of expressed sequence tags (ESTs) are available in public databases for both these genomes and other cereals.

The introduction of new "drought genes" from other species through genetic engineering may provide new and useful variations for drought tolerance. Several candidate genes putatively involved in drought tolerance are listed in the literature and gene databases (Skriver and Mundy, 1990; Bray, 1993; Ingram and Bartels, 1996; Bohnert et al., 2000; Seki et al., 2001). One of these stress-induced genes, *rd29DREB1A,* is induced through an abscisic acid-independent pathway. This gene expresses under conditions of dehydration and cold stress (Yamaguchi-Shinozaki and Shinozaki, 1994). Introduction of the *rd29DREB1A* gene from rice into wheat may control or influence the biochemical pathways important in conferring drought tolerance in wheat. However, to date the evidence remains inconclusive (Pellegrineschi et al., 2002).

Moisture Stress Concluding Remarks

Regardless of the breeding or selection method used to develop new cultivars of wheat, breeders must be conversant with the wider farming system in which these cultivars will be sown. Knowledge of agronomic practices, in particular crop rotations and tillage regimes, will influence plant morphology and disease resistance. In many drier environments farmers have begun adopting conservation tillage to conserve soil moisture and reduce wind erosion. Some debate remains over the existence of tillage × genotype interactions. However, lack of such interactions within the current gene pool does not indicate that interactions will not occur. All such studies have included genotypes developed using conventional tillage, which may have led to loss of genetic variability, important for adaptation to zero tillage. Better emerging genotypes with vigorous early growth may enhance establishment and subsequently grain yield. In conservation farming systems where stubble is retained from the previous cropping cycle, changes in disease patterns are frequent. Root rots and foliar diseases such as yellow spot become increasingly important and must be included in breeding objectives.

BREEDING WHEAT FOR TOLERANCE TO HEAT STRESS

High temperature limits grain yield potential in many wheat-production environments. Periods of heat stress are often of short duration and frequently occur toward the end of the crop cycle, during grain filling and seed maturation. Minimum levels of heat tolerance can be achieved in wheat if breeding, selection, and yield screening are carried out under these conditions in the field. However, in some wheat-growing regions in the world tolerance to heat must be specifically pursued. In such areas temperatures reach very high levels, above 30°C, for extended periods. Most commonly such stressed conditions are encountered during flowering and the post-flowering period in regions of intermediate elevation (500 to 1500 m above sea level); however, significant heat stress may also occur in the lowland tropics (<500 m above sea level) early in the cropping cycle. Apart from the timing of the onset of heat stress, e.g., at emergence, anthesis, and during grain filling, production conditions determine specific breeding needs. First, heat stress may occur in well-watered, irrigated, or high rainfall areas, or may be prevalent in dryland areas where it coincides with drought stress. Second, heat stress may or may not be accompanied by high relative humidity (Kohli et al., 1991).

This section will focus on breeding wheat for these distinct types of heat stress. Expert reviews have been published detailing progress in understanding the mechanisms controlling tolerance to heat stress up to the early 1990s (Acevedo et al., 1991). This section will emphasize literature published between 1990 and 2001.

Geographical Spread of Areas Affected by Heat Stress

The total area of wheat production affected by some form of heat stress is estimated to be 65 to 70 million ha. Of these, 7 million ha are grown under continual heat stress (Reynolds et al., 1994).

Well-Watered, Low-Humidity Environments

Most well-watered, low-humidity areas are irrigated, lack major soil or nutrient problems, and are characterized by low incidence of foliar diseases. More than 20 million ha of wheat grown in these environments periodically suffer heat stress during the cropping cycle. Some of the most important wheat-producing countries representing these conditions are listed here.

- *India:* Wheat in most of the irrigated areas of India regularly experiences early (at or following planting) or late (grain filling) heat stress (Tandon, 1994a). These production areas are found primarily in the north and northwest of the country, in the states of western Uttar Pradesh, Haryana, and Punjab, although representative areas are found in the central and peninsular zones of Madhya Pradesh and Karnataka. The area annually affected by heat stress in India is approximately 5 to 8 million ha.
- *China:* In the northern spring-planted spring wheat region, late heat stress is common, with some 3 to 5 million ha annually affected. The central eastern tier of facultative and winter wheat regularly experiences high temperatures during grain filling, also affecting 3 to 5 million ha. The southern provinces of Guangdong, Fujian, and Hainan grow fall-planted spring wheat and experience high temperatures during the early growth stages (Changjian, 1994). The total area annually affected by heat stress in China is 6 to 10 million ha.
- *Egypt:* The irrigated middle and upper Nile wheat-growing areas are affected by predominately postanthesis heat stress, with temperatures of 39°C common (Ghanem, 1994). The majority of this area is irri-

gated and can be affected by hot winds (Abdelghani et al., 1994). Total wheat area annually affected by heat stress in Egypt is close to 300,000 ha.

- *Sudan:* Delay in land preparation and hence seeding is a primary cause of late heat stress in the irrigated and mechanized production schemes in Sudan (Abdelshafi Ali and Ageeb, 1994). The Gezira Scheme grows 70 percent of the country's wheat, and average temperatures during the three main months of the crop cycle are 24.1°C (December), 23.6°C (January), and 25.4°C (February) (Adam and Ageeb, 1994). However, maximum air temperatures of 38 to 40°C are routinely observed (Ishag et al., 1998). The entire wheat area is periodically affected by heat, totaling some 300,000 ha (Pingali, 1999).

Well-Watered, High-Humidity Environments

Well-watered, high-humidity regions are either irrigated or receive ample rainfall. Disease pressure tends to be high, with specialized pathogens, e.g., *Helminthosporium sativum,* and bacterial diseases constraining production. Soils are frequently nutrient deficient. The total wheat area that experiences heat stress under these conditions is more than 10 million ha. Examples of countries affected by this stress are as follows.

- *India:* The entire eastern section of the Gangetic Plains, e.g., the state of Bihar, suffers from heat stress during much of the crop cycle. The area affected is approximately 6 to 7 million ha (Tandon, 1994a).
- *Bangladesh:* The respective mean temperatures in December, January, February, and March are 19 to 20°C, 18°C, 19 to 20°C, and 24 to 25°C, respectively, and relative humidity in these areas varies between 62 and 87 percent (Razzaque et al., 1994). For the 700,000 ha of wheat grown, 10 to 15 percent is frequently exposed to early heat stress and 45 to 55 percent to terminal heat stress. The incidence of terminal heat stress is exacerbated by late planting due to the delayed harvest of the preceding "aman" paddy rice crop (Samad et al., 1994). On average, at least half the total wheat area is affected by heat annually.
- *China:* Warm, humid, and cloudy conditions, typical of the southern Chinese province of Guangxi (Huang and Zhu, 1994) and the Yangtze valley cause periodic heat stress. Approximately 1 million ha are affected.
- *Nepal:* The low-lying, irrigated Terai region of Nepal experiences warm conditions and high humidity one year in three. Late maturity of the favored rice varieties and difficulties in preparing waterlogged

seedbeds for the following wheat crop delay planting, leading to high temperatures and hot winds during grain filling (Devkota, 1994). More than half the total wheat area (650,000 ha [Pingali, 1999]) is periodically affected by heat stress.

- *Paraguay:* Extreme variation in temperature characterizes the wheat-production areas of Paraguay, where 200,000 ha (Pingali, 1999) are heat affected. It is not uncommon for temperature to fluctuate from 0 to more than 30°C within a matter of days (Pedretti and Kohli, 1991).
- *Thailand:* Although most of the few thousand ha of wheat production in Thailand experience warm conditions, it is particularly in the lowland areas that temperatures restrict plant development in the vegetative phase, with subsequent effects on grain set (Moolsri, 1994). Mean temperatures in the period November to January of 26 to 32°C are common, with variation mostly driven by fluctuating minimum temperatures (Vongburi, 1994). Late heat stress is common when the preceding rice crop is harvested late, causing delay in the wheat planting.

Semiarid Drought-Stressed Environments

Heat stress in semiarid conditions is exacerbated by moisture deficit. Both heat and drought stress may occur at any time during the cropping cycle. The total wheat-growing area in the world affected in this way is more than 35 million ha. The following are representative countries.

- *United States:* Within the Great Plains both the winter wheats in the southern and central regions and the spring wheats in the northern regions are often exposed to high-temperature stress, generally during grain filling. Some 10 million ha are affected.
- *West Asia and North Africa:* Most of the semiarid wheat grown in West Asia and North Africa, from Morocco to Libya, and from Yemen to Iran, suffers from postanthesis heat and drought stress. The total area affected is about 10 million ha.
- *Australia:* The majority of wheat grown in Australia experiences warm conditions during grain filling, with temperatures exceeding 30°C and sometimes approaching 40°C. The area affected covers 6 to 8 million ha.
- *Canada:* Analysis of 50 years of data indicated that durum wheats in Canada might already start to suffer from temperature-induced stress at relatively low temperatures of 20 to 24°C (McCaig, 1996). The total wheat area, mostly bread wheat, but some durum wheat, affected is 6 to 8 million ha.

- *India:* Rainfed crops in the central and peninsular zones are sown in October and rely on residual moisture from the late-summer monsoon rains, when temperatures are still quite high (mean temperatures close to 26°C), but while the water table is still within reach for the germinating seedlings (Tandon, 1994b). Delayed planting coincides with reduced temperatures but also less water in the soil profile. The area planted under these conditions has been declining since the early 1960s, with the expansion of irrigation facilities (Peacock et al., 1994), and is presently estimated at 2 to 3 million ha.

The Future

Global warming is likely to impact agriculture by increasing variability in temperature extremes, thereby reducing wheat yields (Ferris et al., 1998). Although such models also predict elevated CO_2 levels, most show that these increases in CO_2 cannot compensate for the negative effects of an increase in temperature, even of just 1 to 2°C, on yield (Rosenzweig and Tubiello, 1996; Wheeler, Batts, et al., 1996; Batts et al., 1997). However, response may well depend on the variety used (Batts et al., 1998). As a result of global warming, the global spring wheat-producing zone may move farther away from the equator (Frank and Bauer, 1996).

Heat Stress and Plant Development

To make an impact in breeding for heat tolerance, the breeder must have a sound understanding of how heat stress affects the wheat plant. This includes an understanding of the underlying physiological processes, their inheritance, and whether these processes are amenable to artificial selection.

The effect of heat stress on the development and physiological processes of wheat is manifold and can adversely affect germination, emergence, root development, tillering, leaf deployment, biomass production, floret initiation, pollination/fertilization, seed growth, final grain yield, and grain quality, both in a nutritional and industrial quality sense. Sometimes setbacks due to stress in crop development in one phase can be compensated for by recovery in later growth stages, e.g., excess tillering compensating for reduced emergence as heat stress dissipates. However, once the plant has suffered heat stress, it is more likely to be affected by additional stresses, such as drought or disease pressure, which will greatly limit final yield. In addition, even though grain yield may not be greatly reduced in some instances, grain quality can still be severely affected, resulting in significant economic losses.

Germination, Emergence, and Root Development

The lethal, minimum, optimum, and maximum temperatures for wheat, based on 65 literature sources, are listed in Table 12.5 (Porter and Gawith, 1999). In general, wheat is less sensitive to heat stress during the vegetative phase than during the reproductive phase.

The narrower temperature range for root growth compared to the sowing to emergence phase (Table 12.5) indicates greater sensitivity to heat stress during this phenological phase. When a high-temperature treatment (36/29°C day/night regime) was applied during grain filling to two durum wheat varieties, the subsequent germination ability of the harvested seed was reduced in the cultivar Marzak (72 percent germination), while the cultivar Oum-rabia germinated normally (98 percent) (Grass and Burris, 1995a). Seed germination itself was delayed by about 24 h for both varieties relative to seed that matured at control temperatures. Heat-stressed seed of 'Marzak' produced some abnormal seedlings upon subsequent planting. For both varieties, seedling and root dry weight diminished on average by 40 percent and 70 percent, respectively, indicating the sensitivity of early root development to heat stress. The reduction in root weight resulted in an increased shoot/root ratio. Root length and root number were also decreased. Oxygen uptake of heat-stressed seed was reduced, and their mitochondria showed structural damage and reduced activity (Grass and Burris, 1995b).

The optimum temperature for emergence of wheat was shown to be 18.6°C, with temperatures above 20.6°C leading to decreased emergence (Khera and Chaudhary, 1989). When comparing 14 different wheat varie-

TABLE 12.5. Summary of lethal, minimum, optimum, and maximum temperature (in °C) for various processes and phenological phases in wheat.

Processes	$T_{minimum}$	$T_{optimum}$	$T_{maximum}$
Lethal limits	−17.2	—	47.5
Leaf initiation	−1	22	24
Shoot growth	3.0	20.3	> 20.9
Root growth	2.0	< 16.3	> 25.0
Sowing to emergence	3.5	22.0	32.7
Vernalization	−1.3	4.9	15.7
Terminal spikelet	1.5	10.6	> 20.0
Anthesis	9.5	21.0	31.0
Grain filling	9.2	20.7	35.4

Source: Porter and Gawith, 1999.

ties at 14 temperatures ranging from 5 to 40°C, the mean optimum germination temperature was 25°C (Ali et al., 1994). The cultivar Gomam showed little response to increases in germination temperature, while other varieties responded dramatically. Dell'Aquila and Turi (1999) found that heat stress of 40°C before emergence of the radicle of durum wheat resulted in significantly delayed germination. In some instances, they found poorer germination, reduced protein synthesis, and increased development of low-molecular-weight heat-shock proteins (HSPs). Similarly, in spring bread wheat the mean germination time was found to increase and germination percentage to decrease with increasing temperatures (Abayomi and Wright, 1999). These trends were greatly increased when drought stress was simultaneously applied, although significant interactions and varietal differences were observed. Although most studies showed germination to be rapidly affected by heat stress, a study by Hampson and Simpson (1990a), in which temperatures were raised from 10 to 30°C, showed no effect on germination percentage, at least in the absence of osmotic stress. Shoot length and maximum root length of seven-day seedlings were even enhanced at 30°C relative to 10 or 18°C at low osmotic stress (Hampson and Simpson, 1990b). However, when osmotic stress (simulating drought) was increased, germination and development at 30°C were most affected relative to the lower temperatures.

As temperatures are raised from 10 to 20 and then to 30°C, shoots and roots tend to appear sooner (Huang et al., 1991c), the diameter of the central late metaxylem and stele of the root is reduced (Huang et al., 1991b) and the relative conductivity of the seminal roots declines. At 25°C root length and dry matter reach an optimum, which steadily declines as temperatures approach 30°C (Huang et al., 1991c). At 30°C four-week-old seedlings produced 20 percent less root dry weight relative to 20°C, 10 to 40 percent fewer seminal roots, and 20 to 50 percent fewer crown roots (Huang and Taylor, 1993). Heat stress also reduces lateral roots on the seminal roots (Huang et al., 1991a). Even increases of just a few degrees Celsius during development will decrease root weight at anthesis due to reduced carbon partitioning to the roots (Batts et al., 1998).

Emergence Until Booting Phase

Heat stress between emergence and booting shortens the length of this phase and reduces tillering, biomass, and spike fertility. Minimum temperature profiles for leaf initiation, shoot growth, and terminal spikelet development (Table 12.5) are –1 to 3°C; the optimum range is 10.6 to 22°C; and the upper limits vary from >20 to 24°C.

Winter wheats are more affected than spring wheats by temperature changes at a lower absolute level. Even a small temperature increase of just 2°C was shown to decrease the vegetative phase of winter wheat in the United Kingdom by 26 days (Batts et al., 1997). A small rise in temperature can also significantly reduce leaf emergence (Batts et al., 1996). Delgado and colleagues (1994) reported that an increase of 4°C over the ambient temperature reduced leaf duration and photosynthetic capacity in winter wheat.

Temperature increases up to 30°C. Spring wheats appear to have two distinct ranges of temperature where responses are different. Initial increases in temperature may even have beneficial effects. Slafer and Rawson (1994) reported that as temperature rose from 10 to 19°C the time needed to reach terminal spikelet initiation decreased by 40 percent. Temperatures above 19°C lengthened the period again and the total number of leaves or spikelets produced was little affected by temperature. However, the effect of temperature on final tiller number is highly variety dependent (Hucl and Baker, 1990). Heat stress significantly reduced tiller number in the cultivars Pitic 62 and Potam but had little effect on the cultivars Neepawa and Ingal. Warmer temperatures were also found to have little effect on time required to reach terminal spikelet initiation; however, the time from terminal spikelet initiation to anthesis was significantly reduced (Slafer, 1995). Heat stress, while not greatly affecting the number of leaves produced, does increase the rate of the leaf appearance (Slafer and Rawson, 1997). In addition, daylength was found to influence early spike development. Under short daylengths, elevated temperatures delayed early spike development, but such temperatures hastened the process under long days (Rawson and Richards, 1993).

Temperature increases above 30°C. The effects of temperature stress are more dramatic above 30°C. Four-week-old seedlings produce less than half the leaf area and shoot dry weight at 30°C relative to 20°C (Huang and Taylor, 1993). Temperatures of 30°C or more also significantly reduce seed set, although there are significant differences among cultivars (Dawson and Wardlaw, 1989); reduce photosynthesis (Balota and Lichtenthaler, 1999); and reduce biomass (Al-Khatib and Paulsen, 1990). Yucel and colleagues (1992) reported that inhibition of PSII (photosystem II) fluorescence in ten-day-old seedlings exposed to 37°C could be fully reversed at 22°C, if the initial heat treatment did not exceed 16 hours. A two-hour heat shock of 37°C can also result in leakage of cellular contents into the intercellular spaces of seedlings of sensitive genotypes (Vierling and Nguyen, 1992). Seven-day-old seedlings exposed to 42°C for up to three days were found to contain thylakoid proteins that during normal senescence would have been degraded, thus prohibiting recycling of the amino acids (Ferguson et al.,

1993). As temperatures rise from 25 to 45°C, thylakoid lipids within the chloroplasts of seedlings can be degraded by peroxidation, inhibiting electron transport and photosynthetic ability (Mishra and Singhal, 1992). Two possibly distinct temperature domains were identified when studying PSII response to heat stress on 28-day-old seedling leaves (Lu and Zhang, 2000). Moderate heat (30 to 37.5°C) elicited no fundamental changes in the PSII: no real variations were noted (F_v/F_m and Q_p), nor did any decrease prove reversible (Φ_{PSII}, Fv'/Fm'). However, following severe heat (>37.5°C) these changes were large and irreversible.

Raising the temperature from 10 to 25°C led to increased leaf length in Maris Huntsman near-isogenic lines (NILs) fully recessive for the gibberellin (GA)-insensitive Rht height-reducing genes (Pinthus et al., 1989). This effect was less marked in the single-gene Rht1 and Rht2 NILs, and was absent in the Rht3, Rht1+2, and Rht2+3 NILs. The temperature effect was mediated through an influence on GA levels. The higher temperature also halved the duration of the leaf-elongation phase, thereby doubling the rate of leaf elongation (Pinthus and Abraham, 1996).

HSPs could be detected in seven-day-old seedlings following a two-hour exposure to 28 and 31°C temperatures, and in flag leaves of adult plants exposed to 32°C (Hendershot et al., 1992). HSPs develop when temperatures are as little as 10°C above optimum levels and are frequently expressed in the absence of observable heat-stress damage. A specific HSP belonging to the HSP26 group was differentially expressed in heat-tolerant RILs (recombinant inbred lines) and not in the heat-sensitive RILs from the same cross (Joshi et al., 1997). A specific cDNA sequence seems associated with this segregation.

Anthesis and Grain-Filling Phase

The general impact of heat stress at, and following, anthesis is increased floret sterility and seed abortion, reduced seed filling, and, ultimately, decreased grain yield. The temperature profile for anthesis and grain filling (Table 12.5) shows minimum values of 9.2 to 9.5°C, an optimum range of 20.7 to 21°C, and maximum values varying from 31 to 35.4°C.

Winter wheat trials showed that a 1°C increase in average temperature throughout the season will reduce grain yield by 10 to 24 percent, mostly due to a shorter crop cycle (Batts et al., 1997, 1996) and a contracted duration of grain filling (Wheeler, Batts, et al., 1996; Wheeler, Hong, et al., 1996). Total biomass and grain yield decline and the numbers of grains per spike reduce while the rate of gain in grain weight increases. An average temperature increase of 4°C over the crop cycle in the United Kingdom decreased grains per m^2, averaged over two N and two CO_2 treatments, by 30 percent, thousand kernel weight (TKW) by 13 percent, and final yield by 18 percent (Mitchell et al., 1993). Interestingly, under high temperature the TKW under high N actually increased by 30 percent, thus offsetting the reduction in grain per m^2 somewhat. Increased fertility may compensate for heat stress under some circumstances.

Temperature increases up to 30°C. In spring wheat, the midanthesis period was shown to be the most sensitive to heat stress. Generally, grain fertilization, grain set, and eventual yield are reduced (Ferris et al., 1998). As daytime temperature approaches 30°C just after anthesis, the total green-leaf area per stem will decrease (Slafer and Miralles, 1992). Fructan is one of the carbon reserves stored in dry matter in the stem. When temperatures increase, fructan accumulation decreases due to the increased demand from the sink (Bancal and Triboi, 1993).

Temperature increases between 30 and 40°C. Temperatures above 30°C significantly increase damage to the wheat plant. Grain size and weight decrease, although there is a significant genotype × temperature interaction (Wardlaw and Moncur, 1995). Cultivars that maintain kernel weight are those with higher rates of grain filling under heat stress. Applying a 30/25°C regime for ten days following anthesis, relative to a 21/16°C regime, reduced the number of grains per spike by 11 percent (Tashiro and Wardlaw, 1990a). Terminal spikelets proved more vulnerable than central or basal spikelets. Likewise, within spikelets, the terminal florets were more sensitive to heat stress than the basal ones. Al-Khatib and Paulsen (1990) found that plants grown at 32/27°C relative to 22/17°C, suffered a 20 percent drop in kernel weight (30 percent maximum) and a 23 percent in mean yield (more than 40 percent in some entries). Heat stress also reduced the photosynthetic rate. Tolerance to heat is generally correlated with longer duration of photosynthetic activity and a higher photosynthetic rate (Al-Khatib and Paulsen, 1990). If temperature is increased shortly after anthesis to 35°C, then grain number, kernel weight, and yield are severely reduced (Gibson and Paulsen, 1999). However, these authors found that by delaying the temperature increase another five days (15 days postanthesis), the yield losses were considerably lower and were primarily associated with reduction on kernel weight. These observed reductions in kernel weight (Banowetz et al., 1999; Gibson and Paulsen, 1999) are likely to be due to a reduction in soluble starch activity and cytokinin in the grain (Hawker and Jenner, 1993). Elevated temperatures after anthesis (35/30°C) enhance leaf senescence, loss of chloroplast integrity, and thylakoid luminal volume, while decreasing the PSII-mediated electron transport (Xu et al., 1995). Floret sterility can also be induced by 36/31°C temperatures two to three days prior to anthesis and is further exacerbated by high humidity (Tashiro and Wardlaw, 1990b). Fokar, Blum, and Nguyen (1998) observed that grain weight per spike was highly positively correlated with photosynthate stem reserves ($r = 0.96$) when wheat plants were exposed to 38°C. The constitutive ability to store such stem reserves was itself correlated to rate of chlorophyll loss ($r = 0.92$), reflecting increased leaf senescence.

Temperature increases over 40°C. Temperatures over 40°C in practical situations are infrequent and associated damage can be very high. Reductions in grains per spike (97 percent) and eventual yield (98 percent) can be drastic at these very high temperatures (Leithold et al., 1997). However, even at temperatures of 40°C or more, there are significant differences among genotypes for chlorophyll content, chlorophyll stability, cell membrane stability (Sairam et al., 1997), and kernel weight (Stone and Nicolas, 1998b). Stone and Nicolas (1998b) found that the greatest reduction in kernel weight occurred in the first day of exposure to heat stress and the earlier in the grain-filling phase stress was applied the greater the reduction in kernel weight (Stone and Nicolas, 1995b). Varieties also differ in their response to sudden heat shock as opposed to gradual increases in temperature (Stone and Nicolas, 1995a). Slow heat acclimation may protect plants from later sudden heat stress, allowing them to adjust their water status and canopy cooling. Earliness, as determined by date of heading, was found to be highly correlated ($r = 0.75$) with durum wheat grain yield under grain-filling heat-stress temperatures up to 40°C (Rekika et al., 2000). They found that for every day that a variety headed earlier, yield increased by 100 kg.

Isogenic lines in a 'Condor' background carrying the 1BL/1RS translocation expressed higher numbers of grains per spike and spike grain weight at the higher temperature (30/25°C) than isolines not carrying the translocation (Dawson and Wardlaw, 1989).

Field trials show that yield reduction under heat stress is associated with decreased grain number per spike (Shpiler and Blum, 1991; Zhonghu and Rajaram, 1994). Shorter grain-filling periods and lower grain growth rates also reduce eventual yield (Viswanathan and Khanna-Chopra, 2001). Zhonghu and Rajaram (1994) found that seed per spike, grain-filling rate, and final yield were more affected by heat stress than spike number per m^2, grain-filling period kernel weight, and test weight. Trials planted deliberately to experience heat across six global locations identified biomass, grains per m^2, days to anthesis, early ground cover, rate of CO_2 fixation, stomatal conductance, and loss of chlorophyll between anthesis and grain filling to be highly correlated with yield (Reynolds et al., 1994).

Some of the latest wheat varieties have increased values for stomatal conductance (Lu et al., 1998). This results in enhanced evaporative leaf cooling, thereby reducing the effects of high temperature. Fokar, Nguyen, and Blum (1998) found that stomatal conductance was highly correlated with yield ($r = 0.93$), but not photosynthetic rate ($r = 0.02$). Cell membrane thermostability (CMS) also correlated well with yield in various hot locations. The correlation between measurements of CMS carried out on seedlings and adult plants of 56 different spring wheats was very high ($r = 0.92$), and somewhat better than when triphenyl tetrazolium chloride (TTC) assays were used (r = 0.84).

At leaf temperatures during anthesis of 32 to 35°C, mRNA for HSP16.9 was detected (Nguyen et al., 1994). Two cDNA clones have been isolated that are related to the HSP16.8 and HSP16.9 heat-shock proteins in wheat (McElwain and Spiker, 1992). HSP70 mRNA has also been detected in grain endosperm following 40°C heat shock during the grain-filling period (Hurkman et al., 1998).

Grain Development and Quality

The deposition of compounds in the grain under heat stress affects final industrial and end-use quality characteristics.

Starch

Heat stress causes a reduction in starch granule size and reduces amyloplast number (Tester et al., 1995). Short-term exposure to elevated temperatures will decrease soluble starch synthase (SSS) activity, however, this effect is reversible in cooler conditions (Keeling et al., 1993). Longer-term exposure takes longer to reverse, although there are significant differences among varieties (Caley et al., 1990), which translate to differences in endosperm weight. Following extended exposure to heat stress, enzyme activity is lost and is not recoverable.

Starch deposition is reduced at temperatures over 30°C and does not change when plants are moved to cooler conditions (Jenner, 1994). Findings suggest that two distinct temperature response mechanisms may be operating in the wheat plant. It was found that SSS activity in the grain dropped to less than half within one day of exposure to 35°C, with longer exposures having little additional effect (Hawker and Jenner, 1993). This initial quick response is the likely reason why grain weights decline following heat stress, whereas daily five-hour periods of 40°C during grain filling result in reduced starch accumulation and smaller deformed starch granules (Shi et al., 1994).

Protein

Following exposure to heat stress the protein content of the grain is often decreased, although protein percentage increases (Stone and Nicolas, 1998b). This is because loss of grain mass is greater than protein accumulation. When five-day episodes of 40°C temperatures (6 h per day) were spaced throughout the grain-filling period, varieties responded differently in protein synthesis and accumulation, and the phase at which they were most affected (Stone and Nicolas, 1996). Heat reduces the duration of protein accumulation, and therefore the amount accumulated, but does not affect the rate of protein accumulation. Moving plants that were heat shocked (40/16°C) during grain filling to lower temperatures does not allow recovery of the negative effects on protein (Stone, Nicholas, and Wardlaw, 1996). If temperature is increased slowly, the effects of heat stress on protein accumulation are reduced (Stone and Nicolas, 1998a). Genotypes also vary significantly in their protein accumulation response under heat stress (Stone and Nicolas, 1994, 1995c, 1998a).

Gliadin:Glutenin Ratio

Extreme heat stress (>35°C) causes a reduction in the high-molecular-weight (HMW) glutenins and increases gliadin accumulation (Blumenthal et al., 1994; Daniel and Triboi, 2000). The resulting lack of larger aggregates of glutenin causes weakening of dough properties. Three days of exposure to high temperature (maximum 40°C) results in variation in the gliadin:glutenin ratio from –9 to 18 percent (Stone and Nicolas, 1994). It has been proposed that heat-shock elements upstream of the coding regions for gliadins influence or control these effects (Blumenthal et al., 1993).

Industrial Quality

Heat stress causes significant reductions in flour yield and mixing time (Gibson et al., 1998). Both moderate (21 to 30°C) and high (40°C) temperatures decrease flour protein percent and dough strength (Stone et al., 1997). Years in which temperatures during grain filling in farmers' fields were high produced grain that had particularly weak dough properties and poor loaf volumes (Blumenthal, Batey, et al., 1991; Blumenthal, Bekes, et al., 1991). This was accompanied by an increase in protein, in particular the gliadins. Heat shock (40°C) applied at any time during grain filling negatively affected rheological performance (Corbellini et al., 1997). Just five days of stress reduced mixing time by 40 to 60 percent, which was associated with an increase in production of low-molecular-weight (LMW) gliadins and a reduction in the required complex protein aggregates. Varietal responses to heat stress vary significantly. Genotypes with HMW glutenin subunit combination 5+10 were more heat tolerant than those with 2+12.

Temperatures during grain filling of up to 30°C increase dough strength; however, temperatures above this tend to decrease dough strength (Randall and Moss, 1990). HSP70 could not be shown to be directly involved in the degradation of dough properties in heat-stressed wheat, although HPS70 levels in the grain did increase with increasing temperatures (Blumenthal et al., 1998).

Genetic Variability

Various studies, many referenced in the preceding sections, have shown that varieties vary in their response to heat stress. Genetic variability for good performance under heat stress is available within the common bread and durum wheat genepools and among wheats' wild relatives. As discussed earlier, tolerance to heat in agronomic performance is genetically independent from tolerance in terms of grain quality. However, certain factors may influence both. Table 12.6 lists some of the wheat genotypes that differentiate in various heat-stress-related studies. Both tolerant and sensitive types are indicated, organized by the trait investigated.

Reduction in grain weight of varieties maturing at 30/25°C varied from 30 to 60 percent (Wardlaw and Moncur, 1995). Differences between five varieties for reduction in grain weight varied from 39.3 to 58.3 percent, when exposed to an average of 38°C during grain filling (Fokar, Blum, and Nguyen, 1998). When 75 varieties were studied for agronomic (phasic development, yield components) and quality (nitrogen, SDS-soluble protein fractions, noodle swelling power) response to a three-day period postanthesis of 40°C, large genetic variability was observed (Stone and Nicolas, 1994). About 5 percent of the entries expressed a high level of tolerance. Table 12.7 lists some of the entries and their responses (Stone and Nicolas, 1995c). The variety Bass was able to combine good agronomic tolerance with good quality tolerance; in this instance, the relative glutenin concentration actually increased.

Variability for cellular tolerance among 56 spring wheat varieties, measured either as cell membrane thermostability or with the triphenyl tetrazolium chloride assay, showed up to four-fold differences (Fokar, Nguyen, and Blum, 1998). Among wild relatives of wheat, two geographic groups of *Aegilops geniculata* were shown to carry distinct adaptive features to heat (Zaharieva et al., 2001). The accessions collected from moderately warm

TABLE 12.6. Wheat varieties that are tolerant or sensitive in regard to specific traits when exposed to elevated temperatures.

Trait	Heat tolerant	Heat sensitive	Reference
Seed germination after heat stress during grain filling	Oum-rabia	Marzak	Grass and Burris, 1995a
Seed germination	APCB53	Adamello	Dell'Aquila and Turi, 1999
Damage to cellular ultrastructure	M3 *(Triticum monococcum)*	M9 *(Triticum monococcum)*	Vierling and Nguyen, 1992
Decline in protein synthesis	APCB53	Adamello	Dell'Aquila and Turi, 1999
LMW HSP synthesis	APCB53	Adamello	Dell'Aquila and Turi, 1999
Seedling photosynthetic rate	Ventnor	Lancero	Al-Khatib and Paulsen, 1990
Leaf extension	Rht3, Rht1+2, Rht2+3 NILs (Maris Huntsman)	Rht1Rht2Rht3 NILs (Maris Huntsman)	Pinthus et al., 1989
Increased tillering	Potam	Pitic 62	Hucl and Baker, 1990
Membrane stability	HD2285	Raj3077	Sairam et al., 1997
Sensitivity at booting	SUN 9E, ISIS, Hybrid Titan	Olympic, Pinnacle	Dawson and Wardlaw, 1989.
Stem reserves mobilization	V5	V2183	Blum et al., 1994
Photosynthate stem reserves	Danbata	Nacozari	Fokar, Nguyen, and Blum, 1998
Overall adult performance	Ventnor	Lancero	Al-Khatib and Paulsen, 1990
Grain weight	Kalyansona	SUN 9E	Jenner, 1991
TKW tolerance to sudden high heat stress	Egret	Oxley	Stone and Nicolas, 1995b
Yield (main ear)	Naxos	Nacozari	Leithold et al., 1997
Yield (plot)	Avalon	Galahad, Hereward, Mercia	Batts et al., 1998
Stable yield across planting dates	El Neilein	–	Ishag and Mohamed, 1996.
Starch synthesis	Hindi62	PBW154	Viswanathan and Khanna-Chopra, 2001
Protein synthesis	Hindi62	PBW154	Viswanathan and Khanna-Chopra, 2001

Trait	Heat tolerant	Heat sensitive	Reference
Stable glutenin:gliadin ratio and stable TKW	Bass	Dagger, King, Osprey, Oxley, Veery 5	Stone and Nicolas, 1995c
Yield and quality	Lillimur	Oxley	Stone and Nicolas, 1994

TABLE 12.7. Varieties ordered by response to heat stress as regards their TKW and glutenin:gliadin ratio

Response trait	Varieties
Stable TKW	
Increased glutenin:gliadin ratio	Bass
Stable glutenin:gliadin ratio	Egret, Lillimur, Gamenya
Decreased glutenin:gliadin ratio	Hybrid Titan, IW 744, Lachish#156, ME71
Decreased TKW	
Increased glutenin:gliadin ratio	Durati, Sun 9E-16, Trigo 1
Stable glutenin:gliadin ratio	Bodallin, Meering
Decreased glutenin:gliadin ratio	Dagger, King, Osprey, Oxley, Veery#5

Source: Modified after Stone and Nicolas, 1995c.

areas showed high C-isotope discrimination and chlorophyll content and relied more on transpiration as an adaptive mechanism to heat stress. Those from the high-temperature regions showed low C-isotope discrimination and low chlorophyll content, thereby limiting energy load and thus avoiding heat-stress damage. Variability for heat tolerance has also been found among *Triticum dicoccoides* accessions (Rekika et al., 2000). It may be concluded that sufficient genetic variability for heat tolerance is available to allow progress to be made in breeding.

Genetics and Breeding

Breeders have been successful in developing wheat genotypes with good performance under warm to hot conditions, and these have found acceptance among farmers. However, little research has been devoted to studying the inheritance of the mechanisms involved. The previous sections indicate that the mechanisms conferring heat tolerance during the various stages of plant development are many and that the data collected are not always congruent. Only with a thorough understanding of the underlying mechanisms of heat tolerance can progress in breeding be maximized.

Genetics

The heritability of tolerance to high temperature (30/25°C), as measured by reduction in grain weight, is very low ($h^2 = 0.2$) despite a considerable range in genotypic response (Wardlaw, 1994). Additive gene effects were most important in determining tolerance to high levels of heat (37°C) (Porter et al., 1995). Fokar, Nguyen, and Blum (1998) found that gene action was dominant and most of the genetic variation was due to additive genetic variance. These authors calculated broad-sense heritabilities as high as 89 percent. Using durum wheat substitution lines, genes coding for membrane thermostability were shown to be most likely located on chromosomes 3A, 3B, 4A, and 4D in the cultivar Langdon (Sun and Quick, 1991). A study of ditelosomic lines also showed that chromosomes 3, 4, and 7 contain a higher frequency of genes controlling the development of heat-shock proteins (Porter et al., 1989).

Breeding

Developing an understanding of candidate traits that could be used in breeding for heat tolerance was listed as a top research priority a decade ago (Kohli et al., 1991) and remains a high priority today. The effect of acclimation to heat and varietal differences for this characteristic make it imperative that comparisons among varieties for short heat-stress response are made only following a period of gradual increase in temperature (Stone and Nicolas, 1995b, 1998a). This seems appropriate since it reflects the most likely response in the field. Low light levels were shown to increase the reduction in grain weight caused by heat stress (30/25°C) (Wardlaw, 1994). This may confound selection of plants in sowing arrangements that lead to varying levels of shade. Plants with more tillers may also limit the influx of light, resulting in increased sensitivity to heat. Two separate mechanisms in response to heat stress may be operating, one functioning at temperatures below 30°C and one above 30°C (Jenner, 1994). If this is the case, then the breeder may aim to combine both mechanisms in one genetic background.

When heat stress occurs at the beginning of the crop cycle, it is beneficial if good crop establishment can be achieved (Ali et al., 1994). Harvest index (HI) is dramatically reduced by high temperature (Gibson and Paulsen, 1999). Selection of plants with high kernel weight, associated with high HI, which is driven by a longer duration and more stable rate of photosynthetic activity, was proposed by Al-Khatib and Paulsen (1990). The number of grains per spike was also shown to influence yield variation under heat (Shpiler and Blum, 1991) and may provide a more easily measurable selection criterion than HI. Zhonghu and Rajaram (1994) concluded that grains per spike, biomass, HI, and test weight were potential selection criteria for heat tolerance. If there is more variation in rate of grain filling than in duration of grain filling during heat stress, then the former can also be used as a selection tool in breeding (Wardlaw and Moncur, 1995). Grain-filling rate and grain weight per ear were also proposed as breeding objectives for locations where temperatures preanthesis reached 27°C (Mashiringwani and Schweppenhauser, 1992). Epicuticular wax may also improve tolerance to heat by reflecting light and lowering temperature (Ishag and Mohamed, 1996).

The stability of chlorophyll variable fluorescence (F_v) measured several times postanthesis showed some correlation with yield under heat stress (maximum 38.8°C) but could not serve as a single reliable criterion for selection of heat tolerance (Moffatt, Sears, and Paulsen, 1990). Stage of crop development also influenced F_v measurements and significant maternal effects were noted, making it important to choose the appropriate parent to serve as female for this trait in a breeding program (Moffatt, Sears, Cox, and Paulsen, 1990). Additive gene effects for (F_v) were shown to be dominant. The transverse relaxation time (T_2) of water in the leaf, noninvasively measured using NMR (nuclear magnetic resonance), becomes shorter as temperatures increase. Heat-tolerant varieties maintain a higher T_2 (Maheswari et al., 1999).

Selection of genotypes according to their membrane thermostability response corresponded very well to their field performance in warm conditions (Saadala, Quick, and Shanahan, 1990; Shanahan et al., 1990). Membrane thermostability (MT) correlated well (r = 0.60) with yield at five international sites (Reynolds et al., 1994). CMS observations on seedlings of 16 genotypes also correlated well with yield across locations (0.51 to 0.63) (Fokar, Nguyen, and Blum, 1998). Membrane stability determinations on heat-stressed (34°C) seedlings showed a 0.79 correlation with measurements taken on adult plants exposed to the same temperature (Saadalla, Shanahan, et al., 1990). Tests on a subset of eight lines showed the correlations between seedling and adult plants using the CMS or TTC assays for cellular thermotolerance to be high (0.84 to 0.92) (Fokar, Nguyen, and Blum, 1998). Even higher correlations were noted between CMS measurements taken on seedlings and reduction in grain weight, when plants were exposed to very high temperatures (mean of 38°C) (r = –0.98) (Fokar, Blum, and Nguyen, 1998). Reynolds and colleagues (1994), however, found the correlation with MT measured on seedlings was somewhat less (0.46).

Modifications and improvements on the membrane thermostability methodology have been proposed (Tahir and Singh, 1993). In field tests, CMS correlated well with yield and heritability was estimated at 0.74 (Blum et al., 2001). However, the authors warn that CMS should not be used as the sole selection criterion to determine heat tolerance, but it can be used as an initial indication.

One of the most promising tools for selecting for yield under heat stress is canopy temperature depression (CTD) (Amani et al., 1996). It is an integrated approach that measures the effect of stomatal conductance over time. Stomatal conductance was shown to be highly correlated (0.93) to yield in field plots exposed to increasing temperature (Lu et al., 1998). Correlations of up to 0.89 between CTD and yield were observed in field experiments in tropical, low-latitude locations. Postanthesis measurements taken at a warm site in Mexico even correlated well (0.65 to 0.76) with yield in five heat-stress locations around the world (Reynolds et al., 1994). The best time to measure CTD is between noon and 1600 h, and it can be measured either pre- or postanthesis.

Some cultivars, such as the bread wheat 'Halberd,' express a stable response in dough-mixing properties following exposure to high temperatures. These genotypes are good sources of heat tolerance for breeding wheats that maintain good quality under heat (Wrigley et al., 1994). In durum wheat, the reduction in solubility of albumins and globulins following heat stress could be used as a measure of eventual pasta-processing quality (Every et al., 1990).

Since heat tends to decrease the length of the crop cycle, one approach to obtain acceptable yields may be to grow later-maturing lines that are planted early, so grain filling coincides with the coolest month (Ishag et al., 1998). Such materials would tiller more, have higher biomass, and thus produce more yield.

Crosses are made based on detailed information on the parents, which involves a multitude of traits, including tolerance to heat. Segregating populations are then alternated between two distinct sites in Mexico; a high-rainfall, high-altitude site with many foliar diseases and an irrigated, low-rainfall, high-yield-potential site. This alternating process is called "shuttle breeding." No deliberate exposure to either early or late heat stress occurs during this phase. When advanced lines become available for yield testing, they are grown in yield trials planted at dates that will result in either early or late heat stress. Rather than planting in late November or early December, these lines are sown in early October to experience early heat, or between mid January and late March to experience heat stress during flowering and/or grain filling. CTD observations are taken on some of these trials. The top-yielding lines with desired agronomic type are disseminated internationally to collaborators, where the final proof of heat tolerance occurs. Lines found to outperform the commercial varieties are released as new options to farmers.

Conclusions and Recommendations

Heat stress in wheat is a common occurrence in 65 to 70 million ha around the world. It occurs in irrigated and rainfed conditions, as well as at moderate or high relative humidities; it can affect the crop early, late, or throughout the entire crop cycle. Predictions are that heat stress may gain in importance as global warming intensifies. In general, winter wheats start suffering from heat stress to the roots at lower temperatures than spring wheats.

Heat stress above 30°C may double germination time and significantly reduce germination percentage. Less than three-quarters of the normal amount of seminal, crown, and lateral roots may be produced following heat stress, and root conductivity may decline. In conjunction with additional stresses such as drought or nutrient deficiencies these effects become larger.

During the vegetative growth phase, at temperatures of 20 to 24°C, a rise in temperature may have no negative effect on plant growth. However, temperatures of 28°C and higher cause the production of heat-shock proteins before observable heat damage is evident. At about 30°C, leaf mass in seedlings may decrease by half and photosynthesis may drop by a third. A temperature of 30°C at booting may result in pollen sterility. Above 35 to 37°C, sudden heat shocks may damage thylakoids and cause leaking of cellular contents, with electron transport compromised. Exposure to high temperature for more than 16 h may lead to irreversible damage. During the most sensitive reproductive phase, slight seasonal increases in temperatures of just a few degrees will significantly affect the crop cycle in winter wheat, with associated declines in biomass, grains per m^2, and yield.

In spring wheat, temperatures in the upper 20°C range started to cause increased senescence and reduced fertilization. Fructans ceased accumulating in the stems as sink demand increased. As temperatures exceed 30°C, the number of grains per spike drops significantly. Significant reductions in kernel weight and yield may occur, although higher photosynthetic activity during this time may provide some tolerance to heat. Following temperatures of 35°C and above, a return to lower temperatures does not alter the rate of grain filling, resulting in low kernel weight.

Photosynthate stem reserves and their export to the filling grain provide a good measure of tolerance to heat, providing they are not associated with excessive senescence. Early maturity may also be beneficial at these high temperatures, as it provides an escape mechanism. However, evidence is not conclusive as to which is most important in conferring heat tolerance, maintaining grain-filling period or maintaining grain-filling rate. Varieties that withstand heat stress best have high stomatal conductance, providing leaf cooling, and maintain cell membrane thermostability.

Starch synthase activity is reduced as temperatures rise, resulting in reduced starch accumulation and lower kernel weight. At temperatures over 30°C these effects are not reversible, and over 35°C starch synthase activity can be halved within a day. Protein levels may rise slightly up to 25°C and begin declining at higher temperatures. As with other physiological processes, a return to lower temperatures does not reverse earlier damage, although gradual rather than sudden increases in temperature causes less damage. At high temperatures, the HMW glutenins in the grain reduce and the gliadins tend to increase. Varieties with the HMW-glutenin subunit combination 5+10 tend to be more heat tolerant than wheats with other subunits. A rise in temperature up to 35°C tends to have a dough-strengthening effect, while temperatures over 35°C cause dough weakening.

Large genetic diversity within the wheat gene pool is available for most traits considered important in conferring tolerance to heat stress. Additive gene effects control heat tolerance, and heritability estimates varying from low to very high have been reported. The mechanisms of heat tolerance for agronomic and quality performance are largely independently inherited. Wheats carrying the 1BL/1RS translocation may have a higher heat tolerance.

When comparing heat tolerance in a breeding program, it is crucial to take into account the ability of some wheats to acclimate during gradual temperature increases, providing protection against higher levels. It also appears that the physiological plant response when exposed to temperatures up to 35°C is likely to be different than the response above 35°C.

Biomass, grains/spike, harvest index, kernel weight, and test weight are often proposed as morphological selection tools. Membrane thermostability and canopy temperature depression seem especially promising physiological selection tools, expressing high correlations with yield. The latter integrative measure is probably the easier to implement. The expression of heat-shock proteins is not correlated with plant response and is unlikely to be useful in practical breeding.

Heat stress is an important global stress in wheat, the understanding of the underlying mechanisms is quite considerable, sufficient genetic diversity is at hand, and practical breeding tools are available. It would appear that a concerted effort to breed wheats with tolerance to heat stress should bear fruit in the near future.

BREEDING FOR TOLERANCE TO COLD-TEMPERATURE STRESS IN WHEAT

Cold damage occurs in winter wheat primarily during the vegetative phase in winter, when temperatures can fall as low as –20°C. However, low autumn temperatures also damage seedlings and affect stand establishment, and late spring frosts can damage mature foliage, the reproductive organs, and the filling grain. Fall-planted spring wheat is more likely to suffer from late frosts, causing damage to the growing spikes or grain during grain filling. Spring-planted spring wheat can also suffer vegetative damage from late spring frosts, and early fall frosts sometimes cause floret sterility, poor grain set, and shrunken grain.

Cold temperatures are also needed for vernalization-sensitive winter wheats. In this case, cold enables the plant to reach its potential, rather than limiting it. A vernalization requirement is also a mechanism by which varieties can avoid spring frosts. This chapter focuses on the damaging effects of cold temperature on wheat development and explores literature from the past decade. Good reviews have been published on vernalization (Yan and Wallace, 1996) and on cold tolerance more than a decade ago (Guy, 1990). This section will be complementary to these reviews and will attempt to view cold tolerance from a plant breeder's perspective.

Geographical Distribution of Cold-Affected Environments

As latitude increases and at higher elevations, mean temperatures decrease, and variation in minimum and maximum temperatures increases. Daylength also increases at higher latitudes. Five primary wheat-production environments requiring wheat varieties with differing phasic development can be identified.

From the Equator to 30°N or S

Spring wheats during the early fall to early winter period dominate. These wheats carry dominant alleles for the vernalization sensitivity genes on one or more of the three major *Vrn/vrn* loci. They do not require vernalization to initiate the reproductive (flowering) growth phase. They also carry dominant alleles for the photoperiod sensitivity genes on one or both of the two major *Ppd/ppd* loci, making them insensitive to variation in daylength. They are therefore adapted to the conditions of moderate temperature and short days found in the tropics. These wheats are planted in the fall, as summer temperatures at lower elevations tend to be too high for wheat (maximum 30 to 45°C). However, at higher elevations (>1500 m above sea level), they may be planted year-round (e.g., Kenya). Countries where wheat is grown under these conditions include Mexico, Ecuador, Brazil, Morocco, Egypt, southern Iran, south and central Pakistan, India, Ethiopia, central and northern South Africa, south and central China, and Australia.

Between 30 and 35°N or S

Under these conditions spring wheats are fall planted. These wheats carry dominant alleles for the vernalization sensitivity genes on one or more of the three major *Vrn/vrn* loci. The initiation of the reproductive growth phase is therefore independent of vernalization. However, a mild photoperiod response may be beneficial due to slightly longer day lengths; these genotypes may have dominant alleles at one of the *Ppd/ppd* loci rather than both. These wheats mature a few weeks later and can better exploit the longer crop cycle. Countries representative of these production conditions include the United States (Texas), eastern Argentina, Chile, Morocco (Atlas region), Iran, Pakistan (NWFP), South Africa, and Australia (Victoria).

Between 35 and 40°N or S

At these latitudes winters are too cold for spring wheats, so fall-planted winter wheats are grown. These wheats carry recessive alleles for the vernalization sensitivity genes on all three major *Vrn/vrn* loci. They therefore require low temperatures (optimum 5°C) or vernalization to initiate the reproductive growth phase. They need not carry recessive alleles for the photoperiod sensitivity genes on both major *Ppd/ppd* loci; however, some photoperiod response will avoid damage to reproductive organs from late spring frosts. Wheats grown under these conditions will require some early or late cold tolerance. Representative countries are: the United States (Kansas), Chile (south), Spain, South Korea, Turkey (Eskisehir), and New Zealand (north).

Between 40 and 50°N or S

At these latitudes fall-sown winter wheats are grown. These wheats carry recessive alleles for the vernalization sensitivity genes on all three major

Vrn/vrn loci, making initiation of flowering dependent on low vernalizing temperatures (optimum 5°C). At these higher latitudes, varieties should carry recessive alleles for the photoperiod sensitivity genes on both major *Ppd/ppd* loci. Regulation of flowering is dependent on long daylength; this ensures that flowering takes place when the danger of late frosts has passed. Wheats grown under these conditions require considerable cold tolerance. Countries where wheat is grown under these conditions include the United States (Nebraska), Canada (Ontario), France, Romania, China (Beijing), and Turkey (north).

Between 45 to 50 and 60 to 65°N or S

Spring-sown spring wheat is grown at these latitudes because winter temperatures regularly fall below –20°C. These low temperatures are too cold for winter wheat production. The crop cycle is very short and limited to just three to four months. Daylengths are very long, almost 24 h in the most northern latitudes, and the increased radiation partly compensates for the shorter cropping period. These wheats carry dominant alleles for the vernalization sensitivity genes on one or more of the three major *Vrn/vrn* loci, making them true spring wheats, independent of low vernalization temperatures. Most of these wheats carry recessive alleles for the photoperiod sensitivity genes on both major *Ppd/ppd* loci. This requirement prevents the flowering during high-frost-risk periods, thereby protecting reproductive organs from damage. Wheats grown under these conditions require cold tolerance both early and late in the crop cycle. Representative areas include the United States (North Dakota), Canada (most), Sweden, Kazakhstan, Russia, and China (Heilongjiang). The total global area affected regularly by cold in any form is about 80 million ha.

Cold Stress and Plant Development

The breeder must have a sound understanding of the underlying physiological mechanisms of cold tolerance, their inheritance, and the practicalities of applying artificial selection if progress in breeding is to be made. Cold stress affects the wheat plant by adversely affecting germination, emergence, root development, leaf health, biomass production, floret initiation, pollination/fertilization, seed growth, final grain yield, and grain industrial and nutritional quality. Sometimes cold stress in one phase of crop development can be compensated for by recovery in later growth stages (e.g., excess tillering compensating for reduced emergence). Grain yield may sometimes appear unaffected by cold stress and yet the grain quality can be severely reduced, resulting in considerable economic losses.

Emergence and Root Development

Cold hardening or cold acclimation is defined according to Guy (1990) (see Galiba, 1994) as "adjustment of metabolism and basic cellular function to the biophysical constraints imposed by low non-freezing temperature, and the induction of freezing tolerance" (p. 116). Acclimation may be reversed if temperatures rise (Galiba, 1994). The following section summarizes the crop response to cold stress during the initial phase of this hardening period during crop emergence and establishment.

The shoot:root ratio of spring wheat decreases when root zone temperature is 12°C rather than 20°C, and N concentrations increase in both the shoots and roots (Engels, 1994). When the temperature is lowered to an 8/8°C day/night temperature regime, from 20/20°C, relative root-growth rate of winter wheat decreases (White et al., 1991). Evidence also suggests that low root-zone temperatures may reduce levels of P, S, Cu, Mn, and Zn, thereby restricting plant growth (Miyasaka and Grunes, 1997). The relative shoot growth rate (RGR) is also greatly reduced at low temperatures (Equiza et al., 1997). Equiza and colleagues (1997) found that root RGR in winter wheat drops more than in spring wheat at low temperatures, resulting in a smaller root:shoot ratio. At the same time, sugar and fructan levels in the shoots and roots increase in both spring wheat and winter wheat at low temperature. The smaller root system of winter wheat at low temperatures is associated with fewer stomates and thicker, more xeromorphic epidermal leaf cell walls, resulting in lower transpiration. In comparison with spring wheats, winter wheats under cold stress have lower osmotic potential and more root fructans, which may influence cold tolerance (Santoiani et al., 1993). Root dry weight and protein accumulation in winter wheat were similar or slightly higher at 4°C than at 20°C, while both processes were retarded at 4°C in spring wheat (Karimzadeh et al., 2000). Under cold stress, the CO_2 assimilation rate and translocation of photosynthate to the root and the shoot drops significantly. Temperatures of 2°C cause cessation of root growth in winter wheat and reduce the polysaccharide content of the cell walls (Zabotin et al., 1998). The date of emergence is also significantly delayed when soil temperature drops to 5°C (Addae et al., 1991).

Emergence to Booting Phase

The hardening process in wheat occurs prior to booting. Osmotic compounds such as sugars and proline accumulate and work as cryoprotectants during hardening, thereby preventing ice formation between the cells and subsequent damage (Galiba, 1994).Cold hardening has a large effect on the subsequent ability of the plant to withstand freezing temperatures. Acclimation of a winter wheat at 2°C significantly improves plant survival at subzero temperatures, and sugars, lipids, and proteins in the leaves increase (Zhou et al., 1994; Klimov et al., 1999). Dallaire and colleagues (1994) found that hardening at 2°C lowered the maximum freezing tolerance from –3 to –9°C in winter wheat; however, exogenous ABA application could improve freezing tolerance by another 3°C. However, they were unable to show that endogenous ABA levels are increased during the hardening process. Evidence suggests that overly extended cold acclimation may actually reduce the plants ability to withstand below-freezing temperatures (Fowler, Chauvin, et al., 1996). Spring wheats and winter wheats differ in this regard. Spring wheats are not able to maintain low-temperature-induced genes in an up-regulated fashion, thereby reducing their ability to withstand temperatures below 0°C. During acclimation at 2/0°C crown moisture content, crown osmotic potential and leaf water potential decreased more so for winter wheats than for spring wheats (Fu et al., 2000).

Earlier studies postulated a complete linkage between vernalization and cold tolerance. However, recent evidence shows that the end of the vernalization process does not coincide with a "switching off" of low-temperature-tolerance genes (Fowler, Limin, et al., 1996). Low-temperature-tolerance genes even functioned up to the heading stage.

Various proteins increase in absolute amount and relative content during cold hardening, and a rise in proline levels is commonly reported. During acclimation at 5°C, the winter wheat cultivar Kharkov accumulated large amounts of proline (more than fivefold), especially in the crown, and soluble protein (twofold) primarily in the leaves (Charest and Phan, 1990). At the same time, the spring wheat 'Glenlea' hardly responded. Cold hardening, therefore, causes an increase in dry weight and proline content and a decrease in osmotic potential that coincides with increased freezing resistance (Dörffling et al., 1990). Final proline and peak ABA content correlate well with freezing resistance (Machackova et al., 1989).

Apart from proline, a number of other proteins are specifically synthesized upon cold treatment. Polysomes, which are indicative of active protein synthesis, also accumulate during cold acclimation (Perras and Sarhan, 1990). The protein TaADF, previously known as Wcor719, has also been detected in cold-hardened plants (Ouellet et al., 2001). Similarly, glutathione (GSH) was found to increase rapidly in frost-tolerant winter wheat (Kocsy et al., 2000), indicating a probable role in frost tolerance. The TACR7 protein, coded for by the *tacr7* gene and specifically produced at low temperatures (2°C), was shown to be linked to freezing tolerance in wheat (Gana et al., 1997). Accumulation of acidic dehydrin WCOR410, coded for by *wcr410,* is also correlated with the expression of freezing tolerance (Danyluk et al., 1998). This protein concentrates near the plasma membrane of the vascular transition area, and may provide cryoprotection during freezing. The presence of these dehydrin-like proteins appear to influence freezing tolerance in both spring and winter wheats (Fu et al., 2000).

While a shift in the type of proteins synthesized is evident under cold stress, overall protein synthesis is not markedly different to noncold-stressed plants (Lasztity et al., 1999). In some cases gene transcription may take place, but translation may be blocked (Paldi et al., 1996).

Specific enzyme activity, evident in winter wheat, is absent in spring wheat (Hurry et al., 1995). These enzymatic processes form the basis of cold acclimation. These processes increase sugar content, which acts as a cryoprotective agent, allowing basal metabolism to proceed. Winter wheats express higher levels of photosynthesis than spring wheats when exposed for long periods to low temperature. The sucrose/starch ratio also increases significantly in leaves at low temperature (Savitch et al., 2000). However, only winter wheat increases sucrose- and fructan-processing enzymes under cold stress (Larsson, 1989; Tognetti et al., 1990). Clearly, frost-tolerant winter wheats accumulate more water-soluble carbohydrates than sensitive ones (Vagujfalvi et al., 1999).

Comparative studies of *Vrn1+Fr1, vrn1+fr1,* and a rare recombinant showed that the accumulation of sucrose at low temperature was closely associated with the *Vrn1* gene and hence distinct from *Fr1* (Galiba et al., 1997). The chromosome interval between *Vrn 1* and *Fr1* explains up to 80 percent of the variation in winter kill (Storlie et al., 1998).

Evidence suggests that cold hardening is reversible. When temperature is lowered from 23 to 4°C the level of sucrose and sucrose synthase (SS) increases; however, when plants were moved back to 23°C these levels decreased (Crespi et al., 1991). Similar observations have been made for fructan (Jeong and Housley, 1990).

Peroxidases may help protect the frost-tolerant variety against activated oxygen species, which are overproduced at low temperature (5°C), through their antioxidant function (Scebba et al., 1998). Okuda and colleagues (1991) found that hydrogen peroxide levels increase rapidly (threefold) when winter wheat seedlings are exposed to 4°C, indicating leakage of electrons from the electron transport system.

There is considerable variation for freezing tolerance in winter wheat. However, there is a limit, and at 4 cm soil depth temperatures below –20°C for less than 1 h, or of –19°C for at least 8 h, will result in 100 percent winter kill of any wheat (Bauer and Black, 1990).

At –2°C polyamines increase in the roots, crown, and leaves of winter wheats. This accumulation is correlated with frost resistance (Racz et al., 1996). When temperatures are lowered to –4°C, most antioxidant enzymes increase in the roots (Scebba et al., 1999). Nevertheless, it is difficult to relate this to frost resistance in the field (Fedulov, 1996).

Morphological modifications also contribute to frost resistance. When 27 winter wheats and three spring wheats were compared for cold tolerance, the most tolerant three were winter wheats, and these had the smallest guard cell measured at 17°C (Limin and Fowler, 1994). A spring wheat had the largest cell size. Guard cell size correlated with freezing in the field. Smaller cells tend to suffer less from intercellular ice-crystal formation.

Anthesis and Grain Filling Phase

Low temperatures during anthesis can have a devastating effect on yield and cool periods postanthesis can even reduce grain filling and subsequent yield. Generally, winter wheat varieties have more accumulated frost tolerance than spring wheats. Demotes-Mainard and colleagues (1995) reported no loss of pollen viability in winter wheat at temperatures as low as 1.5°C, although a slight reduction in grain set was noted. This is at odds with later work by Subedi, Gregory, and colleagues (1998); these authors found that cold weather (2°C) at anthesis will cause significant floret sterility. In spring wheat, mildly cool temperatures (<10°C) during heading can delay anthesis and increase sterility in cold-susceptible varieties (Subedi, Floyd, and Budhathoki, 1998). Earlier heading varieties tend to suffer cold stress more than later varieties due to a combination of cold escape and inherent cold tolerance. Maintenance of the rate of transpiration, which helps dissipate energy, may also influence the plants' ability to tolerate low temperatures (Subedi et al., 2001). Others have suggested that pubescence on the floral bracts may partially insulate that floret at low temperature (Maes et al., 2001).

Grain Development and Quality

Yield, kernel weight, test weight, protein content, flour yield, starch damage, SDS-sedimentation, wet gluten, Farinograph stability, and loaf volume values can be decreased by temperatures of less than –3°C during grain filling in spring wheat, while grain hardness and ash content increase (Preston et al., 1991). Allen and colleagues (2001) noted that baking quality and falling number also decrease under cold stress. However, the largest grains (>2 mm) were often undamaged and had high germination values. A seed subsample separated by sieving to include only this large grain fraction met all mixing and baking quality requirements.

Genetic Variability

Large differences have been observed in response to cold between frost-tolerant and frost-sensitive varieties. Genetic variability for frost tolerance is available within the common bread and durum wheat gene pools and, to a lesser extent, among wild relatives of wheat. Table 12.8 lists some of the wheat genotypes with differential response to cold stress in various studies. Both tolerant and sensitive types are indicated, organized by the trait that was investigated.

When comparing three winter wheat varieties, the most winter hardy was 'Roughrider' (Bauer and Black, 1990). Comparisons among six varieties at –16°C identified variety Mv9 as having the best survival (98 percent) (Veisz et al., 2001). Lines from Nebraska and Kazakhstan, in general, were very frost tolerant, with more than 90 percent survival at –15°C. Freezing tests at –10 and –14°C identified the cultivar Albidum 114 as the most freezing tolerant of 35 winter varieties (Fedulov et al., 1990).

Bread wheat was more cold tolerant (5-10°C) than durum wheat when 102 mostly winter wheat varieties were compared (Boubaker and Yamada, 1991). The most tolerant durum wheats were 1460, 'Adamello,' 'Creso,' Echiopia Komugi 72-201, 'Kunduru 1149,' 'Kyperounda,' 'Latino,' 'Mindum,' and 'Norba,' with seedling emergence times similar to bread wheat. Some spring wheats also expressed a degree of cold tolerance. When 90 durum wheat landraces were exposed to field frost conditions (minimum of –8.9°C) during vegetative growth and booting stage the entries originating from Syria and Turkey proved most cold tolerant, and the Moroccan accessions the least (Pecetti et al., 1993). Among these entries, flag leaf size was negatively correlated to frost tolerance (r = –0.65), indicating that small leaves contribute to frost tolerance.

The frost-tolerant triticale variety Aubrac tested at –14°C showed a lower level of membrane damage (40.3 percent) than the winter wheat 'Salmone' (56.3 percent) (Rizza et al., 1997). In addition, the spelt wheat

TABLE 12.8. Wheat varieties that are cold tolerant or sensitive in regard to specific traits when exposed to low temperatures.

Trait	Cold tolerant	Cold sensitive	Reference
Emergence time (5 to 10°C)	DW: 1460, Adamello, Creso, Echiopia Komugi 72-201, Kunduru 1149, Kyperounda, Latino, Mindum, and Norba	Beloturka 69, Clowson Long Valley, Montpellier 1, Peru 1, Quedzenote (368)	Boubaker and Yamada, 1991
Proline	Mironovska 808	Slavia	Machackova et al., 1989
TaADF accumulation	Norstar (WW)	Glenlea (SW)	Ouellet et al., 2001
Guaiacol peroxidase activity	Brasilia	Eridano	Scebba et al., 1998
Sterility·	Annapurna-3 (SW)	NL683 (SW)	Subedi, Gregory, et al., 1998
Cold-hardiness index (visual)	Varma	Salut	Larsson, 1989
Freezing tolerance (–10 to –14°C)	Albidum 114	Rusalka	Fedulov et al., 1990
Survival at –16°C	Mv9	Zagrepchanka	Veisz et al., 2001
LT_{50}	Mv-4	Vitka	Racz et al., 1996
LT_{50}	Mv-11-75	Mv-13-74	Paldi et al., 1996
LT_{50}	Norstar (WW)	Glenlea (SW)	Fowler, Limin et al., 1996
LT_{50}	Roughrider	Capelle	Dörffling et al., 1990

varieties Ostro and Rouquin filled their grain better at cool temperatures (13/8°C, relative to 18/13°C) than winter wheat, and may be a useful genetic source for stable grain filling (Ruegger et al., 1990).

A survey to identify new genetic variability among 46 wild relatives of wheat from diverse ecological regions in the western Asian region for cold tolerance was disappointing (Damania and Tahir, 1993). No entry was as tolerant as the check commercial winter wheat cultivar TAM105. The *Triticum urartu* group proved to be the most cold tolerant of the related species. Amphiploids and disomic addition lines involving *Aegilops glaucum* expressed improved cold tolerance (100 percent survival) at –18°C (Sutka, 1994).

Genetics and Breeding

Genetics

Chromosome 5A contains two loci responsible for cold hardiness (–1°C to –11.5°C), one of which is linked to short leaves on hardened plants (*r:* –0.46 to –0.88), while the other is linked to *Vrn1* (Roberts, 1990). Chromosomes 2B, 4B, and 7A also carry genes involved in frost tolerance (Sutka, 1994). At very low temperatures (–11°C) other research has implicated chromosomes 5AL, 3Bl, 5DL, 7AS, 7AL, and 7BL in the inheritance of frost tolerance (Veisz and Sutka, 1993). While the inheritance of frost resistance is additive and may involve several genes, chromosomes 5A and 5D appear to be of major importance (Sutka, 2001). Chromosomes 5A and 5D are implicated in the coding of proline during cold acclimation or hardening (Galiba, 1994), and the frost tolerance gene *Fr1* is completely linked to *Vrn1* on chromosome 5A. While the *Fr1* locus is located on chromosome 5AL and closely linked to *Vrn1,* some recombination (1/68) does occur (Galiba et al., 1995). *Vrn-A1* and *Fr1*, both on 5A, are separated by 2 cM. The homeologous genes *Vrn-D1* and *Fr2* are located on 5DL (Kocsy et al., 2000). The cultivar Cheyenne and the Chinese Spring substitution line CS (Ch 5A) show elevated levels of glutathione (GSH) upon cold acclimation, indicating that chromosome 5A from 'Cheyenne' contributes to GSH production.

Earlier work proposed that most genetic variation for cold tolerance was quantitatively inherited (Galiba, 1994) and that additive gene action is more important than dominant gene action. Nevertheless, Limin and Fowler (1993) found that both additive and dominant genes determine cold tolerance, and Mladenov and colleagues (1998) concluded that the mode of gene action was partial dominance with underlying additive and dominance effects. Others have shown overdominance to be of prime importance in conferring tolerance to "perishing under snow" and in frost tolerance in three winter wheats from the former Soviet Union (Kokurina and Rybakova, 1989). Limin and Fowler (1993) studied synthetic wheats developed from crossing cold-tolerant durum wheats and cold-tolerant *Triticum tauschii* lines and calculated broad-sense heritability estimates of 63 to 70 percent. Surprisingly, no transgressive segregants were noted in crosses between the most cold-tolerant and diverse parents. Hence, the synthetic hexaploids may not contribute any new genetic variability to this trait.

Evidence suggests that winter wheat cytoplasm may offer a slight advantage over spring wheat cytoplasm under cold stress (Sutka et al., 1991). However, subsequent work has not confirmed these findings (Sutka, 1994).

Crosses between in vitro selected and regenerated hydroxyproline lines and the wild-type demonstrated a single-gene segregation ratio for frost tolerance and proline content (Dörffling et al., 1997). The high proline content in homozygous F3 individuals from the cross was directly correlated with impressive frost tolerance. Frost tolerance of the best Hyp-lines was comparable to that of some of the most frost-tolerant winter wheat varieties known.

Breeding

Various authors have shown that heritabilities for cold-stress tolerance are relatively high, indicating that selection should be effective in identifying superior types. However, the most frost-tolerant cultivars, e.g., Minhardi, Kharkov 22MC, were developed almost 100 years ago and continue to be among the best sources of cold tolerance (Limin and Fowler, 1991). It has proven almost impossible to combine resistances, resulting in a genotype superior to both parents. It appears that most genetic variability has already been collected and recombined into available varieties. Nor has much new variability of promise been found in alien species, with the possible exception of *Aegilops cylindrica*.

No reliable, nondestructive, single-plant test is available to measure cold hardiness (Limin and Fowler, 1991). However, a number of morphological traits have been suggested that may be useful. Since no genotype × temperature interaction exists when cold tolerance is compared at 5 to 10°C, the 10°C level may be conveniently used for screening (Boubaker and Yamada, 1991). Although sucrose and fructose correlate well with cold hardiness, simple freezing tests, which are much less expensive, can be used (Larsson, 1989). The freezing test involves exposing young seedlings to a slow decrease in temperature following hardening, down to –15°C, followed by a visual estimation of cold hardiness based on leaf damage. Veisz and colleagues (2001) found that reduction in grain mass of side spikes showed the greatest differentiation among lines exposed to –16°C in a freezing test, and they concluded that this may be useful in breeding for cold tolerance.

Selection for small cell size may be beneficial since these tend to suffer less freezing damage (Limin and Fowler, 1991); selection of small leaves may also contribute to frost tolerance, since flag leaf size is negatively correlated with frost tolerance (Pecetti et al., 1993). When leaf traits at 4°C relative to 17°C were compared, it was found that plants with short, narrow leaves and small cell size were most frost tolerant (Limin and Fowler, 2000). Prostrate growth was also a good indicator of frost tolerance. Limin and Fowler (1994) suggested that selection for smaller guard cells could be used as a cheap selection tool. Pubescence on the glumes can maintain temperatures slightly above those of nonpubescent wheats and may serve as an additional trait contributing to cold tolerance (Maes et al., 2001).

Crop modeling can also help the breeder understand the limitations in the target environment. This approach was used in Australia to avoid frost risk by identifying the optimal range of flowering dates across five regions (Boer et al., 1993). First and last frost events were found to be correlated in such a way that should an early-season frost occur, then the probability of late-season frosts increased significantly. Such information may help farmers choose appropriate varieties and plant breeders in determination of their objectives.

To separate varieties with large differences in frost response, proline can be used as a marker, although its use is limited when varieties show similar behavior (Machackova et al., 1989). ABA tests can distinguish smaller differences; however, the procedure is too laborious for use in a breeding program. Dörffling and colleagues (1993, 1997) showed that it is possible to select stable hydroxyproline-resistant cell lines in vitro that express improved frost tolerance at a reasonable cost.

One of most widely used and recommended methods of selecting better winter hardiness is determination of sucrose and/or fructan levels (Galiba et al., 1997). The total water-soluble carbohydrate content may also function as a selection tool for frost tolerance (Vagujfalvi et al., 1999).

The *tacr7* gene provides low-temperature (2°C) tolerance, and tagging this gene or assaying its product (TACR7) would be useful in breeding (Gana et al., 1997). The *Xwg644* probe linked to the *Vrn1-Fr1* interval has also been used to identify winter-hardy genotypes (Storlie et al., 1998).

Although based on limited studies, synthetic wheats derived from the best donor materials have not yet provided breeders with new genetic variability (Limin and Fowler, 1993). Nevertheless, *Aegilops glaucum* may be a useful source of improved frost tolerance (Sutka, 1994).

At CIMMYT some winter wheats are grown at 2640 m above sea level from the fall (November), through winter and spring, maturing in early summer (July). During the winter period temperatures drop for a number of weeks below –3 to –5°C, allowing almost all introduced winter wheats to properly vernalize. The mean lowest night temperatures recorded during those periods are –6 to –8°C. The location is subtropical and predominately receives summer rainfall; therefore, skies are mostly clear during winter, and daytime temperatures invariably reach 10 to 15°C in the early afternoon. No snow falls at this location. The plots are irrigated regularly from planting in November until late May, when the rainy season begins. Material can suffer some cold damage in the winter when water is withheld; however, under irrigation little cold damage is inflicted.

The breeding sites in Mexico do not provide reliable screening for frost tolerance. CIMMYT has a complete international winter wheat-breeding program based in Turkey in conjunction with Turkish scientists. Breeding and screening for frost tolerance is conducted in Turkey. For spring wheats, CIMMYT's breeders use data provided by collaborators around the world, situated in sites where cold stress regularly occurs, to plan crosses and gain better understanding of the extent of this problem.

Conclusions and Recommendations

Cold damage can occur across the wheat-growing areas of the world. Damage can be observed in spring wheat grown at high elevations near the equator, in the traditional winter wheat regions, and in the high-latitude spring wheat production areas. The total area affected globally is 80 million ha.

Cold hardening or acclimation is a critical adaptive mechanism improving survival under very low temperatures. Shoot and root growth diminishes as night temperatures fall below 10°C, and winter wheat experiences a greater reduction in root growth at low temperatures than spring wheat. At the same time, sugar and fructan levels increase more dramatically in winter wheat than in spring wheat, with osmotic potential dropping in unison. The leaves of winter wheat are generally smaller and transpire less. Cold-hardened wheat expresses double the amount of sugars, lipids, and proteins in the leaves than do nonhardened controls. This enables them to withstand much lower subzero temperatures. Nevertheless, temperatures of –20°C will kill even the most hardened winter wheat.

Proline levels dramatically rise during cold hardening by as much as fivefold, which directly correlates with the ability to withstand freezing temperatures. Several other proteins also increase in level, such as TaADF, glutathione, TACR7, and dehydrins, all of which are associated with a drop in osmotic potential and act as cryoprotectants. Sugar levels also increase during cold stress and fall when plants are subject to warmer conditions. Peroxidases that protect against activated oxygen also increase under cold stress.

Cool conditions before flowering can cause sterility or at least delay anthesis. Temperatures just above 0°C can cause severe sterility, and pubescence on the glumes can give some protection against light frost. Almost all grain-quality parameters are negatively affected by cold damage.

Large genetic differences were noted among varieties, both within spring and winter wheats, for tolerance to cold stress. The frost-tolerance gene *Fr1* is located in chromosome 5AL, close to but not fully linked to

Vrn1. This chromosome is also implicated in higher proline and glutathione levels: Most cold response is based on additive gene action. Heritability estimates are very high, 63 to 98 percent, and no cytoplasmic effects are evident.

Although tolerance sources are known and heritabilities are high, the absolute level of cold tolerance has improved little in 100 years, and little if any new superior variability has been found in wild wheat relatives. A number of morphological selection traits have been proposed, including response to simple freezing tests; small cell size; short, narrow leaves; prostrate growth; and pubescent glumes. Proline and sugar levels are possible physiological selection tools, and markers are also available for the *Vrn1-Fr1* interval on 5AL.

In summary, cold is a globally common stress in wheat, the physiology is quite well-understood, very tolerant lines are known, and selection tools are available. However, genetic variability is not extensive and genes appear common across the present gene pool. Hence, it may be difficult to move beyond current levels. Nevertheless, it is feasible to move existing genes into commercially acceptable varieties, thereby improving their cold-stress tolerance.

BREEDING FOR TOLERANCE TO WATERLOGGING STRESS IN WHEAT

In the twenty-first century, the increasing scarcity of water for agriculture will become a primary limitation as burgeoning urbanization and global warming impact water supplies. Nevertheless, water is one of the most abundant commodities on the globe. The issue facing humanity is clearly not water availability but its proper "harvest" and management. One of the ironies of the debate on water is that the agricultural lands suffering from excess water or waterlogging are increasing, as are those where water is increasingly scarce. Estimates are that about 10 million ha of agricultural land suffer from waterlogging-related crop losses (Boru et al., 2001).

Geographical Distribution of Waterlogging Stress

In many of the irrigated agricultural systems in developing countries, proper drainage systems were never built. These areas suffer from poor water-table management and the associated problems of waterlogging and salinity. The Indo-Gangetic alluvial plains and the irrigated areas in Egypt are examples of irrigated systems affected by poor water management. Waterlogging is a seasonal problem in high-rainfall areas on heavy soils, such as the cracking vertisols of Ethiopia. In the central Ethiopian highlands, summer rainfall saturates the soil profile and precipitation is lost as runoff. Deforestation in central Ethiopia exacerbates the problem, with both water and soil washed from the agricultural lands into lower-lying regions and rivers. Regions in the United Kingdom also suffer from seasonal excess rainfall, which often overpowers the natural drainage capacities of the land. The U.S. agricultural lands bordering the lower regions of the Missouri and Mississippi river systems experience occasional waterlogging as the water table rises.

The problem is somewhat different in the rice-wheat rotation systems in southern, southeastern, and eastern Asia. In some of these regions, wheat is a recent introduction and farmers are less accustomed to its optimum water management. They tend to overirrigate wheat, causing waterlogging damage. Most damage is patchy in nature, with low-lying corners of fields most affected.

Waterlogging Stress and Plant Development

As discussed in previous sections breeders must have a sound understanding of the underlying physiological mechanisms, their inheritance, existing genetic variability, and effectiveness of selection methods. Obviously, the problem can and should be mitigated by installation of proper drainage systems and improved water management. The development of tolerant cultivars in conjunction with these improvements in water management will provide farmers with the best possible outcome. However, engineering solutions are expensive and time-consuming to implement; and genetic variability for waterlogging tolerance in wheat has been identified. Therefore, in the short term, improved genetic materials may lessen the impact of waterlogged soils in some regions.

Waterlogging initially impacts root growth, with subsequent effects on the aboveground biomass. Emergence and crop establishment is badly affected as root development is poor and tillering is reduced; and long-term waterlogging will cause floret sterility. Seed set and growth are depressed and final grain yield reduced. Sometimes, setbacks due to waterlogging early in crop development can be compensated for by recovery in later growth stages, e.g., excess tillering compensating for reduced emergence when waterlogging stress is lifted. Application of fertilizer will also help plant recovery following waterlogging. However, if additional stresses are present, such as salt or pH extremes, then grain yield will be more affected.

Vegetative Phase

Huang and colleagues (1994) observed that waterlogging reduced the number of nodal roots in sensitive genotypes and not at all in the tolerant cultivar Savannah. 'Savannah' showed more conspicuous aerenchyma tissue in the cortex of seminal and nodal roots and superior root weight (Huang et al., 1994, 1997). Other responses to waterlogging are reduction in the root:shoot ratios (Huang and Johnson, 1995), lower root respiration rate, and, in the case of tolerant genotypes, increased sugar transport from shoots to the roots. In one variety, anaerobic conditions were shown to increase the size of the cortical plasmodesmata five- to tenfold, allowing larger molecules such as sugars to be transported cell-to-cell to affected regions of the root (Cleland et al., 1994). In general, during anoxia the amount of sugars transported from the shoots to the roots is reduced by 79 to 97 percent (Waters et al., 1991).

Rooting depth and root volume are reduced by waterlogging (Box, 1991), and in Japanese varieties it was noted that those adapted to conditions of frequent excess soil moisture (southern Japan) developed shallower root systems than those grown in the dryer north (Oyanagi et al., 1993).

Lophopyrum elongatum is a weed adapted to salt marshes that expressed tolerance to waterlogging (0 percent loss in shoot weight) compared with a wheat variety (63 percent loss) (McDonald et al., 2001). Seminal root weight was reduced by 95 to 97 percent in the wheat variety, and by just 19 percent in *Lophopyrum elongatum*. Independent of waterlogging exposure, *L. elongatum* produces internal porosity in the roots, allowing better O_2 movement. Amphiploids developed between the two parents were intermediate at best for waterlogging tolerance and were often similar to the wheat parent.

Ding and Musgrave (1995) found that a reddish-brown coating which tends to form on roots under waterlogged conditions consisted mostly of iron and was strongly negatively correlated with yield. In their study, the internal root area containing aerenchyma was, surprisingly, negatively correlated to yield.

Shoot weight is severely reduced by waterlogging (Huang et al., 1995) as is stomatal conductance (Huang et al., 1994). In waterlogging-sensitive genotypes, all stomatal activity will cease three days after the onset of waterlogging, while stomatal conductance in tolerant cultivars such as Savannah will continue, albeit at a reduced rate (Huang et al., 1997). Leaf water potential and photosynthesis are similarly affected in sensitive genotypes (Huang et al., 1994, 1997). Huang and colleagues (1995) observed that the shoots of sensitive genotypes showed greatly reduced levels of N, P, Mg, and Zn, but increased levels of these in the roots. These changes were considerably less in tolerant materials. Increased levels of root minerals (Fe + Mn + P) are also strongly negatively correlated with yield (Musgrave and Ding, 1998). In addition, deprivation of oxygen in young seedlings causes a rapid increase in the pH level of cellular cytoplasm of the shoot (Menegus et al., 1991).

Waterlogging reduces grain weight and hence yield by decreasing both grains per m^2 and kernel weight (Musgrave, 1994; Musgrave and Ding, 1998). These authors found that grain yield under waterlogged and under nonwaterlogged conditions were not correlated. Melhuish and colleagues (1991) reported that in a 5 $t \cdot ha^{-1}$ environment, yield under waterlogged conditions declined at a rate of 69 $kg \cdot ha^{-1}$ for every day of waterlogging, whereas Gill and colleagues (1992) observed a 26 percent reduction in yield following four to 12 days of flooding on sodic soils, with the flowering stage more sensitive to waterlogging than the tillering stage. When waterlogging occurs at anthesis, the levels of Mn and Zn increase in the leaves, while levels of K, P, and Mg decline in the grain (Stieger and Feller, 1994).

Genetic Variability

Several authors have reported significant variability for waterlogging tolerance among varieties. However, waterlogging tolerance and genetic variability for this trait are not common within the bread and durum wheat genepools. Nevertheless, evidence suggests that sources may be available among wild relatives. Table 12.9 lists some of the wheat genotypes that respond differentially to waterlogging stress reported in various studies. Both tolerant and sensitive types are indicated, organized by the trait that was investigated.

The winter wheat cultivar Coker9877 experienced only a 9 percent loss in yield compared with 69 percent for LA8564 A80-3-1-X (Musgrave and Ding, 1998). The spring wheat 'Ducula' was shown to be very tolerant to waterlogging under northwestern Mexican conditions, expressing only 12 percent leaf chlorosis after 40 days of exposure to stagnant conditions in the field, relative to 71 percent chlorosis in the sensitive check, 'Seri-82' (Boru et al., 2001).

Flooding of sodic soils for 12 days identified the genotypes CSW540-1 and CSW538-2 as the most tolerant (Gill et al., 1992). The most tolerant varieties following flooding on alkaline soils (pH 8.0 to 9.6) for up to 12 days were CSW538-2 and E14-3 (Gill et al., 1993).

Four synthetic wheats expressed high levels of tolerance to waterlogging during seven weeks of exposure (Villareal et al., 1998, 2001). *Lophopyrum elongatum*, a diploid, may provide a new source of tolerance to water-

TABLE 12.9. Wheat varieties that are tolerant or sensitive in regard to specific traits when exposed to waterlogging.

Trait	Waterlogging tolerant	Waterlogging sensitive	Reference
Leaf chlorosis	Ducula	Seri-82	Boru et al., 2001
Stomatal conductance	Savannah	Bayles	Huang et al., 1997
Aerenchyma formation	Savannah	Bayles	Huang et al., 1994
Grain yield	CSW540-1, CSW538-2	E26, HD2329	Gill et al., 1992
Grain yield	Coker9877	LA8564 A80-3-1-X	Musgrave and Ding, 1998

logging, with its constitutive porosity allowing freer O_2 movement (McDonald et al., 2001). Unfortunately, amphidiploids produced and studied to date most closely resemble the wheat parent in response to excess water.

Genetics and Breeding

Genetics

When segregating populations were exposed to 40 days of stagnant waterlogged conditions, the gene action involved was mainly additive (Boru et al., 2001) and no cytoplasmic effects were noted. These authors found that four genes were responsible, with the presence of just two sufficient for a tolerant response. Genotypes also vary with respect to optimal growth stages at which tolerance may be expressed (Sayre et al., 1994). For instance, a variety may be sensitive to waterlogging early in the crop cycle but tolerant to waterlogging later.

Breeding

Early generation selection should be effective in identifying waterlogging-tolerant genotypes in those populations where a few major genes with additive effects segregate (Boru et al., 2001). The presence of a reddish-brown mineral coating of mostly iron was shown to be negatively correlated with yield and may also serve as a selection tool (Ding and Musgrave, 1995). Selection for shallow root systems may provide a measure of escape from anoxia (Oyanagi et al., 1993); however, selection for this character may penalize plants in nonwaterlogged conditions.

At CIMMYT, waterlogging trials were first carried out in the 1970s (Luxmoore et al., 1973) and screening was revived again in the early 1990s (Van Ginkel et al., 1992). During the 1990s, four lines, identified out of 1344 entries tested, were outstanding under extended waterlogging exposure from ten days after emergence until booting. Timing of the waterlogging stress was shown to affect response across genotypes, with some tolerant to early stress and others to late stress (Sayre et al., 1994). Selection for waterlogging tolerance is considered an important objective at CIMMYT (Samad et al., 2001).

Conclusions and Recommendations

Waterlogging stress in wheat is a common occurrence in at least 10 million ha globally, occurring in both irrigated and high-rainfall environments. The number of roots may decrease by up to half in sensitive varieties, with significant loss of root weight. Aerenchyma tissue forms under these conditions, particularly in tolerant genotypes, which allows better O_2 movement. Tolerant lines also transport more sugar to the roots. Shoots are reduced in weight by up to three-quarters in sensitive varieties, with stomatal conductance dropping significantly and, in some cases, completely shutting down. The concentration of minerals in the shoot decreases, but it increases in the roots, particularly in sensitive varieties. Yield losses of up to 69 percent have been reported, with the flowering stage the most sensitive to stress.

Genetic variability for tolerance has been found, involving large differences among varieties; however, its occurrence is relatively rare. Some of the wild relatives of wheat show great promise in this regard. The inheritance of tolerance appears to be relatively simple and additive in nature. However, varieties may differ in their response to the timing of flooding stress. Several indirect selection tools for tolerance have potential. Nevertheless, root number and mass, elevated root-sugar levels, decreased root-mineral levels, aerenchyma formation, and rooting depth remain to be explored.

In summary, waterlogging is a problem that has engineering and breeding solutions, and evidence suggests the problem is increasing in importance. A knowledge base of the physiological mechanisms involved has been established and needs expansion. Genetic variability is available within wheat and among wild relatives. The genetic basis may be simple and response to selection measurable; however, proven and efficient selection tools are lacking.

BREEDING FOR MICRONUTRIENT STRESSES IN WHEAT

Between 30 and 40 million ha planted to wheat experience mineral stress. Soil acidity and alkalinity are the two major causes of micronutrient stress. Reducing the soil chemical constraints for growth by applying amendments is one way of improving yield potential. However, in many instances to do so is impractical and costly. The alternative is to breed or select plants that are tolerant to moderate mineral stresses (Baligar et al., 1990). Furthermore, in some situations a breeding approach can complement agronomic solutions in integrated farming strategies. For example, in the case of the presence of Al toxicity, additional yield increases over those realized by lime application can be achieved by breeding Al-tolerant wheat cultivars (Scott et al., 2001). In wheat, breeding for trace-element stresses has been successful in improving productivity in a number of trials.

To successfully breed for mineral nutritional characteristics, several criteria need to be met (Graham, 1984). First, biological, ecological, and economic considerations dictate that a breeding solution would be better than an agronomic one. Second, it is critical to identify and/or develop a location where genotypes exhibit good expressions of variability for tolerance to nutrient stress. Third, sufficient and heritable variability must exist for the desired characteristic, hopefully not negatively associated with yield (Clark, 1990).

When breeding for nutritional characteristics in wheat, the following definitions are used. An efficient variety has a higher yield than others under low nutrient supply, while a responsive variety has a higher yield under high nutrient supply. A tolerant variety can adapt to a soil that has toxic levels of a nutrient, e.g., B, Al, Mn, or the combined harmful effect of several nutrients.

Much has been published on some of the key nutritional stresses. This section will focus on various aspects involved in the development of micronutrient-efficient and -tolerant varieties, emphasizing the recent information relevant to plant breeding, such as genetic diversity, gene location, gene number, and potential breeding strategies.

Sillanpaa (1990) tested 190 soils in 15 countries and concluded that the most common micronutrient deficiencies were the following (numbers in parenthesis represent the percentage of soils that tested deficient): Zn (49 percent), B (31 percent), Mo (15 percent), Cu (14 percent), and Mn (10 percent). Mo deficiencies are more commonly associated with leguminous crops, since Mo is an important nutrient in the process of symbiotic nitrogen fixation. Therefore, Mo deficiency is of lower importance for wheat.

Zinc Deficiency

Approximately 14 million ha in Central Anatolia, Turkey (Eyupoglu et al., 1994), 8 million ha in Western Australia, and one-third of the agricultural lands in China (White and Zasoski, 1999) have been classified as potentially Zn deficient. There are reports of areas showing Zn deficiency in India (Dwivedi and Tiwari, 1992), Pakistan, Mexico, Nepal, and various other countries (Sillanpaa, 1990). Wheat is a major crop in these countries. Evidence suggests that Zn deficiency has not yet been diagnosed in many areas around the world, particularly in developing countries. Therefore, Zn deficiency is the most important micronutrient stress in wheat systems around the world.

Genetic variability for Zn efficiency has been reported in wheat (Graham et al., 1992; Cakmak, Ekiz, et al., 1997), with durum wheat being less Zn efficient compared to bread wheat in studies conducted in Turkey (Cakmak, Derici, et al., 1997) and Australia (Graham et al., 1992). Rengel and Romheld (2000) used nutrient solution to characterize several Australian bread wheat and durum wheat genotypes. They classified the bread wheat genotypes 'Aroona,' Excalibur, Stilleto, and Trident as tolerant to both Zn deficiency and Fe deficiency, while durum wheats 'Durati' and Yallaroi were sensitive to Zn deficiency and moderate to sensitive to Fe deficiency. They were the first to report on combined tolerance to Zn and Fe deficiency in wheat. Genotypes Excalibur, Stilleto, and Trident come from the same breeding program and have the common parental line 'MEC3' (Sonora64// TZPP/YAQUI54), which is the likely donor of tolerance to Zn deficiency. Cakmak and colleagues (2001) screened durum genotypes for tolerance to Zn deficiency and found significant genetic variation. Cakmak and Braun (2001) reported that the genes controlling Zn efficiency are dominant in bread wheat, and Cakmak, Ekiz, and colleagues (1997) found that rye (cv. King-II) has an exceptionally high Zn efficiency, with genes located on rye chromosomes 1R and 7R. They also showed that the genes controlling Zn efficiency in rye are transferable to wheat and can be used to develop new wheat and triticale varieties with high Zn efficiency. Cakmak and Braun (2001) concluded that high Zn efficiency was associated to several morphological and physiological plant traits. These were root surface area, release of Zn-mobilizing PS from roots, membrane absorption and transport of Zn, and better utilization of Zn at the cellular level. Zinc efficiency in wheat was reported to be independent of copper efficiency (Ascher-Ellis et al., 2001).

A crown rot disease in wheat, caused by the fungus *Fusarium graminearum* Schw. (Group 1), is a widespread problem in chronically Zn-deficient Australian soils. Sparrow and Graham (1988) observed an association between crown rot and Zn deficiency. However, Grewal and colleagues (1996) later established that wheat genotypes more efficient at extracting zinc from low-zinc soils were in general more resistant to infection by this pathogen.

Boron Toxicity

Boron (B) toxicity is common in the arid and semiarid regions of the world and has been recognized as an important problem limiting wheat production in the low-rainfall areas of southern Australia, West Asia, and North Africa. The amelioration of B toxicity through soil modifications, e.g., incorporating gypsum, leaching with water, is, in general, not economically feasible, and breeding tolerant varieties is the best option to minimize yield losses (Nable, 1992).

Genetic variation for tolerance to B toxicity in wheat was observed by Mehrotra and colleagues (1980), Chatterjee and colleagues (1980), and Moody and colleagues (1990). The three loci conferring tolerance to B have been classified *Bo1, Bo2,* and *Bo3* (Paull et al., 1992). Jamjod and colleagues (1992) characterized four genotypes for these loci:

- 'Halberd'—moderately tolerant *(Bo1, Bo2, Bo3),*
- 'Warigal'—moderately sensitive *(bo1, Bo2, Bo3),*
- W1*MMC—sensitive *(bo1, Bo2, bo3),* and
- 'Kenya Farmer'—very sensitive *(bo1, bo2, bo3).*

In durum wheat, Yau and colleagues (1997) screened nine genotypes under greenhouse conditions for tolerance to high levels of B and identified the cultivars Oued Zenati, Omrabi-5, and Gezira-17 as being tolerant; yet they were less tolerant than tolerant bread wheats to high soil B levels. Jefferies and colleagues (2000) showed that RFLP markers on chromosome 7B and 7D were effective in selecting for improved boron tolerance in two different wheat genetic backgrounds. Alleles in the cultivar Halberd on chromosome 7B were associated with the concentration of boron in whole shoots and grain. Manyowa (1989) discovered sources of B tolerance in the tribe *Triticeae* with potentially higher B efficiency compared to hexaploid wheat. These genetic resources include *Secale cereale* cultivars, *Th. bessarabicum, Agropyron elongatum,* and *Aegilops sharonensis.*

Boron Deficiency

Cereals tend to have a lower B requirement than dicotyledonous plants and are, therefore, less affected by B deficiency (Pilbeam and Kirkby, 1983). Clark (1990) suggests that monocotyledonous plants can be separated into two groups according to their sensitivity to B deficiency: (1) those that show early vegetative growth reduction and final death (maize and barley) and (2) those that show deficiency in the reproductive stage (wheat, oat, and rye) (Dugger, 1983). B deficiency causes grain set in wheat to fail, and significant interaction with genotype has been observed. Rerkasem and Jamjod (1997) screened wheat genotypes in low-B soil and sand cultures, and classified them as very sensitive, sensitive, moderately sensitive, moderately tolerant, or tolerant. At very low levels of B, the very sensitive and sensitive genotypes were completely male sterile and set few or no grain, while the tolerant genotypes displayed normal grain set. They mention the following three practical implications based on their findings:

1. Genotypes that are tolerant to low B can overcome the problem of grain-set failure caused by B deficiency.
2. Outcrossing in male sterile B-deficient wheat can cause problems in the maintenance of pure lines in low-B soils.
3. A simple and novel method for hybridization is proposed, in which B deficiency can be used to induce male sterility to facilitate cross hybridization.

Subedi and colleagues (1997) screened genotypes from Nepal, India, Bangladesh, Thailand, China, Pakistan, and CIMMYT Mexico for tolerance to B deficiency in Nepal. About one-third of all the genotypes had <25 percent sterility, and 16 of them suffered >75 percent sterility. The genotypes 'Fang-60,' 'Sonalika,' De Mai 6-22, BUC/FLK/MYNA/VUL, and HDW-234 were identified as highly tolerant to boron deficiency. In contrast, the genotypes BOW/BUC//SERI/THB, 'Glennson,' SW-41, 'Yumai-33,' and 'UHU' were highly sensitive and almost no grains were set in the boron-deficient plots. They concluded that given these large genetic differences, progress in breeding wheat for B-deficient soils in Nepal should be possible.

Copper Deficiency

Copper deficiency is associated with soils high in organic matter and with highly weathered, sandy mineral soils (Murphy and Walsh, 1972).

Copper deficiency delays maturity and reduces straw and grain yield in wheat (Nambiar, 1976). The sensitivity of small grain cereals to Cu deficiency follows the sequence wheat > oat > barley > rye, according to Clark (1990). Rye was considerably more tolerant to Cu deficiency when grown on low-Cu soil than wheat, whereas triticale showed an intermediate response (Graham, 1978b, 1984; Graham et al., 1981; Harry and Graham, 1981). The higher Cu efficiency of rye does result in beneficial Cu nutritional properties and can be transferred to wheat (Graham, 1978a,b). Graham (1984) and Graham and colleagues (1987) could locate the genes for Cu efficiency in rye on the long arm of chromosome 5 (5RL). The presence of this chromosome in four wheat genotypes increased yield by 100 percent when grown on Cu-deficient soil (Graham et al., 1987). Cu efficiency in wheat appears to be independent of zinc efficiency (Ascher-Ellis et al., 2001).

Manganese Deficiency

Manganese (Mn) deficiency is a widespread problem on alkaline soils, particularly for durum wheat, which is more sensitive than either bread wheat or barley. Manganese deficiency is a problem for bread wheat production on highly permeable sandy soils in India that have been cropped with rice for five or more years. Bansal and Nayyar (1998) screened wheat genotypes for Mn efficiency under field conditions in India. They used foliar application of Mn to measure yield response and grouped genotypes as intolerant (PBW 34 and PBW 215), moderately tolerant (PBW 226 and PBW 343), and tolerant (WH 542 and HD 2329) to Mn deficiency. In a recent study, Bansal and Nayyar (2000) found durum wheat genotypes PBW 34 and PDW 233 were less Mn efficient than the bread wheat varieties PBW 373, Raj 3765, and PBW 343. Saberi and colleagues (1999) reported that the mode of inheritance of Mn efficiency was incomplete dominance. They also concluded that Mn efficiency is controlled by two genes. Significant genetic diversity for manganese efficiency is present in the wheat gene pool and additional variability is present in rye (Ascher-Ellis et al., 2001). The rankings of zinc and manganese efficiency tend to be inversely correlated in Australian varieties (Graham, 1990). Zinc-efficient Excalibur had poor manganese efficiency. The cultivars Bayonet, Millewa, and Takari were relatively zinc efficient but were highly inefficient regarding manganese. However, several varieties are efficient (cv. Aroona and Machete) or inefficient (cv. Durati, Kamilaroi, Songlen, Gatcher) for both elements according to Ascher-Ellis and colleagues (2001).

Manganese Toxicity

Manganese toxicity is part of the nutrient-stress complex constraining plant growth in acidic soils. The area affected by Mn toxicity is unknown; however, it is believed to be a less common constraint to crop growth than Al toxicity. Scott and colleagues (1998) screened 164 wheat accessions, predominantly genotypes from South America, Australia, and CIMMYT in a solution culture for tolerance to Mn toxicity. Fourteen genotypes were classified as tolerant; they did not display toxicity symptoms, and their total dry matter yield was not reduced when exposed to Mn in solution culture. In contrast, grain yield was reduced by 87 percent in the very sensitive cultivar Teal. Mn tolerance appeared to be simply inherited and the mechanism related to tolerance of internal tissue to high concentrations of Mn.

Aluminum Toxicity

Acidic soils with high levels of aluminum concentration severely limit plant growth on more than 1.6 billion ha worldwide. Wheat production is most affected in Brazil. In the United States alone, a gradual decline in pH in the Great Plains and the Southwest has led to increased levels of free Al. The deployment of Al-tolerant varieties is a prerequisite for successful crop production in low-pH, Al-toxic soils. Soil amelioration, combined with the production of tolerant varieties, will significantly improve productivity. Soil amelioration by liming generally affects the upper root zone; Al-intolerant genotypes are unable to penetrate deeper soil horizons, particularly during periods of restricted root growth due to heat or drought stress.

Genetic diversity for tolerance to Al toxicity has been found in numerous studies. De Sousa (1998) screened wheat germplasm for aluminum toxicity tolerance under field conditions and identified a number of cultivars with high levels of tolerance to Al: BH 1146, Embrapa15, Embrapa 24, Embrapa 49, IAC 5-Maringa, IAC 18-Xavantes, IAC27-Pantaneiro, IAPAR 46, IAPAR 53, RS 8-Westphalen, Trigo BR 15, Trigo BR 20-Guato, Trigo BR 25, Trigo BR 35, and Trigo BR 41-Ofaie. Aluminum tolerance was traced by pedigree analysis to a small number of landraces introduced to Brazil in the early twentieth century. Therefore, it can be assumed that all Brazilian Al-tolerant cultivars have a common major gene for Al tolerance on the long arm of chromosome 4D.

Hede and colleagues (2001) suggested that it might not be possible to improve Al tolerance over the level present in the 50-year-old Brazilian cultivar BH1146 by exploiting the primary diploid, tetraploid, and hexaploid wheat gene pool. They recommend utilizing the secondary (*Aegilops* spp. and rye) and tertiary (annual and perennial forage grasses) gene pools, which may offer greater potential to enhance Al tolerance in wheat. However, these authors recognize the difficulties and high costs associated with accessing genes from the secondary and tertiary gene pool. Kim and colleagues (2001) confirmed these observations after testing new and old wheat varieties in Brazil. They could not identify genotypes superior to BH1146 for Al tolerance and concluded that, in spite of yield and production increases from breeding efforts in Brazil, Al tolerance per se remained unchanged over the past 50 years. There are, however, potential sources of Al toxicity tolerance in the primary gene pool, such as Barbela, a wheat landrace grown for centuries in acidic soil regions of Portugal, that offer scope for further improving Al tolerance over existing levels (Hede et al., 2001). Ribeiro-Carvalho and colleagues (1997) detected an introgression of rye-origin chromatin in Barbela in cytological studies and later confirmed the location of the segment on wheat chromosome arm 2DL (Ribeiro-Carvalho et al., 2001).

Early studies (Slootmaker, 1974) revealed insignificant variation for Al tolerance in the secondary gene pool *Aegilops* species. However, in a more recent study, Miller and colleagues (1997) suggested that the wild relatives of wheat can provide valuable genetic variation for Al tolerance. One wild species, *Aegilops uniaristata* Vis. ($2n = 2x = 14$, NN) has been shown to confer tolerance to Al toxicity on chromosome 3N. By manipulating the genetic control of homeologous chromosome pairing, chromosome 3N can be recombined with its wheat homeologous chromosome to introduce a smaller alien segment carrying the gene(s) for tolerance. Hede and colleagues (2001) reported that *Aegilops umbellulata* has useful levels of Al tolerance.

Somers and colleagues (1996) and Basu and colleagues (1997) proposed that Al tolerance is simply inherited and controlled by a major dominant gene. Other publications suggest a more complex genetic mechanism involving a minimum of two dominant genes and the existence of minor genes (Camargo, 1981; Rajaram et al., 1991). More recently, Camargo, Ferreira, and Felicio (2000) evaluated wheat seedlings that were derived from F2 populations from crosses among tolerant and sensitive wheat cultivars. They concluded that the tolerance to Al^{3+} toxicity was dominant, and in all crosses the tolerant cultivars differed from the sensitive by one pair of genes. Camargo, Neto, and colleagues (2000) compared the Al tolerance of 17 gamma irradiation mutant lines derived from the aluminum-sensitive wheat cultivar Anahuac with two sensitive (cv. Siete Cerros and Anahuac) and three tolerant cultivars (BH-1146, IAC-60, and IAC-24). The results indicate that tolerance in the induced mutants was due to a single pair of dominant alleles and that these alleles expressed the same degree of tolerance as BH-1146 and IAC-24.

Several reports show that transferring genes from Al-tolerant germplasm into locally adapted wheat and triticale varieties has successfully increased yield potential in acidic soil areas in Australia, Brazil, and the United States. Scott and colleagues (2001) used the Brazilian variety 'Carazinho' as a source for Al tolerance in a recurrent selection backcross program to introduce tolerance into the locally adapted Australian variety 'Egret.' Al tolerance from 'Carazinho' increased the grain yield under acidic soil conditions in the 'Egret' derivatives. Kim and colleagues (2001) compared Al tolerance in wheat, triticale, and rye by measuring root elongation response in Al-containing hydroponic nutrient solutions. Rye was the most tolerant crop and wheat the most sensitive. The triticales from the mid-1970s carrying a 2D(2R) chromosome substitution were the most sensitive among the triticales. More recently developed triticale advanced lines with the full complement of rye chromosomes show the highest degree of Al tolerance and approach or exceed the levels observed in rye. Using the backcross method, Johnson and colleagues (1997) introgressed the genes for Al tolerance from 'Atlas 66' into the wheat cultivars Chisholm and Century. When planted in Al-toxic soil, backcross-derived 'Century' did not have increased grain yield compared to the recurrent parent 'Century,' in contrast to backcross-derived 'Chisholm,' which out-yielded its recurrent parent by 50 to 74 percent. This indicates that the genetic background is of critical importance in the expression of genes for Al-toxicity tolerance.

Salinity/Sodicity

Crop improvement for salinity tolerance should be conducted in conjunction with land management and reclamation strategies including leaching and drainage. Salt-tolerant cultivars would allow expansion of crop production to areas where conventional reclamation procedures are economically or technically limited (James, 1990).

Tetraploid durum wheat (genome AB) is more salt sensitive than hexaploid bread wheat (genome ABD), a feature that restricts its expansion into areas with sodic or saline soils. Salt tolerance in bread wheat is associated with genes on the D genome controlling low Na^+ uptake and enhanced K^+/Na^+ discrimination. The Kna1 locus controlling the K^+/Na^+ discrimination in saline environments in wheat has been mapped on chromosome 4DL (Gorham et al., 1997). Munns and colleagues (1999), in attempting to transfer salt tolerance to durum wheat from AB-genome sources, evaluated the genetic variation for salt tolerance in a wide range of *Triticum turgidum* subspecies (*T. durum, T. carthlicum, T. turgidum, T. turanicum, T. polonicum*). Accessions were screened for low Na^+ uptake and enhanced K^+/Na^+

discrimination by growing seedlings in 150 mM NaCl with supplemental Ca^{2+} and measuring the Na^+ and K^+ accumulated in the leaf blade over ten days. Large and repeatable genetic variation for low Na^+ accumulation and high K^+/Na^+ discrimination was found in the durum subspecies, similar in magnitude to bread wheat. These alien sources offer promise for the improvement of salt tolerance in durum wheat.

Interactions Between Micronutrient Efficiency and Disease Resistance

Interactions between micronutrient deficiencies and root diseases complicate the improvement of wheat for tolerance to nutrient stress tolerance. Manganese-deficient wheat plants are more susceptible to take-all, *Gaeumannomyces graminis* var. *tritici* (Graham and Rovira, 1984; Huber and Wilhelm, 1988), and powdery mildew, *Erysiphe graminis* (Graham, 1990). Zinc deficiency decreases the resistance of wheat to one of the pathogens causing crown rot, *Fusarium graminearum* (Sparrow and Graham, 1988), and to *Rhizoctonia solani* (Thongbai et al., 1993), the causal agent of bare patch.

Marker-Assisted Selection

Ascher-Ellis and colleagues (2001) concluded that in view of the difficulties in developing fast methods for screening for nutrient efficiency (such as seedling selection in pots) and the difficulty in relating in situ tests with field performance, molecular marker-assisted selection will greatly facilitate crop improvement. The breeder can then screen a larger number of lines in earlier generations with enhanced precision. Several markers for key nutrient-related traits are available, including tolerance to B and improved Zn efficiency.

Conclusion

The existing genetic diversity in wheat for efficiency and/or tolerance to the major nutrient stresses suggests that progress in crop improvement is feasible. Reference was made to the available genetic sources of tolerance/efficiency for the different key nutrient stresses that may be useful in crop improvement. Efficiency or tolerance to several nutrient stresses are controlled by single dominant genes, and relatively "easy" genetic gains from conventional breeding can be projected. However, the more complex inheritance of other nutrients and confounding G×E, will slow progress using conventional breeding and will complicate the development and use of molecular markers. Sustainable increases in productivity in areas subject to nutrient stress require better understanding of the key genetic and environmental determinants of tolerance coupled with the integration of improved crop management practices.

REFERENCES

Abayomi, Y.A. and D. Wright (1999). Osmotic potential and temperature effects on germination of spring wheat genotypes (*Triticum aestivum* L.). *Tropical Agriculture* 76:114-119.

Abdelghani, A.M., A.M. Abdelshafi Ali, and E.H. Ghanem (1994). Genetic factors and differential plant adaptations as indicator of wheat potential under heat stress in upper Egypt. In *Wheat in Heat-Stressed Environments: Irrigated, Dry Areas and Rice-Wheat Farming Systems,* D.A. Saunders and G.P. Hettel (Eds.). D.F., Mexico: CIMMYT, pp. 153-157.

Abdelshafi Ali, A.M. and O.A.A. Ageeb (1994). Breeding strategy for developing heat-tolerant wheat varieties adapted to upper Egypt and Sudan. In *Wheat in Heat-Stressed Environments: Irrigated, Dry Areas and Rice-Wheat Farming Systems,* D.A. Saunders and G.P. Hettel (Eds.). D.F., Mexico: CIMMYT, pp. 33-39.

Acevedo, E., M. Nachit, and G. Ortiz-Ferrara (1991). Effects of heat stress on wheat and possible selection tools for use in breeding for tolerance. In *Wheat for the Nontraditional Warm Areas,* D.A. Saunders (ed.). D.F., Mexico: CIMMYT, pp. 401-421.

Adam, H.S. and O.A.A. Ageeb (1994). Temperature analysis and wheat yields in the Gezira Schemein. In *Wheat in Heat-Stressed Environments: Irrigated, Dry Areas and Rice-Wheat Farming Systems,* D.A. Saunders and G.P. Hettel (Eds.). D.F., Mexico: CIMMYT, pp. 143-145.

Addae, P.C., N. Collis-George, and C.J. Pearson (1991). Overriding effects of temperature and soil strength on wheat seedlings under minimum and conventional tillage. *Field Crops Research* 28:103-116.

Ali, Z.I., V. Mahalakshmi, M. Singh, G. Ortiz-Ferrara, and J.M. Peacock (1994). Variation in cardinal temperatures for germination among wheat *(Triticum aestivum)* genotypes. *Annals of Applied Biology* 125:367-375.

Al-Khatib, K. and G.M. Paulsen (1990). Photosynthesis and productivity during high-temperature stress of wheat genotypes from major world regions. *Crop Science* 30:1127-1132.

Allen, H.M., J.K. Pumpa, and G.D. Batten (2001). Effect of frost on the quality of samples of Janz wheat. *Australian Journal of Experimental Agriculture* 41:641-647.

Amani, I., R.A. Fischer, and M.P. Reynolds (1996). Canopy temperature depression association with yield of irrigated spring wheat cultivars in a hot climate. *Journal of Agronomy and Crop Science* 176:119-129.

Ascher-Ellis, J.S., R.D. Graham, G.J. Hollamby, J. Paul, P. Davies, C. Huang, M.A. Pallotta, N. Howes, H. Khabaz-Saberi, S.P. Jefferies, and M. Moussavi-Nik (2001). Micronutrients. In *Application of Physiology in Wheat Breeding,* M.P. Reynolds, J.I. Ortiz-Monasterio, and A. McNab (Eds.). D.F., Mexico: CIMMYT, pp. 219-240.

Babu, C.R., M.P. Safiullah, A. Blum, and H.T. Nguyen (1999). Comparison of measurements methods of osmotic adjustment in rice cultivars. *Crop Science* 39:150-158.

Baligar V.C., R.R. Duncan, and N.K. Fageria (1990). Soil-plant interactions on nutrient use efficiency in plants: An overview. In *Crops As Enhancers of Nutrient Use,* V.C. Baligar and R.R. Duncan (Eds.). New York: Academic Press, pp. 351-373.

Balota, M. and H.K. Lichtenthaler (1999). Red chlorophyll fluorescence as an ecophysiological method to assess the behaviour of wheat genotypes under drought and heat. *Cereal Research Communications* 27:179-187.

Bancal, P. and E. Triboi (1993). Temperature effect on fructan oligomer contents and fructan-related enzyme activities in stems of wheat (*Triticum aestivum* L.) during grain filling. *New Phytologist* 123:247-253.

Banowetz, G.M., K. Ammar, and D.D. Chen (1999). Postanthesis temperatures influence cytokinin accumulation and wheat kernel weight. *Plant, Cell and Environment* 22:309-316.

Bansal, R.L. and V.K. Nayyar (1998). Screening of wheat *(Triticum aestivum)* varieties tolerant to manganese-deficiency stress. *Indian Journal of Agricultural Sciences* 68:66-69.

Bansal, R.L. and V.K. Nayyar (2000). Differential tolerance of durum *(Triticum durum)* and bread wheat *(Triticum aestivum)* varieties to manganese in a manganese deficient field. *Indian Journal of Agricultural Sciences* 70:507-511.

Basu, U., J.L. McDonald, D.J. Archambault, A.G. Good, K.G. Briggs, T. Aung, and G.J. Taylor (1997). Genetic and physiological analysis of double-haploid, aluminum-resistant lines of wheat provide evidence for the improvement of a 23 kD, root exudate polypeptide in mediating resistance. *Plant and Soil* 196:283-288.

Batts, G.R., R.H. Ellis, J.I.L. Morison, P.N. Nkemka, P.J. Gregory, and P. Hadley (1998). Yield and partitioning in crops of contrasting cultivars of winter wheat in response to CO_2 and temperature in field studies using temperature gradient tunnels. *Journal of Agricultural Science* 130:17-27.

Batts, G.R., J.I.L. Morison, R.H. Ellis, P. Hadley, and T.R. Wheeler (1997). Effects of CO_2 and temperature on growth and yield of crops of winter wheats over four seasons. *European Journal of Agronomy* 7:43-52.

Batts, G.R., T.R. Wheeler, J.I.L. Morison, R.H. Ellis, and P. Hadley (1996). Developmental and tillering responses of winter wheat *(Triticum aestivum)* crops to CO_2 and temperature. *Journal of Agricultural Science* 127:23-35.

Bauer, A. and A.L. Black (1990). Stubble height effect on winter wheat in the Northern Great Plains. I. Soil temperature, cold degree-hours, and plant population. *Agronomy Journal* 82:195-199.

Bennetzen, J.L. (2000). Comparative genomics approaches to the study of drought tolerance. In *Molecular Approaches for the Genetic Improvement of Cereals for Stable Production in Water-Limited Environments,* J. Ribaut and D. Poland (Eds.). El Batan, Mexico: CIMMYT, pp. 41-44.

Bennetzen, J.L. and M. Freeling (1997). The unified grass genome: Synergy in synteny. *Genome Research* 7:301-306.

Blum, A. (2000). Toward standard assays of drought resistance in crop plants. In *Molecular Approaches for the Genetic Improvement of Cereals for Stable Production in Water-Limited Environments,* J.M. Ribaut and D. Poland (Eds.). El Batan, Mexico: CIMMYT, pp. 29-35.

Blum, A., N. Klueva, and H.T. Nguyen (2001). Wheat cellular thermotolerance is related to yield under heat stress. *Euphytica* 117:117-123.

Blum, A., B. Sinmena, J. Mayer, G. Golan, and L. Shpiler (1994). Stem reserve mobilisation supports wheat-grain filling under heat stress. *Australian Journal of Plant Physiology* 21:771-781.

Blumenthal, C.S., E.W.R. Barlow, and C.W. Wrigley (1993). Growth, environment and wheat quality: The effect of heat stress on dough properties and gluten proteins. *Journal of Cereal Science* 18:3-21.

Blumenthal, C.S., I.L. Batey, F. Bekes, C.W. Wrigley, and E.W.R. Barlow (1991). Seasonal changes in wheat-grain quality associated with high temperatures during grain filling. *Australian Journal of Agricultural Research* 42:21-30.

Blumenthal, C.S., F. Bekes, I.L. Batey, C.W. Wrigley, H.J. Moss, D.J. Mares, and E.W.R. Barlow (1991). Interpretation of grain quality results from wheat variety trials with reference to high temperature stress. *Australian Journal of Agricultural Research* 42:325-334.

Blumenthal, C.S., P.J. Stone, P.W. Gras, F. Bekes, B. Clarke, E.W.R. Barlow, R. Appels, and C.W. Wrigley (1998). Heat-shock protein 70 and dough-quality changes resulting from heat stress during grain filling in wheat. *Cereal Chemistry* 75:43-50.

Blumenthal, C.S., C.W. Wrigley, I.L. Batey, and E.W.R. Barlow (1994). The heat-shock response relevant to molecular and structural changes in wheat yield and quality. *Australian Journal of Plant Physiology* 21:901-909.

Boer, R., L.C. Campbell, and D.J. Fletcher (1993). Characteristics of frost in a major wheat-growing region of Australia. *Australian Journal of Agricultural Research* 44:1731-1743.

Bohnert, H., R. Fischer, S. Kawasaki, C. Michalowski, H. Wang, J. Yale, and G. Zepeda (2000). Cataloging stress-inducible genes and pathways leading to stress tolerance. In *Molecular Approaches for the Genetic Improvement of Cereals for Stable Production in Water-Limited Environments,* J. Ribaut and D. Poland (Eds.). El Batan, Mexico: CIMMYT, pp.156-161.

Borlaug, N.E. and C.R. Dowswell (1997). The acid lands: One of agriculture's last frontiers. In *Plant-Soil Interactions at Low pH,* A.C. Moniz (Ed.). Campinas, Brazil: Brazilian Soil Science Society, pp. 5-15.

Boru, G., M. van Ginkel, W.E. Kronstad, and L. Boersma (2001). Expression and inheritance of tolerance to waterlogging stress in wheat. *Euphytica* 117:91-98.

Boubaker, M. and T. Yamada (1991). Screening spring wheat genotypes (*Triticum* sp.) for seedling emergence under optimal and suboptimal temperature conditions. *Japanese Journal of Breeding* 41:381-387.
Box, J.E. (1991). The effects of waterlogging on rooting of soft red winter wheat. In *Plant Roots and Their Environment*, Proceedings of an ISSR Symposium held August 21-26, 1988, B.L. McMichael and H.O. Persson (Eds.). Uppsala, Sweden: ISSR, pp. 418-430.
Bray, E.A. (1993). Molecular response to water deficit. *Plant Physiology* 103:1035-1040.
Byerlee, D. and G. Traxler (1995). National and international wheat improvement research in the post green revolution period: Evolution and impacts. *American Journal of Agricultural Economics* 77:268-278.
Byerlee, D. and G. Traxler (1999). Estimation of actual spillovers of national and international wheat improvement research. In *The Global Wheat Improvement System: Prospects for Enhancing Efficiency in the Presence of Spillovers*, M.K. Maredia and D. Byerlee (Eds.). CIMMYT Research Report No. 5. D.F., Mexico: CIMMYT, pp. 46-59.
Cakmak, I. and H.-J. Braun (2001). Genotypic variation for zinc efficiency. In *Application of Physiology in Wheat Breeding*, M.P Reynolds, J.I. Ortiz-Monasterio, and A. McNab (Eds.). D.F., Mexico: CIMMYT, pp. 183-199.
Cakmak, I., R. Derici, B. Torun, I. Tolay, H.-J. Braun, and R. Schlegel (1997). Role of rye chromosomes in improvement of zinc efficiency in wheat and triticale. *Plant and Soil* 196:249-253.
Cakmak, I., H. Ekiz, A. Yilmaz, B. Torun, N. Koleli, I. Gultekin, A. Alkan, and S. Eker (1997). Differential response of rye, triticale, bread wheat and durum wheats to zinc efficiency in calcareous soils. *Plant and Soil* 188:1-10.
Cakmak, O., L. Oztruk, S. Karanlik, H. Ozkan, Z. Kaya, and I. Cakmak (2001). Tolerance of 65 durum wheat genotypes to zinc deficiency in a calcareous soil. *Journal of Plant Nutrition* 24:1831-1847.
Calderini, D.F., M.P. Reynolds, and G.A. Slafer (1999). Genetic gains in wheat yield and main physiological changes associated with them during the 20th century. In *Wheat: Ecology and Physiology of Yield Determination*, E.H. Satorre and G.A. Slafer (Eds.). Binghamton, NY: Food Products Press, pp. 351-377.
Caley, C.Y., C.M. Duffus, and B. Jeffcoat (1990). Effects of elevated temperature and reduced water uptake on enzymes of starch synthesis in developing wheat grains. *Australian Journal of Plant Physiology* 17:431-439.
Camargo, C.E.O. (1981). Wheat Improvement. I. The heritability of tolerance to aluminum toxicity (in Portuguese). *Bragantia* 44:49-64.
Camargo, C.E.O., A.W.P. Ferreira, and J.C. Felicio (2000). Inheritance of aluminum tolerance in wheat hybrid populations. *Pesquisa Agropecuaria Brasileira* 35(3):517-522.
Camargo, C.E.O., A.T. Neto, A.W.P. Ferreira, and J.C. Felicio (2000). Genetic control of aluminum tolerance in mutant lines of the wheat cultivar Anahuac. *Euphytica* 114(1):47-53.
Ceccarelli, S. (1987). Yield potential and drought tolerance of segregating populations of barley in contrasting environments. *Euphytica* 36:265-273.

Ceccarelli, S. (1989). Wide adaptation: How wide? *Euphytica* 40:197-205.

Ceccarelli, S. and S. Grando (1991). Selection, environment and environment sensitivity in barley. *Euphytica* 57:157-167.

Ceccarelli, S., M.M. Nachit, G.O. Ferrara, M.S. Mekni, M. Tahir, J. van Leur, and J.P. Srivastava (1987). Breeding strategies for improving cereal yield and stability under drought. In *Drought Tolerance in Winter Cereals*, J.P. Srivastava, E. Porceddu, E. Acevedo, and S. Varma (Eds.). New York: John Wiley and Sons, pp. 101-131.

Changjian, S. (1994). Wheat breeding objectives, yield testing system, and recent progress in southern China. In *Wheat in Heat-Stressed Environments: Irrigated, Dry Areas and Rice-Wheat Farming Systems*, D.A. Saunders and G.P. Hettel (Eds.). D.F., Mexico: CIMMYT, pp. 224-225.

Charest, C. and C.T. Phan (1990). Cold acclimation of wheat *(Triticum aestivum):* Properties of enzymes involved in proline metabolism. *Physiologia Plantarum* 80:159-168.

Chatterjee, B.N., M. Chatterjee, and N.R. Das (1980). Note on the difference in the response of wheat varieties to boron. *Indian Journal of Agricultural Sciences* 50:796.

Clark, R.B. (1990). Physiology of cereals for mineral nutrient uptake, use, and efficiency. In *Crops As Enhancers of Nutrient Use*, V.C. Baligar and R.R. Duncan (Eds.). London, UK: Academic Press, pp. 132-183.

Cleland, R.E., T. Fujiwara, and W.J. Lucas (1994). Plasmodesmal-mediated cell-to-cell transport in wheat roots is modulated by anaerobic stress. *Protoplasma* 178:81-85.

Cooper, M., N.M. Jensen, B.J. Carroll, I.D. Godwin, and D.W. Podlich (2000). QTL mapping activities and marker-assisted selection for yield in the germplasm enhancement program of the Australian northern wheat improvement program. In *Molecular Approaches for the Genetic Improvement of Cereals for Stable Production in Water-Limited Environments*, J. Ribaut and D. Poland (Eds.). El Batan, Mexico: CIMMYT, pp. 120-127.

Corbellini, M., M.G. Canevar, L. Mazza, M. Ciaffi, D. Lafiandra, and B. Borghi (1997). Effect of the duration and intensity of heat shock during grain filling on dry matter and protein accumulation, technological quality and protein composition in bread and durum wheat. *Australian Journal of Plant Physiology* 24:245-260.

Crespi, M.D., E.J. Zabaleta, H.G. Pontis, and G.L. Salerno (1991). Sucrose synthase expression during cold acclimation in wheat. *Plant Physiology* 96:887-891.

Dallaire, S., M. Houde, Y. Gagne, H.S. Saini, S. Boileau, N. Chevrier, and F. Sarhan (1994). ABA and low temperature induce freezing tolerance via distinct regulatory pathways in wheat. *Plant and Cell Physiology* 35:1-9.

Damania, A.B. and M. Tahir (1993). Heat and cold tolerance in wild relatives and primitive forms of wheat. In *Biodiversity and Wheat Improvement*, A.B. Damania (Ed.). Chichester, UK: Wiley-Sayce, pp. 217-224.

Daniel, C. and E. Triboi (2000). Effects of temperature and nitrogen nutrition on the grain composition of winter wheat: Effects on gliadin content and composition. *Journal of Cereal Science* 32:45-56.

Danyluk, J., A. Perron, M. Houde, A. Limin, B. Fowler, N. Benhamou, and F. Sarhan (1998). Accumulation of an acidic dehydrin in the vicinity of the plasma membrane during cold acclimation of wheat. *The Plant Cell* 10:623-638.

Dawson, I.A. and I.F. Wardlaw (1989). The tolerance of wheat to high temperatures during reproductive growth. III. Booting to anthesis. *Australian Journal of Agricultural Research* 40:965-980.

Delgado, E., R.A.C. Mitchell, M.A.J. Parry, S.P. Driscoll, V.J. Mitchell, and D.W. Lawlor (1994). Interacting effects of CO_2 concentration, temperature and nitrogen supply on the photosynthesis and composition of winter wheat leaves. *Plant, Cell and Environment* 17:1205-1213.

Dell'Aquila, A. and M.D. Turi (1999). Amplification of aging symptoms in two differently thermal-sensitive wheat (*Triticum durum* L.) genotypes by heat-shock: Relationships between germination response and embryo protein patterns. *Seed Science and Technology* 27:467-476.

Demotes-Mainard, S., G. Doussinault, and J.M. Meynard (1995). Effects of low radiation and low temperature at meiosis on pollen viability and grain set in wheat. *Agronomie* 15:357-365.

De Sousa, C.N.A. (1998). Classification of Brazilian wheat cultivars for aluminium toxicity in acid soils. *Plant Breeding* 117:217-221.

Devkota, R.N. (1994). Wheat breeding objectives in Nepal: The national testing system and recent progress. In *Wheat in Heat-Stressed Environments: Irrigated, Dry Areas and Rice-Wheat Farming Systems,* D.A. Saunders and G.P. Hettel (Eds.). D.F., Mexico: CIMMYT, pp. 216-223.

Devos, K.M. and M.D. Gale (1997). Comparative genetics in the grasses. *Plant Molecular Biology* 35:3-15.

Ding, N. and M.E. Musgrave (1995). Relationship between mineral coating on roots and yield performance of wheat under waterlogging stress. *Journal of Experimental Botany* 46:939-945.

Dörffling, K., H. Dörffling, and G. Lesselich (1993). In vitro-selection and regeneration of hydroxyproline-resistant lines of winter wheat with increased proline content and increased frost tolerance. *Journal of Plant Physiology* 142:222-225.

Dörffling, K., H. Dörffling, G. Lesselich, E. Luck, C. Zimmermann, and G. Melz (1997). Heritable improvement of frost tolerance in winter wheat by in vitro-selection of hydroxyproline-resistant proline overproducing mutants. *Euphytica* 93:1-10.

Dörffling, K., S. Schulenburg, G. Lesselich, and H. Dörffling (1990). Abscisic acid and proline levels in cold hardened winter wheat leaves in relation to variety-specific differences in freezing resistance. *Journal of Agronomy and Crop Science* 165:230-239.

Dugger, W.M. (1983). Boron in plant metabolism. In *Inorganic Plant Nutrition,* A. Lauchli and R.L. Bieleski (Eds.). *Encyclopedia of Plant Physiology,* New Series, Vol. 15B. New York: Springer-Verlag, pp. 626-650.

Dwivedi, B.S. and K.N. Tiwari (1992). Effect of native and fertilizer zinc on dry matter yield and zinc uptake by wheat *(Triticum aestivum)* in Udic Ustochrepts. *Tropical Agriculture* 69:357-361.

Engels, C. (1994). Effect of root and shoot meristem temperature on shoot to root dry matter partitioning and the internal concentrations of nitrogen and carbohydrates in maize and wheat. *Annals of Botany* 73:211-219.

Equiza, M.A., J.P. Mirave, and J.A. Tognetti (1997). Differential inhibition of shoot vs. root growth at low temperature and its relationship with carbohydrate accumulation in different wheat cultivars. *Annals of Botany* 80:657-663.

Every, D., C.M. Hayman, and H. Henderson (1990). A rapid, simple test for heat damage in durum wheat. *New Zealand Journal of Crop and Horticultural Science* 18:43-47.

Eyupoglu, F., N. Kurucu, and U. Samsag (1994). Status of plant available micronutrients in Turkish soils. In *Soil and Fertilizer Research Institute* (in Turkish), Annual Report No. R-118 Ankara, Turkey, pp. 25-32.

Fedulov, P., M. Puchkov, N.M. Belikova, and G.D. Nabokov (1990). Change in structure of frost resistance of winter wheat during wintering. *Soviet Agricultural Sciences* 6:1-4.

Fedulov, Y. (1996). Assessment of the frost resistance of winter wheat by chlorophyll delayed fluorescence method. *Cereal Research Communications* 24:425-431.

Feldman, M. (2001). Origin of cultivated wheat. In *The World Wheat Book: A History of Wheat Breeding,* A.P. Bonjean and W.J. Angus (Eds.). Paris, France: Lavoisier Publishing, pp. 3-53.

Ferguson, D.L., K. Al-Khatib, J.A. Guikema, and G.M. Paulsen (1993). Degradation of proteins from thylakoid membranes in senescing wheat leaves at high temperature. *Plant, Cell and Environment* 16:421-428.

Ferris, R., R.H. Ellis, T.R. Wheeler, and P. Hadley (1998). Effect of high temperature stress at anthesis on grain yield and biomass of field-grown crops of wheat. *Annals of Botany* 82:631-639.

Fokar, M., A. Blum, and H.T. Nguyen (1998). Heat tolerance in spring wheat. II. Grain filling. *Euphytica* 104:9-15.

Fokar, M., H.T. Nguyen, and A. Blum (1998). Heat tolerance in spring wheat. I. Estimating cellular thermotolerance and its heritability. *Euphytica* 104:1-8.

Fowler, D.B., L.P. Chauvin, A.E. Limin, and F. Sarhan (1996). The regulatory role of vernalization in the expression of low-temperature-induced genes in wheat and rye. *Theoretical and Applied Genetics* 93:554-559.

Fowler, D.B., A.E. Limin, S.Y. Wang, and R.W. Ward (1996). Relationship between low-temperature tolerance and vernalization response in wheat and rye. *Canadian Journal of Plant Science* 76:37-42.

Frank, A.B. and A. Bauer (1996). Temperature, nitrogen, and carbon dioxide effects on spring wheat development and spikelet numbers. *Crop Science* 36:659-665.

Fu, P., R.W. Wilen, G. Wu, A.J. Robertson, and L.W. Gusta (2000). Dehydrin gene expression and leaf water potential differs between spring and winter cereals during cold acclimation. *Journal of Plant Physiology* 156:394-400.

Galiba, G. (1994). In vitro adaptation for drought and cold hardiness in wheat. *Plant Breeding Reviews* 12:115-162.

Galiba, G., I. Kerepesi, J.W. Snape, and J. Sutka (1997). Location of a gene regulating cold-induced carbohydrate production on chromosome 5A of wheat. *Theoretical and Applied Genetics* 95:265-270.

Galiba, G., S.A. Quarrie, J. Sutka, A. Morgounov, and J.W. Snape (1995). RFLP mapping of the vernalization (Vrn1) and frost resistance (Fr1) genes on chromosome 5a of wheat. *Theoretical and Applied Genetics* 90:1174-1179.

Gana, J.A., F. Sutton, and D.G. Kenefick (1997). cDNA structure and expression patterns of a low-temperature-specific wheat gene tacr7. *Plant Molecular Biology* 34:643-650.

Ghanem, E.H. (1994). Wheat improvement in Egypt with emphasis on heat tolerance. In *Wheat in Heat-Stressed Environments: Irrigated, Dry Areas and Rice-Wheat Farming Systems,* D.A. Saunders and G.P. Hettel (Eds.). D.F., Mexico: CIMMYT, pp. 12-16.

Gibson, L.R., P.J. McCuskey, K.A. Tilley, and G.M. Paulsen (1998). Quality of hard red winter wheat grown under high temperature conditions during maturation and ripening. *Cereal Chemistry* 75:421-427.

Gibson, L.R. and G.M. Paulsen (1999). Yield components of wheat grown under high temperature stress during reproductive growth. *Crop Science* 39:1841-1846.

Gill, B.S., H.C. Sharma, and W.J. Raupp (1985). Evaluation of *Aegilops* species for resistance to wheat powdery mildew, wheat leaf rust, hessian fly, and greenbug. *Plant Diseases* 69:314-316.

Gill, K.S., A. Qadar, and K.N. Singh (1992). Response of wheat *(Triticum aestivum)* genotypes to sodicity in association with waterlogging at different stages of growth. *Indian Journal of Agricultural Sciences* 62:124-128.

Gill, K.S., A. Qadar, and K.N. Singh (1993). Effect of interaction of flooding and alkalinity at various growth stages on grain yield of wheat *(Triticum aestivum). Indian Journal of Agricultural Sciences* 63:795-802.

Gorham, J., J. Bridges, J. Dubcosvsky, J. Dvorak, P.A. Hollington, M.C. Luo, and J.A. Khan (1997). Genetic analysis and physiology of a trait for enhanced K^+/Na^+ discrimination in wheat. *New Phytologist* 137(1):109-116.

Graham, R.D. (1978a). Nutrient efficiency objective in cereal breeding. In *Proceedings of the Eighth International Colloqium, Plant Analysis and Fertilizer Problems,* A.R. Ferguson, R.L. Bieleski, and I.B. Ferguson (Eds.). Auckland, New Zealand: Information Service—New Zealand Department of Scientific and Industrial Research, pp. 165-170.

Graham, R.D. (1978b). Tolerance of triticale, wheat and rye to copper deficiency. *Nature* 271:542-543.

Graham, R.D. (1984). Breeding for nutritional characteristics in cereals. *Advances in Plant Nutrition* 1:57-102.

Graham, R.D. (1990). Breeding wheats for tolerance to micronutrient deficient soil: Present status and priorities. In *Wheat for the Nontraditional Warm Areas,* D.A. Saunders (Ed.). D.F., Mexico: CIMMYT, pp. 315-332.

Graham, R.D., G.D. Anderson, and J.S. Ascher (1981). Absorption of copper by wheat, rye and some hybrid gentoypes. *Journal of Plant Nutrition* 3:679-686.

Graham, R.D., J.S. Ascher, P.A.E. Ellis, and K.W. Shepherd (1987). Transfer to wheat of the copper efficiency factor carried on rye chromosome arm 5RL. *Plant and Soil* 99:107-114.

Graham, R.D., J.S. Ascher, and S.C. Hynes (1992). Selecting zinc-efficient cereal genotypes for soils low in zinc status. *Plant and Soil* 146:241-250.

Graham, R.D. and A.D. Rovira (1984). A role for manganese in the resistance of wheat plants to take-all. *Plant and Soil* 78:441-444.

Grass, L. and J.S. Burris (1995a). Effect of heat stress during seed development and maturation on wheat *(Triticum durum)* seed quality. I. Seed germination and seedling vigor. *Canadian Journal of Plant Science* 75:821-829.

Grass, L. and J.S. Burris (1995b). Effect of heat stress during seed development and maturation on wheat *(Triticum durum)* seed quality. II. Mitochondrial respiration and nucleotide pools during early germination. *Canadian Journal of Plant Science* 75:831-839.

Grewal, H.S., R.D. Graham, and Z. Rengel (1996). Genotypic variation in zinc efficiency and resistance to crown rot disease (*Fusarium graminearum* Schw. Group 1) in wheat. *Plant and Soil* 186:219-226.

Guy, C.L. (1990). Cold acclimation and freezing tolerance: Role of protein metabolism. *Annual Review of Plant Molecular Biology* 41:187-223.

Hampson, C.R. and G.M. Simpson (1990a). Effects of temperature, salt, and osmotic potential on early growth of wheat *(Triticum aestivum)*. I. Germination. *Canadian Journal of Botany* 68:524-528.

Hampson, C.R. and G.M. Simpson (1990b). Effects of temperature, salt, and osmotic potential on early growth of wheat *(Triticum aestivum)*. II. Early seedling growth. *Canadian Journal of Botany* 68:529-553.

Harry, S.P. and R.D. Graham (1981). Tolerance of triticale, wheat and rye to copper deficiency and low and high pH. *Journal of Plant Nutrition* 3:721-730.

Hawker, J.S. and C.F. Jenner (1993). High temperature affects the activity of enzymes in the committed pathway of starch synthesis in developing wheat endosperm. *Australian Journal of Plant Physiology* 20:197-209.

Hede, A.R., B. Skovmand, and J. Lopez-Cesati (2001). Acid soils and aluminum toxicity. In *Application of Physiology in Wheat Breeding,* M.P Reynolds, J.I. Ortiz-Monasterio, and A. McNab (Eds.). D.F. Mexico: CIMMYT, pp. 172-182.

Hendershot, K.L., J. Weng, and H.T. Nguyen (1992). Induction temperature of heat-shock protein synthesis in wheat. *Crop Science* 32:256-261.

Huang, B.R. and J.W. Johnson (1995). Root respiration and carbohydrate status of two wheat genotypes in response to hypoxia. *Annals of Botany* 75:427-432.

Huang, B.R., J.W. Johnson, and D.S. Nesmith (1997). Responses to root-zone CO_2 enrichment and hypoxia of wheat genotypes differing in waterlogging tolerance. *Crop Science* 37:464-468.

Huang, B.R., J.W. Johnson, D.S. Nesmith, and D.C. Bridges (1994). Growth, physiological and anatomical responses of two wheat genotypes to waterlogging and nutrient supply. *Journal of Experimental Botany* 45:193-202.

Huang, B.R., J.W. Johnson, D.S. Nesmith, and D.C. Bridges (1995). Nutrient accumulation and distribution of wheat genotypes in response to waterlogging and nutrient supply. *Plant and Soil* 173:47-54.

Huang, B.R. and H.M. Taylor (1993). Morphological development and anatomical features of wheat seedlings as influenced by temperature and seeding depth. *Crop Science* 33:1269-1273.

Huang, B.R., H.M. Taylor, and B.L. McMichael (1991a). Behavior of lateral roots in winter wheat as affected by temperature. *Environmental and Experimental Botany* 31:187-192.

Huang, B.R., H.M. Taylor, and B.L. McMichael (1991b). Effects of temperature on the development of metaxylem in primary wheat roots and its hydraulic consequence. *Annals of Botany* 67:163-166.

Huang, B.R., H.M. Taylor, and B.L. McMichael (1991c). Growth and development of seminal and crown roots of wheat seedlings as affected by temperature. *Environmental and Experimental Botany* 31:471-477.

Huang, J. and G. Zhu (1994). Wheat breeding objectives and the provincial yield testing system in Guangxi, China. In *Wheat in Heat-Stressed Environments: Irrigated, Dry Areas and Rice-Wheat Farming Systems,* D.A. Saunders and G.P. Hettel (Eds.). D.F., Mexico: CIMMYT, pp. 226-227.

Huber, D.M. and N.S. Wilhelm (1988). The role of manganese in resistance to plant diseases. In *Manganese in Soils and Plants,* R.D. Graham, R.J. Hannam, and N.C. Uren (Eds.). Dordrecht, The Netherlands: Kluwer Academic Publishers, pp. 155-173.

Hucl, P. and R.J. Baker (1990). Effect of seeding depth and temperature on tillering characteristics of four spring wheat cultivars. *Canadian Journal of Plant Science* 70:409-417.

Hurkman, W.J., F.M. DuPont, S.B. Altenbach, A. Combs, R. Chan, C.K. Tanaka, M. Reuveni, and J.E. Bernardin (1998). BiP, HSP70, NDK and PDI in wheat endosperm. II. Effects of high temperature on protein and mRNA accumulation. *Physiologia Plantarum* 103:80-90.

Hurry, V.M., A. Strand, M. Tobiaeson, P. Gardestrom, and G. Oquist (1995). Cold hardening of spring and winter wheat and rape results in differential effects on growth, carbon metabolism, and carbohydrate content. *Plant Physiology* 109:697-706.

Ingram, J. and D. Bartels (1996). The molecular basis of dehydration tolerance in plants. *Annual Review of Plant Physiology and Plant Molecular Biology* 47:377-403.

Ishag, H.M. and B.A. Mohamed (1996). Phasic development of spring wheat and stability of yield and its components in hot environments. *Field Crops Research* 46:169-176.

Ishag, H.M., B.A. Mohamed, and K.H.M. Ishag (1998). Leaf development of spring wheat cultivars in an irrigated heat-stressed environment. *Field Crops Research* 58:167-175.

James, D.W. (1990). Plant nutrient interactions in alkaline and calcareous soils. In *Crops As Enhancers of Nutrient Use,* V.C. Baligar and R.R. Duncan (Eds.). New York: Academic Press, pp. 132-183.

Jamjod, S., C.E. Mann, and B. Rerkasem (1992). Combining ability of the response to boron deficiency in wheat. In *Boron Deficiency in Wheat,* C.E. Mann and B.

Rerkasem (Eds.). CIMMYT Wheat Special Report No. 11. D.F., Mexico: CIMMYT, pp. 83-85.

Jefferies, S.P., M.A. Pallotta, J.G. Paull, A. Karakousis, J.M. Kretschmer, S. Manning, A.K.M.R. Islam, P. Langridge, and K.J. Chalmers (2000). Mapping and validation of chromosome regions conferring boron toxicity tolerance in wheat *(Triticum aestivum). Theoretical and Applied Genetics* 101:767-777.

Jenner, C.F. (1991). Effects of exposure of wheat ears to high temperature on dry matter accumulation and carbohydrate metabolism in the grain of two cultivars. II. Carry-over effects. *Australian Journal of Plant Physiology* 18:179-190.

Jenner, C.F. (1994). Starch synthesis in the kernel of wheat under high temperature conditions. *Australian Journal of Plant Physiology* 21:791-806.

Jeong, B.R. and T.L. Housley (1990). Fructan metabolism in wheat in alternating warm and cold temperatures. *Plant Physiology* 93:902-906.

Johnson, J.P., B.F. Carver, and V.C. Baligar (1997). Productivity in Great Plains acid soils of wheat genotypes selected for aluminum tolerance. *Plant and Soil* 188:101-106.

Joshi, C.P., N.Y. Klueva, K.J. Morrow, and H.T. Nguyen (1997). Expression of a unique plastid-localized heat-shock protein is genetically linked to acquired thermotolerance in wheat. *Theoretical and Applied Genetics* 95:834-841.

Karimzadeh, G., D. Francis, and M.S. Davies (2000). Low temperature-induced accumulation of protein is sustained both in root meristems and in callus in winter wheat but not in spring wheat. *Annals of Botany* 85:769-777.

Keeling, P.L., P.J. Bacon, and D.C. Holt (1993). Elevated temperature reduces starch deposition in wheat endosperm by reducing the activity of soluble starch synthase. *Planta* 191:342-348.

Khera, R. and M.R. Chaudhary (1989). Interactive effect of soil temperature and seeding depth on wheat seedling emergence. *Indian Journal of Agronomy* 34:449-452.

Kim, B.Y., A.C. Baier, D.J. Somers, and J.P. Gustafson (2001). Aluminum tolerance in triticale, wheat and rye. *Euphytica* 120:329-337.

Kirigwi, F., R.G. Sears, M. van Ginkel, G.M. Paulsen, and R. Trethowan (2000). Developing a selection strategy for drought tolerant wheat (*Triticum aestivum* L). Poster presented at the American Society of Agronomy Meetings, Minneapolis, MN, November.

Klimov, S.V., N.V. Astakhova, and T.I. Trunova (1999). Changes in photosynthesis, dark respiration rates and photosynthetic carbon partitioning in winter rye and wheat seedlings during cold hardening. *Journal of Plant Physiology* 155:734-739.

Kocsy, G., G. Szalai, A. Vagujfalvi, L. Stehli, G. Orosz, and G. Galiba (2000). Genetic study of glutathione accumulation during cold hardening in wheat. *Planta* 210:295-301.

Kohli, M.M., C.E. Mann, and S. Rajaram (1991). Global status and recent progress in breeding wheat for the warmer areas. In *Wheat for the Nontraditional, Warm Areas,* D.A. Saunders (Ed.). D.F., Mexico: CIMMYT, pp. 96-112.

Kokurina, E.A. and M.I. Rybakova (1989). Genetic control of resistance to perishing under snow and frost resistance of winter wheat. *Soviet Agricultural Sciences* 10:1-5.

Larsson, S. (1989). Sugar content and membrane lipid modifications in winter wheat (*Triticum aestivum* L.) during cold hardening. *Sveriges Utsadesforenings Tidskrift* 99:143-149.

Lasztity, D., I. Racz, and E. Paldi (1999). Effect of long periods of low temperature exposure on protein synthesis activity in wheat seedlings. *Plant Science* 149:59-62.

Leithold, B., G. Muller, W.E. Weber, and T. Westermann (1997). Investigations on heat tolerance of spring wheat varieties of different origin under growth chamber conditions. *Journal of Agronomy and Crop Science* 179:115-122.

Li Z., L.S. Shen, B. Courtois, and R. Lafitte (2000). Development of near isogenic line (NIL) sets for QTLs associated with drought tolerance in rice. In *Molecular Approaches for the Genetic Improvement of Cereals for Stable Production in Water-Limited Environments,* J. Ribaut and D. Poland (Eds.). El Batan, Mexico: CIMMYT, pp. 103-107.

Limin, A.E. and D.B. Fowler (1991). Breeding for cold hardiness in winter wheat: Problems, progress and alien gene expression. *Field Crops Research* 27:201-218.

Limin, A.E. and D.B. Fowler (1993). Inheritance of cold hardiness in *Triticum aestivum* × synthetic hexaploid wheat crosses. *Plant Breeding* 110:103-108.

Limin, A.E. and D.B. Fowler (1994). Relationship between guard cell length and cold hardiness in wheat. *Canadian Journal of Plant Science* 74:59-62.

Limin, A.E. and D.B. Fowler (2000). Morphological and cytological characters associated with low-temperature tolerance in wheat (*Triticum aestivum* L. em Thell.). *Canadian Journal of Plant Science* 80:687-692.

Lu, C.M. and J.H. Zhang (2000). Heat-induced multiple effects on PSII in wheat plants. *Journal of Plant Physiology* 156:259-265.

Lu, Z.M., R.G. Percy, C.O. Qualset, and E. Zeiger (1998). Stomatal conductance predicts yields in irrigated Pima cotton and bread wheat grown at high temperatures. *Journal of Experimental Botany* (Special Issue) 49:453-460.

Luxmoore, R.J., R.A. Fisher, and L.H. Stolzy (1973). Flooding and soil temperature effects on wheat during grain filling. *Agronomy Journal* 65:361-364.

Machackova, I., A. Hanisova, and J. Krekule (1989). Levels of ethylene, ACC, MACC, ABA and proline as indicators of cold hardening and frost resistance in winter wheat. *Physiologia Plantarum* 76:603-607.

Maes, B., R.M. Trethowan, M.P. Reynolds, M. van Ginkel, and B. Skovmand (2001). The influence of glume pubescence on spikelet temperature of wheat under freezing conditions. *Australian Journal of Plant Physiology* 28:141-148.

Maheswari, M., D.K. Joshi, R. Saha, S. Nagarajan, and P.N. Gambhir (1999). Transverse relaxation time of leaf water protons and membrane injury in wheat (*Triticum aestivum* L.) in response to high temperature. *Annals of Botany* 84:741-745.

Manyowa, N.M. (1989). "The genetics of aluminum, excess boron, copper and manganese stress tolerance in the tribe *Triticeae* and its implications for wheat improvement." PhD thesis, Cambridge University, Cambridge, U.K.

Mashiringwani, N.A. and M.A. Schweppenhauser (1992). Phenotypic characters associated with yield adaptation of wheat to a range of temperature conditions. *Field Crops Research* 29:69-77.

McCaig, T.N. (1996). Temperature and precipitation effects on durum wheat grown in southern Saskatchewan for fifty years. *Canadian Journal of Plant Science* 77:215-223.

McDonald, M.P., N.W. Galwey, P. Ellneskog-Staam, and T.D. Colmer (2001). Evaluation of *Lophopyrum elongatum* as a source of genetic diversity to increase the waterlogging tolerance of hexaploid wheat *(Triticum aestivum)*. *New Phytologist* 151:369-380.

McElwain, E.F. and S. Spiker (1992). Molecular and physiological analysis of a heat-shock response in wheat. *Plant Physiology* 99:1455-1460.

Mehrotra, O.N., R.D.L. Srivastava, and P.H. Mishra (1980). Some observations on the relative tolerance of wheat genotypes to boron. *Indian Agriculture* 24:223-238.

Melhuish, F.M., E. Humphreys, W.A. Muirhead, and R.J.G. White (1991). Flood irrigation of wheat on a transitional red-brown earth. I. Effect of duration of ponding on soil water, plant growth, yield and N uptake. *Australian Journal of Agricultural Research* 42:1023-1035.

Menegus, F., L. Cattaruzza, M. Mattana, N. Beffagna, and E. Ragg (1991). Response to anoxia in rice and wheat seedlings. *Plant Physiology* 95:760-767.

Mian, M.A.R., E.D. Nafziger, F.L. Kolb, and R.H. Teyker (1993). Root growth of wheat genotypes in hydroponic culture and in the greenhouse under different soil moisture regimes. *Crop Science* 33:283-286.

Miller, T.E., N. Igbal, S.M. Reader, A. Mahmood, K.A. Cant, and I.P. King (1997). A cytogenetic approach to the improvement of aluminum tolerance in wheat. *New Physiologist* 137(1):93-98.

Mishra, R.K. and G.S. Singhal (1992). Function of photosynthetic apparatus of intact wheat leaves under high light and heat stress and its relationship with peroxidation of thylakoid lipids. *Plant Physiology* 98:1-6.

Mitchell, R.A.C., V.J. Mitchell, S.P. Driscoll, J. Franklin, and D.W. Lawlor (1993). Effects of increased CO_2 concentration and temperature on growth and yield of winter wheat at two levels of nitrogen application. *Plant, Cell and Environment* 16:521-529.

Miyasaka, S.C. and D.L. Grunes (1997). Root zone temperature and calcium effects on phosphorus, sulfur, and micronutrients in winter wheat forage. *Agronomy Journal* 89:742-748.

Mladenov, N., S. Dencic, M. Kraljevic-Balalic, N. Przulj, and N. Hristov (1998). Genetic analysis of resistance to low temperatures in wheat. *Cereal Research Communications* 26:53-58.

Moffatt, J.M., R.G. Sears, T.S. Cox, and G.M. Paulsen (1990). Wheat high temperature tolerance during reproductive growth. II. Genetic analysis of chlorophyll fluorescence. *Crop Science* 30:886-889.

Moffatt, J.M., R.G. Sears, and G.M. Paulsen (1990). Wheat high temperature tolerance during reproductive growth. I. Evaluation by chlorophyll fluorescence. *Crop Science* 30:881-885.

Moody, D.B., A.J. Rathjen, and B. Cartwright (1990). Yield evaluation of a gene for boron tolerance using backcross-derived lines. In *Proceedings of the Sixth Assembly Wheat Breeding Society of Australia,* L. O'Brien, F.W. Ellison, R.A. Hare, and M.C. Mackay (Eds.). Tamworth, Australia: Wheat Breeding Society of Australia, pp. 117-121.

Moolsri, S. (1994). Wheat breeding program in northern Thailand. In *Wheat in Heat-Stressed Environments: Irrigated, Dry Areas and Rice-Wheat Farming Systems,* D.A. Saunders and G.P. Hettel (Eds.). D.F., Mexico: CIMMYT, pp. 236-238.

Munns, R., R.A. Hare, R.A. James, and G.J. Rebetzke (1999). Genetic variation for improving the salt tolerance of durum wheat. *Australian Journal of Agricultural Research* 51:69-74.

Murphy, L.S. and L.M. Walsh (1972). Correction of micronutrient deficiencies with fertilizer. In *Micronutrients in Agriculture,* J.J. Mortvedt, P.M. Giordano, and W.L. Lindsay (Eds.). Madison, WI: Soil Science Society of America, pp. 347-387.

Musgrave, M.E. (1994). Waterlogging effects on yield and photosynthesis in eight winter wheat cultivars. *Crop Science* 34:1314-1318.

Musgrave, M.E. and N. Ding (1998). Evaluating wheat cultivars for waterlogging tolerance. *Crop Science* 38:90-97.

Nable, R.O. (1992). Mechanism of tolerance to boron toxicity in plants. In *Boron Deficiency in Wheat,* C.E. Mann and B. Rerkasem (Eds.). CIMMYT Wheat Special Report No. 11. D.F., Mexico: CIMMYT, pp. 98-109.

Nambiar, E.K.S. (1976). Genetic differences in the nutrition of cereals. I. Differential responses of genotypes to copper. *Australian Journal of Agricultural Research* 27:453-463.

Nguyen, H.T., C.P. Joshi, N. Klueva, J. Weng, K.L. Hendershot, and A. Blum (1994). The heat-shock response and expression of heat-shock proteins in wheat under diurnal heat stress and field conditions. *Australian Journal of Plant Physiology* 21:857-867.

O'Brien, L. (1982). Victorian wheat yield trend, 1898-1977. *Journal of the Australian Institute of Agricultural Science* 48:162-168.

Okuda, T., Y. Matsuda, A. Yamanaka, and S. Sagisaka (1991). Abrupt increase in the level of hydrogen peroxide in leaves of winter wheat is caused by cold treatment. *Plant Physiology* 97:1265-1267.

Ouellet, F., E. Carpentier, M.J.T.V. Cope, A.F. Monroy, and F. Sarhan (2001). Regulation of a wheat actin-depolymerizing factor during cold acclimation. *Plant Physiology* 125:360-368.

Oyanagi, A., T. Nakamoto, and S. Morita (1993). The gravitropic response of roots and the shaping of the root system in cereal plants. *Environmental and Experimental Botany* 33:141-158.

Paldi, E., I. Racz, and D. Lasztity (1996). Effect of low temperature on the rRNA processing in wheat (*Triticum aestivum* L.). *Journal of Plant Physiology* 148:374-377.

Paull, J.G., R.O. Nable, and A.J. Rathjen (1992). Physiological and genetic control of the tolerance of wheat to high concentrations of boron and implications for plant breeding. *Plant and Soil* 146:251-260.

Peacock, J.M., V. Mahalaksmi, G. Ortiz Ferrara, C.J. Howarth, M. Nachit, and J. Hamblin (1994). Physiologists' approaches to screening for high temperature tolerance in cereals: The Indian experience. In *Wheat in Heat-Stressed Environments: Irrigated, Dry Areas and Rice-Wheat Farming Systems,* D.A. Saunders and G.P. Hettel (Eds.). D.F., Mexico: CIMMYT, pp.158-169.

Pecetti, L., P. Annicchiarico, and G. Kashour (1993). Flag leaf variation in Mediterranean durum wheat landraces and its relationship to frost and drought tolerance and yield response in moderately favourable conditions. *Plant Genetic Resources Newsletter* 93:25-28.

Pedretti R.R. and M.M. Kohli (1991). Wheat production in Paraguay: Trends, major constraints, and potential. In *Wheat for the Nontraditional, Warm Areas,* D.A. Saunders (Ed.). D.F., Mexico: CIMMYT, pp. 84-95.

Pellegrineschi, A., J.M. Ribaut, R. Trethowan, K. Yamaguchi-Shinozaki, and D. Hoisington (2002). Progress in the genetic engineering of wheat for water-limited conditions. *JIRCAS Working Report* 2002:55-60.

Peña, R.J., J. Zarco-Hernandez, and A. Mujeeb-Kazi (1995). Glutenin subunit compositions and bread-making quality characteristics of synthetic hexaploid wheats derived from *Triticum turgidum* × *Triticum tauschii* (Coss.) Schmal crosses. *Journal of Cereal Science* 21:15-23.

Perry, M. and M. D'Antuono (1989). Yield improvement and associated characteristics of some Australian spring wheat cultivars introduced between 1860 and 1982. *Australian Journal of Agricultural Research* 40:457-472.

Perras, M. and F. Sarhan (1990). Polysome metabolism during cold acclimation in wheat. *Plant and Cell Physiology* 31:1083-1089.

Pilbeam, D.J. and E.A. Kirkby (1983). The physiological role of boron in plants. *Journal of Plant Nutrition* 6:563-582.

Pingali, P.L. (Ed.) (1999). *CIMMYT 1998-99 World Wheat Facts and Trends. Global Wheat Research in a Changing World: Challenges and Achievements.* D.F., Mexico: CIMMYT.

Pingali, P.L. and S. Rajaram (1999). Global wheat research in a changing world: Options for sustaining growth in wheat productivity. In *CIMMYT 1998-99 World Wheat Facts and Trends. Global Wheat Research in a Changing World: Challenges and Achievements,* P.L. Pingali (Ed.). D.F., Mexico: CIMMYT, pp.1-18.

Pinstrup-Andersen, P., R. Pandya-Lorch, and M.W. Rosegrant (1999). World food prospects: Critical issues for the early twenty-first century. 2020 Vision Food Policy Report Number 29. Washington, DC: IFPRI.

Pinthus, M.J. and M. Abraham (1996). Effects of light, temperature, gibberellin (GA_3) and their interaction on coleoptile and leaf elongation of tall, semi-dwarf and dwarf wheat. *Plant Growth Regulation* 18:239-247.

Pinthus, M.J., M.D. Gale, N.E.J. Appleford, and J.R. Lenton (1989). Effect of temperature on gibberellin (GA) responsiveness and on endogenous GA1 content of tall and dwarf wheat genotypes. *Plant Physiology* 90:854-859.

Porter, D.R., H.T. Nguyen, and J.J. Burke (1989). Chromosomal location of genes controlling heat shock proteins in hexaploid wheat. *Theoretical and Applied Genetics* 78:873-878.

Porter, D.R., H.T. Nguyen, and J.J. Burke (1995). Genetic control of acquired high temperature tolerance in winter wheat. *Euphytica* 83:153-157.

Porter, J.R. and M. Gawith (1999). Temperatures and the growth and development of wheat: A review. *European Journal of Agronomy* 10:23-36.

Preston, K.R., R.H. Kilborn, B.C. Morgan, and J.C. Babb (1991). Effects of frost and immaturity on the quality of a Canadian hard red spring wheat. *Cereal Chemistry* 68:133-138.

Racz, I., M. Kovacs, D. Lasztity, O. Veisz, G. Szalai, and E. Paldi (1996). Effect of short-term and long-term low temperature stress on polyamine biosynthesis in wheat genotypes with varying degrees of frost tolerance. *Journal of Plant Physiology* 148:368-373.

Rajaram, S. (2001). Prospects and promise of wheat breeding in the twenty-first century. In *Wheat in a Global Environment,* Z. Bedo and L. Lang (Eds.). Dordrecht, The Netherlands: Kluwer Academic Publishers, pp. 37-52.

Rajaram, S., H.J. Braun, and M. van Ginkel (1996). CIMMYT's approach to breeding for drought tolerance. *Euphytica* 92:147-153.

Rajaram, S., M.M. Kohli, and J. Lopez-Cesati (1991). Breeding for tolerance to aluminum toxicity in wheat. In *Plant-Soil Interactions at Low pH,* R.J. Wright, V.C. Baliger, and R.P. Murrman (Eds.). Dordrecht, The Netherlands: Kluwer Academic Publishers, pp. 1019-1028.

Rajaram, S. and M. van Ginkel (2001). Mexico, 50 years of international wheat breeding. In *The World Wheat Book: A History of Wheat Breeding,* A.P. Bonjean and W.J. Angus (Eds.). Paris, France: Lavoisier Publishing, pp. 579-604.

Randall, P.J. and H.J. Moss (1990). Some effects of temperature regime during grain filling on wheat quality. *Australian Journal of Agricultural Research* 41:603-617.

Rathjen, A.J. (1994). The biological basis of genotype × environment interaction: Its definition and management. In *Proceedings of the Seventh Assembly of the Wheat Breeding Society of Australia: Wheat Breeding—into the Second Century,* J. Paull, I.S. Dundas, K.J. Shepherd, and G.J. Hollamby (Eds.). Adelaide, Australia: Wheat Breeding Society of Australia, pp. 13-17.

Rawson, H.M. and R.A. Richards (1993). Effects of high temperature and photoperiod on floral development in wheat isolines differing in vernalisation and photoperiod genes. *Field Crops Research* 32:181-192.

Razzaque, M.A., A.B.S. Hossain, S.M. Ahmed, A.K.D. Sarkar, M. Rahman, C.D. Barma, and R. Amin (1994). Wheat breeding strategies in the rice farming system of Bangladesh. In *Wheat in Heat-Stressed Environments: Irrigated, Dry Areas and Rice-Wheat Farming Systems,* D.A. Saunders and G.P. Hettel (Eds.). D.F., Mexico: CIMMYT, pp. 239-246.

Rekika, D., Y. Kara, I. Souyris, M.M. Nachit, A. Asbati, and P. Monneveux (2000). The tolerance of PSII to high temperatures in durum wheat (*T. turgidum* conv. *durum*): Genetic variation and relationship with yield under heat stress. *Cereal Research Communications* 28:395-402.

Rengel, Z. and V. Romheld (2000). Differential tolerance to Fe and Zn deficiencies in wheat germplasm. *Euphytica* 113:219-225.

Rerkasem, B. and S. Jamjod (1997). Boron deficiency induces male sterility in wheat (*Triticum aestivum* L.) and implications for plant breeding. *Euphytica* 96:257-262.

Reynolds, M.P., M. Balota, M.I.B. Delgado, I. Amani, and R.A. Fischer (1994). Physiological and morphological traits associated with spring wheat yield under hot, irrigated conditions. *Australian Journal of Plant Physiology* 21:717-730.

Ribeiro-Carvalho, C., H. Guedes-Pinto, G. Harrison, and J.S. Heslop-Harrison (1997). Wheat-rye chromosome translocations involving small terminal and intercalary rye chromosome segments in the Portuguese wheat landrace Barbela. *Heredity* 78:539-546.

Ribeiro-Carvalho, C., H. Guedes-Pinto, J.S. Heslop-Harrison, and T. Shwarzacher (2001). Introgression of rye chromatin on chromosome 2D in the Portuguese wheat landrace "Barbela." *Genome* 44:1122-1128.

Richards, R.A., A.G. Condon, and G.J. Rebetzke (2001). Traits to improve yield in dry environments. In *Application of Physiology in Wheat Breeding,* M.P. Reynolds, J.I. Ortiz-Monasterio, and A. McNab (Eds.). D.F., Mexico: CIMMYT, pp. 88–100.

Richards, R.A., G.J. Rebetzke, R. Appels, and A.G. Condon (2000). Physiological traits to improve the yield of rain-fed wheat: Can molecular genetics help? In *Molecular Approaches for the Genetic Improvement of Cereals for Stable Production in Water-Limited Environments,* J. Ribaut and D. Poland (Eds.). El Batan, Mexico: CIMMYT, pp. 54-58.

Rizza, F., P. Baldi, L. Cattivelli, and G. Delogu (1997). Cold hardening in triticale in comparison with rye and wheat. *Cereal Research Communications* 25:947-954.

Roberts, D.W.A. (1990). Identification of loci on chromosome 5A of wheat involved in control of cold hardiness, vernalization, leaf length, rosette growth habit, and height of hardened plants. *Genome* 33:247-259.

Rosegrant, M.W., M.A. Sombilla, R.V. Gerpacio, and C. Ringler (1997). "Global food markets and US exports in the twenty-first century." Paper presented at the Illinois World Food and Sustainable Agriculture Program conference, Meeting the Demand for Food in the Twenty-First Century: Challenges and Opportunities, May 28, University of Illinois, Urbana-Champaign.

Rosenzweig, C. and F.N. Tubiello (1996). Effects of changes in minimum temperature on wheat yields in the central US: A simulation study. *Agricultural and Forest Meteorology* 80:215-230.

Ruan, Y., J. Gilmore, and T. Conner (1998). Toward *Arabidopsis* genome analysis: Monitoring expression profiles of 1400 genes using cDNA microarrays. *Plant Journal* 15:821-823.

Ruegger, A., H. Winzeler, and J. Nosberger (1990). Dry matter production and distribution of 14C-assimilates of spelt (*Triticum spelta* L.) and wheat (*Triticum aestivum* L.) as influenced by different temperatures before and during grain filling. *Journal of Agronomy and Crop Science* 165:110-120.

Saadalla, M.M., J.S. Quick, and J.F. Shanahan (1990). Heat tolerance in winter wheat. II. Membrane thermostability and field performance. *Crop Science* 30:1248-1251.

Saadalla, M.M., J.F. Shanahan, and J.S. Quick (1990). Heat tolerance in winter wheat. I. Hardening and genetic effects on membrane thermostability. *Crop Science* 30:1243-1247.

Saberi, H.K., R.D. Graham, and A.J. Rathjen (1999). Inheritance of manganese efficiency in durum wheat. *Journal of Plant Nutrition* 22:11-21.

Sairam, R.K., P.S. Deshmukh, and D.S. Shukla (1997). Tolerance of drought and temperature stress in relation to increased antioxidant enzyme activity in wheat. *Journal of Agronomy and Crop Science* 178:171-177.

Samad, A., C.A. Meisner, M. Saifuzzaman, and M. van Ginkel (2001). Waterlogging tolerance. In *Application of Physiology in Wheat Breeding,* M.P. Reynolds, J.I. Ortiz-Monasterio, and A. McNab (Eds.). D.F., Mexico: CIMMYT, pp. 136-144.

Samad, M.A., M.M. Rahman, M.R. Amin, Z.I. Sarkar, and M. Islam (1994). Selecting traits for breeding heat-tolerant wheat varieties. In *Wheat in Heat-Stressed Environments: Irrigated, Dry Areas and Rice-Wheat Farming Systems,* D.A. Saunders and G.P. Hettel (Eds.). D.F., Mexico: CIMMYT, pp. 379-381.

Santoiani, C.S., J.A. Tognetti, H.G. Pontis, and G.L. Salerno (1993). Sucrose and fructan metabolism in wheat roots at chilling temperatures. *Physiologia Plantarum* 87:84-88.

Savitch, L.V., T. Harney, and N.P.A. Huner (2000). Sucrose metabolism in spring and winter wheat in response to high irradiance, cold stress and cold acclimation. *Physiologia Plantarum* 108:270-278.

Sayre, K.D., M. van Ginkel, S. Rajaram, and I. Ortiz-Monasterio (1994). Tolerance to waterlogging losses in spring bread wheat: Effect of time of onset on expression. In *Annual Wheat Newsletter,* Colorado State University, 40:165-171.

Scebba, F., L. Sebastiani, and C. Vitagliano (1998). Changes in activity of antioxidative enzymes in wheat *(Triticum aestivum)* seedlings under cold acclimation. *Physiologia Plantarum* 104:747-752.

Scebba, F., L. Sebastiani, and C. Vitagliano (1999). Protective enzymes against activated oxygen species in wheat (*Triticum aestivum* L.) seedlings: Responses to cold acclimation. *Journal of Plant Physiology* 155:762-768.

Scott, B.J., J.A. Fisher, and B.R. Cullis (2001). Aluminium tolerance and lime increase wheat yield on the acidic soils of central and southern New South Wales. *Australian Journal of Experimental Agriculture* 41:523-532.

Scott, B.J., J.A. Fisher, and L.J. Spohr (1998). Manganese tolerance in a range of wheat genotypes. *Communications in Soil and Plant Analysis* 29:219-235.

Seki, M., M. Narusaka, H. Abe, M. Kasuga, K. Yamaguchi-Shinozaki, Y. Carninci, P. Hayashizaki, and K. Shinozaki (2001). Monitoring the expression pattern of 1300 *Arabidopsis* genes under drought and cold stresses by using a full-length cDNA microarray. *The Plant Cell* 13:61-72.

Shanahan, J.F., I.B. Edwards, J.S. Quick, and J.R. Fenwick (1990). Membrane thermostability and heat tolerance of spring wheat. *Crop Science* 30:247-251.

Shi, Y.C., P.A. Seib, and J.E. Bernardin (1994). Effects of temperature during grain-filling on starches from 6 wheat cultivars. *Cereal Chemistry* 71:369-383.
Shpiler, C. and A. Blum (1991). Heat tolerance for yield and its components in different wheat cultivars. *Euphytica* 51:257-263.
Sillanpaa, M. (1990). Micronutrient assessment at the country level: An international study. FAO Soils Bulletin 63. Rome, Italy: Food and Agricultural Organization of the United Nations, in cooperation with the Finish International Development Agency (FINNIDA).
Singh, A.J. and D. Byerlee (1990). Relative variability in wheat yields across countries and over time. *Journal of Agricultural Economics* 41:21-32.
Singh, R.P., J. Huerta-Espino, S. Rajaram, and J. Crossa (2001). Grain yield and other traits of tall and dwarf isolines of modern bread and durum wheats. In *Wheat in a Global Environment,* Z. Bedo and L. Lang (Eds.). Dordrecht, The Netherlands: Kluwer Academic Publishers, pp. 579-584.
Singh, R.P., S. Rajaram, A. Miranda, J. Huerta-Espino, and E. Autrique (1998). Comparison of two crossing and four selection schemes for yield, yield traits, and slow rusting resistance to leaf rust in wheat. *Euphytica* 100:35-43.
Skovmand, B., M.P. Reynolds, and I.H. Delacy (2001). Searching genetic resources for physiological traits with potential for increasing yield. In *Application of Physiology in Wheat Breeding,* M.P. Reynolds, J.I. Ortiz-Monasterio, and A. McNab (Eds.). D.F., Mexico: CIMMYT, pp. 17-28.
Skriver, K. and J. Mundy (1990). Gene expression in response to abscisic acid and osmotic stress. *The Plant Cell* 2:503-512.
Slafer, G.A. (1995). Wheat development as affected by radiation at two temperatures. *Journal of Agronomy and Crop Science* 175:249-263.
Slafer, G.A. and D.J. Miralles (1992). Green area duration during the grain filling period of an Argentine wheat cultivar as influenced by sowing date, temperature and sink strength. *Journal of Agronomy and Crop Science* 168:191-200.
Slafer, G.A. and H.M. Rawson (1994). Does temperature affect final numbers of primordia in wheat? *Field Crops Research* 39:111-117.
Slafer, G.A. and H.M. Rawson (1997). Phyllochron in wheat as affected by photoperiod under two temperature regimes. *Australian Journal of Plant Physiology* 24:151-158.
Slootmaker, A.L.J. (1974). Tolerance to high soil acidity in wheat related species, rye, and triticale. *Euphytica* 23:505-513.
Somers, D.J., K.G. Briggs, and J.P. Gustafson (1996). Aluminum stress and protein synthesis in near isogenic lines of *Triticum aestivum* differing in aluminum tolerance. *Physiologia Plantarum* 97:694-700.
Sparrow, D.H. and R.D. Graham (1988). Susceptibility of zinc-deficient wheat plants to colonization by *Fusarium graminearum* Schw. Group 1. *Plant and Soil* 112:261-266.
Stieger, P.A. and U. Feller (1994). Nutrient accumulation and translocation in maturing wheat plants grown on waterlogged soil. *Plant and Soil* 160:87-95.
Stone, P.J., P.W. Gras, and M.E. Nicolas (1997). The influence of recovery temperature on the effects of a brief heat shock on wheat. 3. Grain protein composition and dough properties. *Journal of Cereal Science* 25:129-141.

Stone, P.J. and M.E. Nicolas (1994). Wheat cultivars vary widely in their responses of grain yield and quality to short periods of post-anthesis heat stress. *Australian Journal of Plant Physiology* 21:887-900.

Stone, P.J. and M.E. Nicolas (1995a). Comparison of sudden heat stress with gradual exposure to high temperature during grain filling in two wheat varieties differing in heat tolerance. 1. Grain growth. *Australian Journal of Plant Physiology* 22:935-944.

Stone, P.J. and M.E. Nicolas (1995b). Effect of timing of heat stress during grain filling on two wheat varieties differing in heat tolerance. 1. Grain growth. *Australian Journal of Plant Physiology* 22:927-934.

Stone, P.J. and M.E. Nicolas (1995c). A survey of the effects of high temperature during grain filling on yield and quality of 75 wheat cultivars. *Australian Journal of Agricultural Research* 46:475-492.

Stone, P.J. and M.E. Nicolas (1996). Effect of timing of heat stress during grain filling on two wheat varieties differing in heat tolerance. 2. Fractional protein accumulation. *Australian Journal of Plant Physiology* 23:739-749.

Stone, P.J. and M.E. Nicolas (1998a). Comparison of sudden heat stress with gradual exposure to high temperature during grain filling in two wheat varieties differing in heat tolerance. II. Fractional protein accumulation. *Australian Journal of Plant Physiology* 25:1-11.

Stone, P.J. and M.E. Nicolas (1998b). The effect of duration of heat stress during grain filling on two wheat varieties differing in heat tolerance: Grain growth and fractional protein accumulation. *Australian Journal of Plant Physiology* 25:13-20.

Stone, P.J., M.E. Nicolas, and I.F. Wardlaw (1996). The influence of recovery temperature on the effects of a brief heat shock on wheat. 2. Fractional protein accumulation during grain growth. *Australian Journal of Plant Physiology* 23:605-616.

Storlie, E.W., R.E. Allan, and M.K. Walker-Simmons (1998). Effect of the Vrn1-Fr1 interval on cold hardiness levels in near-isogenic wheat lines. *Crop Science* 38:483-488.

Subedi, K.D., C.B. Budhathoki, and M. Subedi (1997). Variation in sterility among wheat (*Triticum aestivum* L.) genotypes in response to boron deficiency in Nepal. *Euphytica* 95(1):21-26.

Subedi, K.D., C.N. Floyd, and C.B. Budhathoki (1998). Cool temperature-induced sterility in spring wheat (*Triticum aestivum* L.) at high altitudes in Nepal: Variation among cultivars in response to sowing date. *Field Crops Research* 55:141-151.

Subedi, K.D., M.J. Gooding, and P.J. Gregory (2001). Cultivar variation in boron accumulation and grain set in wheat under the influence of cold temperature. *Annals of Applied Biology* 138:97-101.

Subedi, K.D., P.J. Gregory, R.J. Summerfield, and M.J. Gooding (1998). Cold temperatures and boron deficiency caused grain set failure in spring wheat (*Triticum aestivum* L.). *Field Crops Research* 57:277-288.

Sun, Q.X. and J.S. Quick (1991). Chromosomal locations of genes for heat tolerance in tetraploid wheat. *Cereal Research Communications* 19:431-437.

Sutka, J. (1994). Genetic control of frost tolerance in wheat (*Triticum aestivum* L.). *Euphytica* 77:277-282.

Sutka, J. (2001). Genes for frost resistance in wheat. *Euphytica* 119:167-172.

Sutka, J., A.J. Worland, and O.I. Maystrenko (1991). Slight effect of the cytoplasm on frost resistance in wheat *(Triticum aestivum). Cereal Research Communications* 19:311-317.

Tahir, M. and M. Singh (1993). Assessment of screening techniques for heat tolerance in wheat. *Crop Science* 33:740-744.

Tandon, J.P. (1994a). National testing system, breeding strategies, and recent progress in wheat research for warmer area rice farming systems in India. In *Wheat in Heat-Stressed Environments: Irrigated, Dry Areas and Rice-Wheat Farming Systems,* D.A. Saunders and G.P. Hettel (Eds.). D.F., Mexico: CIMMYT, pp. 251-256.

Tandon, J.P. (1994b). Wheat cultivation, research organization, and production technology in the hot dry regions of India. In *Wheat in Heat-Stressed Environments: Irrigated, Dry Areas and Rice-Wheat Farming Systems,* D.A. Saunders and G.P. Hettel (Eds.). D.F., Mexico: CIMMYT, pp.17-23.

Tashiro, T. and I.F. Wardlaw (1990a). The effect of high temperature at different stages of ripening on grain set, grain weight and grain dimensions in the semi-dwarf wheat 'Banks'. *Annals of Botany* 65:51-61.

Tashiro, T. and I.F. Wardlaw (1990b). The response to high temperature shock and humidity changes prior to and during the early stages of grain development in wheat. *Australian Journal of Plant Physiology* 17:551-561.

Tester, R.F., W.R. Morrison, R.H. Ellis, J.R. Piggott, G.R. Batts, T.R. Wheeler, J.I.L. Morison, P. Hadley, and D.A. Ledward (1995). Effects of elevated growth temperature and carbon dioxide levels on some physicochemical properties of wheat starch. *Journal of Cereal Science* 22:63-71.

This, D. and B. Teulat-Merah (2000). Toward a comparative genomics of drought tolerance in cereals: Lessons from a QTL analysis in barley. In *Molecular Approaches for the Genetic Improvement of Cereals for Stable Production in Water-Limited Environments,* J. Ribaut and D. Poland (Eds.). El Batan, Mexico: CIMMYT, pp. 92-96.

Thongbai, P., R.J. Hannam, R.D. Graham, and M.J. Webb (1993). Interaction between zinc nutritional status of cereals and *Rhizoctonia* root rot severity. I. Field observations. *Plant and Soil* 153:207-214.

Tognetti, J.A., G.L. Salerno, M.D. Crespi, and H.G. Pontis (1990). Sucrose and fructan metabolism of different wheat cultivars at chilling temperatures. *Physiologia Plantarum* 78:554-559.

Trethowan, R.M., J. Crossa, M. van Ginkel, and S. Rajaram (2001). Relationships among bread wheat international yield testing locations in dry areas. *Crop Science* 41:1461-1469.

Trethowan, R.M., M. van Ginkel, and S. Rajaram (2002). Progress in breeding for yield and adaptation in global drought affected environments. *Crop Science* 42:1141-1446.

Vagujfalvi, A., I. Kerepesi, G. Galiba, T. Tischner, and J. Sutka (1999). Frost hardiness depending on carbohydrate changes during cold acclimation in wheat. *Plant Science* 144:85-92.

Valkoun, J. (2001). Wheat pre-breeding using wild progenitors. In *Wheat in a Global Environment,* Z. Bedo and L. Lang (Eds.). Dordrecht, The Netherlands: Kluwer Academic Publishers, pp. 699-707.

Van Ginkel, M., S. Rajaram, and M. Thijssen (1992). Waterlogging in wheat: Germplasm evaluation and methodology development. In *Proceedings of the Seventh Regional Wheat Workshop for Eastern, Central and Southern Africa,* D.G. Tanner and W. Mwangi (Eds.). Nakuru, Kenya: CIMMYT, pp. 115-124.

Veisz, O., H.J. Braun, and Z. Bedo (2001). Plant damage after freezing, and the frost resistance of varieties from the facultative and winter wheat observation nurseries. *Euphytica* 119:179-183.

Veisz, O. and J. Sutka (1993). Ditelosomic analysis of frost resistance in wheat (cv. Chinese spring). *Cereal Research Communications* 21:263-267.

Vierling, R.A. and H.T. Nguyen (1992). Heat-shock protein gene expression in diploid wheat genotypes differing in thermal tolerance. *Crop Science* 32:370-377.

Villareal, R.L., O. Banuelos, K. Sayre, M. Mujeeb-Kazi, and M. van Ginkel (1998). Screening of synthetic bread wheats *(Triticum turgidum* × *Aegilops tauschii)* for waterlogging tolerance. In *Agronomy Abstracts.* Annual Meeting of the American Agronomy Society, 90: October 18-22, Madison, Wisconsin: p.159.

Villareal, R.L., A. Mujeeb-Kazi, E. Del Toro, J. Crossa, and S. Rajaram (1994). Agronomic variability in selected *triticum turgidum* × *T. tauschii* synthetic hexaploid wheats. *Journal of Agronomy and Crop Science* 173:307-317.

Villareal, R.L., K. Sayre, O. Banuelos, and A. Mujeeb-Kazi (2001). Registration of four synthetic hexaploid wheat *(Triticum turgidum/Aegilops tauschii)* germplasm lines tolerant to waterlogging. *Crop Science* 41:274.

Viswanathan, C. and R. Khanna-Chopra (2001). Effect of heat stress on grain growth, starch synthesis and protein synthesis in grains of wheat (*Triticum aestivum* L.) varieties differing in grain weight stability. *Journal of Agronomy and Crop Science* 186:1-7.

Vongburi, K. (1994). Using phasic development differences in wheat to overcome heat stress in northern Thailand. In *Wheat in Heat-Stressed Environments: Irrigated, Dry Areas and Rice-Wheat Farming Systems,* D.A. Saunders and G.P. Hettel (Eds.). D.F., Mexico: CIMMYT, pp. 257-264.

Wardlaw, I.F. (1994). The effect of high temperature on kernel development in wheat: Variability related to pre-heading and post-anthesis conditions. *Australian Journal of Plant Physiology* 21:731-739.

Wardlaw, I.F. and L. Moncur (1995). The response of wheat to high temperature following anthesis. I. The rate and duration of kernel filling. *Australian Journal of Plant Physiology* 22:391-397.

Waters, I., P.J. Kuiper, E. Watkin, and H. Greenway (1991). Effects of anoxia on wheat seedlings. *Journal of Experimental Botany* 42:1427-1435.

Wheeler, T.R., G.R. Batts, R.H. Ellis, P. Hadley, and J.I.L. Morison (1996). Growth and yield of winter wheat (*Triticum aestivum* L.) crops in response to CO_2 and temperature. *Journal of Agricultural Science, Cambridge* 127:37-48.

Wheeler, T.R., T.D. Hong, R.H. Ellis, G.R. Batts, J.I.L. Morison, and P. Hadley (1996). The duration and rate of grain growth, and harvest index, of wheat (*Triticum aestivum* L.) in response to temperature and CO_2. *Journal of Experimental Botany* 47:623-630.

White, J.G. and R.J. Zasoski (1999). Mapping soil micronutrients. *Field Crops Research* 60:11-26.

White, P.J., H.D. Cooper, D.T. Clarkson, M.J. Earnshaw, and B.C. Loughman (1991). Effects of low temperature on growth and nutrient accumulation in rye (*Secale cereale*) and wheat (*Triticum aestivum*). *Annals of Botany* 68:23-31.

Wrigley, C.W., C. Blumenthal, P.W. Gras, and E.W.R. Barlow (1994). Temperature variation during grain filling and changes in wheat-grain quality. *Australian Journal of Plant Physiology* 21:875-885.

Xu, Q., A.Q. Paulsen, J.A. Guikema, and G.M. Paulsen (1995). Functional and ultrastructural injury to photosynthesis in wheat by high temperature during maturation. *Environmental and Experimental Botany* 35:43-54.

Yamaguchi-Shinozaki, K. and K. Shinozaki (1994). A novel *cis*-acting element in an *Arabidopsis* gene is involved in responsiveness to drought, low-temperature, or high-salt stress. *The Plant Cell* 6:251-264.

Yan, W. and D.H. Wallace (1996). A model pf photoperiod × temperature interaction effects on plant development. *Critical Reviews in Plant Sciences* 15:63-96.

Yau, S.K., M. Nachit, and J. Ryan (1997). Variation in growth, development and yield of durum wheat in response to high soil boron. 2. Differences between genotypes. *Australian Journal of Agricultural Research* 48:951-957.

Yucel, M., J.J. Burke, and H.T. Nguyen (1992). Inhibition and recovery of photosystem II following exposure of wheat to heat shock. *Environmental and Experimental Botany* 32:125-135.

Zabotin, A.I., T.S. Barisheva, O.A. Zabotina, I.A. Larskaya, V.V. Lozovaya, G. Beldman, and A.G.J. Voragen (1998). Alterations in cell walls of winter wheat roots during low temperature acclimation. *Journal of Plant Physiology* 152:473-479.

Zaharieva, M., E. Gaulin, M. Havaux, E. Acevedo, and P. Monneveux (2001). Drought and heat responses in the wild wheat relative *Aegilops geniculata* Roth: Potential interest for wheat improvement. *Crop Science* 41:1321-1329.

Zhonghu, H. and S. Rajaram (1994). Differential responses of bread wheat characters to high temperature. *Euphytica* 72:197-203.

Zhou, B.L., K. Arakawa, S. Fujikawa, and S. Yoshida (1994). Cold-induced alterations in plasma membrane proteins that are specifically related to the development of freezing tolerance in cold-hardy winter wheat. *Plant and Cell Physiology* 35:175-182.

Chapter 13

Breeding for Abiotic Stress Tolerance in Barley

Sakti Jana
Ron W. Wilen

INTRODUCTION

The cultivation of barley is nearly as old as the origin of agriculture, dating back over 10,000 BP (recent reviews by Fischbeck, 2002; Molina-Cano et al., 2002). Since its domestication in the Near East fertile crescent (Zohary and Hopf, 1994), and possibly in two other regions of the world (Zohary, 1999; Molina-Cano et al., 2002), this neolithic crop has spread over a large part of the world from Siberia to Chile. The crop now covers over 64 million ha of cultivated land (FAO, 2003). In the Near East center of diversity, barley is grown primarily as a winter cereal. Much of the 5 million ha of land under barley in this region is subject to one or more stresses. In these regions, barley is typically grown on marginal land, leaving more favorable cropland to wheat cultivation (Ceccarelli, 1984). Because of frequent drought and high temperatures in the growing season, genetic adaptation to adverse growing conditions has been an important determining factor for barley's success as a major field crop.

Adaptation to stress conditions may be considered synonymous to crop evolution (Blum, 1988). These conditions may be biotic (e.g., diseases and insect pests) or abiotic (e.g., physical, chemical, and nutritional stresses). Barley is capable of rapidly organizing adaptive multilocus genetic structures in response to its growing conditions (Allard, 1988). This capacity is conducive to its adaptation to a wide range of agroclimatic conditions around the world. Adaptation to abiotic stresses was a key factor in the evolution of barley and its spread to many varied, and often very harsh, environments.

GENETIC ARCHITECTURE OF BARLEY

Barley is a predominantly self-pollinated species with the natural outcrossing rate rarely exceeding 1 percent. Conscious hybridization within the species, and with its immediate evolutionary progenitor, wild barley (*Hordeum spontaneum* Koch), is simple. It is a diploid organism with seven large chromosomes. Over 200 genetic markers have been located on the barley chromosomes (review by Kleinhofs and Han, 2002), and over 1000 genes and 500 translocation stocks of barley are known. In addition, many natural and induced variants are available. Barley has well-developed linkage maps that include both morphological and molecular markers. Molecular markers are also a promising tool for identifying quantitative trait loci (QTL) in barley and may help to identify and map QTLs for response to abiotic stresses.

The genetic architecture of barley seems to be conducive to its genetic adaptation to extreme environments. Tolerance to environmental stresses such as drought and salinity is known to have multigenic inheritance, mediated by a large number of genes distributed throughout the seven chromosomes of cultivated barley. Epistasis, genotype-environment interaction, and large environmental effects complicate breeding efforts to increase the stress tolerance of barley. These difficulties can be partially alleviated by targeting stress-responsive characters with high heritability. The use of genetic maps involving molecular markers and QTLs offers promise to enhance the genetic potential of barley to abiotic stress resistance (see review by Cattevelli et al., 2002).

SELECTION CRITERIA FOR TOLERANCE TO ENVIRONMENTAL STRESSES

Blum (1983, 1988) described seven environmental stresses that limit the growth and productivity of agricultural crops: drought, heat, chilling, freezing, mineral deficiency, mineral toxicity, and salinity. Although barley encounters all of these stresses in its worldwide cultivation range, the most predominant stresses for barley are drought and heat. Compared with most other winter cereals, barley has relatively high drought and salinity tolerance, but it is known to be less cold tolerant than is its close evolutionary associate, wheat. A number of physiological characters have been used as selection criteria for plant response to water stress, including photosynthetic CO_2 resistance, internal CO_2 concentration, chlorophyll a and b content, chlorophyll a/b ratio, leaf water loss, leaf water content, specific leaf weight, leaf area, canopy temperature, plant dry weight, and water-use efficiency (WUE). Of these, WUE has been long regarded as the most useful indicator of drought response, and, C_{13}/C_{12} discrimination has been applied as an indicator of WUE with varying degrees of success. An efficient selection criterion should (1) have strong association with response to stress, (2) exhibit simple inheritance and high heritability, and (3) be easy to genetically manipulate.

The most direct criterion for assessing response to stresses is yield. Because the underlying genetic mechanisms for both grain and biomass yield are complex with significant environmental influence, the use of simply inherited morphological and physiological traits associated with yield is more attractive. Obviously, the success of indirect selection for stress tolerance using morphophysiological characters as markers depends on the strength of relationship of such markers with plant response to stresses under consideration. However, selection based on several markers is likely to be more effective than selection based on single markers. Shannon (1985) observed that the use of phenotypic markers was not successful in breeding for salt tolerance and suggested that absolute salt tolerance could be improved by increasing both absolute yield and relative salt tolerance.

Despite its complexity, grain yield is often used as an indicator of response of cereal crops to environmental stresses, although in some harsh environments total biomass yield may be a more appropriate indicator of stress tolerance than grain yield (Ceccarelli and Grando, 1996). Selection in low-input environments has been found to be an efficient strategy for improving grain yield and yield stability of barley in low-input arid environments (Ceccarelli et al., 1998). Ceccarelli and colleagues (1998) suggest the most effective way to achieve genetic advance for low-yielding and, presumably, stressful environments is to select promising genotypes in representative low-yielding conditions. However, the difficulty with yield is that phenotypic expression and, consequently, genetic variation for grain or biomass yield is suppressed in stressful environments. Complex inheritance and low heritability limit the effectiveness of selection for grain yield or total biomass yield under environmental stresses (Blum, 1983, 1988). For this reason, the use of associated characters that have simple inheritance and high heritability for response to stresses is an attractive alternative.

SELECTION TECHNIQUES FOR ENHANCING TOLERANCE TO ENVIRONMENTAL STRESSES

Of the major cereal crops, barley exhibits a relatively high degree of resistance to abiotic stresses. However, further improvements would be useful to expand the area that could be cultivated with this crop. Improved drought tolerance would be beneficial in the Middle East and Central Asia, improved salt tolerance would allow expansion of acreage on saline soils, and improved winterhardiness would permit the growth of winter varieties in the Northern Great Plains region of Canada and the United States. Within current genotypes of cultivated barley there is little chance of increasing the extremes of the stress tolerance that already exists. Improvements can be made in regional varieties by introduction of resistance from cultivars developed in other regions, but major increases will need the introduction of germplasm from outside the standard breeding material.

There are two basic approaches toward improving crop productivity under environmental stresses: modifying the environment to a more favorable condition for crop growth and development, and the genetic improvement of the crop to increase its ability to grow well and produce an economic yield under the given environmental constraints (Blum, 1983). Epstein (1977) argued that the latter is the more realistic approach toward maintaining and enhancing crop productivity in saline soils. However, this argument also holds for other types of environmental stresses such as drought, cold, and high temperature. The presence of adequate genetic variation is basic to the genetic improvement for stress tolerance in barley, as in other field crops. The basic steps in developing stress-tolerant barley cultivars are (1) acquisition of adequate germplasm, (2) identification of reliable screening mechanisms, (3) screening of germplasm for tolerance of the relevant stress, and (4) incorporation of genes responsible for stress tolerance into otherwise desirable genotypes by conventional breeding, or a suitable combination of conventional and more recent methods, such as double haploid production, mutation breeding, and molecular breeding.

Molecular marker-assisted selection has received considerable attention and popularity since the early 1990s. This indirect selection procedure is being used to increase tolerance to mineral toxicity (e.g., boron and aluminum) and deficiency (e.g., manganese) in barley (Jefferies et al., 2000). The progress of this work indicates that it is an efficient method of selection for response to abiotic stresses mediated by relatively few genes with low heritability. Molecular marker-assisted selection for stress tolerance is an attractive complementary approach to conventional breeding when the linked markers have simple inheritance and high heritability.

Selection for stress tolerance would be easier if the cultivars that yield well under nonstress (favorable) conditions also yield well under saline (stressful) conditions. This has been found to be true for barley grown in saline and nonsaline soils in Alberta, Canada (Bole and Wells, 1979), and California (Richards et al., 1987).

Another difficulty of genetic improvement of stress tolerance is that plant response to abiotic stresses is usually controlled by many genes and is significantly modified by the environment (review in Blum, 1988). Because physiological characters are often found to have relatively high heritability (Mahon, 1983; King, 1991), genetic improvement of crop plants for physiological traits associated with stress tolerance should be an attractive and attainable option in plant breeding.

Specific adaptation to a target environment (macroenvironment) has been considered to be the most rational approach to breeding for stress resistance, especially drought resistance. The rationale for subjecting a population to the specific stress is that genotypes that display the best performance under these conditions are the best adapted and thus will yield the highest under these conditions. However, the range of microenvironments that exist within stressful environments are far more numerous than in a nonstressed environment (Ceccarelli, 2000). Thus, while macroenvironment selection techniques may identify genotypes that will perform well in the presence of a uniform stress, these genotypes may not be widely adapted to the range of microenvironments encountered in local fields. There will be a significant adaptive cost to plants selected under uniform conditions, and maximum yield will not be as high as those of populations that have adapted to local growing conditions. To our knowledge, plant breeding programs to increase stress tolerance in barley have paid little attention to the utilization of populational buffering to address the problem of microenvironmental heterogeneity in stress environments.

Ideotype Breeding

As stress tolerance usually has a metabolic cost (Gale and Zeroni, 1985), characters imparting tolerance to a high level of stress, such as soil salinity, may not be conducive to high productivity (Tal, 1985). In addition, there may be different levels of synergy among the "stress-tolerance" characters in varying conditions. The overall metabolic cost of incorporating these attributes into a barley plant may be prohibitive. Ideotype breeding essentially aims at utilizing favorable multicharacter (or multilocus) associations. For successful utilization of ideotype breeding, the breeder must have not only a list desirable characters but also a thorough knowledge of synergistic interactions among multiple stress-related characters within the target range of stress.

Male-Sterility-Facilitated Recurrent Selection

Ramage (1980) proposed a scheme to develop broad-based, genetically heterogeneous populations to breed stress tolerance in plants. Originally proposed for increasing salt tolerance, R. T. Ramage and his associates in Tucson, Arizona, successfully applied this selection scheme, described as male-sterility-facilitated recurrent selection, to develop barley cultivars that performed well under low water-use regimes. The program involved selecting large numbers of both male-fertile and male-sterile plants from a population, segregating for male sterility and the desired traits. The selected plants were intercrossed using the male-sterile plants as females. The crossed seeds were harvested from the female plants, bulked, and grown to produce the F_1 generation. Bulk-harvested seeds from the F_1 (the F_2 generation) provided the population for the next cycle of selection. The selection was continued for several cycles under water-limiting conditions, and the resulting bulk population was released as a composite cross of barley. Drought-tolerant cultivars were selected from the genetically heterogeneous composite cross. This method was used to produce drought-tolerant barley cultivars for the semiarid southwestern United States.

Jana (1999) proposed a scheme for population improvement that can be easily adapted to small grain cereals, such as barley. The scheme involves the following key steps:

1. assessment of genetic variation,
2. systematic and reliable screening of germplasm collections for each stress,
3. identification of sources of resistance,
4. production of heterogeneous populations by carefully planned hybridization to produce genetically broad-based bulk populations, such as composite crosses, and
5. en masse propagation of genetically heterogeneous bulk populations under moderate to high levels of stress, periodic genetic survey, followed by identification and isolation of novel genotypes.

This scheme is a modified version of the male-sterility-facilitated recurrent selection of R. T. Ramage (1980). It emphasizes the development of broad-based germplasm pools by intercrossing salt-tolerant lines of diverse origin, followed by mass propagation and selection under saline conditions. These genetically heterogeneous germplasm pools, such as composite crosses, are propagated en masse under stress conditions and enriched by introduction of promising lines of wild progenitor species. The selection and introgression are continued in successive generations, and the resulting populations are used in breeding for salt tolerance. Although the scheme was originally proposed for improving salt tolerance, this dynamic population improvement scheme may be applied for developing genetically enriched breeding populations for other environmental stresses, such as drought and cold.

Participatory Plant Breeding

Since the early 1990s, participatory plant breeding has received considerable attention and popularity. The best-known example of participatory barley breeding comes from Syria, where barley breeders under the leadership of Professor S. Ceccarelli compared the relative frequencies of lines with high grain yield identified under four different selection schemes: (1) decentralized participatory, (2) decentralized nonparticipatory, (3) centralized participatory, and (4) centralized nonparticipatory (Ceccarelli, 2000; Ceccarelli et al., 2000). In the first scheme, farmers in different locations selected barley lines according to their individual preference of phenotypes (decentralized participatory). Measured in terms of high-yielding lines selected after one selection cycle, decentralized selection was superior to centralized selection, and selection by farmers was superior to selection by breeders in the decentralized scheme. The experiment was especially significant for plant breeding for stress resistance because, "in general, farmers' ability to identify the highest yielding lines is higher in stressful environments than in relatively favorable environments" (Ceccarelli et al., 2000, p. 103). The results are encouraging and consistent with the observations of Ceccarelli and Grando (1991) and Ceccarelli and colleagues (1998) that barley genotypes developed through repeated selection in low-yielding, marginal conditions are more likely to be more productive in low-yielding environments than those developed by selecting in high-yielding, favorable conditions.

It is important to note that decentralized participatory plant breeding is essentially a modified version of the basic processes involved in the development and maintenance of landraces that have been followed by peasant farmers in traditional agricultural systems in the Near East centers of diversity of cultivated barley for millennia. Peasant farmers in this region practiced phenotypic selection in successive generations, which permitted progressive genetic improvement of their field crops, with emphasis on adaptation to local conditions (specific adaptation). The process proved highly successful for producing locally adapted landraces, while maintaining a high level of genetic diversity, both within and between landrace populations. Like the ancient method of landrace development, selection, and propagation, decentralized participatory plant breeding holds considerable promise for barley improvement for harsh environments, as has been shown by Ceccarelli and colleagues (2000).

Genetically Heterogeneous Populations

Barley is grown in diverse agroclimates, many of which are characterized by wide spatial and temporal variation in the level of stress. Significant variation in stress levels in a single barley field is a common occurrence in stressed environments, especially in drought- and salt-affected fields. A successful barley cultivar, therefore, must possess genetic potential for adaptation to a wide range of microenvironmental differences commonly encountered in representative stressed crop fields. Some evidence suggests that heterozygosity, as well as genetic heterogeneity, impart populational buffering in predominantly self-pollinated species such as barley by synergistic intergenotype interactions. Landrace populations of barley with inherent genetic heterogeneity may be better buffered against microenvironmental heterogeneity than genetically homogeneous cultivars (Ceccarelli and Grando, 2000). The role of heterogeneous populations, for example, cultivar mixtures and composite crosses, in increasing the stability and productivity of barley in stress environments needs to be fully explored.

SOURCES OF VARIATION FOR BARLEY IMPROVEMENT UNDER STRESS CONDITIONS

The primary source of genes for improving stress resistance in barley is the gene pool of cultivated barley. This gene pool consists of modern and obsolete cultivars, composite crosses, and other broad-based, genetically heterogeneous populations (mass reservoirs), landrace populations, and wild relatives. Cultivated barley is widely believed to have been domesticated from the diploid wild barley, *H. spontaneum* C. Koch. Sometimes considered a subspecies of *H. vulgare* L., wild barley has been frequently cited and used in the genetic improvement of barley, including resistance to biotic and abiotic stresses. Wild barley grows in a wide range of habitats ranging from desert to high rainfall areas, in extreme heat to cold, and in many locations under saline conditions, and it contains genetic information that could be useful in increasing the stress tolerance of barley.

Several research groups, notably of S. Ceccarelli in Syria, R. P. Ellis in Scotland, and E. Nevo in Israel, have considered wild barley *(H. spontaneum)* to be a valuable genetic resource for barley improvement. *Hordeum agriocrithon,* known to be a secondary derivative of natural hybridization between six-rowed barley and *H. spontaneum,* grows well under harsh conditions on marginal lands and may also be a potentially rich genetic resource.

Barley landraces collected from very low rainfall areas in Syria proved to be the most promising sources of drought tolerance (Ceccarelli et al., 1998). Extrapolating from the assessment of genetic variation in wild and cultivated landrace populations of barley of the fertile crescent by Holwerda and colleagues (1986) and Jana and Pietrzak (1988), one would expect that the amount of genetic variation for tolerance to abiotic stresses in cultivated barley may be just as high as in wild barley.

Wild barley has been used to increase tolerance of barley to a number of abiotic stresses, such as drought, salinity, low temperature, and waterlogging (Ellis, 2002). In its centers of primary diversity, landrace populations of cultivated barley have evolved side by side with wild barley, with occasional introgression. These populations have been exposed to both conscious, the equivalent of present-day participatory plant breeding, and natural selection, which have allowed the organization of multilocus gene complexes. These coadapted gene complexes can be very useful for barley breeding for harsh environments.

BREEDING FOR STRESS TOLERANCE

It has been suggested that modern barley cultivars contain less than 40 percent of the alleles contained within the wild barley populations (Ellis et al., 2000). In the 10,000 years of cultivation, breeders have removed much of the genetic variability that existed in the wild species and greatly narrowed the genetic base of barley (Tanksley and McCouch, 1997). However, while the genetic variability in modern cultivars has been reduced and many potential sources of stress tolerance may exist in the wild genotypes, the wild genotypes also contain genetic information that can be detrimental to the development of yield in barley varieties. Methods are therefore required to distinguish the desired genetic information introduced in these crosses and eliminate as easily as possible the unwanted alleles.

Several studies have identified morphological and developmental traits, such as root growth, canopy closure, and flowering time, that correlate with stress tolerance (van Oosterom and Acevedo, 1992a,b; Ellis et al., 2000). These morphological markers can serve as an early selection criteria; however, a much faster and potentially more reliable method of selection is the use of molecular marker technology. Although the ultimate goal of marker-assisted selection programs is to identify single markers linked to the phenotypic, biochemical, or physiological trait of interest, given the multigenic nature of stress tolerance, the possibility of using a single marker in a breeding program is highly unlikely. However, barley genetic diversity has been studied using a variety of marker techniques, including RFLPs (restriction fragment length polymorphisms) (Saghai-Maroof et al., 1984), RAPDs (random amplified polymorphic DNA) (Noli et al., 1997; Volis et al., 2001), AFLPs (amplified fragment length polymorphisms) (Russell et al., 1997), and SSRs (simple sequence repeats) (Becker and Heun, 1995; Russell et al., 1997). This work has contributed to the rapid development of linkage maps for each of the seven barley chromosomes. This information can be applied to breeding for enhanced stress resistance.

The presence of these developing maps has led to the assignment of several QTLs for a variety of traits that could potentially impact on stress tolerance. Teulat and colleagues (1997) identified several stable QTLs correlating with water content and osmotic adjustment in barley that could be linked to drought tolerance. QTLs for saline germination (Mano and Takeda, 1997), winter survival (Hayes et al., 1993), heading and vernalization (Pan et al., 1994), and frost resistance and vernalization (Galiba et al., 1995) have been reported. However, there have been no reports utilizing these developed QTL in the development of a new stress-tolerant variety. Recently, an ambitious program was initiated to develop a functional genetic map for cold resistance using SSR markers. A cross between a cold-resistant winter barley and a frost-sensitive spring cultivar was used as the mapping population (Pecchioni et al., 2000).

Mutation Breeding

Mutation breeding has been applied to barley for a large number of traits, but only recently has improved stress tolerance been a target of this approach. Baha'a Eldin (2000) generated a population of mutants through gamma irradiation and, within this population, identified a line (A-4806) that showed superior grain yield in large-scale field trials in southern Syria when compared with the parent cultivar, Arabic Aswad. The line had reduced grain-protein content and, surprisingly, somewhat reduced freezing tolerance. It is expected that A-4806 will be released as a local variety in southern Syria in the very near future. Improved drought and salt tolerance were the primary objectives of a research program that utilized gamma-irradiated barley mutants derived from the cultivar Kaya as the starting germplasm (Toker et al., 1999). Mutants were screened for cuticular wax, photosynthetic capacity, and foliar salt adsorption. Eight lines were identified for further study in field trials in Turkey for improved salt and drought resistance.

Mutation breeding has not been reported to improve cold tolerance. However, recent developments in other species would indicate that this avenue of research might be worth pursuing. A constitutive freezing-tolerant mutant was identified in a mutant screen of *Arabidopsis* (Xin and Browse, 1998), and attempts are being made to identify homologous mutations in *Brassica* oilseed crops.

Haploid Breeding

Production of haploid barley has been used as a breeding tool for more than 30 years (Symko, 1969; Kasha and Reinbergs, 1972). Although the potential of using doubled haploid technology to improve abiotic stress resistance is very good, the application of this technology has not been extensive to date. A haploid population was produced from a cross between a winter and spring barley. The progeny were then screened for winter and spring growth habit and frost tolerance (Dunn et al., 1991). The populations were then used to determine the relative expression of putative cold-regulated genes in barley. A similar approach was taken by Crosatti and colleagues (1996), although to develop their doubled haploid (DH) population they utilized the F_1 of a cross between wild barley and a commercial cultivar to increase the variability. Within the collection of DH plants generated, a large degree of variation was noted in the resistance to cold. Doubled haploids were also used to screen for drought-resistant plants in a program developed to increase barley production in Syria (Mayer et al., 1995).

Genetic Engineering

Barley was the last major cereal to be successfully transformed, and stable transformation has occurred only within the past decade. The first reports of stable, fertile transgenic barley plants (Ritala et al., 1994; Wan and Lemaux, 1994) used particle bombardment. Transformed barley has now been produced through electroporation (Salmenkallio-Marttila et al., 1995) and *Agrobacterium*-mediated infection (Tingay et al., 1997). Most of the research has been focused on basic studies, such as improved transformation frequencies and promoter activity analysis. The few applied projects have been limited to the introduction of thermostable β-glucanases for improved malting stability (Nuutila et al., 1999), seed quality traits (Zhang et al., 2000), and disease resistance (Wang et al., 2001).

Although no transformation work has been reported for enhanced abiotic stress tolerance, significant work toward this objective has been completed. A large number of genetic studies have identified genes associated with increased drought, cold, or salt in barley. The number of genes that have been identified to be associated with some form of abiotic stress in barley numbers greater than 50 (for a recent listing of stress-induced genes, see

Cattivelli et al., 2002). Many of these genes have been mapped to specific locations within the barley genome, and this information, in combination with molecular marker technologies, can be used to improve the stress tolerance of new varieties.

Potential for Improving Salt Tolerance

Barley is one of the most salt tolerant of the small grain crops; however, yield could be improved and cultivation expanded if an increase in salt tolerance could be realized. The primary mechanisms that exist for enhanced tolerance are reduced uptake or sequestration into vacuoles. Barley plants are able to tolerate moderate saline conditions, owing to the ability to elevate their cell osmolarity through the accumulation of compounds such as glycinebetaine (GB). Nakamura and colleagues (1996) showed that GB differentially accumulates in young versus old leaves in response to salt stress, and as a result Na^+ ions preferentially accumulate in the older leaves, thus permitting continued growth of the plant. The peroxisomal betaine aldehyde dehydrogenase *(BADH)* gene has been isolated from barley, and when introduced into rice increased salt and drought tolerance (Kishitani et al., 2000). Whether overexpression or altered expression of the *BADH* gene in barley will further enhance salt tolerance remains to be determined.

The second major mechanism of salt tolerance is sequestration of ions to the vacuole. The Na^+/H^+ antiporter is involved in the removal of Na^+ from the cytoplasm. A comparative study between rice and barley showed the antiporter gene in barley was induced in the presence of salt and vacuolar salt concentrations increased (Fukuda et al., 1998). The expression of a tonoplast Na^+/H^+ antiporter gene in either *Arabidopsis* or *Brassica napus* increased the salt tolerance of the transgenic plants (Apse et al., 1999; Zhang et al., 2001). The fact that the overexpression of the *Arabidopsis* Na^+/H^+ antiporter gene in *Arabidopsis* led to increased tolerance would indicate that there is room to further increase the salt tolerance of barley through this mechanism.

Potential for Increased Drought Tolerance

A common link between salt and drought stress is the increased accumulation of cell osmolytes in response to the stress. The *BADH* gene could also be used to potentially increase drought tolerance as well as salt tolerance. The *HVA1* gene from barley was used to transform rice plants. Expression of this gene, which encodes a late embryogenesis-abundant (LEA)-type protein, resulted in enhanced drought tolerance in the transgenic rice lines created (Xu et al., 1996). LEA proteins are thought to help stabilize cell macromolecules during desiccation and are particularly abundant in mature seeds.

Perhaps the most studied stress-associated proteins in barley are the dehydrins. At least 13 members in the barley dehydrin gene family are distributed over four chromosomes (Choi et al., 1999). Within this gene family, the expression of individual members has been correlated with cold, drought, or salt tolerance, or in some cases with all three stresses. Dehydrin proteins have been identified in numerous other species and their accumulation also correlated with increased abiotic stress tolerance (Close et al., 1989; Robertson et al., 1994; Arora and Wisniewski, 1996; Hara et al., 2001). Although dehydrins are well documented for their association with enhanced stress tolerance, surprisingly no papers of which we are aware show the use of a dehydrin gene to increase abiotic stress resistance in a transgenic system. Marginal increases in freezing and drought tolerance were observed in transgenic canola and flax plants overexpressing the *dhn4* gene from barley (L. V. Gusta et al., unpublished).

Potential for Increased Cold Tolerance

Spring barley cultivars have been bred to behave as short-season crops and there is generally little problem with spring or early fall frosts, which are damaging to many other crops in most regions where barley is currently cultivated. However, an expansion of the area suitable for barley cultivation is possible with increased cold tolerance. The genetic potential to develop cultivars that can withstand low temperatures is probably already present, but understanding how to exploit that potential is key to success. One area of interest is to understand how genes associated with stress are regulated. A number of studies have examined the temperature requirements necessary for plants to respond to temperature stress. Crosatti and colleagues (1996) used a population derived from a wild barley/barley cross and found a correlation between the temperature at which the cold-regulated genes were induced and the overall cold tolerance of the line. Some of our work with wheat and canola has also shown this correlation (unpublished data).

In contrast to our ability to successfully grow barley as a short-season spring crop in most cool regions, a major increase in cold tolerance of winter barley cultivars will be necessary to expand the acreage devoted to this crop. Improving winter hardiness will require increased resistance to winter desiccation at the extreme low temperatures common to winters on the

Northern Great Plains. Fu and colleagues (2000) demonstrated that dehydrin gene expression differed between winter and spring barley with dehydrin transcripts persisting into early winter in the winter types. However, they also showed that of the three winter cereals analyzed, rye, wheat, and barley, the level of dehydrin transcripts was lowest in the barley plants. Barley is also the least hardy of the three cereals compared. Although overexpression of the barley *dhn4* gene in canola and flax gave only marginal increases in stress tolerance, these were both spring crops, and increasing the accumulation of this protein in winter barley may increase the level of winter hardiness. The work of Crosatti and colleagues (1996) that demonstrated the ability of barley plants to initiate cold-regulated gene expression at higher temperature was related to cold tolerance and may be key to increasing winter survival. Wilen and colleagues (1998) also showed that the more hardy the cereal, the sooner the genes associated with cold tolerance were induced in the plant.

A second approach to increasing winter hardiness is through the alleviation of oxidative stress. Drought and cold both result in the overproduction of free radicals in the chloroplast, which in turn causes damage to cell membranes, proteins, and DNA. One of the major mechanisms through which the cell counteracts this problem is the removal of free radicals through the activity of a number of enzymes, including superoxide dismutase (SOD). The expression of either an Fe SOD or a Mn SOD resulted in improved winter survival of transgenic alfalfa (McKersie et al., 1999, 2000). When the expression of two wheat SOD genes was examined, a correlation with both cold acclimation and relative hardiness of winter cereals was noted by Wu and colleagues (1999). Introduction of these genes into canola, potato, and flax increased cold, drought, and heat tolerance of the transgenic plants (Unpublished research by L. V. Gusta and associates). Thus, the expression of SOD genes appears to serve as a general mechanism of abiotic stress tolerance and may be applicable to increasing the winterhardiness of barley.

The last approach that needs to be considered is the introduction of genes encoding transcription factors. A family of genes encoding transcription factors called cold-binding factor (CBF) have been identified in *Arabidopsis*. The presence of these proteins in the cell regulates the expression of numerous genes responsive to cold temperature and ultimately freezing tolerance. Gilmour and colleagues (2000) overexpressed *CFB3* in *Arabidopsis* and the transgenic plants that were shown to be constitutively freezing tolerant. Similar genes will ultimately be isolated in cereal crops and could be used in a similar fashion.

Potential for Improving Heat Tolerance

Heat is most detrimental during flowering. Most current cultivars to date utilize an avoidance strategy. By seeding early in cool-temperature regions or by growing winter types in other areas, the crop has generally completed flowering prior to the onset of elevated temperatures. However, unseasonably early periods of high temperature will cause yield losses. With global warming predicted to increase the likelihood of earlier and higher summer temperatures, an increase in heat tolerance would be desirable. Most work has focused on heat-shock proteins (HSPs), and several HSPs have been isolated in barley (Marmiroli et al., 1993; Walther-Larsen et al., 1993). The promoter from a low-molecular-weight barley HSP gene *(HvHsp18)* has been isolated and found to convey heat-inducible expression of reporter genes in transgenic tobacco (Raho et al., 1996). This information will allow researchers the opportunity to express genes that will enhance heat tolerance under only high-temperature conditions. As to candidate genes, for improving heat tolerance, the only HSPs that have been shown to be required for thermotolerance are members of the HSP100 family. HSP101 is absolutely required for thermotolerance in *Arabidopsis* (Gurley, 2000; Hong and Vierling, 2000). Whether barley has a HSP101-like protein is unknown, but the overexpression of this gene may result in increased heat resistance. Other candidate genes for improving heat tolerance include genes associated with free-radical damage repair.

FUTURE TRENDS

Advancements in improving barley stress tolerance are likely to be incremental. The methods for achieving this goal will probably utilize a combination of traditional and novel techniques. The development of genetic maps in rice and other cereal crops will assist breeders in determining the physical location of major QTLs to be identified in any crosses for stress tolerance. Additional application of marker technology will decrease the time and monetary investment in breeding for new traits. The increased use of haploid breeding techniques has a large potential for barley improvement, particularly if used in combination with intra- and interspecific crosses with wild and landrace barleys. Some recent developments in using mutagenesis as a method of increasing genetic diversity indicate that this technique has some potential to increase stress tolerance. Finally, genetic engineering could be a means to achieve increased stress tolerance. Numerous examples exist in other species where the introduction of novel genes, or the altered expression of endogenous genes, results in improved stress resistance.

REFERENCES

Allard, R.W. (1988). Genetic changes associated with the evolution of adaptedness in cultivated plants and their wild progenitors. *Journal of Heredity* 79:225-238.

Apse, M.P., G.S. Aharon, W.A. Snedden, and E. Blumwald (1999). Salt tolerance conferred by overexpression of a vacuolar Na^+/H^+ antiport in *Arabidopsis*. *Science* 285:1256-1258.

Arora, R. and M. Wisniewski (1996). Accumulation of a 60-kD dehydrin protein in peach xylem tissues and its relationship to cold acclimation. *HortScience* 31: 923-925.

Baha'a Eldin, J. (2000). A-4806 barley mutant for Syrian dry land. In *Barley Genetics VIII, Proceedings of the Eighth International Barley Genetics Symposium,* Volume III, S. Logue (Ed.). Glen Osmond, South Australia: Department of Plant Science of Adelaide University, pp. 139-141.

Becker, J. and M. Heun (1995). Barley microsatellites: Allele variation and mapping. *Plant Molecular Biology* 27:835-845.

Blum, A. (1983). Breeding crop varieties for stress environments. *CRC Critical Reviews in Plant Sciences* 12:199-238.

Blum, A. (1988). *Plant Breeding for Stress Environments.* Boca Raton, FL: CRC Press.

Bole, J.B. and S.A. Wells (1979). Dryland soil salinity: Effect on the yield and yield components of 6-row barley, 2-row barley, wheat and oats. *Canadian Journal of Soil Science* 59:11-17.

Cattivelli, L., P. Baldi, C. Crosatti, M. Grissi, G. Vale, and A.M. Stanca (2002). Genetic basis of barley physiological response to stressful conditions. In *Barley Science: Recent Advances from Molecular Biology to Agronomy of Yield and Quality,* G.A. Slafer, J.L. Molina-Cano, R. Savin, and I. Ramagosa (Eds.). Binghamton, NY: Food Products Press, pp. 307-360.

Ceccarelli, S. (1984). Utilization of landraces and *H. spontaneum* in barley breeding for dry areas. *Rachis* 3:8-11.

Ceccarelli, S. (2000). Decentralized participatory plant breeding: Adapting crops to environments and clients. In *Barley Genetics VIII, Proceedings of the Eighth International Barley Genetics Symposium,* S. Logue (Ed.). Glen Osmond, South Australia: Department of Plant Science of Adelaide University, pp. 159-166.

Ceccarelli, S. and S. Grando (1991). Selection, environment and environmental sensitivity in barley. *Euphytica* 57:157-167.

Ceccarelli, S. and S. Grando (1996). Drought as a challenge for the plant breeder. *Plant Growth Regulation* 20:149-155.

Ceccarelli, S. and S. Grando (2000). Barley landraces from the Fertile Crescent: a lesson for plant breeders. In *Genes in the Field,* S. Brush (Ed.). Boca Raton, FL: Lewis Publishers, pp. 51-76.

Ceccarelli, S., S. Grando, and A. Impiglia (1998). Choice of selection strategy in breeding barley for stress environments. *Euphytica* 103:307-318.

Ceccarelli, S, S. Grando, R. Tutwiler, J. Baha, A.M. Martini, H. Salahieh, A. Goodchild, and M. Michael (2000). A methodological study of participatory barley breeding. I. Selection phase. *Euphytica* 111:91-104.
Choi, D.W., B. Zhu, and T.J. Close (1999). The barley (*Hordeum vulgare* L.) dehydrin multigene family: Sequences, allele types, chromosome assignments, and expression characteristics of 11 Dhn genes of cv Dicktoo. *Theoretical and Applied Genetics* 98:1234-1247.
Close, T.J., A.A. Kortt, and P.M. Chandler (1989). A cDNA-based comparison of dehydration-induced proteins (dehydrins) in barley and corn. *Plant Molecular Biology* 13:95-108.
Crosatti, C., E. Nevo, A.M. Stanca, and L. Cattivelli (1996). Genetic analysis of the accumulation of COR14 proteins in wild *(Hordeum spontaneum)* and cultivated *(Hordeum vulgare)* barley. *Theoretical and Applied Genetics* 93:975-981.
Dunn, M.A., M.A. Hughes, L. Zhan, R.S. Pearce, A.S. Quigley, and P.L. Jack (1991). Nucleotide sequence and molecular analysis of the low temperature induced cereal gene, BLT4. *Molecular General Genetics* 229:389-394.
Ellis, R.P. (2002). Wild barley as a source of genes for crop improvement. In *Barley Science: Recent Advances from Molecular Biology to Agronomy of Yield and Quality,* G.A. Slafer, J.L. Molina-Cano, R. Savin, and I. Ramagosa (Eds.). Binghamton, NY: Food Products Press, pp. 65-83.
Ellis, R.P., B.P. Forster, D. Robinson, L.L. Handley, D.C. Gordon, J.R. Russell, and W. Powell (2000). Wild barley: A source of genes for crop improvement in the twenty-first century? *Journal of Experimental Botany* 51:9-18.
Epstein, E. (1977). Genetic potentials for solving problems of mineral stress: Adaptation of crops to salinity. In *Plant Adaptation to Mineral Stress in Problem Soils,* M.J. Wright (Ed.) Ithaca, NY: Cornell University, pp. 73-84.
FAO (2003). Annual crop statistics, FAO Web site: <http://apps.fao.org/page/collections>.
Fischbeck, G. (2002). Contribution of barley to agriculture: A brief overview. In *Barley Science: Recent Advances from Molecular Biology to Agronomy of Yield and Quality,* G.A. Slafer, J.L. Molina-Cano, R. Savin, and I. Ramagosa (Eds.). Binghamton, NY: Food Products Press, pp. 1-14.
Fu, P., R.W. Wilen, G.H. Wu, A.J. Robertson, and L.V. Gusta (2000). Dehydrin gene expression and leaf water potential differs between spring and winter cereals during cold acclimation. *Journal of Plant Physiology* 156:394-400.
Fukuda, A., Y. Yazaki, T. Ishikawa, S. Koike, and Y. Tanaka (1998). Na^+/H^+ antiporter in tonoplast vesicles from rice roots. *Plant and Cell Physiology* 39:196-201.
Gale, J. and M. Zeroni (1985). The cost to plants of different strategies of adaptation to stress and the alleviation of stress by increasing assimilation. *Plant and Soil* 89:57-67.
Galiba, G., S.A. Quarrie, J. Sutka, A. Morgounov, and J.W. Snape (1995). RFLP mapping of the vernalization (Vrn1) and frost resistance (Fr1) genes on chromosome 5A of wheat. *Theoretical and Applied Genetics* 90:1174-1179.
Gilmour, S.J., A.M. Sebolt, M.P. Salazar, J.D. Everard, and M.F. Thomashow (2000). Overexpression of the *Arabidopsis* CBF3 transcriptional activator mim-

ics multiple biochemical changes associated with cold acclimation. *Plant Physiology* 124:1854-1865.

Gurley, W.B. (2000). HSP101: A key component for the acquisition of thermotolerance in plants. *Plant Cell* 12:457-460.

Hara, M., S. Terashima, and T. Kuboi (2001). Characterization and cryoprotective activity of cold-responsive dehydrin from *Citrus unshiu*. *Journal of Plant Physiology* 158:1333-1339.

Hayes, P.M., T. Blake, T.H.H. Chen, S. Tragoonrung, F. Chen, A. Pan, and B. Liu (1993). Quantitative trait loci on barley (*Hordeum vulgare* L.) chromosome 7 associated with components of winterhardiness. *Genome* 36:66-71.

Holwerda, B.C., S. Jana, and W.C. Crosby (1986). Chloroplast and mitochondrial DNA variation in *Hordeum vulgare* and *Hordeum spontaneum*. *Genetics* 114:1271-1291.

Hong, S.W. and E. Vierling (2000). Mutants of *Arabidopsis thaliana* defective in the acquisition of tolerance to high temperature stress. *Proceedings of the National Academy of Science USA* 97:4392-4397.

Jana, S. (1999). Some recent issues on the conservation of crop genetic resources in developing countries. *Genome* 42:562-569.

Jana, S. and L.N. Pietrzak (1988). Comparative assessment of genetic variability in wild and cultivated barley in a center of diversity. *Genetics* 119:981-990.

Jefferies, S.P., H. Raman, M.A. Pallotta, and A.R. Barr (2000). Genetic approach to ameliorating mineral stresses in barley. In *Barley Genetics VIII, Proceedings of the Eighth International Barley Genetics Symposium,*, S. Logue (Ed.). Glen Osmond, South Australia, Department of Plant Science of Adelaide University, pp. 253-259.

Kasha, K.J. and E. Reinbergs (1972). The haploid technique in barley breeding. *Master Brewers Association of America, Technical Quarterly* 9:128-130.

King, J. (1991). *The Genetic Basis of Plant Physiological Processes*. New York: Oxford University Press.

Kishitani, S., T. Takanami, M. Suzuki, M. Oikawa, S. Yokoi, M. Ishitani, A.M. lvarez-Nakase, T. Takabe, and T. Takabe (2000). Compatibility of glycinebetaine in rice plants: Evaluation using transgenic rice plants with a gene for peroxisomal betaine aldehyde dehydrogenase from barley. *Plant, Cell and Environment* 23:107-114.

Kleinhofs, A. and F. Han (2002). Molecular mapping of the barley genome. In *Barley Science: Recent Advances from Molecular Biology to Agronomy of Yield and Quality*, G.A. Slafer, J.L. Molina-Cano, R. Savin, and I. Ramagosa (Eds.). Binghamton, NY: Food Products Press, pp. 31-53.

Mahon, J.D. (1983). Limitation to the use of physiological variability in plant breeding. *Canadian Journal of Plant Science* 63:11-21.

Mano, Y. and Y. Takeda (1997). Mapping quantitative trait loci for salt tolerance at germination and the seedling stage in barley (*Hordeum vulgare* L.). *Euphytica* 94:263-272.

Marmiroli, N., A. Pavesi, G. Di-Cola, H. Hartings, G. Raho, M.R. Conte, and C. Perrotta (1993). Identification, characterization, and analysis of cDNA and genomic sequences encoding two different small heat shock proteins in *Hordeum vulgare*. *Genome* 36:1111-1118.

Mayer, M., A. Gland, S. Ceccarelli, and H.H. Geiger (1995). Comparison of doubled haploid lines and F_2 bulks for the improvement of barley in the dry areas of north Syria. *Plant Breeding* 114:45-49.

McKersie, B.D., S.R. Bowley, and K.S. Jones (1999). Winter survival of transgenic alfalfa overexpressing superoxide dismutase. *Plant Physiology* 119:839-847.

McKersie, B.D., J. Murnaghan, K.S. Jones, and S.R. Bowley (2000). Iron-superoxide dismutase expression in transgenic alfalfa increases winter survival without a detectable increase in photosynthesis oxidative stress tolerance. *Plant Physiology* 122:1427-1437.

Molina-Cano, J.L., E. Igartua, A.M. Casas, and M. Moralejo (2002). New views on the origin of cultivated barley. In *Barley Science: Recent Advances from Molecular Biology to Agronomy of Yield and Quality,* G.A. Slafer, J.L. Molina-Cano, R. Savin, and I. Ramagosa (Eds.). Binghamton, NY: Food Products Press, pp. 15-21.

Nakamura, T., M. Ishitani, P. Harinasut, A.M. Nomura, T. Takabe, and T. Takabe (1996). Distribution of glycinebetaine in old and young blades of salt stressed barley plants. *Plant and Cell Physiology* 37:873-877.

Noli, E., S. Salvi, and R. Tuberosa (1997). Comparative analysis of genetic relationships in barley based on RFLP and RAPD markers. *Genome* 40:607-616.

Nuutila, A.M., A. Ritala, R.W. Skadsen, L. Mannonen, and V. Kauppinen (1999). Expression of fungal thermotolerant endo-1,4-beta-glucanase in transgenic barley seeds during germination. *Plant Molecular Biology* 41:777-783.

Pan, A., P.M. Hayes, F. Chen, T.H.H. Chen, T. Blake, S. Wright, I. Karsi, and Z. Bedo (1994). Genetic analysis of the components of winterhardiness in barley (*Hordeum vulgare* L.). *Theoretical and Applied Genetics* 89:900-910.

Pecchioni, N., E. Francia, P. Faccioli, L. Cattivelli, and A.M. Stanca (2000). A new SSRs-COR genes function map derived from a "winter x malt" barley doubled haploid population. In *Barley Genetics VIII, Proceedings of the Eighth International Barley Genetics Symposium,* Volume III, S. Logue (Ed.). Glen Osmond, South Australia: Department of Plant Science of Adelaide University, p. 90.

Raho, G., E. Lupotto, H. Hartings, P.A. Della-Torre, C. Perrotta, and N. Marmiroli (1996). Tissue-specific expression and environmental regulation of the barley Hvhsp17 gene promoter in transgenic tobacco plants. *Journal of Experimental Botany* 47:1587-1594.

Ramage, R.T. (1980). Genetic methods to breed salt tolerance in plants. In *Genetic Engineering for Osmoregulation,* D.W. Rains, R.C. Valentine, and A. Hollaender (Eds.). New York: Plenum, pp. 311-318.

Richards, R.A., C.W. Dennett, C.O. Qualset, E. Epstein, J.D. Norlyn, and M.D. Winslow (1987). Variation in yield of grain and biomass in wheat, barley and Triticale in a salt-affected field. *Field Crops Research* 15:277-287.

Ritala, A., K. Aspegren, U. Kurten, M. Salmenkallio-Marttila, L. Mannonen, R. Hannus, V. Kauppinene, T. Teeri, and T.M. Enari (1994). Fertile transgenic barley by particle bombardment of immature embryos. *Plant Molecular Biology* 24:317-325.

Robertson, A.J., A. Weninger, R.W. Wilen, P. Fu, and L.V. Gusta (1994). Comparison of dehydrin gene expression and freezing tolerance in *Bromus inermis* and *Secale cereale* grown in controlled environments, hydroponics, and the field. *Plant Physiology* 106:1213-1216.

Russell, J.R., J.D. Fuller, M. Macaulay, B.G. Hatz, A. Jahoor, W. Powell, and R. Waugh (1997). Direct comparison of levels of genetic variation among barley accessions detected by RFLPs, AFLPs, SSRs and RAPDs. *Theoretical and Applied Genetics* 95:714-722.

Saghai-Maroof, M.A., K.M. Soliman, R.A. Jorgensen, and R.W. Allard (1984). Ribosomal DNA spacer-length polymorphisms in barley: Mendelian inheritance, chromosomal location and population dynamics. *Proceedings of the National Academy Science USA* 81:8014-8018.

Salmenkallio-Marttila, M., K. Aspegren, S. Akerman, U. Kurten, L. Mannonen, A. Ritala, T.H. Teeri, and V. Kauppinen (1995). Transgenic barley (*Hordeum vulgare* L.) by electroporation of protoplasts. *Plant Cell Reports* 15:301-304.

Shannon, M.C. (1985). Principles and strategies in breeding for high salt tolerance. *Plant and Soil* 89:227-241.

Symko, S. (1969). Haploid barley from crosses of *Hordeum bulbosum* (2*x*) × *Hordeum* vulgare (2*x*). *Canadian Journal of Cytology and Genetics* 11:602-608.

Tal, M. (1985). Genetics of salt tolerance in higher plants: Theoretical and practical considerations. *Plant and Soil* 89:199-226.

Tanksley, S.D. and S.R. McCouch (1997). Seed banks and molecular maps: Unlocking genetic potential from the wild. *Science* 227:1063-1066.

Teulat, B., P. Monneveux, J. Wery, C. Borries, I. Souyris, A. Charrier, and D. This (1997). Relationships between relative water content and growth parameters under water stress in barley: A QTL study. *New Phytologist* 137:99-107.

Tingay, S., D. McElroy, R. Kalla, S. Fieg, M.B. Wang, S. Thornton, and R. Brettell (1997). *Agrobacterium tumefaciens*-mediated barley transformation. *Plant Journal* 11:1369-1376.

Toker, C., J. Gorham, and M.I. Cagirgan (1999). Assessment of response to drought and salinity stresses of barley (*Hordeum vulgare* L.) mutants. *Cereal Research Communications* 27:411-418.

Van Oosterom, E.J. and E. Acevedo (1992a). Adaptation of barley (*Hordeum vulgare* L.) to harsh Mediterranean environments. I. Morphological traits. *Euphytica* 62:1-14.

Van Oosterom, E.J. and E. Acevedo (1992b). Adaptation of barley (*Hordeum vulgare* L.) to harsh Mediterranean environments. II. Apical dominance, leaf and appearance. *Euphytica* 62:15-27.

Volis, S., B. Yakubov, I. Shulgina, D. Ward, V. Zur, and S. Mendlinger (2001). Tests for adaptive RAPD variation in population genetic structure of wild barley, *Hordeum spontaneum* Koch. *Biological Journal of Linneaus Society* 74:289-303.

Walther-Larsen, H., J. Brandt, D.B. Collinge, and H. Thordal-Christensen (1993). A pathogen-induced gene of barley encodes a HSP90 homologue showing striking similarity to vertebrate forms resident in the endoplasmic reticulum. *Plant Molecular Biology* 21:1097-1108.

Wan, Y.C. and P.G. Lemaux (1994). Generation of large numbers of independently transformed fertile barley plants. *Plant Physiology* 104:37-48.

Wang, M.B., D.C. Abbott, N.M. Upadhyaya, J.V. Jacobsen, and P.M. Waterhouse (2001). *Agrobacterium tumefaciens*-mediated transformation of an elite Australian barley cultivar with virus resistance and reporter genes. *Australian Journal of Plant Physiology* 28:149-156.

Wilen, R.W., P. Fu, A.J. Robertson, and L.V. Gusta (1998). Comparison of vernalized and non vernalized winter cereals subjected to cold hardening conditions. In *Plant Cold Hardiness,* P.H. Li and T.H. Chen (Eds.). New York: Plenum, pp. 191-201.

Wu, G.H., R.W. Wilen, A.J. Robertson, and L.V. Gusta (1999). Isolation, chromosomal localization and differential expression of mitochondrial Mn superoxide dismutase and chloroplastic Cu/Zn superoxide dismutase genes in wheat *(Triticum aestivum). Plant Physiology* 120:513-520.

Xin, Z. and J. Browse (1998). Eskimo-1 mutants of *Arabidopsis* are constitutively freezing-tolerant. *Proceedings of the National Academy of Science USA* 95:7799-7804.

Xu, D., X. Duan, B. Wang, B. Hong, T.H.D. Ho, and R. Wu (1996). Expression of a late embryogenesis abundant protein gene, HVA1, from barley confers tolerance to water deficit and salt stress in transgenic rice. *Plant Physiology* 110:249-257.

Zhang, H.X., J.N. Hodson, J.P. Williams, and E. Blumwald (2001). Engineering salt-tolerant *Brassica* plants: Characterization of yield and seed oil quality in transgenic plants with increased vacuolar sodium accumulation. *Proceedings of the National Academy of Science USA* 98:12832-12836.

Zhang, Y., H. Darlington, J. Napier, H.D. Jones, M.R. Davey, N.G. Halford, P.A. Lazzeri, and P.R. Shewry (2000). Expression of γ-zein, a maize storage protein in barley seed. In *Barley Genetics VIII, Proceedings of the Eighth International Barley Genetics Symposium,* Volume III, S. Logue (Ed.). Glen Osmond, South Australia: Department of Plant Science of Adelaide University, pp. 31-33.

Zohary, D. (1999). Monophyletic vs. polyphyletic origin of the crops on which agriculture was founded in the Near East. *Genetic Resources and Crop Evolution* 46:133-142.

Zohary, D. and M. Hopf (1994). *Domestication of Plants in the Old World,* Second Edition. Oxford, UK: Oxford Science Publications, Clarendon Press.

Chapter 14

Breeding for Abiotic Stress Tolerance in Rice

Glen B. Gregorio
Gloria S. Cabuslay

INTRODUCTION

The Green Revolution of the 1970s resulted in remarkable increases in rice production. However, since then, the rate of production in most rice-growing countries has slowed down (Hossain, 1994) and has now reached a plateau. Contributing factors include a higher population growth rate and the conversion of some highly productive rice lands for industrial and residential purposes. These developments pushed rice cultivation to less productive lands, including saline- and drought-prone areas.

Millions of hectares in the humid regions of South and Southeast Asia are technically suited for rice production but are left uncultivated or give very poor yields because of salinity and abiotic stresses. In rice, only 1.5 $t{\cdot}ha^{-1}$ yield is obtained in rainfed-lowland ecosystems (Wade et al., 1995), where environmental stresses range from a lack of water to excessive water supply. Associated soil constraints in these areas are salinity, alkalinity, iron toxicity, phosphorus deficiency, and organic and acid sulfate conditions (Mackill et al., 1996). On the other hand, a range of 3 to 9 $t{\cdot}ha^{-1}$ is achievable in irrigated areas (IRRI, 1997). To narrow the yield gap between areas susceptible to abiotic stresses and irrigated fields, multiple abiotic stress tolerance traits should be considered aside from high-yielding ability and disease resistance.

For inland salinity brought about by poor-quality groundwater or irrigation, associated stresses are alkalinity, phosphorus and zinc deficiency, and boron toxicity. In this type of soil, rice could not be grown without a supply of good-quality irrigation water. For coastal salinity, on the other hand, tidal intrusion is the source of salinity in the soil. In addition, crops grow under rainfed conditions and planting is timed with the occurrence of heavy rains, which flush out some of the salts from the soil. This type of soil is usually associated with (1) acidity, where deficiency in phosphorus and zinc and toxicity in iron and aluminum are prevalent; (2) acid sulfate conditions, where deficiency in phosphorus and zinc and toxicity in aluminum and sulfides are also present; and (3) peatiness, where phosphorus and zinc deficiency are always present, with accompanying iron and organic acid toxicities.

Progress in breeding for salinity tolerance and its related abiotic stresses has been slow owing to (1) limited knowledge concerning the genetics of tolerance, (2) the involvement of several complex tolerance mechanisms, (3) inadequate screening techniques, and (4) low selection efficiency. High genotype × environment interaction adds to the difficulty of identifying consistently superior genotypes (Cabuslay et al., 1999). Recent progress in germplasm improvement, screening techniques, genetic inheritance, and application of molecular marker technology for tolerance of salinity and related abiotic stresses at IRRI will accelerate breeding progress.

GENETIC VARIABILITY AND SOURCES OF TOLERANCE

Salinity Tolerance

Some degree of varietal tolerance for salinity, drought, submergence, Al and Fe toxicity, P and Zn deficiency, and elongating ability is available in rice germplasm. Since the start of the Genetic Evaluation and Utilization (GEU) program at IRRI in 1969, 138,000 rice genotypes have been screened. Of these, 17 percent had acceptable tolerance at electrical conductivity (EC) of 10 $dS{\cdot}m^{-1}$ at the seedling stage. These entries come from germplasm collections and elite breeding lines from various breeding programs.

Salt-tolerant indica rice varieties seem to have originated or been selected in coastal areas of India (states of Kerala and West Bengal) and Bangladesh. An evaluation of 250 traditional varieties collected from Orissa and Tamil Nadu in India revealed that there are also some salt-tolerant types in the southeastern coastal areas of India. Among the tolerant varieties identified, the cultivars NV, Pat, Solla, and DH were found to have tolerance similar to that of well-known 'Pokkali.' Similar tests done with coastal rice varieties of Indonesia, Thailand, and Vietnam identified three highly tolerant cultivars: Ketumbar (Indonesia), Khao Seetha (Thailand), and Soc Nau (Vietnam). For tolerance mechanisms, the newly identified donors seemed to be similar to 'Pokkali.' However, the cultivars Pat, Ketumbar, and Soc Nau have much better grain quality characteristics than 'Pokkali.'

Traditional varieties showed good tolerance for abiotic stresses (Table 14.1). Cultivars Pokkali, Cheriveruppu, Nona Bokra, SR26B, Damodar, and Getu are tolerant of salinity but possess poor agronomic characters; they are tall and photosensitive, have poor grain quality, and are low yielding. These salt-tolerant varieties are also tolerant of P deficiency but are very sensitive to Zn deficiency and are very sensitive to pests and diseases. If these tolerance traits are properly combined through breeding, multiple stress tolerance can be developed. However, salt tolerance and desired plant type should be selected independently (Garcia et al., 1995). Selection for low Na content (salt-tolerance parameter) preserves variation for desirable agronomic plant types, but early selection for agronomic characters strongly depletes the population of potentially salt-tolerant genotypes.

Attempts have also been made to identify other sources of salt tolerance. Wild species of *Oryza* and the African cultivated variety *Oryza glaberrima* were screened, but were all highly susceptible (Table 14.2) (Akbar et al., 1987). However, some wild rice *(O. rufipogon)* from Sri Lanka shows salinity tolerance. A distant relative of *Oryza (Porteresia coarctata)*, a halophyte, needs high salinity (EC of 12 to 18 $dS \cdot m^{-1}$) to produce grains (Table 14.3). At lower salinity, it produces almost no yield. Halophytes such as this one may not be a good source of tolerance in breeding programs. Aside from their difficulty in producing a fertile F_1, they need extremely high salinity to grow. In saline-prone areas, especially in coastal lands, salinity may not be present or may occur at lower levels during particular cropping seasons. What is needed are salt-tolerant varieties that could produce high yields even under low salinity.

Drought Tolerance

In this chapter, drought tolerance refers to the ability of the plant to grow and complete its life cycle under water-limited conditions. In rice, drought at the vegetative stage decreases yield via a reduction in leaf area and tiller number. At the reproductive stage, heading and, consequently, crop maturity is delayed. Poor panicle exsertion leads to production of sterile spikelets (IRRI, 1980) and reduction in grain quality (IRRI, 1979).

Since the 1970s, traditional rice varieties such as 'Khao Dawk Mali 105' (KDML 105), 'Nam Sagui 19' (NSG 19), and improved variety 'Mahsuri' continue to be used as tolerant checks during field and greenhouse screening for drought tolerance. They also served as donors of tolerance genes, resulting in progenies that perform well under water-deficit conditions. NSG 19, a short duration variety, and KDML 105, an aromatic photoperiod-sensitive variety, are grown in drought-prone areas of Thailand. In 1994, 500 F_2

TABLE 14.1. Characteristics of saline-tolerant traditional varieties.

Traditional varieties	Plant height (cm)	Days to flowering	Disease	Salinity	Alkalinity	Fe toxicity	Zn deficiency	P deficiency
Pokkali	165-176	72-104	S[a]	T	T	S	T	T
Nona Bokra	126-156	77-112	S	T	S	T	S	T
SR26B	141-151	81-107	S	T	S	T	S	T
Damodar	119-139	70-89	S	T	T	S	T	T
Cheriviruppu	147-157	78-92	S	T	T	T	T	T
Gatu	127-136	71-95	S	T	T	S	T	T

[a]S = sensitive, T = tolerant.

TABLE 14.2. Survival of *Oryza* species at 6800 ppm (EC 12 $dS \cdot m^{-1}$) salinity, IRRI phytotron.

***Oryza* species or variety**	**Accession no.**	**Genome**	**Seedlings tested (no.)**	**Survival (%)**
O. glaberrima	100855	AA	25	16.0
O. nivara	102163	AA	21	14.3
O. punctata	101434	BB	60	8.3
O. officinalis	100896	CC	88	5.7
O. minuta	101125	BBCC	55	12.7
O. latifolia	100885	CCDD	46	13.0
O. australiensis	100882	EE	54	1.9
O. brachyantha	100115	FF	43	0
Nona Bokra			50	100

Source: Akbar et al. (1987). Used by permission of the International Rice Research Institute.

pedigree lines coming from KDML 105 were evaluated and released at six sites in northeast Thailand (IRRI, 1995). 'Mahsuri,' on the other hand, was a breeding line from the cultivar cross Taichung 65/Mayang Ebos 80*2 and was released in 1965 (Mackill et al., 1996). 'Taichung 65' was a japonica variety, while 'Mayang Ebos 80' was a tall indica variety from Malaysia.

SCREENING METHODS FOR DROUGHT TOLERANCE

Drought-prone environments occur in two different ecosystems, namely, the upland and rainfed ecosystems. Upland areas tend to have aerobic soils, which offer little resistance to root penetration (Mackill et al., 1996). Rainfed lowland sites, on the other hand, are characterized by a shallow plow layer and the presence of a hardpan, which restricts root growth. Thus, plants that are adapted to upland areas may not perform well under rainfed sites and vice versa. Different tolerance mechanisms might operate under these two ecosystems, which warrants the use of different screening methods.

Field Testing

IRRI (International Rice Research Institute) developed a protocol for varietal screening for drought tolerance (IRRI, 1979) which had been extensively used for a number of years. Dry-season field screening was con-

TABLE 14.3. Effects of salinity on growth duration, straw and grain yields, and yield components of *Porteresia coarctata* tateoke.

Salinity level ($dS \cdot m^{-1}$)	Days to flower	Straw dry weight (g)	No. of panicles	No. of spikelets	Spikelet fertility (%)	Grain yield of five plants (g)
Control	102	4.2	6	50.4	5.5	0.3
EC 12	99	14.4	9	112.2	41.8	4.9
EC 18	144	12.9	9	119.5	37.4	4.6
EC 24	157	11.3	6	65.2	7.4	0.4
LSD (0.05)		4.1	4	50.2	8.8	2.2

ducted at IRRI and in farmers' fields, and included drought-tolerant and susceptible checks. Visual scoring based on leaf rolling (Table 14.4) and leaf drying (Table 14.5) was done following the IRRI (1996b) Standard Evaluation System (SES) for rice to assess damage. At the vegetative stage, plants were scored for drought tolerance at 0.2, 0.5, and 1.0 MPa soil moisture tension (SMT) at the 20 cm soil depth. Plots were sprinkler irrigated after the 1.0 MPa SMT scoring and scored for drought recovery (see Table 14.6) 15 days later.

Screening cultivars/lines that could perform well under rainfed condition was done by draining the field 50 days after seeding and allowing to dry for three weeks to one month, to simulate a prolonged drought. Fields representing a control treatment were kept wet by frequent irrigation.

Visual score was found to be positively correlated ($r = 0.66$, $P < 0.01$) with leaf water potential, i.e., plants showing less injury from drought had

TABLE 14.4. Scoring system for drought reaction (leaf rolling) at the vegetative stage.

Score	Description
0	Leaves healthy
1	Leaves start to fold (shallow V-shape)
3	Leaves folding (deep V-shape)
5	Leaves fully cupped (U-shape)
7	Leaf margins touching (O-shape)
9	Leaves tightly rolled

TABLE 14.5. Scoring system for drought reaction (leaf drying) at the vegetative stage.

Score	Description
0	No symptoms
1	Slight tip drying
3	Tip drying extended up to one-fourth length in most leaves
5	One-fourth to one-half of all leaves fully dried
7	More than two-thirds of all leaves fully dried
9	All plants apparently dead

TABLE 14.6. Scoring system for recovery after water stress.

Score	Description (percent plants recovered)
1	90-100
3	70-89
5	40-69
7	20-39
9	0-19

high leaf water potential. Therefore, maintenance of plant water potential is related to the performance of the rice plant.

Greenhouse Screening

Dry-season field screening can be conducted in the greenhouse, with the advantage of having a more controlled soil-moisture regime (IRRI, 1977). In early 1976, a special greenhouse for screening rice for drought tolerance was constructed in IRRI. The greenhouse had 12 concrete tanks, each measuring 6.8 × 3.5 × 1.35 m. The tanks contained 1 m deep upland soil placed on a 0.35 m deep sand-and-gravel drainage bed. Irrigation could be subsurface, by a perforated system laid in the gravel, or supplied on the soil surface, simulating rainfall. Soil moisture was monitored by using tensiometers and electric resistance blocks. A single tank could accommodate 66 entries and two check varieties. Results of the greenhouse screening were similar to those of dry-season field screening. There was close agreement between check varieties of both tests.

MECHANISMS OF DROUGHT TOLERANCE

The earliest response of plants to drought is stomatal closure. This is a strategy to minimize moisture loss via transpiration. However, prolonged stomatal closure reduces photosynthesis, therefore, decreasing growth. Leaf temperature increases lead to other stresses such as photoinhibition, leading to oxidative stress. Thus, the plant needs to utilize other mechanisms to be able to adapt to water deficit.

Deep Root and Other Morphological Characters

Growing deep roots allows plants to access water in the deeper soil horizon. Having thick roots enables plants to penetrate through soil barriers such as the hardpan in rainfed lowland areas. Thus, 'IR52,' a semidwarf rice cultivar developed at IRRI, was released in 1980 in the Philippines as a rainfed lowland rice suited to drought-prone conditions (IRRI, 1983). It had consistently high drought-tolerance scores at the vegetative stage, presumably due to a deeper and more prolific root system, which could have enabled the roots to extract more water during drought stress.

Path coefficient analysis involving 34 rice cultivars with diverse reactions to field drought in the reproductive phase showed that thick- and deep-rooted plants are more tolerant of field drought (IRRI, 1990). Moderate tillering ability, short and narrow leaves, and intermediate plant height also contribute to a more stable yield in drought-prone environments.

Osmotic Adjustment

Some plants are able to minimize water loss from tissues by lowering the water potential of the cell without accompanying loss in turgor. This is made possible by production of compatible solutes such as sugars, ammonium compounds such as proline, glycinebetaine, and low-molecular-weight proteins such as LEAs (late embryogenesis abundant) and dehydrins. These solutes are also produced during salt and cold stresses, where plants are subjected to osmotic stress that occurs during water deficit. Among the compatible solutes accumulated during stress, proline appears to be the most widely distributed (Delauney and Verma, 1993). In plants, proline biosynthesis occurs via either the glutamate or the ornithine pathways. However, in osmotically stressed plant cells, glutamate is the primary precursor for proline biosynthesis (Hu et al., 1992).

A quaternary ammonium compound that is accumulated during osmotic stress, as occurs during drought and salinity, is glycinebetaine (betaine). However, this compound is synthesized only in a number of halophytes and in a few crop plants, including barley and beet (Gibon et al., 1997). In rice seedlings, exogenous application of betaine did not confer tolerance to salt stress (Lin and Kao, 1995). Thus, this solute could have inhibitory effects on higher plants that do not accumulate it. Using rape leaf discs, Gibon and colleagues (1997) showed that, indeed, betaine could be a noncompatible organic solute. Results suggested that absorption of betaine could have inhibitory effects both at the chloroplast level, such as suppression of Rubisco (ribulose-1,5-bisphosphate carboxylase oxygenase) large subunit synthesis, and at the mitochondrial level, such as decrease of the TTC (2,3,5-triphenyltetrazolium chloride)-reducing activity.

Glycinebetaine is synthesized in chloroplasts of higher plants via the pathway: choline → betaine aldehyde → glycinebetaine (Rathinasabapathi et al., 1994; Ishitani et al., 1995). These reactions are catalyzed by choline monooxygenase and betaine aldehyde dehydrogenase (BADH), respectively. Why some plants could not accumulate betaine could be attributed partly to lack of either a certain reactant or a catalyst in the biosynthetic pathway. Tobacco plants are unable to accumulate betaine during osmotic stress because of lack of these two enzymes involved in its biosynthesis (Rathinasabapathi et al., 1994). Targeting choline monooxygenase and BADH in transgenic plants could enable nonaccumulators to accumulate betaine during abiotic stress. Rathinasabapathi and colleagues (1994) demonstrated that BADH targeted to the chloroplast in leaves of transgenic tobacco plants converted supplied betaine aldehyde to betaine. Hayashi and colleagues (1997) transformed *Arabidopsis thaliana* with a *codA* gene, which encodes for choline oxidase, the enzyme that converts choline into betaine. Transformed plants accumulated betaine and enhanced their tolerance to salt and cold stress. Therefore, despite the concern aired by Gibon and colleagues (1997) regarding betaine being incompatible with the leaf tissues of non-betaine-accumulating species, it is possible to engineer betaine accumulation into these species and increase their tolerance to environmental stresses.

Dehydrins are a family of proteins that generally accumulate in plants in response to abiotic stresses, including water deficit, cold temperature, and application of abscisic acid (ABA) (Close et al., 1995). The dehydrins are also called LEA group II or RAB (responsive to ABA) proteins. Seven dehydrin genes have been reported in *Arabidopsis,* namely, *COR47, ERD14, LT130, RAB18, PAP310,* and *DHNXERO1* (Nylander et al., 2001).

Chemical Signals

It is widely held that during water stress a reduction in leaf turgor initiates stomatal responses. However, under stress conditions, stomata can close in the absence of measurable reductions in leaf-water potential (ψ_w) or turgor (Saab and Sharp, 1989). This indicates that shoots can sense reduced water availability at the roots, independent of their own water status. Transport of chemical signals originating from the drying roots to the leaves have been implicated in the regulation of stomatal aperture as an adaptation to drought. Split root experiments by Gowing and colleagues (1990) had shown the involvement of a positive root signal in inhibition of leaf initiation, leaf expansion, and leaf transpiration rate, with no significant differences detected in leaf water status between the partially dried and the well-watered plants.

Abscisic Acid

ABA is the most widely investigated chemical signal. It is a plant hormone that promotes stomatal closure, thereby reducing transpiration. Interest in ABA was spawned by the need to develop antitranspirants that would prevent excessive water loss. This is of economic importance, especially in semiarid regions (Jones and Mansfield, 1970). Then known as "inhibitor β," ABA was shown to increase in detached, wilted wheat leaves (Wright, 1969). The increase in inhibitor content was dependent on duration, severity of the stress, and temperature. "Inhibitor β" was identified as abscisic acid by Milborrow (1967).

Additional experiments by Wright (1977), using excised wheat leaves, showed a sigmoid relationship between ABA level and decreasing leaf water potential values. Capacity to biosynthesize ABA in leaves during stress conditions has been shown to confer drought resistance in some species (Larque-Saavedra and Wain, 1976). In addition, ABA application has been shown to mimic various effects of water stress on plants. Exogenous ABA applied to excised first leaves of wheat and barley led to induction of stomatal closure and inhibition of transpiration (Mittelheuser and van Steveninck, 1969). Application of ABA to the cut base of a detached leaf was found by Kriedemann and colleagues (1972) to elicit rapid response, with stomata closing beginning at 3 min from application, in the case of corn.

The significant role of ABA in stomatal control was further confirmed using dehydrated, intact plants. Exogenous application to the leaf surface caused marked stomatal closure in *Xanthium pennsylvanicum* (Jones and Mansfield, 1970). Blake and Ferrell (1977) showed a close relationship between soil water potential and ABA, with ABA returning to normal levels within six days after rewatering. Significant correlation was reported between xylem ABA and stomatal behavior under controlled environmental conditions (Khalil and Grace, 1993).

Although ABA is also produced in the leaves and the hormone can be remobilized from mature to immature leaves, more studies point to the roots as the source of ABA during signal transduction. Cornish and Zeevaart (1985) reported that ABA accumulated in cut roots and in roots of girdled plants, proving that ABA was synthesized in the roots and not just transported from the shoots. Root ABA content was also shown by Zhang and

Davies (1989) to be a sensitive indicator of soil water status around the roots. In this study, ABA of the xylem sap also increased. Lebreton and colleagues (1995) compared the coincidence of QTL for ABA content and stomatal conductance, and found xylem ABA, rather than the whole-leaf ABA content, more likely to have a regulatory role on stomatal conductance. Results also showed the importance of the rooting behavior in regulating stress responses, especially in determining xylem ABA contents. Drought resistance was associated with having a greater number of nodal roots, leading to more ABA translocated through the xylem to the stomata, thus lowering stomatal conductance and delaying the onset of severe dehydration.

Further understanding of the role of ABA in stomatal regulation is provided by experiments utilizing ABA-deficient genotypes. In the absence of specific inhibitors of ABA synthesis, these mutants make it possible to modify endogenous ABA contents (Quarrie, 1982). The wilty tomato mutants *flacca, sitiens,* and *notabilis* are deficient in ABA production, thus their stomata resist closure (Tal and Nevo, 1973). All three mutants reverted to the normal phenotype upon daily applications of ABA. Similarly, a droopy mutant of potato had higher stomatal conductances than those of normal plants. Three days after ABA application onto the surface of several leaves of droopy plants, conductánces became similar to those of non-droopy plants (Quarrie, 1982). Cohen and Bray (1990) demonstrated that three mRNAs which accumulate in wilted wild-type tomato leaves were not expressed in wilted ABA-deficient mutant leaves. This inability was reversed with exogenously applied ABA. More recently, symptoms of ABA deficiency in transgenic tobacco were generated by blocking ABA with intracellularly expressed, specific single-chain variable fragment (scFv) antibodies (Artsaenko et al., 1995). The wilty phenotype of transgenic plants is a consequence of the failure of the stomata to close. Similar techniques of modifying ABA biosynthesis further demonstrate that endogenous ABA regulates gene expression.

Increase in ABA during stress also induces accumulation of ABA-responsive genes, indicating that ABA controls the accumulation of specific mRNAs and proteins. One such gene is RAB 21, which is induced in ABA-treated tissues and in rice plants subjected to water stress (Mundy and Chua, 1988). Accumulation of RAB 21 mRNA during stress suggests that expression of the gene is induced during dehydration and that, under these conditions, ABA protects cells by acting as a signal for the initiation of these responses.

Cytokinin

Roots also produce other physiological substances, such as cytokinins, which may also function as chemical signals during soil drying. Cytokinin is a plant hormone that promotes stomatal conductance. Working on maize seedlings, Blackman and Davies (1985) divided the root system between two containers. Part of the root system was subjected to water stress while the rest was adequately supplied with water. Results suggested that cytokinin, rather than ABA, from roots controlled stomatal responses during soil drying. Reduction in transport of cytokinin from roots to leaves during stress provided a signal for the shoot to modulate its stomatal aperture in reaction to an impending limitation in water supply.

GENETICS OF TOLERANCE

Salinity Tolerance

Genetic studies at IRRI indicated that both additive and dominance effects are important in the inheritance of almost all characters associated with salt tolerance (Mishra et al., 1990; Gregorio and Senadhira, 1993; Lee, 1995). At the seedling stage, characters such as shoot length, Na^+ and K^+ content in the shoots, and dry weights of shoots and roots showed a highly significant magnitude of additive effects. The estimates of heritability in these characters were low. These high predictable values may be used as criteria for selecting salt-tolerant lines at the seedling stage. Characters at the maturing stage such as plant height and yield per plant showed a highly significant magnitude of additive effects, suggesting the greater importance of additive gene action in the inheritance of these characters (Moeljopawiro and Ikehashi, 1981; Akbar et al., 1986; Mishra et al., 1990).

Narayanan and colleagues (1990) reported the result of their 6 × 6 diallel study and showed the nonsignificance of additive effects for grain yield per plant under normal conditions. Under saline stress conditions, the additive effects became highly significant. The magnitude of grain yield expression along with other characters was substantially reduced because of salinity. From this, it was inferred that part of the grain yield due to additive action was less affected by salinity. It was then argued that in a variety with more additive genes for grain yield the suppression of yield would be relatively less under salinity stress. Most studies have used a partial diallel analysis and had large environmental interactions that could diffuse or mask the phenotypic differences in tolerance.

The study by Gregorio and Senadhira (1993) using a nine-parent complete diallel analysis showed that both additive and dominance gene effects govern salinity tolerance, shown as low Na-K ratio in shoots. The trait exhibits overdominance and is controlled by at least two groups of genes that exhibit dominance. Environmental effects were large and narrow-sense heritability was low (19.18 percent). These findings suggest that (1) early-generation breeding populations must be large and replicated and (2) selection for tolerance must be done in a later generation and under controlled conditions to minimize environmental effects. This further suggests the use of modified bulk and single-seed descent (SSD) as breeding methods to develop salinity-tolerant rice varieties.

A genetic combining ability study showed the significance of both general combining ability (GCA) and specific combining ability (SCA) effects in the inheritance of salinity tolerance in rice. Moderately tolerant parents (e.g., IR4595-4-1-13 and IR98884-54-3-1E-P1) were the best combiners to produce salt-tolerant progenies. Most of the best specific cross combinations for salt tolerance had susceptible but high-yielding parents, such as IR28 and IR29. The presence of reciprocal effects among crosses necessitates the use of high-yielding susceptible parents as males in a hybridization program. Further, the large heterotic effects suggest the potential of a hybrid rice program for salt-affected lands.

Elongation Ability

The physiological mechanisms and inheritance of elongation ability in deepwater rice are fairly well understood. Deepwater rice can withstand excess water by elongating (leaf sheath, leaf blades, and internodes) to keep its foliage above the water level. It accumulates starch before flooding and uses this starch for rapid elongation when the plant is submerged.

Floating rice varieties in South Asia, such as 'Jalmagna,' differ from their counterparts in Southeast Asia ('Baisbish' and 'Rayada 16-3') in terms of elongation ability. The South Asian types elongate faster. Genetic analysis showed overdominance in the F_1 of all crosses, indicating a heterotic response of the genes governing elongation. Transgression was always positive with Jalmagna, whereas both positive and negative transgressions were observed in crosses involving cultivars Baisbish and Rayada 16-3. These suggest differences in minor genes among the two types. In 'Jalmagna,' the differences led to complementary gene action. Crosses between the two types could produce faster elongation than the better parent. In a separate 7 × 7 diallel analysis, 'Jalmagna' was the best combiner. Thus, elongation ability can be combined with derived genotypes (IRRI, 1996a).

Submergence Tolerance

The genetics of submergence tolerance was studied by analyzing the segregation pattern of the F_2 generation. Results showed that FR13A, Kurkaruppan, and Thavalu possess the same genes for submergence tolerance, whereas Goda Heenati has a recessive gene that is nonallelic to the dominant genes of 'FR13A,' 'Kurkaruppan,' or 'Thavalu' (IRRI, 1995). Another genetic study showed that a single dominant gene (Mishra et al., 1996) governed submergence tolerance. Previous studies showed that the submergence tolerance gene of 'FR13A' and 'Kurkaruppan' is different from that of 'Goda Heenati.'

PREBREEDING RESEARCH TO IMPROVE DONOR GERMPLASM

In the mid-1970s, attempts to transfer salinity tolerance from traditional varieties to high-yielding cultivars were not successful. Several high-yielding lines developed at IRRI using traditional salt-tolerant parents such as 'Nona Bokra,' 'Pokkali,' 'SR26B,' and 'Kalarata' do not possess the level of tolerance found in the donors (Table 14.7). Difficulty in transferring the tolerance trait is attributed to donors having too many undesirable traits, most of them linked to salinity tolerance (Table 14.1). Thus, their full level of tolerance has not been transferred to an improved plant type. Prebreeding activities were given much attention to improve the agronomic characters of the tolerant donors.

Somaclonal Variants of 'Pokkali'

The 'Pokkali' type of rice variety is highly salt tolerant and commonly grows in coastal areas of Kerala, India. It is a traditional, tall, photoperiod-

TABLE 14.7. Moderate salt-tolerant lines developed at IRRI and their sources.

Improved salt-tolerant lines	Source of tolerance
IR4630-22-2-5-1-3	Pokkali
IR4595-4-1-13	Pokkali
IR9884-54-3	Nona Bokra
IR10198-66-2	Nona Bokra
IR10167-129-3-4	Kalanata 1-24
IR10206-29-2-1	SR26B

sensitive rice variety that is susceptible to lodging and has low tillering and long, broad, dark, and droopy leaves. The leaves senesce quickly after flowering. It has a red pericarp and poor grain quality. 'Pokkali' is one of the few rice varieties that produce vigorously growing seedlings. It is highly tolerant of salinity, but its yielding ability is low.

'Pokkali' and other tolerant donors underwent cell culture for induction of somaclonal variation. Somaclonal variants of 'Pokkali' with improved agronomic traits were identified. Table 14.8 shows the traits for which the variants differed significantly from 'Pokkali.' The variant isolated (TCCP 266-2-49-B-B-3) had desirable levels of all the characteristics tested. Its level of tolerance for salinity remains as high as that of the donor. It has vigorous growth and, unlike the parent 'Pokkali,' is semidwarf, a trait essential in increasing yield potential without lodging. The variant had a white pericarp, which is a distinct advantage. There was also improvement in cooked rice quality, with the semidwarf variant having medium consistency. These characteristics make the somaclonal variant TCCP 266-2-49-B-B-3 superior to 'Pokkali' as a donor of salt tolerance in hybridization programs. It has become a popular donor parent and has produced new, high-yielding, salinity-tolerant lines, some of which have already been released as varieties (Table 14.9).

TABLE 14.8. Major agronomic differences and salinity tolerance between 'Pokkali' and its somaclonal variant.

Trait	Pokkali	TCCP266-2-49
Leaves	Droopy	Erect
Days to flowering	PS (72-104)	82
Plant height	Tall	Semidwarf
Lodging (SES)[a]	9	1
Pericarp	Red	White
Gel consistency	83	59
Salinity tolerance (SES)		
Nutrient culture		
test 1	2	2
test 2	1	1
test 3	1	1
Greenhouse	3	3
Field	3	3

[a]According to IRRI's Standard Evaluation System for rice (IRRI 1996b), with slight modification (Gregorio et al., 1997). On a scale of 1 to 9, where 1 = highly tolerant and 9 = highly susceptible.

TABLE 14.9. New salt-tolerant lines developed at IRRI and their donor sources.

Improved varieties/lines	Source of tolerance
IR65185-3B-8-3-2	TCCP266-2-49-2B-3
IR65195-3B-2-3	TCCP266-2-49-2B-3
IR52713-2B-8-2B-1-2	Pokkali
IR63291-B-3R-8-3	TCCP266-2-49-2B-3
IR61920-3B-22-2-1	Wagwag
IR72132-AC6-1	TCCP266, IR51491-AC10
IR58443-6B-10-3	AT69-5
IR69997-AC1,2,3,4	TCCP266-2-49-2B-3
IR63307-4B-4-3	TCCP266-2-49-2B-3

F_1 *Anther Culture (AC)-Derived Lines*

Development of a promising salinity-tolerant line through hybridization and selection could take eight to ten years. This is aggravated by the fact that, in areas where salinity or other stresses are found, only one cropping season per year is possible; thus, generation of stable lines takes much longer. This period could be substantially shortened by applying the F_1 anther culture technique.

Some high-yielding, salt-tolerant, anther culture-derived lines have been generated at IRRI. It took only three years for the tolerant lines to be isolated. These AC-derived lines were IR51500-AC11-1, IR51500-AC17, IR51485-AC6534-4, IR72132-AC6-1, IR69997-AC1, IR69997-AC2, IR69997-AC3, and IR69997-AC4. Most of these had been used as donor parents in breeding programs in Bangladesh, Dominican Republic, Egypt, Mexico, Myanmar, Philippines, and Thailand (IRSTON, 1991). IR51500-AC11-1 was released as a salt-tolerant variety in the Philippines with the name PSBRc50 or 'Bicol'. This is the first time an AC-derived line from an indica-indica cross had been released as a variety, and this was also the first variety to be recommended for cultivation in adverse environments. IR51500-AC17 and IR51485-AC6534-4, named CSR21 and CSR28, respectively, were identified as commercial varieties in India. Both varieties were derived from F_1 anther culture and were recommended for cultivation on degraded saline-alkaline soils.

Combined Salinity and Submergence Tolerance, and Salinity and Cold Tolerance

Most coastal areas are prone to flash floods or submergence. Therefore, varieties with good tolerance to both salinity and submergence are required. Eleven elite lines coming from the cross FR13A × IR17494-32 were identified. The best among them were IR42598-4B-12-1-2 and IR42598-4B-15-1-3, both tolerant of salinity and submergence (ten to twelve days underwater).

The demand for salinity-tolerant rice varieties having cold tolerance is high in coastal areas of North and South Korea, China, Egypt, and some European countries where japonica rice is grown. Salt- and cold-tolerant rice lines were identified from multiple crosses of japonica, salt-tolerant indicas, and 'Moroberekan.'

ACHIEVEMENTS IN GERMPLASM IMPROVEMENT

IRRI's abiotic stress breeding program is not designed to produce varieties for release but to develop improved germplasm that will be used by national breeding programs. National agricultural research experiment stations (NARES) are in need of promising lines that can be used as donor parents or for release as varieties after national or regional yield tests produce favorable results. Table 14.10 shows the types of salinity-tolerant breeding lines required by NARES. These lines should have a wide range of stress tolerance with different growth duration combinations. This requires more diverse breeding populations and would necessitate pyramiding traits for tolerance of different stresses to be able to produce specific tolerance combinations.

TABLE 14.10. Types of salt-tolerant varieties required by national agricultural research and extension systems.

	Soil condition		
Growth duration	**Saline**	**Saline acid**	**Saline alkaline**
Early (115 d)	+	−	+
Medium (130 d)	+	+	+
Late (145 d)	+	+	−
Moderately photosensitive	+	+	−
Highly photosensitive	+	+	−

Note: Pest and disease resistance, grain quality, and climatic stresses (cold, submergence, etc.) were not considered.

Tolerance of Salinity and Problem Soils

In IRRI's breeding program, the modified bulk method and rapid generation advance (modified single-seed descent) were used. Early generation (F_2 to F_5) breeding materials were evaluated for salinity tolerance in several saline fields; population sizes were large and selection pressure was mild. Rigorous plant selection for salinity tolerance was done in the advanced generations (F_6 to F_7) under severe salinity conditions. Selection for good plant type and seedling tolerance was done in the IRRI field and under controlled conditions in the phytotron, respectively (Gregorio et al., 1997).

Various mechanisms of salinity tolerance in rice have been identified (Yeo and Flowers, 1986 ; Akita and Cabuslay, 1990). Integration of all these physiological mechanisms for overall salinity tolerance is the desirable long-term goal to increase the level of tolerance in rice. Donors having these salinity-tolerance mechanisms have been identified from improved rice lines and traditional varieties in South and Southeast Asia. Hybridizations were done to produce improved salinity tolerance by pyramiding the different mechanisms. Based on the genetic component analysis of some donors, both the additive (fixable) and nonadditive (nonfixable) genes significantly contributed to the inheritance of the Na exclusion mechanism. The best combining donors for Na exclusion and tissue tolerance mechanisms were also identified.

Fifty-nine advanced lines developed in IRRI are listed in Table 14.11, some of which were advanced using the rapid generation advance (RGA) facility. All these were selected based on seedling vigor, survival, Na exclusion, and agronomic traits. All showed good tolerance for salinity at the

TABLE 14.11. Promising lines with pyramided salinity-tolerance mechanisms.

No. of lines	Designation	Parentage[a]
9	IR68652-3B-series	IR20/Pokkali
1	IR69587-3R-P-4-1	IR24//IR20/Pokkali
9	IR69588-4R-P-series	IR26//IR20/Pokkali
9	IR70865-B-P-series	IR50184-3B-18-2B-1//IR20/IR24
2	IR70866-B-P-series	IR63731-1-1-1-4-3//IR20/IR24
14	IR70868-B-P-series	IR63731-1-1-1-4-3//IR20/IR26
7	IR70869-B-P-series	Pokkali//IR20/IR24
8	IR70870-B-P-series	Pokkali//IR20/IR26

[a]IR20 for tissue tolerance, IR24 and IR26 for leaf-to-leaf compartmentation, others for Na exclusion, low shoot Na, seedling vigor, and survival.

seedling and reproductive stages. These lines had seed increased for distribution to different hot spots/target sites in India.

Seven advanced lines of the cross IR68652 were tested for tissue tolerance, and two had a tissue tolerance level equal to or better than that of IR20, the donor for the trait. These lines demonstrated that tissue tolerance could be combined with other tolerance traits and with high-yielding ability.

The advanced bulk populations or lines developed are evaluated in target environments of different countries to select mutants adaptable to their conditions. Promising lines are identified and further tested in their respective national or regional multilocational yield trials. IRRI follows up their selections and nominates lines to the International Rice Stress Tolerance Observational Nursery (IRSTON) of the International Network for Genetic Evaluation of Rice (INGER) for global distribution.

Deepwater Rice (Elongating Ability)

Germplasm improvement focuses on prebreeding research on flood tolerance and providing improved breeding materials to NARES. It develops crosses involving traditional deepwater rice (DWR) and tropical japonicas, aiming to produce elongating plant types with sturdy stems and long, heavy panicles. By 1995, IRRI's collaborative effort focused mainly on assisting lead national research centers in widening the gene pool. IRRI generates advanced bulk populations in the dry season and characterizes advanced lines. The advanced breeding materials produced are given to the lead centers and NARES for evaluation.

Prototypes of the new DWR plant type developed at IRRI were tested for yield in the dry and wet seasons. IR11141-6-1-4 was grown in the dry season at 80 to 90 cm water depth and was compared with IR72 grown in normal irrigated conditions. The experiment was repeated in the next wet season, with four more new deepwater-plant-type lines, two improved DWR lines, and a traditional plant type (FR13A). IR8 and IR74 were added to the irrigated rice variety set. Although DWR is not usually planted in the dry season, the 1996 dry-season experiment showed that the new DWR plant type is capable of as high a yield as modern irrigated rice. The water depth of 80 to 90 cm seemed to have no effect on yield. The results were similar in the wet season. The average yield of five new deepwater-plant-type lines was 4.9 $t \cdot ha^{-1}$ (flooded) compared with 4.7 $t \cdot ha^{-1}$ (normal field) of the irrigated varieties. Yields of the new deepwater plant type were about double those of the traditional plant type, and the new deepwater line IR62364-2B-10-2-2 yielded 6 $t \cdot ha^{-1}$. These results demonstrate that the new plant type yields the same as irrigated rice grown in deepwater areas (80 to 90 cm) (IRRI, 1998).

Very deepwater (>100 cm) elite lines from *O. sativa/O. rufipogon* crosses were developed. These 52 lines were derived from the cross among IR60601 (FR13A/*O. rufipogon*), IR60608 (IR42/*O. rufipogon*), and IR64588 (IR11141-6-1-4/*O. rufipogon*/2*IR11141-6-1-4). They elongate rapidly under increasing water level, are high tillering, have dense and nonshattering panicles, and have erect, broad, green leaves.

Since 1994, the Prachinburi Rice Research Center (PRC) and the Huntra Experiment Station (HTA) in Thailand have been the lead centers for evaluation and selection of breeding materials. The Rice Research Institute of Thailand produced two new plant type lines, HTA FR 810442-B-7-1 and HTA FR84038-B-5-0-1, which yield 5.0 $t \cdot ha^{-1}$ at 100 cm water depth. In contrast, the local check variety LPT 123 yields 3.7 $t \cdot ha^{-1}$.

Submergence Tolerance

Previous studies showed that the submergence-tolerance genes of 'FR13A' and 'Kurkaruppan' are different from that of 'Goda Heenati.' With the long-term objective of pyramiding these genes to enhance tolerance, two types of crosses were made in 1992. One cross was made to transfer the submergence-tolerance gene from traditional donors to an improved plant-type background. The crosses were IR74/FR13A, IR74/Kurkaruppan, and IR74/Goda Heenati. The other type of cross was made between IR40931-26-3-3-5 and IR31142-14-1-1-3-1-12 (IR66036), both moderately tolerant of submergence. The improved advanced lines developed from these crosses were tested for submergence tolerance, and five highly submergence-tolerant lines with improved plant type were isolated from the cross IR74/Goda Heenati, six from IR74/Kurkaruppan, and 14 from IR74/FR13A. Lines from IR74/Goda Heenati will be crossed with others to pyramid the different genes. Eight lines of the cross IR66036 were more tolerant than the parents, indicating the accumulation of genes from different sources of tolerance. More than 500 elite submergence-tolerant lines with improved plant types and varying tolerance for other biotic and abiotic stresses were identified and categorized. These lines are available for national testing and are good sources of tolerance.

A replicated yield trial in both submerged (13 days under water) and normal field conditions was conducted in the 2000 dry season. Most of the ten submergence-tolerant lines with modern plant types had yields comparable with that of IR72 (3.5 $t \cdot ha^{-1}$) under normal irrigated field conditions and much higher than that of traditional check FR13A (1.2 $t \cdot ha^{-1}$). However, under submerged conditions, modern varieties IR72 and IR42 did not survive after 13 days of submergence, thus producing no yield. All other submergence-tolerant entries survived and produced yields. One of the new lines yielded 3.4 $t \cdot ha^{-1}$, more than double the yield of FR13A (1.5 $t \cdot ha^{-1}$). This yield trial demonstrates the possibility of increasing the yield of submergence-prone rice by genetic improvement.

TOWARD THE DEVELOPMENT OF MOLECULAR MARKER-ASSISTED SELECTION (MAS)

In breeding rice for tolerance of salinity and adverse environments, multiple-stress-tolerance traits have been emphasized. However, selection efficiency for these stresses using conventional selection is low. Thus, achieving multiple tolerance traits requires gene pyramiding through molecular marker-assisted breeding.

Rapid and reliable techniques have been developed at IRRI for major abiotic stresses, such as elongation ability, and tolerance for salinity, submergence, iron toxicity, aluminum toxicity, and phosphorus deficiency. However, those techniques have limited capacity, are expensive, and are suitable for only basic studies of genetics and physiological mechanisms. Research at IRRI has focused on developing MAS techniques that are efficient and inexpensive. Six distinct steps are involved:

1. generate mapping populations from crosses between selected parents,
2. phenotype the mapping population,
3. genotype to tagged genes and quantitative trait loci (QTL) analysis,
4. do fine mapping to identify molecular markers linked to important genes,
5. convert molecular markers to usable and inexpensive polymerase chain reaction (PCR)-based markers, and
6. test MAS and apply it to NARES breeding programs.

Progress in developing MAS for salinity tolerance and related abiotic stresses is shown in Figure 14.1. Donor germplasm has been identified and mapped. Most of the important stress-tolerance traits were tagged, and we are now at the fine-mapping stage. Application of MAS to national breeding programs is the ultimate goal.

The permanent mapping population (F_8 recombinant inbred lines, RILs) developed at IRRI and its mapping status are shown in Table 14.12. Contrasting parents (highly tolerant and highly sensitive) were identified for each abiotic stress using available screening techniques. The contrasting

Trait	Develop mapping population	Develop phenotype technique	Phenotype population	Study genetics	Tagged genes/ QTLs	Fine-map	Converted to PCR-based MAS	Test to PCR-based MAS	Demo. to NARES
Photosensitivity									
Salinity (Na-K ratio)									
P efficiency									
Submergence									
Elongation									
Al-toxicity tolerance									
Fe-toxicity tolerance									
Zn in grain Zn efficiency									
High Fe (grain)									
Salinity (Na exclusion)									
Salinity tissue tolerance									

FIGURE 14.1. Progress in the development of marker-aided selection (MAS) for tolerance of salinity-related stresses. Lightly shaded portions have already been done by IRRI, and those with darker shading are ongoing activities. Used by permission of the International Rice Research Institute.

TABLE 14.12. Permanent mapping populations (F_8 RILS) and anchor markers available.

Trait	Cross	Population size	Mapping status[a]	No. of markers
Salinity				
Na-K ratio	IR29 × Pokkali	275	Mapped AFLP	172
Tissue tolerance	IR4630 × IR15324	118	Phenotyping	
Na exclusion	AGAMI × IR63352-AC202	204	AFLP	216
Submergence	IR74 × FR13A	252	Mapped AFLP	202
Elongation	IR74 × Jalmagna	247	Mapped AFLP	201
P deficiency	IR20 × IR55178-3B-9-3	285	Mapped AFLP/RFLP/SSR	217
Al toxicity	IR1552 × Azucena	190	Mapped AFLP/RFLP	207
Fe toxicity	IR64 × Azucena	135 (DH)[b]	Mapped RFLP/RAPD	175
Zn deficiency	IR26 × Madhukar	250	Mapped SSR	

[a]AFLP = amplified fragment length polymorphism, RFLP = restriction fragment length polymophisms, SSR = simple sequence repeat, RAPD = random amplification of polymorphic DNA.
[b]DH = doubled haploid.

parents were crossed and RILs developed by single-seed descent in the RGA facility at IRRI. RILs were phenotyped by the respective screening method and in collaboration with IRRI's genome mapping laboratory. QTL mapping was done by selective genotyping. Major genes and QTLs for tolerance of salinity, P and Zn deficiency, submergence, Al and Fe toxicity, and elongation ability were mapped as shown in Table 14.13.

Salinity Tolerance

The genes governing tolerance in 'Pokkali' were mapped through the use of F_8 RILs derived from the cross between 'Pokkali' and salinity-sensitive variety IR29. Because of the strong interaction between the salinity tolerance trait and environmental factors, the phenotyping was done in highly controlled environmental conditions in the IRRI phytotron.

Phenotyping was done at the seedling stage using salinized nutrient solution (EC = 12 $dS{\cdot}m^{-1}$). Eighty RILs were selected from the extremes of the population in response to salinity and were used to construct an amplified fragment length polymorphism (AFLP) map consisting of 206 markers. A major gene for salinity tolerance was mapped on chromosome 1. It explained 64 to 80 percent of the phenotypic variation.

QTLs governing high K absorption, low Na absorption, and low Na-K ratio were mapped to five chromosomes (Table 14.13). For high K absorption, QTLs were detected on chromosomes 4 and 12, whereas for low Na absorption, there were QTLs on chromosomes 1, 3, and 10. For Na-K ratio, three QTLs were mapped to chromosomes 1, 10, and 12. A common major QTL was detected on chromosome 1 for the three traits putatively associated with salinity tolerance (IRRI, 1997). In a different mapping population (Agami × IR63352-AC202), a major QTL was also mapped on the same segment of chromosome 1. Fine mapping using near-isogenic lines (NILs) is in progress.

TABLE 14.13. Identified QTLs for abiotic stresses in rice.

Trait	Major QTL	Minor QTLs
Salinity	Ch 1	Ch 3, 4, 10, 12
Submergence	Ch 9	Ch 6, 7, 11, 12
Elongation	Ch 1	Ch 4, 5, 10, 12
Phosphorus deficiency	Ch 12	Ch 1, 6, 9
Fe-toxicity tolerance	Ch 1	Ch 2, 7
Al-toxicity tolerance	Ch 1	Ch 12

Submergence Tolerance

RILs for submergence tolerance derived from the cross IR74 (sensitive)/FR 13A (tolerant) were screened by subjecting 50-day-old seedlings to complete submergence until IR74 was dead. Two screening tests of 250 RILs were conducted in artificial ponds at IRRI and at the Huntra Rice Experiment Station in Thailand. A third test at IRRI used concrete tanks in a greenhouse. All tests produced similar results.

Survival ranged from zero in IR74 to 96.7 percent in FR13A. About 20 percent of the RILs had the same survival rate as IR74, but less than 10 percent of the lines scored as high as FR13A. RILs with a survival rate of >70 percent and <10 percent were selected for AFLP analysis.

The AFLP linkage map constructed from the selected genotypes had a total map length of 1756 cM. The average interval size was 8.5 cM, with the smallest value of 5.5 cM for chromosome 6 and the largest of 10.5 cM for chromosome 9. Based on screening data, a major gene for submergence tolerance was mapped near AFLP markers P1/M3-6, P2/M7-7, and P3/M4-10 on chromosome 9. In addition, QTLs were located on chromosomes 6, 7, 11, and 12. Each of these four QTLs explained more than 19 percent of total phenotypic variation. The QTL on chromosome 11 was in the vicinity of the genes *Adh1* and *Adh2* for alcohol dehydrogenase. Because alcohol dehydrogenase is believed to be associated with submergence tolerance, the QTL on chromosome 11 might correspond to the *Adh* genes.

Elongation Ability

RILs for elongation ability were developed from the cross IR74 (little or no internode elongation)/Jalmagna (a fast-elongating, traditional cultivar adapted to water as deep as 4 m). The 247 RILs developed were phenotyped for internode elongation in field tests with fast-rising floodwater and slow-rising floodwater. The maximum water depth in all tests was 120 cm. Results were used to select 41 internode-elongating lines and 41 nonelongating lines.

AFLP analysis of the 82 selected lines was used to construct a linkage map of 2203.55 cM and a major gene governing internode elongation was mapped on chromosome 1 between markers P2M6-7 and P1M6-2. It explained 85.7 percent of the phenotypic variation for the trait. QTLs were located on chromosomes 1, 4, 5, 6, and 10 with respective phenotypic variance of 22.7, 18.1, 19.4, 16.9, and 20.4 percent.

Phosphorus Deficiency

IRRI uses two methods to screen rice for tolerance for P deficiency. In both methods, genotypes are grown in P-deficient and P-sufficient conditions and relative tillering ability (tiller number in P-deficient conditions/tiller number in P-sufficient conditions × 100) is measured 28 to 30 days after seeding (DAS) to determine the level of tolerance.

A greenhouse method induces P deficiency by reducing the P concentration in culture solution from 10.0 to 0.5 ppm. Tolerant genotypes detected by this method are considered internally efficient or to have a low P requirement. Because the physiological mechanisms of P-deficiency tolerance are not known, the strategy is to map the genes using RILs developed.

Greenhouse screening was used to identify contrasting parents for internal efficiency. They were IR20 (efficient) and IR55178-3B-9-3 (inefficient). A total of 285 F_8 RILs were developed from IR20/IR55178-3B-9-3. The F_8s were phenotyped using the same screening technique (relative tillering ability), and 42 efficient and 42 inefficient lines were identified for AFLP analysis. Based on qualitative data, a major gene for P-deficiency tolerance was mapped on chromosome 12. The availability of some restriction fragment length polymophism (RFLP) markers in this region allowed further analysis and linkage was found between RFLP markers RG9 and RG241 with a P-deficiency-tolerance gene. RG9 was only 5.5 cM away from the gene (IRRI, 1998).

QTL analysis revealed that P-deficiency tolerance using relative tillering ability as a measurement has important QTLs. The first QTL flanked simple sequence repeat (SSR) marker R69 and OSR 32 on chromosome 12. The second QTL was mapped on chromosome 6, closely linked to AFLP markers E3/M8-3 and E1/M7-7.

Zn Deficiency

A mapping population (F_8 RIL) from a cross of Madhukar (Zn-deficiency-tolerant) and IR20 (Zn-deficiency-sensitive) was generated at IRRI. A greenhouse method of phenotyping was used to identify contrasting RILs. Pregerminated seeds of each RIL were sown in pots with Zn-deficient soil (0.04 kg Zn/ha) with a pH of 7.8. After 45 days, Zn-deficiency symptoms were visually evaluated based on IRRI's Standard Evaluation System (IRRI, 1996b). Selective genotyping (93 RILs) was done using microsatellite markers to locate QTLs. QTL analysis demonstrated that Zn-deficiency tolerance was located on chromosome 5 flanking microsatellite markers RM167 and RM87, explaining 61.9 percent of the variation.

Fe Toxicity

Inheritance of Fe toxicity is a complex quantitative trait, but a symptom such as leaf bronzing is a straightforward indicator. Tissue tolerance for Fe toxicity in rice was considered to be an important parameter for genetic studies. Wu and colleagues (1998) mapped these genes using the doubled-haploid (DH) population developed at IRRI. A total of 135 DH-derived lines from the cross of Azucena (Fe-toxicity-tolerant japonica variety) and IR64 (sensitive indica variety) were phenotyped in a culture solution with Fe-toxic conditions (250 mg·liter^{-1}) maintained at pH 4.5. Results revealed that Fe-toxicity tolerance using the leaf-bronzing index (LBI) as a measurement is located on two loci of chromosome 1. The first locus was flanked by RG146 and RG345, which explained 31 percent of the variation, and the second locus was flanked by RG810 and RG331, where both LBI and Fe concentration in the shoot were located. The first locus was also associated with the gene loci for the Fe-induced enzyme involved in the ascorbate-specific H_2O_2 scavenging system, glutathione reductase activity, and dehydroascorbate concentration, which were also detected on chromosome 7 between RG773 and RG769. Another QTL was detected on chromosome 2, flanking PalI and RZ58, for Fe concentration in the shoot.

Al Toxicity

Toxic concentrations of Al are generally found in acid soils (pH<5.0). A reduction in the amount of available P at low soil pH increases Al-toxicity stress. A severe inhibition of root growth is a direct effect of Al on plants (Lu-Ning and Hou-Tian, 1989). This reduction in root growth restricts water and nutrient uptake and leads to poor growth and yield. Mapping populations for tagging Al-toxicity tolerance were developed at IRRI. IR28 and IR1552 are highly sensitive to Al-toxicity, whereas Azucena is tolerant. IR28/Azucena and IR1552/Azucena crosses were used to develop RILs. Most lines of the IR28/Azucena cross showed high spikelet sterility, making them unsuitable as a mapping population. The F_8 RIL produced from the IR1552/Azucena cross showed no abnormality, thus making it suitable as a mapping population. The DH population of IR64/Azucena was also suitable for mapping Al-toxicity tolerance because IR64 is also sensitive to the stress. The main advantage of the F_8 RIL is that IR1552 is more sensitive to Al stress than IR64. Wu and colleagues (2000) used the F_8 RIL of the IR1552/Azucena population developed at IRRI to map the genes for Al-toxicity tolerance using relative root length (RRL) as a parameter. RRL was measured after two and four weeks in a nutrient solution with 1 mM of Al and a control at pH 4.0. Three QTLs on chromosome 1 and 12 were detected after two weeks of stress, and one significant QTL flanked RG323 and RZ801 on chromosome 1, explaining 19 percent of the total variation. These flanking markers were also associated with Fe-toxicity tolerance (Wu et al., 1998). A QTL expressed on chromosome 12 was detected after both two and four weeks of Al stress. It was linked to RG9, which was also reported to be linked with the major QTL for P deficiency. These relationships of the genes for the stresses suggest a greater possibility of developing rice lines with multiple-stress tolerance.

A breeding program for tolerance of salinity and related abiotic stresses in rice should emphasize developing MAS techniques that can increase selection efficiency and permit the simultaneous selection of several traits to deal with abiotic stresses.

REFERENCES

Akbar, M., K.K. Jena, and D.V. Seshu (1987). Salt tolerance in wild rices. *International Rice Research Newsletter* 12(5):15.

Akbar, M., G.S. Khush, and D. HilleRisLambers (1986). Genetics of salt tolerance in rice. In *Proceedings of the International Rice Genetics Symposium.* Los Baños, Philippines: IRRI, pp. 399-409.

Akita, S. and G.S. Cabuslay (1990). Physiological basis of differential response to salinity in rice cultivars. *Plant and Soil* 123:277-294.

Artsaenko, O., M. Peisker, U.Z. Nieden, U. Fiedler, E.W. Weiler, K. Muntz, and U. Conrad (1995). Expression of a single-chain Fv antibody against abscisic acid creates a wilty phenotype in transgenic tobacco. *The Plant Journal* 8:745-750.

Blackman, P.G. and W.J. Davies (1985). Root to shoot communication in maize plants of the effects of soil drying. *Journal of Experimental Botany* 36:39-48.

Blake, J. and W.K. Ferrell (1977). The association between soil and xylem water potential, leaf resistance, and abscisic acid content in droughted seedlings of Douglas-fir *(Pseudotsuga menziesii). Physiologia Plantarum* 39:106-109.

Cabuslay, G., O. Ito, and A. Alejar (1999). Genotypic differences in physiological responses to water deficit in rice. In *Genetic Improvement of Rice for Water-Limited Environments,* O. Ito, J. O'Toole, and B. Hardy (Eds.). Proceedings of the Workshop on Genetic Improvement of Rice for Water-Limited Environments, December 1-3, 1998, Los Baños, Philippines: International Rice Research Institute, pp. 99-116.

Close, T.J., N.C. Meyer, and J. Radik (1995). Nucleotide sequence of a gene encoding a 58.5-kilodalton barley dehydrin that lacks a serine tract. *Plant Physiology* 107:289-290.

Cohen, A. and E.A. Bray (1990). Characterization of three mRNAs that accumulate in wilted tomato leaves in response to elevated levels of endogenous abscisic acid. *Planta* 182:27-33.

Cornish, K. and J.A.D. Zeevaart (1985). Abscisic acid accumulation by roots of *Xanthium strumarium* L. and *Lycopersicon esculentum* Mill. in relation to water stress. *Plant Physiology* 79:653-658.

Delauney, A.J. and D.P.S. Verma (1993). Proline biosynthesis and osmoregulation in plants. *The Plant Journal* 4:215-223.

Garcia, A., D. Senadhira, T.J. Flowers, and A.R. Yeo (1995). The effects of selection for sodium transport and of selection for agronomic characteristics upon salt resistance in rice (*Oryza sativa* L.). *Theoretical and Applied Genetics* 90:1106-1111.

Gibon, Y., M.A. Bessieres, and F. Larher (1997). Is glycinebetaine a non-compatible solute in higher plants that do not accumulate it? *Plant, Cell and Environment* 20:329-340.

Gowing, D.J.G., W.J. Davies, and H.G. Jones (1990). A positive root-sourced signal as an indicator of soil drying in apple, *Malus* × *domestica* Borkh. *Journal of Experimental Botany* 41:1535-1540.

Gregorio, G.B. and D. Senadhira (1993). Genetic analysis of salinity tolerance in rice. *Theoretical and Applied Genetics* 86:333-338.

Gregorio, G.B., D. Senadhira, and R.D. Mendoza (1997). Screening rice for salinity tolerance. *IRRI Discussion Paper Series No. 22*. Manila, Philippines: International Rice Research Institute.

Hayashi, H., H. Alia, L. Mustardy, P. Deshnium, M. Ida, and N. Murata (1997). Transformation of *Arabidopsis thaliana* with the *codA* gene for choline oxidase: Accumulation of glycinebetaine and enhanced tolerance to salt and cold stress. *The Plant Journal* 12:133-142.

Hossain, M. (1994). A note on constraints to growth in rice production in the Philippines. Paper presented at the conference on Rice Research Prioritization in Asia, February 15-22. Los Baños, Philippines: IRRI.

Hu, C.A., A.J. Delauney, and D.P.S. Verma (1992). A bifunctional enzyme (Δ-pyrroline-5-carboxylate synthetase) catalyzes the first two steps in proline biosynthesis in plants. *Proceedings of the National Academy of Science* 89:9354-9358.

IRRI (International Rice Research Institute) (1977). *Annual Report for 1976*. Manila, Philippines: IRRI.

IRRI (International Rice Research Institute) (1979). *Annual Report for 1978*. Manila, Philippines: IRRI.

IRRI (International Rice Research Institute) (1980). *Annual Report for 1979*. Manila, Philippines: IRRI.

IRRI (International Rice Research Institute) (1983). *Annual Report for 1981*. Manila, Philippines: IRRI.

IRRI (International Rice Research Institute) (1990). *Annual Report for 1989*. Manila, Philippines: IRRI.

IRRI (International Rice Research Institute) (1995). *Program Report for 1994*. Manila, Philippines: IRRI.

IRRI (International Rice Research Institute) (1996a). *Program Report for 1995*. Manila, Philippines: IRRI.

IRRI (International Rice Research Institute) (1996b). *Standard Evaluation System Manual.* Los Baños, Philippines: IRRI.

IRRI (International Rice Research Institute) (1997). *Rice Almanac,* Second Edition. Manila, Philippines: IRRI with Centro Internacional de Agricultura Tropical and West Africa Rice Development Association.

IRRI (International Rice Research Institute) (1998). *Program Report for 1997.* Manila, Philippines: IRRI.

IRSTON, 1991. Final report of the Fifteenth International Rice Salinity Tolerance Observation Nursery (Fourteenth IRSTON) 1990. In *Final Report of the 1990 INGER Nurseries.* Los Baños, Philippines: IRRI.

Ishitani, M., T. Nakamura, S.Y. Han, and T. Takabe (1995). Expression of the betaine aldehyde dehydrogenase gene in barley in response to osmotic stress and abscisic acid. *Plant Molecular Biology* 27:307-315.

Jones, R.J. and T.A. Mansfield (1970). Suppression of stomatal opening in leaves treated with abscisic acid. *Journal of Experimental Botany* 21:714-719.

Khalil, A.A.M. and J. Grace (1993). Does xylem sap ABA control the stomatal behavior of water-stressed sycamore (*Acer pseudoplatanus* L.) seedlings. *Journal of Experimental Botany* 44:1127-1134.

Kriedemann, P.E., B.R. Loveys, G.L. Fuller, and A.C. Leopold (1972). Abscisic acid and stomatal regulation. *Plant Physiology* 49:842-847.

Larque-Saavedra, A. and T.A. Wain (1976). Studies on plant growth-regulating substances. XLII. Abscisic acid as a genetic character related to drought tolerance. *Annals of Applied Biology* 83:291-297.

Lebreton, C., V. Lazic-Jancic, A. Steed, S. Pekic, and S.A. Quarrie (1995). Identification of QTL for drought responses in maize and their use in testing causal relationships between traits. *Journal of Experimental Botany* 46:853-865.

Lee, K.S. (1995). "Variability and genetics of salt tolerance in japonica rice (*Oryza sativa* L.)." PhD Thesis, University of the Philippines, Los Baños, Philippines.

Lin, C.C. and C.H. Kao (1995). NaCl stress in rice seedlings: Effects of L-proline, glycinebetaine, L- and D-asparagine on seedling growth. *Biologia Plantarum* 37:305-308.

Lu-Ning, and Hou-Tian (1989). Effects of aluminum on physiological functions of rice seedlings. *Acta Botanica Sinica* 31:847-853.

Mackill, D.J., W.R. Coffman, and D.P. Garrity (1996). *Rainfed Lowland Rice Improvement.* Manila, Philippines: IRRI.

Milborrow, B.V. (1967). The identification of (+)-Abscisin II [(+)-Dormin] in plants and measurement of its concentrations. *Planta* 76:93-113.

Mishra, B., M. Akbar, and D.V. Seshu (1990). Genetic studies on salinity tolerance in rice toward better productivity in salt-affected soils. Paper presented at the Rice Research Seminar, July 12, 1990, Los Baños, Philippines: IRRI.

Mishra, S.B., D. Senadhira, and N.L. Manigbas (1996). Genetics of submergence tolerance in rice (*O. sativa* L.). *Field Crops Research* 46:177-181.

Mittelheuser, C.J. and R.F.M. van Steveninck (1969). Stomatal closure and inhibition of transpiration induced by (RS)-Abscisic acid. *Nature* 221:281-282.

Moeljopawiro, S. and H. Ikehashi (1981). Inheritance of salt tolerance in rice. *Euphytica* 30:291-300.

Mundy, J. and N. Chua (1988). Abscisic acid and water-stress induce the expression of a novel rice gene. *The EMBO Journal* 7:2279-2286.

Narayanan, K.K., S. Krishnaraj, and S.R. Sree Rangaswamy (1990). Genetic analysis for salt tolerance in rice. In *Proceedings of the Second International Rice Genetics Symposium*, May 14-18, 1990, Los Baños, Philippines: IRRI.

Nylander, M., J. Svensson, E. Tapio Palva, and B.V. Welin (2001). Stress-induced accumulation and tissue-specific localization of dehydrins in *Arabidopsis thaliana*. *Plant Molecular Biology* 45:263-279.

Quarrie, S.A. (1982). Droopy: A wilty mutant of potato deficient in abscisic acid. *Plant, Cell and Environment* 5:23-26.

Rathinasabapathi, B., K.F. McCue, D.A. Gage, and A.D. Hanson (1994). Metabolic engineering of glycine betaine synthesis: Plant betaine aldehyde dehydrogenases lacking typical transit peptides are targeted to tobacco chloroplasts where they confer betaine aldehyde resistance. *Planta* 193:155-162.

Saab, I.N. and R.E. Sharp (1989). Non-hydraulic signals from maize roots in drying soil: Inhibition of leaf elongation but not stomatal conductance. *Planta* 179:466-474.

Tal, M. and Y. Nevo (1973). Abnormal stomatal behavior and root resistance, and hormonal imbalance in three wilty mutants of tomato. *Biochemical Genetics* 8:291-300.

Wade, L.J., S. Sarkarung, C.G. McLaren, A. Guhey, B. Quader, C. Boonwite, S.T. Amarante, A.K. Sarawgi, A. Haque, D. Harnpichitvitaya, et al., (1995). Genotype × environment interaction and selection methods for identifying improved rainfed lowland rice genotypes. In *Fragile Lives in Fragile Ecosystems*, Proceedings of the International Rice Research Conference, February 13-17, 1995, Los Baños, Philippines: IRRI, pp. 885-900.

Wright, S.T.C. (1969). An increase in the "inhibitor-β" content of detached wheat leaves following a period of wilting. *Planta* 86:10-20.

Wright, S.T.C. (1977). The relationship between leaf water potential (ψ leaf) and the levels of abscisic acid and ethylene in excised wheat leaves. *Planta* 134:183-189.

Wu, P., B. Hu, C.Y. Liao, J.M. Zhu, Y.R. Wu, D. Senadhira, and A. Paterson (1998). Characterization of tissue tolerance to iron by molecular markers in different lines of rice. *Plant and Soil* 203:217-226.

Wu, P., C.Y., Liao, B. Hu, K.K. Yi, W.Z. Jin, J.J. Ni, and C. He (2000). QTLs and epistasis for aluminum tolerance in rice (*Oryza sativa* L.). *Theoretical and Applied Genetics* 100:1295-1303.

Yeo, A.R. and T.J. Flowers (1986). Salinity resistance in rice and a pyramiding approach to breeding varieties for saline soils. In *Plant Growth, Drought, and Salinity*, N.C. Turner and J.B. Passioura (Eds.). Melbourne, Australia: CSIRO, pp. 161-173.

Zhang, J.H. and W.J. Davies (1989). Sequential response of whole plant water relations to prolonged soil drying and the involvement of xylem sap ABA in the regulation of stomatal behavior of sunflower plants. *New Phytologist* 113:167-174.

Chapter 15

Breeding for Abiotic Stress Tolerance in Maize

Hans-Werner Koyro
Bernhard Huchzermeyer

INTRODUCTION

It is forecasted that by 2050 the world population will increase from the current level of approximately 6 million people to more than 8 billion. Feeding this population will require an increase in food production, especially in developing countries. For the foreseeable future, conventional agriculture will be our primary response, with cereal grains playing a pivotal role (Hoisington et al., 1999). In contrast to the enormous gains in plant productivity, the extension of irrigated agriculture, poor local irrigation practices, and the intensive use of water resources, combined with high evaporative rates in arid and semiarid regions, have inevitably given rise to the problem of salinity in the soil and in underground water (Koyro and Huchzermeyer, 2002b). Every year, 6 million ha of arable land are lost in developing countries and the economic existence of more than 1 billion people in 100 countries is threatened by the expansion of the deserts.

Improvement of salinity and drought tolerance of crop plants such as maize and rice by genetic means has an important but largely unfulfilled role in modern agricultural development. However, maize and rice will account for almost 80 percent of the cereal import requirement of developing countries. The International Food Policy Research Institute has predicted that by the year 2020 almost 60 percent of the world's maize consumption will be in developing countries. The annual global demand will be close to 775 million tons for maize.

All breeding aspects of different stresses in maize cannot be reviewed in a single chapter without major omissions, because there are many different biotic, such as pathogens, and abiotic stressors to be mentioned. The challenge posed by the world's expanding demand for food and the outstanding importance of food production in developing countries in arid or semiarid regions led this review to focus on plant water deficit with several different stressors such as drought, salinity, and mineral deficiency.

Optimistic discussions of future salt and drought tolerance have invoked the unrealistic vision of crop plants such as maize growing as xero- or halophytes on severely salt-affected land or being irrigated with sea water (Winicov, 1998). For the discussion of recent developments in stress-tolerance breeding in maize, "improved resistance to water deficit" will imply continued growth under conditions where a sensitive genotype shows severely inhibited growth or dies.

Multigenic Responses to Abiotic Stresses

Environmental factors that impose water-deficit stress, such as salinity, drought, and temperature extremes, place major limits on plant productivity (Boyer, 1982; Cushman and Bohnert, 2000). To reduce these limitations in the face of a growing world population, more stress-tolerant crops must be developed (Kush, 1999). At the genetic level, drought and salinity tolerance have been considered to be quantitative traits with influence on maximal plant yield and productivity (Foolad and Jones, 1993). The improvement of tolerance to drought and salinity generally leads to a lowering of the potential yield under normal conditions. Therefore, the use of plants with these characteristics is likely to be restricted to areas of limited arable lands.

Improvement in salt and drought tolerance in crops has been elusive because both abiotic stresses lead to multigenic responses (Winicov and Bastola, 1997). However, diverse genes that are induced and repressed by dehydration, such as oxidative stress defenses, also respond to salt stress and some, but not all, respond to cold stress (Shinozaki and Yamaguchi-Shinozaki, 1996; Winicov, 1998). The commonality of responses indicates similar functions of these gene products for plants under stress conditions involving water deficit.

There are some very basic physiological observations, but the underlying biochemical events are a network of interacting reactions rather than an effect brought about by a single enzyme. For instance, when maize encounters water deficits, there is a decline in photosynthesis per plant. This can be due to a reduction in light interception as leaf expansion is reduced or as leaves senescence, and a reduction in C fixation per unit leaf area as stomates close or as photooxidation damages the photosynthetic mechanism. Leaf senescence in general is associated with higher oxidative stress and a decline in antioxidant activity toward maturity. Mn-SOD has a major role in the scavenging of superoxide radicals during maize leaf senescence (Prochazkova et al., 2001). The accumulation of abscisic acid may enhance survival but reduce productivity (Leung and Giraudat, 1998; Mugo et al., 2000).

Genes that are induced in a stress-specific manner increase the complexity of engineering tolerance traits in plants. In addition, during salt stress, different short-term and long-term responses contribute to salt tolerance. This problem becomes even more complicated by the fact that quantitative trait loci (QTL) that are linked to tolerance at one stage of development can differ from those linked to tolerance at other stages (Foolad, 1999).

PHYSIOLOGICAL ASPECTS RELEVANT TO STRESS-TOLERANCE BREEDING

The sense of applying genetic engineering depends completely on the understanding of the complexity and interplay of tolerance mechanisms. In no known example are physiological data sufficiently informative to completely describe the tolerance mechanisms of a species. It is extremely helpful to concentrate all efforts in salt- and drought-tolerance breeding in maize on the four major constraints of plant growth on saline substrates (Koyro and Huchzermeyer, 2002b): (1) water deficit, (2) restriction of CO_2 uptake, (3) ion toxicity, and (4) nutrient imbalance. Some physiological aspects of stress-tolerance breeding in maize are presented in the following section.

Special Physiological Examinations

Although plants are exposed to many types of environmental stresses, they respond to these stresses generally at morphological, anatomical, cellular, and molecular levels (Winicov, 1998; Zhang et al., 1999). One major reason for the lack of understanding of stress tolerance or sensitivity to plant water deficit, including drought or salinity, is missing data about factors such as compartmentation; homeostasis, including selective ion uptake and transfer across membranes; and metabolic flux (Niu et al., 1995; Holmberg and Bülow, 1998; Koyro and Huchzermeyer, 1999). The understanding of biochemical mechanisms in physiological or metabolic adaptation to plant water deficit or other abiotic stress at the cellular level provides potential candidate genes for engineering improved tolerance (Winicov and Bastola, 1997). Physiological or metabolic adaptation to abiotic stress at the cellular level are the main responses amenable to molecular analysis (Shinozaki and Yamaguchi-Shinozaki, 1997). A possibility to find such limiting factors is the study of the relations inside single cells such as (1) the compartmentation between cytoplasm and vacuole, (2) the selective transfer across membranes, and (3) the distribution of elements in different cell types or along a diffusion zone in a root apoplast.

The analysis of the distribution of elements or compatible solutes (compartmentation) in a distinct cell type can be used to answer several questions such as:

1. Is there a compartmentation and a division of labor in cytoplasm and vacuole?
2. Is it necessary to look in more detail to toxic effects caused by high cytoplasmic sodium or chlorine concentrations?
3. Is there a compatibility of the metabolism with high NaCl concentrations in the symplast (cytoplasm)?
4. Is there a high ionic imbalance in the symplast (cytoplasm)?

However, without such knowledge, genetic engineering cannot have a complete and direct target. The complexity of stress adaptation and possible models of action were listed by Cushman and Bohnert (2000), knowing that they could not supply a complete list. As these traits are better understood, the underlying genes can be isolated and used to identify corresponding genes in the genetic resources.

Adaptation to Osmotic and Ionic Stress

As a rule, includer species, which accumulate large quantities of inorganic ions in their leaves for osmotic adjustment, must protect their enzymes in the cytoplasm and in the chloroplasts by strict compartmentation. In maize-breeding concepts it may be expected that the ability to protect cellular membranes and enzymes from stress and to recover from water deficits will also enhance the crop's capacity to survive and produce grain (Zhang et al., 1999). In turn, the ability of the plant to capture water from a drying soil through deeper or more thorough root exploration or through osmotic adjustment should stabilize and increase the flow of assimilates to developing grains. A report by Lemcoff and colleagues (1998) links osmotic adjustment to stable grain yield under the transient stresses that are often encountered in wide-area testing of modern maize hybrids. In addition, based on the relation between cell volume and osmotic potential, various methods have been proposed to quantify osmotic adjustment capacity (Zhang et al., 1999), mainly as a basis for the development of molecular strategies for the improvement of plant-stress resistance through manipulating osmotic adjustment.

Certain molecules are produced or accumulated to detectable levels during water deficits, suggesting a protective role for the plants. Polyamines, for example, are thought to play a role in plant growth and development, especially under stressful growing conditions (Ben-Hayyim et al., 1994). Many of the changes due to dehydratation stress in plants include the accumulation of a variety of sugars, proline, and glycinebetaine in addition to changes in protein levels (Bartels and Nelson, 1994; Ingram and Bartels, 1996; Verslues and Sharp, 1999; Zinselmeier et al., 1999). These changes are associated with osmotic adjustment and the protection of membranes from damage as cell contents desiccate.

Proline dehydrogenase (P-DH), the first enzyme in the proline oxidation pathway, was shown to be down-regulated in maize during drought stress but not affected by ABA treatments, suggesting an ABA-independent drought-mediated regulation (Dallmier and Stewart, 1992). Proline was postulated also to have a protective function similar to polyamines during water deficits, but the exact mechanism is unknown. These data suggest that P-DH could be included as a target for screening of water-deficit resistance in maize.

Glycinebetaine is an important osmolyte, and its genetics have been studied by scientists through the physiological-genetic approach. In maize, Yang and colleagues (1995) developed near-isogenic F_8 pairs of glycinebetaine-containing (wild-type) and glycinebetaine-deficient lines. The pairs of lines differed for alternative alleles at a single locus (wild-type *Bet1*, mutant *bet1*). This gene affects glycinebetaine biosynthesis at the level of choline oxidation (Rhodes et al., 1993). These data suggest that choline oxidation and proline dehydrogenating could be included as a target for screening of water-deficit resistance in maize.

Determining Yield

Many of the mechanisms that favor survival have little value in enhancing grain yield under that level of plant water deficit, including salinity, that varies spatially and across seasons. Hybrids usually give higher yields than varieties under drought, with heterosis acting as an important source of stress tolerance (Blum, 1997). Obviously, yield potential is important in determining yield under moderate stress. It becomes much less so when yields fall below 50 to 60 percent of potential, when stress-adaptive secondary traits assume a real significance (Bänziger and Lafitte, 1997). For maize, the critical developmental period of determining grain yield centers on flowering and early seed development (Claassen and Shaw, 1970; Shaw, 1977). Numerous studies have examined components of stress response during the flowering stage of development (Bolaños and Edmeades, 1996; Zinselmeier et al., 1999, 2000). A recent study showed that a steady flow of photoassimilates plays an important role in both ovule and seed development (Zinselmeier et al., 1999). Interruption of this flow due to dehydration causes reduction in key carbon metabolic enzyme activities.

Grain yield is normally highly correlated with the kernel number per unit area and per plant rather than with weight per kernel (Bolaños and Edmeades, 1996; Edmeades et al., 1999; Andrade et al., 1999). Factors affecting grain set under drought, therefore, are of special interest in commercial breeding endeavors. Because maize is an out-crossing species, pollen must move from the anthers at the top of the plant to the exposed silks of the same and the surrounding plants. This process is risky because pollen and the delicate stigmatic tissue are exposed to a desiccating environment. One universal phenomenon observed when maize flowers are under drought is the delay of silking in relation to pollen shed, giving rise to the anthesis-silking interval (ASI), whose duration is highly correlated with kernel set (DuPlessis and Dijkhuis, 1967; Edmeades et al., 2000). Under such conditions, pollen can arrive after it has desiccated, when silks have withered or senesced, or after ovaries have exhausted their starch reserves (Bassetti and Westgate, 1993a,b; Saini and Westgate, 2000; Zinselmeier et al., 2000).

The Influence of the Root System on Maize Yield Under Stress

Root systems obviously play an important role in water acquisition for plants and are a significant component of tolerance to water-deficit stress (Barker and Varughese, 1992; Weerathaworn et al., 1992; McCully, 1999). In rice, some groups have shown the significance of certain root attributes that contribute to higher yields in low-water-stress-prone environments and identified the corresponding quantitative trait loci (Hemamalini et al., 2000; Zheng et al., 2000). Others have made progress in using markers in breeding modifications of root attributes in rice that correlate well with improved tolerance to dry environments (Mackill et al., 1999). It is apparent from such studies, including the recurrent selection described previously, that shifting root biomass from the upper soil zone to deeper regions would enhance the chances of encountering new water sources. This shift in biomass would need to occur in response to drying soils.

An attempt has been made to show a relationship between early seedling root development and the effects of drought on yield (Bolaños et al., 1993). It is likely that vigorous root growth may come at a cost to grain production despite the improved advantage of water acquisition in dry soils. Defining early seedling traits correlating with stress tolerance would benefit the selection process in breeding programs by reducing input resources and time to develop new germplasm.

Storage of Assimilates

Selection for a high and sustainable stomatal conductance during the period of stem elongation is among the current breeding strategies. Increasing assimilate allocation to the reproductive primordial so as to establish a large potential sink should also indirectly increase total crop photosynthesis. Evidence in the major grain crops suggests that by anthesis the capacity for photosynthesis is not limiting during grain filling. To use this surplus capacity, it was suggested that carbon and nitrogen partitioning to the reproductive meristem be increased so as to establish a high potential grain number and the potential for a large grain size. It is then expected that additional photosynthesis will follow by an extended leaf-area duration (Richards, 2000).

The flux of currently formed assimilates to the developing ear is a powerful determinant of ear growth, suggesting that stored reserves in the stem and ear shank have little to do with reproductive success (Westgate and Boyer, 1985; Schussler and Westgate, 1995). Despite the long history of germplasm improvement by selfing under conditions of high plant density and occasional exposure to heat and drought, delayed silking (increasing ASI) under severe drought stress can still be evident and thus negatively affect grain yield. However, it has to be taken into account that the storage capacity of developing maize grains is established during the early stage of endosperm development by the process of cell division and cell enlargement. This way, metabolic capacity and volume of storage compartment are created (Setter and Flannigan, 2001). Any kind of abiotic stress subsequent to fertilization, in the initial period of grain development, inhibits these early events and results in a decrease of final grain yield (Commuri and Jones, 1999; Ober and Sharp, 1994).

During early endosperm development, the mitotic phase is followed by a developmental transition, after which cells undergo repeated rounds of S phase without entering mitosis (Setter and Flannigan, 2001). During this phase, mRNAs for starch pathway enzymes as well as storage proteins can be detected at high concentrations (Ober and Sharp, 1994). These mRNA concentrations correlate with cell size among different cereal genotypes subjected to environmental stress (Lemontey et al., 2000; Setter and Flannigan, 2001). With respect to maize kernel growth, it has been shown that the stage in which the number of endosperm cells is established is extremely sensitive to stress. However, during starch accumulation, maize is relatively tolerant to drought and salt stress (Ouattar et al., 1987). With respect to breeding stress-tolerant maize, the difference in sensitivity has to be elucidated in more detail.

Other studies demonstrated that acid invertase activity is affected by drought stress and thus the assimilate flux to the developing ear is inhibited (Schussler and Westgate, 1995; Zinselmeier et al., 1999, 2000). The insoluble form of acid invertase activity correlated well with ovary dry-mass accumulation and final kernel number between well-watered and water-stressed plants. These results suggest that increased acid invertase activity is important for the continuing development of ovaries just prior to pollination and kernels after pollination (Zinselmeier et al., 2000).

The Intrinsic Regulation of Photosynthetic Capacity and Transport

Because maize evolved in the tropics, where nitrogen is inherently deficient, natural selection has favored the early uptake of N from the soil, its storage in leaves as photosynthetic enzymes, and its subsequent remobilization to developing grains during grain filling. When the grain sink size is increased by selection, this can sometimes accelerate foliar senescence as the demands for N by the grains exceed capacity for uptake from a dry soil (Chapman and Edmeades, 1999). In this context, it must be mentioned that a strict control of photosynthesis exists in all plant species. In response to high assimilate availability, sinks are elaborated and new ones are developed. On the other hand, an inhibition of building new sink capacity results in an accumulation of sugars in the source leaves and in a feedback inhibition, particularly via cytokinins, of photosynthesis genes in the developing leaves. The interaction between carbon-to-nitrogen ratio, photosynthetic capacity, and grain sink size is complex and far from being understood in detail, but it is obvious that grain set is a crucial step on the path to grain yield (Paul and Foyer, 2001).

As summarized by Richards (2000), the grain yield of cereals has almost doubled during the past 70 years. Surprisingly for geneticists, but in total agreement with basic concepts of biochemistry, there has been no change in the rate of photosynthesis per unit leaf area. However, total photosynthesis has increased as a result of an increase in leaf area and leaf-area duration. As leaf development is crucial for grain yield, many attempts have been made to identify traits to monitor growth performance and to predict, during plant growth, the extent of any positive or adverse effect on growth and development of leaves. An approach to elucidate the biochemical mechanisms controlling leaf growth is to combine spatial analysis of expansion rate and measurement of candidate enzyme activities (Palmer and Davies, 1996). In maize leaves, the evaporative demand and cell water deficit independently affect the elongation rate. Both tissue expansion rate and cell division rate are inhibited to the same extent by soil water deficit (Tardieu et al., 2000).

Substantial opportunities to continue to improve crop yield genetically using conventional breeding practices still remain. Selectable traits have to be identified as early indicators of increasing biomass under favorable conditions. Opportunities exist to alter crop duration and the timing of crop development to match these better to radiation, temperature, and water supply during crop growth. The importance of any trait depends on the environment in which the crop is grown.

Increases in crop photosynthesis through breeding are also likely to come via indirect means. Photosynthesis is known to be inhibited, for instance in saline habitats, by adverse cytosolic NaCl concentrations. Therefore, monitoring salt flux into and out of leaves is among the most-often measured parameters, and it has been found that maize is different from other crops. Approximately 13 to 36 percent of the Na^+ and Cl^- imported into the maize leaves through the xylem were exported by the phloem (Lohaus et al., 2000). This high phloem translocation rate of ions found in maize may be due to the higher carbon translocation rate observed for C_4 plants than for C_3 plants. On the other hand, inhibiting effects of internal salt were not reduced by rain, and extremely low ion leaching was measured with maize leaves compared with other plant species. Until now, no breeding concept has focused directly on aspects of salt transport within the shoot.

The Concept of ABA Signaling

It is important to note in this context that in modern farming, maize plants are grown close to one another. Therefore, the roots of individual plants compete with neighboring ones for water as well as nutrients. Moreover, there is an interacting network for ABA. The hormone is synthesized in leaves as well as in roots. It may move freely from plant to soil and from there to a neighboring plant. It can also move rapidly through the plant in both the xylem and the phloem and will partition between different compartments in different tissues largely as a function of pH (Sauter et al., 2001). These ABA translocations not only bring about interactions of neighboring plants with respect to the regulation of transpiration and photosynthesis but also affect root nutrient uptake. It was found that ABA regulated the membrane potential of root cells and that this regulation is consistent with the hypothesis that ABA-induced K^+ accumulation in roots is mediated by K^+ channels (Roberts and Snowman, 2000). ABA is produced to high levels in response to water deficit, including salinity (Ingram and Bartels, 1996; Leung and Giaudat, 1998). Stomatal closure is one obvious result of the chain of events but, in the long run, these interactive effects eventually lead to leaf-surface reduction through limitation of nitrogen uptake, bringing about reduced plant growth under drought (Bahrun et al., 2002).

Exogenous ABA treatments can also affect the expression of dehydration-regulated genes, suggesting that ABA acts as a signal transducer for response to water-deficit stress (Leung and Giraudat, 1998). However, other drought-induced genes do not respond to ABA treatments, suggesting an ABA-independent pathway and a greater complexity in the drought-response phenomenon (Gosti et al., 1995; Shinozaki and Yamaguchi-Shinozaki, 1996). In *Arabidopsis,* numerous genes that respond to dehydration stress were identified and have been categorized as "responsive to dehydration" (rd), and "early response to dehydration" (erd) (Yamaguchi-Shinozaki et al., 1992). Moreover, the potential existence of four ABA-independent and one ABA-dependent pathway has been described (Yamaguchi-Shinozaki et al., 1992). The transferability of these results to maize needs to be studied, and the corresponding genes need to be identified.

PLANTING REGIME AND SITE-SPECIFIC EFFECTS ON PLANT PERFORMANCE

Since hybrids are planted at ever-higher population densities, the stresses imposed by high density merit examination (Modarres et al., 1998; Tollenaar and Wu, 1999). Higher planting densities obviously create greater competition for input resources such as solar radiation capture, nutrients, and water acquisition (Scheiner et al., 2000). Increasing planting densities leads to less available water per plant and thus an acceleration of the occurrence of water-stress symptoms as dry spells lengthen (Tollenaar and Wu, 1999).

In many breeding projects it has been pointed out very clearly that it is most important to test newly produced germplasm for productivity at different sites. It was obvious that the yield could differ remarkably at different sites, especially when the plots were located in different countries. This observation was explained in terms of different patterns of local stresses. Recently, it was pointed out that in many cases such observations may be due to symbiotic interactions between the maize root system and microorganisms in the soil. A prerequisite for such interactions is a matching receptor pattern on the surface of both roots and microorganisms. Besides nutrient allocation, the supply of stress-reducing components may be among the benefits maize can receive from soil symbionts. The population of microorganisms greatly varies among different locations, and it eventually changes with the supply of fertilizers and, even more drastically, under drought and salt stress (Lieth, personal communication).

Maize improvements in grain yield have been due to a variety of factors, one of which is greater tolerance to environmental stresses. Drought-stress resistance is among these factors. For maize, the period during pollination and early grain filling is more sensitive to water stress than are the preflowering and late grain-filling stages. To achieve efficient selection for drought tolerance, the selection procedure must include managed environments where the timing, duration, and intensity of water-deficit stress can be controlled.

GENETIC RESOURCES

Plant breeding depends on the correct combination of specific alleles at the ~50,000 genetic loci present in a plant's genome. Obviously, access to a wide range of genetic diversity is essential for the success of any breeding program. The gene banks of the Consultative Group on International Agricultural Research hold nearly 600,000 of the estimated 6 million accessions stored globally (Hoisington et al., 1999). Although ~50,000 accessions of maize exist in germplasm banks around the world (Ayad et al., 1980; Goodman, 1983), most of these have never been adequately evaluated for useful traits (Hoisington et al., 1999). These resources evolved an assortment of alleles needed for resistance and tolerance to the diseases, pests, and abiotic stressful environments found in many developing countries. The International Center for Maize and Wheat Improvement has a mandate for improving the productivity and sustainability of maize and wheat in developing countries. The genetic resource of this center alone contains 18,000 accessions of American maize. However, reasons for low utilization include, as mentioned previously, lack of physiological data and poor linkages between gene banks and breeders. On a global basis only about 5 percent of the available maize germplasm is used commercially (Hoisington et al., 1999). The resulting lack of genetic diversity and the widespread deployment of genetically uniform varieties lead to increased susceptibility to diseases, pests, or abiotic stresses.

It seems to be likely that the significant variation of exotic germplasm has a high potential to improve many quality traits of the germplasm in use and can help to ensure a yield plateau by increased tolerance to drought, other abiotic stresses, or pathogens. Wild relatives of maize, such as *Tripsacum floridanum* or *Zea mexicana,* are an excellent source for novel characteristics improving drought, salinity, pest, or disease resistance (Zscheischler et al., 1990). The utilization of regenerated plants after selection of salt-tolerant cells in cultures has successfully been undertaken with alfalfa and rice (Winicov, 1998). However, this method does not provide ready identification of the affected genes without further studies. The application of molecular techniques ultimately should result in the isolation of critical genes of these species. Applications for in vitro genetic manipulation of maize somatic cell tissue cultures already have been established (Armstrong, 1994). Fertile transgenic plants are reproducibly obtained in some laboratories through either direct DNA delivery into protoplasts or particle bombardment of intact, cultured cells. The successful establishment and development of such molecular techniques to more effectively transfer the traits of stress tolerance, for example against the parasitic weed *Striga* sp., or rust disease, drought, and salinity, to maize varieties or to other crop species are of tremendous importance, especially for tropical agriculture (Zscheischler et al., 1990; Pryor, 1994).

The main goals of breeders are to generate hybrids with higher grain-yield potential, better grain-yield stability, and improved grain traits for end users (Duvick, 1997). One special demand is to produce new hybrids with enhanced tolerance to stressful growing conditions. Such plant material would more easily overcome the problem of variations for available water within a field. This implies a need for a reasonable level of drought tolerance in all maize hybrids, since farmers typically grow only one hybrid in any given field (Kitchen et al., 1999).

Despite its complexities, nearly all of the breeding programs include yield as a selection parameter. However, a better understanding of the components that contribute to higher yields should improve the selection process. Many of the recent improvements seem to involve greater resistance to biotic and abiotic stresses rather than improved yield production under ideal growing conditions. It is generally accepted that modern hybrids show an increased level of stress tolerance that counters the potential water limitations with significantly improved levels of productivity (Castleberry et al., 1984; Duvick, 1997; Tollenaar and Wu, 1999).

THE CLASSICAL APPROACH FOR STRESS-TOLERANCE BREEDING

A classical approach to improve the genetic variation of maize tropical germplasm for drought tolerance began in 1975. Selected populations underwent up to eight cycles of recurrent selections in rain-free crop seasons at Tlaltizapán and Obregón, Mexico, where the timing and intensity of drought stress can be regulated by irrigation. Improvements in performance under drought have averaged 126 $kg{\cdot}ha^{-1}$/cycle. Similar gains were also observed under low N supply, suggesting a common mechanism of tolerance to stress at flowering. These improved sources also exhibited tolerance to drought when transferred to other regions in the world. The superiority of local hybrids under stress conditions, however, made the introgression of Mexican sources into local germplasm necessary. This, in combination with screening under managed stress levels, has led more recently to maize hybrids with superior and stable performance across a wide range of growing environments in Africa. The development of an open-pollinated variety that shows a 35 percent superiority over commercial hybrid controls under moderate to severe levels of water stress has been reported (Bänziger, Pixley, et al., 2000).

Gains in yield were largely due to reduced barrenness under drought and not to increased biomass production (Bolaños and Edmeades, 1993; Edmeades et al., 1999). Anthesis date became slightly earlier. Root biomass in the top 50 cm of the soil declined by 35 percent, but there was no change in any trait indicative of plant water status (Bolaños, et al., 1993). Recurrent selection also reduced the final number of spikelets per ear by 2.5 percent per cycle in one population (Edmeades et al., 1993). It is generally accepted that this change is associated with selection for early vigorous silking under drought. Thus, fewer spikelets were formed, but these grew more rapidly and were ultimately more successful in forming grain, especially under conditions of drought at the time of flowering. A dominant symptom of these changes was a reduction in ASI. The QTLs for this trait have been identified and used in marker-assisted backcrossing (Ribaut et al., 1999). It may be mentioned that the mechanism leading to improved yield under drought also appears to hold under low N supply.

Traits for Classical Maize Breeding Concepts

Grain yield under stressed conditions is usually the primary trait for selection. A suitable secondary trait is (1) genetically associated with grain yield under stress, (2) highly heritable, (3) stable and feasible to measure, and (4) not associated with yield loss under ideal growing conditions (Edmeades et al., 2002). Very few proposed secondary traits meet these criteria. It has been shown that the use of secondary traits plus yield improved selection gains for maize yield under low N supply by 20 percent compared with selection for yield alone, with the gains increasing as N deficiency intensified (Bänziger and Lafitte, 1997). Key secondary traits under drought are reduced barrenness, ASI, stay green, and leaf rolling under drought (Bänziger, Edmeades, et al., 2000). Other traits such as changes in the root system will be worth adding only when they have been field-tested and have met these criteria. The apparent increase in stress tolerance in modern maize germplasm has been attributed to a number of factors (Duvick, 1984, 1997; Tollenaar and Wu, 1999), including (1) the occurrence of heat and drought in nurseries with no available irrigation, (2) the use of high plant densities during hybrid line development, (3) large-scale, broad-area testing that includes the use of winter nurseries, and (4) the use of stable, high-yielding progenitors to form the next crop of parental inbreed lines.

GENETIC ENGINEERING STRATEGIES

In contrast to traditional breeding and marker-assisted selection programs, the direct introduction of a small number of genes by genetic engineering seems to be a more attractive and rapid approach to improve stress tolerance (Cushman and Bohnert, 2000). The use of innovative plant breeding and, more recently, molecular engineering led to enormous gains in plant productivity and resistance to a variety of pests and diseases. Traditional breeding strategies that have attempted to utilize genetic variation arising from parental germplasm, interspecific or intergeneric hybridization, induced mutations, and somaclonal variation of cell and tissue cultures have met with only limited success for drought or salt resistance (Flowers and Yeo, 1995; Koyro and Huchzermeyer, 1999). Plants regenerated from cells from cell or tissue culture, selected for improving salinity tolerance, have not shown an unequivocal increase in drought or salinity tolerance (Winicov, 1998; Winicov and Bastola, 1999). Only very few plant introductions with improved stress resistance under field conditions have resulted. This may be due to the limitation of traditional approaches by the complexity of stress-tolerance traits and the low generic variance of yield components (Ribaut et al., 1996, 1997; Frova, Caffulli, and Pallavera, 1999; Frova, Krajewski, et al., 1999).

Engineering-Compatible Osmolytes

Engineered overexpression of biosynthetic enzymes for osmoprotectants, scavengers of reactive oxygen species, sodium/proton antiporters, components of stress-signaling pathways, stress-adaptive proteins, and biochemical "pathway" engineering involving multiple targets for salinity-stress tolerance by improving ion homeostasis are among the approaches to improving stress tolerance (Bohnert and Sheveleva, 1998; Holmberg and Bülow, 1998; Noctor and Foyer, 1998; Pardo et al., 1998; Apse et al., 1999; McNeil et al., 1999; Winicov and Bastola, 1999).

A large number of genes inducible under stress conditions were identified after physiological studies at the cellular level, for example, by gene expression and proteomics, and classified in groups (after Winicov, 1998):

1. photosynthesis,
2. structural components (wall and membrane),
3. osmoprotectants and molecular chaperones,
4. water-channel proteins,
5. ion transport,
6. oxidative stress defenses,
7. detoxifying enzymes,
8. proteinases,
9. proteins involved in signaling, and
10. transcription factors.

After stress-induced gene activation has been demonstrated for a plant, the results should be tested in transgenic plants for their effect on stress resistance or tolerance. This system has been demonstrated for maize by Saneoka and colleagues (1995). It has been shown that the increased accumulation of the compatible solute glycinebetaine in maize leaves correlates with the improvement of salt and drought tolerance (Saneoka et al., 1995) and with *Bet1* gene copy number.

Cataloging of Complex Abiotic Stress Responses

Nonetheless, marker-associated selection of specific secondary traits that are indirectly related to yield, such as osmotic adjustment, membrane stability, or physiological tolerance indices, might prove increasingly useful as the resolution of the genetic and physical chromosome maps on the major crops improves (Frova, Caffulli, and Pallavera, 1999; Frova, Krajewski, et al., 1999; Zhang et al., 1999; Cushman and Bohnert, 2000). The analysis of drought tolerance depended mainly on broad-area testing, which often used data from more than 300 different environments (Jensen and Cavaleri, 1983; Bruce et al., 2002). It appears that maize is generally susceptible to drought around the flowering period, and variation in this susceptibility could give rise to genotype × drought interactions (Shaw, 1977).

Marker-assisted breeding methods are becoming increasingly useful in the development of new germplasm with improved stress tolerance (Hoisington et al., 1996; Quarrie et al., 1999; Hemamalini et al., 2000). To clarify the genetic network involved in stress tolerance, key agronomic traits, such as grain yield, need to be categorized into individual components to reduce complex analyses (Jacobs and Pearson, 1991; Modarres et al., 1998; Tollenaar and Wu, 1999).

In maize it is helpful to consider both the number and size of the kernel generated on the ear, since these are determined at different stages in the life cycle of the crop (Bolaños and Edmeades, 1996; Edmeades et al., 1999; Andrade et al., 1999). Factors affecting kernel development thus become the main research focus in improving yield and yield stability in germplasm development. The target genes that play central roles in developing kernels can also be used to screen for variations in a population showing a continuum of sensitivity to drought stress. Likewise, transgenic approaches can exploit the appropriate modifications of target genes, ultimately to improve drought tolerance (Bartels and Nelson, 1994; Ingram and Bartels, 1996; Xu et al., 1996). Some methods of identifying key genes rely on physiological and biochemical examinations, whereby certain pathways that are rate limiting under stress conditions can be easily identified (Bartels and Nelson, 1994; Leung and Giraudat, 1998).

Engineering Stress Perception and Signaling

Water-deficit stress by drought, salinity, or low temperature is sensed by an unknown set of receptors that initiate a signal cascade with or without ABA and thus affect gene expression of dehydratation-responsive genes through activation of a set of particular transcription factors. Several of the genes induced in these pathways have been identified and include the *rd29B* and *bZIP* transcription factors that regulate rd29B, AREB1, 2, and 3 (Uno et al., 2000).

In *Arabidopsis,* cold- and drought-affected gene expression has been monitored using cDNA microarray technology (Seki et al., 2001). Forty-four cDNA clones were shown to increase twofold or greater at the mRNA level after 2 h of drought stress. A subset of these genes also shows mRNA level modulation after 2 h of cold stress. This suggests similarities in the signaling to gene expression control of both stresses (Yamaguchi-Shinozaki et al., 1992; Seki et al., 2001) and they can also be expected for salt stress.

Mutants have been identified that are especially sensitive to drought effects and may illuminate some of the underlying mechanisms of the stress response and tolerance. For example, several of the nonallelic mutations designated "wilty" in maize exhibit chronic wilting, mostly on the top leaves, under mild drought stress (Neuffer et al., 1997). These mutants do not seem to be affected by ABA sensing, and it has been concluded that the corresponding genes control vascular element development (Rock and Ng, 1999). It is possible that modern maize lines differ more subtly in vascular developments that regulate water flow and contribute to differential tolerance to water stress. Mutations that affect ABA sensing and response in maize and other model plants may also uncover classes of genes whose variations in a population could provide gene targets for associating with tolerant lines (McCarty, 1995; Leung and Giraudat, 1998).

Methods Currently Used

To identify less obvious gene networks that are responding to stress, more massively parallel and sensitive methods are necessary. The advent of whole genomic-related technology is providing the necessary tools to identify the key gene networks that respond to drought stress and relating their regulation to adaptive events occurring during stress. Differential display was one of the earliest methods of parallel screening for differences in the levels of cDNA fragments generated from mRNA isolated from samples between experimental treatments (Liang and Pardee, 1992). This technology depends on semirandom polymerase chain reaction (PCR) using degenerate primers and gel-based separation of the amplified cDNA fragments.

Other methods were subsequently developed using polynucleotides bound to miniature solid supports (Gene-Chips, cDNA microarrays, etc.) hybridized with complex probes either as one or both treatments simultaneously (Schena et al., 1995; Chee et al., 1996). Despite the robustness of this new technology, such methods can rely only on available clones or sequence information. Comprehensive sequence information, however, is just now becoming readily available.

Alternative methods, such as open-ended technologies, use modifications and improvements of the original differential display technology. One such method has been shown to be useful in identifying the regulation of known and novel genes associated with particular phenotypic traits (Bruce et al., 2000). In order to improve the identification of candidate genes, many groups work on the production of a comprehensive expressed sequence tag database for important crops (Gai et al., 2000). Other, more high-throughput methods such as genome map-based and functional genomic approaches can provide indications of less obvious pathways that are nonetheless just as important (Bruce et al., 2000; Seki et al., 2001). The combination of such methods can be expected to reduce the high number of candidate genes to a few target genes that can be handled in experimental approaches.

The Functions of Transcription Factors

Numerous studies have shown that transcription factors are important in regulating plant responses to environmental stress. However, specific functions for most of the genes encoding transcription factors are unclear. Transcription factors potentially controlling downstream gene expression in stress-signal transduction pathways were identified by observed activation and repression of the genes after certain stress treatments. Among 43 transcription factor genes that are induced during senescence, 28 of them also are induced by stress treatment. This finding suggests extensive overlap responses to these stresses. Although further validation is needed, the correlative results available provide a vast amount of information that guide hypothesis-driven research to elucidate the molecular mechanisms involved in transcriptional regulation and signaling networks in plants (Chen et al., 2002).

Probing breeding populations that show improved traits during environmental stress with genomic tools is expected to uncover important gene networks involved in drought and salinity resistance. The cDNA microarray technique has been used to discover maize genes expressed in the ear that change due to water deficit during flowering. The cDNA for selections of maize genes corresponding to cell cycle, some stress responses, carbohydrate metabolism, and ABA-responsive genes were included on the microarrays. Three different patterns of cell-cycle gene expression have been found this way: (1) an increase in mRNA levels during the period of stress, (2) a decrease in mRNA during the stress followed by an increase during subsequent recovery, and (3) an increase in mRNA only during the recovery period. These results suggest distinct functions for members of the cell-cycle-regulation gene family during dehydration and recovery. Similar patterns of expression in response to water-deficit stress were also noted for other genes. These results suggest further analysis of this type.

IDENTIFYING POTENTIAL STRESS-TOLERANT LINES IN MAIZE

Newer genomic-related tools are being used to facilitate the selection of tolerant lines. These tools include associated DNA markers, known orthologous stress-related genes, and newly identified genes based on more conventional physiological and biochemical approaches. RNA profiling experiments to date have identified key target genes that respond to stress inputs and are targets for screening variations in the breeding population. Ultimately, approaches described in this chapter will provide better knowledge of the nature of improved hybrid performance in stressed environments and generate faster and cheaper breeding germplasm for drought tolerance.

Molecular genetics can provide higher resolution of the genome of any species, allowing precise gene identification before attempted transfers (Hoisington et al., 1999). Sequencing of randomly selected cDNA clones or expressed sequence tags (ESTs) are one way toward cataloging genetically complex abiotic stress responses (Schena et al., 1995; Cushman and Bohnert, 2000). Large-scale EST were started for maize. The sequences of EST 3' UTRs are of great value in distinguishing members of gene families as potential sources of IDPs (indel polymorphisms are a new class of codominant, allele-specific, genetic markers suitable for high-throughput analyses) and for producing the gene-specific probes needed for the mapping array (Kehoe et al., 1999). Up to now, more than 20,000 ESTs have been isolated by the maize genome project, part of the plant research program initiated in 1998 and of a national plant-genome research initiative established by the Office of Science and Technology Policy of the United States. The long-term goal of this program is to understand the structure, organization, and function of plant genomes important to agriculture, the environment, energy, and health. More details about this project are provided in the selected publications related to the *Zea mays* database (ZmDB) and the Maize Gene Discovery Project (Gai et al., 2000; Walbot and Petrov, 2001; Xing and Brendel, 2001; Bennetzen et al., 2002; Fernandes et al., 2002).

However, even sequences of the chromosomes of maize are very large and complex in structure. There are approaches generating DNA sequences of approximately 100 kbp segments for the maize genome, isolated as clones in bacterial artificial chromosome vectors. The aim of these approaches is to analyze the sequences for gene content and organization, and for the presence and nature of other types of DNAs.

Maize can control gene expression at the level of chromatin. Chromatin is the proteinaceous material that together with DNA comprises chromosomes. A key requirement for the expression of genes in chromosomes is that chromatin be remodeled and RNA polymerases can have access to the DNA. Projects are in progress with the overall goal to identify and functionally analyze most, if not all, of the several hundred genes in maize that contribute to chromatin-level control of gene expression.

A further possibility of improving maize to react to abiotic stresses is modification at the place of the centromeres. The centromeres are long stretches of DNA that, along with a set of proteins known as kinetochore proteins, are responsible for ensuring that chromosomes are accurately segregated and passed on to the next generation. Several projects are well under way to use artificial centromeres to create fully autonomous artificial chromosomes. Such artificial chromosomes could be used as vehicles for carrying agronomically important genes and could greatly improve existing methods for transferring genes into maize plants.

Proteome Analysis of Maize

Beside genomics, approaches with proteomics will be necessary to explain the genome sequence information and to access the modifications of composition and quantity of proteins as well as protein-ligand interactions relevant to abiotic stress response (Thiellement et al., 1999; Porubleva and Chitnis, 2000; Zivy and de Vienne, 2000; de Vienne et al., 2001). Proteomics is becoming a necessity in plant biology for deciphering the function and role of the genes that are or will be sequenced. Several applications of the proteomic tools have been developed in recent years, such as characterization of individuals or lines, estimation of genetic variability within and between populations, establishment of genetic distances that can be used in phylogenetic studies, characterization of mutants, and localization of the genes encoding the revealed proteins. A precise quantification of the variation in amounts of proteins was leading to the concept of "protein quantity loci" which, combined with the "quantitative trait loci" approach, results in testable hypotheses regarding the role of "candidate proteins" in the metabolism under abiotic stress or phenotype under study. The accumulation of data regarding genomic and cDNA sequencing needs to be connected to the protein databases currently developed in maize.

Another area of interest connected to proteomics is reversible protein phosphorylation, a key component of virtually every regulated biological process and with potentially high importance for a complex abiotic stress response. Approximately 5 percent of the genes in higher plants are directly involved in signaling by protein phosphorylation. Projects are running with the goal to mine maize sequence databases to gain new insights into the structural and functional relationships of the genes and proteins involved in posttranslation modification by phosphorylation (protein kinase and phosphatase).

Further studies in the same context are improving the understanding of how plants control gene expression. This knowledge seems to be essential for understanding how plants grow and develop and how they respond and adapt to the environment. Quality and yield improvements in crops continue to depend on applying new genetic insights and tools such as those presented previously. The research could have a direct impact on genetic improvement of maize and could also be applicable to many other important crop plants.

Salinity-Stress-Specific Transcripts

Only some initial attempts to identify stress-specific transcripts have been conducted in glycophytic plant species exposed to water deficits due to drought or NaCl salinity. As part of the gene discovery effort in maize, EST collections have been established from salt-stressed roots and shoots. However, although all plant species on earth are based on halophytic ancestors, and adaptation mechanisms could be largely conserved among glycophytes, present halophytic organisms still have working regulations accounting for their ability to withstand severe osmotic or ionic stress. *Aster tripolium* possesses, in contrast to nonhalophytic relatives, a Na^+-sensing system that down-regulates K^+ uptake in guard cells in response to high salinity (Véry et al., 1998). *Beta vulgaris* ssp. *maritima* expresses, in contrast to its relatives, a high selective and effective compartmentation of potentially toxic and essential ions (Koyro and Huchzermeyer, 2004a). Comparative studies of leaf tissues of well-watered and salinity-stressed halophytes have shown that the stressed plants expressed approximately 15 percent more functionally unknown genes than unstressed plants, supporting the notion that genes related to salinity stress are underrepresented in current gene-bank databases (Miyazaki et al., 1999; Bohnert et al., 2001; Dietz et al., 2001). Similar results were found for drought-resistant plant species, accumulating compatible solutes such as sucrose or proline, that are involved in the osmotic adaptation in the vacuole or in the protection of the metabolically more active compartment, the cytoplasm (Koyro and Huchzermeyer, 2004a).

Potential of Mutants

Plants integrate stress responses in the context of development and a complex architecture of tissues, each with different functions and, therefore, potentially differential sensitivity to different environmental stresses. In addition, every single cell consists of different compartments, showing that the understanding of cellular responses also depends on single-cell relations. Currently, surveys of stress-responsive gene expression cannot reach the target because their resolution is not high enough to permit comparisons between single cells and their compartments, thereby permitting only the identification of differences and similarities in expression pattern or gene components that contribute to abiotic stress tolerance at a tissue or plant level. Mutant screens for stress-hypersensitive maize may lead to the discovery of single-cell-related structural and signaling components that are important for stress tolerance in the near future.

Limitations of Including Genetic Resources

Even if useful characteristics can be identified, the transfer of them to cultivated species is often difficult and time is a major factor. Complete transfer can take years and there is often the additional problem that more than the desired gene is transferred. For genetic resources to be a major factor on stress-tolerance breeding, new methods must be developed. Physiological measures of various parameters are essential for a more accurate determination of new sources of useful characteristics.

The highest potential impacts of the use of genetic resources are in the areas of biotic and abiotic stress resistances and tolerances. At present, relatively few of the genes isolated and studied by plant molecular biologists have been mapped. Their routine integration onto molecular maps will facilitate the development of plant resistance through breeding and molecular approaches.

Alternative Approach and Future Attempts

An alternative approach makes use of the plants that already have the requisite level of high productivity at high external salinity levels or low soil water potentials, the halophytes and xerophytes (Lieth, 1999). This would refresh the discussion about crop plants growing on severely salt-affected land or being irrigated with sea water (Winicov, 1998). However, the research with cash crop halophytes is only at the beginning, and production efficiency needs to be improved. On the basis that genetic variability exists within crop species, it seems promising to introduce important traits of resistance to water deficit into crop species from their wild relatives through interspecific hybridization, for example, *Beta vulgaris* ssp. *vulgaris* (sugar beet) and *Beta vulgaris* ssp. *maritima* (sea beet). These are valuable tools available for screening and breeding.

During recent years, gene-modification technology has been applied to obtain genetically improved crop plants (Yeo, 1998; Winicov and Bastola, 1999). However, genetic as well as molecular biological methods should be tested according to their possible applications. The selection and breeding for water-deficit tolerance and specific salinity tolerance are important issues for traditional agricultural production systems, especially in semiarid and arid regions. It is conceivable that an improvement of the utilization potential can be reached in the near future by gene transfer of profitable features to a xerophyte or halophyte or transfer of tolerance-enhancing features to a sensitive genotype (Winicov, 1998). Many international organizations dealing with desertification problems are now looking for halophytes as one means to solve certain problems (Lieth and Moschenko, 2002). Recent studies have shown that a relatively small number of QTLs may govern complex physiological characters. After isolation and cloning of genes from a sensitive cultivated plant, responsible, for example, for yield-enhancing characteristics, the gene could be transferred into a xerophyte or halophyte. Alternatively, the transfer of genes responsible for salinity or drought tolerance could occur from a xerophyte or halophyte to a sensitive crop.

The use of halophytes is already leading to the development of several seawater-tolerant crops (Lieth and Moschenko, 2002). This chapter has shown that the genetic resources of maize also have a high potential to contribute to the rate of yield increase, especially in more marginal lands and in harsher environments. However, a rapidly expanding base of information on strategies in plant adaptation to stress is important to achieve improved tolerance. The biotechnological techniques or resources alone will not suffice; rather, the effective and complementary use of all technological tools and materials is required (Hoisington et al., 1999).

REFERENCES

Andrade, F.H., C. Vega, S. Uhart, A. Cirilo, M. Cantarero, and O. Valentinuz (1999). Kernel number determination in maize. *Crop Science* 39:453-459.

Apse, M.P., G.S. Aharon, W.A. Snedden, and E. Blumwald (1999). Salt tolerance conferred by overexpression of a vacuolar Na+/H+ antiport in *Arabidopsis*. *Science* 285:1256-1258.

Armstrong, C.L. (1994). Regeneration of plants from somatic cell cultures: Applications for in vitro genetic manipulation. In *The Maize Handbook,* M. Freeling and V. Walbot (Eds.). New York: Springer-Verlag, pp. 663-670.

Ayad, G., J. Troll, and T.J. Esquinas-Alcazat (1980). *Directory of Germplasm Collections: III Cereals. 2. Maize.* Rome, Italy: International Board for Plant Genetic Resources.

Bahrun, A., C.R. Jensen, F. Asch, and V.O. Mogenese (2002). Drought-induced changes in xylem pH, ionic composition, and ABA concentration act as early signals in field-grown maize (*Zea mays* L.). *Journal of Experimental Botany* 53:251-263.

Bänziger, M., G.O. Edmeades, D. Beck, and M. Beuon (2000). *Breeding for Drought and Nitrogen Stress Tolerance in Maize: From Theory to Practice.* El Batan, Mexico CIMMYT.

Bänziger, M. and H.R. Lafitte (1997). Efficiency of secondary traits for improving maize for low-nitrogen target environments. *Crop Science* 37:1110-1117.

Bänziger, M., K.V. Pixley, B. Vivek, and B.T. Zambezi (2000). *Characterization of Elite Maize Germplasm Grown in Eastern and Southern Africa: Results of the 1999 Regional Trials Conducted by CIMMYT and the Maize and Wheat Improvement Research Network for SADC (MWIRNET).* Harare, Zimbabwe: CIMMYT.

Barker, T.C. and G. Varughese (1992). Combining ability and heterosis among eight complete spring hexaploid triticale lines. *Crop Science* 32:340-344.

Bartels, D. and D. Nelson (1994). Approaches to improve stress tolerance using molecular genetics. *Plant, Cell and Environment* 17:659-667.

Bassetti, P. and M.E. Westgate (1993a). Senescence and receptivity of maize silks. *Crop Science* 33:275-278.

Bassetti, P. and M.E. Westgate (1993b). Water deficit affects receptivity of maize silks. *Crop Science* 33:279-282.

Ben-Hayyim, G., J.P. Damon, J. Martin-Tanguy, and D. Tepfer (1994). Changing root system architecture through inhibition of putrescine and feruloyl putrescine accumulation. *FEBS Letters* 342:145-148.

Bennetzen, J.L., V.L. Chandler, and P. Schnable (2002). National Science Foundation-Sponsored Workshop Report. Maize Genome Sequencing Project. *Plant Physiology* 127:1572-1578.

Blum, A. (1997). Constitutive traits affecting plant performance under stress. In *Developing Drought and Low-N Tolerant Maize,* G.O. Edmeades, M. Bänziger, H.R. Mickelsen, and C.B. Peiia-Valdivia (Eds.). El Batan, Mexico: CIMMYT, pp. 131-135.

Bohnert, H.J., P. Ayoubi, C. Borchert, R.A. Bressan, R.L. Burnap, J.C. Cushman, M.A. Cushman, M. Deyholos, R. Fischer, D.W. Galbraith, et al. (2001). A genomics approach toward salt stress tolerance. *Plant Physiology and Biochemistry* 39:295-311.

Bohnert, H.J. and E. Sheveleva (1998). Plant stress adaptations: Making metabolism move. *Current Opinion in Plant Biology* 9:214-219.

Bolaños, J. and G.O. Edmeades (1993). Eight cycles of selection for drought tolerance in lowland tropical maize. I. Responses in grain yield, biomass and radiation utilization. *Field Crops Research* 31:233-252.

Bolaños, J. and G.O. Edmeades (1996). The importance of the anthesis-silking interval in breeding for drought tolerance in tropical maize. *Field Crops Research* 48:65-80.

Bolaños, J., G.O. Edmeades, and L. Martinez (1993). Eight cycles of selection for drought tolerance in lowland tropical maize. III. Responses in drought-adaptive physiological and morphological traits. *Field Crops Research* 31:269-286.

Boyer, J.S. (1982). Plant productivity and environment. *Science* 218:443-448.

Bruce, W.B., G.O. Edmeades, and T.C. Barker (2002). Molecular and physiological approaches to maize improvement for drought tolerance. *Journal of Experimental Botany* 53:13-25.

Bruce, W.B., O. Folkerts, C. Garnaat, O. Crasta, B. Roth, and B. Bowen (2000). Expression profiling of maize flavonoid pathway genes controlled by estradiol-inducible transcription factors CRC and P. *Plant Cell* 12:65-80.

Castleberry, R.M., C.W. Crum, and C.F. Krull (1984). Genetic improvements of US maize cultivars under varying fertility and climatic environments. *Crop Science* 24:33-36.

Chapman, S.C. and G.O. Edmeades (1999). Selection improves drought tolerance in tropical maize populations. II. Direct and correlated responses among secondary traits. *Crop Science* 39:1315-1324.

Chee, M., R. Yang, E. Hubbell, A. Berno, X.C. Huang, D. Stern, J. Einkler, D.J. Lockart, M.S. Morris, and S.P. Fodor (1996). Accessing genetic information with high-density DNA arrays. *Science* 274:610-614.

Chen, W., N.J. Provart, J. Glazebrook, F. Katagiri, H.-S. Chang, T. Eulgem, F. Mauch, S. Luan, G. Zou, S.A. Whitman, et al. (2002). Expression profile matrix of *Arabidopsis* transcription factor genes suggests their putative functions in response to environmental stresses. *The Plant Cell* 14:559-574.

Claassen, M.M. and R.H. Shaw (1970). Water deficit effects on corn. II. Grain components. *Agronomy Journal* 62:652-655.

Commuri, P.D. and R.J. Jones (1999). Ultrastructural characterization of maize (*Zea mays* L.) kernels exposed to high temperature during endosperm cell division. *Plant Cell and Environment* 22:375-385.

Cushman, J.C. and H.J. Bohnert (2000). Genomic approaches to plant stress tolerance. *Current Opinion in Plant Biology* 3:117-124.

Dallmier, K.A. and C.R. Stewart (1992). Effect of exogenous abscisic acid on proline dehydrogenase activity in maize (*Zea mays* L.). *Plant Physiology* 99:762-764.

De Vienne, D., B. Bost, J. Fievet, M. Zivy, and C. Dillmann (2001). Genetic variability of proteome expression and metabolic control. *Plant Physiology and Biochemistry* 39:271-283.

Dietz, K.J., N. Tavakoli, C. Kluge, T. Mimura, S.S. Sharma, G.C. Harris, A.N. Chardonnens, and D. Golldack (2001). Significance of the V-type ATPase for the adaptation to stressful growth conditions and its regulation on the molecular and biochemical level. *Journal of Experimental Botany* 52:1969-1980.

DuPlessis, D.P. and F.J. Dijkhuis (1967). The influence of time lag between pollen shedding and silking on the yield of maize. *South African Journal of Agricultural Science* 10:667-674.

Duvick, D.N. (1984). Genetic contributions to yield gains of US hybrid maize, 1930 to 1980. In *Genetic Contributions to Yield Gains of Five Major Crop Plants,* W.R. Fehr (Ed.). Madison, WI: CSSA, ASA, pp. 1-47.

Duvick, D.N. (1997). What is yield? In *Developing Drought and Low-N Tolerant Maize,* G.O. Edmeades, M. Bänziger, H.R. Mickelsen, and C.B. Pefia-Valdivia (Eds.). El Batan, Mexico: CIMMYT, pp. 332-334.

Edmeades, G.O., J. Bolanos, S.C. Chapman, H.R. Lafitte, and M. Bänziger (1999). Selection improves drought tolerance in tropical maize populations. I. Gains in biomass, grain yield and harvest index. *Crop Science* 39:1306-1315.

Edmeades, G.O., J. Bolanos, A. Elings, J.-M. Ribaut, M. Bänziger, and M.E. Westgate (2000). The role and regulation of the anthesis-silking interval in maize. In *Physiology and Modeling Kernel Set in Maize,* M.E. Westgate and K.J. Boote (Eds.). Madison, WI: CSSA, pp. 43-73.

Edmeades, G.O., J. Bolanos, M. Hernandez, and S. Beuon (1993). Causes for silk delay in lowland tropical maize. *Crop Science* 33:1029-1035.

Edmeades, G.O., M. Cooper, H.R. Lafitte, C. Zinselmeier, J.-M. Ribaut, J.E. Habben, C. Uiffler, and M. Bänziger (2002). Abiotic stresses and staple crops. In *Proceedings of the Third International Crop Science Congress,* Hamburg, Germany: CABI, pp. 18-23.

Fernandes, J., V. Brendel, X. Gai, S. Lal, V.L. Chandler, R.P. Elumalai, D.W. Galbraith, E.A. Pierson, and V. Walbot (2002). Comparison of RNA Expression Profiles Based on Maize Expressed Sequence. Tag Frequency Analysis and Micro-Array Hybridization. *Plant Physiology* 128:896-910.

Flowers, T.J. and A.R. Yeo (1995). Breeding for salinity tolerance in crop plants: Where next? *Australian Journal of Plant Physiology* 22:875-884.

Foolad, M.R. (1999). Comparison of salt tolerance during seed germination and vegetative growth in tomato by QTL mapping. *Genome* 42:727-734.

Foolad, M.R. and R.A. Jones (1993). Mapping salt-tolerance genes in tomato *(Lycopersicon esculentum)* using trait-based marker analysis. *Theoretical and Applied Genetics* 87:184-192.

Frova, C., A. Caffulli, and E. Pallavera (1999). Mapping quantitative trait loci for tolerance to abiotic stresses in maize. *Journal of Experimental Zoology* 282:164-170.

Frova, C., P. Krajewski, N. di Fonzo, M. Villa, and M. Sari-Gorla (1999). Genetic analysis of drought tolerance in maize by molecular markers. 1. Yield components. *Theoretical and Applied Genetics* 99:280-288.

Gai, X., S. Lal, L. Xing, V. Brendel, and V. Walbot (2000). Gene discovery using the maize genome database ZmDB. *Nucleic Acids Research* 28:94-96.

Goodman, M.M. (1983). *Report of the 1983 Plant Breeding Research Forum.* Ames, IA: Pionneer Hi-Bred Int.

Gosti, F., N. Bertauche, N. Vartanian, and J. Giraudat (1995). Abscisic acid-dependent and -independent regulation of gene expression by progressive drought in *Arabidopsis thaliana. Molecular and General Genetics* 246:10-18.

Hemamalini, G.S., H.E. Shashidhar, and S. Hittalmani (2000). Molecular marker assisted tagging of morphological and physiological traits under two contrasting moisture regimes at peak vegetative stage in rice (*Oryza sativa* L.). *Euphytica* 112:69-78.

Hoisington, D., C. Jiang, M. Khairallah, J.-M. Ribaut, M. Bohn, A. Melchinger, M. Willcox, and D. Gonzalez-de-Leon (1996). QTL for insect resistance and drought tolerance in tropical maize: Prospects for marker assisted selection. *Symposium of the Society for Experimental Biology* 50:39-44.

Hoisington, D., M. Khairallah, T. Reeves, J.-M. Ribaut, B. Skovmand, S. Taba, and M. Warburton (1999). Plant genetic resources: What can they contribute toward increased crop productivity? *Proceedings of the National Academy of Science, USA* 96:5937-5943.

Holmberg, N. and L. Bülow (1998). Improving stress tolerance in plants by gene transfer. *Trends in Plant Science* 3:61-66.

Ingram, J. and D. Bartels (1996). The molecular basis of dehydration tolerance in plants. *Annual Review of Plant Physiology and Plant Molecular Biology* 47:377-403.

Jacobs, B.C. and C.J. Pearson (1991). Potential yield of maize, determined by rates of growth and development of ears. *Field Crops Research* 27:281-298.

Jensen, S.D. and A.J. Cavalieri (1983). Drought tolerance in US maize. *Agricultural Water Management* 7:223-236.

Kehoe, D.M., P. Villand, and S. Somerville (1999). DNA microarrays for studies of higher plants and other photosynthetic organisms. *Trends in Plant Science* 4: 38-41.

Kitchen, N.R., K.A. Sudduth, and S.T. Drummond (1999). Soil electrical conductivity as a crop productivity measure for claypan soils. *Journal of Production Agriculture* 12:607-617.

Koyro, H.-W. and B. Huchzermeyer (1999). Salt and drought stress effects on metabolic regulation in maize. In *Handbook of Plant and Crop Stress*, M. Pessaraki (Ed.). New York: Marcel Dekker, pp. 843-878.

Koyro, H.-W. and B. Huchzermeyer (2004a). Ecophysiological mechanisms leading to salinity tolerance: Screening of cashcrop halophytes. *Recent Research Developments in Plant Science* 1:187-207.

Koyro, H.-W. and B. Huchzermeyer (2004b). Ecophysiological needs of the potential biomass crop *Spartina townsendii* Grov. *Tropical Ecology* 45:123-139.

Kush, G.S. (1999). Green revolution: Preparing for the 21st century. *Genome* 42:646-655.

Lemcoff, J.H., C.A. Chimenti, and A.E. Davezac (1998). Osmotic adjustment in maize (*Zea mays* L.): Changes with ontogeny and its relationship with phenotypic stability. *Journal of Agronomy and Crop Science* 180:241-247.

Lemontey, C., C. Mousset-Declas, and N. Munier-Jolain (2000). Maternal genotype influences pea seed size by controlling both mitotic activity during early embryogenesis and final endoreduplication level/cotyledon cell size in mature seed. *Journal of Experimental Botany* 51:167-175.

Leung, J. and J. Giraudat (1998). Abscisic acid signal transduction. *Annual Review of Plant Physiology and Plant Molecular Biology* 49:199-222.

Liang, P. and A.B. Pardee (1992). Differential display of eukaryotic messenger RNA by means of the polymerase chain reaction. *Science* 257:967-971.

Lieth, H. (1999). Development of crops and other useful plants from halophytes. In *Halophytes Uses in Different Climates, Ecological and Ecophysiological Studies,* H. Lieth, M. Moschenko, M. Lohmann, H.-W. Koyro, and A. Hamdy (Eds.). Leiden, The Netherlands: Backhuys Publishers, pp. 1-18.

Lieth, H. and M. Moschenko (2002). *Halophyte Uses in Different Climates IV. Cashcrop Halophytes for Future Halophytes Growers.* Leiden, The Netherlands: Backhuys Publishers.

Lohaus, G., M. Hussmann, K. Pennewiss, H. Schneider, J.J. Zhu, and B. Sattelmacher (2000). Solute balance of a maize (*Zea mays* L.) source leaf as affected by salt treatment with special emphasis on phloem retranslocation and ion leaching. *Journal of Experimental Botany* 51:1721-1732.

Mackill, D.J., U.T. Nguyen, and J. Zhang (1999). Use of molecular markers in plant improvement programs for rainfed lowland rice. *Field Crops Research* 64:177-185.

McCarty, C. (1995). Genetic control and integration of maturation and germination pathways in seed development. *Annual Review of Plant Physiology and Plant Molecular Biology* 46:71-93.

McCully, M.E. (1999). Roots in soil: Unearthing the complexities of roots and their rhizospheres. *Annual Review of Plant Physiology and Plant Molecular Biology* 50:695-718.

McNeil, S.D., M.L. Nuccio, and A.D. Hanson (1999). Betaines and related osmoprotectants. Targets for metabolic engineering of stress resistance. *Plant Physiology* 120:945-949.

Miyazaki, S., R. Koga, H.J. Bohnert, and T. Fukuhara (1999). Tissue- and environmental response-specific expression of 10 PP2C transcripts in *Mesembryanthemum crystallinum. Molecular and General Genetics* 261:307-316.

Modarres, A.M., R.I. Hamilton, M. Dijak, L.M. Dwyer, D.W. Stewart, D.E. Mather, and D.L. Smith (1998). Plant population density effects on maize inbred lines grown in short-season environments. *Crop Science* 38:104-108.

Mugo, S.N., M. Bänziger, and G.O. Edmeades (2000). Prospects of using ABA in selection for drought tolerance in cereal crops. In *Proceedings of Workshop on Molecular Approaches for the Genetic Improvement of Cereals for Stable Production in Water-Limited Environments.* El Batan, Mexico: CIMMYT, pp. 73-78.

Neuffer, M.G., E.H. Coe, and S.R. Wessler (1997). *Mutants of Maize.* Cold Spring Harbor, NY: Cold Spring Harbor Laboratory Press.

Niu, X., R.A. Bressan, P.M. Hasegawa, and J.M. Prado (1995). Ion homeostasis in NaCl stress environments. *Plant Physiology* 109:735-742.

Noctor, G. and C.H. Foyer (1998). Acsorbate and glutathione: Keeping active oxygen under control. *Annual Review of Plant Physiology and Plant Molecular Biology* 49:249-279.

Ober, E.S. and R.E. Sharp (1994). Proline accumulation in maize (*Zea mays* L.) primary roots at low water potentials. *Plant Physiology* 105:981-987.

Ouattar, S., R.J. Jones, and R.K. Crookston (1987). Effect of water deficit during grain filling on the pattern of maize kernel growth and development. *Crop Science* 27:726-730.

Palmer, S.J. and W.J. Davies (1996). An analysis of relative elemental growth rate, epidermal cell size and xyloglucan endotransglycosylase activity through the growing zone of ageing maize. *Journal of Experimental Botany* 52:1383-1400.

Pardo, J.M., M.P. Reddy, S. Yand, A. Maggio, A.-H. Huh, T. Matsumoto, M.A. Coca, M. Paino-DÚrzo, and H. Koiwa (1998). Stress signaling through Ca^{2+}/calmodulin-dependent protein phosphatase calcineurin mediates salt adaptation in plants. *Proceedings of the National Academy of Science, USA* 95:9681-9686.

Paul, M.J. and C.H. Foyer (2001). Sink regulation of photosynthesis. *Journal of Experimental Botany* 52:1383-1400.

Porubleva, L. and P.R. Chitnis (2000). Proteomics: A powerful tool in the post-genomic era. *Indian Journal of Biochemistry and Biophysics* 37:360-368.

Prochazkova, D., R.K. Sairam, G.C. Srivastava, and D.V. Singh (2001). Oxidative stress and antioxidant activity as the basis of senescence in maize leaves. *Plant Science* 161:765-771.

Pryor, T. (1994). Maize and *Puccinia sorghi:* A system for the study of the genetic and molecular basis of host-pathogen interactions. In *The Maize Handbook,* M. Freeling and V. Walbot (Eds.). New York: Springer-Verlag, pp. 286-290.

Quarrie, S.A., V. Lazic-Janic, D. Kovacevic, A. Steed, and S. Pekic (1999). Bulk segregant analysis with molecular markers and its use for improving drought resistance in maize. *Journal of Experimental Botany* 50:1299-1306.

Rhodes, D., W.J. Yang, Y. Samaras, K. Wood, C.C. Bonham, J.C. Rhodes, and B. Burr (1993). Map locations of genes conferring glycinebetaine and trigonelline accumulation in maize. *Plant Physiology* 102:160.

Ribaut, J.-M., G.O. Edmeades, F.J. Betran, C. Jiang, and M. Bänziger (1999). Marker-assisted selection for improving drought tolerance in tropical maize. In *Genetic Improvements of Rice for Water-Limited Environments,* J. O'Toole, O. Ito, and B. Hardy (Eds.). Los Baños, Philippines: IRRI, pp. 193-209.

Ribaut, J.-M., D.A. Hosington, J.A. Deitsch, C. Jiang, and D. Gonzalez-de-Leon (1996). Identification of quantitative trait loci under drought conditions in tropical maize. 1. Flowering parameters and the anthesis-silking interval. *Theoretical and Applied Genetics* 92:905-914.

Ribaut, J.-M., C. Jiang, D. Gonzalez-de-Leon, G.O. Edmeades, and D.A. Hosington (1997). Identification of quantitative trait loci under drought conditions in tropical maize. 2. Yield components and marker-assisted selection strategies. *Theoretical and Applied Genetics* 94:887-914.

Richards, R.A. (2000). Selectable traits to increase crop photosynthesis and yield of grain crops. *Journal of Experiment Botany* 51:447-458.

Roberts, S.K. and B.N. Snowman (2000). The effect of ABA on channel-mediated K^+ transport across higher plant roots. *Journal of Experimental Botany* 51:1585-1594.

Rock, C.D. and P.P.F. Ng (1999). Dominant wilty mutants of *Zea mays (Poaceae)* are not impaired in abscisic acid perception or metabolism. *American Journal of Botany* 86:1796-1800.

Saini, H.S. and M.E. Westgate (2000). Reproductive development in grain crops during drought. *Advances in Agronomy* 68:59-96.

Saneoka, H., C. Nagasaka, D.T. Hahn, W.J. Yang, G.S. Premachandra, R.J. Joly, and D. Rhodes (1995). Salt tolerance of glycinebetaine-deficient and glycinebetaine-containing maize lines. *Plant Physiology* 107:631-638.

Sauter, A., W.J. Davies, and W. Hartung (2001). The long-distance abscisic acid signal in the droughted plant: The fate of the hormone on its way from root to shoot. *Journal of Experimental Botany* 52:1991-1997.

Scheiner, J.D., F.H. Gutierrez-Boem, and R.S. Lavado (2000). Root growth and phosphorus uptake in wide- and narrow-row soybeans. *Journal of Plant Nutrition* 23:1241-1249.

Schena, M., D. Shalon, R.W. Davis, and P.O. Brown (1995). Quantitative monitoring of gene expression patterns with a complementary DNA microarray. *Science* 270:467-470.

Schussler, J.R. and M.E. Westgate (1995). Assimilate flux determines kernel set at low water potential in maize. *Crop Science* 35:1074-1080.

Seki, M., M. Narasuka, H. Abe, M. Kasuga, K. Yamaguchi-Shinozaki, P. Caminci, Y. Hayashizaki, and K. Shinozaki (2001). Monitoring the expression pattern of 1300 *Arabidopsis* genes under drought and cold stresses by using a full-length cDNA microarray. *The Plant Cell* 13:61-72.

Setter, T.L. and B.A. Flannigan (2001). Water deficit inhibits cell division and expression of transcripts involved in cell proliferation and endoreduplication in maize endosperm. *Journal of Experimental Botany* 52:1401-1408.

Shaw, R.H. (1977). Water use and requirements of maize: A review. In *Agrometeorology of the Maize (Corn) Crop.* Geneva, Switzerland: World Meteorological Organization, pp. 198-208.

Shinozaki, K. and K. Yamaguchi-Shinozaki (1996). Molecular responses to drought and cold stress. *Current Opinion in Biotechnology* 7:161-167.

Shinozaki, K. and K. Yamaguchi-Shinozaki (1997). Gene expression and signal transduction in water-stress response. *Plant Physiology* 115:327-334.

Tardieu, F., M. Reymond, P. Hamard, C. Granier, and B. Muller (2000). Spatial distributions of expansion rate, cell division rate and cell size in maize leaves: A synthesis of the effects of soil water status, evaporative demands and temperature. *Journal of Experimental Botany* 51:1505-1514.

Thiellement, H., N. Bahrman, C. Damerval, C. Plomion, M. Rossignol, V. Santoni, D. de Vienne, and M. Zivy (1999). Proteomics for genetic and physiological studies in plants. *Electrophoresis* 20:2013-2026.

Tollenaar, M. and J. Wu (1999). Yield improvement in temperate maize is attributable to greater stress tolerance. *Crop Science* 39:1597-1604.

Uno, Y., T. Furihata, H. Abe, R. Yoshida, K. Shinozaki, and K. Yamaguchi-Shinozaki (2000). *Arabidopsis* basic leucine zipper transcription factors involved in an abscisic acid-dependent signal transduction pathway under drought

and high-salinity conditions. *Proceedings of the National Academy of Science, USA* 97:11632-11637.
Verslues, P.E. and R.E. Sharp (1999). Proline accumulation in maize (*Zea mays* L.) primary roots at low water potentials. II. Metabolic source of increased proline deposition in the elongation zone. *Plant Physiology* 119:1349-1360.
Véry, A.A., M.F. Robinson, T.A. Mansfield, and D. Sanders (1998). Guard cell cation channels are involved in Na^+-induced stomatal closure in a halophyte. *Plant Journal* 14:509-521.
Walbot, V. and D.A. Petrov (2001). Gene galaxies in the maize genome. *Proceedings of the National Academy of Science, USA* 98:8163-8164.
Weerathaworn, P., A. Soldati, and P. Stamp (1992). Anatomy of seedling roots of tropical maize (*Zea mays* L.) cultivars at low water supply. *Journal of Experimental Botany* 43:1015-1021.
Westgate, M.E. and J.S. Boyer (1985). Carbohydrate reserves and reproductive development at low leaf water potentials in maize. *Crop Science* 25:762-769.
Winicov, I. (1998). New molecular approaches to improving salt tolerance in crop plants. *Annals of Botany* 82:703-710.
Winicov, I. and D.R. Bastola (1997). Salt tolerance in crop plants: New approaches through tissue culture and gene regeneration. *Physiologia Plantarum* 19:435-449.
Winicov, I. and D.R. Bastola (1999). Transgenic overexpression of the transcription factor Alfin1 enhances expression of the endogenous *MsPRP2* gene in alfalfa and improves salinity tolerance of the plants. *Plant Physiology* 120:473-480.
Xing, L. and V. Brendel (2001). Multi-query sequence BLAST output examination with MuSeqBox. *Bioinformatics* 17:744-745.
Xu, D., X. Duan, B. Wang, B. Hong, T.D.H. Ho, and R. Wu (1996). Expression of a late embryogenesis abundant protein gene, *HVA1,* from barley confers tolerance to water deficit and salt stress in transgenic rice. *Plant Physiology* 110:249-257.
Yamaguchi-Shinozaki, K., K. Koizumi, S. Urao, and K. Shinozaki (1992). Molecular cloning and characterization of 9 cDNAs for genes that are responsive to desiccation in *Arabidopsis thaliana*: Sequence analysis of one cDNA clone encodes a putative transmembrane channel protein. *Plant Cell Physiology* 33:217-224.
Yang, W.J., A. Nadolska-Orczyk, K.V. Wood, D.T. Hahn, P.J. Rich, A.J. Woods, H. Saneoka, G.S. Premachandra, C.C. Bonham, J.C. Rhodes, et al. (1995). Near isogenic lines of maize differing in glycinebetaine. *Plant Physiology* 107:621-630.
Yeo, A.R. (1998). Molecular biology of salt tolerance in the context of whole-plant physiology. *Journal of Experimental Botany* 49:915-929.
Zhang, J.X., H.T. Nguyen, and A. Blum (1999). Genetic analysis of osmotic adjustment in crop plants. *Journal of Experimental Botany* 50:291-302.
Zheng, H.G., R.C. Babu, M.S. Pathan, L. Ali, N. Huang, B. Courtois, and H.T. Nguyen (2000). Quantitative trait loci for root-penetration ability and root thickness in rice: Comparison of genetic backgrounds. *Genome* 43:53-61.
Zinselmeier, C., J.E. Habben, and M.E. Westgate (2000). Carbohydrate metabolism in setting and maize ovaries. In *Modeling Kernel Set in Maize,* M.E. Westgate and K.J. Boote (Eds.). Madison, WI: CSSA, pp. 1-13.

Zinselmeier, C., B.R. Jeong, and J.S. Boyer (1999). Starch and the control of kernel number in maize at low water potentials. *Plant Physiology* 121:25-36.

Zivy, M. and D. de Vienne (2000). Proteomics: A link between genomics, genetics and physiology. *Plant Molecular Biology* 44:575-580.

Zscheischler, J., M.C. Estler, W. Staudacher, F. Groß, G. Burgstaller, H. Streyl, and T. Rechmann (1990). *Handbuch Mais. Umweltgerechter Anbau, wirtschaftliche Verwertung*. Frankfurt, Germany: DLG-Verlags-GmbH.

Chapter 16

Breeding for Abiotic Stress Tolerance in Oilseed Crops

David A. Dierig

INTRODUCTION

There were 309 million metric tons of oilseeds, which include soybeans, cottonseed, peanuts, sunflowers, and rapeseed produced worldwide in 2001 (USDA, 2001). The world exports of oilseeds were valued at $8.6 billion in the year 2000, with soybeans accounting for more than half of this production. Other significant annual oilseed crops included canola, flaxseed, safflower, linseed, corn oil, and castor (USDA, 2001). These oils have edible uses in baking or frying fats, margarine, and salad or cooking oils, as well as industrial applications in paints, varnishes, plasticizers, resins, lubricants, greases, and coatings, for cosmetic uses, and pharmaceuticals.

Worldwide population growth has increased the demands on agriculture. Scarcity of agricultural lands lost to urban sprawl has increased land costs and forced less costly marginal lands with harsher microclimates into production. Countries with lower land and production inputs maintain a competitive advantage in the international market for major oilseeds. As a result, it is not only critical for yields to increase, but they must also occur in more stressful environments. Most high-yielding varieties are not sustainable in such areas. Development of genotypes possessing desirable genes able to confer abiotic stress resistance in crops is necessary for improvement.

A number of environmental conditions inhibit the growth and development of plants. Drought, salinity, deficits or excesses of plant nutrients, excesses of metals, temperature extremes, and exposure to pollutants are some of these. Pollutants may include pesticides, detergents, lubricants, and fuels entering the environment. Dudal (1976) estimated that only 10 percent of the world's arable lands could be categorized as free of stress, with drought and salt stress impacting agriculture the most of any abiotic stress (Ashraf, 1994). Flowers and colleagues (1977) reported that the total area of countries so far mapped for saline lands is more than 3×10^6 km^2, which does not include areas with major deserts. At least another million square km have not yet been mapped. Secondary salinization adds to the gravity of the problem where once-productive agricultural lands are going out of use due to poor-quality irrigation water. Since no economically feasible technological improvements for this situation are available for use in the near future, tolerance to stress must be developed.

Methodologies to improve a plant's yielding capacity under abiotic stresses have changed in the recent past. Plant breeding has expanded from traditional selection of plants based on morphology and physiology, or the recombination of favorable traits through hybridization, to include the use of molecular markers in marker-assisted selection and genetic transformations, including the identification, characterization, and manipulation of gene sequences.

There are more than 200 seed oils, composed of various fatty acids, produced as storage oils in higher plants (Somerville and Somerville, 1999). Synthesis of fatty acids up to the oleic acid (C18:1) chain length is confined to proplastids in developing oilseeds. Subsequent modification of the oleic acid occurs outside the plastid, generally on the endoplasmic reticulum, which is also the intracellular location of the triacylglycerol formation (John, 1992). These fatty acids differ primarily in the presence of double bonds, hydroxyl or epoxy groups, triple bonds, or secondary modification of functional groups at various carbons along the fatty acyl chain (Lee et al., 1998).

GENETIC ARCHITECTURE OF OILSEED CROPS

A key component in understanding the crop's genetic architecture is the development of genetic maps. Some have debated whether these maps have provided pertinent information for breeders when based only on cytological markers and mutations, as in the past (Lee, 1995). As they have become more integrated with relevant DNA markers, maps have become more useful in providing plant breeders and geneticists with a perspective of the genetic inheritance of various traits. They may also contribute essential information in developing crop improvement schemes or modifying plant breeding methods (Lee, 1995).

Sequencing an entire genome of selected plants such as *Arabidopsis* has played an important role in providing information and in gene discovery. Although *Arabidopsis* is not agronomically significant, it offers important advantages for basic research as a model crop, such as a relatively small nuclear genome, a short generation time, and a large number of seeds produced each generation (Anonymous, 2000). It is also relevant in this discussion because *Arabidopsis* is an oilseed and related taxonomically to other agronomically important oilseed crops. The sequenced regions include 115.4 Mb of 125 Mb total genome, which was sequenced in the year 2000 (The Arabidopsis Genome Initiative, 2000). There are extensive genetic and physical maps of all five chromosomes and a large number of mutant lines and genomic resources available (http://www.ncbi.nlm.nih.gov). The differences between *Arabidopsis* and other oilseed crops with larger genome sizes are primarily due to amplification of interspersed repetitive sequences (Somerville and Somerville, 1999). No evidence indicates that plants with larger amounts of DNA per cell have a substantially greater number of functional genes compared to those with relatively small amounts (Somerville and Somerville, 1999). The colinearity between *Arabidopsis* and *Brassica* crops has been studied by Gale and Devos (1998). They found from cross mapping the genomes of *Arabidopsis* and oilseed rape, which are estimated to be 10 million years apart in evolutionary terms, that in the diploid *Brassica* crops the genome is essentially triplicated. The DNA content of oilseed rape is at 480 Mb, which is more than three times that of *Arabidopsis*. Many of the major genes, such as flowering time and vernalization, indicate the similarities between *Arabidopsis* and *Brassica* crops (Osborn et al., 1997). It is still unclear how close *Arabidopsis* is to other non-*Brassica* oilseed crops.

Ten to 12 percent of the genes on each of the five *Arabidopsis* chromosomes are classified as having a plant defense function (The Arabidopsis Genome Initiative, 2000). Other genes with functions such as transport and signaling make up 4 to 11 percent on each chromosome, with less than half of the functions still unclassified. It will be interesting to compare such statistics to the genomes of other oilseeds.

Partial sequencing of genes has provided major targets for engineering stress tolerance (Cushman and Bohnert, 2000). This has been accomplished in randomly selected cDNA clones or expressed sequence tags (ESTs). This is a rapid way to establish an inventory of expressed genes where single-pass sequences of 300 to 500 bp are determined from one or both ends of randomly chosen cDNA clones (Cushman and Bohnert, 2000). The sequences have a relatively high error rate but contain enough information to identify the corresponding gene (Bouchez and Höfte, 1998). Extensive EST collections already exist for *Arabidopsis* (Bevan et al., 1999) and soybean (Walbot, 1999) but are lacking for other nonmajor oilseed crops.

Alternative approaches to EST collections in assessing gene expression patterns have been developed, such as serial analysis of gene expression (SAGE) (Bertelsen and Valculescu, 1998). These quantify the occurrence of a large number of transcripts of a population by sequencing multiple tags within a single clone. Nuclear expressed sequence tags (NEST) are another option (Macas et al., 1998). This combines fluorescence-assisted nucleus sorting and cDNA generated from RNA of isolated nuclei using a cell-specific promoter.

Stress-resistant crop varieties can be associated either with marginal or low-performance areas of production or with high-input and high-value production systems. There are many examples of genes that respond to the numerous types of abiotic stresses such as drought, salt, and cold (Hwang and Goodman, 1995), indicating a similar function of genes in detoxification and cell maintenance for plants under water stress (Winicov, 1998).

McCue and Hanson (1990) listed four levels of change in the plant's genetic architecture to tolerate stress, which include developmental, structural, physiological, and biochemical levels. The limitations in exploiting plants for tolerance at these levels are in the lack of necessary information on the infrastructure of the genetic architecture or for genetic analysis.

A change at the developmental level for stress tolerance is a complex process. Delaying the time of flowering or a reduction in the vegetative phase can involve many genes due to the interplay of plant products controlling the trait. McCue and Hanson (1990) suggested that intervention of the earlier or later timing of flowering is predictable in the amount of stress that will be altered, but change in stress resistance has been ineffective at this level. Some progress has transpired over the past ten years. Thomashow (1998) reported a strategy involving several genes controlling overexpression of late embryogenesis-abundant (LEA) biosynthetic enzymes transferred for cold acclimation. LEA proteins have played a role in desiccation tolerance in seed development and in the response to drought, salinity, and cold stress (Winicov, 1998).

The structural level may or may not be as complex. Some of the plant's structural changes can be the result of a single gene change, such as dwarfing traits (Mahon, 1983), while other traits are more complex, such as root patterns for more efficient water uptake. Very and colleagues (1998) discuss the role of guard cells in regulating salt tolerance and breeding for increased stomatal apertures. Guard cells of *Aster tripolium* respond to high salt concentrations by down-regulating K^+ uptake, preventing excessive accumulation of Na^+. Nonhalophytic plants without this sensitivity might respond similarly if bred for larger stomatal apertures. However, reports of this have not yet appeared in the literature.

The physiological level includes traits such as ion transport, a temperature tolerance mechanism, oxidative stress, signaling proteins, osmoprotectants, and carbon metabolism (Winicov, 1998). McCue and Hanson (1990) state that, in general, this level is less complex to alter than the developmental or structural levels. At the biochemical level, many examples of mutations of enzymes or proteins confer stress tolerance. Liu and Zhu (1998) reported the cloning of the *SOS3* gene encoding for a calcineurin B-like protein critical for stress tolerance in *Arabidopsis*. Salt overly sensitive (SOS) qualities helped elucidate important signaling-pathway components of tolerance. Success in engineering biochemical stress-resistance traits has generally been limited by the lack of understanding of metabolic flux, compartmentation, and function (Cushman and Bohnert, 2000).

Quantitative trait loci (QTLs) and marker-assisted selection have still proven useful in indirectly selecting for traits such as osmotic adjustment (Zhang et al., 1999). QTLs that are linked to tolerance at one developmental stage can differ from those linked to tolerance at other stages. Foolad (1999) found in most cases that QTLs linked to tolerance measured during seed germination differed from those measured during vegetative growth.

SOURCES OF GENETIC VARIATION

The existence of genetic variability is a prerequisite for any breeding program. Most crop species are produced outside of their native localities. Recent changes in germplasm exchange brought on by the Convention for Biological Diversity in 1993 (Zedan, 2002) have made plant explorations more difficult and has limited our abilities to freely obtain germplasm from native sources. This limits the ability of breeding programs to incorporate new or exotic germplasm accessions.

An important criterion in the breeding of stress tolerance is the availability of sufficient genetic variation for the characteristic within the germplasm base. This may include not only the cultivated species but also wild progenitors of the target species. Interspecific crosses can be used where there is limited variation for tolerance. Some closely related species may possess tolerance to some abiotic stress but may have inferior agronomic characteristics. Although valuable germplasm may be available, the transfer from the wild to cultivated species is often difficult and may require ovule culture or protoplast fusion to overcome barriers. Strategies for utilizing wild or exotic germplasm were described by Tanksley (1993). Mutagenesis is another option to produce the desired variability when not available from existing germplasm or the use of tissue culture for somaclonal variation.

Open-pollinated crops are more likely to have substantial plant-to-plant variation that can be utilized for stress tolerance. Variation from intraspecific cross has been reported widely in salt tolerance (Yeo, 1998). Variation may be lacking in some species due to a narrowing genetic base through continued selection pressure. Alleles for stress tolerance may be lost or occur only at a low frequency. Lack of variation may also be due to poor representation of available germplasm. Either accessions being screened do not adequately represent germplasm currently available in a collection such as the National Plant Germplasm System (NPGS), or the collection does not adequately represent what is available in the wild and requires further expansion of the germplasm base.

Salt-tolerant plants have high intercellular salt levels. Plants use three strategies for maintaining low Na^+ concentration: sodium exclusion, compartmentation, and secretion. Sodium is transported out of the cell by the operation of plasma-membrane-bound Na^+/H^+ antiports (Blumwald et al., 2000; also see Zhang et al., 2001). It can also be excluded by actively moving ions across the tonoplast into the vacuole, removing the potentially harmful ions from the cytosol.

BREEDING FOR TOLERANCE

Traditional approaches in breeding for abiotic stresses attempt to utilize genetic variation from intraspecific, interspecific, or intergeneric hybridization; mutations; and somaclonal variation. These approaches have been successful in improving stress tolerance by different selection schemes. This has been the backbone of modern agriculture over the years and has resulted in numerous improvements. This approach, although hampered by the complexity of traits and traits with low genetic variances, is a cost-effective means of crop improvement. With the development of molecular markers for many oilseed crops, major QTLs controlling or regulating stress tolerance have been identified (Quarrie, 1996). Marker-assisted selection of secondary traits related to a component of yield or trait for tolerance has also proven successful in breeding for stress tolerance. When a specific set of molecular markers are found to be associated with the primary trait being selected that is difficult to phenotype, the secondary markers are used to screen a population. This also has the effect of selecting against linkage with undesirable traits. Closely spaced genetic maps can be produced for crops with appropriate mapping population of plants. These populations can also be scored and statistically analyzed for variation of the quantitative traits to identify the minimum number and location of the genes regulating those traits or QTL (Lebreton et al., 1995).

In cases such as osmotic adjustment, it is considered impractical to breed for improvement by traditional methods. Osmotic adjustment is too tedious to measure since it requires all genotypes to be at the same rate of dehydration and plant water deficit (Zhang et al., 1999). In segregating generations, this becomes impossible because of the quantity of measurements. Technologies such as molecular markers and genetic transformations have allowed manipulation of this trait. Osmotic adjustment has been associated with increased yield under drought stress in *Brassica,* since the solutes causing the adjustment can substitute for water in some biochemical processes (Kumar et al., 1984; also see Zhang et al., 1999).

Jamaux and colleagues (1997) used QTL analysis to improve osmotic adjustment in sunflower and examined relative water loss of two lines that contrasted in their rates of desiccation. The two cDNA libraries were screened, which led to the isolation of a clone showing high-sequence homology with a mammalian amino-acid transporter. Restriction fragment length polymorphism (RFLP) analysis and the probe were able to differentiate between the two lines. It is important to note that the same level of stress needs to be maintained for all genotypes in mapping QTLs for osmotic adjustment because solute accumulation in a given genotype is a function of leaf desiccation and can lead to erroneous results (Nguyen et al., 1997).

Zhang and colleagues (1999) described a common backcross breeding procedure coupled with marker-assisted selection of QTLs to incorporate a tolerance trait, such as osmotic adjustment, and developed near-isogenic lines (NILs) along the way. The NILs are used to fine map the QTL. An F1 produced from the QTL parent and the other parent from an elite line were backcrossed to the recurrent, elite-line parent. The BC_1F_1-generation plants were screened with identified markers linked to the QTL. Those that showed linkage were then backcrossed again to the recurrent parent. This whole procedure was carried out for several generations. At the end of this procedure, plants carrying the targeted QTL were selfed and taken one more generation for seed multiplication. This breeding scheme could be adapted to any open-pollinated crop.

Several criteria must be met before establishing a breeding program for tolerance to an identified abiotic stress. Devine (1982) established criteria for edaphic adaptations, and some of these can be applied in breeding for other stresses. First, the problem must be severe and widespread enough to merit investment of resources. This has become especially critical with biotechnological approaches for developing tolerant crops because the method can be very costly. This is also tied in to the economics of the oilseed market. When a vast market is available as is the case for soybeans, there is more room to invest in resources for problems such as these. In other cases, as with minor crops or the development of new crops, the oils may be competing against an imported oil by a narrow margin. Therefore, the market may not be able to sustain the investment.

In vitro cellular selection and the exploitation of somaclonal variation produced salt-tolerant lines of linseed (Rowland et al., 1995). One of the salt-tolerant lines had increased heat tolerance, improved low-temperature germination, and other positive attributes, and was registered as a variety (Rowland et al., 1989). In this case, the strategy was to evaluate over 11,000 plants regenerated from callus cultures of a number of Canadian linseed varieties. Eventually, 3,100 lines were field evaluated in single rows and approximately 25 percent selected for a replicated trial the following year. About 33 percent were carried on to a second replicated trial. Winicov (1996) states that early problems encountered with this method were due to prolonged selection on NaCl in culture and lack of secondary screening of an adequate number of plants.

In some cases, the problem can be amenable to alleviation more economically by means other than breeding. Salts may be managed by a leaching-irrigation management practice. Fertilization or liming may be a more practical approach. Techniques must be available to assay plant response to the particular stress. It is also necessary to distinguish between environmental interactions. An example of this is provided in the study by Magistad and colleagues (1943) on the effect of climatic conditions in modifying crop growth at different salt concentrations.

SELECTION CRITERIA AND TECHNIQUES

Plants with particular traits of interest have been selected and characterized to determine whether the trait was the result of the genetic contribution. The selection criteria are usually based on plant morphology or physiology. In the more recent past, selection has also been based on traits associated with molecular markers. Most of the applications with molecular markers for selection have utilized nuclear DNA, although organellar DNA has also been used (Bretting and Widrlechner, 1995). Organellar DNA is more highly conserved and, as a result, less polymorphic than nuclear DNA. The most useful markers are

1. highly heritable,
2. highly polymorphic,
3. ubiquitous throughout the plant genome,
4. selective neutral,
5. codominant inherited,
6. inexpensive, and
7. nonlabor intensive (Bretting and Widrlechner, 1995).

A comprehensive review of the status of DNA markers in crop improvement is found in Lee (1995).

Smith and colleagues (1994), in an attempt to simplify selection of accessions with increased likelihood of salt tolerance, tried to relate climatological information and geographical data as criteria. They evaluated 630 plant introductions from the NPGS in the greenhouse under saline and nonsaline conditions, relating tolerance to the type of climate at the collection site. Rumbaugh and Pendery (1990) were able to accomplish this in selecting for salt tolerance at germination by grouping accessions according to passport information. Agronomic practices, especially irrigation, interfered with the ability to predict salt tolerance based on this type of information in the former study.

A screening protocol was developed by Kumar and colleagues (1999) for thermotolerant sunflower lines selected on the basis of temperature-induction response. Mild heat treatment elicits a heat-shock response leading to the immediate induction of a set of new proteins, or the overexpression of existing proteins, known as heat-shock proteins (HSPs), that play a key role in survival at extreme temperatures. This was done by screening an open-pollinated population of seedlings to a gradually increased temperature to induce HSPs. Surviving plants were selected and cross-pollinated. Antibodies were used to detect HSP90 and HSP104 in the leaf tissue. This same HSP101 was also isolated in soybean (Lee et al., 1994) and in *Arabidopsis* (Schirmer et al., 1994). The protein AtHSP101 from *Arabidopsis* was 43 percent identical to HSP104 from yeast and, when expressed in yeast, protected it from severe heat stress.

Spehar and Galwey (1997) found a relationship between root length of soybeans grown in solution and plants stressed due to low calcium as a nondestructive method of screening. Areas characterized by acidity and low levels of calcium result in moisture stress due to shallow root systems and reduction in yields. They found that varieties had a different response of elongation to Ca levels and were able to be classified into sensitive varieties and efficient varieties. In addition, they suggested that genetic improvement to low calcium should also improve aluminum tolerance. A similar screening method was found for aluminum tolerance in soybean by examining root growth. Aluminum tolerance was first reported in soybean in the 1960s (Armiger et al., 1968). Campbell and Carter (1990) found good genotypic correlation between root growth of 40-day-old hydroponic plants for Al tolerance. Foy and colleagues (1993) observed a high correlation between Al tolerance of soybean expressed as shoot and root growth in soil of 37-day-old plants based on tolerance ratings. Bianchi-Hall and colleagues (1998) identified QTLs associated with Al tolerance from RFLP marker loci, indicating that marker-facilitated selection may be useful in breeding for tolerance in soybeans.

Carter and Rufty (1993) also reported on PI 416937's ability to maintain leaf tissue turgidity during drought stress. Hudak and Patterson (1995) found that tolerance in this PI did not depend upon reduced leaf area, as in other crops with drought-resistant morphology. It was characterized by relatively few, very large leaves along with greater root mass, volume, and surface area. These examples are based on selection for agronomic characters such as survival under various stresses. Noble and Rogers (1992) suggested that the use of physiological mechanisms responsible for tolerance such as ion exclusion, ion accumulation, and compatible solute production has been largely nonexistent. With the progress made in the past ten years from molecular biology, this has begun to change.

Whether to select under stress conditions and narrowly rely on the landrace genotypes has been debated over the years. Ceccarelli and colleagues (1992) proposed selecting in a stressful environment to minimize genotype × environment interaction. Braun and colleagues (1992) suggested evaluating genotypes under optimum conditions that maximize heritability. Simmonds (1991) contends that no conclusive data have been presented to demonstrate that heritabilities are generally lower in stress than in optimum environments. Many other authors such as Falconer (1990), Calhoun and colleagues (1994), and Igartua (1995) have emphasized the importance of testing and selecting in a range of environments, including nonstress ones because it translates into improved gains in yield. The traditional method of selecting only under drought conditions and narrowly relying on the landrace genotype does not move yield beyond those traditionally obtained and does not provide the farmer with a bonus yield in the "fat years" (Rajaram et al., 1996). This option, however, is more costly than the other two options. Hohls (2001) used computer simulations to compare yield improvements between a stress and nonstress environment and found that if the goal is yield improvement, selection should be conducted directly in the stress environment. Selection should be conducted in the nonstress environment only if the genetic correlation between environments is high and the heritability of yield is higher in that environment.

ACHIEVEMENTS IN CROP IMPROVEMENT IN THE PAST FEW DECADES

New technologies have developed in genetic engineering, some of which were mentioned in prior sections: large-scale EST preparation and expression profiling, microarray analysis identifying target genes whose expression is stress regulated, and the generation of mutants in stress-response pathways (Bohnert et al., 2001). A suitable bioinformatics program is required to produce unique sequences for these large-scale DNA sequencing projects to be successful (Selkov et al., 1998). The sequencing of species such as *Arabidopsis,* corn, and rice has led to information on the frequency and type of transcribed genes obtained from organs, tissues, and cells during development. Changes in expression levels over time can be more easily observed under various stresses. Bohnert and colleagues (2001) referred to another impact in molecular genetics that has occurred in the past decade. Saturation mutagenesis for the loss or gain of a plant function has provided proof of concepts of experiments on a large scale.

Shannon and Noble (1990) reported that few varieties have been selected for improved tolerance and an even smaller number of released varieties exist that have been specifically bred for improved tolerance. Since this time, few stress-resistant varieties associated with high yield have become available. The lack of progress may likely be due to the multigenic nature of stress-sensitive and stress-tolerant phenotypes (Bohnert and Jensen, 1996). As a result of sequencing entire organisms during the past 15 years, progress has been made in the discovery of genes that could be essential components of stress tolerance. Plant stress in oilseed crops is still not well understood at the genome level. Even with the newly identified genes related to tolerance, both traditional breeding strategies and new variations of strategies that are adapted to technologies such as marker-assisted selection and QTLs are still needed.

One of the most obvious achievements in the past decade has been the wealth of information generated and made available from the large-scale sequencing projects and the high-throughput molecular analysis projects. The information has become better defined in areas such as genetic function and, as a result, is becoming more available for plant breeders. As different genes that are a part of some biochemical pathway or signaling pathway related to stress tolerance are discovered, they have been engineered into crop species. One example of this is the *AtNHX1* gene, which, when overexpressed in *Arabidopsis* or *Brassica napus,* allowed plants to grow, flower, and produce seed in high levels of salt (Apse et al., 1999; Zhang et al., 2001). It is interesting that moving a single trait such as this particular protein into the rapeseed genotype significantly improved salt tolerance. This suggests that multigenes may not always be necessary to produce a salt-tolerant variety. Similar achievements in screening for mutants have been made, one of which is for the SOS phenotypes in *Arabidopsis* (Zhu et al., 1998). These phenotypes all have different responses but have produced a genetic model for salt-tolerance mechanisms.

GENETIC ENGINEERING FOR CROP IMPROVEMENT

The direct introduction of a small number of genes by genetic engineering has provided an alternative to selection breeding schemes and a complement to it. The discovery of novel genes and the determination of their expression patterns in response to abiotic stress have improved stress adaptation of crops in recent years. Genetic engineering of crops relies on the transfer of one or several genes that encode either biochemical pathways or endpoints of signaling pathways that are controlled by a constitutively active promoter (Cushman and Bohnert, 2000).

There have been many examples of this in the recent literature. As stated previously, the engineering of *Brassica napus* to overexpress a vacuolar Na^+/H^+ antiport from *Arabidopsis thaliana* was described by Zhang and colleagues (2001). In a greenhouse experiment, the engineered plants grew, flowered, and produced seed in 200 mM NaCl. Growth was correlated to levels of the engineered protein AtNHX1. There was a sixfold increase of proline content in the engineered plants. The wild-type all died under these conditions. Seed oil was unaffected in either quality or quantity from the engineered and control plants, indicating the AtNHX1 protein enhanced accumulation of Na^+ and allowed the engineered plants to mitigate the toxic effects. *Arabidopsis* plants were also engineered to overexpress this gene by Apse and colleagues (1999). The control plants were, in general, growth inhibited when watered with increased NaCl-containing solution up to 200 mM NaCl. Engineered plants were unaffected in plant development and seed production.

Another example is the *codA* gene from the bacteria *Arthrobacter globiformis* for biosynthesis of glycinebetaine to improve salt stress, which was inserted in *Brassica juncea* cv. Pusa jaikisan (Prasad et al., 2000). Plants from the mustard family lack the potential to synthesize glycinebetaine (McCue and Hanson, 1990). This was the first successful introduction and expression of a bacterial gene for choline oxidase into an oilseed crop. Previous reports were restricted to transformation of either *Nicotiana* or *Arabidopsis* (Prasad et al., 2000). Improvements to salt-stress tolerance of these young seedlings was attributed to increased levels of total sugars in the leaves, contributing to osmotic adjustment and stabilization of macromolecules.

Overexpression of a subset of a cold-regulated binding factor 1 (CBF1) has led to enhanced freezing tolerance (Jaglo-Ottosen et al., 1998; see also Winicov, 1998). In this case, some adaptative response provided increased tolerance by CBF1 turning on the cold-regulated genes. Numerous other examples of genetic engineering of oilseeds continue to surface with new technologies that have become available. The bioinformatics, gene discovery projects, mapping, and QTLs have all greatly increased our ability to introduce new genes into crops that target stress tolerance. However, this has not replaced but has enhanced plant breeding.

LIMITATIONS AND FUTURE PROSPECTS

Michelmore (2000) states that the new technologies we have experienced in the field of breeding and genetics will cause shifts in the existing paradigms for experimentation. Hypothesis-driven research will shift to a period of descriptive studies that have begun to generate large amounts of data. New hypotheses will be derived from examining large databases, resulting in computer-simulated experimentation prior to actual experimentation (Michelmore, 2000). The field of genomics is still in the infancy stage but appears to be growing at a rapid pace. As costs decrease and technology is improved, more crops will also be sequenced and, hopefully, complex traits will be described and manipulated. Many limitations must still be faced in developing new, more tolerant oilseed crops. Poor understanding of the genetic, physiological, morphological, and environmental effects have reduced the progress of breeding for stress tolerance.

Great gains have been made in the genomics of stress tolerance with the large-scale sequencing projects (The Arabidopsis Genome Initiative, 2000). This has resulted in mapping of genes based on segregation analysis at low density for some oilseed crops. Soybeans and rapeseed already have high-density maps. Identifying single-gene traits, engineering them into and overexpressing them in plants, has occurred and has resulted in improved tolerance to abiotic stress. Inferences have been made regarding function of genes between homologs to determine similarities and taxonomic relations. Some of this discovery has been applied to major oilseed crops.

In the future, other crop species will be sequenced, and it is likely that resequencing will occur for genetic variants and the functions of these genes will be elucidated. Michelmore (2000) states that as more species are sequenced and the extent of synteny has been established for each species it will become possible to predict the position of some, but not all, of the genes of a genome. Resistance genes may be located in less stable regions or be less conserved. In several species, resistance was either telomeric or close to the centromere region of the chromosome (Han et al., 1999). Mapping will be improved as other crops are sequenced. Soybeans will be the first of the oilseeds (Michelmore, 2000). Integration of QTL mapping with genomic sequence data will speed the process for finding candidate genes for tolerance. Breeding practices such as genotype × environment interaction and progeny testing will follow once strategies for introgression of traits or QTLs are accomplished.

It is obvious from the literature that in engineering stress tolerance, most of the studies have been with model crops such as *Arabidopsis*. Now the focus must switch to agronomically relevant oilseed crops. Field testing of these crops will not only verify what has been found in the laboratory but also determine the effect of the trait in an agronomic environment. Greater success will be achieved in the future as we obtain a better understanding of the metabolic pathways, the various mechanisms such as osmotic, desiccation, and temperature tolerance, and their signaling pathways in oilseeds. The next step will be to move from model crops to crops of agronomical importance.

REFERENCES

Anonymous (2000). Analysis of the genome sequence of the flowering plant *Arabidopsis thaliana*. *Nature* 408:796-815.

Apse, M.P., G.S. Aharon, W.A. Snedden, and E. Blumwald (1999). Salt tolerance conferred by overexpression of a vacuolar Na^{+} / H^{+} antiport in *Arabidopsis*. *Science* 285:1256-1258.

Arabidopsis Genome Initiative (2000). Analysis of the genome sequence of the flowering plant *Arabidopsis thaliana*. *Nature* 408:796-815.

Armiger, W.H., C.D. Foy, A.L. Fleming, and B.E. Caldwell (1968). Differential tolerance of soybean varieties to an acid soil high in exchangeable aluminum. *Agronomy Journal* 60:67-70.

Ashraf, M. (1994). Breeding for salinity tolerance in plants. *Critical Reviews in Plant Science* 13:17-42.

Bertelsen, A.H. and V.E. Valculescu (1998). High throughput gene expression analysis using SAGE. *Drug Discovery Today* 3:152-159.

Bevan M., I. Bancroft, H.W. Mewes, R. Martienssen, and R. McCombie (1999). Clearing a path through the jungle: Progress in *Arabidopsis* genomics. *Bioessays* 21:110-120.

Bianchi-Hall, C.M., T.E. Carter, T.W. Rufty, C. Arellano, H.R. Boerma, D.A. Ashley, and J.W. Burton (1998). Heritability and resource allocation of aluminum tolerance derived from soybean PI 416937. *Crop Science* 38:513-522.

Blumwald, E., G.S. Aharon, and P. Apse (2000). Sodium transport in plant cells. *Biochimica Biophysica Acta* 1465:140-151.

Bohnert, H., P. Ayoubi, C. Borchert, R.A. Bressan, R.L. Burnap, J.C. Cushman, M.A. Cushman, M. Deyholos, R. Fischer, D.W. Galbraith, et al. (2001). A genomics approach toward salt stress tolerance. *Plant Physiology and Biochemistry* 39:295-311.

Bohnert, H.J. and R.G. Jensen (1996). Strategies for engineering water-stress tolerance in plants. *Trends in Biotechnology* 14:89-97.

Bouchez, D. and H. Höfte (1998). Functional genomics in plants. *Plant Physiology* 118:725-732.

Braun, H., W.H. Pfeiffer, and W.G. Pollmer (1992). Environments for selecting widely adapted spring wheat. *Crop Science* 32:1420-1427.

Bretting, P.K. and M.P. Widrlechner (1995). Genetic markers and plant genetic resource management. In *Plant Breeding Reviews*, J. Janick (Ed.). New York: John Wiley & Sons, pp. 11-86.

Calhoun, D.S., C. Gebeyehu, A. Miranda, S. Rajaram, and M. van Ginkel (1994). Choosing evaluation environments to increase grain yield under drought conditions. *Crop Science* 34:673-678.

Campbell, K.A.G. and T.E. Carter Jr. (1990). Aluminum tolerance in soybean. I. Genotypic correlation and repeatability of solution culture and greenhouse screening methods. *Crop Science* 30:1049-1054.

Carter, T.E. and T.W. Rufty (1993). Soybean plant introductions exhibiting drought and aluminum tolerance. In *Adaptation of Food Crops to Temperature and Water Stress: International Symposium Proceedings*, C.G. Kuo (Ed.). Taipei: Taiwan Asian Vegetable Research and Development Center, pp. 335-346.

Ceccarelli, S., S. Grando, and J. Hamblin (1992). Relationship between barley grain yield measured in low- and high-yielding environments. *Euphytica* 64:49-58.

Cushman, J.C. and H.J. Bohnert (2000). Genomic approaches to plant stress tolerance. *Current Opinion in Plant Biology* 3:117-124.

Devine, T.E. (1982). Genetic fitting of crops to problem soils. In *Breeding Plants for Less Favorable Environments*, M.N. Christiansen and C.F. Lewis (Eds.). New York: John Wiley Press, pp. 143-173.

Dudal, R. (1976). Inventory of major soils of the world with special reference to mineral stress. In *Plant Adaptations to Mineral Stress*, Wright, M.J. (Ed.). Ithaca, NY: Cornell University Agricultural Experimental Station, pp. 3-14.

Falconer, D.S. (1990). Selection in different environments: Effects on environmental sensitivity (reaction norm) and on mean performance. *Genetical Research* 56:57-70.

Flowers, T.J., P.F. Troke, and A.O. Yeo (1977). The mechanism of salt tolerance in holophytes. *Annual Review of Plant Physiology* 28:89-121.

Foolad, M.R. (1999). Comparison of salt tolerance during seed germination and vegetative growth in tomato by QTL mapping. *Genome* 42:727-734.

Foy, C.D., T.E. Carter Jr., J.A. Duke, and T.E. Devine (1993). Correlation of shoot and root growth and its role in selection for aluminum tolerance in soybean. *Journal of Plant Nutrition* 16:305-325.

Gale, M.D. and K.M. Devos (1998). Plant comparative genetics after 10 years. *Science* 282:656-658.

Han, F., A. Kilian, J.P. Chen, D. Kudrna, B. Steffenson, K. Yamamoto, T. Matsumoto, T. Sasaki, and A. Kleinhofs (1999). Sequence analysis of a rice BAC covering the syntenous barley Rpg1 region. *Genome* 42:1071-1076.

Hohls, T. (2001). Conditions under which selection for mean productivity, tolerance to environmental stress, or stability should be used to improve yield across a range of contrasting environments. *Euphytica* 120: 235-245.

Hudak, C.M. and R.P. Patterson (1995). Vegetative growth analysis of a drought-resistant soybean plant introduction. *Crop Science* 35:464-471.

Hwang, I. and H.M. Goodman (1995). An *Arabidopsis thaliana* root specific kinase homolog is induced by dehydration, ABA, and NaCl. *The Plant Journal* 8:37-43.

Igartua, E. (1995). Choice of selection environment for improving crop yields in saline areas. *Theoretical and Applied Genetics* 91:1016-1021.

Jaglo-Ottosen, K.R., S.J. Gilmour, D.G. Zarka, O. Schabenberger, and M.F. Thomashow (1998). *Arabidopsis* CBF1 overexpression induces COR genes and enhances freezing tolerance. *Science* 280:104-106.

Jamaux, I., A. Steinmetz, and E. Belhassen (1997). Looking for molecular and physiological markers of osmotic adjustment in sunflower. *New Phytologist* 137:117-127.

John, P. (1992). *Biosynthesis of Major Crop Products*. New York: Wiley and Sons Press.

Kumar, A., P. Singh, D.P. Singh, H. Singh, and H.C. Sharma (1984). Differences in osmoregulation in *Brassica* species. *Annals of Botany* 54:537-541.

Kumar, G., B.T. Krishnaprasad, M. Savitha, R. Gopalakrishna, K. Mukhopadhyay, G. Ramamohan, and M. Udayakumar (1999). Enhanced expression of heat-shock proteins in thermo-tolerant lines of sunflower and their progenies selected on the basis of temperature-induction response. *Theoretical and Applied Genetics* 99:359-367.

Lebreton, C., V. Lazic-Jancic, A. Steed, S. Pekic, and S.A. Quarrie (1995). Identification of QTL for drought responses in maize and their use in testing causal relationships between traits. *Journal of Experimental Botany* 46:853-865.

Lee, M. (1995). DNA markers and plant breeding programs. In *Advances in Agronomy,* D.L. Sparks (Ed.). San Diego, CA: Academic Press, Volume 55, pp. 265-344.

Lee, M., M. Lenman, A. Banás, M. Bafor, S. Singh, M. Schweizer, R. Nilsson, C. Lijenberg, A. Dahlqvist, P. Gummeson, et al. (1998). Identification of non-heme diiron proteins that catalyze triple bond and epoxy group formation. *Science* 280:915-918.

Lee, Y.R., R.T. Nagao, and J.L. Key (1994). A soybean 101-kD heat shock protein complements a yeast HSP104 deletion mutant in acquiring thermotolerance. *The Plant Cell* 6:1887-1897.

Liu, J. and J.-K. Zhu (1998). A calcium sensor homolog required for plant salt tolerance. *Science* 280:1943-1945.

Macas, J., G.M. Lambert, D. Dolezel, and D.W. Galbraith (1998). Nuclear expressed sequence tag (NEST) analysis: A novel means to study transcription through amplification of nuclear RNA. *Cytometry* 33:460-468.

Magistad, O.C., A.D. Ayers, C.H. Wadleigh, and H.G. Gauch (1943). Effect of salt concentration, kind of salt, and climate on plant growth in sand cultures. *Plant Physiology* 18:151-166.

Mahon, J.D. (1983). Limitations to the use of physiological variablity in plant breeding. *Canadian Journal of Plant Science* 63:11-21.

McCue, K.F. and A.D. Hanson (1990). Drought and salt tolerance: Toward understanding application. *Trends in Biotechnology* 8:358-362.

Michelmore, R. (2000). Genomic approaches to plant disease resistance. *Current Opinion in Plant Biology* 3:125-131.

Nguyen, H.T., R.C. Babu, and A. Blum (1997). Breeding for drought resistance in rice: Physiology and molecular genetics considerations. *Crop Science* 37:1426-1434.

Noble, C.L. and M.E. Rogers (1992). Arguments for the use of physiological criteria for improving the salt tolerance in crops. *Plant and Soil* 146:99-107.

Osborn, T.C., C. Kole, I.A.P. Parkin, A.G. Sharpe, M. Kuiper, D.J. Lydiate, and M. Treck (1997). Comparison of flowering time genes in *Brassica rapa, B. napus* and *Arabidopsis thaliana*. *Genetics* 146:1123-1129.

Prasad, K.V.S.K., P. Sharmila, P.A. Kumar, and P.P. Saradhi (2000). Transformation of *Brassica juncea* (L.) Czern. with bacterial *codA* gene enhances its tolerance to salt stress. *Molecular Breeding* 6:489-499.

Quarrie, S.A. (1996). New molecular tools to improve the efficiency of breeding for increased drought resistance. *Plant Growth Regulation* 20:167-178.

Rajaram, S., H.J. Braun, and M. van Ginkel (1996). CIMMYT's approach to breed for drought tolerance. *Euphytica* 92:147-153.

Rowland, G.G., A. McHughen, and R.S. Bhatty (1989). Andro flax. *Canadian Journal Plant Science* 69:911-913.

Rowland, G.G., A. McHughen, R.S. Bhatty, S.L. MacKenzie, and D.C. Taylor (1995). The application of chemical mutagenesis and biotechnology to the modification of linseed (*Linum usitatissimum* L.). *Euphytica* 85:317-321.

Rumbaugh, M.D. and B.M. Pendery (1990). Germination salt resistance of alfalfa *Medicago sativa* L. germplasm in relation to subspecies and centers of diversity. *Plant and Soil* 124:47-51.

Schirmer, E.C., S. Lindquist, and E. Vierling (1994). An *Arabidopsis* heat shock protein complements a thermotolerance defect in yeast. *The Plant Cell* 6:1899-1909.

Selkov, E.J., Y. Grechkin, N. Mikhailova, and E. Salkov (1998). MPW: The metabolic pathways database. *Nucleic Acids Research* 26:43-45.

Shannon, M.C. and C.L. Noble (1990). Genetic approaches for developing economic salt tolerant crops. In *Agricultural Salinity Assessment and Management*. New York: American Society of Civil Engineers Press, pp. 161-185.

Simmonds, N.W. (1991). Selection of local adaptation in a plant breeding programme. *Theoretical and Applied Genetics* 82:363-367.

Smith, S.E., D.W. Johnson, D.M. Conta, and J.R. Hotchkiss (1994). Using climatological, geographical, and taxonomic information to identify sources of mature-plant salt tolerance in alfalfa. *Crop Science* 34:690-694.

Somerville, C. and S. Somerville (1999). Plant functional genomics. *Science* 285:380-383.

Spehar, C.R. and N.W. Galwey (1997). Screening soybeans [*Glycine max* (L.) Merill.] for calcium efficiency by root growth in low-Ca nutrient solution. *Euphytica* 94:113-117.

Tanksley, S.D. (1993). Mapping polygenes. *Annual Review of Genetics* 27:205-233.

Thomashow, M.F. (1998). Plant cold acclimation: Freezing tolerance genes and regulatory mechanisms. *Annual Review of Plant Physiology and Plant Molecular Biology* 50:571-599.

USDA (2001). *Oil Crops, Situation and Outlook Yearbook.* Washington, DC: USDA Economic Research Service.

Very, A.A., M.F. Robinson, T.A. Mansfield, and D. Sanders (1998). Guard cell cation channels are involved in Na^+-induced stomatal closure in a halophyte. *Plant Journal* 14:509-521.

Walbot, V. (1999). Genes, genomes, genomics. What can plant biologists expect from the 1998 national science foundation plant genome research program? *Plant Physiology* 119:1151-1155.

Winicov, I. (1996). Characterization of rice (*Oryza sativa* L.) plants regenerated from salt-tolerant cell lines. *Plant Science* 113:105-111.

Winicov, I. (1998). New molecular approaches to improving salt tolerance in crop plants. *Annals of Botany* 82:703-710.

Yeo, A.R. (1998). Molecular biology of salt tolerance in the context of whole-plant physiology. *Journal of Experimental Botany* 49:915-929.

Zedan, H. (2002). *Quarterly Report on the Administration of the Convention on Biological Diversity.* Montreal, Canada: Convention on Biodiversity.

Zhang, H.X., J.N. Hodson, J.P. Williams, and E. Blumwald (2001). Engineering salt tolerant *Brassica* plants: Characterization of yield and seed oil quality in transgenic plants with increased vacuolar sodium accumulation. *Proceedings of the National Academy of Science* 98:12832-12836.

Zhang, J., H.T. Nguyen, and A. Blum (1999). Genetic analysis of osmotic adjustment in crop plants. *Journal of Experimental Botany* 50:291-302.

Zhu, J.K., J. Liu, and L. Xiong (1998). Genetic analysis of salt tolerance in *Arabidopsis*: Evidence for a critical role of potassium nutrition. *The Plant Cell* 10:1181-1191.

Chapter 17

Breeding for Abiotic Stress Tolerance in Cotton

Daryl T. Bowman
Randy Wells

GENETIC ARCHITECTURE OF COTTON

Over 90 percent of the world's cotton is upland cotton or *Gossypium hirsutum* L. (Brubaker et al., 1999); this chapter will concentrate on this one species. A passing reference will be given to *Gossypium barbadense* L., since it is the second most important cotton species; this species includes Sea Island, Egyptian, and Pima cottons, along with extra-long staple cottons of other countries such as India, Peru, Sudan, and the former Soviet Union. Both of these species are allotetraploids with $2n = 52$ chromosomes, and both have "A" and "D" subgenomes. Allotetraploids are native only to the New World, although *G. hirsutum* is grown extensively throughout the world. The A subgenome is characteristic of the Old World diploid cottons ($2n = 26$), *G. arboreum* L. and *G. herbaceum* L. The D subgenome is characteristic of the New World diploid cottons, of which there are at least a dozen wild species. The A subgenome consists of relatively large chromosomes, while the D subgenome has relatively small chromosomes. It is theorized that the tetraploids resulted from natural hybridization of an Old World and a New World diploid.

Genetically, the allotetraploids behave as normal diploids (Kimber, 1961). As such there have been nearly 200 qualitative alleles identified from the tetraploids (Percy and Kohel, 1999). Recent evidence suggests that *G. hirsutum* may actually be an octaploid (David Stelly, personal communication). Nevertheless, the diploid-like behavior of the allotetraploids affords possibilities for genetic improvements without the complications of polyploidy.

SOURCES OF GENETIC VARIATION

The primary gene pool for variation for the allotetraploids consists of the other recognized tetraploids. For *G. hirsutum,* it would be *G. barbadense, G. tomentosum, G. mustelinum,* and *G. darwinii* (Table 17.1). Only *G. hirsutum* and *G. barbadense* produce fiber, and these have the most variability, probably because most of the exploration efforts have been directed toward these two. *G. darwinii, G. tomentosum,* and *G. mustalinum* have limited variability, thought to be due to their limited geographic distribu-

TABLE 17.1. List of *Gossypium* species grouped by germplasm pool.

Species	Genome	Species	Genome
Primary Pool		Tertiary Pool	
hirsutum	AD_1	*sturtianum*	C_1
barbadense	AD_2	*robinsonii*	C_2
tomentosum	AD_3	*bickii*	G_1
mustelinum	AD_4	*australe*	G
darwinii	AD_5	*nelsonii*	G
Secondary Pool		*costulatum*	K
herbaceum	A_1	*cunninghamii*	K
arboreum	A_2	*enthyle*	K
anomalum	B_1	*exgiuum*	K
triphylium	B_2	*nobile*	K
capitis-viridis	B_3	*pilosum*	K
trifurcatum	B?	*populifolium*	K
longicalyx	F_1	*pulchellum*	K
thurberi	D_1	*rotundifolium*	K
armouranium	$D_{2\text{-}1}$	*G. sp. nov.*	K
harknessi	$D_{2\text{-}2}$	*stocksii*	E_1
davidsonii	$D_{3\text{-}d}$	*somalense*	E_2
klotzchianum	$D_{3\text{-}k}$	*areysianum*	E_3
aridum	D_4	*incanum*	E_4
raimondii	D_5	*benadirense*	E
gossypiodes	D_6	*bricchetti*	E
lobatum	D_7	*vollesenii*	E
trilobum	D_8		
laxum	D_9		
turneri	D_{10}		
schwendimanii	D_{11}		

Source: From Stewart, 1994.

tion. *G. tomentosum* is limited geographically to the Hawaiian Islands, *G. darwinii* to the Galapagos Archipelago, and *G. mustalinum* to northeastern Brazil.

The primary gene pool by definition does not pose a major obstacle to hybridization; crosses are fairly easily made and viable seed is the normal result. However, when interspecific crosses are made, there appears to be less crossing over (gene exchange) than in intraspecific crosses. As a result, the genotypes in the subsequent segregating generations tend to revert to one of the parents. Thus, large segregating populations and cycles of intercrossing are required to transfer desired alleles between species in the primary gene pool.

Many of the allotetraploid wild species and *G. hirsutum* landraces are daylength sensitive and require a short daylength (<12 h) to flower. Jack McCarty's race stock conversion program (USDA-ARS, Mississippi State) has made accessible to cotton breeders more than 600 different race stocks of *G. hirsutum*. Within this group, bollweevil (*Anthonomus grandis* Boh.) resistance (McCarty et al., 1985); root-knot nematode *(Meloidogyne incognita)* resistance (Shepherd et al., 1988); and genes for boll size, seed size, fiber fineness, fiber length, fiber elongation, and fiber strength (McCarty and Jenkins, 1992) are potentially useful to the breeder.

Other examples of useful traits gleaned from the primary gene pool include bacteria (*Xanthamonas campestris* pv. *malvacerum*), blight-resistance genes from *G. hirsutum* and *G. barbadense* (Endrizzi et al., 1985), resistance to Fusarium (*Fusarium oxysporum* f.sp. *vasinfectum*) and Verticillium wilt *(Verticillium dahliae)* from *G. hirsutum* and *G. darwinii* (Bell, 1984), and nectariless genes from *G. tomentosum* (Meyer and Meyer, 1961).

Most commercial breeders utilize only the elite germplasm within the primary gene pool; thus variability within *G. hirsutum* species is underutilized except by public breeders. U.S. cotton breeders typically use in-house material as their primary parental source, with public germplasm the second most-used parental source (Van Esbroeck et al., 1998). However, in recent years the most successful cultivars in the United States have included more public germplasm than any other source. One can understand the reluctance of breeders to use any exotic germplasm, since exchange of genes usually involves large blocks of genes, thus resulting in genetic drag of less-desirable alleles from exotic germplasm. These unfavorable linkages require many years of crossing and intercrossing to break, and most breeders are hesitant to devote major resources to this effort.

The secondary gene pool for allotetraploid cotton is the group of diploid species of the same genome (A and D) and those diploids with genomes of close homeology with the A genome (Table 17.1). The latter includes the B and F diploids. Some of the D genome diploids such as *G. davidsonii, G. klotzschianum,* and *G. gossypiodes* have lethal genes conditioning death of hybrid seedlings when crossed with allotetraploids, but this has been overcome (Lee, 1981). Techniques have been discovered that improve the chances of success; these include ovule culture (Stewart and Hsu, 1977, 1978) and hormone application (Altman, 1988). Traditionally, diploids in the secondary gene pool are crossed with an allotetraploid to produce a sterile triploid. The chromosome number is doubled with colchicine to produce a hexaploid that is then backcrossed to the allotetraploid until a 4 *n* number is reached. This is a cumbersome approach with a very low success rate.

A novel approach to this problem of transferring genes from a diploid species to an allotetraploid was first employed by Beasley (1940). A synthetic allotetraploid is produced by crossing either A, B, or F genome species with a D genome species. The chromosome number of the resulting hybrid is doubled to produce a synthetic allotetraploid that is then crossed with a natural allotetraploid. Subsequently, the new hybrid is backcrossed to the natural allotetraploid while screening for the desired trait, if known.

A third approach involves the use of synthesized hexaploids (Stewart, 1994). Genetic hexaploid stocks were created by crossing an allotetraploid with a D genome diploid and doubling the chromosome number. The resulting hexaploid can be used to cross with a diploid species in the secondary gene pool. After this final cross, it is just a matter of backcrossing to the cultivated allotetraploid.

Bacterial blight resistance has been extracted from the secondary gene pool, specifically *G. arboreum, G. herbaceum,* and *G. anomalum* (Endrizzi et al., 1985). Meyer (1975) created cytoplasmic male sterility using the cytoplasm from *G. harknessii,* and Meyer (1975) transferred a glabrous trait from *G. armourianum.*

The tertiary gene pool comprises all other wild diploid species. A complete listing taken from Stewart (1994) is found in Table 17.1. Basically, this group consists of the E genome species of Africa, and the C, G, and K genome species from Australia. This group will require extraordinary efforts to extract genetic material usable in a cultivated allotetraploid. The direct approach for gene transfer via hybridization, creation of a triploid, doubling the chromosome number, and then backcrossing to the cultivated allotetraploid will require extensive efforts. Viable seed are not assured, thus requiring some form of embryo rescue. Recombination is greater in the hexaploid state but at the expense of fertility (Stewart, 1994). Other attempts are summarized by Stewart (1994). No record of successful transfer of a genetic trait from this gene pool has been published.

Outside the traditional germplasm pool lies the "alpha omega" pool as described by Stewart (1994). This gene pool includes all other life forms and the genetic material associated with those life forms. It has been made possible only recently through the advent of molecular biology. This germplasm pool has been tapped for insect resistance via *Bacillus thuringiensis* (Perlak et al., 1990) and herbicide resistance (Bayler et al., 1992). Currently, only monogenic (qualitative) traits are being explored for possible commercialization.

CROP PRODUCTION

Cotton production is primarily found in the world between latitudes 30° and 37° north. The United States, parts of the former Soviet Union, and the People's Republic of China produce two-thirds of the world's cotton in this region. The remainder of the world cotton crop is produced between 30° north and 30° south latitude with a small portion planted above 37° north latitude (Bell and Gillham, 1989).

Cotton, being a subtropical plant, requires a minimum number of growing days or heat units to produce the crop. When there are no limitations of nutrients or moisture, there is a direct relationship between growing days and yield, which is shown in Figure 17.1. Typically, heat units for cotton

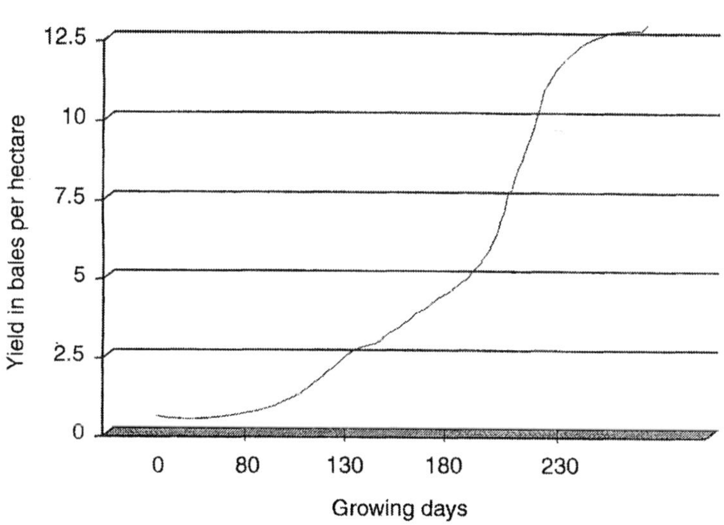

FIGURE 17.1. Relationship between yield and growing days with no nutrient or moisture limitations. *Source:* From B. A. Waddle (1984), Crop growing practices. In R. J. Komel and C. P. Lewis, *Cotton* (pp. 233-263). Madison, WI: American Society of Agronomy. Used by permission of ASA.

production are calculated on the basis of 60°F or 15.6°C, and sometimes on a basis of 55°F or 13°C. A cotton crop requires a minimum of 2500 heat units based on 15.6°C, or more than 160 days above 15°C (Waddle, 1984). Closely related to heat units is solar energy. Too much cloudy weather will negatively impact yield via a reduction in photosynthate production, fruit abortion, and poor seed set. Areas with greater than 50 percent cloudiness would not be suitable for cotton production (Waddle, 1984).

Rainfall distribution is critical to successful cotton production. A cotton crop requires a minimum of 50 cm of moisture and becomes more productive as moisture increases, up to a critical point at which excess moisture could actually lower yields. Relative humidity above 90 percent for extended lengths of time could reduce boll set (Waddle, 1984). Thunderstorms in the summer affect yield by rendering the pollen useless, so boll set as well as seed set are reduced (Lloyd, 1920). About half of world cotton production is under irrigation due to unfavorable soils and lack of natural rainfall.

Soil type is not absolutely critical; essentially any soil type that can grow row crops can grow cotton (Waddle, 1984). Soils are used for nutrients, water, oxygen, and anchorage for roots. Cotton, like most plants, cannot tolerate anaerobic conditions for long periods of time; therefore, poorly drained soils must be managed to prevent this from happening. It is the moisture-holding capacity of the soil that differentiates the production practices used to manage the crop. Silt-loam soils hold water better than sandy soils that may require supplemental water. Clay soils have excellent water-holding capacity but are subject to anaerobic conditions when saturated.

ABIOTIC STRESSES

Probably the most noticeable abiotic stress is drought. In some cases the opposite extreme occurs and drowning takes place, but this is so rare, except for certain soil types, that it is not necessarily worthy of becoming a breeding objective. Drought can affect the crop from planting to boll maturation. Nonuniform soil moisture at planting can result in uneven stands. Drought is most critical from flowering until boll maturation.

Subsoil acidity and the associated toxicities have received attention in Louisiana, where they are a major problem on the Macon Ridge (Kennedy and Jones, 1991). Root growth may be inhibited, and shoot growth is affected at certain pH levels.

Tolerance to both low and high temperatures is desired. Low temperatures shortly after planting and during early seedling growth is common in the northern fringes of the cotton belt. Extremely high temperatures can cause flowers to become sterile in sensitive genotypes; there are genetic differences for heat tolerance. Some breeders make it a standard practice to screen promising lines in the lower elevations of Arizona for heat tolerance.

Some soils have salinity problems, particularly if they are irrigated on a regular basis. The Aridosols of the former Soviet Union and the People's Republic of China fit this category. Smog has become a major factor in cotton production for areas such as the lower San Joaquin Valley of California. Smog blocks solar energy and lowers yield; it may also adversely affect yield via a change in the chemical composition of the atmosphere.

BREEDING TECHNIQUES

Drought

Although few cotton cultivars specifically identified as having drought resistance have been released (see Table 17.2), work toward this goal has been reported. The Texas MAR program has developed cultivars that have outperformed many of their own lines under drought conditions (Thaxton and El-Zik, 2000). 'Tamcot Pyramid' and 'Tamcot Luxor' looked impressive in their trials. The previous year they had reported on progress in breeding for drought resistance by examining relative performance in irrigated

TABLE 17.2. Identified drought-tolerant phenotypes.

Cultivar/Line	Reference	Cultivar/Line	Reference
Tamcot CD3H	Cook and El-Zik, 1993	4521	Penna et al., 1998
SRT1	Singh, 1993	Mac 13	McDaniel, 2000
JK125-2-42	Singh, 1993	Mac 95	McDaniel, 2000
Indore 2	Singh et al., 1996	CNPA Precoce 2	Pereira et al., 1998
170-C02	Singh et al., 1996	CNPA 7H	Pereira et al., 1998
Siokra L-23	Nepomuceno et al., 1998	CNPA Precoce 1	Pereira et al., 1998
T-1521	Nepomuceno et al., 1998	TCH1002	Rajeswari, 1995
IAC-13-1	Penna et al., 1998	LH900	Rajeswari, 1995
IAC-RM4-SM5	Penna et al., 1998	L886	Rajeswari, 1995
Minas Sertaneja	Penna et al., 1998	T25	Peterschmidt and Quisenberry, 1981
Acala 1517E-1	Penna et al., 1998		

versus nonirrigated locations (Thaxton and El-Zik, 1999). They have not reported screening for any specific trait or character for drought resistance in the breeding process. Quisenberry and Roark (1976) reported genetic variability for shoot growth under dryland and irrigated conditions. The *G. hirsutum* racestocks T147, T292, and T133 were not negatively impacted by dryland conditions when compared to irrigated conditions. It is not always easy to decide on the approach to use in breeding for drought resistance. The timing of drought affects cultivars differently. An early-season drought impacts determinate cultivars more than indeterminate cultivars (Quisenberry and Roark, 1976).

Breeding for drought resistance is complicated by the very complex nature of the resistance mechanisms. Stomatal behavior influences leaf temperature, CO_2 intake, and transpiration rate, which in turn affect the response of plants during a drought. Avoiding the detrimental effects of drought by closing stomates during the daytime may result in reduced growth when water is not limiting. Since the timing of drought cannot always be predicted, it may be difficult to make breeding progress for improved drought tolerance and yield at the same time. Another mechanism for avoiding the detrimental effects of drought is extracting relatively more water from the soil via a more extensive root system; this may be advantageous where soil water is extractable but not otherwise. However, a combination of mechanisms, if possible, may be best for viable drought resistance with consequential yield stability.

Quisenberry and colleagues (1981) reported that drought resistance, as expressed in dry-matter yield, was the result of early growth under adequate soil moisture. The implication is that selection for drought resistance should occur under optimum soil moisture. Later, Quisenberry and colleagues (1985) stated that one should look at leaf turgidity under dry field conditions to select for drought tolerance. Infrared thermometry was then suggested as a method of screening germplasm for water-use efficiency by selecting plants with cooler leaves (Hatfield et al., 1987). This was confirmed by Singh and colleagues (1992), who also stated that one should look for cultivars with lower leaf temperatures that are accompanied by a lower transpiration rate along with a higher photosynthetic rate and stomatal conductance. Stomatal conductance was positively associated with drought tolerance and heat tolerance; in most instances, heat and drought occur simultaneously. Cook and El-Zik (1992) concluded that one needed to improve seedling vigor, increase early root growth, and lower root-to-shoot ratio in order to improve drought tolerance.

Heat

Tolerance to heat has been achieved in the Pima cottons via changes in stomatal conductance (Kittock et al., 1988; Percy et al., 1996). There is significant variability for heat tolerance (Malik et al., 1999; Hassan et al., 2000). At the same time, various approaches are suggested for identifying heat-tolerant genotypes in the breeding process. A simple method was advocated by Rodreguez-Garay and Barrow (1988); they used the gametophytic stage by subjecting cotton pollen to 35°C for 15 h at 100 percent relative humidity. They proved that pollen carrying a dominant heat-tolerant allele would survive to generate a pollen tube that would penetrate and eventually fertilize the ovules, while heat-sensitive pollen are less likely to generate a pollen tube.

Taking a biotech approach, Trolinder and Shang (1991) exposed cell suspension cultures to temperature/time treatments that served to select cultures in later cycles that were more tolerant to the treatments. Another nonmorphological approach to heat tolerance is the production of heat-shock proteins. Fender and O'Connell (1989) found it difficult to distinguish heat-sensitive proteins between heat-tolerant and heat-sensitive genotypes. Yang and colleagues (1997) concluded that the optimum temperature to induce heat-shock proteins was 42°C, while Fender and O'Connell (1989) reported an optimum temperature of 45°C. Either way, it appears that extreme heat, atypical of most cotton-growing areas, is needed to generate heat-shock proteins, and this may not be a practical approach to breeding for heat tolerance.

Heat tolerance was measured by Ronde and colleagues (1995) using triphenyltetrazolium chloride in a viability assay to distinguish heat-tolerant 'Letaba' from heat-sensitive 'Acala 1517-70' and 'Sclati'. They compared absorbances at 485 nm between control plants at 29°C and heat-acclimated plants at 40°C. Lu and Zeigler (1994) suggested using gas-exchange analysis to select for heat resistance based on their findings with improved lint yields in the Pima cottons. A later report (Lu et al., 1997) based on this research showed that genetic variation for stomatal conductance was higher in a *G. hirsutum* cultivar (Deltapine 90) than in the advanced Pima cultivars (Pima S-6 and Pima S-7). They suggested that improvement in heat tolerance could be made in Pima cottons through introgression of upland cotton alleles.

Cold

Although there is no known cultivar of *G. hirsutum* with pronounced cold tolerance, there is genetic variability for root and shoot growth inhibition (Bradow, 1991). Abdullaev and Kiktev (1985) have reported cold tolerance in *G. gossypioides, G. thurberi,* and *G. raimundii*. John Green of the United States (Green and Roberts, 2001) has selected for cold tolerance in a segregating population, allowing for earlier-than-normal planting.

Techniques for screening for cold tolerance range from measuring electrical conductivity of the leachate from seed, to counting germination in both warm and cold conditions, to measuring seedling root-length inhibition under suboptimal temperatures. Schulze and colleagues (1996) suggested measuring electrical conductivity from seed after exposure to chilling during the imbibition period. Duesterhaus and colleagues (2000) chilled cotton seed at 5°C for 6 h, planted the seed in sand for 14 days at 30°C, and then divided the emergence rate by the warm germination percentage. This they called the imbibitional emergence percentage. They also examined metabolic cold tolerance by planting the seed in sand kept at 18°C for 21 days, then divided the emergence rate by the warm germination percentage. Genotypes with excellent cold tolerance would have both imbibitional emergence percentage and metabolic cold tolerance above 80 percent.

Bauer and Bradow (1996) measured seedling root-length inhibition after subjecting four-day-old seedlings to 15, 20, 25, or 30°C for six days. There were genotypic differences for this trait, with two cultivars (Dp. 5690 and Dp. 90) having the same root length at all temperatures. The Texas MAR program has taken the approach of selecting genotypes that germinate but have a reduced rate of radicle growth under cool (13.3°C) conditions for eight days (Bird, 1982). The theory is that rapid growth under cool conditions will subject the young seedlings to adverse stress, which may be avoided by delaying emergence.

Salt

Cotton is classified as being relatively tolerant to salt by Downton (1984). However, within the genus some genotypes are more tolerant than others (Läuchli et al., 1981; Gossett et al., 1994) (Table 17.3). Cotton is most vulnerable to excess salt at the seedling stage, when salt stress results in a greater loss in yield than at any later time in development (Khan et al., 1995a). Various techniques have been used to screen for salt tolerance. Ledbetter (1987) used salt concentrations ranging from –1.2 to –1.8 MPa

TABLE 17.3. Salt-tolerant cotton cultivars.

Cultivar/Line	Reference
MNH-93	Malik and Makhdum, 1987
ROELCA	Lira-Saldivar and Hernandez-Rosales, 1988
1656-52-36	Lira-Saldivar and Hernandez-Rosales, 1988
Paymaster 404	Lira-Saldivar and Hernandez-Rosales, 1988
NIAB-78	Khan et al., 1995b
Dhy 286	Hebbara et al., 1996
Laxmi	Hebbara et al. 1996
Sairvottam	Hebbara et al. 1996
Z407	Leidi and Saiz 1997
Acala 1517-88	Gossett et al., 1994
Acala 1517-SR2	Gossett et al., 1994
Paymaster 303	Läuchli et al., 1981
Acala SJ-2	Läuchli et al., 1981

NaCl to successfully differentiate cotton seedling emergence. Lira-Saldivar and Hernandez-Rosales (1988) used a lower osmotic potential of –0.7 MPa and successfully differentiated three salt-tolerant lines. Wang and colleagues (1991) recognized an increasing response of cotyledon, hypocotyls, and radicle explants to salinity concentrations. Gossett and colleagues (1994) used 150 mM NaCl to confirm the salt tolerance of greenhouse plants of two Acala cultivars. Ye and Liu (1994) compared the use of table salt containing iodine and pure NaCl in screening for salt tolerance and suggested 0.6 to 0.9 percent concentration of table salt, noting that NaCl had a more discriminating effect than regular table salt. Shen-FaFu and colleagues (1997) suggested using 100 or 150 mM NaCl to screen for salt tolerance and indicated that salt tolerance could be observed in the pollen via germination in the salt solution.

Air Pollution

Air pollution, e.g., ozone, has received very little attention. Temple (1990) studied the effect of determinance on susceptibility to ozone (O_3). The short-season cultivars were more vulnerable to this pollutant than the full-season cultivars. This was due to timing of flowering and high O_3 concentrations, and the ability of full-season cultivars to cope with early season fruit loss.

Soil Acidity and Toxic Metals

Subsoil acidity and the associated toxicities have been studied by the USDA at Beltsville, Maryland (Foy et al., 1967), and by scientists at Louisiana State University (Kennedy et al., 1987). There is genetic variability for tolerance to both aluminum (Foy et al., 1967, 1980) and manganese (Foy et al., 1969, 1981) (Table 17.4).

Manganese toxicity can occur at a soil pH that is too high for Al toxicity, and Mn toxicity causes visible symptoms only in the plant foliage and does not directly affect the roots (Foy et al., 1981). Attempts were made by Kennedy and Jones (1991) to devise a reliable screening method for detecting Mn toxicity in cotton genotypes. They concluded that one could measure peroxidase activity as a reliable quantitative method of identifying leaf damage from Mn toxicity.

Genotypes tolerant to aluminum are not necessarily tolerant to manganese and vice versa (see Table 17.4) (Foy et al., 1967, 1969). There was not a high positive correlation between area of development and either manganese or aluminum tolerance (Foy et al., 1967, 1969). Genotypes developed in the acid soils of the southeast United States did not necessarily show tolerance to aluminum and manganese, whereas some genotypes developed in areas where soils are not known for being acidic did show tolerance.

TABLE 17.4. Cotton genotypes with tolerance to acid, aluminum, or manganese toxic soil.

Aluminum		Manganese	
Genotype	**Reference**	**Genotype**	**Reference**
Pima S-2	Foy et al., 1980	C 310	Foy et al., 1981
Acala 4-42	Foy et al., 1980	73-307	Foy et al., 1981
La. DASS 5194	Foy et al., 1980	La. DSIS 125.3	Foy et al., 1981
La. DASS 5187	Foy et al., 1980	La. DASB 12609	Foy et al., 1981
Stoneville 213	Foy et al., 1980	Rex Smoothleaf	Foy et al., 1969
Delcott 277	Foy et al., 1980	Gregg	Foy et al., 1969
McNair 612	Foy et al., 1980	Acala 1517D	Foy et al., 1969
Dixie King III	Foy et al., 1980		
Pima S-5	Kennedy et al., 1986		
Stoneville 825	Kennedy et al., 1986		
Acala 44-10	Foy et al., 1967		

GENETIC ENGINEERING FOR IMPROVEMENT OF ABIOTIC STRESS TOLERANCE

Genetic engineering opens up a vast gene pool for crop improvement, making virtually all life forms potential sources of genetic variability. Already, alien genes have been incorporated into the cotton genome and commercialized. Over half of the U.S. cotton crop in 2001 contained alien genes that were engineered into the cotton genome (USDA, 2001). These include resistance to glyphosate, bromoxcynil, and lepidopteran insects. Work is also progressing on developing resistance to the boll weevil. There is potential for resistance to nematodes. Molecular biology approaches are being used to improve cold tolerance in cotton (Haigler et al., 2000), one of the rare instances where abiotic stress is being addressed. Identification of useful alleles is the first step in engineering improvement of abiotic stress tolerance, and this requires phenotypic screening for the desired alleles. Movement of desired traits from exotic sources will be slow and tedious.

LIMITATIONS AND FUTURE PROSPECTS

Very few breeders put any effort into breeding for abiotic stress, other than as a final screen prior to commercial release. A notable exception is Dr. Robert McDaniel of the University of Arizona. Indirectly, the USDA Pima breeders in Arizona had been improving abiotic stress tolerance by selecting for improved lint yield. Most of the research in abiotic stress has been performed by physiologists, and rightfully so.

Abiotic stress tolerance will probably be a quantitative trait, which means that breeding progress may be slow and incremental. Variability exists within the *G. hirsutum* species for some abiotic stresses. The native habitat of many of the wild species would suggest that variability exists for heat, drought, and salt tolerance. Cold tolerance may exist in some genomes of cotton, but one would not expect pressure from natural selection to retain those alleles. Variability does exist in *G. hirsutum* for subsoil acidity and associated toxicities, which may or may not have existed in ancestral lines of cotton.

As the yield plateau is reached by breeders, then attention will be given to genetic mechanisms that may enhance yield via tolerance to abiotic stresses. Molecular biology is most likely to help the breeder via marker-assisted selection, whereby major QTLs may be tracked in the breeding process.

REFERENCES

Abdullaev, A.A. and M.M. Kiktev (1985). Donors of resistance in cotton to relatively low temperatures. *Doklady Akademii Nauk Uzbekskoi SSR* 10:52-53.

Altman, D.W. (1988). Exogenous hormone applications at pollination for in vitro and in vivo production of cotton interspecific hybrids. *Plant Cell Reports* 7:257-261.

Bauer, P.J. and J.M. Bradow (1996). Cotton genotype response to early-season cold temperatures. *Crop Science* 36:1602-1606.

Bayler, C., N. Trolinder, C. Ray, M. Morgan, J.E. Quisenberry, and D.W. Ow (1992). Engineering 2,4-D resistance into cotton. *Theoretical and Applied Genetics* 83:645-649.

Beasley, J.O. (1940). The production of polyploids in *Gossypium*. *Journal of Heredity* 31:39-48.

Bell, A.A. (1984). Cotton protection practices in the USA and world. Section B: Disease. In *Cotton, Agronomy Monograph 24,* R.J. Kohel and C.F. Lewis (Eds.), Volume 24. Madison, WI: ASA, pp. 288-309.

Bell, T.A. and F.E. Gillham (1989). *World of Cotton*. Washington, DC: ContiCotton, EMR.

Bird, L.S. (1982). Multi-adversity (disease, insects, and stresses) resistance (MAR) in cotton. *Plant Disease* 66:172-176.

Bradow, J.M. (1991). Cotton cultivar responses to suboptimal postemergent temperatures. *Crop Science* 31:1595-1599.

Brubaker, C.L., F.M. Bourland, and J.F. Wendel (1999). The origin and domestication of cotton. In *Cotton: Origin, History, Technology, and Production,* C.W. Smith and J.T. Cothren (Eds.). New York: John Wiley & Sons, pp. 3-31.

Cook, C.G. and K.M. El-Zik (1992). Cotton seedling and first-bloom plant characteristics: Relationships with drought-influenced boll abscission and lint yield. *Crop Science* 32:1464-1467.

Cook C.G. and K.M. El-Zik (1993). Fruiting and lint yield of cotton cultivars under irrigated and non-irrigated conditions. *Field Crops Research* 33:411-421.

Downton, W.J.S. (1984). Salt tolerance of food crops: Prospectives for improvements. *CRC Critical Reviews in Plant Sciences* 1:183-201.

Duesterhaus, B., N. Hopper, J. Gannaway, and T.D. Valco (2000). A screening test for the evaluation of cold tolerance in cottonseed germination and emergence. In *Proceedings Beltwide Cotton Conference.* Memphis, TN: National Cotton Council of America, pp. 596-599.

Endrizzi, J.E., E.L. Turcotte, and R.J. Kohel (1985). Genetics, cytology, and evolution of *Gossypium*. *Advances in Genetics* 23:271-375.

Fender, S.E. and M.A. O'Connell (1989). Heat shock protein expression in thermotolerant and thermosensitive lines of cotton. *Plant Cell Reports* 8:37-40.

Foy, C.D., W.H. Armiger, A.L. Fleming, and C.F. Lewis (1967). Differential tolerance of cotton varieties to an acid soil high in exchangeable aluminum. *Agronomy Journal* 59:415-418.

Foy, C.D., A.L. Fleming, and W.H. Armiger (1969). Differential tolerance of cotton varieties to excess manganese. *Agronomy Journal* 61:690-694.

Foy, C.D., J.E. Jones, and H.W. Webb (1980). Adaptation of cotton genotypes to an acid, aluminum toxic soil. *Agronomy Journal* 72:833-839.

Foy, C.D., H.W. Webb, and J.E. Jones (1981). Adaptation of cotton genotypes to an acid, manganese toxic soil. *Agronomy Journal* 73:107-111.

Gossett, D.R., E.P. Millhallon, and M.C. Lucus (1994). Antioxidant response to NaCl stress-tolerant and salt-sensitive cultivars of cotton. *Crop Science* 34:706-714.

Green, J.M. and F.L. Roberts (2001). Cold tolerant cotton for ultra early planting. In *Proceedings of the Beltwide Cotton Conference.* Memphis, TN: National Cotton Council of America, pp. 30-31.

Haigler, C.H., W. Cai, L.K. Martin, J. Tummala, R. Anconetani, J.G. Gannaway, G.J. Jividen, and A.S. Holladay (2000). Mechanisms by which fiber quality and fiber and seed weight can be improved in transgenic cotton growing under cool night temperatures. In *Proceedings of the Beltwide Cotton Conference.* Memphis, TN: National Cotton Council of America, p. 483.

Hassan, I.S., S.A. El Shaarawy, and H.B. Abou Tour (2000). Response of some Egyptian genotypes to varied climatic measurements over varied environments. In *Proceedings of the Beltwide Cotton Conference.* Memphis, TN: National Cotton Council of America, pp. 553-560.

Hatfield, J.L., J.E. Quisenberry, and R.E. Dilbeck (1987). Use of canopy temperatures to identify water conservation in cotton germplasm. *Crop Science* 27:269-273.

Hebbara, M., S.G. Patil, M.V. Manjunatha, and R.K. Gupta (1996). Performance of cotton (*Gossypium* species) genotypes under different salinity and water-table conditions. *Indian Journal of Agricultural Sciences* 8:446-454.

Kennedy, C.W., M.T. Ba, A.G. Caldwell, R.L. Hutchinson, and J.E. Jones (1987). Differences in root and shoot growth and soil moisture extraction between cotton cultivars in an acid subsoil. *Plant and Soil* 101:241-246.

Kennedy, C.W. and J.E. Jones (1991). Evaluating quantitative screening methods for manganese toxicity in cotton genotypes. *Journal of Plant Nutrition* 14:1331-1339.

Kennedy, C.W., W.C. Smith Jr., and M.T. Ba (1986). Root cation exchange capacity of cotton cultivars in relation to aluminum toxicity. *Journal of Plant Nutrition* 9:1123-1133.

Khan, A.N., R.H. Qureshi, N. Ahmad, and A. Rashid (1995a). Response of cotton cultivars to salinity at various growth development stages. *Sarhad Journal of Agriculture* 11:729-731.

Khan, A.N., R.H. Qureshi, N. Ahmad, and A. Rashid (1995b). Selection of cotton cultivars for salinity tolerance at seedling stage. *Sarhad Journal of Agriculture* 11:153-159.

Kimber, G. (1961). Basis of the diploid-like meiotic behaviour of polyploid cotton. *Nature* 191:98-100.

Kittock, D.L., E.L. Turcotte, and W.C. Hofmann (1988). Estimation of heat tolerance improvement in recent American Pima cotton cultivars. *Journal of Agronomy and Crop Science* 161:305-309.

Läuchli, A., L.M. Kent, and J.C. Turner (1981). Physiological responses of cotton genotypes to salinity. In *Proceedings of the Beltwide Cotton Conference*. Memphis, TN: National Cotton Council of America, p. 40.

Ledbetter, C.A. (1987). Heritability of salt tolerance during germination and emergence in short-staple cotton (*Gossypium hirsutum* L.). *Dissertation Abstracts International, Science and Engineering* 47:113.

Lee, J.A. (1981). Genetics of D_3 complementary lethality in *Gossypium hirsutum* and *G. barbadense*. *Crop Science* 72:299-300.

Leidi, E.O. and J.F. Saiz (1997). Is salinity tolerance related to Na accumulation in Upland cotton *(Gossypium hirsutum)* seedlings? *Plant and Soil* 190:67-75.

Lira-Saldivar, R.H. and S.P. Hernendez-Rosales (1988). Characterization of salt tolerance in genotypes of cotton. *Informes-de-Investigaciones-PRONAPA* 299-317.

Lloyd, F.E. (1920). Environmental changes and their effect upon boll-shedding in cotton. *Annals of the New York Academy of Sciences* 29:1-131.

Lu, Z.M., J.W. Chen, R.G. Percy, and E. Zeiger (1997). Photosynthetic rate, stomatal conductance and leaf area in two cotton species (*Gossypium barbadense* and *Gossypium hirsutum*) and their relation with heat resistance and yield. *Australian Journal of Plant Physiology* 24:693-700.

Lu, Z.M. and E. Zeiger (1994). Selection for higher yields and heat resistance in pima cotton has caused genetically determined changes in stomatal conductances. *Physiologia Plantarum* 92:273-278.

Malik, M.N., F.I. Chaudhry, and M.I. Makhdum (1999). Cell membrane thermostability as a measure of heat-tolerance in cotton. *Pakistan Journal of Scientific and Industrial Research* 42:44-46.

Malik, M.N. and M.I. Makhdum (1987). Salinity tolerance of cotton cultivars (*G. hirsutum* L.) at germination. *The Pakistan Cotton* 31:171-174.

McCarty, J.C. Jr. and J.N. Jenkins (1992). Cotton germplasm: Characteristics of 79 day-neutral primitive race accessions. Mississippi Agricultural and Fertility Experiment Station Technical Bulletin 184.

McCarty, J.C., Jr., J.N. Jenkins, and W.L. Parrott (1985). Notice of release of two germplasm lines of cotton with resistance to the boll weevils, *Anthonomous grandis* Boh. *Crop Science* 26:1088-1092.

McDaniel, R.G. (2000). Genetic manipulation of cotton leaf stomatal density to enhance drought tolerance. In *Proceedings of the Beltwide Cotton Conference*. Memphis, TN: National Cotton Council of America, pp. 562-564.

Meyer, J.R. and V.G. Meyer (1961). Origin and inheritance of nectariless cotton. *Crop Science* 1:167-169.

Meyer, V.G. (1975). Male sterility from *Gossypium harknessii*. *Journal of Heredity* 66:23-27.

Nepomuceno, A.L., D.M Oosterhuis, and J.M. Stewart (1998). Physiological responses of cotton leaves and roots to water deficit induced by polyethylene glycol. *Environmental and Experimental Botany* 40:29-41.

Penna, J.C.V., L.M. Verhalen, M.B. Kirkham, and R.W. McNew (1998). Screening cotton genotypes for seedling drought tolerance. *Genetics and Molecular Biology* 21:545-549.

Percy, R.G. and R.J. Kohel (1999). Qualitative genetics. In *Cotton: Origin, History, Technology, and Production,* C.W. Smith and J.T. Cothren (Eds.). New York: John Wiley & Sons, pp. 319-360.

Percy, R.G., Z. Lu, J.W. Radin, E.L. Turcotte, and E. Zeiger (1996). Inheritance of stomatal conductance in cotton *(Gossypium barbadense). Physiologia Plantarum* 96:389-394.

Pereira, J.R., P.D. Fernandes, and N.E. de Macedo Beltrao (1998). Germination of genotypes of upland cotton (*Gossypium hirsutum* L. r. *latifolium* H.) under water stress. *Revista de Oleaginosase Fibrosas* 2:41-51.

Perlak, F.J., R.W. Deaton, T.A. Armstrong, R.L. Fuchs, S.R. Sims, J.T. Greenplate, and D.A. Fischhoff (1990). Insect resistant cotton plants. *Biotechnology* 8:939-943.

Peterschmidt, N.A. and J.E. Quisenberry (1981). Plant water status among cotton genotypes. In *Proceedings of the Beltwide Cotton Conference.* Memphis, TN: National Cotton Council of America, pp. 43-44.

Quisenberry, J.E., W.R. Jordan, B.A. Roark, and D.W. Frynear (1981). Exotic cottons as genetic sources for drought resistance. *Crop Science* 21:889-895.

Quisenberry, J.E. and B. Roark (1976). Influences of indeterminate growth habit on yield and water-use efficiency in upland cotton. *Crop Science* 16:762-765.

Quisenberry, J.E., C.W. Wendt, J.D. Berlin, and B.L. McMichael (1985). Potential for using leaf turgidity to select drought tolerance in cotton. *Crop Science* 25:294-299.

Rajeswari, V.R. (1995). Evaluation of cotton genotypes for drought tolerance under rainfed conditions. *Annals of Plant Physiology* 9:109-112.

Rodreguez-Garay, B. and J.R. Barrow (1988). Pollen selection for heat tolerance in cotton. *Crop Science* 28:857-859.

Ronde, J.A., A. Mescht, and W.A. Creas (1995). The biochemical responses of six cotton cultivars to heat stress. *South African Journal of Science* 91:363-366.

Schulze, D.H., N.W. Hopper, J.R. Gannaway, and G.M. Jividen (1996). Laboratory tests used to screen for chilling tolerance in cotton genotypes. In *Proceedings of the Beltwide Cotton Conference.* Memphis, TN: National Cotton Council of America, pp. 1240-1243.

Shen-FaFu, Yi Yin-Cheng, YuanJie Ju, Feng Zhen Lu, Li Chen, T.T. Shen, and LiYi Yih (1997). Screening of whole plants and pollen grains of cotton for salt tolerance. *Acta Agronomica Sinica* 23:620-625.

Shepherd, R.L., J.C. McCarty Jr., J.N. Jenkins, and W.L. Parrott (1988). Registration of twelve nonphotoperiodic lines with root-knot nematode resistant primitive cotton germplasm. *Crop Science* 28:868-869.

Singh, D. (1993). Effect of moisture stress on physiological parameters and yield in cotton. *Journal of Cotton Research and Development* 7:256-258.

Singh, S.B., R.G. Dani, S. Devendra, and D. Singh (1992). Improving drought resistance in cotton: A review. *Coton-et-Fibres-Tropicales* 47:289-300.

Singh, S.B., S. Devendra, S.S. Navaganan, and D. Singh (1996). Variation in physio-morphological characters related to drought tolerance in cotton (*Gossypium* species). *Indian Journal of Agricultural Sciences* 66:357-359.

Stewart, J. McD. (1994). Potential for crop improvement with exotic germplasm and genetic engineering. In *Proceedings of the World Cotton Research Conference*. Brisbane, Australia, February 14-17, 1994, G.A. Constable and N.W. Forrester (Eds.). Melbourne, Australia: CSIRO, pp. 313-327.

Stewart., J. McD. and C.L. Hsu (1977). In-ovule embryo culture and seedling development of cotton (*Gossypium hirsutum* L.). *Planta* 137:113-117.

Stewart, J. McD. and C.L. Hsu (1978). Hybridization of diploid and tetraploid cotton through *in ovulo* embryo culture. *Crop Science* 18:404-408.

Temple, P.J. (1990). Growth form and yield responses of four cotton cultivars to ozone. *Agronomy Journal* 82:1045-1050.

Thaxton, P.M. and K.M. El-Zik (1999). Superior new MAR cotton germplasm for drought, productivity, and quality. In *Proceedings of the Beltwide Cotton Conference*. Memphis, TN: National Cotton Council of America, pp. 470-471.

Thaxton, P.M. and K.M. El-Zik (2000). Comparative characteristics of recent and prospective TAMCOT cultivar releases. In *Proceedings of the Beltwide Cotton Conference*. Memphis, TN: National Cotton Council of America, pp. 521-524.

Trolinder, N.L. and X.M. Shang (1991). In vitro selection and regeneration of cotton resistant to high temperature stress. *Plant Cell Reports* 10:448-452.

USDA (2001). *Cotton Varieties Planted 2001 Crop*. Memphis, TN: USDA Agricultural Marketing Service, Cotton Division.

Van Esbroeck, G.A., D.T. Bowman, D.S. Calhoun, and O.L. May (1998). Changes in the genetic diversity of cotton in the USA from 1970 to 1995. *Crop Science* 38:33-37.

Waddle, B.A. (1984). Crop growing practices. In *Cotton, Agronomy Monograph 24*, R.J. Kohel and C.F. Lewis (Eds.). Madison, WI: ASA, pp. 233-263.

Wang. Z.Z., Z.G. Shen, Y. Ding, Z.N. Ye, K.Q. Li, L.F. Hao, D.Z. Guo, and D.L. Zhang (1991). A preliminary report of research on the salt tolerance response and screening of cotton explants. *China Cottons* 1:16, 18.

Yang, G.Z., X.R. Zhang, X.N. Sun, W.J. Li, and G.Z. Yang (1997). Preliminary study on cotton heat shock proteins. *Journal of Huozhong Agricultural University* 16:18-25.

Ye, W.W. and J.D. Liu (1994). The effect of NaCl and table salt on the germination of cotton seed. *China Cottons* 21:14-15.

Chapter 18

Breeding for Abiotic Stress Tolerance in Tomato

Majid R. Foolad

GENETIC ARCHITECTURE OF TOMATO

The cultivated tomato, *Lycopersicon esculentum* Mill., a fruit that is almost universally treated as a vegetable, belongs to the nightshade family Solanaceae, which is in Division Magnoliophyta, Class Magnoliopsida, Subclass Asteridae, Order Solanales, and Suborder Solanineae. The family Solanaceae consists of 96 genera and about 2800 species in the three subfamilies Solanoideae, Cestroideae, and Solanineae (Nee et al., 1991). The genus *Lycopersicon,* which is in the subfamily Solanoideae, is one of the smallest genera in the Solanaceae and the closest to the genus *Solanum* (nightshade). Originally, Linnaeus included the tomato in *Solanum;* however, in 1754 Miller separated tomatoes and designated the genus *Lycopersicon* (Warnock, 1988). Some researchers today wish to treat *Lycopersicon* under the genetic heading of *Solanum* (Spooner et al., 1993). There are nine known species in the genus *Lycopersicon,* including the cultivated tomato and its wild form, *L. esculentum* var. *cerasiforme* (Dun.) Gray, and eight wild species: *L. pimpinellifolium* (Jusl.) Mill.; *L. cheesmanii* Riley; *L. chmielewskii* Rick, Kes., Fob., & Holle; *L. chilense* Dun.; *L. parviflorum* Rick, Kes., Fob., & Holle; *L. peruvianum* (L.) Mill.; *L. hirsutum* Humb. & Bonpl.; and *L. pennellii* (Corr.) D'Arcy (Rick, 1976a, 1979b). All species are native to western South America, between Ecuador and Chile, although *L. esculentum* var. *cerasiforme* (wild cherry) is also found in Mexico, Central America, and other parts of South America. The natural habitat of the *Lycopersicon* is highly variable, from very dry to very wet and from coastal areas to mountainous elevations of more than 3300 m (Warnock, 1988). This diversity in natural habitat has undoubtedly contributed to the great variation that can be found in *Lycopersicon.*

All species within *Lycopersicon* produce perfect, hermaphrodite, pentamerous flowers. However, a complete range of mating systems is found, from the almost completely autogamous *L. cheesmanii* and *L. parviflorum* to obligately outcrossed, self-incompatible biotypes of *L. chilense, L. hirsutum, L. peruvianum,* and *L. pennellii*. Self-fertility with various degrees of facultative outcrossing is found in *L. chmielewskii, L. esculentum, L. pimpinellifolium,* and the self-compatible biotypes of *L. hirsutum* and *L. pennellii* (Rick, 1982). The nine species within *Lycopersicon* have been grouped into two intracrossable, inter-incrossable groups, or complexes: the "*esculentum* complex," including *L. esculentum, L. pimpinellifolium, L. cheesmanii, L. chmielewskii, L. parviflorum, L. hirsutum,* and *L. pennellii,* and the "*peruvianum* complex," including *L. peruvianum* and *L. chilense* (Rick, 1979a). The barrier between the two complexes can be broken only by the application of embryo culture and succeeds when the member of *esculentum* complex is used as the pistillate parent.

Only *L. esculentum* has become a domesticated crop, although *L. pimpinellifolium* is casually planted as an ornamental and its tiny red berries (diameter ≈ 1 cm) are edible. Domestication of *L. esculentum* was accompanied by a transition from exerted to inserted stigmas and consequent change from facultative outcrossing to enforced autogamy (Rick, 1976b). As a result of autogamy, most accessions within the cultivated species are essentially pure lines. It is widely accepted that *L. esculentum* var. *cerasiforme* (fruit diameter = 1.5 to 3.0 cm) is the immediate progenitor of the cultivated tomato, although *L. pimpinellifolium* is also a likely candidate (Jenkins, 1948; Rick, 1976b). The wild cherry and currant tomatoes *(L. pimpinellifolium)* produce shiny, bright red berries, whereas the Galapagos tomatoes *(L. cheesmanii)* produce dull purplish berries (diameter ≈ 1 cm). The rest of the wild species produce small (diameter = 1.0 to 1.5 cm), inedible green fruit.

Lycopersicon esculentum is generally broadly defined as the common fresh-market and processing varieties of tomato together with the wild cherry tomato. All variants within *L. esculentum* are reciprocally interfertile with each other and with the two pigmented, closely related wild species, *L. pimpinellifolium* and *L. cheesmanii*. *Lycopersicon esculentum* is also crossable with all other wild species, although in most cases only unilaterally as the pistillate parent, and in crosses with *L. peruvianum* and *L. chilense* only with the application of embryo culture (Rick, 1979a). It is especially fortunate for tomato improvement that the cultivated species can be hybridized with, and thereby receive genes from, every other species in the genus.

All *Lycopersicon* species are herbaceous perennials, although the cultivated tomato is almost universally grown as an annual (Rick, 1976b). In the wild, they grow as short-lived perennials, if enough water is available and they are not killed by frost. However, all can be grown to flowering and fruiting in five months or less. All species are diploid and are similar in chromosome number ($2n = 2x = 24$) and morphology. Most species are indeterminate in growth habit; however, the cultivated tomato can be either determinate or indeterminate. The tomato has become an important crop worldwide in a relatively short period of less than one century; it is the second most consumed vegetable per person worldwide after potato and the second most economically important vegetable in the United States after lettuce. Furthermore, tomato ranks first among all fruits and vegetables as a source of vitamins and minerals in the U.S. diet. The significance of tomato to health is only recently becoming apparent. The fruit contains different antioxidant carotenoids, in particular lycopene, which gives the tomato its red color. Consumption of lycopene has been connected with the prevention of prostate cancer and other degenerative diseases (Chen et al., 1999).

SOURCES OF GENETIC VARIATION

The cultivated tomato *(L. esculentum)* has a narrow germplasm base, largely because of several population bottlenecks, in the forms of founder events and natural and artificial selections, that occurred during domestication and evolution of modern cultivars (Rick, 1976b). Although higher levels of variability are found in primitive cultivars in the native regions of tomato, only approximately 5 percent of the total genetic variation within *Lycopersicon* can be found within *L. esculentum* (Miller and Tanksley, 1990; Rick and Fobes, 1975). Therefore, genes for many desirable agricultural characteristics, including environmental stress tolerance, are not found within the cultivated species. Fortunately, however, the related wild species of tomato are a rich source of desirable genes and characteristics for tomato crop improvement, and all can be hybridized with *L. esculentum,* albeit with different degrees of difficulty (Rick, 1976a, 1979a; Rick et al., 1987). The species with the greatest genetic variability are *L. chilense, L. hirsutum, L. peruvianum,* and *L. pennellii,* whereas the least-variable species is *L. cheesmanii* (Miller and Tanksley, 1990; Breto et al., 1993). The exotic resources have been extensively utilized in tomato breeding programs for improving desirable agricultural characteristics, in particular disease resistance and, to a lesser extent, insect resistance and fruit quality. In fact, tomato is a prime example of a cultivated plant that has benefited significantly from the introgression of genes from wild and exotic taxa. The major world collections of tomato accessions include the Asian Vegetable Research and Development Center (AVRDC) at Tainan, Taiwan, China; the United States Department of Agriculture (USDA) North Central Regional Plant Introduction Station at Ames, Iowa; the USDA Plant Genetic Resources Unit at Geneva, New York; and the C. M. Rick Tomato Genetics Resource Center, University of California, Davis.

Although thus far only a superficial assessment of the extent of the genetic variation for environmental stress tolerance within *Lycopersicon* has been made, some accessions with significant tolerance to different abiotic stresses, including salinity, drought, and extreme temperatures, have been identified. These accessions have been utilized for characterization of the physiological and genetic bases of abiotic stress tolerance as well as for developing stress-tolerant tomato plants, as described in this chapter. In addition to the wild species of *Lycopersicon,* a few species within the genus *Solanum* have been shown to be rich sources of desirable genes, in particular for environmental stress tolerance; some of these species are rather closely related to the tomato and might be useful for tomato improvement (Rick, 1979a, 1988; Rick et al., 1986, 1987, 1990). These species include *S. lycopersicoides* Dun. and *S. rickii* Corr., which are more closely related, and *S. juglandifolium* Dun. and *S. ochranthum* Dun., which are more distantly related. These four species belong to section *Petota,* subsection *Potatoe,* series *Juglandifolia*. In this chapter, the existing variation in *Lycopersicon* in relation to environmental stress tolerance and the recent advancements in genetics of and breeding for stress tolerance in tomato are reviewed.

CROP PRODUCTION AND VARIOUS STRESSES

Although the cultivated tomato is widely adapted to different climates, ranging from the tropics and subtropics to a few degrees from the arctic circle, its growth and development is rather sensitive to different environmental stresses, including salinity, drought, excessive moisture, extreme temperatures, and environmental pollution. In common with those for other major crop species, most tomato breeding programs have largely focused on the development of cultivars with high yield potential in favorable, i.e., nonstress, environments, although significant progress also has been made in breeding for disease resistance. Such efforts have been extremely successful in improving the efficiency of crop production per unit area and have resulted in a significant increase in total agricultural production (Duvick, 1986). For example, between the 1920s and 1990s, the average yield per unit area of processing tomatoes in the United States increased by sevenfold (Warren, 1998). However, it is estimated that only 10 percent of the world's arable land is free of stress (Dudal, 1976). Extreme temperatures, drought, soil salinity, and nutrient imbalances, including mineral toxicities and deficiencies, are among the major environmental constraints to crop production worldwide. With the rapid increase in human population and a greater need for food production, and with an increasing diminution in natural resources and arable lands, greater efforts must be devoted to increasing crop productivity in stressful agricultural environments as well as bringing marginal lands under cultivation. The genetic improvement of plant stress tolerance for crop production under stressful environments is an economic solution (Blum, 1988). Toward this goal, within the past few decades considerable research has been conducted and significant information accumulated regarding the physiology and genetics of tomato tolerance to different environmental stresses. Furthermore, some tomato breeding lines and cultivars with improved resistance/tolerance to different stresses have been developed. In this chapter, the current information on tomato response to environmental stresses, including salt, drought, cold, and heat stress, and the available genetic resources for tolerance breeding are reviewed and the prospects for developing commercially acceptable, stress-tolerant cultivars are discussed.

Salt Stress

Salinity is an increasingly important environmental constraint to crop production in the arid and semiarid regions of the world (Boyer, 1982; Tanji, 1990; Rowley, 1993). Regardless of the cause, whether ion toxicity, water deficit, and/or nutritional imbalance, high salinity in the root zone severely impedes normal tomato growth and development, resulting in reduced crop productivity or total crop failure. Of the total 14 billion ha of land available on earth, 6.5 billion ha are estimated to be arid and semiarid, of which about 1 billion ha are natural saline soils. It is estimated that worldwide about 20 percent of cultivated lands and 33 percent of irrigated agricultural lands are afflicted by high salinity (Epstein et al., 1980; Flowers et al., 1986; Tanji, 1990; Szabolcs, 1992; Francois and Maas, 1994). Furthermore, the salinized areas are increasing at a rate of 10 percent annually (Syverstein et al., 1989; Tanji, 1990; Kalaji and Pietkiewica, 1993; Szabolcs, 1994). Low precipitation, high surface evaporation, weathering of native rocks, irrigation with saline water, and poor cultural practices are among the major contributors to the increasing salinity. Although tomato is grown in a wide variety of climates, its production is concentrated in a few warm and rather dry areas where soil salinity is a limiting factor to production. For example, more than 30 percent of world tomato production is from countries around the Mediterranean Sea and about 20 percent from California (FAO, 1995). Although these areas provide the optimal climates for tomato production, salinity can be a serious problem.

To minimize the deleterious effects of high soil/water salinity, two major approaches must be considered simultaneously to achieve sustainable crop production (Epstein et al., 1980). The first approach is technological implementation of large engineering schemes for reclamation, drainage, and irrigation with high-quality water. This approach has been employed successfully in some areas; however, the associated costs are high and it often provides only a temporary solution to the problem. The second approach entails biological strategies focused upon the exploitation or development of plants capable of tolerating high soil/water salinity. This approach includes diversifying cropping systems to include crops that are known to be salt tolerant (e.g., by crop substitution), exploiting wild or feral species that are adapted to saline environments (e.g., by domestication), and genetically modifying domesticated crops by breeding and selection for improved salt tolerance. Crop substitution has been largely practiced since the beginning of agriculture, and it is still one of the easiest and most practical strategies to deal with salinity when growing field crops (Shannon, 1996). However, most vegetable crops, including tomato, are generally more sensitive to salinity than field crops. Thus, in the past few decades, the idea of developing salt-tolerant tomato plants that can grow and produce economic yield under salinity stress has gained significant interest.

Commercial cultivars of tomato are generally sensitive to salinity at all stages of plant development, including during seed germination, vegetative growth, and reproduction, and as a result their growth and economic yield are substantially reduced under salt stress (Maas, 1986; Jones et al., 1988; Bolarin et al., 1993). However, genetic resources for salt tolerance have been identified largely within the related wild species of tomato. Attempts to find sources of genes for salt tolerance in tomato were first made by Lyon (1941), who suggested that salt tolerance of the cultivated tomato might be improved by introgression of genes from *L. pimpinellifolium,* the most closely related wild species of tomato. Later investigations resulted in the identification of a few other salt-tolerant accessions within *L. esculentum* and the related wild species, including *L. pimpinellifolium, L. peruvianum, L. cheesmanii, L. hirsutum,* and *L. pennellii* (Tal, 1971; Rush and Epstein, 1976; Phills et al., 1979; Tal and Shannon, 1983; Sarg et al., 1993; Foolad and Lin, 1997b; Jones, 1986b). However, it is expected that more salt-tolerant accessions can be found within the wild species of tomato if a thorough screening of the available germplasm is made.

A significant finding in tomato (Jones and Qualset, 1984; Asins, Breto, Cambra, and Carbonell, 1993; Foolad and Lin, 1997a; Foolad, 1999) and other plant species (Ashraf and McNeilly, 1988; Johnson et al., 1992) has been that salt tolerance at each stage of plant development is often not correlated with tolerance at other developmental stages. Furthermore, it has been demonstrated that, in tomato, salt tolerance is generally increased with plant age (Bolarin et al., 1993), again similar to findings in other plant species such as barley (*Hordeum* spp.), corn (*Zea mays* L.), rice (*Oryza sativa* L.), and wheat (*Triticum* spp.) (Maas, 1986). Therefore, salt tolerance at each developmental stage in tomato has often been evaluated separately for the assessment of tolerance and the identification, characterization, and utilization of useful genetic components. In the paragraphs following, recent findings on the salt tolerance/sensitivity of tomato during different developmental stages are briefly reviewed.

Salt Tolerance During Seed Germination and Seedling Emergence

The commercial cultivars of tomato are most vulnerable to salinity stress at the seed germination and early seedling growth stages (Cook, 1979; Jones, 1986a; Maas, 1986; Foolad and Jones, 1991). At these stages, tomato exhibits sensitivity to even low concentrations of salts (~75 mM NaCl) (Foolad and Lin, 1997b; Jones 1986b; Cuartero and Fernandez-Munoz, 1999). However, surface soils may have salinities several-fold that of the subsoil, presenting a serious problem at the germination stage. High salinity delays the onset, reduces the rate and final percentage of germination, and increases the dispersion of germination events. This sensitivity has important biological and applied significance. The costly operations of greenhouse seedling production and transplantation into the field have encouraged many tomato producers to grow direct-seeded crops. Furthermore, the dependence upon mechanization in modern cultivation systems and the use of costly hybrid seed requires rapid, uniform, and complete germination. Improving the uniformity and rapidity of seed germination responses under saline conditions would contribute significantly to the efficiency of stand establishment in tomato. Genetic resources for salt tolerance during seed germination have been identified within the cultivated and related wild species of tomato, including *L. pennellii, L. pimpinellifolium,* and *L. peruvianum* (Jones, 1986b; Foolad and Lin, 1997b; Cuartero and Fernandez-Munoz, 1999). Some salt-tolerant accessions have been used to investigate the genetic basis of salt tolerance during seed germination in tomato, as briefly described in the section Genetics of, and Breeding for, Tolerance to Different Stresses.

Salt Tolerance During Vegetative Growth and Reproduction

For tomato production under saline conditions, salt tolerance during the vegetative stage is more important than salt tolerance during seed germination, because most tomato crops are established by seedling transplantation rather than direct seeding in the field. Furthermore, salt tolerance during the vegetative stage may also be more important than salt tolerance during reproduction, i.e., flowering and fruit set, because tomato tolerance of salinity generally increases with plant age and plants are usually most tolerant at maturation (Bolarin et al., 1993). For example, during the flowering and fruiting stages, tomato plants can withstand salt concentrations that can kill them at the seedling stage. Furthermore, there is a positive correlation between tomato yield and plant size under salt stress (Bolarin et al., 1993; Pasternak et al., 1979), indicating the importance of salt tolerance during the vegetative stage.

Most commercial cultivars of tomato are moderately sensitive to salt stress during the vegetative stage (Foolad and Lin, 1997b; Maas, 1986; Tal and Shannon, 1983). At low concentrations of salt (EC = 3 to 5 $dS \cdot m^{-1}$), tomato growth is mainly restricted by nutritional imbalances, as nutrients become the limiting factor under such conditions (Cuartero and Fernandez-Munoz, 1999). At moderate to high levels of salt (EC = 6 $dS \cdot m^{-1}$), in addition to nutrient imbalances, osmotic effects and ion toxicity contribute to reductions in growth. Phenotypic variation for salt tolerance during the vegetative stage in tomato has been identified within the cultivated (Cuartero et al., 1992; Sarg et al., 1993; Foolad, 1997) and related wild species, including *L. peruvianum* (Tal and Gavish, 1973), *L. pennellii* (Dehan and Tal, 1978; Saranga et al., 1991; Perez-Alfocea et al., 1994; Cano et al., 1998), *L. cheesmanii* (Rush and Epstein, 1976; Asins, Breto, Cambra, and Carbonell, 1993), and *L. pimpinellifolium* (Bolarin et al., 1991; Cuartero et al., 1992; Asins, Breto, Cambra, and Carbonell, 1993; Foolad and Chen, 1999). Although much of this phenotypic variation has not been genetically verified yet, it can potentially be useful for developing tomato cultivars with enhanced salt tolerance.

Much less research has been conducted on tomato salt tolerance during reproduction than during earlier stages. For example, little effort has been devoted to determine tomato pollen viability or stigma receptivity, and/or the ability of the plant to produce flowers and set fruit under salt stress. This may be due in part to a higher salt tolerance generally observed during reproduction than during earlier stages in tomato. Adams and Ho (1992) reported that increasing salinity to 10 $dS \cdot m^{-1}$ did not significantly affect fruit set in tomato, which was reduced only at 15 $dS \cdot m^{-1}$. Grunberg and colleagues (1995) reported that salinity does not affect tomato pollen viability;

however, the number of pollen grains per flower decreases with the duration of salinity, being only about 30 percent of the normal 70 days after the beginning of salinization. The suggestion of a higher salt tolerance during reproduction, however, is in contrast with a report that several tomato genotypes that were capable of growing adequately under moderate saline conditions failed to produce any fruit (Asins, Breto, and Carbonell, 1993).

In a recent study, 13 tomato accessions from three different species were grown under both saline (300 mM NaCl + 30 mM $CaCl_2$; equivalent to ~28 $dS \cdot m^{-1}$) and control (no salt) conditions, and their pollen production and in vitro pollen germination were examined (S. Prakash and M. R. Foolad, unpublished data). For most accessions, there was no significant reduction in pollen production per flower in response to salt stress. Pollen from both salt-grown and control-grown plants was cultured at different salt concentrations, including 0, 0.2, 0.4, and 0.8 percent NaCl, and evaluated for percentage germination after 4 or 8 h of incubation. In all accessions, pollen germinability was lower in the salt than in the control (no salt) treatment, and the reduction was greater at higher (0.8 percent) than lower (0.2 percent) salt concentrations. However, in most accessions, in vitro pollen germinability of salt-grown plants was generally higher than the control-grown plants at high concentrations of salt (0.4 and 0.8 percent NaCl), suggesting that pollen salt tolerance in tomato was increased by growing plants under salt stress.

The cultivated tomato is considered "moderately sensitive" to salinity, as measured by the reduction in fruit yield under salt stress compared with control conditions (Maas, 1990). Fruit yield of tomato cultivars starts decreasing when the EC of the saturated soil extract exceeds 2.5 $dS \cdot m^{-1}$ (Maas, 1990; Saranga et al., 1991), although there are reports of higher thresholds for yield reduction in tomato (Adams, 1991). A 10 percent reduction in fruit yield is expected per additional $dS \cdot m^{-1}$ beyond the threshold level (Saranga et al., 1991). The major cause of yield reduction in tomato under low to moderate levels of salinity (EC = 3 to 9 $dS \cdot m^{-1}$) is the reduction in the average fruit size and not a reduction in fruit number (vanIeperen, 1996). A 10 percent reduction in fruit size is caused following irrigation with 5 to 6 $dS \cdot m^{-1}$ water, a 30 percent reduction with 8 $dS \cdot m^{-1}$, and about 50 percent reduction at 9 $dS \cdot m^{-1}$ (Cuartero and Fernandez-Munoz, 1999). Thus, small-fruited cultivars, including cherry tomatoes, may be grown under low to moderate levels of salt stress more successfully than large-fruited cultivars. This is because the relative importance of reduction in fruit size is less in smaller fruits (Caro et al., 1991). However, at higher levels of salinity or prolonged exposure to salinity, a reduction in the total number of fruits per plant, mainly due to a reduction in the number of trusses per plant, is the major cause of yield reduction in tomato (van-

Ieperen, 1996; Cuartero and Fernandez-Munoz, 1999). Thus, under high levels of salt stress, significant yield reduction is expected for both large-fruited and small-fruited cultivars. Furthermore, because upper inflorescences are more sensitive to salinity (Cuartero and Fernandez-Munoz, 1999), when breeding for salt tolerance in tomato, it would be better to develop cultivars with compact plant type and short life cycle, in which only four to six trusses are produced and harvested.

The potential of tomato wild species as sources of salt tolerance during reproduction, i.e., flower and fruit development and fruit yield, has not been assessed critically, mainly because most of the wild accessions are self-incompatible and/or produce very small fruits that cannot be easily compared with those of the cultivated species. Furthermore, many of the wild accessions require specific growing conditions to complete their life cycles, in particular day length, and thus cannot be grown and compared with the cultivated tomato under standard growing conditions. However, progenies from early or late generations derived from interspecific crosses with the cultivated tomato have been used for salt-tolerance studies, as described in the section Genetics of, and Breeding for, Tolerance to Different Stresses.

Drought Stress

Drought, defined as the occurrence of a substantial water deficit in the soil or in the atmosphere, is an increasingly important constraint to crop productivity and yield stability worldwide. It is by far the leading environmental stress in agriculture in many areas of the world, including Asia, Africa, the Middle East, and America. In the United States, for example, up to 45 percent of the land surface is subject to continuous or frequent water stress (Boyer, 1982; Tanji, 1990), and a drought occurs somewhere in the country every year, costing billions of dollars in damage to crops and businesses (Ross and Lott, 2000). The worldwide losses in yield due to drought stress probably exceed the losses from all other causes combined (Blum, 1988; Kramer, 1980; Schonfeld et al., 1988).

Most crop plants, including tomato, are sensitive to drought stress throughout the ontogeny of the plant, from seed germination to harvest (Hsiao, 1973). Plant response to drought stress can be generally classified into three categories: drought escape or avoidance, dehydration avoidance or postponement, and dehydration tolerance (Kramer, 1983; Blum, 1988). Drought escape includes situations where plants with short growth cycles and early maturity avoid experiencing the drought. Breeding for drought escape involves developing cultivars with very early maturity so that, by the time drought occurs, the plant has already completed its life cycle. Advantages and disadvantages of plants with a short growth cycle under conditions of water stress and under normal conditions have been reviewed elsewhere (Reitz, 1974; Jordan et al., 1983; Saeed and Francis, 1983; Blum, 1988). Extremely late cultivars may also escape the drought.

Dehydration avoidance is defined as the ability of the plant to retain a relatively higher level of "hydration" during the period of water stress (Blum, 1988). In this concept, the plant protects its various growth-related physiological, biochemical, and metabolic processes from the external water stress. A common measure of dehydration avoidance is the maintenance of a higher tissue-water or turgor potential under conditions of water stress. Thus, osmotic adjustment, as a means for retaining a higher turgor at a given tissue-water potential, is a component of dehydration avoidance and is an example of avoidance at the cell level. Osmotic adjustment is usually obtained by the production and accumulation of compatible organic solutes, such as amino acids, glycine betaine, sugars, proline, and ectoine in the cytoplasm. The various mechanisms employed by different species to avoid dehydration have been described elsewhere (Levitt, 1980).

When the tissue is not protected by any of the avoidance mechanisms, cells lose turgor and dehydrate, resulting in various cellular physicochemical injuries, as described by Hsiao and Bradford (1983). Complete loss of free water will result in desiccation or dehydration. Although our understanding of the effects of dehydration at the cellular level and on various plant processes at higher levels of plant organization, e.g., photosynthesis and respiration, has been improving, particularly in light of the application of new molecular techniques, it is far from complete and there are many unanswered questions (Blum, 1988; McCue and Hanson, 1990; Bohnert et al., 1995; Bohnert and Jensen, 1996; Richards, 1996; Shinozaki and Yamaguchi-Shinozaki, 1996, 1997; Bray, 1997; Zhu et al., 1997; Lutfor-Rahman, 1998). What is known, however, is that different genotypes exhibit different responses to cellular and whole-plant strains caused by dehydration, and that there are different levels of dehydration tolerance. It should also be noted that characteristics of the three categories of plant response to drought stress, drought escape, dehydration avoidance, and dehydration tolerance, are not generally independent of one another, and some plants may exhibit characteristics of more than one category (Blum, 1988).

Most commercial cultivars of tomato are sensitive to drought stress during different stages of plant development; however, genotypic variation for drought tolerance is found within the cultivated tomato (Wudiri and Henderson, 1985) and related wild species, including *L. chilense, L. pennellii, L. pimpinellifolium,* and *L. esculentum* var. *cerasiforme* (Yu, 1972; Rick, 1973, 1978, 1979b, 1982; Richards and Phills, 1979; Martin et al., 1989; Pillay and Beyl, 1990). These wild species are native to coastal deserts of western South America, including the Galapagos Islands, which are nearly rainless for long periods between the occasional El Niño episodes, when warm ocean water generates heavy rains. Some accessions within these species grow where condensation of dew at night and fog drip are the main sources of moisture. They are also remarkably capable of recovering from brief wilting. Rana and Kalloo (1990) evaluated 150 lines of cultivated and wild species of tomato under water-deficit conditions and identified a few selections within *L. esculentum* and a few accessions of *L. pimpinellifolium* and *L. chilense* with various drought-tolerance attributes. However, insufficient effort has been devoted to the characterization of the physiology or genetics of this variation in tomato to warrant its use in breeding programs. In the following paragraphs, recent findings on the drought tolerance/sensitivity of tomato during different developmental stages are briefly reviewed.

Drought Tolerance During Seed Germination and Seedling Emergence

Although it may not be related to plant performance during later stages, the ability of the tomato seed to germinate rapidly and uniformly under drought stress is a desirable trait for direct seeding in the field. Large areas of land in tomato production are established by sowing seed directly into the field instead of using transplants (Liptay and Schopfer, 1983). Successful establishment of direct-seeded crops depends on successful germination and emergence. Most commercial cultivars of tomato are sensitive to drought stress during seed germination; however, sources of tolerance have been identified within the related wild species of tomato, including *L. pennellii* and *L. pimpinellifolium* (M. R. Foolad et al., unpublished data). Recently, one accession of *L. pimpinellifolium* was used in crosses with *L. esculentum* to examine the genetic basis of drought tolerance during seed germination in tomato.

Drought Tolerance During Vegetative Growth and Reproduction

Potential sources of genes for drought tolerance during vegetative growth and later stages in tomato have been identified within the related wild species, mostly from among accessions native to dry habitats. For example, accessions have been identified within *L. chilense* and *L. pennellii* with various drought-tolerance attributes (Rick, 1973, 1978, 1979b, 1982). Different tolerance indices (TIs) have been employed to characterize the physiological and/or genetic bases of drought tolerance in tomato, including dry weight of shoot and root, root length, root morphology, leaf rolling, flower and fruit set, fruit weight, yield, water-use efficiency (WUE), recovery after rewatering, stomatal resistance, plant survival, leaf water potential, leaf osmotic potential, osmoregulation, oxidative damages, transpiration rate, photosynthetic rate, enzyme (e.g., superoxide dismutase and Rubisco) activities, and pollen viability (Richards and Phills, 1979; Blum, 1988; Martin and Thorstenson, 1988; Rana and Kalloo, 1989; Pillay and Beyl, 1990; Cohen et al., 1991; Kalloo, 1991; Lutfor-Rahman, 1998). For example, in a germplasm screening study, tomato cultivar Saladette tolerated water stress as determined by a small reduction in fruit set compared with other cultivars, which was attributed to its ability to roll up its leaves under a high evaporative demand and thereby maintain a high leaf water potential (Wudiri and Henderson, 1985). Rick (1978) suggested that the physiological basis for drought tolerance in *L. chilense* might be related to its deep, vigorous root system. Cultivar Red Rock performed better than other cultivars under drought conditions, and its drought resistance was similarly attributed to its deep and more vigorous root system (Stoner, 1972). Drought tolerance in the *L. pimpinellifolium* was also attributed to root length (Rana and Kalloo, 1989). Conversely, the drought-tolerant *L. pennellii* accession LA716 has a limited root system, and the basis for its drought tolerance is most likely due to its ability to conserve moisture during periods of limited rainfall. LA716 has been characterized as having a greater WUE under drought stress than *L. esculentum,* as measured by g DW produced per kg of water consumed (Martin and Thorstenson, 1988). A higher WUE in this accession than in *L. esculentum* was attributed to different characteristics of this accession, including smaller leaf conductance due to fewer and smaller stomata, longer trichomes, lower chlorophyll content and Rubisco activity per unit leaf area, and larger mesophyll cell surface exposed to intercellular air space (Martin et al., 1999). Although WUE may be a good indicator of drought tolerance in tomato, its measurement under field condition is extremely difficult. Thus, attempts have been made to determine whether in tomato there is a relationship between WUE and stable carbon isotope discrimination (Δ), a measure of the proportion of ^{13}C relative to ^{12}C in plant organic matter, which is easier to measure when dealing with large number of plants (Martin et al., 1999). A recent study suggested that WUE in progeny of crosses between *L. esculentum* and *L. pennellii* LA716 could be increased by selecting for low Δ; however, this could lead to the selection of small plants, an agriculturally undesirable characteristic (Martin et al., 1999). The authors suggested that the small plant size could be corrected by conventional breeding following selection for drought tolerance, although no such effort has yet been reported.

Cold Stress

The cultivated tomato originated and was domesticated in the tropical and subtropical regions of South and Central America (Rick, 1975). However, commercial tomato production is extensive in many temperate regions of the world, where low temperatures occur during at least part of the growing season. Temperatures in the range of 0 to 15°C (chilling temperatures) injure many crops of tropical origin, including tomatoes. Most commercial cultivars of tomato are sensitive to low temperatures during all stages of plant development, including seed germination, the vegetative stage, and reproduction (Patterson et al., 1978; Lyons et al., 1979; Graham and Patterson, 1982; Patterson, Mutton, et al., 1987). This sensitivity limits the geographic distribution of tomato cultivation and the time of year for planting. In addition, the spring soil temperatures in temperate regions, which are often below 15°C, may restrict direct seeding of tomato crops (Liptay et al., 1982). Low-temperature sensitivity also affects greenhouse production of tomato, as cold-sensitive cultivars may require expensive greenhouse heating throughout the life of the plant. One approach to minimizing deleterious effects of cold stress is to develop cold-tolerant tomato cultivars.

The search for genetic diversity for temperature adaptation logically leads one to examine extreme environments for adapted cultivars or ecotypes. The potential for this geographic approach in identifying tomato genotypes tolerant to low temperatures was first examined by Thompson (1970) and Patterson and colleagues (1978) and reviewed by Vallejos (1979). Genetic resources for cold tolerance have been identified within the cultivated and related wild species of tomato (Patterson et al., 1978; Vallejos, 1979; Scott and Jones, 1982; Patterson and Payne, 1983; Patterson, 1988; Wolf et al., 1986; Foolad and Lin, 2000; Henk-Venema et al., 2000). These resources have been employed to investigate the physiology and genetics of low-temperature tolerance in tomato and are potentially useful for developing cultivars with improved cold tolerance.

Cold Tolerance During Seed Germination and Seedling Emergence

Germination rate of tomato seed decreases progressively as the temperature of the germination medium is reduced from 25 to 10°C, and is inhibited below 10°C (Scott and Jones, 1982). Low temperatures (10 to 15°C) delay the onset, reduce the rate, and increase the dispersion of germination events. As a result, under such conditions, many seeds do not germinate or germinate so sporadically that plants grow unevenly, delaying plant establishment and leading to variability in crop maturation. Most commercial cultivars of tomato are sensitive to low temperatures during seed germination; however, considerable genetic variation has been identified within the cultivated and related wild species. Genetic variability for cold tolerance during seed germination in *Lycopersicon* was first reported by Smith and Millet (1964) and has since been the subject of numerous studies. For example, Scott and Jones (1982) evaluated a total of 37 accessions of the cultivated and wild species of tomato and identified one accession of *L. chilense* that germinated better at 10°C than PI120256, the fastest-germinating *L. esculentum* accession known at that time. Additional accessions exhibiting rapid seed germination at 10°C were also identified within *L. peruvianum* and *L. hirsutum*. In a later study, Scott and Jones (1985a) evaluated seed germination of three fast-germinating *L. esculentum* accessions, PI120256, PI174263, and PI341988, and a slow-geminating tomato breeding line, T3, at temperatures of 6 to 20°C. They concluded that rapid germination of these plant introductions at low temperatures might not be due to cold tolerance, but to seed characteristics that promoted rapid seed germination. Furthermore, early seedling emergence of the same three plant introductions under low temperatures was attributed to their rapid seed germination rather than rapid hypocotyl development. More recently, Foolad and Lin (2000) evaluated germination cold tolerance of 31 tomato accessions (cultivars, breeding lines, and plant introductions) representing six *Lycopersicon* species. They identified significant phenotypic variation for the ability of the seed to germinate rapidly under low temperatures. Several accessions were identified within *L. pimpinellifolium* and *L. hirsutum* with high CT during seed germination.

Generally, accessions germinating rapidly at low temperatures also exhibit fast seedling growth under such conditions (Scott and Jones, 1985a). A few accessions within *L. chilense,* e.g., LA460, *L. peruvianum,* e.g., PI126435 and PI127832, were identified with the characteristic of fast hypocotyl growth under low temperatures (Scott and Jones, 1986). The identified phenotypic variation in these and other studies should potentially be useful for improving low-temperature germination and hypocotyl growth of commercial cultivars of tomato. To facilitate breeding for cold tolerance during seed germination in tomato, several researchers have examined the genetic basis of this seed-related characteristic using a number of low-temperature-tolerant accessions of tomato, as briefly reviewed in the section Genetics of, and Breeding for, Tolerance to Different Stresses. The physiological basis of low-temperature seed germination in tomato has been described elsewhere (Simon, 1979; Maluf and Tigchelaar, 1982; Dahal et al., 1990; Leviatov et al., 1994, 1995).

Cold Tolerance During Vegetative Growth and Reproduction

Chilling tolerance during the vegetative stage in tomato has been defined as the ability of the plant to resist damage below 10°C but above freezing temperature (Lyons, 1973). Went (1957) determined that the cultivated tomato preferred a growth temperature about 7°C higher than the cultivated potato, *Solanum tuberosum* L., which, although closely related to tomato, is much less susceptible to chilling injury. Although both species originated from the Andean region of South America (Jenkins, 1948; Correll, 1962), the cultivated tomato originated from the low altitudes whereas the cultivated potato came from the cooler, higher elevations. Plants that regularly experience temperatures below 10°C in their native habitat would be expected to be comparatively chilling resistant. Near the equator, night temperatures regularly are below 10°C only at altitudes of 2000 m or more. High-altitude wild species of tomato have not contributed to the ancestry of the cultivated tomato, which may explain why the cultivated tomato is less tolerant to chilling than the potato.

A green-fruited related wild species of tomato, *L. hirsutum,* grows naturally over a wide range of altitudes, from sea level in Ecuador to 3300 m in Peru (Rick, 1973). Patterson and colleagues (1978) and Patterson and Payne (1983) evaluated several accessions of this wild species and reported that chilling tolerance was greatest in those that originated in the higher altitudes, e.g., LA1363 and LA1777, and reduced in accessions from low altitudes, e.g., LA407. In these studies, the authors used either a survival test (survival of seedlings after chilling at 0°C for four to seven days) or a night-chill test (growing seedlings at day/night temperatures of 20/0°C for several days) to evaluate chilling tolerance of tomatoes. Wolf and colleagues (1986) evaluated the cold tolerance of several high-altitude accessions of wild tomato species, including *L. hirsutum* (LA1363, LA1777), *L. chilense* Dun. (LA1969, LA1971), and *Solanum lycopersicoides* Dun. (LA1964) during vegetative growth and compared that with the cold tolerance of the *L. esculentum* breeding line UC82B. These researchers used chlorophyll fluorescence (Walker et al., 1990; Bruggemann and Linger, 1994), electrolyte leakage (Van-De-Dijk et al., 1985), and plastochron index (Coleman and Greyson, 1976) as evaluation criteria and reported that the high-altitude wild accessions were all more tolerant to low temperatures than UC82B. Recently, by producing a cybrid between *L. hirsutum* LA1777 (source of the cytoplasm) and *L. esculentum* cv. Large Red Cherry (source of nucleus), Henk-Venema and colleagues (2000) demonstrated that the low-temperature tolerance of LA1777 was not due to genes in the cytoplasm. The cybrid was as cold sensitive as the *L. esculentum* parent, as determined by various growth- and photosynthesis-related characteristics under suboptimal temperatures (day/night 16/14°C). Vallejos and colleagues (1983) reported that a high-altitude accession of *L. hirsutum* had less growth reduction under a low-temperature (day/night 12/5°C) compared to a high-temperature regime (25/18°C) than a low-altitude *L. hirsutum* or a *L. esculentum* breeding line. The high-altitude *L. hirsutum* accession was capable of a faster growth rate and faster rate of leaf production and acquired greater biomass than the *L. esculentum* and the low-altitude *L. hirsutum* accessions. Miltan and colleagues (1986) reported that the dry-weight accumulation and rate of leaf-area increase were less adversely affected by low temperatures in the wild species *L. hirsutum* and *L. chilense* than in *L. esculentum*. In contrast, Raison and Brown (1989) evaluated cold tolerance of high (LA1777; 3200 m), mid (LA1625, 1500 m), and low (LA1361, 50 m) altitudinal ecotypes of *L. hirsutum* based on inhibition of photosynthesis at low temperatures (day/night 15/5°C) and observed that the three tomato ecotypes were similar with respect to critical temperature for chilling-induced photoinhibition (between 10 and 15°C) and the rate of their response to chilling stress, and thus were similarly sensitive to chilling. Photosynthesis, however, is only one component of response to cold stress.

An important attribute for selecting genotypes adapted to low temperatures is root growth. Scott and Jones (1986) evaluated hypocotyl and root elongation of four wild accessions of *L. chilense* and *L. peruvianum* and two control genotypes of *L. esculentum* under low (10°C) and control (20°C) temperatures. Hypocotyl growth rates of the wild accessions were less inhibited at 10°C relative to 20°C than were either of the cultivated accessions, suggesting that these wild accessions had greater cold tolerance than the cultivated genotypes for early seedling growth. In a more recent study, 31 accessions of the cultivated and related wild species of tomato were evaluated for cold tolerance during the vegetative stage to identify gene resources that could be potentially useful for improving chilling tolerance of modern tomato cultivars (Foolad and Lin, 2000). In this study, plant vegetative growth was evaluated under two day/night temperature regimes of 12/5°C (cold stress) and 25/18°C (control) with 12 h photoperiod and a photon flux of 350 $\mu mol \cdot m^{-2} \cdot s^{-1}$ in growth chambers. Cold tolerance during the vegetative stage was defined as the ratio of shoot dry weight under cold stress to shoot dry weight under control, and referred to as vegetative stage tolerance index (TI_{VS}). Across accessions, TI_{VS} ranged from 0.12 to 0.39, indicating the presence of genotypic variation for cold tolerance. Cold tolerance during the vegetative stage was independent of plant vigor, as judged by the absence of a significant correlation ($r = 0.14$, $P > 0.05$) between TI_{VS} and dry weight under control. Several accessions were identified that exhibited considerable cold tolerance, including *L. hirsutum* PI127826, LA1777, and LA386, and *L. esculentum* PI174263 and PI120256. Several of the identified accessions have been used to examine the genetic basis of cold tolerance during the vegetative stage in tomato, as briefly reviewed in the next section. Potential sources of low-temperature tolerance for tomato breeding also have been identified in *Solanum* species, in particular *S. lycopersicoides* (Wolf et al., 1986; Rick, 1988; Walker et al., 1990; R. Chetelat, UC Davis, CA, personal communication).

The cultivated tomato sets fruit poorly at or below 10°C (Charles and Harris, 1972), and this has been attributed mainly to a reduced viability of pollen (Picken, 1984), poor release of pollen (Maisonneuve, 1982), and failure of pollen production or pollination (Picken, 1984) under low temperatures. Generally, in tomatoes grown under low temperatures, anther dehiscence and pollination are poor and the pollen tube grows very slowly before effecting fertilization (Dempsey, 1970). Other important processes involved in fruit setting in tomato, such as the position of stigma in the anther cone, the formation of fertile ovules, or the early development of the embryo, are not adversely affected by chilling temperatures (Fernandez-Munoz and Cuartero, 1991; Fernandez-Munoz et al., 1995a). Exposure of immature flowers of *L. esculentum* to repeated night temperatures below 10°C resulted in infertile pollen grains (Patterson, Mutton, et al., 1987). However, high-altitude *L. hirsutum* accessions LA1363, LA1393, and LA1777 were found to produce functional pollen under such conditions (Maisonneuve, 1983; Patterson, Paull, and Graham, 1987). Pollen of *L. hirsutum* accession LA1366 was shown to be able to germinate at 6°C (Patterson, Paull, and Graham, 1987). Similarly, pollen germination of *L. hirsutum* accession LA1777 was less inhibited at 5°C than pollen from a cultivated tomato. Furthermore, the frequency of *L. hirsutum* gametes contributing to hybrid zygote formation was doubled when controlled fertilizations with pollen mixtures of *L. esculentum* and *L. hirsutum* occurred at day/night temperatures of 12/6°C as compared to crosses with the same mixtures at 24/19°C (Zamir et al., 1981). The results of the latter study suggested competitive ability of high-altitude *L. hirsutum* pollens at low temperatures. Furthermore, this ability has been shown to be under genetic control and heritable (as described in the section Genetics of, and Breeding for, Tolerance to Different Stresses). In another study, evaluation of pollen fertility and pollen shedding of 170 accessions from eight different *Lycopersicon* species indicated that accessions from *L. hirsutum, L. peruvianum,* and *L. pennellii* generally produced fertile pollen below 10°C and shed pollen satisfactorily (Fernandez-Munoz et al., 1995b). Lyakh (1992) demonstrated that pollen of some accessions of *L. pennellii* and *L. hirsutum* fertilized more efficiently than that of *L. esculentum* at 10°C. In other germplasm-screening studies, several *L. esculentum* lines were also identified with the ability to produce flower and set fruit at low temperatures, including the cultivars Early North, PI205040, PI280597, Cold Set, Precoce, Apedice, Montfavet, and Supermarmande (Kemp, 1968; Philouze and Maisonneuve, 1979; Gautam et al., 1981). In conclusion, results from the different studies indicate the presence of potential genetic resources within the cultivated and related wild species for improving low-temperature tolerance of commercial cultivars of tomato. The utility of such genetic resources for improving tomato cold tolerance has been examined in a few studies.

Heat Stress

Although tomato plants can grow in a wide range of climatic conditions, their vegetative and reproductive growths are severely impeded at high temperatures, resulting in reduced yield and fruit quality (Abdul-Baki, 1991; Dane et al., 1991; Scott, 1993; Wessel-Beaver and Scott, 1992). Generally, when the ambient temperature exceeds 35°C, seed germination, seedling and vegetative growth, flowering and fruit set, and fruit ripening are adversely affected in tomato (Kalloo, 1991). This high-temperature sensitivity is particularly important in areas with tropical or subtropical climates. In such climates, heat stress may become a major factor limiting or preventing field production of tomatoes. Although tomato plants are sensitive to high temperatures during all stages of plant development, flowering and fruit set are the most sensitive stages. Fruit set is partially interrupted at day/night temperatures above 26/20°C and is severely interrupted above 35/26°C (Rudich et al., 1977; George et al., 1984; Stevens and Rudich, 1987; Berry and Rafique-Uddin, 1988). In most cultivars of tomato, exposure to daytime temperature of 40°C for only 4 h during the reproductive phase prevents fruit set (Charles and Harris, 1972; Rudich et al., 1977). Moore and Thomas (1952) reported that fruit set in tomatoes was low when the average maximum temperature was above 32°C and the average minimum temperature was above 21°C. Villareal (1978) defined heat tolerance in tomato as the ability to set fruits under night temperatures not lower than 21°C. Other researchers, however, have argued that day and night temperatures may not affect tomato fruit set independently, and that diurnal mean temperature is a better predictor of plant response to high temperature. Accordingly, the sensitivity to high temperature should be determined as the daily mean temperature above which tomato fruit set is reduced (Peet and Bartholemew, 1996; Peet et al., 1997, 1998; Peet and Willits, 1998). By examining different day/night temperature regimes, Peet and colleagues (1997, 1998) and demonstrated that percent fruit set in tomato decreased as mean daily temperature rose above 25°C and was severely impeded at 29°C. These authors proposed that the primary parameter affecting fruit set in tomato was mean daily temperature, with day temperature having a secondary role.

Processes that are reported to be adversely affected by high temperature in tomato include meiosis in the microspore and megaspore mother cells (Kinet and Peet, 1997); amount of pollen produced (El-Ahmadi and Stevens, 1979a; Stevens and Rudich, 1987); pollen dehiscence and release (El-Ahmadi and Stevens, 1979a); pollination (Charles and Harris, 1972; Shelby et al., 1978); pollen germination and tube growth (Weaver and Timm, 1989); ovule viability (Kinet and Peet, 1997); stigma and style position (Charles and Harris, 1972; El-Ahmadi and Stevens, 1979a); number of pollen grains retained by the stigma; fertilization and postfertilization processes; and growth of the endosperm, proembryo, and fertilized embryo (Kinet and Peet, 1997; Peet et al., 1998). However, the most noticeable morphological effect of high temperature is the production of exserted style and stigma, which may prevent self-pollination (Rick and Dempsey, 1969). Poor fruit set at high temperature has also been associated with low levels of carbohydrates and poor production of growth regulators in plant sink tissues (Kinet and Peet, 1997). Furthermore, high-temperature effects on fruit set depend on the stage of flower development when exposed (Iwahori, 1965) and the genotype (Rudich et al., 1977; Shelby et al., 1978; El-Ahmadi and Stevens, 1979a,b; Shen and Li, 1982). Growth chamber and greenhouse studies suggest that high temperature is most deleterious when flowers are first visible, and sensitivity continues for 10 to 15 days (Calvert, 1969). Reproductive phases most sensitive to high temperature are during meiosis (eight to nine days before anthesis) and fertilization (one to three days after anthesis) (Iwahori, 1966). Although both male and female gametophytes are sensitive to high temperature, with response varying with genotype (Shelby et al., 1978; El-Ahmadi and Stevens, 1979a,b), ovules are generally less heat sensitive than pollen (Charles and Harris, 1972; Peet et al., 1998; Rudich et al., 1977). While most researchers have concluded that poor fruit set at high temperature cannot be attributed to a single factor, decreases in pollen germination and/or pollen tube growth are among the most commonly reported factors. Pollen germination and pollen tube length exhibited large positive correlations with fruit set at high temperatures (Weaver and Timm, 1989), although, according to Charles and Harris (1972), pollen tube growth was not a deciding factor for fruit set at high temperatures. Thus, pollen viability, which generally decreases as the ambient air temperature exceeds 32°C (Shelby et al., 1978), has been suggested as an indirect selection criterion for heat tolerance (Weaver and Timm, 1989). Levi and colleagues (1978) reported that heat-tolerant cultivars often exhibited higher pollen viability than heat-sensitive cultivars under high temperature.

A strong positive correlation has been observed between fruit set and yield under high temperature (El-Ahmadi and Stevens, 1979b; Abdul-Baki, 1991; Wessel-Beaver and Scott, 1992). Therefore, evaluation of tomato germplasm to identify sources of heat tolerance has regularly been accomplished by screening for flowering and fruit set under high temperature (El-Ahmadi and Stevens, 1979b; Berry and Rafiq-Uddin, 1988; Abdul-Baki, 1991). High-temperature conditions also induce flower abscission and the incidence of fruit disorders. Major fruit disorders under high temperature include cracks, blossom-end rot, deterioration of fruit color, watery tissue, and small, immature fruit, which all may cause reductions in marketable yield (Charles and Harris, 1972; Scott et al., 1986; Abdul-Baki, 1991; Johjima, 1995). Production of viable seed also is often reduced under high temperature and thus high seed set has been reported as an indication of heat tolerance (Charles and Harris, 1972; El-Ahmadi and Stevens, 1979b; Berry and Rafique-Uddin, 1988; Abdul-Baki, 1991). Moreover, nonreproductive processes in tomato that are affected by heat stress include photosynthesis efficiency (Bar-Tsur et al., 1985), assimilate translocation (Tanaka et al., 1974), mesophyll resistance (Stevens and Rudich, 1987), and disorganization of cellular membrane (Chen et al., 1982). However, general criteria most often used for screening tomatoes for heat tolerance include fruit set, yield, fruit quality, and seed production.

Wide genotypic variation has been observed in the cultivated and related wild species of tomato for the effect of high temperature on pollen and ovule production and viability, pollen dehiscence, pollination effectiveness (Shelby et al., 1978), style elongation (Rudich et al., 1977; El-Ahmadi and Stevens, 1979b), splitting of the antheridial cone and stigma exsertion (Levi et al., 1978), and the ability to set fruit (Stoner and Otto, 1975; Rudich et al., 1977; Berry and Rafique-Uddin, 1988; Abdul-Baki, 1991; Dane et al., 1991; Abdul-Baki and Stommel, 1995). However, a definite association between each of these characteristics and heat tolerance has not always been revealed (Charles and Harris, 1972; El-Ahmadi and Stevens, 1979b; Lohar and Peat, 1998). Rudich and colleagues (1977) reported that in heat-tolerant lines stigma position was generally near the staminal cone, whereas in heat-sensitive lines the style was elongated and exserted. Lohar and Peat (1998) suggested that the use of stigma exsertion as a criterion for selecting against fruit set at high temperatures might be misleading and should be avoided. El-Ahmadi and Stevens (1979b) reported that for an optimal heat-tolerance response, a cultivar must exhibit a combination of essential characteristics under high temperature, including high number of flowers per plant, absence of stigma exsertion, high pollen production, ovule viability, and substantial fruit and seed set. However, none of the heat-tolerant cultivars studied by these researchers possessed all of the model traits, and different factors have been associated with fruit set in different tolerant cultivars.

Two common and undesirable characteristics of tomato genotypes with high fruit-setting ability under high temperature include production of small fruit and poor leaf coverage (Dane et al., 1991; El-Ahmidi and Stevens, 1979b; Hanna and Hernandez, 1982; Rudich et al., 1977; Scott, 1993; Scott et al., 1986, 1997; Shelby et al., 1978; Villareal, 1978; Wessel-Beaver and Scott, 1992). The former characteristic is most likely due to the adverse effect of high temperature on the production of auxins in the fruit and the latter is due to the highly reproductive nature of the heat-tolerant plants. Dane and colleagues (1991) evaluated fruit set ability of 47 tomato genotypes under high temperatures in the field and/or in the greenhouse and concluded that small-fruited genotypes were generally less affected by heat than large-fruited cultivars, consistent with many other reports on the negative correlations between heat tolerance and fruit size (e.g., Wessel-Beaver and Scott, 1992). However, despite all the complexities of heat tolerance and the difficulties encountered during transferring of the tolerance, heat-tolerant inbred lines and hybrid cultivars with commercial acceptability have been developed and released.

Seed germination in tomato is also greatly affected by high temperature. Tomato seed germinates best between 20 and 25°C; however, germination is reduced at 30°C, highly inhibited at 35°C (Jaworski and Valli, 1965; Thompson, 1974; Lorenz and Maynard, 1988; Coons et al., 1989), and totally inhibited at 39 to 39.5°C (Thompson et al., 1977). However, there is significant genotypic variation in response to germination under high temperature, with total germination ranging from 0 percent to 95 percent at or above 35°C (Berry, 1969; Thompson, 1974; El-Hassan, 1978; Taylor et al., 1982; Coons et al., 1989). The ability to germinate under high-temperature conditions is important for fall production in the tropics and subtropics, where seeding is required during late summer when soil temperature often exceeds 35°C (Coons et al., 1989).

GENETICS OF, AND BREEDING FOR, TOLERANCE TO DIFFERENT STRESSES

Salt Tolerance

During the past few decades, the identified salt-tolerant germplasm has been used to characterize the genetic basis of salt tolerance and to develop tomatoes with improved salt tolerance during different developmental stages. Some notable progresses are briefly reviewed here.

Salt Tolerance During Seed Germination and Seedling Emergence

To characterize the inheritance of salt tolerance during seed germination in tomato, parental, filial, and backcross populations (total of ten generations) of a cross between a salt-sensitive tomato breeding line (UCT5) and a salt-tolerant *L. esculentum* plant introduction (PI174263) were evaluated for rate of seed germination under salt stress. Generation means analysis indicated that the ability of tomato seed to germinate rapidly under salt stress was genetically controlled with a narrow-sense heritability (h^2) of 0.75 ± 0.03 (Foolad and Jones, 1991). These results were confirmed in a different study using regression analysis and F_2:F_3 and F_3:F_4 progeny of the same intraspecific cross (Foolad and Jones, 1992). Both investigations suggested that tomato salt tolerance during seed germination was genetically controlled, with additivity being the major genetic component, and that it could be improved by directional phenotypic selection. However, in a later study, the effectiveness of phenotypic selection to improve tomato seed germination under salt stress was examined using F_2, F_3, and F_4 progeny populations of a cross between UCT5 and PI174263 (Foolad, 1996b). The results indicated that directional phenotypic selection for rapid seed germination under salt stress significantly improved progeny salt tolerance; estimates of realized h^2 for salt tolerance during seed germination ranged between 0.67 and 0.76 (Foolad, 1996b). The overall results from these studies suggested that salt tolerance during seed germination in tomato was most likely controlled by a few major genes with additive effects and that it could be improved by directional phenotypic selection. However, to better understand the genetic basis of and improve the efficiency of selection for salt tolerance during seed germination in tomato, a few quality trait loci (QTL) mapping experiments were carried out during the 1990s and determined the number, genomic location, and individual effects of QTLs affecting this trait.

Seed of an F_2 population of a cross between a salt-sensitive tomato breeding line (UCT5) and a salt-tolerant accession (LA716) of *L. pennellii* was subjected to salt stress, and the two extreme tails (salt tolerant and salt sensitive) of the response distribution were selected (Foolad and Jones, 1993). The selected individuals in each tail were subjected to marker analysis, and marker allele frequency differences between the two tails were determined. A trait-based marker analysis, also known as selective genotyping or distributional extreme analysis, which determines the significance of the marker allele frequency differences between selected tails of a response distribution (Lebowitz et al., 1987; Darvasi and Soller, 1992), was used to identify marker-linked QTLs. Five genomic locations were identified on chromosomes 1, 3, 7, 8, and 12 bearing QTLs for salt tolerance during seed germination in this population (Foolad and Jones, 1993). The validity of these QTLs was examined in a few subsequent investigations, using various populations of the same (UCT5 × LA716) or different interspecific crosses, including BC_1, BC_1S_1 (self-progeny of BC_1), and recombinant inbred line (RIL) populations of a cross between a *L. esculentum* breeding line (NC84173) and a salt-tolerant accession (LA722) of *L. pimpinellifolium* (Foolad and Chen, 1998; Foolad et al., 1997, 1998b) (M. R. Foolad et al., unpublished data). In all of these studies, larger populations and larger numbers of molecular markers than in the original study were used. In addition, different mapping strategies, including trait-based and marker-based (standard interval mapping) analyses, were employed. These studies validated most of the previously identified QTLs and detected a few new QTLs on chromosomes 2 and 9. The combined results supported the previous suggestion that salt tolerance during seed germination in tomato was a quantitative trait controlled by more than one gene. Notably, however, in all of these studies it was determined that salt tolerance was controlled by a few QTLs with major effects and several QTLs with smaller effects. Furthermore, these studies suggested the absence of significant epistatic interactions among the identified QTLs. The demonstration that a few QTLs with major and independent effects determined most of the variation for salt tolerance during seed germination in tomato was consistent with the previous finding that this trait could be effectively improved by directional phenotypic selection (Foolad, 1996b). The results also suggest a good prospect for improving this trait by marker-assisted selection (MAS), although no such attempt has been made yet.

A comparison of QTLs identified for salt tolerance during seed germination in different populations (e.g., BC_1S_1 and RILs) of the same cross (*L. esculentum* × *L. pimpinellifolium*) indicated that some QTLs were stable across populations/generations whereas others were population-specific. A comparison of QTLs identified in different populations, including those derived from *L. esculentum* × *L. pennellii* (Foolad and Chen, 1998; Foolad et al., 1997) and *L. esculentum* × *L. pimpinellifolium* crosses (Foolad et al., 1998b) (M. R. Foolad, unpublished data), indicated that some QTLs were conserved across species whereas others were species specific. The combined results indicate that, in comparison with salt tolerance at later stages of plant development, salt tolerance during seed germination in tomato is less affected by variation in genetic and environmental backgrounds and thus it should be feasible to transfer this trait to commercial cultivars of tomato by phenotypic selection or MAS. However, because in most cases tolerance genes (QTLs) are often found within the wild species of tomato, a combination of phenotypic selection and MAS may be more useful. Furthermore, because in most cases more than one gene resource is utilized during the life of a breeding project, when breeding for salt tolerance it might be possible to incorporate QTLs from different resources by MAS.

In many saline soils the concentration of salts varies across the soil horizon, ranging from low to moderate and high (Richards and Dennett, 1980). A successful cultivar would be one that exhibits salt tolerance in a wide range of salt stress and whose performance would not decline in the absence of salts. From a practical point of view, however, selection under different levels of salt-stress may not always be feasible in a breeding program. Thus, it is essential to examine relationships among responses to different levels of salts and determine whether there is a critical salt concentration at which selections should be made. Several investigations have been conducted to examine relationships among germination responses under different salt-stress levels in tomato. For example, evaluation of 56 tomato genotypes (commercial lines, plant introductions, and wild accessions) for seed germination at 75 mM (low), 150 mM (intermediate), and 200 mM (high) salt indicated generally genotypes that germinated rapidly at the low-salt concentration also germinated rapidly at the moderate or the high-salt concentration (Foolad and Lin, 1997b). Linear correlation analysis indicated the presence of a strong phenotypic relationship ($r = 0.90$, $P < 0.01$) between germination at 75 mM and 150 mM salt. The results suggested that the same genes most likely control the rate of tomato seed germination under different levels of salt stress. This suggestion was later confirmed genetically by an analysis of response and correlated response to selection for salt tolerance, where selections were made separately under 100 mM (low), 150 mM (intermediate), and 200 mM (high) salt, and progeny responses were examined at all three salt concentrations (Foolad, 1996b). In that study, selection for rapid seed germination at any salt-stress level resulted in progeny with enhanced germination rate at all three (100, 150, and 200 mM) salt-stress levels, regardless of what salt concentration was used during the selection process. A genetic correspondence (genetic correlation) of up to 100 percent between germination at different salt-stress levels was observed. The results suggested that similar or identical genes with additive effects contributed to rapid seed germination at different salt-stress levels. This suggestion was also consistent with the finding that in F_2 populations of a cross between a salt-tolerant accession (LA716) of *L. pennellii* and a salt-sensitive tomato cultivar (UCT5) identical QTLs contributed to salt tolerance during seed germination at different salt-stress levels (Foolad and Jones, 1993). The combined results indicate that to develop tomato cultivars with improved seed germination at diverse salt-stress levels, it is sufficient to conduct selection and breeding at a single salt-stress level. However, because the rate of tomato seed germination at a moderate level of salt stress (150 mM salt) was highly correlated with that at both a low (100 mM salt) and a high salt-stress level (200 mM salt), it was suggested that selection and breeding for salt tolerance should be performed at an intermediate salt-stress level (Foolad, 1996b).

Although the QTLs for salt tolerance during seed germination in tomato have been identified and verified, the genetic nature of these QTLs or the physiological mechanism(s) that they modulate have not been determined. However, based on the current knowledge of the genetics and physiology of salt tolerance during seed germination, some speculations can be made as to the role of these QTLs. Salt tolerance during seed germination is a measure of the seed's ability to withstand the effects of high concentrations of salts in the medium. Excessive salt depresses the external water potential, making water less available to the seed. Slower seed germination under salt stress than in nonstress conditions, however, could be due to osmotic and/or ionic effects of the saline germination medium. Physiological investigations to distinguish between these two types of effect have been very limited. However, accumulating evidence in different crop species suggests that low water potential of the external medium, rather than ion toxicity effects, is the major limiting factor to germination under salt stress (Kaufman, 1969; Ungar, 1978; Bliss et al., 1986; Haigh and Barlow, 1987; Bradford, 1995), although a few reports have indicated otherwise (Choudhuri, 1968; Younis and Hatata, 1971; Redmann, 1974). Furthermore, in a recent investigation, germination responses of eight tomato genotypes, including salt-tolerant or salt-sensitive accessions of *L. esculentum* and *L. pimpinellifolium,* were evaluated in isoosmotic (water potential = –700 kPa or 15 $dS \cdot m^{-1}$) medium of NaCl, $MgCl_2$, KCl, $CaCl_2$, sorbitol, sucrose, or mannitol (J. R. Hyman and M. R. Foolad, unpublished data). Comparison of germination rates in the salt-stress treatments (NaCl, $MgCl_2$, KCl, and $CaCl_2$) with those in osmotic-stress treatments (mannitol and sorbitol) indicated that all genotypes responded similarly to these two types of stresses. In addition, comparison of germination rates among the salt-stress treatments indicated that the different types of salt in the germination medium generally similarly affected germination of all genotypes. The results confirmed the previous suggestion that the delay in the germination of tomato seed under salt stress was mainly due to osmotic rather than ion toxicity effects. The overall results indicated that salt tolerance during seed germination of the tomato genotypes examined was in fact an adaptation to germinate rapidly under low water potential and that there might be little or no detrimental ionic effects during seed germination in tomato (J. R. Hyman and M. R. Foolad, unpublished data).

Anatomically, the tomato seed comprises a seedcoat (testa) that encloses a curved filiform embryo and an endosperm that practically fills the lumen of the seed not occupied by the embryo (Esau, 1953). For germination to occur, the hydraulic extension force of the embryo must exceed the opposing force of the seed coat and the living endosperm tissues at the placental end of the seed (Hegarty, 1978; Liptay and Schopfer, 1983; Bradford, 1986; Groot and Kourssen, 1987). Thus, embryo genotype was suggested to play a major role in determining the time to germination of tomato seed under nonstress or stress conditions (Liptay and Schopfer, 1983). According to this hypothesis, differences in salt sensitivity of tomato seeds during germination reside either in the osmotic potential or pressure potential of the germinating embryo. However, osmotic stress can also negatively affect seed imbibition and thus retard (or prevent) weakening of the restrictive forces of the endosperm and testa, resulting in reduced rate, or inhibition, of germination (Liptay and Schopfer, 1983; Groot and Karssen, 1987; Dahal et al., 1990). Thus, the rate of seed germination may be influenced by the physical, chemical, and genetic composition of the embryo, endosperm, and/or testa. The identified QTLs for salt tolerance during seed germination in tomato could therefore affect germination rate by affecting the vigor of the germinating embryo; the variation in the thickness of the endosperm; the physical and permeability properties of the endosperm cell walls; the time of onset or rate of activity of enzymes which modify the properties of the endosperm cell wall; the release of gibberellin by the embryo, which is necessary for endosperm weakening; the base water potential required for seed germination; the hydrotime constant (Bradford, 1995); the rate of metabolic activities in the embryo or endosperm under osmotic stress; osmoregulation during germination; or any other physiological or metabolic processes that are essential for the initiation of germination. However, isolation, characterization, and comparison of functional genes that facilitate rapid seed germination under salt stress would be necessary to determine the actual role(s) of the identified QTLs. Nonetheless, irrespective of the physiological mechanism(s) of salt tolerance during seed germination in tomato, the identified QTLs could potentially be useful for improving tomato salt tolerance during seed germination using MAS and breeding.

Salt Tolerance During Vegetative Growth and Reproduction

Most of the earlier research on tomato salt tolerance during the vegetative stage was focused on the physiology of the plant response to salt stress. During the past few decades, however, a greater emphasis has been placed on the genetics of salt tolerance, with the goal of facilitating the development of tomato cultivars with improved tolerance. About 20 years ago, Epstein proposed the exploitation of gene resources within the wild *Lycopersicon* species to increase salt tolerance of the cultivated tomato (Rush and Epstein, 1976). However, at that time, there was insufficient information regarding the genetic control of salt tolerance in tomato to warrant such endeavor. Rush and Epstein (1981a) hybridized a salt-tolerant accession (LA1401) of *L. cheesmanii* with a salt-sensitive tomato cultivar (Walter) and produced filial (F_1 and F_2) and backcross (BC) progeny. Selections for salt tolerance in the F_2 and BC generations resulted in progeny with enhanced salt tolerance compared with 'Walter,' suggesting that salt tolerance of LA1401 was heritable and could be transferred to the cultivated tomato by hybridization and selection. Saranga and colleagues (1992) developed BC_1 and BC_1S_1 populations of a cross between a salt-sensitive tomato processing line (M82) and a salt-tolerant accession (LA716) of *L. pennellii* and evaluated them for salt tolerance under field conditions. Evaluations were based on total dry matter and total fruit yield under saline conditions as well as total dry matter under salt relative to a control (no salt). Estimates of h^2 for these traits were moderate (0.3 to 0.45), suggesting that salt tolerance of the cultivated tomato could be improved by selection for dry matter and yield parameters under salinity using LA716 as a gene resource. By evaluating F_2 progeny of a cross between a salt-sensitive cultivated tomato and a salt-tolerant accession of *L. pimpinellifolium* under salt stress, Asins, Breto, and Carbonell (1993) concluded that total fruit yield and total fruit number were useful selection criteria for improving tomato salt tolerance; estimates of broad-sense h^2s for these two traits were 0.53 and 0.73, respectively. Furthermore, in a greenhouse hydroponics study, using parental, filial, and backcross generations (total of ten populations) of an intraspecific cross between a salt-sensitive tomato breeding line (UCT5) and a salt-tolerant primitive cultivar (PI174263), Foolad (1996a) determined that growth under salt stress relative to the control, the most widely used index in physiological investigation of salt tolerance in tomato, was under additive genetic control and could be an excellent selection criterion for improving tomato salt tolerance. In none of the latter three studies, however, was any empirical selection made to verify the suggestion that salt tolerance of tomato could be improved by directional phenotypic selection under saline conditions.

Nonetheless, these and other studies (Bolarin et al., 1991; Foolad, 1996a) have suggested that shoot growth under salinity relative to a control, also known as relative growth under salt stress, should be the best indicator of salt tolerance in tomato.

It is generally agreed that direct selection for salt tolerance under field conditions is difficult because of the confounding effects of numerous other environmental factors (Richards, 1983; Tal, 1985; Yeo and Flowers, 1990). A suggested approach to improve the efficiency of selection for salt tolerance has been the adoption of new selection criteria based on genetic knowledge of physiological processes that limit crop production under saline conditions (Tal, 1985; Flowers and Yeo, 1988, 1997; Yeo and Flowers, 1990). Physiological criteria that have been suggested as potential indicators of salt tolerance in tomato include tissue water potential, tissue ion content, K^+/Na^+ ratio, osmoregulation, succulence, and water-use efficiency (WUE). Whether these physiological parameters are good indicators of salt tolerance in tomato or if there is variation in these responses must be determined before the question of their genetic controls can be addressed. Several researchers have investigated the relationship between tomato salt tolerance and some of these physiological responses and speculated on their utility as indirect selection criteria (Tal and Gavish, 1973; Tal et al., 1979; Asins, Breto, and Carbonell, 1993; Saranga et al., 1993; Foolad, 1997; Cuartero and Fernandez-Munoz, 1999). A few of these studies and their findings are briefly reviewed here.

Physiological investigations have indicated that, generally, most of the salt-tolerant genotypes within the cultivated tomato and the closely related wild species *L. pimpinellifolium* exhibit a glycophytic response to salinity. That is, the exclusion of toxic ions, in particular Na^+, at the root or shoot level and the synthesis and accumulation of compatible organic compounds, e.g., sugars and amino acids, in the cell for osmoregulation (Caro et al., 1991; Cuartero et al., 1992; Bolarin et al., 1993; Perez-Alfocea et al., 1993; Foolad, 1997; Santa-Cruz et al., 1998). In contrast, salt-tolerant accessions within the tomato wild species *L. pennellii, L. cheesmanii,* and *L. peruvianum* generally exhibit a halophytic response to salinity, in which osmotic adjustment is achieved by uptake of inorganic ions from the soil and compartmentalization in the cell vacuoles in the leaf or other plant organs (Sacher et al., 1983; Tal and Shannon, 1983; Bolarin et al., 1991; Perez-Alfocea et al., 1994). However, differential accumulation of ions in the leaf tissue has not always been identified as a major factor in determining tomato salt tolerance or salt sensitivity. Analysis of BC_1 and BC_1S_1 populations of a cross between a salt-sensitive *L. esculentum* (M82) and a salt-tolerant *L. pennellii* (LA716) indicated that tissue ion content was not likely to provide an efficient selection criterion for salt tolerance, as no direct relationship was observed (Saranga et al., 1992). On the other hand, analysis of the relationship between salt tolerance and leaf ion compositions in the cultivated and three wild species of tomato prompted Saranga and colleagues (1993) to conclude that dry-matter production under salt stress was positively correlated with K^+/Na^+ ratio in the stem and negatively correlated with the Cl^- concentration in leaves and stems. The authors suggested that tissue ion content and ion selectivity were good selection criteria for breeding for salt tolerance in tomato. Potassium selectivity over Na^+ was also reported as a good indicator of salt tolerance in a study of several genotypes of the cultivated and wild species of tomato (Cuartero et al., 1992). Further studies of wild species of tomato, including *L. peruvianum* (Tal, 1971), *L. cheesmanii* (Rush and Epstein, 1981b), and *L. pimpinellifolium*, *L. hirsutum*, and *L. pennellii* (Bolarin et al., 1991), related elevated concentrations of Na^+ in the leaf tissue to the plant salt tolerance. In contrast, no relationship was observed between tissue ion content and plant salt tolerance, as determined by survival under salt stress, in BC_1 and BC_1S_1 populations of a cross between a moderately salt-sensitive tomato breeding line (NC84173) and a salt-tolerant accession (LA722) of *L. pimpinellifolium* (Foolad and Chen, 1999). Furthermore, other studies suggested that the ability to regulate Na^+ concentration in the leaf tissue was more closely correlated with salt tolerance than Na^+ concentration per se (Sacher et al., 1983), and that the distribution of Na^+ in young and mature leaves were an important part of such regulation (Shannon et al., 1987). The overall conclusion from the various studies is that tissue ion content per se may not be a universal indicator of salt tolerance across tomato genotypes.

It has frequently been reported that in salt-tolerant genotypes of tomato which exhibit a glycophytic response to salt stress, as ion concentration increases beyond a threshold level, the exclusion mechanism fails at the root and/or shoot level; further increases in ion concentration in the root zone would result in declining growth and gradual plant death (Perez-Alfocea et al., 1993; Foolad, 1997). Thus, salt-tolerant tomato genotypes bearing an exclusion mechanism may be useful for cultivation only in regions with low to moderate levels of salt. At higher salinity levels, genotypes that exhibit a halophytic-type response to salt stress may be more advantageous. Unfortunately, however, many salt-tolerant wild accessions of tomato that exhibit a halophytic response to salt stress often have the undesirable characteristic of slow growth rate (Foolad, 1996a; Tal, 1997). Although such accessions may survive high levels of salinity, they often grow extremely slowly, a trait that is highly undesirable for tomato cultivation. Whether this association is due to pleiotropic effects of the same genes or linkage between different genes affecting growth rate and salt tolerance is unknown. Furthermore, it has not been determined whether this association can be eliminated in segregating progeny derived from crosses between slow-growing, salt-tolerant wild accessions and fast-growing, salt-sensitive cultivated types of tomato. If it can be determined, the salt-tolerant wild accessions would be potentially useful for salt-tolerance breeding in tomato. Several studies in tomato and in other plant species, however, have suggested that genes contributing to plant vigor are different from those conferring salt tolerance (Forster et al., 1990; Foolad, 1996a). When breeding tomatoes for efficient production under saline conditions, genes for both plant vigor and salt tolerance are important. Recent transgenic approaches have resulted in the development of tomato plants with improved salt tolerance and vigorous growth, which are described in the section Genetic Engineering for Crop Improvement.

Research to determine the genetic basis of ion accumulation/exclusion in tomato plants grown under salt stress has been very limited. In one study, analysis of the parental, filial, and backcross generations of an intraspecific cross between a salt-sensitive tomato breeding line (UCT5) and a salt-tolerant primitive cultivar (PI174263) indicated that growth under salt stress was positively correlated with leaf Ca^{2+} content and negatively correlated with leaf Na^{+} content (Foolad, 1997). Generation means analysis of these populations indicated that, under salt stress, accumulation of both Na^{+} and Ca^{2+} in the leaf was genetically controlled, with additivity being the major genetic component. Tissue ion concentration was therefore suggested as a useful selection criterion when breeding for improved salt tolerance of tomato using 'PI174263' as a genetic resource (Foolad, 1997).

In spite of these and other studies, there is still no consensus on what might be the best physiological or morphological characteristic(s) that should be used as indirect selection criteria when breeding tomatoes for salt tolerance. Most likely a combination of different characteristics must be considered if salt-tolerant genotypes with commercial value are anticipated. This, by itself, indicates the complexity of salt tolerance and the need for identifying better approaches for characterizing genetic bases of tolerance components and facilitating breeding for salt tolerance. Recent advances in molecular marker technology, QTL mapping, MAS, and genetic transformation have provided some promising approaches.

Significant progress has recently been made in mapping QTLs for salt tolerance during the vegetative stage and reproduction in tomato. In one study, a BC_1S_1 population of a cross between a moderately salt-sensitive tomato breeding line (NC84173) and a salt-tolerant accession of *L. pimpinellifolium* (LA722) was evaluated for survival under salt stress using an aerated hydroponics system (Foolad and Chen, 1999). The two parents were distinctly different in salt tolerance; while 80 percent of LA722 survived two weeks after the final salt concentration (700 mM NaCl + 70 mM

$CaCl_2$; equivalent to ~64 dS·m^{-1}) was reached, only 25 percent of NC84173 remained alive within that period. The BC_1S_1 population, consisting of 119 families, exhibited a continuous variation for salt tolerance, with survival rate ranging from 9 percent to 94 percent across families and with a mean of 51 percent. Interval mapping identified five putative QTLs for salt tolerance on tomato chromosomes 1, 3, 5, and 9, with individual effects ranging between 5.7 percent and 17.7 percent and with combined effects of 46 percent of the total phenotypic variation under salt stress (Foolad and Chen, 1999). All QTLs had the positive QTL alleles from the salt-tolerant *L. pimpinellifolium* parent. The results supported the previous suggestion that salt tolerance during the vegetative stage in tomato is controlled by more than one gene (Foolad, 1996a, 1997). However, the involvement of only a few major QTLs, which accounted for a large portion of the total phenotypic variation, suggested the utility of MAS for improving tomato salt tolerance using LA722 as a gene resource. Analyses of leaf ion content, including Na^+, K^+, Mg^{2+}, Ca^{2+}, Cl^-, NO_3^-, SO_4^{2-} and PO_4^{3-}, in the BC_1S_1 population indicated the absence of a correlation between salt tolerance and tissue ion content in this population. Furthermore, despite the presence of significant variation among BC_1S_1 families in concentration of the various ions, no major QTL was identified for tissue ion content under salt stress.

Using a different BC_1 population of the same *L. esculentum* × *L. pimpinellifolium* cross, a selective genotyping approach was employed to verify the previously identified QTLs and possibly identify new QTLs for salt tolerance during the vegetative stage (Foolad et al., 2001). From a population of 792 BC_1 plants, 37 (4.7 percent of the total) that exhibited the highest salt tolerance were selected, grown to maturity, and self-pollinated to produce BC_1S_1 seeds. The 37 selected BC_1S_1 families and 119 nonselected (random) BC_1S_1 families were evaluated for salt tolerance and their performances were compared. A realized h^2 of 0.46 was obtained for salt tolerance during the vegetative stage, consistent with a previous estimate of h^2 for this trait obtained from an intraspecific cross of tomato (Foolad, 1996a). The 37 selected BC_1 plants and the 119 nonselected BC_1 plants were subjected to RFLP analysis using 115 markers, and marker allele frequencies determined. A comparison of marker allele frequency differences between the selected and nonselected populations detected five genomic regions on chromosomes 1, 3, 5, 6, and 11 bearing QTLs for salt tolerance (Foolad et al., 2001). Except for one, all QTLs had positive alleles contributed from the salt-tolerant *L. pimpinellifolium* parent. Three of the five QTLs were at the same locations as those identified in the previous study (Foolad and Chen, 1999). Only one of the major QTLs that was identified in the previous study was not detected in this study (Foolad et al., 2001). The high level of consistency between the results of the two studies indicated the genuine nature of the detected QTLs and their potential usefulness for salt-tolerance breeding using MAS. In each of the two studies described, a few BC_1S_1 families were identified with most or all of the identified QTLs and with a salt tolerance comparable to that of the salt-tolerant *L. pimpinellifolium* accession. These families should be particularly useful for salt-tolerance breeding in tomato.

In a more recent study, 145 F_9 RILs of the same *L. esculentum* × *L. pimpinellifolium* cross were evaluated in replicated trials for salt tolerance during the vegetative stage (plant survival under salt stress). The RILs were genotyped for 129 RFLP and 62 resistance gene analog (RGA) markers, covering 1505 cM of tomato genome with an average marker distance of 7.9 cM. Interval analysis identified seven QTLs for salt tolerance during the vegetative stage on tomato chromosomes 3, 4, 5, 7, 8, 9, and 12 (M. R. Foolad et al., unpublished data). The QTLs detected on chromosomes 3, 5, and 9 were the same as those identified in the previous studies (Foolad and Chen, 1999; Foolad et al., 2001) and exhibited larger effects than the newly identified QTLs on chromosomes 4, 7, 8, and 12. The overall results from these three studies indicate that the stable QTLs on chromosomes 3, 5, and 9 should be useful for introgression into the cultivated tomato via MAS to improve tomato salt tolerance during the vegetative stage.

Comparatively little research has been done to identify QTLs for salt tolerance during reproduction in tomato. Using an F_2 population of a cross between a salt-sensitive *L. esculentum* cultivar and a salt-tolerant accession of *L. pimpinellifolium,* and by using only 14 genetic markers, Breto and colleagues (1994) identified a few QTLs that exhibited association with fruit yield, fruit number, and/or fruit size under salt stress. However, because of the extreme difference in fruit size between the parents of this F_2 population, QTL identification was most likely confounded by the effects of genes controlling fruit size and, thus, the identified QTLs should be considered with caution and should be verified in more advanced generations before using for MAS. Similar studies were conducted in F_2 populations of different crosses between *L. esculentum* and either *L. pimpinellifolium* or *L. cheesmanii,* and several other QTLs were identified for the same fruit-related traits under salt stress (Monforte et al., 1996, 1997, 1999). However, again, large morphological and physiological differences between parental lines of these populations, including differences in flowering habits, maturity time, fruit size and number, and total fruit yield, would have adversely affected the power of the experiments in detecting true QTLs for salt tolerance. These QTLs also should be validated in more advanced populations, where such confounding effects are eliminated or minimized, before using for MAS.

Comparison of Salt Tolerance in Different Stages of Plant Development

A plant's response to salinity may vary with growth stage. Knowledge of the genetic relationship between salt tolerance at different plant stages is necessary to expedite breeding efforts to develop cultivars with enhanced tolerance throughout the ontogeny of the plant. Early studies in different plant species had suggested the absence of phenotypic relationships among different developmental stages with regard to plant salt tolerance (Abel and MacKenzie, 1963; Greenway and Munns, 1980; Johnson et al., 1992). Recently, however, systematic approaches were taken to examine the phenotypic and genetic relationships between salt tolerance during seed germination and later stages in tomato. In one study, an F_4 population of a cross between a tomato breeding line (UCT5) with salt sensitivity during both seed germination and the vegetative stage, and a primitive tomato cultivar (PI174263) with salt tolerance during both of these stages, was evaluated for salt tolerance independently during seed germination and the vegetative stage. Although there was significant variation among F_4 families in salt tolerance during both stages, there was no significant phenotypic correlation ($r = -0.10$, $P > 0.05$) between the ability of the seed to germinate rapidly and the ability of the plant to grow under salt stress (Foolad and Lin, 1997a). In a second approach, to examine the genetic correlation between salt tolerance during seed germination and the vegetative stage, selection was made for rapid seed germination under salt stress in an F_2 population of the same cross, and the selected F_3 progeny were evaluated for salt tolerance separately during seed germination and the vegetative stage (Foolad and Lin, 1997a). Selection for salt tolerance during seed germination significantly improved germination salt tolerance of the F_3 progeny, indicating that the selection was effective. However, the selection for salt tolerance during seed germination did not affect salt tolerance of the F_3 progeny during the vegetative stage, suggesting that the genetic and physiological mechanisms that contributed to salt tolerance during seed germination in these genetic materials were different from those conferring salt tolerance during the vegetative stage (Foolad and Lin, 1997a). Although these studies strongly suggested the absence of a relationship between salt tolerance during seed germination and the vegetative stage, this relationship was further examined by QTL analysis.

Using a BC_1S_1 population of a cross between a tomato breeding line (NC84173, salt sensitive during both seed germination and the vegetative stage) and a *L. pimpinellifolium* accession (LA722, salt tolerant during both stages), QTLs for salt tolerance during both seed germination and the vegetative stage were independently identified and compared. The results indicated that in most cases the locations of QTLs for salt tolerance during seed germination were different from the locations of QTLs for salt tolerance during the vegetative stage (Foolad, 1999). This study suggested the involvement of different genes (QTLs) controlling salt tolerance during seed germination and the vegetative stage in these genetic materials. A similar study was recently conducted using 145 F_9 RILs of a cross between NC84173 and LA722 (M. R. Foolad et al., unpublished data). The RILs were evaluated in replicated trials for salt tolerance during seed germination and the vegetative stage (plant survival under salt stress). The results indicated the absence of a significant phenotypic correlation between the rate of seed germination and the plant survival under salt stress across the RILs. The RILs were genotyped for 129 RFLP and 62 RGA markers. Marker analysis identified QTLs for salt tolerance during seed germination on chromosomes 2, 3, 4, 8, 9, and 12, and QTLs for salt tolerance during the vegetable stage on chromosomes 3, 4, 5, 7, 8, 9, and 12. Mostly different QTLs were detected for salt tolerance during seed germination and the vegetative stage in this population, suggesting the absence of a genetic relationship between salt tolerance in these two stages of plant development.

The overall results from these investigations indicate that, in tomato, salt tolerance during seed germination is generally independent of salt tolerance during the vegetative stage. This conclusion is consistent with earlier reports that salt tolerance of young tomato plants did not correlate with that of mature plants (Shannon et al., 1987) and that ranking of salt-tolerant tomato genotypes based on vegetative characteristics in mature plants differed from the ranking based on fruit yield (Caro et al., 1991). Absence of genetic relationships among different developmental stages with regard to salt tolerance have also been reported in other plant species, including alfalfa (*Medicago sativa* L.) (Johnson et al., 1992), wheat (*Triticum aestivum* L.) (Ashraf and McNeilly, 1988), triticale (*Triticale hexaploide* Lart.) (Norlyn and Epstein, 1984), and slender wheatgrass [*Elymus trachycalus* spp. *Trachycalus* (Link) Malte] (Pearen et al., 1997). These findings indicate that when breeding for improved salt tolerance, each stage of plant development must be evaluated separately for the assessment of tolerance and the identification, characterization, and utilization of useful genetic components. However, the identification of QTLs for salt tolerance at different developmental stages can facilitate simultaneous or sequential introgression of QTLs for tolerance and the development of cultivars with improved tolerance at all desirable stages. Furthermore, the finding that in tomato only a few major QTLs account for a large portion of the total phenotypic variation for salt tolerance at each developmental stage suggests that MAS for salt tolerance should be feasible, providing the opportunity to develop tomato germplasm with enhanced salt tolerance at different stages.

Drought Tolerance

A complementary approach to agricultural systems currently employed to minimize losses incurred by water stress is to develop, via genetic means, drought-tolerant cultivars with the ability to escape, avoid, and/or tolerate effects of water stress. Development of such plants would have a lasting economic impact on crop production worldwide. However, despite many decades of research on drought tolerance, drought continues to be a major challenge to plant breeders. This is in part due to the complexity of the trait. Accumulating evidence suggests that plant response to drought stress is controlled by the function of many genes and physiological mechanisms (Blum, 1988; Zhu et al., 1997; Subudhi et al., 2000; Zhang, Zheng, et al., 2001) and is varied depending on the influence of innumerable other environmental factors (Ceccarelli and Grando, 1996; Richards, 1996). Selection and breeding for drought tolerance is also difficult because tolerance appears to be a developmentally regulated, stage-specific phenomenon (Blum, 1988; Ludlow and Muchow, 1990; Richards, 1996; Mitchell et al., 1998). Tolerance at one stage of plant development may not be related to tolerance at other developmental stages. Each stage, which may be considered as a separate trait, may require a different screening method. Furthermore, no reliable screening procedure is known that can be effectively and efficiently used to identify drought-tolerant plants at different developmental stages. These and other complexities have led to a limited success in developing drought-tolerant plants or improving crop yields in dry environments.

Although genetic resources for tolerance to drought stress have been identified within the cultivated and related wild species of tomato, very limited research has been done on the physiology or genetics of this variation (Pillay and Beyl, 1990; Kahn et al., 1993; Martin et al., 1999) to warrant its use in breeding programs to develop drought-tolerant tomatoes. This is unlike the extensive research that has been done on drought tolerance in many other plant species, including rice (*Oryza sativa* L.; Nguyen et al., 1997; Zhang, Zheng, et al., 2001), maize (*Zea mays* L.; Ribant et al., 1997), sorghum (*Sorghum bicolor* L. Moench; Subudhi et al., 2000), and lettuce (*Lactuca sativa* L.; Johnson et al., 2000). Furthermore, comparatively much less research has been done on tomato tolerance to drought stress than its tolerance to other abiotic stresses, such as salinity and extreme temperatures. In the following paragraphs, the current genetic information on drought tolerance of the tomato during different developmental stages is reviewed.

Drought Tolerance During Seed Germination and Seedling Emergence

To determine the h^2 of drought tolerance during seed germination in tomato, a BC_1 population (N = 1000) of a cross between a germination drought-tolerant *L. pimpinellifolium* accession (LA722) and a drought-sensitive tomato breeding line (NC84173; maternal and recurrent parent) was evaluated for seed germination under drought stress (14 percent PEG, $\psi_w = -680$ kPa), and the most rapidly germinating seeds (first 3 percent germinated) were selected (M. R. Foolad et al., unpublished data). The 30 selected BC_1 individuals were grown to maturity and self-pollinated to produce BC_1S_1 progeny seeds. Twenty of the 30 selected BC_1S_1 progeny families were evaluated for germination under drought stress and their average performance was compared with that of a nonselected BC_1S_1 population of the same cross. Results indicated that selection for rapid seed germination under drought stress was effective and significantly improved progeny seed germination rate under drought stress (selection gain = 19.6 percent). A realized h^2 of 0.41 was obtained for drought tolerance during seed germination in this population. The results suggested that drought tolerance during seed germination in tomato was genetically controlled and could be improved by directional phenotypic selection.

To further characterize the genetic basis of drought tolerance during seed germination in tomato, two QTL studies were recently carried out. In one study, a trait-based marker analysis, using a BC_1 population of a cross between NC84173 and drought-tolerant *L. pimpinellifolium* accession LA722, detected four QTLs on chromosomes 1, 4, 8, 9, and 12 for drought tolerance during seed germination (M. R. Foolad et al., unpublished data). The results indicated that drought tolerance during seed germination in tomato was a quantitative trait, controlled by more than one gene. A few BC_1S_1 families were identified with most or all of the QTLs and with a drought tolerance comparable to that of LA722. These families should be useful for developing germination drought-tolerant tomato lines using MAS. In a second study, 145 F_9 RILs of the same cross were evaluated for germination rate under drought stress and, by using a composite interval mapping analysis, several QTLs for drought tolerance during seed germination were identified on tomato chromosomes 1, 2, 3, 4, 8, 9, and 12 (M. R. Foolad et al., unpublished data). The results of this study were highly consistent with those of the previous study, which used a BC_1 population of the same cross. The combined results indicated the presence of stable QTLs for drought tolerance during seed germination across tomato populations derived from the NC84173 × LA722 cross, suggesting the usefulness of these QTLs for improving tomato seed germination under drought stress by MAS. However, the stability of these QTLs across tomato populations and in other interspecific crosses using LA722 should be examined before considering them for MAS and transferring of drought tolerance to the cultivated tomato.

Drought Tolerance During Vegetative Growth and Reproduction

Very limited research has been done to characterize the genetic control of, or to develop tomatoes with improved, drought tolerance during the vegetative stage or reproduction. To facilitate selection for low Δ ($^{13}C/^{12}C$ discrimination), three QTLs associated with this trait were previously identified using F_3 and BC_1S_1 progeny of a cross between *L. esculentum* UC82B and *L. pennellii* LA716 (Martin et al., 1989). However, it has not been determined whether selection for these QTLs would increase WUE in tomato. Other related research on genetics of tomato drought tolerance includes the identification of several genes or mRNA whose expressions are elevated in response to drought stress. For example, four tomato genes that were drought induced, *le4, le16, le25,* and *le20,* were identified and characterized in *L. esculentum* (Cohen et al., 1991; Plant et al., 1991; Kahn et al., 1993). It was further determined that the increase in the expression of these genes occurred after a longer period of water deficit in *L. pennellii* than in *L. esculentum,* although these genes did not appear to be responsible for drought tolerance in *L. pennellii* (Kahn et al., 1993).

Cold Tolerance

As described earlier, genetic resources for cold-tolerance breeding during different stages of plant development have been identified within the cultivated and related wild species of tomato. There are several advantages in developing tomatoes with cold-tolerance attributes. When grown under suboptimal temperatures, cold-tolerant plants could exhibit improved earliness, adaptability, water use, and yield of high-quality fruits. Cold-tolerant genotypes can be planted earlier in the season and harvested earlier than cold-sensitive genotypes, when the crop may have higher economic value. Cold-tolerant plants could also be grown in the field for longer periods of time, i.e., extended growing season, thus the final yield may be higher than with cold-sensitive plants. Furthermore, production may be improved because early plantings avoid high temperatures typically observed during midsummer in many climates. Moreover, cold tolerance may result in water savings in areas with a Mediterranean climate, because early plant growth could make better use of early-season rains and water available in the root zone.

The physiological aspects of cold tolerance in tomato and other plant species have been extensively studied and reviewed (Lyons, 1973; Lyons et al., 1979; Graham and Patterson, 1982; Vallejos et al., 1983; Parkin et al., 1989; Guy, 1990; Alberdi and Corduera, 1991; Leviatov et al., 1994, 1995; Bohnert et al., 1995; Bradford, 1995; Keller and Steffen, 1995). Comparatively, less research has been conducted to characterize genetic controls of cold tolerance in tomato at the whole-plant and cellular levels. In this section, recent findings on genetics of cold tolerance during different developmental stages in tomato are briefly reviewed and the prospects for developing cold-tolerant cultivars through the use of various techniques are contemplated.

Cold Tolerance During Seed Germination and Seedling Emergence

Earlier studies on cold tolerance during seed germination in tomato attributed the genetic control of this trait to one recessive gene (Cannon et al., 1973), additive genes (Whittington and Fierlinger, 1972), or three to five genes (Ng and Tigchelaar, 1973; DeVos et al., 1981) when using different germplasms. Broad-sense h^2 was estimated at 0.97 and narrow-sense h^2 at 0.67 to 0.69 (Ng and Tigchelaar, 1973). These estimates were independently confirmed by DeVos and colleagues (1981). In a recent study, a low temperature fast-germinating *L. esculentum* accession (PI120256) and a low-temperature, slow-germinating tomato cultivar (UCT5) and their reciprocal F_2, F_3, and BC_1 progeny (total of ten populations) were evaluated for germination at low (11 ± 0.5°C) and normal (control; 20 ± 0.5°C) temperature regimes (Foolad and Lin, 1998). Weighted least-square regression analysis indicated that under cold stress most of the variation in germination time was due to additive genetic effects; dominance and epistatic interactions were not significant. Partitioning of the total genetic variance into those attributable to the effects of embryo, endosperm, testa, and the cytoplasm indicated that additive effects of endosperm could account for 80 percent of the total genetic variance for germination under cold stress. These results supported the previous physiological studies that had suggested the significant role of the endosperm in determining the rate of germination in tomato seed (Groot and Karssen, 1987; Haigh and Barlow, 1987).

To examine the response to selection for improved cold tolerance during seed germination in tomato, an F_2 population of a cross between UCT5 and PI120256 was evaluated under a low temperature (11 ± 0.5°C) regime and the fastest-germinating seeds (the first 5 percent germinated) were selected (Foolad and Lin, 1998). The selected F_2 progeny were grown to maturity and self-pollinated to produce F_3 seed. The selected F_3 progeny were evaluated for seed germination under the low temperature and their mean performance compared with that of a nonselected F_3 population of the same cross. The results indicated that selection was effective and significantly improved germination performance of the progeny; a realized h^2 of 0.74 was obtained for cold tolerance during seed germination (Foolad and Lin, 1998). This study indicated that the rate of tomato seed germination under cold stress was genetically controlled, with additivity being the major genetic component, and that it could be improved by directional phenotypic selection. A recent similar study, using interspecific BC_1 and BC_1S_1 progeny of a cross between a low-temperature, fast-germinating *L. pimpinellifolium* accession (LA722) and a low-temperature, slow-germinating tomato cultivar (NC84173), suggested a more moderate realized heritability ($h^2 = 0.43$) for cold tolerance during seed germination in tomato (Foolad et al., 2003).

To improve low-temperature seed germination of processing tomatoes, breeders have used *L. esculentum* lines screened at 10°C, including PI120256, PI174263, and derived lines PI341985 and PI341988 (Cannon et al., 1973; Ng and Tigchelaar, 1973; DeVos et al., 1981; Scott and Jones, 1985a,b) or wild accessions from high altitudes screened for high germination at 6 to 7°C, including LA1363 and LA1777 *(L. hirsutum)* and LA 460 *(L. chilense)* (Patterson et al., 1978; Patterson and Payne, 1983). Several processing breeding lines have been released in Europe, and field tests with early direct seeding confirmed an advantage of three to five days in terms of emergence, with lower variability than with standard cultivars (Damidaux and Martinez, 1992). Taken all together, it seems that, due to its moderate-to-high heritability, low-temperature tolerance during seed germination in tomato can be relatively easily transferred to commercial cultivars by hybridization and phenotypic selection. However, more recently, molecular mapping techniques have been employed to examine in more detail the genetic basis of and the prospect for improving this trait through MAS.

The number, chromosomal locations, and genetic effects of QTLs contributing to cold tolerance during seed germination in tomato have been determined in different interspecific tomato populations. A BC_1S_1 population of a cross between breeding line NC84173 (cold sensitive) and *L. pim-*

pinellifolium accession LA722 (cold tolerant during seed germination) was evaluated for the rate of seed germination under a low-temperature regime. Interval mapping analysis detected a few QTLs on tomato chromosomes 1 and 4 with significant effects on cold tolerance during seed germination (Foolad et al., 1998a). The *L. pimpinellifolium* accession had favorable QTLs on chromosome 1, and NC84173 had favorable QTLs on chromosome 4. The percentage of phenotypic variation explained (PVE) by individual QTLs ranged from 11.9 percent to 33.4 percent. Multilocus analysis indicated that the cumulative action of all significant QTLs accounted for 43.8 percent of the total phenotypic variance. Digenic epistatic interactions were evident between only two QTL-linked markers and two QTL-unlinked markers. Transgressive phenotypes were observed in the direction of cold sensitivity. The validity of the identified QTLs was further examined in two different populations of the same interspecific cross.

A BC_1 population (N = 1000) was evaluated for the rate of seed germination under cold stress, and the most rapidly germinating seeds (the first 3 percent germinated) were selected as the cold-tolerant individuals. The selected BC_1 plants were subjected to molecular marker analysis, using 119 RFLP markers covering the 12 tomato chromosomes with an average marker distance of 9.7 cM. A distributional extreme analysis resulted in the identification of QTLs for cold tolerance on chromosomes 1, 4, 8, 9, and 12 (L. P. Zhang, S. Prakash, and M. R. Foolad, unpublished data). In a different study, seeds of 145 F_9 RILs of the NC84173 × LA722 cross were evaluated for germination under cold stress and the time to 50 percent germination (T50) for each RIL was determined. The RILs were also subjected to marker analysis using 129 RFLP and 62 RGAs and a molecular linkage map was constructed. Interval mapping analysis detected QTLs for cold tolerance on tomato chromosomes 1, 2, 3, 7, 8, 9, and 12 (M. R. Foolad et al., unpublished data). The latter two studies validated all of the QTLs that were identified in the first study (Foolad et al., 1998a) and further detected a few new QTLs. The combined results supported the suggestion that cold tolerance during seed germination in tomato was a quantitative trait controlled by more than one gene. Comparison of QTLs in different populations (e.g., BC_1, BC_1S_1, and RILs) of the same *L. esculentum* × *L. pimpinellifolium* cross indicated that most QTLs were stable across populations/generations, whereas a few QTLs were population specific. Although no attempt has been made to improve cold tolerance during seed germination in tomato via MAS, the consistency of QTLs across populations suggests that such an attempt might be fruitful. However, before engaging with such an attempt, the stability of these QTLs should be examined in other interspecific crosses of tomato using LA722 as a source of cold tolerance.

Cold Tolerance During Vegetative Growth and Reproduction

A high-altitude ecotype of *L. hirsutum,* which was able to develop the first true leaf when grown at a night temperature of 0°C, was hybridized with a cold-sensitive *L. esculentum* cultivar (pistillate and recurrent parent), and F_1 and BC_1 progeny were produced (Patterson and Payne, 1983). Subsequent selections for survival under a night temperature of 0°C in the BC_1 and its self-progeny indicated that the cold-tolerance attributes of the *L. hirsutum* accession could be transferred to *L. esculentum* through hybridization and selection. Using the BC_1 population of an interspecific cross between a cold-sensitive *L. esculentum* line and a cold-tolerant *L. hirsutum* accession, Vallejos and Tanksley (1983) identified three QTLs responsible for growth at low temperatures. In a more recent study, parental and reciprocal F_1, F_2, F_3, and BC_1 progeny (total of 12 populations) of a cross between a cold-sensitive tomato breeding line (UCT5) and a cold-tolerant primitive cultivar (PI120256) were grown under two day/night temperature regimes of 15/10°C (cold stress) and 25/15°C (control) (Foolad and Lin, 2001a). Plants were evaluated for shoot dry weight under cold stress and for TI, measured as the ratio of dry weight under cold stress to dry weight under control conditions. Shoot dry weight was reduced in all genotypes in response to cold stress, however, PI120256 exhibited the highest tolerance (TI = 90.5 percent) and UCT5 the lowest (TI = 38.9 percent). The TIs of the filial and backcross progeny were intermediate to the parents, indicating that cold tolerance of PI120256 was inherited to the progeny. Across generations, there was a positive correlation (r = 0.76, P <0.01) between dry weight under cold stress and dry weight under control, suggesting that growth under cold stress was influenced by plant vigor. However, the absence of a significant correlation between dry weight under control and the TI (r = 0.47, P >0.05) and, in contrast, the presence of a significant positive correlation (r = 0.92, P <0.01) between dry weight under stress and the TI suggested that plant vigor was not a determining factor in the expression of cold tolerance in PI120256 and its progeny. Generation means analyses of dry weight under cold stress and TI indicated that the variation among generations was genetically controlled, with additive effects accounting for most of the variation (≈90 percent for TI). There were no significant dominance effects, and epistatic effects were minor and involved only additive × additive interactions. The results suggested that the inherent cold tolerance of PI120256 would be useful for improving low-temperature tolerance of commercial cultivars of tomato. The overall results from the different studies suggest that cold tolerance during the vegetative stage of commercial cultivars of tomato can potentially be improved using the cold-tolerant accessions within the cultivated and related wild species of tomato as gene resources, although no commercial cultivar with improved cold tolerance has yet been released.

Using parental and F_3 progeny of a cross between *L. esculentum* and a high-altitude, cold-tolerant accession of *L. hirsutum,* Patterson, Mutton, and colleagues (1987) demonstrated that some F_3 plants were able to produce and liberate pollen of normal appearance at low temperatures, suggesting inheritance of the cold-tolerance characteristics from *L. hirsutum.* Fernandez-Munoz and colleagues (1995a) investigated low-temperature pollen viability of parental, F_1, F_2, and BC progeny of a *L. esculentum* × *L. pennellii* cross and concluded that this trait was polygenic and heritable. In segregating progeny derived from crosses between *L. esculentum* and *L. pimpinellifolium,* the ability to set fruit under low temperatures was determined to be controlled by recessive factors (Kalloo and Banerjee, 1990). The cold tolerance of a hybrid between the sensitive *L. esculentum* cv. Sub Arctic Maxi and tolerant *S. lycopersicoides* was suggested to be due to dominant nuclear genes (Kamps et al., 1987). The available information in the literature on tomato cold tolerance during reproduction suggests that sources of genetic variation exist within the related wild species of tomato and that tomato cold tolerance can be potentially improved by hybridization and selection. However, further research is needed to determine the number and genomic locations of QTLs or major genes in order to facilitate the development of cold-tolerant tomatoes via MAS or genetic transformation.

Comparison of Cold Tolerance Among Developmental Stages

Low-temperature tolerance is needed at all stages of plant development, including seed germination, to facilitate early-season direct seeding in temperate regions, the vegetative stage, to extend the location and the range of growing season, reproduction, to ensure flower production and fruit set at low temperatures, and, in the fruit, to facilitate postharvest storage at low temperatures. When searching for genetic resources for cold tolerance, it is desirable to identify genotypes that exhibit cold tolerance throughout the life cycle of the plant. However, cold tolerance at one stage of plant development may not be correlated with cold tolerance at other stages, and there might not be accessions bearing tolerance at all desirable stages (Herner and Kemps, 1983). Patterson and colleagues (1978) studied altitudinal ecotypes of *L. hirsutum* and reported variable tolerance to low temperatures at the whole-plant level, as described previously. However, equivalent germination rates of *L. hirsutum,* native to 3100 m altitude in Peru, occurred at 3°C lower temperature than *L. esculentum* cv. Rutgers. This marginal performance suggests that cold tolerance during the vegetative stage is not necessarily a good indication of cold tolerance during other stages, including seed germination. Comparison of different accessions of *L. hirsutum* indicated that the ability of the seed to germinate quickly at low temperatures was not related to the ability of the pollen to germinate at low temperatures (Patterson et al., 1978; Patterson, 1988). Similarly, screening *L. hirsutum* accessions for cold tolerance indicated that tolerance at the seedling stage was not necessarily correlated with tolerance at later stages, including pollen development (Patterson, Paull, and Graham, 1987). Lack of positive relationships between cold tolerance during seed germination and reproduction (pollen development and germination) and between seedling growth and reproduction were also reported in other studies (Patterson et al., 1978; Maisonneuve and Den-Jigs, 1984; Maisonneuve et al., 1986; Patterson, Paull, and Graham, 1987). In contrast, it was demonstrated that *L. hirsutum* accessions that were cold tolerant in the sporophytic stage (e.g., seedling growth) also exhibited greater cold tolerance in the gametophytic stage, i.e., pollen germination and growth (Zamir et al., 1981). Similarly, it was shown that the inhibition of root elongation at low temperatures was less in plants of the crosses in which pollen was used from low-temperature-adapted *L. hirsutum* plants than from plants of normal temperatures (Zamir and Gadish, 1987). It was also shown that the frequency of two isozyme markers from high-altitude *L. hirsutum* LA1777 were favored in the backcross progeny of *L. esculentum* pollinated at a low temperature with pollen from the hybrid *L. esculentum* × *L. hirsutum* (Zamir et al., 1982). However, pollen selection at a low temperature was not efficient for improving tomato cold tolerance during sporophytic stage (Maisonneuve et al., 1986).

Two recent studies examined in more detail the phenotypic and genetic relationships between cold tolerance during seed germination and the vegetative stage in tomato. In one study, the phenotypic relationship was investigated by evaluating 31 tomato accessions (cultivars, breeding lines, and plant introductions), representing six *Lycopersicon* spp., for cold tolerance during both stages (Foolad and Lin, 2000). Seed germination was evaluated under two temperature regimes of 11 ± 0.5°C (cold stress) and 20 ± 0.5°C (control) in darkness. Cold tolerance during seed germination was defined as the inverse of the ratio of germination time under cold stress to germination time under control conditions, and it was called germination tolerance index (TI_G). Across accessions, TI_G ranged from 0.15 to 0.48, indicating the presence of notable genotypic variation for cold tolerance during seed germination. Vegetative growth was evaluated in growth chambers with day/night temperatures of 12/5°C (cold stress) and 25/18°C (control) and a 12 h photoperiod of 350 $\mu mol \cdot m^{-2} \cdot s^{-1}$ (photosynthetic photon flux). Cold tolerance during the vegetative stage was defined as the ratio of shoot dry weight under cold stress to shoot dry weight under control and called vegetative stage TI (TI_{VS}). Across accessions, TI_{VS} ranged from 0.12 to 0.39, indicating the presence of notable genotypic variation for cold tolerance during the vegetative stage. Cold tolerance during the vegetative stage was independent of plant vigor, as judged by the absence of a significant correlation ($r = 0.14$, $P > 0.05$) between TI_{VS} and dry weight under control. Furthermore, cold tolerance during the vegetative stage was independent of cold tolerance during seed germination, as judged by the absence of a significant rank correlation ($r_R = 0.14$, $P > 0.05$) between TI_{VS} and TI_G. However, a few accessions were identified with cold tolerance during both seed germination and the vegetative stage (Foolad and Lin, 2000). The results indicated that for cold-tolerance breeding in tomato, each stage of plant development would have to be evaluated and selected for independently.

To determine the genetic relationship between cold tolerance during seed germination and the vegetative stage, an F_2 population of a cross between *L. esculentum* PI120256, cold tolerant during both seed germination and the vegetative stage, and UCT5, cold sensitive during both stages, was evaluated for germination under cold stress, and the most cold-tolerant individuals (the first 5 percent germinated) were selected (Foolad and Lin, 2001b). Selected F_2 individuals were grown to maturity and self-pollinated to produce F_3 families, referred to as the selected F_3 population. The selected F_3 population was evaluated for cold tolerance separately during seed germination and the vegetative stage, and its performance was compared with that of a nonselected F_3 population of the same cross. The results indicated that selection for cold tolerance during seed germination significantly improved cold tolerance of the progeny during seed germination, and a realized h^2 of 0.75 was obtained for this trait. However, selection for cold tolerance during seed germination did not affect plant cold tolerance during the vegetative stage. There was no significant difference between the selected and nonselected F_3 populations in either absolute cold tolerance, defined as shoot FW under cold stress, or relative cold tolerance, defined as shoot FW under cold as a percentage of control. The results indicated that, in PI120256, cold tolerance during seed germination was genetically independent of cold tolerance during the vegetative stage (Foolad and Lin, 2001b).

The overall information suggests the absence of relationships among cold tolerance in different development stages in tomato, including seed germination, the vegetative stage, and reproduction. Thus, to develop tomato cultivars with improved cold tolerance throughout the ontogeny of the plant, selection protocols that include all relevant developmental stages are necessary. In this regard, the use of molecular markers and MAS may be desirable for achieving this goal.

Heat Tolerance

Development of tomato cultivars with improved fruit set and yield under high temperature is desired for tomato production in regions where the temperature during part of the growing season reaches 35°C (Scott et al., 1986; Stevens and Rudich, 1987; Scott, 1993). However, breeding for heat tolerance in tomato is a difficult task, in part due to the complexity of the trait (Charles and Harris, 1972; Rudich et al., 1977; Levi et al., 1978; El-Ahmadi and Stevens, 1979b; Kuo et al., 1979; Dane et al., 1991; Aung, 1979) and its low to moderate h^2 (El-Ahmadi and Stevens, 1979a; Villareal and Lai, 1979; Scott et al., 1986). It is especially difficult to obtain heat-tolerant lines with large fruit, as most heat-tolerant genotypes tend to have smaller fruit than is commercially acceptable (Villareal, 1978; Villareal and Lai, 1979; Scott, 1993; Scott et al., 1997). It is also difficult to obtain heat-tolerant lines with good vine coverage, as most heat-tolerant genotypes exhibit poor leaf cover (Scott et al., 1997; Scott, 1993). Furthermore, heat-tolerant genotypes often exhibit significant G × E interactions (Rudich et al., 1977). To develop widely adapted lines with acceptable horticultural characteristics in a breeding program, utilization of several genotypes capable of fruit setting at high temperatures in different environments and in combinations with genotypes with large fruit and good vine growth has been recommended.

To estimate h^2 for heat tolerance and produce heat-tolerant tomato lines, Wessel-Beaver and Scott (1992) developed a synthetic population of tomato from polycrosses among seven selected genotypes (cultivars and breeding lines) and evaluated the population for heat tolerance in two different locations, Florida and Puerto Rico, during one summer. Single-location h^2s were high for percent fruit set (0.74 to 0.77), yield (0.65 to 0.81), and fruit weight (0.89 to 0.97). Across-location h^2 was low for yield (0.14), intermediate for fruit set (0.60), and high for fruit weight (0.92). G × E interaction was most important for yield and least important for fruit weight. Large genetic correlations ($r = 0.71$ to 0.74) were observed between yield and fruit set under high temperatures in both locations (Wessel-Beaver and Scott, 1992). El-Ahmadi and Stevens (1979a) reported the inheritance of fruit set under high temperature of heat-tolerant genotypes to be due to additive and dominant gene actions with moderate h^2. Hanna and colleagues (1982) found additive gene action to be more important than nonadditive gene action for percent fruit set, percent flower drop, and percent undeveloped ovaries under high field temperatures. In a diallel analysis using several heat-tolerant and heat-sensitive tomato genotypes, Dane and colleagues (1991) determined that pollen fertility and fruit set under high field temperatures were primarily under additive gene controls. However, evaluating parental and hybrid progeny of crosses between several genotypes with various levels of fruit-set ability under high temperature, Scott and colleagues (1986) reported the presence of dominance for high fruit set under high temperature and recommended the development and use of hybrid cultivars with improved heat tolerance. A more recent study, including parental and hybrid progeny of crosses among several heat-tolerant and heat-sensitive genotypes, confirmed the presence of dominance for fruit-set ability under high temperature (Scott et al., 1997). The advantage of the dominant nature of high-temperature fruit-setting ability is that commercial, heat-tolerant, fresh-market tomato cultivars can feasibly be developed by making hybrids between heat-tolerant and heat-sensitive parents. In this approach, defects of the heat-tolerant parent, such as disease susceptibility, small fruit, and poor vine coverage, could be at least partially corrected, if the right heat-sensitive parent is used for the hybrid production. Thus, Scott and colleagues (1997) strongly recommended the use of hybrids for the production of commercially acceptable, heat-tolerant, fresh-market tomato cultivars.

A review of the literature on heat tolerance breeding in tomato indicates that breeding progress for improving tomato fruit set at high temperatures in combination with desirable fresh-market characteristics such as large fruit size and good vine coverage has been feasible but difficult (Scott et al., 1986; Dane et al., 1991; Scotgt, 1993; Scott et al., 1997). Almost all tomato genotypes with good fruit-setting ability under high temperature are small-fruited (Rudich et al., 1977; Shelby et al., 1978; Villareal, 1978; El-Ahmidi and Stevens, 1979b; Hanna and Hernandez, 1982; Scott et al., 1986; Dane et al., 1991; Wessel-Beaver and Scott, 1992), a characteristic that is not desirable for fresh-market tomato production. It has been difficult to transfer heat tolerance to large-fruited cultivars without compromising the fruit size. Furthermore, almost all of the heat-tolerant genotypes have inadequate leaf cover due to the highly reproductive nature of the plants (Scott et al., 1997). However, despite these and other difficulties, several tomato breeding lines and hybrid cultivars with improved heat tolerance and rather acceptable horticultural characteristics have been developed and released for commercial use (Scott et al., 1989, 1995, 1997; Hanna et al., 1992; Scott, 2000). The use of such cultivars has already resulted in dramatic increases in tomato yield in areas with high day or night temperatures during the growing season, such as that in Florida (Scott et al., 1997). In most cases, some compromise had to be made in the fruit size and/or fruit-setting ability under high temperatures. Recently, however, J. W. Scott (personal communication) at the University of Florida has been able to develop heat-tolerant tomato cultivars with medium to large fruit size and with a more vigorous vine and better leaf coverage. He has overcome some of the problems of small fruit size by crossing heat-tolerant inbred lines with large-fruited, heat-sensitive inbreds and producing heat-tolerant hybrids with commercially acceptable fruit size (Scott et al., 1997; Scott, 2000). Scott has strongly advocated the use of hybrids for developing commercially acceptable heat-tolerant cultivars (Scott and Angell, 1998). When producing hybrids, heat-sensitive parents should be used that can complement the undesirable characteristics of the heat-tolerant parents. For example, heat-tolerant inbred lines with medium-size fruit can be crossed with extremely large-fruited, heat-sensitive lines to produce heat-tolerant hybrids with commercially acceptable fruit size. This approach has been successful in developing heat-tolerant hybrid cultivars for production in Florida (Scott, 2000; Scott et al., 1997). Furthermore, most of the heat-tolerant varieties developed to date have a jointed pedicel, since the development of heat-tolerant, large-fruited, jointless inbreds has been extremely difficult. Most recently, Scott was able to develop heat-tolerant, large-fruited, jointless inbred lines and hybrid tomato cultivars, although such genotypes do not perform as well as the equivalent genotypes with jointed pedicel (Scott, 2000; Scott et al., 1997). The available heat-tolerant genotypes are currently being used extensively to develop further commercially acceptable tomato cultivars with enhanced heat tolerance.

GENETIC ENGINEERING FOR CROP IMPROVEMENT

In tomato, transgenic approaches to improve crop stress tolerance have been successfully employed to develop plants with tolerance to salt stress. Such technologies, however, have not yet been applied to improving tomato plants with tolerance to other environmental stresses.

A plant's response to salt stress involves the functions of many genes that lead to a wide variety of biochemical and physiological changes. These include expression of genes that facilitate compartmentation of toxic ions in the vacuole, activation of detoxification enzymes, synthesis of late-embryogenesis-abundant (LEA) proteins, and accumulation of low-molecular-weight organic compounds, the latter collectively known as compatible solutes or osmolytes. During the past several years, genetic engineering approaches have been employed to produce transgenic plants with enhanced tolerance to various abiotic stresses, including salt stress, by overexpression of genes controlling different tolerance-related physiological mechanisms (Bajaj et al., 1999; Serrano et al., 1999; Rontein et al., 2002). For example, plants have been engineered with genes encoding enzymes that enhance the synthesis of compatible solutes, such as mannitol (Thomas et al., 1995), glycinebetaine (Lilius et al., 1996), proline (Zhu et al., 1997), and polyamines (Galston et al., 1997), all contributing to osmotic adjustment and improving plant stress tolerance (Rathinasabapathi, 2000; Rontein et al., 2002). Compatible solutes may also contribute to stress tolerance through other functions, such as protection of enzyme and membrane structure and scavenging of radical oxygen species (Shen et al., 1997; Bohnert and Shen, 1999; Rathinasabapathi, 2000). Transgenic plants have also been produced with overexpression of different vacuolar antiports that facilitate exclusion of toxic ions from the cell cytosol (Apse et al., 1999; Serrano et al., 1999; Zhang and Blumwald, 2001; Zhang, Hodson, et al., 2001). Furthermore, transgenic plants have been developed with increased expression of detoxification enzymes that reduce oxidative stress (Tanaka et al., 1999). Although in almost all cases plant growth and stress treatments were carried out in controlled conditions, and in many cases the increased tolerance has been rather marginal, the transgenic approach has provided opportunities for a better understanding of mechanisms leading to stress tolerance. The preliminary results have been particularly encouraging for enabling scientists to better understand the effects of single-gene transfers to plants.

Efforts to develop tomatoes with enhanced salt tolerance using transgenic approaches have been rather limited. However, a significant advancement was recently reported by Zhang and Blumwald (2001), who developed transgenic tomato plants overexpressing *AtNHX1*, a single-gene controlling vacuolar Na^+/H^+ antiport protein introduced from *Arabidopsis thaliana*. The overexpression of this gene was previously shown to increase salt tolerance in *Arabidopsis* (Apse et al., 1999). Transgenic tomato plants overexpressing this gene were able to grow, set flowers, and produce fruit in the presence of 200 mM NaCl in greenhouse hydroponics, whereas the nontransgenic (control) plants did not survive the saline conditions (Zhang and Blumwald, 2001). The transgenic plants acquired truly a halophytic response to salt stress; that is, they accumulated salts in the cell and sequestered them in the vacuole. As indicated before, the normal response of the cultivated tomato to salinity is of a glycophytic type, that is, excluding salt from the cell at the plasma membrane in the root and/or in the shoot. This was the first reported example of a single-gene transformation in any crop species that resulted in such a significant enhancement in plant salt tolerance. According to this report, under high-salinity conditions, transgenic tomato plants accumulated very high concentrations of Na^+ and Cl^- in the leaves. The overproduction of the vacuolar Na^+/H^+ antiport protein enhanced the ability of the transgenic plants to sequester Na^+ in their vacuole, averting its toxic effects in the cell cytosol. At the same time, Na^+ was used to maintain an osmotic balance to drive water into the cells, thus being able to use saline water for cell expansion and growth. Furthermore, there were only minimal increases in concentrations of Na^+ and Cl^- in the fruit, a great horticultural advantage for commercial production of such transgenic plants. The low Na^+ content of the fruit was attributed to the ability to maintain a high cytosolic K^+/Na^+ concentration ratio along the symplastic pathway in the transgenic plants. The results indicated that the enhanced accumulation of Na^+, mediated by the vacuolar Na^+/H^+ antiport, allowed the transgenic plants to ameliorate the toxic effects of Na^+. More recently, the transfer and overexpression of the same *AtNHX1* gene into canola (rape seed), *Brassica napus,* resulted in salt-tolerant transgenic plants that were able to grow, flower, and produce seeds in the presence of 200 mM NaCl (Zhang, Hodson, et al., 2001). Although the transgenic tomato and canola plants are yet to be evaluated for salt tolerance under field conditions, these findings suggest the potential for producing salt-tolerant plants using the transgenic approach.

It should be noted that the transformation technology for improving plant stress tolerance is still at its infancy, and the success to date represents only a beginning. There is no report to date of any field studies testing the performance of transgenic plants under salt-stress conditions. Much more work is needed to gain a better understanding of the genetic, biochemical, and physiological basis of plant salt tolerance. Future knowledge of other tolerance components and the identification and cloning of responsible genes may allow transformation of plants with multiple genes and production of highly stress-tolerant transgenic plants. With the current transformation technology, it is becoming possible to transfer multiple genes, which may act synergistically and additively to improve plant stress tolerance.

LIMITATIONS AND FUTURE PROSPECTS

Most commercial cultivars of tomato are sensitive to environmental stresses, including salinity, drought, and extreme temperatures. Gene resources for tolerance to most environmental stresses are not found within the cultivated species of tomato. This is in part due to the limited genetic variation within this species, which resulted from the occurrence of several population bottlenecks during tomato's domestication and evolution. Fortunately, however, an almost inexhaustible supply of genetic diversity exists within the wild species of *Lycopersicon,* which can be exploited for tomato breeding. Although thus far only a superficial assessment of the extent of the genetic variation for environmental stress tolerance has been made within *Lycopersicon,* several accessions with substantial tolerance to different stresses have been identified. However, it is expected that additional sources of tolerance can be identified if a thorough screening of the wild tomato germplasm is conducted. Such screening is highly necessary, in particular in search of tolerance to drought stress.

During the past few decades, the known sources of tolerance in tomato have been used rather extensively in physiological studies, and considerable knowledge has been accumulated as to the physiological mechanisms and cellular bases of stress tolerance. Although our knowledge in this area is far from complete, it is much greater than our understanding of the genetic basis of stress tolerance in tomato. Limited research has been conducted to characterize the genetic controls of stress tolerance in tomato, and this has significantly restricted the development of tomato cultivars with improved stress tolerance. It has only been during the past several years that, with the advent of the new tools of plant molecular biology and with a better understanding of the importance of increasing crop tolerance to environmental stresses, more emphasis has been placed on the genetic and molecular basis of stress tolerance in tomato. These investigations have determined that abiotic stress tolerance in tomato is generally a complex trait, controlled by more than one gene, and highly influenced by environmental variation. It has also been determined that stress tolerance in tomato is a developmentally regulated, stage-specific phenomenon, similar to findings in other plant species. Tolerance at one stage of plant development is often not correlated with tolerance at other developmental stages. Furthermore, at each developmental stage usually more than one physiological mechanism or genetic factor contributes to tolerance, and different mechanisms may be involved in different genotypes. Some tolerance components for various environmental stresses have been identified in tomato and their genetic controls determined. Many controlling genes or QTLs with major effects have been identified and/or cloned. It seems that with the recent interest in and emphasis on plant stress tolerance, the prospect for developing tomatoes with enhanced tolerance to different environmental stresses is rapidly improving.

Prospects for Developing Salt-Tolerant Tomatoes

Genetic resources for tolerance to salt stress have been identified within the cultivated and related wild species of tomato. Considerable knowledge has been accumulated regarding the physiological mechanisms and cellular bases of plant salt tolerance, and notable progress has been made in characterizing genetic controls of tolerance during different stages of plant development. Heritabilities for salt tolerance during different stages, including seed germination, the vegetative stage, and reproduction (fruit yield), have been determined and controlling QTLs identified. Significant progress also has been made in employing the transgenic approaches for a better understanding of tomato salt tolerance and for developing salt-tolerant plants. Although no commercial cultivar of tomato with proven field tolerance to salinity has yet been released through the use of either the conventional protocols of plant genetics and breeding, MAS, or transgenic approaches, the prospect for such development in the near future is very good. Furthermore, with the current knowledge of the physiological and genetic bases of salt tolerance and with the availability of molecular genetic techniques, coupled with our understanding of the necessity of developing salt-tolerant tomato genotypes, it would not be unexpected to witness tomato cultivars with field salt tolerance in the near future. It is anticipated that the importance of breeding tomatoes for salt tolerance, using either conventional protocols or the contemporary biotechnologies, will increase in the future. This has been happening in many other crop species. Currently, a few research groups in the United States and worldwide are attempting to produce salt-tolerant tomato cultivars.

Prospects for Developing Drought-Tolerant Tomatoes

Currently, physiological and/or genetic information on tomato drought tolerance is too limited to warrant development of drought-tolerant cultivars through plant breeding. This might be in part due to the lack of known useful sources of drought tolerance in tomato. Many investigations of drought tolerance in tomato have employed a single accession (LA716) of *L. pennellii* as a gene resource. However, due to various undesirable characteristics of this accession, including its extremely slow growth rate under drought stress, its usefulness for breeding tomatoes with drought tolerance has not been proven. Although this accession can survive long periods of dryness, it does not appear to be a highly desirable resource for tomato breeding. Considering the normal climatic conditions for growing tomatoes, where short periods of drought may occur intermittently throughout the growing season, it seems that the ability of the plant to survive transient periods of water stress and to recover rapidly upon reavailability of water is far more important than the ability to survive long-term water stress. In other words, the main focus of any breeding program for drought tolerance must be on improving tomatoes for dehydration avoidance. It is extremely important that larger germplasm-screening experiments be carried out in search of additional genetic resources for drought-tolerance breeding in tomato. Furthermore, selection criteria for breeding for drought tolerance in tomato are less clear than those for breeding for other stresses, including salt stress, cold stress, and heat stress. More research is needed to identify and validate useful selection criteria for breeding for drought tolerance in tomato, including agronomic, physiological, and biochemical characteristics. Similar to that for other abiotic stresses, the identification and utilization of molecular markers associated with different tolerance-related physiological, morphological, or agronomic criteria might be an efficient way of improving drought tolerance in tomato. Moreover, transgenic approaches have been employed in other plant species to increase drought tolerance, and more and more drought tolerance-related genes or proteins are being discovered (Kavi-Kishor et al., 1995; Grover et al., 1999; Serrano et al., 1999). Transgenic approaches have also been successful in developing tomatoes with improved salt tolerance, and there is no reason that such attempts would not be successful in developing tomatoes with improved drought tolerance. Therefore, a combination of different germplasm resources, characteristics, tolerance criteria, and technologies have to be employed at different stages of the breeding process if successful tomato cultivars for production under water stress are expected. In summary, more physiological and genetic research on tomato drought tolerance is necessary before we will witness the first drought-tolerant tomato cultivar in the market.

Prospects for Developing Cold-Tolerant Tomatoes

Genetic resources for cold tolerance during different stages of plant development, including seed germination, the vegetative stage, and reproduction, have been identified within the cultivated and related wild species of tomato, in particular *L. hirsutum*. Significant progress has been made in characterizing the physiological mechanisms and metabolic aspects of cold tolerance, and some knowledge has been gained as to the genetic controls of cold tolerance during different stages of plant development in tomato. However, progress in developing tomatoes with improved cold tolerance has been slow, in part due to the insufficient understanding of the genetic basis of cold tolerance and the limited breeding effort that has been devoted to this goal. A better understanding of the genetics of cold tolerance in tomato, in particular at the cellular and molecular levels, is needed to facilitate the identification and transferring of tolerance components from cold-tolerance accessions of *L. hirsutum* into the cultivated tomato. Research to identify QTLs with major effects on cold tolerance may also be useful in the short term for improving tomato low-temperature tolerance. Compared with past genetics and breeding efforts, more emphasis must be placed on characterizing and utilizing the available germplasm resources for developing tomato cultivars with improved cold tolerance. Significant progress has been made in molecular genetic basis of cold tolerance in many other plant species, in particular *A. thaliana*. Several genes, proteins, enzymes, and other compounds have been identified with direct or indirect effects on low-temperature tolerance. The search for potential tolerance components, including genes and proteins, could go beyond the limits of species within *Lycopersicon*, and it could include other genera, such as model plants and microbial organisms. Utilization of such genes or compounds in tomato using the available transgenic approaches may become a useful method for improving tomato cold tolerance.

Prospects for Developing Heat-Tolerant Tomatoes

Germplasm resources with heat tolerance attributes have been widely identified within *Lycopersicon* species, the physiological and morphological components of tolerance/sensitivity have been determined, and the genetic basis of heat tolerance at the whole-plant level has been discerned. During the past several years, many breeding lines and cultivars have been developed with improved fruit-setting ability under high temperature. Many of these lines exhibit some undesirable horticultural characteristics, in particular small fruit size, low yield, poor vine coverage, and jointed pedicel. However, many years of tomato research and breeding efforts at the University of Florida have resulted in the development of new, high-yielding breeding lines and hybrid cultivars with medium to large size fruit, high fruit-setting ability, jointless pedicel, and other heat-tolerance attributes. Currently, further efforts are being made to develop heat-tolerant tomato cultivars with larger fruit size, higher yield, and more vigorous vine for production in tropical and subtropical environments. Judging from the recent advances in this area, it would not be unexpected to witness newer cultivars, in particular hybrids with excellent fruit-setting ability under high temperature and acceptable horticultural characteristics, in the market in the near future. However, very little research has been done on molecular genetic basis of heat tolerance in tomato. This line of research, as well as research on the relationship between tolerances to high and low temperatures during different stages of plant development, should be emphasized in future.

REFERENCES

Abdul-Baki, A.A. (1991). Tolerance of tomato cultivars and selected germplasm to heat stress. *Journal of the American Society for Horticultural Science* 116:1113-1116.

Abdul-Baki, A.A. and J.R. Stommel (1995). Pollen viabililty and fruit set of tomato genotypes under optimum- and high-temperature regimes. *HortScience* 30:115-117.

Abel, G.H. and A.J. Mackenzie (1963). Salt tolerance of soybean varieties (*Glycine max* L. Merrill.) during germination and later growth. *Crop Science* 3:159-161.

Adams, P. (1991). Effects of increasing the salinity of the nutrient solution with major nutrients or sodium chloride on the yield, quality and composition of tomatoes grown in rockwool. *Journal of Horticultural Science* 66:201-207.

Adams, P. and L.C. Ho (1992). The susceptibility of modern tomato cultivars to blossom-end rot in relation to salinity. *Journal of Horticultural Science* 67:827-839.

Alberdi, M. and L.J. Corduera (1991). Cold acclimation in plants. *Phytochemistry* 30:3177-3184.

Apse, M.P., G.S. Aharon, W.A. Snedden, and E. Blumwald (1999). Salt tolerance conferred by overexpression of a vaculolar Na^+/H^+ antiport in *Arabidopsis*. *Science* 285:1256-1258.

Ashraf, M. and T. McNeilly (1988). Variability in salt tolerance of nine spring wheat cultivars. *Journal of Agronomy and Crop Science* 160:14-21.

Asins, M.J., M.P. Breto, M. Cambra, and E.A. Carbonell (1993). Salt tolerance in *Lycopersicon* species. I. Character definition and changes in gene expression. *Theoretical and Applied Genetics* 86:737-743.

Asins, M.J., M.P. Breto, and E.A. Carbonell (1993). Salt tolerance in *Lycopersicon* species. II. Genetic effects and a search for associated traits. *Theoretical and Applied Genetics* 86:769-774.

Aung, L.H. (1979). Temperature regulation of growth and development of tomato during ontogeny. In *Proceedings of the First International Symposium on Tropical Tomato*. Tainan, Taiwan: Asian Vegetables Research and Development Center, pp. 79-93.

Bajaj, S., J. Targolli, L.F. Liu, T.H.D. Ho, and R. Wu (1999). Transgenic approaches to increase dehydration-stress tolerance in plants. *Molecular Breeding* 5:493-503.

Bar-Tsur, A., J. Rudich, and B. Bravdo (1985). High temperature effects on CO_2 gas exchange in heat-tolerant and sensitive tomatoes. *Journal of the American Society for Horticultural Science* 110:582-586.

Berry, S.Z. (1969). Germinating response of the tomato at high temperature. *HortScience* 4:218-219.

Berry, S.Z. and M. Rafique-Uddin (1988). Effect of high temperature on fruit set in tomato cultivars and selected germplasm. *HortScience* 23:606-608.

Bliss, R.D., K.A. Platt-Aloia, and W.W. Thomson (1986). Osmotic sensitivity in relation to salt sensitivity in germinating barley seeds. *Plant, Cell and Environment* 9:721-725.

Blum, A. (1988). *Plant Breeding for Stress Environments*. Boca Raton, FL: CRC Press.

Bohnert, H.J. and R.G. Jensen (1996). Strategies for engineering water-stress tolerance in plants. *Trends in Biotechnology* 14:89-97.

Bohnert, H.J., D.E. Nelson, and R.G. Jensen (1995). Adaptation to environmental stresses. *The Plant Cell* 7:1099-1111.

Bohnert, H.J. and B. Shen (1999). Transformation and compatible solutes. *Scientia Horticulturae* 78:237-260.

Bolarin, M.C., F.G. Fernandez, V. Cruz, and J. Cuartero (1991). Salinity tolerance in four wild tomato species using vegetative yield-salinity response curves. *Journal of the American Society for Horticultural Science* 116:286-290.

Bolarin, M.C., F. Perez-Alfocea, E.A. Cano, M.T. Estan, and M. Caro (1993). Growth, fruit yield, and ion concentration in tomato genotypes after pre- and post-emergence salt treatments. *Journal of the American Society for Horticultural Science* 118:655-660.

Boyer, J.S. (1982). Plant productivity and environment. *Science* 218:443-448.

Bradford, K.J. (1986). Manipulation of seed water relations via osmotic priming to improve germination under stress conditions. *HortScience* 21:1105-1112.

Bradford, K.J. (1995). Water relations in seed germination. In *Seed Development and Germination,* J. Kigel and G. Galili (Eds.). New York: Marcel Dekker, pp. 351-396.

Bray, E.A. (1997). Plant responses to water deficit. *Trends in Plant Science* 2:48-54.

Breto, M.P., M.J. Asins, and E.A. Carbonell (1993). Genetic variability in *Lycopersicon* species and their genetic relationships. *Theoretical and Applied Genetics* 86:113-120.

Breto, M.P., M.J. Asins, and E.A. Carbonell (1994). Salt tolerance in *Lycopersicon* species: III. detection of quantitative trait loci by means of molecular markers. *Theoretical and Applied Genetics* 88:395-401.

Bruggemann, W. and P. Linger (1994). Long-term chilling of young tomato plants under low light. IV. Differential responses of chlorophyll fluorescence quenching coefficients in *Lycopersicon* species of different chilling sensitivity. *Plant and Cell Physiology* 35:585-591.

Calvert, A. (1969). Flower initiation and development in the tomato. *National Agricultural Advisory Service Quarterly Review* 70:79-88.

Cannon, O.S., D.M. Gatherum, and W.G. Miles (1973). Heritability of low temperature seed germination in tomato. *HortScience* 8:404-405.

Cano, E.A., F. Perez-Alfocea, V. Moreno, M. Caro, and M.C. Bolarin (1998). Evaluation of salt tolerance in cultivated and wild tomato species through in vitro shoot apex culture. *Plant Cell, Tissue and Organ Culture* 53:19-26.

Caro, M., V. Cruz, J. Cuartero, M.T. Estan, and M.C. Bolarin (1991). Salinity tolerance of normal-fruited and cherry tomato cultivars. *Plant and Soil* 136:249-255.

Ceccarelli, S. and S. Grando (1996). Drought as a challenge for the plant breeder. *Plant Growth Regulation* 20:149-155.

Charles, W.B. and R.E. Harris (1972). Tomato fruit-set at high and low temperatures. *Canadian Journal of Plant Science* 52:497-506.

Chen, F.Q., M.R. Foolad, J. Hyman, D.A. St. Clair, and R.B. Beelman (1999). Mapping of QTLs for lycopene and other fruit traits in a *Lycopersicon esculentum* × *L-pimpinellifolium* cross and comparison of QTLs across tomato species. *Molecular Breeding* 5:283-299.

Chen, H., Z.Y. Shen, and P.H. Li (1982). Adaptability of crop plants to high temperature stress. *Crop Science* 22:719-725.

Choudhuri, G.N. (1968). Effect of soil salinity on germination and survival of some steppe plants in Washington. *Ecology* 49:465-471.

Cohen, A., A.L. Plant, M.S. Moses, and E.A. Bray (1991). Organ-specific and environmentally regulated expression of two absicisic acid-induced genes of tomato. *Plant Physiology* 97:1367-1374.

Coleman, W.K. and R.I. Greyson (1976). The growth and development of the leaf in tomato *(Lycopersicon esculentum)*. I. The plastochron index, a suitable basis for description. *Canadian Journal of Botany* 54:2421-2428.

Cook, R.E. (1979). Patterns of juvenile morbidity and recruitment in plants. In *Topics in Plant Population Biology,* O.T. Solbrig, S. Jain, G.B. Johnson, and P.H. Raven (Eds.). Los Angeles, CA: Columbia University Press, pp. 207-301.

Coons, J.M., R.O. Kuehl, N.F. Oebker, and N.R. Simons (1989). Germination of eleven tomato phenotypes at constant or alternating high temperatures. *HortScience* 24:927-930.

Correll, D.S. (1962). *The Potato and Its Wild Relatives.* Renner, TX: Texas Research Foundation.

Cuartero, J. and R. Fernandez-Munoz (1999). Tomato and salinity. *Scientia Horticulturae* 78:83-125.

Cuartero, J., A.R. Yeo, and T.J. Flowers (1992). Selection of donors for salt-tolerance in tomato using physiological traits. *New Phytologist* 121:63-69.

Dahal, P., K.J. Bradford, and R.A. Jones (1990). Effects of priming and endosperm integrity on seed germination rates of tomato genotypes. I. Germination at suboptimal temperature. *Journal of Experimental Botany* 41:1431-1440.

Damidaux, R. and J. Martinez (1992). Tomato cold resistance: Present status and future trends. *Acta Horticulturae* 301:73-86.

Dane, F., A.G. Hunter, and O.L. Chambliss (1991). Fruit set, pollen fertility, and combining ability of selected tomato genotypes under high-temperature field conditions. *Journal of the American Society for Horticultural Science* 116:906-910.

Darvasi, A. and M. Soller (1992). Selective genotyping for determination of linkage between a marker locus and a quantitative trait locus. *Theoretical and Applied Genetics* 85:353-359.

Dehan, K. and M. Tal (1978). Salt tolerance in the wild relatives of the cultivated tomato: Responses of *Solanum pennellii* to high salinity. *Irrigation Science* 1:71-76.

Dempsey, W.H. (1970). Effect of temperature on pollen germination and tube growth. *Report of the Tomato Genetics Cooperative* 20:15-16.

DeVos, D.A., J.R.R. Hill, R.W. Helper, and D.L. Garwood (1981). Inheritance of low temperature sprouting ability in F_1 tomato crosses. *Journal of the American Society for Horticultural Science* 106:352-355.

Dudal, R. (1976). Inventory of major soils of the world with special reference to mineral stress. In *Plant Adaptation to Mineral Stress in Problem Soils,* M.J. Wright (Ed.). Ithaca, NY: Cornell University Agricultural Experiment Station, pp. 3-14.

Duvick, D. (1986). Plant breeding: Past achievement and expectations for the future. *Economic Botany* 40:289-294.

El-Ahmadi, A.B. and M.A. Stevens (1979a). Genetics of high-temperature fruit set in the tomato. *Journal of the American Society for Horticultural Science* 104:691-696.

El-Ahmadi, A.B. and M.A. Stevens (1979b). Reproductive responses of heat-tolerant tomatoes to high temperatures. *Journal of the American Society for Horticultural Science* 104:686-691.

El-Hassan, G.W. (1978). Inheritance studies of high temperature germination of tomato. *Egyptian Journal of Horticulture* 5:29-41.

Epstein, E., J.D. Norlyn, D.W. Rush, R.W. Kingsbury, D.B. Kelly, G.A. Gunningham, and A.F. Wrona (1980). Saline culture of crops: a genetic approach. *Science* 210:399-404.

Esau, K. (1953). *Plant Anatomy*. New York: John Wiley.

FAO (1995). *Production Yearbook 1994*. Rome, Italy: FAO.

Fernandez-Munoz, R. and J. Cuartero (1991). Effects of temperature and irradiance on stigma exsertion, ovule viability and embryo development in tomato. *Journal of Horticultural Science* 66:395-401.

Fernandez-Munoz, R., J.J. Gonzalez-Fernandez, and J. Cuartero (1995a). Genetics of the viability of pollen grain produced at low temperatures in *Lycopersicon* Mill. *Euphytica* 84:139-144.

Fernandez-Munoz, R., J.J. Gonzalez-Fernandez, and J. Cuartero (1995b). Viability of pollen tolerance to low temperatures in tomato and related wild species. *Journal of Horticultural Science* 70:41-49.

Flowers, T.J., M.A. Hajibagheri, and N.C.W. Clipson (1986). Halophytes. *Quarterly Review of Biology* 61:313-337.

Flowers, T.J. and A.R. Yeo (1988). Salinity and rice: A physiological approach to breeding for resistance. In *The International Congress of Plant Physiology*. New Delhi, India: Society for Plant Physiology and Biochemistry, pp. 953-959.

Flowers, T.J. and A.R. Yeo (1997). Breeding for salt resistance in plants. In *Strategies for Improving Salt Tolerance in Higher Plants*, P.K. Jaiwal, R.P. Singh, and A. Gulati (Eds.). Enfield, NH: Science Publishers, Inc., pp. 247-264.

Foolad, M.R. (1996a). Genetic analysis of salt tolerance during vegetative growth in tomato, *Lycopersicon esculentum* Mill. *Plant Breeding* 115:245-250.

Foolad, M.R. (1996b). Response to selection for salt tolerance during germination in tomato seed derived from P.I. 174263. *Journal of the American Society for Horticultural Science* 121:1006-1011.

Foolad, M.R. (1997). Genetic basis of physiological traits related to salt tolerance in tomato, *Lycopersicon esculentum* Mill. *Plant Breeding* 116:53-58.

Foolad, M.R. (1999). Comparison of salt tolerance during seed germination and vegetative growth in tomato by QTL mapping. *Genome* 42:727-734.

Foolad, M.R. and F.Q. Chen (1998). RAPD markers associated with salt tolerance in an Interspecific cross of tomato (*Lycopersicon esculentum* × *L. pennellii*). *Plant Cell Reports* 17:306-312.

Foolad, M.R. and F.Q. Chen (1999). RFLP mapping of QTLs conferring salt tolerance during vegetative stage in tomato. *Theoretical and Applied Genetics* 99:235-243.

Foolad, M.R., F.Q. Chen, and G.Y. Lin (1998a). RFLP mapping of QTLs conferring cold tolerance during seed germination in an interspecific cross of tomato. *Molecular Breeding* 4:519-529.

Foolad, M.R., F.Q. Chen, and G.Y. Lin (1998b). RFLP mapping of QTLs conferring salt tolerance during germination in an interspecific cross of tomato. *Theoretical and Applied Genetics* 97:1133-1144.

Foolad, M.R. and R.A. Jones (1991). Genetic analysis of salt tolerance during germination in *Lycopersicon. Theoretical and Applied Genetics* 81:321-326.

Foolad, M.R. and R.A. Jones (1992). Parent-offspring regression estimates of heritability for salt tolerance during germination in tomato. *Crop Science* 32:439-442.

Foolad, M.R. and R.A. Jones (1993). Mapping salt-tolerance genes in tomato (*Lycopersicon esculentum*) using trait-based marker analysis. *Theoretical and Applied Genetics* 87:184-192.

Foolad, M.R. and G.Y. Lin (1997a). Absence of a relationship between salt tolerance during germination and vegetative growth in tomato. *Plant Breeding* 116:363-367.

Foolad, M.R. and G.Y. Lin (1997b). Genetic potential for salt tolerance during germination in *Lycopersicon* species. *HortScience* 32:296-300.

Foolad, M.R. and G.Y. Lin (1998). Genetic analysis of low temperature tolerance during germination in tomato, *Lycopersicon esculentum* Mill. *Plant Breeding* 117:171-176.

Foolad, M.R. and G.Y. Lin (2000). Relationship between cold tolerance during seed germination and vegetative growth in tomato: Germplasm evaluation. *Journal of the American Society for Horticultural Science* 125:679-683.

Foolad, M.R. and G.Y. Lin (2001a). Genetic analysis of cold tolerance during vegetative growth in tomato, *Lycopersicon esculentum* Mill. *Euphytica* 122:105-111.

Foolad, M.R. and G.Y. Lin (2001b). Relationship between cold tolerance during seed germination and vegetative growth in tomato. Analysis of response and correlated response to selection. *Journal of the American Society for Horticultural Science* 126:216-220.

Foolad, M.R., T. Stoltz, C. Dervinis, R.L. Rodriguez, and R.A. Jones (1997). Mapping QTLs conferring salt tolerance during germination in tomato by selective genotyping. *Molecular Breeding* 3:269-277.

Foolad, M.R., P. Subbiah, C. Kramer, G. Hargrave, L.P. Zhang, and G.Y. Lin (2003). Genetic relationships between cold, salt and drought tolerance during seed germination in an interspecific cross of tomato. *Euphytica* 130:199-206.

Foolad, M.R., L.P. Zhang, and G.Y. Lin (2001). Identification and validation of QTLs for salt tolerance during vegetative growth in tomato by selective genotyping. *Genome* 44:444-454.

Forster, B.P., M.S. Phillips, T.E. Miller, E. Baird, and W. Powell (1990). Chromosome location of genes controlling tolerance to salt (NaCl) and vigor in *Hordeum vulgare* and *H. chilense*. *Heredity* 65:99-107.

Francois, L.E. and E.V. Maas (1994). Crop response and management on salt-affected soils. In *Handbook of Plant Crop Stress*. M. Pessarakli (Ed.). New York: Marcel Dekker, pp. 149-181.

Galston, A.W., R. Kaur-Sawhney, T. Altabella, and A.F. Tiburcio (1997). Plant polyamines in reproductive activity and response to abiotic stress. *Botanica Acta* 110:197-207.

Gautam, R.R., G. Kalloo, and B.S. Dhankhar (1981). Evaluation of tomato genotypes for fruit set under low temperature condition. *Haryana Journal of Horticultural Science* 10:81-85.

George, W.L.J., J.W. Scott, and W.E. Splittstoesser (1984). Parthenocarpy in tomato. In *Horticultural Review*, Volume 6, J. Janick (Ed.). Westport, CT: AVI Publishing Co., pp. 65-84.

Graham, D. and B.R. Patterson (1982). Responses of plants to low, nonfreezing temperatures: Proteins, metabolism, and acclimation. *Annual Review of Plant Physiology* 33:347-372.

Greenway, H. and R. Munns (1980). Mechanism of salt tolerance in nonhalophytes. *Annual Review of Plant Physiology* 31:149-190.

Groot, S.P.C. and C.M. Karssen (1987). Gibberellins regulate seed germination in tomato by endosperm weakening: A study with gibberellin-deficient mutant. *Planta* 171:525-531.

Grover, A., C. Sahi, N. Sanan, and A. Grover (1999). Taming abiotic stresses in plants through genetic engineering: Current strategies and perspectives. *Plant Science* 143:101-111.

Grunberg, K., R. Fernandez-Muñoz, and J. Cuartero (1995). Growth, flowering, and quality and quantity of pollen of tomato plants grown under saline conditions. *Acta Horticulturae* 412:484-489.

Guy, C.L. (1990). Cold acclimation and freezing stress tolerance: Role of protein metabolism. *Plant Molecular Biology* 41:187-223.

Haigh, A.H. and E.W.R. Barlow (1987). Water relations of tomato seed germination. *Australian Journal of Plant Physiology* 14:485-492.

Hanna, H.Y., A.J. Adams, and L.L. Black (1992). LHT24 heat-tolerant tomato breeding line. *HortScience* 27:1337.

Hanna, H.Y. and T.P. Hernandez (1982). Response of six tomato genotypes under summer and spring weather conditions in Louisiana. *HortScience* 17:758-759.

Hanna, H.Y., T.P. Hernandez, and K.L. Koonce (1982). Combining ability for fruit set, flower drop, and underdeveloped ovaries in some heat-tolerant tomatoes. *HortScience* 17:760-761.

Hegarty, T.W. (1978). The physiology of seed hydration and dehydration, and the relation between water stress and the control of germination: A review. *Plant, Cell and Environment* 1:101-119.

Henk-Venema, J., M. Eekhof, and P.R. Van-Hasselt (2000). Analysis of low-temperature tolerance of a tomato *(Lycopersicon esculentum)* cybrid with chloroplasts from a more chilling-tolerant *L. hirsutum* accession. *Annals of Botany* 85:799-807.

Herner, R.C. and T. Kemps (1983). Chilling tolerance of wild tomato species. In *Fourth Tomato Quality Workshop,* Volume VEC-83-1, University of Florida, pp. 26-37.

Hsiao, T.C. (1973). Plant responses to water stress. *Annual Review of Plant Physiology* 24:519-570.

Hsiao, T.C. and K.J. Bradford (1983). Physiological consequences of cellular water deficits. In *Limitations to Efficient Water Use in Crop Production,* H.M. Taylor, W.R. Jordan, and T.R. Sinclair (Eds.). Madison, WI: American Society of Agronomy, pp. 227-265.

Iwahori, S. (1965). High temperature injuries in tomato. IV. Development of normal flower buds and morphological abnormalities of flowers treated with high temperature. *Journal of the Japanese Society for Horticultural Science* 34:33-41.

Iwahori, S. (1966). High temperature injuries in tomato. V. Fertilization and development of embryo with special reference to abnormalities caused by high temperature. *Journal of the Japanese Society for Horticultural Science* 35:55-62.

Jaworski, C.A. and V.J. Valli (1965). Tomato seed germination and plant growth in relation to soil temperatures and phosphorus levels. *Proceedings of the Florida State Horticultural Society* 77:177-183.

Jenkins, J.A. (1948). The origin of the cultivated tomato. *Economic Botany* 21:379-392.

Johjima, T. (1995). Inheritance of heat tolerance of fruit coloring in tomato. *Acta Horticulturae* 412:64-70.

Johnson, D.W., S.E. Smith, and A.K. Dobrenz (1992). Genetic and phenotypic relationships in response to NaCl at different developmental stages in alfalfa. *Theoretical and Applied Genetics* 83:833-838.

Johnson, W.C., L.E. Jackson, O. Ochoa, R. van Wijik, J. Peleman, D.A.S. Clair, and R.W. Michelmore (2000). Lettuce, a shallow-rooted crop, and *Lactuca serriola,* its wild progenitor, differ at QTL determining root architecture and deep soil water exploitation. *Theoretical and Applied Genetics* 101:1066-1073.

Jones, R.A. (1986a). The development of salt-tolerant tomatoes: Breeding strategies. *Acta Horticulturae* 190:101-114.

Jones, R.A. (1986b). High salt-tolerance potential in *Lycopersicon* species during germination. *Euphytica* 35:576-582.

Jones, R.A., M. Hashim, and A.S. El-Beltagy (1988). Developmental responsiveness of salt-tolerant and salt-sensitive genotypes of *Lycopersicon.* In *Arid Lands: Today and Tomorrow,* E. Whitehead, F. Hutchison, B. Timmeman, and R. Varazy (Eds.). Boulder, CO: Westview Press, pp. 765-772.

Jones, R.A. and C.O. Qualset (1984). Breeding crops for environmental stress tolerance. In *Application of Genetic Engineering to Crop Improvement,* G.B. Collins and J.F. Petolino (Eds.). The Hague, The Netherlands: Nijihoff/Junk, pp. 305-340.

Jordan, W.R., W.A.J. Dugas, and P.J. Shouse (1983). Strategies for crop improvement for drought prone regions. *Agricultural Water Management* 7:281-299.

Kahn, T.L., S.E. Fender, E.A. Bray, and M.A. O'Connell (1993). Characterization of expression of drought- and abscisic acid-regulated tomato genes in the drought-resistant species *Lycopersicon Pennellii*. *Plant Physiology* 103:597-605.

Kalaji, M.H. and S. Pietkiewica (1993). Salinity effects on plant growth and other physiological processes. *Acta Physiologia Plantarum* 15:89-124.

Kalloo, G. (1991). Breeding for environmental resistance in tomato. In *Genetic Improvement of Tomato*, G. Kalloo (Ed.). Berlin, Germany: Springer-Verlag, pp. 153-165.

Kalloo, G. and M.K. Banerjee (1990). Low temperature fruit set attribute in cultivated variety of tomato derived from *Lycopersicon pimpinellifolium*. In Eucarpia Tomato 90: Eleventh Meeting of the Eucarpia Tomato Working Group, Malaga, Spain, pp. 99-103.

Kamps, T.L., T.G. Isleib, R.C. Herner, and K.C. Sink (1987). Evaluation of techniques to measure chilling injury in tomato. *HortScience* 22:1309-1312.

Kaufman, M.R. (1969). Effects of water potential on germination of lettuce, sunflower, and citrus seeds. *Canadian Journal of Botany* 47:1761-1764.

Kavi-Kishor, P.B., Z. Hong, G.H. Miao, C.A. Hu, and D.P.S. Verma (1995). Overexpression of delta-1-pyrroline-5-carbosxylate synthetase increases proline production and helps maintain osmotic potential in transgenic plants during water stress. *Plant Physiology* 108:1387-1394.

Keller, E. and K.L. Steffen (1995). Increased chilling tolerance and altered carbon metabolism in tomato leaves following application of mechanical stress. *Physiologia Plantarum* 93:519-525.

Kemp, G.A. (1968). Low-temperature growth response of the tomato. *Canadian Journal of Plant Science* 48:281-286.

Kinet, J.M. and M.M. Peet (1997). Tomato. In *The Physiology of Vegetable Crops*, H.C. Wien (Ed.). Wallingford, UK: CAB International, pp. 207-258.

Kramer, P.J. (1980). *Water Relations of Plants*. New York: Academic Press.

Kramer, P.J. (1983). *Water Relations of Plants*. New York: Academic Press.

Kuo, C.G., B.W. Chon, M.H. Chou, C.L. Tsai, and T.S. Tsay (1979). Tomato fruitset at high temperature. In *Proceedings of the First International Symposium in Tropical Tomato*. Tainan, Taiwan: Asian Vegetable Research and Development Center, pp. 94-109.

Lebowitz, R.J., M. Soller, and M. Beckmann (1987). Trait-based analysis for the detection of linkage between marker loci and quantitative loci in crosses between inbred lines. *Theoretical and Applied Genetics* 73:556-562.

Levi, A., H.D. Rabinowitch, and N. Kedar (1978). Morphological and physiological characters affecting flower drop and fruit set of tomatoes at high temperatures. *Euphytica* 27:211-218.

Leviatov, S., O. Shoseyov, and S. Wolf (1994). Roles of different seed components in controlling tomato seed germination at low temperature. *Scientia Horticulturae* 56:197-206.

Leviatov, S., O. Shoseyov, and S. Wolf (1995). Involvement of endomannanase in the control of tomato seed germination under low-temperature conditions. *Annals of Botany* 76:1-6.

Levitt, J. (1980). *Responses of Plants to Environmental Stresses: Water, Salt and Other Stresses*. New York: Academic Press.

Lilius, G., N. Holmberg, and L. Bulow (1996). Enhanced NaCl stress tolerance in transgenic tobacco expressing bacterial choline dehydrogenase. *Biotechnology* 14:177-180.

Liptay, A., E.F. Bolton, and V.A. Dirks (1982). A comparison of field-seeded and transplanted tomatoes grown on a clay soil. *Canadian Journal of Plant Science* 62:483-487.

Liptay, A. and P. Schopfer (1983). Effect of water stress, seed coat restraint, and abscisic acid upon different germination capabilities of two tomato lines at low temperature. *Plant Physiology* 73:935-938.

Lohar, D.P. and W.E. Peat (1998). Floral characteristics of heat-tolerant and heat-sensitive tomato (*Lycopersicon esculentum* Mill.) cultivars at high temperature. *Scientia Horiculturae* 73:53-60.

Lorenz, O.A. and D.N. Maynard (1988). *Knott's Handbook for Vegetable Growers.* New York: Wiley.

Ludlow, M.M. and R.C. Muchow (1990). A critical evaluation of traits for improving crop yields in water-limited environments. *Advances in Agronomy* 43:107-153.

Lutfor-Rahman, S.M. (1998). Eco-physiological study on tomato drought tolerance. In *Division of Environmental Science and Technology.* Kyoto, Japan: Kyoto University, p. 80.

Lyakh, V.A. (1992). Competence of pollen of wild species and cultivar of tomato to affect fertilization at low temperature. *Sexual Plant Reproduction* 5:128-130.

Lyon, C.B. (1941). Responses of two species of tomatoes and the F_1 generation to sodium sulphate in the nutrient medium. *Botanical Gazette* 103:107-122.

Lyons, J.M. (1973). Chilling injury in plants. *Annual Review of Plant Physiology* 24:445-466.

Lyons, J.M., D. Graham, and J.K. Raison (1979). *Low Temperature Stress in Crop Plants: The Role of the Membrane.* New York: Academic Press. p. 565.

Maas, E.V. (1986). Salt tolerance of plants. *Applied Agricultural Research* 1:12-26.

Maas, E.V. (1990). Crop salt tolerance. In *Agricultural Salinity Assessment and Management,* K.K. Tanji (Ed.). ASCE Manuals and Reports on Engineering No. 71, New York, pp. 262-304.

Maisonneuve, B. (1982). Effect of low night temperature on a collection of varieties of tomato *(L. esculentum).* I. Study of fruit production and fertilizing quality of the pollen. *Agronomie* 2:443-451.

Maisonneuve, B. (1983). Cold resistance of *Lycopersicon hirsutum* pollen. *Report of the Tomato Genetics Cooperative* 33:8.

Maisonneuve, B. and A.P.M. Den-Jijs (1984). In vitro pollen germination and tube growth of tomato (*Lycopersicon esculentum* Mill.) and its relation with plant growth. *Euphytica* 33:833-840.

Maisonneuve, B., N.G. Hogenboom, and A.P.M. Den-Jijs (1986). Pollen selection in breeding tomato (*Lycopersicon esculentum* Mill.) for adaptation to low temperature. *Euphytica* 35:983-992.

Maluf, W.R. and E.C. Tigchelaar (1982). Relationship between fatty acid composition and low-temperature seed germination in tomato. *Journal of the American Society for Horticultural Science* 107:620-623.

Martin, B., J. Nienhuis, and G. King (1989). Restriction fragment length polymorphisms associated with water use efficiency in tomato. *Science* 243:1725-1728.

Martin, B., C.G. Tauer, and R.K. Lin (1999). Carbon isotope discrimination as a tool to improve water-use efficiency in tomato. *Crop Science* 39:1775-1783.

Martin, B. and Y.R. Thorstenson (1988). Stable carbon isotope composition (delta ^{13}C), water use efficiency and biomass productivity of *Lycopersicon esculentum, Lycopersicon pennellii,* and the F_1 hybrid. *Plant Physiology* 88:213-217.

McCue, K.F. and A.D. Hanson (1990). Drought and salt tolerance: Towards understanding and application. *Trends in Biotechnology* 8:358-362.

Miller, J.C. and S.D. Tanksley (1990). RFLP analysis of phylogenetic relationships and genetic variation in the genus *Lycopersicon. Theoretical and Applied Genetics* 80:437-448.

Miltan, O., D. Zamir, and J. Rudich (1986). Growth rates of *Lycopersicon* species at low temperatures. *Zeitschrift für Pflanzenzuchtung* 96:193-199.

Mitchell, J.H., D. Siamhan, M.H. Wamala, J.B. Risimeri, E. Chinyamakobvu, S.A. Henderson, and S. Fukai (1998). The use of seedling leaf death score for evaluation of drought resistance of rice. *Field Crops Research* 55:129-139.

Monforte, A.J., M.J. Asins, and E.A. Carbonell (1996). Salt tolerance in *Lycopersicon* species. IV. Efficiency of marker-assisted selection for salt tolerance improvement. *Theoretical and Applied Genetics* 93:765-772.

Monforte, A.J., M.J. Asins, and E.A. Carbonell (1997). Salt tolerance in *Lycopersicon* species. V. Does genetic variability at quantitative trait loci affect their analysis? *Theoretical and Applied Genetics* 95:284-293.

Monforte, A.J., M.J. Asins, and E.A. Carbonell (1999). Salt tolerance in *Lycopersicon* spp. VII. Pleiotropic action of genes controlling earliness on fruit yield. *Theoretical and Applied Genetics* 98:593-601.

Moore, E.L. and W.O. Thomas (1952). Some effects of shading and parachlorophenoxy-acetic acid on fruitfulness of tomatoes. *Proceedings of the American Society for Horticultural Science* 60:289-294.

Nee, M., J.G. Hawkes, R.N. Lester, and N. Estrada (1991). *Solanaceae III: Taxonomy, Chemistry, Evolution.* Kew, UK: The Royal Botanic Gardens.

Ng, T.J. and E.C. Tigchelaar (1973). Inheritance of low temperature seed sprouting in tomato. *Journal of the American Society for Horticultural Science* 98:314-316.

Nguyen, H.T., R.C. Babu, and A. Blum (1997). Breeding for drought resistance in rice: Physiology and molecular genetics considerations. *Crop Science* 37:1426-1434.

Norlyn, J.D. and E. Epstein (1984). Variability in salt tolerance of four Triticale line at germination and emergence. *Crop Science* 24:1090-1092.

Parkin, K.L., A. Marangoni, R.L. Jackman, R.Y. Yada, and D.W. Stanley (1989). Chilling injury: A review of possible mechanisms. *Journal of Food Biochemistry* 13:127-153.

Pasternak, D., M. Twersky, and Y. DeMallach (1979). Salt resistance in agricultural crops. In *Stress Physiology in Crop Plants,* H. Mussell and R.C. Staples (Eds.). New York: Wiley, pp. 127-142.

Patterson, B.D. (1988). Genes for cold resistance from wild tomatoes. *HortScience* 23:794, 947.

Patterson, B.D., D. Graham, and R. Paull (1979). Adaptation to chilling: Survival, germination, respiration and protoplasmic dynamics. In *Low Temperature Stress in Crop Plants: The Role of the Membrane,* J.M. Lyons, D. Graham, and J.K. Raison (Eds.). New York: Academic Press, pp. 25-35.

Patterson, B.D., L. Mutton, R.E. Paull, and V.Q. Nguyen (1987). Tomato pollen development: Stages sensitive to chilling and a natural environment for the selection of resistant genotypes. *Plant Cell and Environment* 10:363-368.

Patterson, B.D., R.E. Paull, and D. Graham (1987). Adaptation to chilling: Survival, germination, respiration and protoplasmic dynamics. In *Low Temperature Stress in Crops Plants: The Role of Membrane,* J.J. Lyons, D. Graham, and J.K. Raison (Eds.). New York: Academic Press, pp. 25-35.

Patterson, B.D., R.E. Paull, and R.M. Smillie (1978). Chilling resistance in *Lycopersicon hirsutum* Humb. and Bonpl., a wild tomato with a wide altitudinal distribution. *Australian Journal of Plant Physiology* 5:609-617.

Patterson, B.D. and L.A. Payne (1983). Screening for chilling resistance in tomato seedlings. *HortScience* 18:340-341.

Pearen, J.R., M.D. Pahl, M.S. Wolynetz, and R. Hermesh (1997). Association of salt tolerance at seedling emergence with adult plant performance in slender wheatgrass. *Canadian Journal of Plant Science* 77:81-89.

Peet, M.M. and M. Bartholemew (1996). Effect of night temperature on pollen characteristics, growth, and fruit set in tomato. *Journal of the American Society for Horticultural Science* 121:514-519.

Peet, M.M., S. Sato, and R.G. Gardner (1998). Comparing heat stress effects on male-fertile and male-sterile tomatoes. *Plant, Cell and Environment* 21:225-231.

Peet, M.M. and D.H. Willits (1998). The effect of night temperature on greenhouse grown tomato yields in warm climate. *Agricultural and Forest Meteorology* 92:191-202.

Peet, M.M., D.H. Willits, and R.G. Gardner (1997). Response of ovule development and post-pollen production processes in mail-sterile tomatoes to chronic, subacute high temperature stress. *Journal of Experimental Botany* 48:101-111.

Perez-Alfocea, F., M.T. Estan, M. Caro, and M.C. Bolarin (1993). Response of tomato cultivars to salinity. *Plant and Soil* 150:203-211.

Perez-Alfocea, F., G. Guerrier, M.T. Estan, and M.C. Bolarin (1994). Comparative salt responses at cell and whole-plant levels of cultivated and wild tomato and their hybrid. *Journal of Horticultural Science* 69:639-644.

Phills, B.R., N.H. Peck, G.E. McDonald, and R.W. Robinson (1979). Differential responses of *Lycopersicon* and *Solanum* species to salinity. *Journal of the American Society for Horticultural Science* 104:349-352.

Philouze, J. and B. Maisonneuve (1979). Breeding tomatoes for their ability to set fruit at low temperatures. *Review of Plant Pathology* 58:54-64.

Picken, A.J.F. (1984). A review of pollination and fruit set in the tomato (*Lycopersicon esculentum* Mill.). *Journal of Horticultural Science* 59:1-13.

Pillay, I. and C. Beyl (1990). Early responses of drought-resistant and -susceptible tomato plants subjected to water stress. *Journal of Plant Growth Regulation* 9:213-219.

Plant, A.L., A. Cohen, M.S. Moses, and E.A. Bray (1991). Nucleotide sequence and spatial expression pattern of drought- and ABA-induced gene for tomato. *Plant Physiology* 97:900-906.

Raison, J.K. and M.A. Brown (1989). Sensitivity of altitudinal ecotypes of the wild tomato *Lycopersicon hirsutum* to chilling injury. *Plant Physiology* 91:1471-1475.

Rana, M.K. and G. Kalloo (1989). Morphological attributes associated with the adaptation under water deficit conditions in tomato (*L. esculentum* Mill.). In *Twelfth Eucarpia Congress 1989,* Volume 15. Goettingen, Germany: Vortrage Pflanzenzucht, pp. 23-27.

Rana, M.K. and G. Kalloo (1990). Evaluation of tomato genotypes under drought conditions (Abstr.). In *Twenty-Third International Horticultural Congress,* Firenze, Italy. Wageningen, The Netherlands: ISHS.

Rathinasabapathi, B. (2000). Metabolic engineering for stress tolerance: Installing osmoprotectant synthesis pathways. *Annals of Botany* 86:709-716.

Redmann, R.E. (1974). Osmotic and specific ion effects on the germination of alfalfa. *Canadian Journal of Botany* 52:803-808.

Reitz, L.P. (1974). Breeding for more efficient water-use: Is it real or a mirage? *Agricultural Meteorology* 14:3-8.

Ribaut, J.M., C. Jiang, D. Gonzalez-de-Leon, G.O. Edmeades, and D.A. Hoisington (1997). Identification of quantitative trait loci under drought conditions in tropical maize. 2. Yield components and marker-assisted selection strategies. *Theoretical and Applied Genetics* 94:887-896.

Richards, M.A. and B.R. Phills (1979). Evaluation of *Lycopersicon* species for drought tolerance (Abstr.). *HortScience* 14:121.

Richards, R.A. (1983). Should selection for yield in saline regions be made on saline or non-saline soils? *Euphytica* 32:431-438.

Richards, R.A. (1996). Defining selection criteria to improve yield under drought. *Plant Growth Regulation* 20:157-166.

Richards, R.A. and C.W. Dennett (1980). Variation in salt concentration in a wheat field. *Soil and Water* 44:8-9.

Rick, C.M. (1973). Potential genetic resources in tomato species: Clues from observation in native habitats. In *Genes, Enzymes, and Populations,* Volume 2, A.M. Srb (Ed.). New York: Plenum, pp. 255-269.

Rick, C.M. (1975). The tomato. In *Handbook of Genetics,* Volume 2. R.C. King (Ed.). New York: Plenum Press, pp. 247-280.

Rick, C.M. (1976a). Natural variability in wild species of *Lycopersicon* and its bearing on tomato breeding. *Genetica Agraria* 30:249-259.

Rick, C.M. (1976b). Tomato *Lycopersicon esculentum* (Solanaceae). In *Evolution of Crop Plants,* N.W. Simmonds (Ed.). London, UK: Longman, pp. 268-273.

Rick, C.M. (1978). The tomato. *Scientific American* 23:76-87.

Rick, C.M. (1979a). Biosystematic studies in *Lycopersicon* and closely related species of *Solanum*. In *The Biology of the Solanaceae,* J.C. Hawkes, R.N. Lester, and A.D. Skelding (Eds.). New York: Academic Press, pp. 1-27.

Rick, C.M. (1979b). Potential improvement of tomatoes by controlled introgression of genes from wild species. In *Proceedings of a Conference on Broadening the Genetic Base of Crops,* A.C. Zeven and A.M. van Harten (Eds.). Wageningen, The Netherlands: Pudoc, pp. 167-173.

Rick, C.M. (1982). The potential of exotic germplasm for tomato improvement. In *Plant Improvement and Somatic Cell Genetics,* I.K. Vasil, W.R. Scowcroft, and K.J. Freys (Eds.). New York: Academic Press, pp. 1-28.

Rick, C.M. (1988). Tomato-like nightshades: Affinities, autoecology and breeders' opportunities. *Economic Botany* 42:145-154.

Rick, C.M. and W.H. Dempsey (1969). Position of the stigma in relation to fruit setting of the tomato. *Botanical Gazette* 130:180-186.

Rick, C.M., J.W. DeVerna, and R.T. Chetelat (1990). Experimental introgression to the cultivated tomato from related wild nightshades. In *Horticultural Biotechnology,* A.B. Bennett and S.D. O'Neill (Eds.). New York: Wiley-Liss, pp. 19-30.

Rick, C.M., J.W. DeVerna, R.T. Chetelat, and M.A. Stevens (1986). Meiosis in sesquidiploid hybrids of *Lycopersicon esculentum* and *Solanum lycopersicoides. Proceedings of the National Academy of Sciences of the United States of America* 83:3580-3583.

Rick, C.M., J.W. DeVerna, R.T. Chetelat, and M.A. Stevens (1987). Potential contribution of wild crosses to improvement of processing tomatoes. *Acta Horticulturae* 200:45-55.

Rick, C.M. and J.F. Fobes (1975). Allozyme variation in the cultivated tomato and closely related species. *Bulletin of the Torrey Botanical Club* 102:376-384.

Rontein, D., G. Basset, and A.D. Hanson (2002). Metabolic engineering of osmoprotectant accumulation in plants. *Metabolic Engineering* 4:49-56.

Ross, R. and N. Lott (2000). *A Climatology of Recent Extreme Weather and Climate Events,* Volume 2002. U.S. Department of Commerce. NOAA/NESDIS, National Climatic Data Center.

Rowley, G. (1993). Multinational and national competition for water in the Middle East: Towards the deepening crisis. *Journal of Environmental Management* 39:187-197.

Rudich, J., E. Zamski, and Y. Regev (1977). Genotypic variation for sensitivity to high temperature in tomato: Pollination and fruit set. *Botanical Gazette* 138:448-452.

Rush, D.W. and E. Epstein (1976). Genotypic responses to salinity: Differences between salt-sensitive and salt-tolerant genotypes of the tomato. *Plant Physiology* 57:162-166.

Rush, D.W. and E. Epstein (1981a). Breeding and selection for salt tolerance by the incorporation of wild germplasm into a domestic tomato. *Journal of the American Society for Horticultural Science* 106:699-704.

Rush, D.W. and E. Epstein (1981b). Comparative studies on the sodium, potassium, and chloride relations of a wild halophytic and domestic salt-sensitive tomato species. *Plant Physiology* 68:1308-1313.

Sacher, R.F., R.C. Staples, and R.W. Robinson (1983). Ion regulation and response of tomato to sodium chloride: A homeostatic system. *Journal of the American Society for Horticultural Science* 108:566-569.

Saeed, M. and C.A. Francis (1983). Yield stability in relation to maturity in grain sorghum. *Crop Science* 23:683-687.

Santa-Cruz, A., F. Perez-Alfocea, M. Caro, and M. Acosta (1998). Polyamines as short-term salt tolerance traits in tomato. *Plant Science* 138:9-16.

Saranga, Y., A. Cahaner, D. Zamir, A. Marani, and J. Rudich (1992). Breeding tomatoes for salt tolerance: Inheritance of salt tolerance and related traits in interspecific populations. *Theoretical and Applied Genetics* 84:390-396.

Saranga, Y., D. Zamir, A. Marani, and J. Rudich (1991). Breeding tomatoes for salt tolerance: Field evaluation of *Lycopersicon* germplasm for yield and dry matter production. *Journal of the American Society for Horticultural Science* 116:1067-1071.

Saranga, Y., D. Zamir, A. Marani, and J. Rudich (1993). Breeding tomatoes for salt tolerance: variation in ion concentration associated with response to salinity. *Journal of the American Society for Horticultural Science* 118:405-408.

Sarg, S.M.H., R.G. Wyn-Jones, and F.A. Omar (1993). Salt tolerance in the Edkawy tomato. In *Toward the Rational Use of High Salinity Tolerant Plants*. Volume 2, H. Lieh and A. Al-Masoom (Eds.). Dordrecht, The Netherlands: Kluwer Academic Publishers, pp. 177-184.

Schonfeld, M.A., B.D. Carver, and D.W. Mornhinweg (1988). Water relations in wheat as drought resistance indicators. *Crop Science* 28:526-531.

Scott, J.W. (1993). Breeding tomatoes for resistance to high temperatures, biotic and abiotic diseases. *Horticultura Brasileira* 11:167-170.

Scott, J.W. (2000). Fla. 7771, a medium-large, heat tolerant, jointless-pedicel tomato. *HortScience* 35:968-969.

Scott, J.W. and F.F. Angell (1998). Tomato. In *Hybrid Cultivar Development*, S.S. Banga and S.K. Banga (Eds.). New Delhi, India: Narosa Publishing House, pp. 451-475.

Scott, J.W., H.H. Bryan, and L.J. Ramos (1997). High temperature fruit setting ability of large-fruited, jointless pedicel tomato hybrids with various combinations of heat-tolerance. *Proceedings of the Florida State Horticultural Society* 110:281-284.

Scott, J.W., S.M. Olson, H.H. Bryan, T.K. Howe, P.J. Stoffella, and J.A. Bartz (1989). Solar set—A heat tolerant, fresh market tomato hybrid. *Florida Agricultural Experiment Station Circular* S-359.

Scott, J.W., S.M. Olson, T.K. Howe, P.J. Stoffella, J.A. Bartz, and H.H. Bryan (1995). 'Equinox' heat-tolerant hybrid tomato. *HortScience* 30:647-648.

Scott, J.W., R.B. Volin, H.H. Bryan, and S.M. Olson (1986). Use of hybrids to develop heat tolerant tomato cultivars. *Proceedings of the Florida State Horticultural Society* 99:311-315.

Scott, S.J. and R.A. Jones (1982). Low temperature seed germination of *Lycopersicon* species evaluated by survival analysis. *Euphytica* 31:869-883.

Scott, S.J. and R.A. Jones (1985a). Cold tolerance in tomato. I. Seed germination and early seedling growth of *Lycopersicon esculentum*. *Physiologia Plantarum* 65:487-492.

Scott, S.J. and R.A. Jones (1985b). Quantifying seed germination responses to low temperatures: Variation among *Lycopersicon* spp. *Environmental and Experimental Botany* 25:129-137.

Scott, S.J. and R.A. Jones (1986). Cold tolerance in tomato. II. Early seedling growth of *Lycopersicon* spp. *Physiologia Plantarum* 66:659-663.

Serrano, R., F.A. Culiañz-Macia, and V. Moreno (1999). Genetic engineering of salt and drought tolerance with yeast regulatory genes. *Scientia Horticulturae* 78:261-269.

Shannon, M.C. (1996). New insights in plant breeding efforts for improved salt tolerance. *HortTech* 6:96-98.

Shannon, M.C., J.W. Gronwald, and M. Tal (1987). Effects of salinity on growth and accumulation of organic and inorganic ions in cultivated and wild tomato species. *Journal of the American Society for Horticultural Science* 112:416-423.

Shelby, R.A., W.H. Greenleaf, and C.M. Paterson (1978). Comparative floral fertility in heat tolerant and heat sensitive tomatoes. *Journal of the American Society for Horticultural Science* 103:778-780.

Shen, B., R.G. Jensen, and J.J. Bohnert (1997). Mannitol protects against oxidation by hydroxyl radicals. *Plant Physiology* 115:527-532.

Shen, Z.Y. and P.H. Li (1982). Heat adaptability of the tomato. *HortScience* 17:924-925.

Shinozaki, K. and K. Yamaguchi-Shinozaki (1996). Molecular responses to drought and cold stress. *Current Opinion in Biotechnology* 7:161-167.

Shinozaki, K. and K. Yamaguchi-Shinozaki (1997). Gene expression and signal transduction in water-stress response. *Plant Physiology* 115:327-334.

Simon, W.W. (1979). Seed germination under low temperatures. In *Low Temperature Stress in Crop Plants: The Role of the Membrane*, J.M. Lyons, D. Graham, and J.K. Raison (Eds.). New York: Academic Press, pp. 37-45.

Smith, P.G. and R.H. Millet (1964). Germinating and sprouting responses of the tomato at low temperatures. *Proceedings of the American Society for Horticultural Science* 84:480-484.

Spooner, D.M., G.J. Anderson, and R.K. Jansen (1993). Chloroplast DNA evidence for the interrelationships of tomatoes, potatoes and pepinos (Solanaceae). *American Journal of Botany* 80:676-688.

Stevens, M.A. and J. Rudich (1987). Genetic potential for overcoming physiological limitations on adaptability, yield, and quality of the tomato. *HortScience* 13:673-679.

Stoner, A.K. (1972). Merit, Red Rock and Potomac-tomato varieties adapted to mechanical harvesting. USDA Product Research Report. Washington, DC: USDA.

Stoner, A.K. and D.E. Otto (1975). A greenhouse method to evaluate high temperature setting ability in the tomato. *HortScience* 10:264-265.

Subudhi, P.K., D.T. Rosenow, and H.T. Nguyen (2000). Quantitative trait loci for the stay green trait in sorghum (*Sorghum bicolor* L. Moench): Consistency across genetic backgrounds and environments. *Theoretical and Applied Genetics* 101:733-741.

Syverstein, J.P., B. Boman, and D.P.H. Tucker (1989). Salinity in Florida citrus production. *Proceedings of the Florida State Horticultural Society* 102:61-64.

Szabolcs, I. (1992). Salinization of soils and water and its relation to desertification. *Desertification Control Bulletin* 21:32-37.

Szabolcs, I. (1994). Soil salinization. In *Handbook of Plant and Crop Stress,* M. Pessarakli (Ed.). New York: Marcel Dekker, pp. 3-11.

Tal, M. (1971). Salt tolerance in the wild relatives of the cultivated tomato: Responses of *Lycopersicon esculentum, L. peruvianum,* and *L. esculentum minor* to sodium chloride solution. *Australian Journal of Agricultural Research* 22:631-638.

Tal, M. (1985). Genetics of salt tolerance in higher plants: Theoretical and practical considerations. *Plant and Soil* 89:199-226.

Tal, M. (1997). Wild germplasm for salt tolerance in plants. In *Strategies for Improving Salt Tolerance in Higher Plants,* P.K. Jaiwal, R.P. Singh, and A. Gulati (Eds.). Enfield, NH: Science Publishers, pp. 291-320.

Tal, M. and U. Gavish (1973). Salt tolerance in the wild relatives of the cultivated tomato: Water balance and abscisic acid in *Lycopersicon esculentum* and *L. peruvianum* under low and high salinity. *Australian Journal of Agricultural Research* 24:353-361.

Tal, M., A. Katz, H. Heikin, and K. Dehan (1979). Salt tolerance in the wild relatives of the cultivated tomato: Proline accumulation in *Lycopersicon esculentum* Mill., *L. peruvianum* Mill., and *Solanum pennellii* Cor. treated with NaCl and polyethylene glycol. *New Phytologist* 82:349-355.

Tal, M. and M.C. Shannon (1983). Salt tolerance in the wild relatives of the cultivated tomato: Responses of *Lycopersicon esculentum, L. cheesmanii, L. peruvianum, Solanum pennellii* and F_1 hybrids to high salinity. *Australian Journal of Agricultural Research* 10:109-117.

Tanaka, A., K. Fujita, and K. Kikuchi (1974). Nutrio-physiological studies on the tomato plant: Photosynthetic rates of individual leaves in relation to the dry matter production in plants. *Soil Science and Plant Nutrition* 20:173-183.

Tanaka, Y., T. Hibino, Y. Hayashi, A. Tanaka, S. Kishitani, T. Takabe, S. Yokota, and T. Takabe (1999). Salt tolerance of transgenic rice overexpressing yeast mitochondrial Mn-SOD in chloroplasts. *Plant Science* 148:131-138.

Tanji, K.K. (1990). Nature and extent of agricultural salinity. In *Agricultural Salinity Assessment and Management,* K.K. Tangi (Ed.). New York: American Society of Civil Engineers, pp. 1-13.

Taylor, A.G., J.E. Motes, and M.B. Kirkham (1982). Germination and seedling growth characteristics of three tomato species affected by water deficit. *Journal of the American Society of the Horticultural Science* 107:282-285.

Thomas, J.C., M. Sepahi, B. Arndall, and H.J. Bohnert (1995). Enhancement of seed germination in high salinity by engineering mannitol expression in *Arabidopsis thaliana. Plant, Cell and Environment* 18:801-806.

Thompson, K., J.P. Grime, and G. Mason (1977). Seed germination in response to diurnal fluctuations of temperature. *Nature (London)* 267:147-149.

Thompson, P.A. (1970). Characterization of the germination response to temperature of species and ecotypes. *Nature* 225:827-831.

Thompson, P.A. (1974). Characterization of the germination responses to temperatures of vegetable crops. I. Tomatoes. *Scientia Horticulturae* 2:35-54.

Ungar, I.A. (1978). Halophyte seed germination. *The Botanical Review* 44:233-264.

Vallejos, C.E. (1979). Genetic diversity of plants for response to low temperature and its potential use in crop plants. In *Low-Temperature Stress in Crop Plants,* J.M. Lyons, D. Graham, and J.K. Raison (Eds.). New York: Academic Press, pp. 473-489.

Vallejos, C.E., J.M. Lyons, R.W. Breidenbach, and M.F. Miller (1983). Characterization of a differential low-temperature growth response of two species of *Lycopersicon*: The plastochron as a tool. *Planta* 159:487-496.

Vallejos, C.E. and S.D. Tanksley (1983). Segregation of isozyme markers and cold tolerance in an interspecific backcross of tomato. *Theoretical and Applied Genetics* 66:241-247.

Van-De-Dijk, S.J., J.A. Maris, and P.R.V. Hasselt (1985). Genotypic variation in chilling-induced leakage of electrolytes from leaf tissue of tomato (*Lycopersicon esculentum* Mill.) in relation to growth under low-energy conditions. *Journal of Plant Physiology* 120:39-45.

Van-Ieperen, W. (1996). Effects of different day and night salinity levels on vegetative growth, yield and quality of tomato. *Journal of Horticultural Science* 71:99-111.

Villareal, R.L. (1978). Screening for heat tolerance in the genus *Lycopersicon. HortScience* 13:479-481.

Villareal, R.L. and S.H. Lai (1979). Development of heat-tolerant tomato varieties in the tropics. In *Proceedings of the First International Symposium on Tropical Tomato,* R. Cowell (Ed.). Tainan, Taiwan: Asian Vegetable Research and Development Center, pp. 188-200.

Walker, M.A., D.M. Smith, K.P. Pauls, and B.D. McKersie (1990). A chlorophyll fluorescence screening test to evaluate chilling tolerance in tomato. *HortScience* 25:334-339.

Warnock, S.J. (1988). A review of taxonomy and phylogeny of the genus *Lycopersicon. HortScience* 23:669-673.

Warren, G.F. (1998). Spectacular increases in crop yields in the twentieth century. *Weed Technology* 12:752-760.

Weaver, M.L. and H. Timm (1989). Screening tomato for high-temperature tolerance through pollen viability tests. *HortScience* 24:493-495.

Went, F.W. (1957). *The Experimental Control of Plant Growth.* New York: Ronald Press Co.

Wessel-Beaver, L. and J.W. Scott (1992). Genetic variability of fruit set, fruit weight, and yield in a tomato population grown in two high-temperature environments. *Journal of the American Society of Horticultural Science* 117:867-870.

Whittington, W.J. and P. Fierlinger (1972). The genetic control of time to germination in tomato. *Annals of Botany* 36:873-880.

Wolf, S., D. Yakir, M.A. Stevens, and J. Rudich (1986). Cold temperature tolerance of wild tomato species. *Journal of the American Society of Horticultural Science* 111:960-964.

Wudiri, B.B. and D.W. Henderson (1985). Effects of water stress on flowering and fruit set in processing tomatoes. *Scientia Horticulturae* 27:189-198.

Yeo, A.R. and T.J. Flowers (1990). Screening of rice (*Oryza sativa* L.) genotypes for physiological characters contributing to salinity resistance, and their relationship to overall performance. *Theoretical and Applied Genetics* 79:377-384.

Younis, A.F. and M.A. Hatata (1971). Studies on the effects of certain salts on germination, on growth of root, and on metabolism. I. Effects of chlorides and sulphates of sodium, potassium, and magnesium on germination of wheat grains. *Plant and Soil* 13:183-200.

Yu, T.T. (1972). "The genetics and physiology of water usage in *Solanum pennellii* Corr. and its hybrids with *Lycopersicon esculentum* Mill." PhD Thesis, University of California, Davis, California.

Zamir, D. and I. Gadish (1987). Pollen selection for low temperature adaptation in tomato. *Theoretical and Applied Genetics* 74:545-548.

Zamir, D., S.D. Tanksley, and R.A. Jones (1981). Low temperature effect on selective fertilization by pollen mixtures of wild and cultivated tomato species. *Theoretical and Applied Genetics* 59:235-238.

Zamir, D., S.D. Tanksley, and R.A. Jones (1982). Haploid selection for low temperature tolerance of tomato pollen. *Genetics* 101.

Zhang, H.X. and E. Blumwald (2001). Transgenic salt-tolerant tomato plants accumulate salt in foliage but not in fruit. *Nature Biotechnology* 19:765-768.

Zhang, H.X., J.N. Hodson, J.P. Williams, and E. Blumwald (2001). Engineering salt-tolerant *Brassica* plants: Characterization of yield and seed oil quality in transgenic plants with increased vacuolar sodium accumulation. *Proceedings of the National Academy of Sciences of the United States of America* 98:12832-12836.

Zhang, J., H.G. Zheng, A. Aarti, G. Pantuwan, T.T. Nguyen, J.N. Tripathy, A.K. Sarial, S. Robin, R.C. Babu, and B.D. Nguyen (2001). Locating genomic regions associated with components of drought resistance in rice: Comparative mapping within and across species. *Theoretical and Applied Genetics* 103:19-29.

Zhu, J.K., P.M. Hasegawa, and R.A. Bressan (1997). Molecular aspects of osmotic stress in plants. *CRC Critical Reviews in Plant Sciences* 16:253-277.

Index